Lecture Notes in Computer Science 5916

Susanne Boll Qi Tian Lei Zhang
Zili Zhang Yi-Ping Phoebe Chen (Eds.)

Advances in Multimedia Modeling

16th International
Multimedia Modeling Conference, MMM 2010
Chongqing, China, January 6-8, 2010
Proceedings

Volume Editors

Susanne Boll
University of Oldenburg
Oldenburg, Germany
E-mail: susanne.boll@informatik.uni-oldenburg.de

Qi Tian
University of Texas at San Antonio
San Antonio, TX, USA
E-mail: qitian@cs.utsa.edu

Lei Zhang
Microsoft Research Asia
Beijing, China
E-mail: leizhang@microsoft.com

Zili Zhang
Southwest University
Beibei, Chongqing, China
E-mail: zhangzl@swu.edu.cn

Yi-Ping Phoebe Chen
Deakin University
Burwood, Victoria, Australia
E-mail: phoebe.chen@deakin.edu.au

Library of Congress Control Number: 2009941456

CR Subject Classification (1998): H.2.4, H.5.1, H.2.8, H.3.3, H.3.5

LNCS Sublibrary: SL 3 – Information Systems and Application, incl. Internet/Web and HCI

ISSN 0302-9743
ISBN-10 3-642-11300-1 Springer Berlin Heidelberg New York
ISBN-13 978-3-642-11300-0 Springer Berlin Heidelberg New York

springer.com

Printed in Germany

Typesetting: Camera-ready by author, data conversion by Scientific Publishing Services, Chennai, India
Printed on acid-free paper SPIN: 12822843 06/3180 5 4 3 2 1 0

Preface

The 16th international conference on Multimedia Modeling (MMM2010) was held in the famous mountain city Chongqing, China, January 6–8, 2010, and hosted by Southwest University. MMM is a leading international conference for researchers and industry practitioners to share their new ideas, original research results and practical development experiences from all multimedia related areas.

MMM2010 attracted more than 160 regular, special session, and demo session submissions from 21 countries/regions around the world. All submitted papers were reviewed by at least two PC members or external reviewers, and most of them were reviewed by three reviewers. The review process was very selective. From the total of 133 submissions to the main track, 43 (32.3%) were accepted as regular papers, 22 (16.5%) as short papers. In all, 15 papers were received for three special sessions, which is by invitation only, and 14 submissions were received for a demo session, with 9 being selected. Authors of accepted papers come from 16 countries/regions. This volume of the proceedings contains the abstracts of three invited talks and all the regular, short, special session and demo papers. The regular papers were categorized into nine sections: 3D modeling; advanced video coding and adaptation; face, gesture and applications; image processing; image retrieval; learning semantic concepts; media analysis and modeling; semantic video concepts; and tracking and motion analysis. Three special sessions were video analysis and event recognition, cross-X multimedia mining in large scale, and mobile computing and applications.

The technical program featured three invited talks, parallel oral presentation of all the accepted regular and special session papers, and poster sessions for short and demo papers. The three distinguished keynote speakers were: Shi-Kuo Chang from Pittsburgh University, Shipeng Li from Microsoft Research Asia, and Hartmut Neven from Google, Inc.

The success of MMM2010 was assured by team efforts from the sponsors, organizers, reviewers, and participants. We would like to thank the special session Co-chairs, Nicu Sebe and Tao Mei, as well as the demo Co-chairs, Berna Erol and Meng Wang. The special and demo sessions at MMM2010 enriched the technical program. We would like to acknowledge the contribution of the individual Program Committee members and thank the external reviewers. Thanks to the Publicity Co-chairs, Liang-Tien Chia, Benoit Huet, and Li Tao, Local Organizing Chair, Boahua Qiang, Publication Chair, Guoqiang Xiao, US Liaison, Jiebo Luo, Asian Liaison, Tat-Seng Chua, and Webmaster, Ming Tang, for their great efforts. Our sincere gratitude goes to the participants and all authors of the submitted papers.

We are grateful to our sponsors: Chongqing Science and Technology Committee and Southwest University. The in-kind support from the Faculty of Computer and Information Science at Southwest University is also much appreciated.

We wish to express our gratitude to the Springer team directed by Alfred Hofmann for their help and cooperation.

January 2010

Susanne Boll
Qi Tian
Lei Zhang
Zili Zhang
Yi-Ping Phoebe Chen

Organization

MMM2010 was hosted and organized by the Faculty of Computer and Information Science, Southwest University, China. The conference was held at Haiyu Hotspring Hotel, Chongqing, January 6–8, 2010.

Conference Committee

Steering Committee	Yi-Ping Phoebe Chen (Deakin University)
	Tat-Seng Chua (National University of Singapore)
	Tosiyasu L. Kunii (University of Tokyo)
	Wei-Ying Ma (Microsoft Research Asia)
	Nadia Magnenat-Thalmann (University of Geneva)
	Patrick Senac (ENSICA, France)
Conference Co-chairs	Yi-Ping Phoebe Chen (Deakin University)
	Zili Zhang (Southwest University)
Program Co-chairs	Susanne Boll (University of Oldenburg)
	Qi Tian (University of Texas at San Antonio)
	Lei Zhang (Microsoft Research Asia)
Special Session Co-chairs	Nicu Sebe (University of Amsterdam)
	Tao Mei (Microsoft Research Asia)
Demo Co-chairs	Berna Erol (Ricoh California Research Center)
	Meng Wang (Microsoft Research Asia)
Local Organizing Chair	Baohua Qiang (Southwest University)
Publication Chair	Guoqiang Xiao (Southwest University)
Publicity Co-chairs	Liang-Tien Chia (Nanyang Technological University)
	Benoit Huet (Institut Eurecom)
	Li Tao (Southwest University)
US Liaison	Jiebo Luo (Kodak Research Lab)
Asian Liaison	Tat-Seng Chua (National University of Singapore)
European Liaison	Susanne Boll (University of Oldenburg)
Webmaster	Ming Tang (Southwest University)

Program Committee

Alan Hanjalic	Delft University of Technology, The Netherlands
Alexander Hauptmann	Carnegie Mellon University, USA
Andreas Henrich	University of Bamberg, Germany
Andruid Kerne	Texas A&M University, USA

Ansgar Scherp	University of Koblenz, Germany
Belle Tseng	NEC Laboratories, USA
Brigitte Kerherve	University of Quebec, Canada
Cathal Gurrin	Dublin City University, Ireland
Cees Snoek	University of Amsterdam, The Netherlands
Cha Zhang	Microsoft Research
Chabane Djeraba	University of Sciences and Technologies of Lille, France
Changsheng Xu	NLPR, Chinese Academy of Science, China
Chong-Wah Ng	City University of Hong Kong
Christian Timmerer	University of Klagenfurt, Austria
Dacheng Tao	Hong Kong Polytechnic University
Dong Xu	Nanyang Technological University, Singapore
Ebroul Izquierdo	Queen Mary University London, UK
Edward Chang	Google
Fernando Pereira	Technical University of Lisbon, Portugal
Franziska Frey	Rochester Institute of Technology, USA
Gamhewage Chaminda De Silva	HP Labs, Japan
Gareth Jones	Dublin City University, Ireland
Harald Kosch	Passau University, Germany
Hari Sundaram	Arizona State University, USA
James Wang	Penn State University, USA
Jane Hunter	University of Queensland, Australia
Jianmin Li	Tsinghua University, China
Jiebo Luo	Kodak R&D Labs, USA
Jinghui Tang	National University of Singapore
Keith van Rijsbergen	University of Glasgow, UK
Lynda Hardman	CWI, The Netherlands
Marcel Worring	University of Amsterdam, The Netherlands
Martha Larson	Technical University of Delft, The Netherlands
Masashi Inoue	Yamagata University, Japan
Meng Wang	Microsoft Research Asia
Michael Lew	Leiden University, The Netherlands
Mohan Kankanhalli	National University of Singapore
Mor Naaman	Rutgers University, USA
Nicu Sebe	University of Amsterdam, The Netherlands
Noel O'Connor	Dublin City University, Ireland
Patrick Schmitz	Ludicrum Enterprises, USA
Raghavan Manmatha	University of Massachusetts, USA
Amherst Rainer Lienhart	University of Augsburg, Germany
Raphael Troncy	CWI, The Netherlands
Roger Zimmermann	National University of Singapore

External Reviewers

Table of Contents

Invited Talks

Regular Papers

3D Modeling

Advanced Video Coding and Adaptation

Face, Gesture and Applications

Image Processing

Image Retrieval

Learning Semantic Concepts

Media Analysis and Modeling

Semantic Video Concepts

Tracking and Motion Analysis

Special Session Papers

Video Analysis and Event Recognition

Cross-X Multimedia Mining in Large Scale

Mobile Computing and Applications

Short Papers

Demo Session Papers

Slow Intelligence Systems
(Extended Abstract)

Shi-Kuo Chang

Department of Computer Science
University of Pittsburgh, Pittsburgh, PA 15260 USA
chang@cs.pitt.edu

Abstract. In this talk I will introduce the concept of slow intelligence. Not all intelligent systems have fast intelligence. There are a surprisingly large number of intelligent systems, quasi-intelligent systems and semi-intelligent systems that have slow intelligence. Such slow intelligence systems are often neglected in mainstream research on intelligent systems, but they are really worthy of our attention and emulation. I will discuss the general characteristics of slow intelligence systems and then concentrate on evolutionary query processing for distributed multimedia systems as an example of artificial slow intelligence systems.

S. Boll et al. (Eds.): MMM 2010, LNCS 5916, p. 1, 2010.

Media 2.0 – The New Media Revolution?
(Extended Abstract)

Shipeng Li

Media Computing Group
Microsoft Research Asia
Shipeng.Li@microsoft.com

Abstract. With the rapid development of Web 2.0 and cloud computing concept and applications, there are many unprecedented web-based multimedia applications are emerging today and they pose many new challenges in multimedia research. In this talk, I first summarize the common features of the new wave of multimedia applications which I call Media 2.0. I use 5 D's to describe Media 2.0 principles, namely, Democratized media life cycle; Data-driven media value chain; Decoupled media system; Decomposed media contents; and Decentralized media business model. Then I explain what the implications of Media 2.0 to multimedia research are and how we should choose our research topics that could make big impacts. Finally, I use example research projects ranging from media codecs, media systems, media search and media related advertisement from MSRA to demonstrate the ideas I have talked about. I hope these ideas and principles could inspire the audience to come up with new media 2.0 research topics and applications in the future.

S. Boll et al. (Eds.): MMM 2010, LNCS 5916, p. 2, 2010.

Designing a Comprehensive Visual Recognition System
(Extended Abstract)

Hartmut Neven

Google Inc.
neven@google.com

Abstract. Computer vision has made significant advances during the last decade. Many capabilities such as the detection of faces or the recognition of rigid textured objects such as landmarks are now working to very satisfying levels. Across the various products and services offered by Google we are interested in analyzing an image crawled on the web in all its aspects. When designing such a comprehensive system it becomes obvious however that important abilities are still lacking. One example is object class recognition that scales to thousands or even millions of classes. Another area where we are still facing obstacles is the reliable recognition of objects that have little surface texture and which are largely contour defined. Even a seemingly simple task such as reading text in a photo is still lacking the accuracy we need. The talk describes our efforts in designing a large scale image recognition system that can analyze any given image on the web with respect to many dimensions. We report on the recognition disciplines in which we made good progress but we also call out areas which still require additional work to reach production ready solutions.

S. Boll et al. (Eds.): MMM 2010, LNCS 5916, p. 3, 2010.

Surface Reconstruction from Images Using a Variational Formulation

Liuxin Zhang and Yunde Jia

Beijing Laboratory of Intelligent Information Technology,
School of Computer Science, Beijing Institute of Technology,
Beijing 100081 P.R. China
{zhangliuxin,jiayunde}@bit.edu.cn

Abstract. In this paper, we present a new approach to recovering the surface of a 3D object from multiple calibrated images. The method is based on a variational formulation which defines an energy functional where both silhouette and texture consistency constraints are included. We cast the surface reconstruction problem as an optimization of the energy functional amenable for minimization with an Euler-Lagrange driven evolution. Starting from an initial surface, the reconstruction result can be obtained at the end of the evolution. Compared to the traditional reconstruction methods using a variational framework, our method can be easily implemented by simple finite difference scheme and is computationally more efficient. The final result of our method is not sensitive to where the initial surface has started its evolution.

1 Introduction

This paper considers the problem of reconstructing the highly realistic 3D model of a real-world object from multi-view images in which the camera intrinsic and extrinsic parameters have been previously obtained. It is one of the most fundamental and extensively studied problems in computer vision. According to the geometrical representation of the 3D object, the state-of-art solutions for this long-studied problem can be roughly classified into two categories: (1) dense stereo methods that recover depth maps with respect to an image plane, and (2) volumetric methods that represent the volume directly, without any reference to an image plane.

A review of the dense stereo methods can be found in [1]. The 3D object is represented as a set of depth maps, one for each input view. The advantage of this 2D representation is its convenience because it avoids resampling the geometry on a 3D domain. But these solutions can only represent depth maps with a unique disparity per pixel, i.e., depth is a function of image point. Reconstructing complete objects requires further processing to merge multiple depth maps [2,3,4,5], which makes the final models suffer from a large number of holes in area of uncertainty. Another drawback of these methods is that the surface smoothness is defined on image disparities or depths and, hence, is viewpoint dependent. So, if a different view is chosen as the reference image the results may be different.

S. Boll et al. (Eds.): MMM 2010, LNCS 5916, pp. 4–14, 2010.

The volumetric methods work on discretized 3D space directly and represent the object as a set of voxels. The space carving methods [6,7] and their variants [8,9] progressively remove inconsistent voxels from an initial volume. The graph cut methods [10,11,12,13,14,15] recover the surface based on exact global optimization of surface photo-consistency. The surface evolution methods [16,17,18,19,20] obtain the optimal surface via gradient descent evolution. We refer readers to [21] for a more thorough discussion of the other volumetric techniques. In these methods, multiple viewpoints can be easily integrated and surface smoothness is free from the viewpoints.

Our work is inspired by Faugeras and Keriven [16], which introduced a variational framework implemented by level set methods [22] for surface reconstruction. The basic idea is to represent surface as the zero level set of an implicit function defined in a higher dimension, usually referred as the *level set function*, and to evolve the level set function according to a partial differential equation (PDE). This evolution PDE is derived from the problem of minimizing a certain energy functional defined as the integral of a data fidelity criterion on the unknown surface. However, the numerical implementation of the PDE in [16] is complicated and requires simplification by dropping some terms. The final result of [16] tends to be sensitive to where the level set function has started its evolution. In contrast, we formulate the energy functional directly in the level set domain as the integral of the data fidelity criterion on the whole 3D space. The minimization of our energy functional leads naturally to the evolution PDE of the level set function and we can use all the terms in the PDE to implement the evolution without dropping any of them. The final result of our method is independence from the initialization. Besides, our variational energy functional also consists of a penalizing term [23] which keeps the evolving level set function close to a signed distance function. As a result, the costly re-initialization procedure of the level set function, which is necessary in the numerical implementation of [16], can be completely eliminated. We can use very simple finite difference scheme rather than the complex upwind scheme to implement the evolution PDE. Compared with Faugeras et al. [16], our method allows more efficient numerical schemes.

The rest of the paper is organized as follows: Section 2 gives the background on the camera model and the level set method. Section 3 describes how surface reconstruction can be formulated as an optimization of a variational energy functional. We introduce the numerical implementation of the proposed method in Section 4. The experimental results and conclusions are discussed in Section 5 and Section 6 respectively.

2 Background

2.1 Camera Model

The perspective (or pinhole) camera is used throughout this paper. A point in 3D space with a coordinate vector $\mathbf{x} \in \Re^3$ is projected to the image point with a coordinate vector $\mathbf{m} \in \Re^2$ according to

$$\lambda \begin{bmatrix} \mathbf{m} \\ 1 \end{bmatrix} = \mathbf{A}[\mathbf{R} \quad \mathbf{t}] \begin{bmatrix} \mathbf{x} \\ 1 \end{bmatrix} \tag{1}$$

where λ is an arbitrary scale factor; $[\mathbf{R} \quad \mathbf{t}]$, called the extrinsic parameters, is the rotation and translation which relates the world coordinate system to the camera coordinate system; $\mathbf{A}$ is a 3×3 matrix called the camera intrinsic matrix. Giving corresponding image points in several views, it is possible to calculate both the intrinsic and extrinsic parameters of multiple cameras using the multi-camera calibration techniques [24,25,26]. In our work, we assume that all the views have already been well calibrated within the global world coordinate system, i.e., for each point in the world space, it is possible to determine the coordinates of its projection onto each view.

2.2 Level Set Representation

The level set method for evolving implicit surfaces was introduced by Osher and Sethian [22]. Let $\mathbf{x}$ be a point in the open set $\Omega \subset \Re^3$. The time dependent surface $S(t)$ is implicitly represented as the zero level set of a Lipschitz smooth function $\phi(\mathbf{x}, t)$: $\Omega \times \Re_+ \rightarrow \Re$ as

$$S(t) = \{\mathbf{x} | \phi(\mathbf{x}, t) = 0, \mathbf{x} \in \Omega\}, \tag{2}$$

where ϕ is commonly referred to as the *level set function*. The sets $\{\mathbf{x}|\phi(\mathbf{x}) < 0, \mathbf{x} \in \Omega\}$ and $\{\mathbf{x}|\phi(\mathbf{x}) > 0, \mathbf{x} \in \Omega\}$ are called the *interior* and the *exterior* of S respectively. Using the definition above, the outward unit normal $\mathbf{n}$ and the mean curvature κ are given by

$$\mathbf{n} = \frac{\nabla \phi}{|\nabla \phi|} \quad \text{and} \quad \kappa = \nabla \cdot \frac{\nabla \phi}{|\nabla \phi|}. \tag{3}$$

One example frequently used to construct a level set function ϕ is the signed distance function, where the additional requirement $|\nabla \phi| = 1$ is imposed. Differentiating $\phi(\mathbf{x}, t) = 0$ with respect to t gives

$$\frac{\partial \phi}{\partial t} + \mathbf{v} \cdot \nabla \phi = 0 \Leftrightarrow \frac{\partial \phi}{\partial t} + v_n |\nabla \phi| = 0, \tag{4}$$

where $\mathbf{v} = d\mathbf{x}/dt$ and $v_n = \mathbf{v} \cdot \mathbf{n}$ is the velocity normal to the surface. This PDE is solved for evolving the surface S according to some derived velocity $\mathbf{v}$.

3 Variational Energy Functional

3.1 Photo-Consistency Energy Term

As do most other approaches, we assume that the surface of a 3D object is nearly Lambertian, i.e., elements belonging to this surface should have a consistent appearance in the viewpoints observing them (*be photo-consistent*). Under this assumption, the photo-consistency energy cost of an element can be defined as

$$\mathbf{A}(\mathbf{x}, \mathbf{n}) = \frac{1}{T} \sum_{i,j=1}^{m} \mathbf{A}_{ij}(\mathbf{x}, \mathbf{n}), \tag{5}$$

where $(\mathbf{x}, \mathbf{n})$ is an infinitesimal element located at point $\mathbf{x}$ and having unit outward normal $\mathbf{n}$, T is the number of items in the summation, m is the number of visible views to the current element, and $\mathbf{A}_{ij}$ between visible views i and j is based on the normalized cross correlation (NCC):

$$\mathbf{A}_{ij}(\mathbf{x}, \mathbf{n}) = 1 - NCC(\mathbf{x}, \mathbf{n}, i, j). \tag{6}$$

$\mathbf{A}_{ij}(\mathbf{x}, \mathbf{n})$ ranges between 0 (best correlation) and +2 (worst). The term $NCC(\mathbf{x}, \mathbf{n}, i, j)$ is the normalized cross correlation of the projections of the element $(\mathbf{x}, \mathbf{n})$ in views i and j. We direct readers to [27] for more details on how to compute this term. The overall surface photo-consistency energy cost is then calculated by integrating the elements' costs over the surface S:

$$E_P(S) = \iint_S \mathbf{A}(\mathbf{x}, \mathbf{n}) dS. \tag{7}$$

Faugeras and Keriven [16] choose the surface S that minimizes Eq. (7) as the final reconstruction result. They use a gradient descent evolution to solve this problem. However, the integral in Eq. (7) is only calculated on surface S, and therefore during the evolution, the velocity $\mathbf{v}$ is derived from the unknown surface S. For this reason, the final result of [16] tends to be rather sensitive to where the surface S has started its evolution. Here we propose the photo-consistency energy term using a variational level set formulation:

$$E_P(\phi) = \iiint_\Omega \mathbf{A}(\mathbf{x}, \mathbf{n}) \delta(\phi) |\nabla \phi| dx dy dz, \tag{8}$$

where δ is the univariate Dirac function and the surface S is the zero level set of ϕ. By introducing the term $\delta(\phi)|\nabla\phi|$ [28] in the integrand, the integral in Eq. (8) can be done on the whole 3D space Ω. Since Eq. (8) is directly formulated in the level set domain, the minimization of it leads naturally to the evolution of ϕ. We use the Euler-Lagrange equation to solve this problem.

3.2 Computing Visibility

Summation in Eq. (5) is only computed for those elements which are visible to the two concerned views. Thus, estimating $\mathbf{A}(\mathbf{x}, \mathbf{n})$ requires the step of computing the visibility status of the elements for all views. We use an implicit ray tracing technique [29] to solve the visibility problem, and then Eq. (5) can be rewritten as

$$\mathbf{A}(\mathbf{x}, \mathbf{n}) = \frac{1}{T} \sum_{i,j=1}^{n} \chi(i, \mathbf{x}) \chi(j, \mathbf{x}) \mathbf{A}_{ij}(\mathbf{x}, \mathbf{n}), \tag{9}$$

where n is the number of all views and $\chi(v, \mathbf{x})$ is a characteristic function that denotes the visibility status of the element $\mathbf{x}$ from the viewpoint v:

$$\chi(v, \mathbf{x}) = \begin{cases} 1 & \text{if } \mathbf{x} \text{ is visible to the viewpoint } v, \\ 0 & \text{if } \mathbf{x} \text{ is invisible to the viewpoint } v. \end{cases} \quad (10)$$

3.3 Regularization Energy Term

It has been demonstrated in [6] that in the absence of texture, different scenes can be consistent with the same set of color images. Therefore, surface reconstruction based solely on photo-consistency is an ill-posed problem. To regularize it, we augment the energy functional with a regularization term:

$$E_R(\phi) = \iiint_{\Omega} \mathbf{B}(\mathbf{x})\delta(\phi)|\nabla\phi| dxdydz, \quad (11)$$

where $\mathbf{B}(\mathbf{x})$ is some volume potential corresponding to the prior tendency for point $\mathbf{x}$ to belong or not to the reconstruction. Here we construct $\mathbf{B}(\mathbf{x})$ based on the silhouette consistency constraint:

$$\mathbf{B}(\mathbf{x}) = \begin{cases} \rho & \text{if } \mathbf{x} \text{ projects to the foreground point for all images,} \\ +\infty & \text{others,} \end{cases} \quad (12)$$

where ρ is some small value which is kept constant throughout our experiments.

3.4 Penalizing Energy Term

The level set function can develop shocks, very sharp and/or flat shape during the evolution [28]. In order to maintain stable evolution and ensure usable results, it is crucial to keep the evolving level set function close to a signed distance function. Re-initialization, a technique for periodically re-initializing the level set function to a signed distance function, has been extensively used in traditional level set methods. However, from the practical viewpoints, the re-initialization process can be quite complicated, expensive, and have subtle side effects. Here, we use the integral [23]

$$P(\phi) = \iiint_{\Omega} \frac{1}{2}(|\nabla\phi| - 1)^2 dxdydz \quad (13)$$

as a metric to characterize how close a function ϕ is to a signed distance function in Ω. This metric plays a key role in our variational energy functional as a penalizing term. It will force the level set function ϕ to be close to a signed distance function in the evolution process, and the costly re-initialization procedure can be completely eliminated.

3.5 The Total Energy Functional

Combining all the energy terms Eqs. (8), (11) and (13) leads to the total energy functional for the surface reconstruction:

$$\begin{aligned} E(\phi) =& \alpha P(\phi) + \beta E_P(\phi) + E_R(\phi) \\ =& \alpha \iiint_\Omega \frac{1}{2}(|\nabla\phi| - 1)^2 dxdydz + \beta \iiint_\Omega \mathbf{A}(\mathbf{x}, \mathbf{n})\delta(\phi)|\nabla\phi| dxdydz \\ &+ \iiint_\Omega \mathbf{B}(\mathbf{x})\delta(\phi)|\nabla\phi| dxdydz. \end{aligned} \quad (14)$$

The second and third energy term drive the zero level set toward the object surface, while the first energy term penalizes the deviation of ϕ from a signed distance function during its evolution. All the terms in the final energy functional are weighted by constants α, β and 1.

3.6 Energy Minimization Framework

The first variation of the functional $E(\phi)$ in Eq. (14) can be given by

$$\frac{\partial E}{\partial \phi} = -\alpha(\Delta\phi - \kappa) - \beta\delta(\phi)(\nabla\mathbf{A}\cdot\mathbf{n} + \mathbf{A}\cdot\kappa) - \delta(\phi)(\nabla\mathbf{B}\cdot\mathbf{n} + \mathbf{B}\cdot\kappa), \quad (15)$$

where Δ is the Laplacian operator, κ and $\mathbf{n}$ are the mean curvature and unit normal of the object surface respectively, which is calculated by Eq. (3). Therefore, the function ϕ that minimizes this functional satisfies the Euler-Lagrange equation $\partial E/\partial\phi = 0$. The steepest descent process for minimization of the functional $E(\phi)$ is the following gradient flow:

$$\frac{\partial \phi}{\partial t} = \alpha(\Delta\phi - \kappa) + \beta\delta(\phi)(\nabla\mathbf{A}\cdot\mathbf{n} + \mathbf{A}\cdot\kappa) + \delta(\phi)(\nabla\mathbf{B}\cdot\mathbf{n} + \mathbf{B}\cdot\kappa). \quad (16)$$

According to Eq. (12), $\mathbf{B}(\mathbf{x})$ can be considered as a constant and satisfies $\nabla\mathbf{B} = 0$. Then Eq. (16) is rewritten as

$$\frac{\partial \phi}{\partial t} = \alpha(\Delta\phi - \kappa) + \beta\delta(\phi)(\nabla\mathbf{A}\cdot\mathbf{n} + \mathbf{A}\cdot\kappa) + \delta(\phi)\cdot\mathbf{B}\cdot\kappa. \quad (17)$$

This gradient flow is the evolution equation of the level set function ϕ and $\partial\phi/\partial t$ denotes derivative with respect to the evolution time t. The reconstruction result can be obtained from the zero level set of ϕ when this evolution is over.

4 Numerical Implementation

In order to implement Eq. (17) numerically, a uniform Cartesian grid described as $\{(x_i, y_j, z_k)|1 \leq i \leq m, 1 \leq j \leq n, 1 \leq k \leq l\}$ in a rectangular domain Ω_d in 3D space is used to discretize the computational domain Ω, and all of the functions and quantities are defined on this grid. Practically, a slightly smooth approximation of the Dirac function $\delta(\phi)$ in Eq. (17), rather than the traditional step form, is used [30]:

$$\delta_\varepsilon(\phi) = \frac{1}{\pi} \cdot \frac{\varepsilon}{\varepsilon^2 + \phi^2}, \tag{18}$$

where $\varepsilon > 0$ is a small constant. It gives a globally positive approximation to $\delta(\phi)$. Using this approximation, Eq. (17) will have the tendency to obtain a global minimum of $E(\phi)$. One of the reasons is that, the Euler-Lagrange equation acts only locally, on a few level surfaces around $\phi = 0$ using the original Dirac function [23,28]. While by Eq. (18), the evolution equation can act on all level sets, especially stronger on the zero level set, but not locally. In this way, our final evolution result is not sensitive to the initial value of ϕ.

Because of the penalizing term introduced by Eq. (14), we no longer need the upwind scheme [28] as in the traditional level set methods to implement the level set evolution. Instead, all the spatial partial derivatives $\partial\phi/\partial x$, $\partial\phi/\partial y$ and $\partial\phi/\partial z$ in Eq. (17) are approximated by the central difference, and the temporal partial derivative $\partial\phi/\partial t$ is approximated by the forward difference. Thus, the approximation of Eq. (17) can be simply rewritten as

$$\frac{\phi_{i,j,k}^{n+1} - \phi_{i,j,k}^{n}}{\Delta t} = f(\phi_{i,j,k}^{n}), \tag{19}$$

where Δt is the time step, $\phi_{i,j,k}^{n}$ is the value of ϕ on the grid node (x_i, y_j, z_k) at the time n and $f(\phi_{i,j,k}^{n})$ is the approximation of the right hand side in Eq. (17). Then, the evolution equation of ϕ can be expressed as

$$\phi_{i,j,k}^{n+1} = \phi_{i,j,k}^{n} + \Delta t \cdot f(\phi_{i,j,k}^{n}). \tag{20}$$

Once the initial level set function ϕ^0 has been given, we can constantly update ϕ through Eq. (20) until the evolution is over.

5 Experimental Results

In this section we discuss some surface reconstruction results obtained by our technique. The system used for all the experiments was a Windows-based Intel Core2 Duo with 2GB of RAM and running at 2.20GHz. We use the *Middlebury* datasets [21,31] as a calibrated sequence of input images. The data consists of two objects, a dinosaur and a temple (see Fig. 1a), and three different sets of input images for each one, with viewpoints forming a sparse ring, a full ring, and a full hemisphere around the object. Our experiments are performed on the *DinoSparseRing* and *TempleSparseRing* inputs. A list of parameters used in the experiments is shown in Table 1 and the initial zero level set function is taken as a sphere. Fig. 1(b), (c), and (d) illustrate some of the initial zero level sets (shown in blue) along with the reconstructed surfaces (shown in red). Our approach can successfully build the surfaces regardless of the initial level set function.

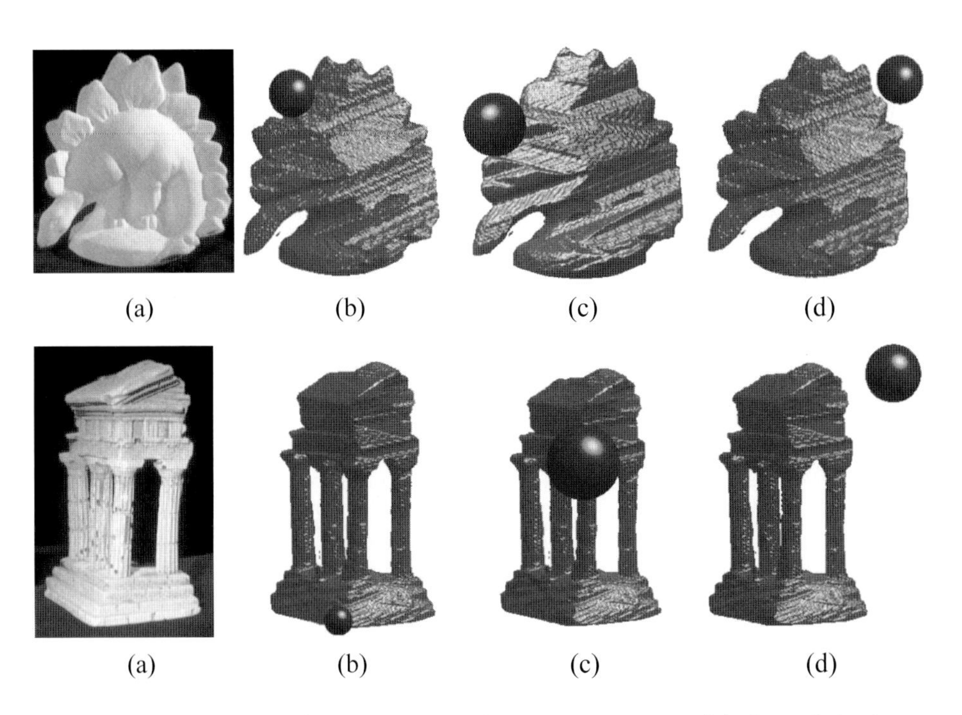

Fig. 1. Dino and temple reconstructions from 16 viewpoints. (a) One of the input images of the dino and temple model (640×480 resolution). (b,c,d) The reconstruction results of dino and temple (shown in red) with several different initializations (shown in blue).

Table 1. Summary of parameters used in the experiments

Dataset	Images	α	β	ρ	ε	Δt	Grid accuracy
DinoSparseRing	16	0.04	3.0	0.1	0.05	5.0	0.0005m
TempleSparseRing	16	0.01	5.0	0.1	0.03	3.0	0.0005m

Table 2. Numerical evaluation of the proposed method for the Middlebury Datasets

Dataset	Grid resolution	Vertices	Triangles	Iterations	Total time
DinoSparseRing	$151 \times 180 \times 152$	58737	117345	4	20mins
TempleSparseRing	$211 \times 326 \times 156$	164793	330977	6	35mins

In our experiments, the most time-consuming part is the photo-consistency cost calculation based on the NCC score. We use an NCC window size of 5×5 and it takes between 5 and 20 minutes to compute this cost for every time evolution depending on numbers of images used and grid resolution. Using a significant larger time step Δt, the final surface can be achieved only in 4 to 6 times evolution for a total time of 20 to 40 minutes. Table 2 shows the quantitative results of the reconstructed models.

6 Conclusions and Future Work

In this paper, we have presented a new variational formulation for reconstructing the 3D surface of an object from a sequence of calibrated images. In our method, the costly re-initialization procedure of the level set function is completely eliminated, and the evolution process of the level set function is easily and efficiently implemented by using very simple finite difference scheme. Our main advantage over other reconstruction methods relying on variational scheme is the insensitiveness to the initialization. Experiments have demonstrated the effectiveness of our approach. Future work will consist of improving the efficiency of the photo-consistency measure and the reconstruction results. Note that level set methods are used only for representing closed curves and surfaces. Therefore, our approach is not yet equipped to treat open surfaces. This limitation will also be dealt with in our future work.

Acknowledgements

This work was partially supported by the Natural Science Foundation of China (90920009) and the Chinese High-Tech Program (2009AA01Z323).

References

1. Scharstein, D., Szeliski, R.: A taxonomy and evaluation of dense two-frame stereo correspondence algorithms. International Journal of Computer Vision 47(1), 7–42 (2002)
2. Bradley, D., Boubekeur, T., Heidrich, W.: Accurate multi-view reconstruction using robust binocular stereo and surface meshing. In: IEEE Computer Society Conference on Computer Vision and Pattern Recognition (2008)
3. Goesele, M., Curless, B., Seitz, S.M.: Multi-view stereo revisited. In: IEEE Computer Society Conference on Computer Vision and Pattern Recognition, vol. 2, pp. 2402–2409 (2006)
4. Merrell, P., Akbarzadeh, A., Wang, L., Mordohai, P., Frahm, J.M., Yang, R., Nister, D., Pollefeys, M.: Real-time visibility-based fusion of depth maps. In: 11th IEEE International Conference on Computer Vision (2007)
5. Strecha, C., Fransens, R., Gool, L.V.: Combined depth and outlier estimation in multi-view stereo. In: IEEE Computer Society Conference on Computer Vision and Pattern Recognition, vol. 2, pp. 2394–2401 (2006)
6. Kutulakos, K.N., Seitz, S.M.: A theory of shape by space carving. International Journal of Computer Vision 38(3), 307–314 (2000)
7. Seitz, S.M., Dyer, C.R.: Photorealistic scene reconstruction by voxel coloring. In: IEEE Computer Society Conference on Computer Vision and Pattern Recognition, pp. 1067–1073 (1997)

8. Slabaugh, G., Culbertson, B., Malzbender, T., Stevens, M.: Methods for volumetric reconstruction of visual scenes. International Journal of Computer Vision 57(3), 179–199 (2004)
9. Zeng, G., Paris, S., Quan, L., Sillion, F.: Progressive surface reconstruction from images using a local prior. In: 10th IEEE International Conference on Computer Vision, vol. 2, pp. 1230–1237 (2005)
10. Kolmogorov, V., Zabih, R.: Multi-camera scene reconstruction via graph cuts. In: Heyden, A., Sparr, G., Nielsen, M., Johansen, P. (eds.) ECCV 2002. LNCS, vol. 2352, pp. 82–96. Springer, Heidelberg (2002)
11. Lempitsky, V., Boykov, Y., Ivanov, D.: Oriented visibility for multiview reconstruction. In: Leonardis, A., Bischof, H., Pinz, A. (eds.) ECCV 2006. LNCS, vol. 3953, pp. 226–238. Springer, Heidelberg (2006)
12. Sinha, S.N., Mordohai, P., Pollefeys, M.: Multi-view stereo via graph cuts on the dual of an adaptive tetrahedral mesh. In: 11th IEEE International Conference on Computer Vision (2007)
13. Sormann, M., Zach, C., Bauer, J., Karner, K., Bischof, H.: Watertight multi-view reconstruction based on volumetric graph-cuts. In: 15th Scandinavian Conference on Image Analysis, pp. 393–402 (2007)
14. Tran, S., Davis, L.: 3D surface reconstruction using graph cuts with surface constraints. In: Leonardis, A., Bischof, H., Pinz, A. (eds.) ECCV 2006. LNCS, vol. 3952, pp. 219–231. Springer, Heidelberg (2006)
15. Vogiatzis, G., Estean, C.H., Torr, P.H.S., Cipolla, R.: Multi-view stereo via volumetric graph-cuts and occlusion robust photo-consistency. IEEE Transactions on Pattern Analysis and Machine Intelligence 29(12), 2241–2246 (2007)
16. Faugeras, O., Keriven, R.: Variational principles, surface evolution, PDEs, level set methods, and the stereo problem. IEEE Transactions on Image Processing 7, 336–344 (1999)
17. Gargallo, M., Prados, E., Sturm, P.: Minimizing the reprojection error in surface reconstruction from images. In: 11th IEEE International Conference on Computer Vision (2007)
18. Keriven, R.: A variational framework for shape from contours. Tech. rep., CERMICS-2002-221b (2002)
19. Lhuillier, M., Quan, L.: Surface reconstruction by integrating 3D and 2D data of multiple views. In: 9th IEEE International Conference on Computer Vision, vol. 2, pp. 1313–1320 (2003)
20. Zaharescu, A., Boyer, E., Horaud, R.: TransforMesh: A topology-adaptive mesh-based approach to surface evolution. In: Yagi, Y., Kang, S.B., Kweon, I.S., Zha, H. (eds.) ACCV 2007, Part II. LNCS, vol. 4844, pp. 166–175. Springer, Heidelberg (2007)
21. Seitz, S.M., Curless., B., Diebel., J., Scharstein, D., Szeliski, R.: A comparison and evaluation of multi-view stereo reconstruction algorithms. In: IEEE Computer Society Conference on Computer Vision and Pattern Recognition, vol. 1, pp. 519–526 (2006)
22. Osher, S., Sethian, J.A.: Fronts propagating with curvature dependent speed: algorithms based on Hamilton-Jacobi formulations. Journal of Computational Physics 79(1), 12–49 (1988)
23. Li, C.M., Xu, C.Y., Gui, C.F., Fox, M.D.: Level set evolution without re-initialization: a new variational formulation. In: IEEE Computer Society Conference on Computer Vision and Pattern Recognition, vol. 1, pp. 430–436 (2005)

24. Ueshiba, T., Tomita, F.: Plane-based calibration algorithm for multi-camera systems via factorization of homography matrices. In: 9th IEEE International Conference on Computer Vision, vol. 2, pp. 966–973 (2003)
25. Unal, G., Yezzi, A.: A variational approach to problems in calibration of multiple cameras. In: IEEE Computer Society Conference on Computer Vision and Pattern Recognition, vol. 1, pp. 172–178 (2004)
26. Wang, L., Wu, F.C., Hu, Z.Y.: Multi-camera calibration with one-dimensional object under general motions. In: 11th IEEE International Conference on Computer Vision (2007)
27. Alejandro, T., Kang, S.B., Seitz, S.: Multi-view multi-exposure stereo. In: 3rd International Symposium on 3D Data Processing, Visualization, and Transmission, pp. 861–868 (2006)
28. Osher, S., Fedkiw, R.: Level set methods and dynamic implicit surfaces. Springer, Heidelberg (2002)
29. Tsai, R., Cheng, L.T., Burchard, P., Osher, S., Sapiro, G.: Dynamic visibility in an implicit framework. UCLA CAM Report 02-06 (2002)
30. Ari, R.B., Sochen, N.: Variational stereo vision with sharp discontinuities and occlusion handling. In: 11th IEEE International Conference on Computer Vision (2007)
31. The middlebury multi-view stereo page, http://vision.middlebury.edu/mview/

Layer-Constraint-Based Visibility for Volumetric Multi-view Reconstruction

Yumo Yang, Liuxin Zhang, and Yunde Jia

Beijing Laboratory of Intelligent Information Technology,
School of Computer Science, Beijing Institute of Technology,
Beijing 100081 P.R. China
{yangyumo,zhangliuxin,jiayunde}@bit.edu.cn

Abstract. Visibility estimation is one of the most difficult problems in multi-view reconstruction using volumetric approaches. In this paper, we present a novel approach called layer-constraint-based visibility (LCBV) to estimating visibility. Based on the layered state of a scene and photo-consistency constraint, this method can determine the more accurate visibility for every point in a scene. We use LCBV in multi-view reconstruction using volumetric graph cuts and obtain satisfactory results on both synthetic and real datasets. We also discuss quantitative error analysis to judge visibility techniques.

1 Introduction

The basic principle in volumetric multi-view reconstruction is to find a classification for all the voxels within a discrete volume whether they belong to the surface of the 3D object or not. There are many volumetric techniques like Space Carving [1], level set [2] and volumetric graph cuts [3,4,5,6]. These methods obtain a uniform 3D representation; in particular, graph cuts are able to find a global optimization [7].

Volumetric approaches usually use the photo-consistency measure of the voxel within the volume to evaluate how consistent would be the reconstructed surface at the voxel. The only requirement to compute the photo-consistency is that visibility is available. However, accurate visibility estimation is very difficult for reconstruction. To estimate the true visibility of some surface point, one needs to know the true scene geometry and vice versa. To solve this chicken-and-egg problem, previous methods compute approximation of visibility. Some papers [8,9] develop probabilistic formulations with the visibility reasoning to evaluate visibility. Hernandez et al. [8] present probabilistic visibility in which the visibility of a point depends only on the probabilistic depth measurements of sensors along optical rays that go through the point. This approach is both computationally and memory efficient, but it is not computed independently. There are also some simple and independent visibility techniques, such as state-based visibility [10] and oriented visibility [11] as shown in Fig. 1a and 1b respectively. State-based visibility uses the current estimate of the geometry to predict visibility for every point on that surface globally. In detail, a point is considered visible

S. Boll et al. (Eds.): MMM 2010, LNCS 5916, pp. 15–24, 2010.

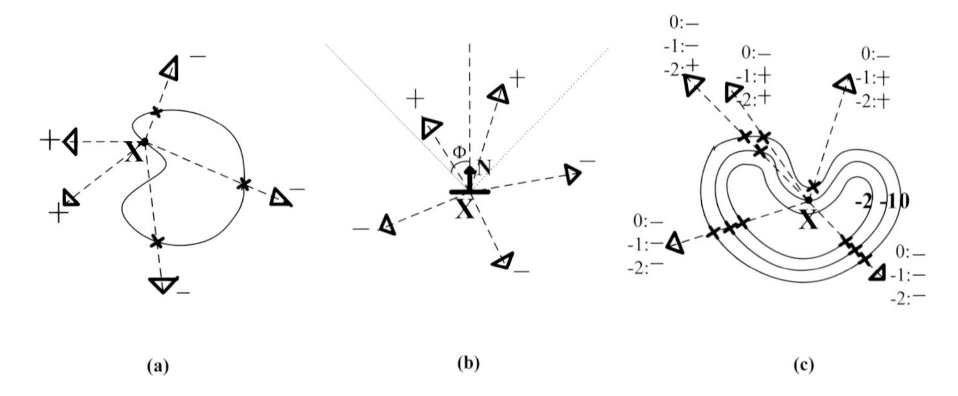

Fig. 1. Three approaches to visibility reasoning. (a) state-based visibility: the visibility for X on its surface is estimated globally. (b) oriented visibility: a local patch (X, N) is considered visible from the viewpoints within a predefined angle from the normal direction N. (c) layer-constraint-based visibility: the layered state and photo-consistency constraint are used to estimate the visibility information for all the voxels within the volume globally. Signs 0, -1, -2 denote layer 0, layer -1, and layer -2 respectively.

to a viewpoint if it is not occluded with current scene configuration. This global technique is often used in the approaches based on iterative process [1,2], which is not guaranteed to converge to the global minimum. Oriented visibility is a local algorithm which infers the visibility of a patch by using the patch position and the normal direction. Concretely, the patch is considered visible to a viewpoint if and only if it locates in front of this viewpoint according to a predefined angle from the normal direction. This method is simple and independent from initialization but, because of unknown the normal directions of the inner points, the accurate visibility for every point inside the volume is not determined. In many papers [4,5,6,12] with this technique, the visibility for the inner point is approximated with the visibility of the closest surface point to it.

In this paper, we propose a novel visibility approach called layer-constraint-based visibility (LCBV) as shown in Fig. 1c. Based on the layered state of the scene and photo-consistency constraint, LCBV can determine the more accurate visibility information for every voxel in the volume in volumetric multi-view reconstruction. Moreover, a quantitative error analysis is presented to compare the visibility methods in detail.

Our solution is inspired by the work of Zhang et al. [10] and Jonathan et al. [13]. Zhang et al. [10] used state-based visibility on the iterated surface which is regarded as the zero of level set. Jonathan et al. [13] divided the scene space into a set of depth layers and the visibility is derived by testing for occluders both on the depth layer and base surface, however, they did not reason visibility in detail. We combine both methods and divide visual hull [14] into several virtual layers like zero level sets. Then state-based visibility is extended to the virtual layers with photo-consistency constraint.

Contrast to the traditional visibility methods such as state-based visibility and oriented visibility, the main benefit of layer-constraint-based visibility is that the visibility information of the points within the volume can be determined accurately. Theoretically, we propose a quantitatively analysis on computing the visibility information for the points within the base surface, while the traditional methods only give a qualitative analysis on the visibility for the points within the base surface. For example, in the traditional visibility techniques, the visibility information for the inner points is approximated with the visibility of the closest surface point to it. In terms of application, the key advantage of LCBV is its ability to produce the better intermediate result to be used in multi-view reconstruction.

Intuitively, the accurate computation of the photo-consistency of the right point (the point is on the true surface) is more important than the computation of the photo-consistency of the wrong point (the point is not on the true surface) in reconstruction. LCBV can infer the more accurate visibility information for these right points to benefit producing a better intermediate result in multi-view reconstruction. According to LCBV, we also give an energy functional which is minimized by graph cuts.

The rest of the paper is organized as follows. Section 2 introduces our layer-constraint-based visibility approach. In Section 3, we describe an energy function which is minimized by graph cuts in multi-view reconstruction. Section 4 presents a significant error analysis and experimental results on synthetic dataset and produces several significant reconstruction results on both synthetic and real datasets. We discuss the paper's main contributions and the future work in Section 5.

2 Layer-Constraint-Based Visibility

In this section, we describe the layer-constraint-based visibility (LCBV) for evaluating the more accurate visibility information for the points within the volume in volumetric multi-view reconstruction. This algorithm is divided into two steps: one step is to develop a method called layered visibility (LV) for computing visibility roughly. The next is to use photo-consistency constraint to obtain the accurate visibility.

2.1 Layered Visibility for Volumetric Reconstruction

We assume that the whole scene is located within a bounding volume $\Omega \in \Re^3$. Let $I_1, \ldots, I_N$ denote the sequence of input calibrated images and $S_1, \ldots, S_N$ denote the foreground silhouettes obtained from these input images. The visual hull VH, the maximal shape consistent with $S_1, \ldots, S_N$, is considered as the initial space. Similar to the work [13], we divide visual hull VH space into a set of virtual layers $V = \{l_0, \ldots, l_D\}$. Each voxel in the virtual layers forms a node in an undirected graph and edges are defined in the graph to link adjacent voxels with a capacity corresponding to the energy function for minimization.

The first layer l_0 corresponds to an external set of voxels and the final layer l_D an internal set, each connected to a source s and sink t node respectively with infinite edge capacity as illustrated in Fig. 2a [13]. We compute LV on multiple layers and modify the function ϕ in [10] to define a new function $\varphi : \Re^3 \rightarrow \Re$ in Ω, which represents the scene layers as $[-D, 0]$ ($-D$ represents the D-th layer in visual hull VH space)of φ, with the following properties:

$$\begin{cases} \varphi(\mathbf{X}) > 0 & \text{for } \mathbf{X} \text{ outside } V \\ \varphi(\mathbf{X}) < -D & \text{for } \mathbf{X} \text{ inside } V \\ \varphi(\mathbf{X}) = 0 & \text{for } \mathbf{X} \in l_0 \subset V \\ \vdots & \quad \vdots \quad \vdots \\ \varphi(\mathbf{X}) = -D & \text{for } \mathbf{X} \in l_D \subset V \end{cases}$$

where φ is similar to a level set function; the layers in V are similar to zeros of level set function; $\mathbf{X}$ denotes a voxel in the bounding volume. The values of φ are easy to derive from layered state.

We use ray tracing [10] to multiple layers and determine in which layer the point in the volume is visible. Let $\mathbf{V_0}$ denote a viewpoint in Ω, then a point $\mathbf{X_1}(\mathbf{X_1} \neq \mathbf{V_0}, \mathbf{X_1} \in \Omega)$ is considered invisible to the viewpoint $\mathbf{V_0}$ for layer $l_i(i = 0, 1, \ldots, D)$ if it lies inside the layer l_i or if it is occluded by another point $\mathbf{X_2}(\mathbf{X_2} \in l_j, j \in [i, \ldots, D])$ in the radial direction $\overrightarrow{\mathbf{X_1 V_0}}$ (Fig. 2b). Under this condition, we shoot a ray towards the viewpoint, and check whether there is a point being inside the layer in this ray. We can determine visibility for a point simply by looking at the value of function φ. That is, $\mathbf{X}$ is inside the layers

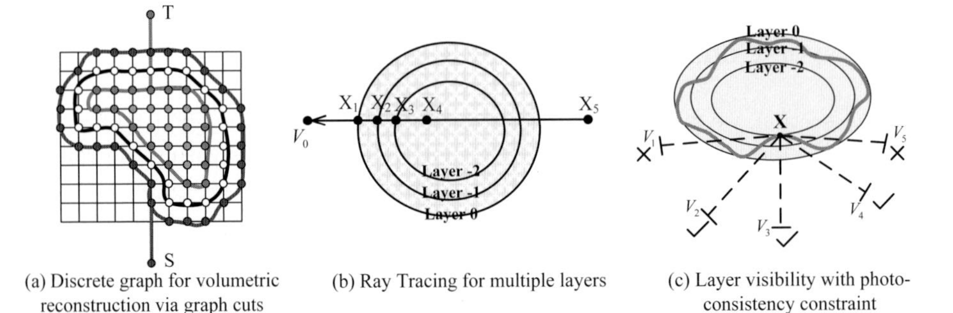

(a) Discrete graph for volumetric reconstruction via graph cuts
(b) Ray Tracing for multiple layers
(c) Layer visibility with photo-consistency constraint

Fig. 2. 3D reconstruction via volumetric graph cuts, ray tracing and photo-consistency constraint. (a) A discrete graph is constructed for volumetric reconstruction via graph cuts, and outer (blue line) and inner (red line) layers are connected to source S and sink T respectively. (b) A point is considered visible for some layer when it is not occluded by the same or inner layer, e.g. $\mathbf{X_1}$ is visible for layer 0; $\mathbf{X_2}$ is invisible for layer 0 but visible for layer -1; $\mathbf{X_3}$ is invisible for layer 0, -1 but visible for layer -2; $\mathbf{X_4}$ and $\mathbf{X_5}$ are invisible for three layers. (c) Red line denotes the true surface. According to layered visibility, point $\mathbf{X}$ is considered visible for v_1, v_2, v_3, v_4 and v_5, but in fact, $\mathbf{X}$ is invisible for v_1 and v_5. After imposing photo-consistency constraint, the last result is that $\mathbf{X}$ is visible for v_2, v_3 and v_4 (v_1 and v_5 are excluded).

when $\varphi(\mathbf{X}) < -D$, outside the layers when $\varphi(\mathbf{X}) > 0$, and on some layer when $\varphi(\mathbf{X}) = -i, i \in \{0, \ldots, D\}$.

2.2 Photo-Consistency Constraint

As discussed in Section 2.1, we approximately evaluate from which views the voxel on some layer is visible. In this part, we obtain the final accurate visibility by imposing photo-consistency constraint as shown in Fig. 2c. The photo-consistency criteria is not used in the traditional visibility techniques, but usually used in the optimization in multi-view reconstruction after computing the visibility. In this paper, we impose photo-consistency constraint on computing the visibility to improve the results. Basically, if a point is on the true surface, it has high photo-consistency. In contrast, if a point has high photo-consistency, it is not always on the true surface. For example, under weak texture, even points very far from the surface may have a strong photo-consistency. However, we mainly consider that accurately computing the visibility information of the right points (see Section 1), since the visibility information of the wrong points (see Section 1) has little influence on multi-view reconstruction.

In this paper, the normalized cross-correlation (NCC) is used as a photo-consistency metric. Let VC denotes the set of all views, VB denotes the set of the views for which the voxel is visible after computing layered visibility and VE denotes the set of the views for which the voxel is visible after using photo-consistency constraint. Obviously, we have $VB \subset VC, VE \subset VC$. Assume, to begin with, after computing layered visibility (see Section 2.1) we obtain the rough visibility information for the voxels and we have $VB = \{v_1, v_2, v_3, v_4, v_5\}$, v_1,v_2,v_3,v_4,v_5 $\in$ VC,VE $=$ Φ. Views v_1, v_2, v_3, v_4, v_5 array in order, and the middle view is chosen as reference view so that its retinal plane is close to parallel to the tangent surface through the voxel with little distortion. Then the closer to the reference the other views are, the more accurate they are.

We first choose the middle value v_3 in the set VB as reference view and turn left or right from v_3 to search a value, e.g. the first value v_2, each time alternately. If such the NCC computed by views v_2 and v_3 is above some threshold W, that means the voxel is invisible for view v_2 and the program is end ($VE = \Phi$). Otherwise, we put v_2 into the set $VE(VE = \{v_3, v_2\})$. Next, v_4 is searched out and the mean NCC computed by views v_2, v_3, v_4 is obtained again to determine whether v_4 is put into the set VE or not. Until all the values in the set VB are searched, we acquire a final accurate visibility set VE. The algorithm excludes inexact views in VB through imposing photo-consistency constraint and produces the more accurate set of views. The results of the synthetic experiment prove that layered visibility with photo-consistency constraint, namely layer-constraint-based visibility (LCBV), is more accurate than oriented visibility.

3 Volumetric Multi-view Reconstruction

Most of multi-view reconstruction methods relying on the traditional visibility techniques need a good initial scene surface to evaluate the visibility information for the points on or in the initial surface [4,5,6]. If the current scene state is far from the true state, e.g. concavity, the traditional visibility methods will approximate the true visibility with significant errors. Therefore, the traditional visibility algorithms are not thought to be good choice in the situation when no good initialization is given. In this section, we deduce an energy functional for multi-view reconstruction based on layer-constraint-based visibility. Compared with other methods, the main benefit of our reconstruction is that concavity can be better recovered.

3.1 Volumetric Graph Cuts

The energy functional is generally defined in terms of a data term E_{data} that imposes photo-consistency and a regularization term E_{smooth} introducing spatial smoothness [13]. Following Section 2.1, we obtain a layered configuration (see Section 2.1). In contrast with the photo-consistency constraint used in visibility (see Section 2.2), here a new photo-consistency metric $\rho(u)$ used in optimization is computed for each node u and edges are constructed between nodes (u, v) using capacities corresponding either to the data term or smoothness term of the cost function.

$$E = E_{data} + E_{smooth}. \tag{1}$$

Data edges are introduced between nodes in adjacent layers (l_i, l_{i+1}) using photo-consistency score at layer $-i$. Smoothing edges are introduced between adjacent nodes within each layer l_i using an average of the photo-consistency. Edge capacities are normalized by edge length and a relative weight $o < k < 1$ controls the degree of smoothing in reconstruction (similar to [13]),

$$E_{data}(u, v) = 1 - k\frac{\rho(u)}{\|x(u) - x(v)\|}, u \in l_i, v \in l_{i+1}, 0 < k < 1, \tag{2}$$

$$E_{smooth}(u, v) = 1 - k\frac{\rho(u) + \rho(v)}{2\|x(u) - x(v)\|}, u, v \in l_i, 0 < k < 1, \tag{3}$$

where $x(u), x(v)$ are the coordinates of node u and v respectively; the photo-consistency score for node u, $\rho(u)$ is the normalized cross correlation (NCC) between the pairs of local image patches that node u projects to in the different views:

$$\rho(u) = \sum_{C_i, C_j \in Vis(u)} NCC(p(C_i, u), p(C_j, u)), \tag{4}$$

where $Vis(u)$ notes the visibility for node u, and $Vis(u)$ is evaluated with layer-constraint-based visibility; C_i, C_j note two camera centers; $p(C_i, u)$ is the local image patch around the images of u in the i -th image I_i.

The global minimization of the energy functional corresponding to an optimal surface is derived from computing the max flow/min cut method [15].

4 Experimental Results

Because visibility results are just intermediate results in multi-view reconstruction and are difficult to be represented by graphic or image, to judge whether a visibility method is better or not is rather difficult. Most of the papers judge the visibility in the indirect way that the final reconstruction results determine whether the visibility technique is better or not. We propose a direct quantitative analysis on the visibility methods by the data form. We demonstrate the performance of our approaches on synthetic dataset, and give a significant error analysis (Table 1). We define two types of errors:

(1)E_{r_1} which denotes the wrong rate of set VE (Section 2) relative to set VT (real visibility set). If there exists a camera $X(X \in VE, X \notin VT)$, the set VE is considered wrong, namely the visibility for the point is wrong.

$$E_{r_1} = \frac{M_{VE}}{N_{VE}} \times 100\%, \tag{5}$$

where M_{VE} denotes the number of the points which have the wrong visibility in V, and N_{VE} denotes the number of all the points in V. Note that there is a problem about E_{r_1}, that is, if $VE \subset VT, VE \neq VT$, then VE is considered right, because of little influence on the photo-consistency, but not exact. Therefore another type of error is presented.

Table 1. Error analysis for the visibility. E_{r_1} denotes wrong rate of set VE(section 2) relative to VT . E_{r_2} denotes the similar degree of VE and VT. Theshold W is 0.3.

Error	E_{r_1}	E_{r_2}(viewpoint)
Oriented visibility	36.56%	4.177
Our method	26.76%	1.237

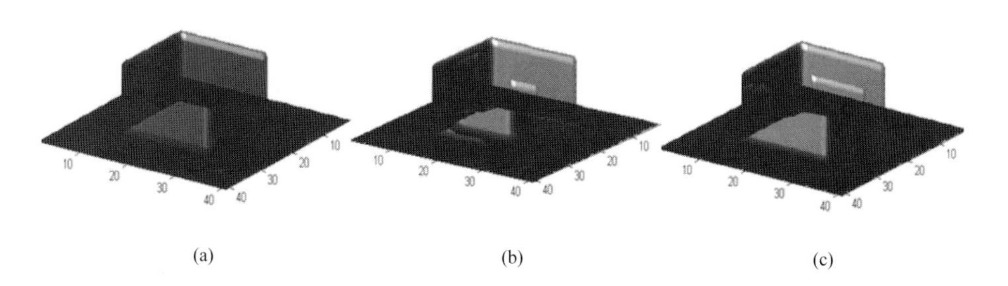

Fig. 3. A synthetic scene contains a non-closed surface with a large cubic concavity. (a) the true surface. (b) the reconstruction result using oriented visibility. (c) the reconstruction result using layer-constraint based visibility.

Table 2. Quantitative analysis on the accuracies for reconstructions using oriented visibility and our method. Accuracy and completeness are similar to the work in [16].

Reconstruction	Accuracy (mm)	Completeness
Oriented visibility	0.209996	70.29%
Our method	0.169690	78.84%

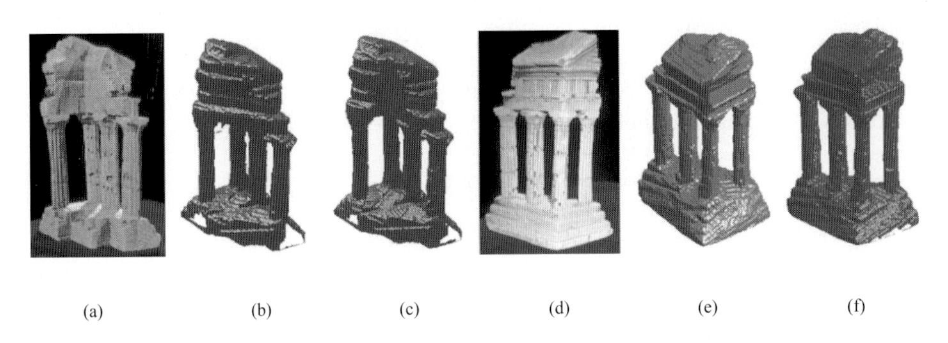

Fig. 4. Reconstruction for temple in Middlebury datasets. (a,d) the true temple images (16 views). (b,e) visual hull. (c,f) the reconstruction result using our method.

(2)E_{r_2} which denotes the similar degree of VE and VT, then

$$E_{r_2} = \frac{\sum |VT - VE|}{N}, \text{when} VE \subset VT, \tag{6}$$

where N denotes the number of investigated voxels. Viewpoint is used as unit of measure, e.g. E_{r_2}=1.900viewpoint, that means when $VE \subset VT$, the average number of the viewpoints which can see the point but not be computed right is 1.9.

Note that the synthetic scene is a non-closed surface with a large cubic concavity which makes it a challenging test (Fig. 3a). 21 viewpoints in line face to the concave surface. We compare the results of our method and oriented visibility to the true visibility (we can get it) respectively on the synthetic image. Table 1 quantitatively analyses the two types of errors on the visibility methods. Fig. 3b,3c show the 3D reconstruction results on the synthetic images using both the methods. Table 2 quantitatively analyses the accuracy and completeness of reconstructions on the synthetic image. The results present that our method is better than oriented visibility in multi-view reconstruction. Similar to the work in [16], we use an accuracy threshold of 80% and a completeness threshold of 0.1mm.

In Fig. 4, we apply our visibility algorithm to reconstruct to the temple (16 images) in Middlebury datasets [17] and acquire a promising result.

5 Conclusion and Future Work

In this paper, a novel visibility method called layer-constraint-based visibility (LCBV) for volumetric multi-view reconstruction has been proposed to determine the accurate visibility for every voxel within the volume. Using this method, an energy functional which can be minimized by graph cuts is presented. Moreover, quantitative error analysis between visibility approaches is first developed. Above are our main contributions. In the future work, we will modify the photo-consistency constraint and improve the results on real datasets. Our results in real datasets will be improved and submitted to Middlebury. Because we put emphasis on the study of the visibility, the reconstruction results via volumetric graph cuts are not perfect. We will also improve graph cuts optimization.

Acknowledgements

This work was partially supported by the Natural Science Foundation of China (90920009) and the Chinese High-Tech Program (2009AA01Z323).

References

1. Kutulakos, K.N., Seitz, S.M.: A theory of shape by space carving. International Journal of Computer Vision 38(3) (2000)
2. Faugeras, O., Keriven, R.: Variational principles, surface evolution, PDEs, level set methods, and the stereo problem. IEEE Transactions on Image Processing 7 (1998)
3. Zeng, G., Paris, S., Quan, L., Sillion, F.: Progressive surface reconstruction from images using a local prior. In: ICCV (2005)
4. Vogiatzis, G., Torr, P., Cipolla, R.: Mult-view stereo via volumetric graph-cuts. In: CVPR (2005)
5. Tran, S., Davis, L.: 3D surface reconstruction using graph cuts with surface constraints. In: Leonardis, A., Bischof, H., Pinz, A. (eds.) ECCV 2006. LNCS, vol. 3952, pp. 219–231. Springer, Heidelberg (2006)
6. Ladikos, A., Benhimane, S., Navab, N.: Multi-view reconstruction using narrow-band graph-cuts and surface normal optimization. In: BMVC (2008)
7. Kolmogorov, V., Zabih, R.: What energy functions can be minimized via graph cuts? In: Heyden, A., Sparr, G., Nielsen, M., Johansen, P. (eds.) ECCV 2002. LNCS, vol. 2352, pp. 65–81. Springer, Heidelberg (2002)
8. Hernandez, C., Vogiatzis, G., Cipolla, R.: Probabilistic visibility for multi-view stereo. In: CVPR (2007)
9. Boykov, Y., Lempitsky, V.: From photohulls to photoflux optimization. In: BMVC (2006)
10. Liuxin, Z., Yumo, Y., Yunde, J.: State-based visibility for 3D reconstruction from multiple views. In: ECCV Workshop (2008)
11. Lempitsky, V., Boykov, Y., Ivanov, D.: Oriented visibility for multiview reconstruction. In: Leonardis, A., Bischof, H., Pinz, A. (eds.) ECCV 2006. LNCS, vol. 3953, pp. 226–238. Springer, Heidelberg (2006)

12. Esteban, C., Schmitt, F.: Silhouette and stereo fusion for 3D object modeling. Computer Vision and Image Understanding 96(3) (2004)
13. Starck, J., Hilton, A., Miller, G.: Volumetric stereo with silhouette and feature constraints. In: BMVC (2006)
14. Laurentini, A.: The visual hull concept for silhouette-based image understanding. TPAMI 16(2) (1994)
15. Boykov, Y., Kolmogorov, V.: An experimental comparison of min-cut/max-flow algorithms for energy minimization in vision. PAMI 26(9) (2004)
16. Seitz, S., Curless, B., Diebel, J., Scharstein, D., Szeliski, R.: A comparison and evaluation of multi-view stereo reconstruction algorithms. In: CVPR (2006)
17. The middlebury multi-view stereo page, `http://vision.middlebury.edu/mview/`

Two Stages Stereo Dense Matching Algorithm for 3D Skin Micro-surface Reconstruction

Qian Zhang and TaegKeun Whangbo

Department of Computer Science, Kyungwon University,
Sujung-Gu, Songnam, Kyunggi-Do, Korea
aazhqg@hotmail.com,
tkwhangbo@kyungwon.ac.kr

Abstract. As usual, laser scanning and structured light projection represent the optical measurement technologies mostly employed for 3D digitizing of the human body surface. The disadvantage is higher costs of producing hardware components with more precision. This paper presents a solution to the problem of in vivo human skin micro-surface reconstruction based on stereo matching. Skin images are taken by camera with 90mm lens. Micro skin images show texture-full wrinkle and vein for feature detection, while they are lack of color and texture contrast for dense matching. To obtain accurate disparity map of skin image, the two stages stereo matching algorithm is proposed, which combines feature-based and region-based matching algorithm together. First stage a triangular mesh structure is defined as prior knowledge through feature-based sparse matching. Region-based dense matching is done in corresponding triangle pairs in second stage. We demonstrate our algorithm with active skin image data and evaluate the performance with pixel error of test images.

Keywords: Stereo vision, Dense matching, Feature-based, Region-based, Skin reconstruction.

1 Introduction

In medical analysis both image-based and modeling-based skin analysis, 3D skin model is very popular and flexible involving computer assisted diagnosis for dermatology, topical drug efficacy testing for the pharmaceutical industry, and quantitative product comparison for cosmetics. Quantitative features of skin surface are the significant but difficult task. In current methods, information has been limited to the visual data such as images, video, etc. The 3D features of skin can make the computer-based visual inspection more accurate and vivid. Our research focuses on human skin micro-surface reconstruction.

As we know, skin surface is a complex landscape influenced by view direction and illumination direction [1]. That means we can take skin surface as a type of texture, but this texture is strongly affected by the light and view direction, and even the same skin surface looks totally different. Some special hardware and techniques (e.g., photometric stereo, reflectance measurement setup) are required to support research. To reduce the impact of these characteristics and also to improve skin data accuracy,

S. Boll et al. (Eds.): MMM 2010, LNCS 5916, pp. 25–34, 2010.

micro skin images are considered in our 3D skin reconstruction. When we observe the skin surface from microcosmic point of view, light projected onto skin surface becomes on part of the skin color with tiny difference from narrow baseline stereo. We will analyze these influences in the paper.

Stereo dense matching between skin image pairs is another difficulty, when we recover skin shape through binocular stereo vision. Corner detection is feasible for feature selection in micro skin images with texture-full wrinkle and vein. It is possible to extract the basic skin structure called as grid texture which is generated by wrinkle and pores. In sparse matching, RANSAC algorithm [2] is a good tool to refine matched points. If we take these sparse matched points as correspondence seeds and find a semi-dense or dense map through seed-growing algorithm, there are many wrong matching because of absence of color and texture contrast in skin images. Even we set the tight constraints to seed-growing and only 2~3 times the number of seed data were matched. We couldn't find the dense map. In region based matching algorithm, energy function was designed and comparison window size was chosen to complete a dense matching through energy function minimization. Graph-cut optimization [3], dynamic programming [4], and region growing [5] are the famous algorithms in this filed. But these methods all rely on a disparity-based formulation. The max disparity will be known as a prerequisite. In our research, we combined the feature based and region based algorithm together to design a two stages matching algorithm to find a dense map. First stage a triangular mesh structure is defined as prior knowledge through feature-based sparse matching. Region-based dense matching is done in corresponding triangle pairs in second stage. It works better than only relying on feature based algorithm and only region based algorithm.

The remainder of this paper is organized as follows. After reviewing related work in Section 2, our matching algorithm is detailed in Section 3, and also the theoretical basis is explained in this section. Experiment results with vivo skin images and the evaluation are shown in Section 4. We conclude our research in Section 5.

2 Related Works

There are several 3D data acquisition techniques with satisfied performance for microscopic objects, small, medium and large objects. These techniques include laser scanning techniques, shape from stereo, and shape from video [6] or shading [7], and so on. Laser scanning techniques are based on a system with a laser source and an optical detector. The laser source emits light in the form of a line or a pattern on the objects surface and the optical detector detects this line or pattern on the objects. Through well-known triangulation algorithm the system is able to extract the geometry of the objects. The most important advantage of laser scanners is high accuracy in geometry measurement. Also it has some disadvantages. First geometry can be extracted without any texture information. Second the high cost of hardware components includes laser, the light sensor and the optical system. Third it is practically impossible to stay immobile for some seconds scanning, such as breath and wink. The technique shape from stereo is the extrapolation of as much geometry information as possible from only a pair of photographs taken from known angles and relative positions, which is the simulation of human visual system. Calibration is critical in terms of achieving accurate measurements. The method can either be fully automated or manually operated. Advantages of this method are the ability to capture both geometry and texture, the low cost and portability. A disadvantage of the method is its low resolution [8]. In our case, the image resolution of region of interest on the skin is 516*468, while on the same region there are only about 800 data obtained by VIVID 9i laser scanner. Even with high precision sparse stereo matching and camera calibration, we can deduce the difference significantly. In our research, we want to obtain more 3d data by dense matching.

3D reconstruction method based on multi-view stereo matching is widely applied for culture heritage, building and other scene reconstruction [9] [10] [11]. The user acquires the images by moving one camera or two cameras together around an object or scene. Many researchers have been interested in this problem and have proposed different processing pipeline, such as space carving [12], level sets [13], and a discrete labeling formulation through global optimization using graph-cut, belief propagation, dynamic programming, and so on. Fan [14] proposed a mesh-based coding method for surface modeling. They obtain a discrete disparity map using block-based hierarchical disparity estimation and then model the obtained disparity map using the Delaunay triangulation on a set of nodes. If the disparity map is not smooth enough, or if some disparity discontinuities exist, a large number of triangulation nodes must be placed very densely around the noise area to reduce the error. Quan [15] proposed a 3D surface reconstruction algorithm both with and without the epipolar constraint in which images play a symmetric role. The idea is based on seed grow contiguous component in disparity space. The greediness algorithm may cause a complete failure in the repetitive texture in the scene, such as our case, the skin texture images. Later, Jac ech [16] proposed a fast and accurate matching algorithm to find a semi-dense disparity map. Our motivation is from their researches to yield a dense matching. We construct sparse matching nodes for triangulation, and region based global optimization is applied for dense matching within each segmented region.

3 Matching Algorithm

3.1 Overview of Proposed Algorithm

We outline our stereo matching processing in figure 1 shown as follows. Our matching algorithm is suit for rectified image pairs. The rectification algorithm is referred to paper [17]. There are two stages included in our pipeline:

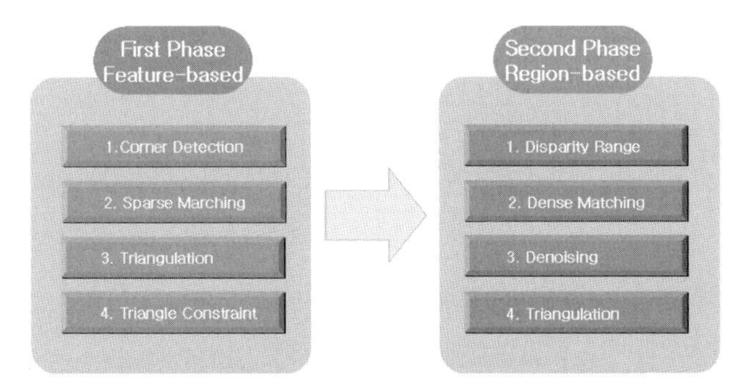

Fig. 1. Overview of the proposed stereo matching algorithm

First stage, triangle constraint is constructed through feature-based sparse matching. We take corners on the skin textures as feature, combined with image gradient, and epipolar constrains for sparse matching. After Delaunay triangulations, we found triangle constraint which is the preparation for dense matching.

Second stage, region-based dense matching is done after disparity formulation. In order to speed up matching process, we use a mask for each triangle region.

3.2 Theoretical Basis

This part refers to paper [18]. Suppose we have a continuous non self-occluding surface. To preserves the topology, each image must have the same topology as the surface does. The disparity gradient is defined as $\frac{|r_L - r_R|}{1/2|r_L + r_R|}$. Vector r_L is in left image, and its corresponding vector r_R is in right image. In the paper, there is a poof that a disparity gradient limits of less than 2 implies that the matches between the two images preserve the topology of the surface.

Suppose p_i, p_j are matched points, and q_i, q_j are another matched points. d_p and d_q are disparity of p and q respectively. $dis_{p_j q_j}$ is the distance between p and q in the second image. Illustration is shown in figure 2.

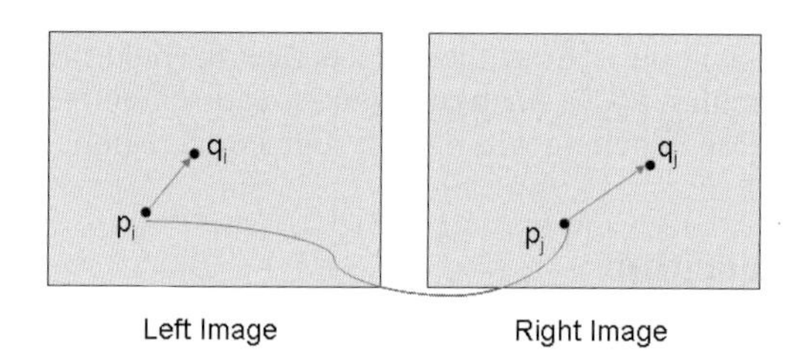

Fig. 2. Illustration of disparity gradient limits in which pixels p and q are in left image and right image

The disparity gradient is defined as follows:

$$\begin{aligned} G &= \frac{|(p_i - q_i) - (p_j - q_j)|}{1/2|(p_i - q_i) + (p_j - q_j)|} \\ &= \frac{|(p_i - p_j) - (q_i - q_j)|}{1/2|(p_i - p_j) - (q_i - q_j) + 2(p_j - q_j)|} \\ &= \frac{|d_p - d_q|}{|1/2(d_p - d_q) + dis_{p_j q_j}|} \end{aligned} \tag{1}$$

Because of the disparity gradient limit, let's suppose $G \le k < 2$ [17], then

$$\begin{aligned} |d_p - d_q| &\le k|1/2(d_p - d_q) + dis_{p_j q_j}| \\ &\le k/2|d_p - d_q| + k \cdot dis_{p_j q_j} \end{aligned} \tag{2}$$

Then, formulation (2) can be simplified

$$\left|d_p - d_q\right| \le \frac{2k}{2-k} dis_{p_j q_j} \tag{3}$$

So the disparity difference between two pixels in the image that taken from a non-occluding surface has the relationship with the distance of the two corresponding points. That means we can estimate the neighboring pixel's disparity through the known-disparity.

3.3 First Stage: Feature-Based Sparse Matching for Triangulation

The inputs to corner detection consist of a pair of rectified stereo images. In skin image, objects are pores and wrinkles which cross each other and form the basic skin structure called grid texture. The feature between two different view images is the basic skin grid. In the matching part, our algorithm for matching point is based on the corner detection. Harris detector was used in the corner detection step and RANSAC algorithm to compute correspondence set to estimate the 2D homography and the (inlier) correspondences. We summarize the processing as follows:

1. Corner detection.
2. Choose window size w for correlation matching.
3. Look for the candidate matching points along the corresponding epipolar line.
4. Calculate the similarity [19] between the candidate matching points. Select the matched points.

With this strategy, only prominent corners in texture-rich areas are extracted. By parameter setting for Harris detector, we can control the density of triangulation. In order to improve the matching accuracy, we can change the searching area size. Parameter w is closer maximum disparity in global images, and accuracy is higher.

A Delaunay triangulation helps us to build the triangular map. It is similar to image segmentation as a prior for stereo matching algorithm. In our work, skin images are taken under narrow base line system considering light influence. Compared with the three triangle vertices' disparities in each region, disparity range belongs to $\{0, \max(V_1, V_2, V_3)\}$ for each to be matched triangle reasonably.

3.4 Second Stage: Region-Based Dense Matching

Along corresponding epipolar line, regional correlation stereo matching will be developed. We define an error energy function for region correlation criterion.

$$Error(d) = \frac{1}{9}\sum_{y=1}^{3}\sum_{c=1}^{3}(l(x, y+d, c) - r(x, y, c))^2 \tag{4}$$

Where, d is the disparity which is in disparity range from *0* to *Max*, c means the RGB color format and its value taken of *{1, 2, 3}*, and *l*, *r* represent the left image and right image. We have known the disparity of three vertexes in each matching area. Set the matched points as the seed. If the error is less than the threshold, associate the point into the region, and then choose the next points. If none points was selected, then give it up for next matching region.

4 Experiments

4.1 Skin Images Acquisition

The taken distance of test images is controlled in 15~20cm for clear skin pictures with surface texture details. In the close distance, long focus lens is necessary. For controlling two cameras to take photos at the same time, computer is necessary with camera SDK improvement. Our system is constituted by two Cannon 30D cameras with Tamron long focus lens 90mm. It performs on the personal computer with Intel Core 2Duo 2.66GHz CPU and 2G memory.

4.2 Experiment Results and Evaluation

The skin image pairs from human calf skin are shown in figure 3 with resolution 509*386. The real skin area is about 4mm*3mm. The biggest disparity value of pixel is 72. After sparse matching, triangulation is done in first stage. The result is shown in figure 4 with rectified images, in which red little circles on the two images represent the corresponding points after sparse matching, and number of matched points is 276. After we do triangulation, the triangle number is 201. Then disparity map is shown in figure 5. Image 5.a is result with our proposed algorithm, and image 5.b is the result with region based algorithm. Both of them are original data without filtering. There are many noises in 5.b, and disparity 5.a is the smoothing of 5.b. Total running time of matching is 102.70 second. The total number of matched points is 108,287. Suppose camera intrinsic matrix is 3×3 identity matrix, and we obtain 3D data points by self-calibration processing shown in figure 6 with a box corner and texture mapping effect.

Sparse matching result is related the system's precision. Table 1 lists the sparse matching results with the increase of detected corners. The accuracy of dense matching compared with manual measurement result is shown in table 2. We select sparse matching points according to figure 4.a manually. For dense matching, we select points in each triangle region randomly and find their matching points manually. Here number of 500 and 1000 points are considered respectively.

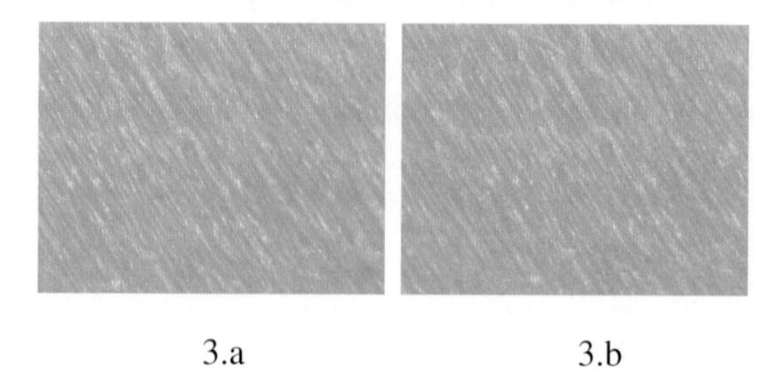

3.a 3.b

Fig. 3. Calf skin. Image 3.a is from left camera and image 3.b is from right camera.

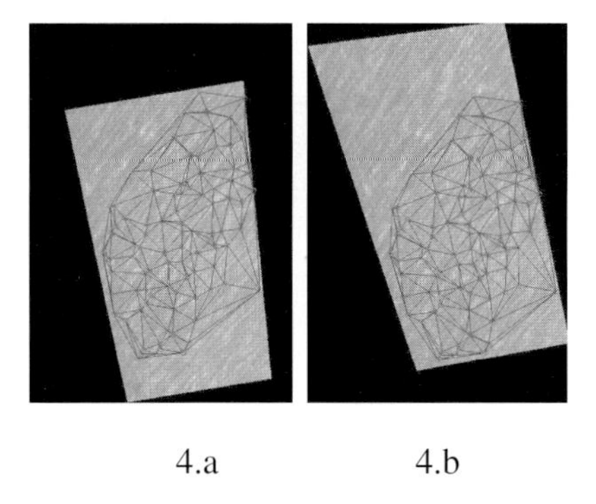

4.a 4.b

Fig. 4. Feature-based matching and triangulation result

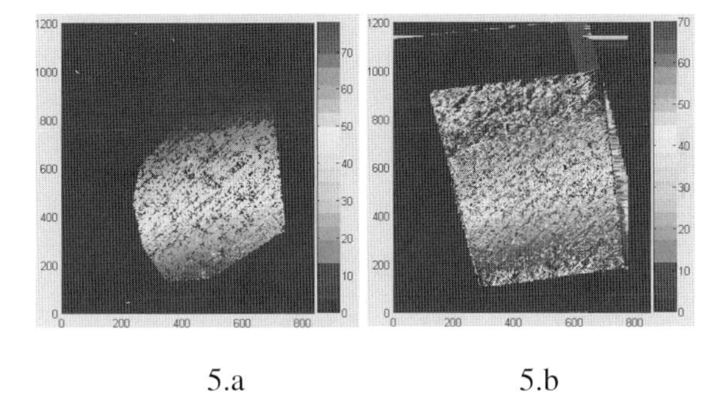

5.a 5.b

Fig. 5. Disparity map of calf skin images with proposed algorithm and region based algorithm respectively

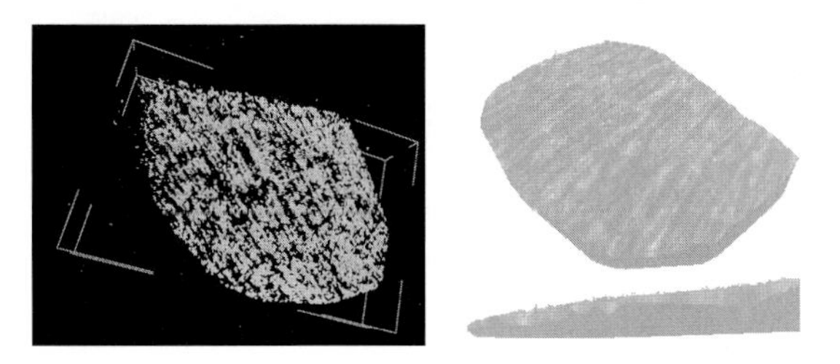

Fig. 6. 3D calf skin data and its visual effect after texture mapping

Table 1. The accuracy comparison of sparse matching in terms of pixel

Numberof corners	Number of matching points	Number of wrong matching points	Accuracy
886	171	0	100%
1340	276	0	100%
2476	480	5	98.96%

Table 2. The evaluation of the proposed algorithm in terms of pixel

Matching stage		Our algorithm		Region growing algorithm	
		Mean error	Max error	Mean error	Max error
Sparse matching: 276 matching points		0.48	1.25		
Dense matching	500 matching points	0.74	4.30	2.23	20.75
	1000 matching points	0.83	6.20	2.46	25.50

We show another stereo matching result in figure 7, in which pairs of image are taken from human thumb back skin. Image resolution is 516*468. The number of sparse matched points is 28. After we do triangulation, the triangle number is 43. Limited by the camera lens performance, some part of the thumb image is blurred. Our algorithm works in the clear region to keep the high accuracy.

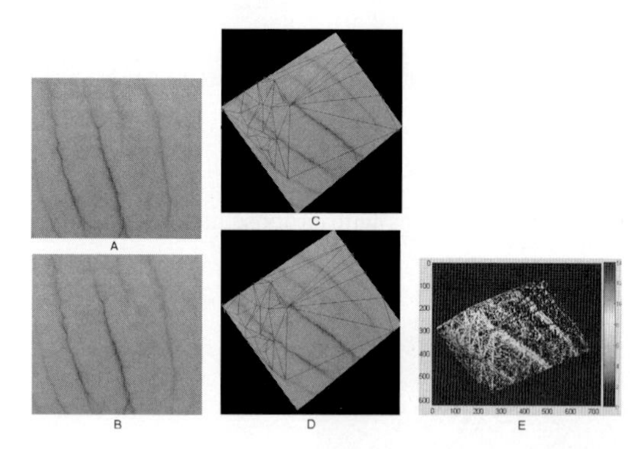

Fig. 7. Thumb skin Image. A, B on the left are original images. C, D in the middle are feature-based matching and triangulation result respectively. E is the finial disparity map.

5 Conclusion and Future Works

The human skin has the own specification properties, such as non-rigid object, micro skin texture structure. In the paper, we proposed a solution for micro skin surface reconstruction without 3D scanner. The solution for 3D reconstruction comes from the computer vision camera model and projective model. In order to reduce the light influence, we take images under narrow base line system. The two stages matching algorithm combined feature-based and region-based matching method has proposed for dense matching with higher accuracy. Considering disparity gradient limitation, we add the disparity constraints to each matching triangle region reasonably.

There are still some problems in our research. Skin wrinkle and texture with non-rigid property, we couldn't obtain the same skin state by time changing. For texture not well distributed skin, the triangular segmentation can be refined and output more matched points. Next step, to improve the performance, we will add more skin images for multi-view reconstruction.

Acknowledgement

This research was supported by the Ministry of Culture & Tourism and Korea Culture & Content Agency, Seoul, Korea, under Supporting Project of Culture Technology Institute.

References

1. Cula, O.-G., Dana, K.-J., Murphy, F.-P., Rao, B.-K.: Skin Texture Modeling. International Journal of Computer Vision 62(1-2), 97–119 (2005)
2. Hartley, R., Zisserman, A.: Multiple View Geometry in Computer Vision. Cambridge University Press, New York (2000)
3. Kolmogorov, V., Zabih, R.: Multi-camera Scene Reconstruction via Graph Cuts. In: Heyden, A., Sparr, G., Nielsen, M., Johansen, P. (eds.) ECCV 2002. LNCS, vol. 2352, pp. 82–96. Springer, Heidelberg (2002)
4. Veksler, O.: Stereo Correspondence by Dynamic Programming on a Tree. In: IEEE Conference on Computer Vision and Pattern Recognition, vol. 2, pp. 384–390. IEEE Press, New York (2005)
5. Wei, Y., Quan, L.: Region-Based Progressive Stereo Matching. In: IEEE Conference on Computer Vision and Pattern Recognition, vol. 1, pp. 106–113. IEEE Press, New York (2004)
6. Chiuso, A., Jin, H., Favaro, P., Soatto, S.: MFm: 3-D Motion and Structure from 2-D Motion Causally Integrated over Time Implementation. In: Vernon, D. (ed.) ECCV 2000. LNCS, vol. 1843, pp. 734–750. Springer, Heidelberg (2000)
7. Zhang, R., Tsai, P.-S., Cryer, J.-E., Shah, M.: Shape from Shading: a Survey. IEEE Transactions on Pattern Analysis and Machine Intelligence 21(8), 690–706 (1999)
8. Pavlidis, G., Koutsoudis, A., Arnaoutoglou, F., Tsioukas, V., Chamzas, C.: Methods for 3D Digitization of Cultural Heritage. Journal of Cultural Heritage (8), 93–98 (2007)
9. Pollefeys, M., Gool, L.-V., Vergauwen, M., Verbiest, F., Cornelis, K., Tops, J., Koch, R.: Visual Modeling with a Hand-Held Camera. International Journal of Computer Vision 59(3), 207–232 (2004)
10. Michael, G., Noah, S., Brian, C., Hugues, H., Steven, M.-S.: Multi-View Stereo for Community Photo Collections. In: Proceedings of ICCV, pp. 1–8 (2007)
11. Pollefeys, M., Nister, D., Frahm, J.-M., Akbarzadeh, A., Mordohai, P., Clipp, B., Engels, C., Gallup, D., Kim, S.-J., Merrell, P., Salmi, C., Sinha, S., Talton, B., Wang, L., Yang, Q., Stewenius, H., Yang, R., Welch, G., Towles, H.: Detailed Real-Time Urban 3D Reconstruction From Video. International Journal of Computer Vision 78(2), 143–167 (2008)
12. Kutulakos, K., Seitz, S.: A Theory of Shape by Space Carving. In: Proceedings of ICCV, pp. 307–314 (1999)

13. Faugeras, O., Keriven, R.: Variational Principles, Surface Evolution, PDE's, Level Set Methods and The Stereo Problem. IEEE Transactions on Image Processing 7(3), 336–344 (1998)
14. Fan, H., Ngan, K.-N.: Disparity Map Coding Based on Adaptive Triangular Surface Modeling. Signal Process.: Image Commun. 14(2), 119–130 (1998)
15. Lhuillier, M., Quan, L.: Matching Propagation for Image-Based Modeling and Rendering. IEEE Transactions on PAMI 27(3), 1140–1146 (2002)
16. Cech, J., Sara, R.: Efficient Sampling of Disparity Space for Fast and Accurate Matching. In: IEEE Conference on Computer Vision and Pattern Recognition, pp. 1–8 (2007)
17. Pollefeys, M., Koch, R., Gool, L.-V.: A Simple and Efficient Rectification Method for General Motion. In: Proceedings of ICCV, pp. 496–501 (1999)
18. Trivedi, H.-P., Lloyd, S.-A.: The Role of Disparity Gradient in Stereo Vision. Perception 14(6), 685–690 (1985)
19. Aschwanden, P., Guggenbuhl, W.: Experimental Results from A Comparative Study on Correlation Type Registration Algorithms. In: Forstner, Ruwiedel (eds.) Robust computer vision, pp. 268–282 (1992)

Safe Polyhedral Visual Hulls

Guojun Chen[1,2], Huanhuan Su[1], Jie Jiang[1], and Wei Wu[2]

[1] College of Computer and Communication Engineering,
China University of Petroleum, Dongying, China
[2] State Key Laboratory of Virtual Reality Technology and Systems,
Beihang University, Beijing, China
cgj@vrlab.buaa.edu.cn

Abstract. The visual hull is used to 3D object modeling in real-time. In this paper, we propose a novel method to remove phantom volumes produced in the polyhedral visual hull (PVH) construction. This method calculates the visual hull by using polygon intersection. By making full use of the information PVH supplied for image planes, we finally obtain the safe polyhedral visual hull (SPVH). Unlike the method based image method, according to the construction process of the PVH, this paper presents a fast ray-volume intersection based on geometry method. Experiment results show that, without adding more cameras, this method can provide view-independent SPVH model with phantom volumes removed, which can be used in the 3D modeling and interactive system.

Keywords: Polyhedral visual hulls, safe zones, polyhedral division, Shape-From-Silhouette.

1 Introduction

Recovering shapes of three dimension object from acquired real images with geometric topology and texture information in complicate scene has long been an important research subject in computer vision, virtual reality and computer simulation. Several methods have been proposed to solve this problem, among which the method of Visual Hull (VH) or Shape-From-Silhouette (SFS) shows the best approximate shape of an object according to the silhouettes of images acquired with multi cameras in interactive rate. Silhouette was first considered by Baumgart [1] who computed the polyhedral shape approximations by intersecting the silhouette cones. The geometric concept of visual hull was later defined by Laurentini [2], and has been widely used in some modeling and interactive systems during recent years.

There are many algorithms to compute the visual hull, which can be generally separated into two categories: volume based approaches and surfaced based approaches. The former was based on space discrete voxel cells to compute the approximation of visual hull, but with a poor precision to complexity trade-off. The latter was then proposed to solve the problem. The image-based visual hull (IBVH) was presented to render view-dependent image, but with no intact model [3]. The exact view-dependent visual hull (EVDVH) improved the precision of model but with

S. Boll et al. (Eds.): MMM 2010, LNCS 5916, pp. 35–44, 2010.

the same problem [4]. The polyhedral visual hull produced a polygon surface model of visual hull, a view-independent approximation, with good quality and speed [5]. Later Franco proposed a method with more precision and robustness but less stability and more complexity [6].

Almost all the visual hull construction algorithms used volume intersection techniques to construct objects by silhouettes which reflected shape information of objects. Nevertheless, when there are complicated objects with self-occlusion, or multi objects with inter-occlusion, the silhouettes only contain part of information of objects in some scenes, the reconstructed visual hull may exhibit phantom volumes, which are part of intersection of visual cones but do not represent objects in the scene. In this paper, we propose a new method to solve the inaccuracy problem above.

Adding more cameras was one of the simplest methods to remove phantom volumes; however, in virtue of the restricted placement of cameras, this is not a guaranteed solution [7], [8]. A method with temporal filtering was proposed to remove phantom volumes by a threshold on size, however, the small real objects would be wrongly removed by the lower threshold value, furthermore, this method does not work well when it is in the self-occlusion situation [9]. In most of the free viewpoints video synthesize based on visual hull method, consistent color or texture information were used in the removal of phantom volumes by color interpretation, however, it is useless for the scene that contains repetitive texture or similar color across a region [10].

The algorithms above could not completely remove phantom volumes in visual hull. The method in this paper follows the recent work which can remove all phantom volumes but cannot produce explicit model based on the EVDVH method [11], [12]. Our approach solves the problem on the basis of the PVH method, produces an accurate model called safe polyhedral visual hull (SPVH), with phantom volumes be removed.

This paper is organized as follows. Section 2 outlines the PVH. Section 3 describes the SPVH, followed by the implementation and experiment results in Section 4. Finally, Section 5 gives the conclusion and future work.

2 Polyhedral Visual Hull

2.1 Visual Hull

Visual hull is the maximal approximation of the surface shape of objects by the intersection of visual cones technology using infinite cameras. In fact, we can only compute visual hull with the limited numbers of images captured from a definite amount of cameras.

Suppose that there are N cameras positioned around the object O in the scene. They are calibrated with C_i being the center of camera i respectively, $\prod_i:R^3\rightarrow R^2$ represents the perspective projection used to map 3D coordinates to image coordinates and the inverse projection $\prod_i^{-1}:R^2\rightarrow R^3$ is used to map image coordinates to 3D coordinates (here result of inverse projection is on the specified plane). Let I_i and S_i denote the image and corresponding silhouette of the object O obtained from camera i. For C_i, the volume VH_i is defined as all the rays starting at C_i and passing through interior points on S_i, thus the visual hull of the object O is expressed as equation (1).

$$VH = \bigcap_{i=1}^{N} VH_i \quad . \tag{1}$$

that is, the intersection of visual cones. Fig. 1 shows the 2D situation of the visual hull with three cameras.

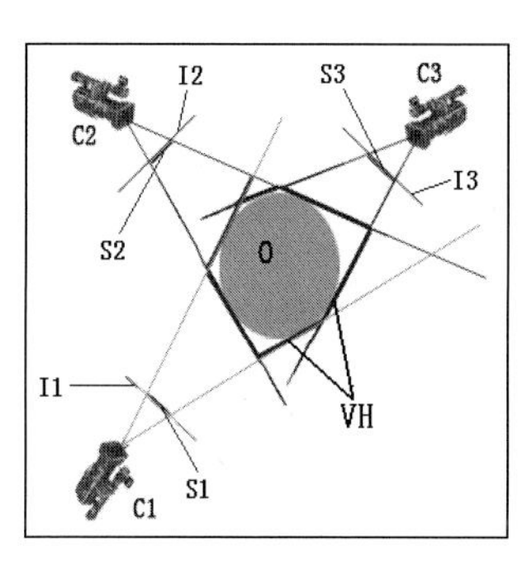

Fig. 1. Visual Hull (bold lines)

2.2 PVH

In fact, the boundary points of silhouettes define the shape of VH absolutely. Thus, we can describe the boundary of silhouettes B_i as approximate polygon which contains M vertices and edges E_{ij}. Each edge is defined by two neighborhood points of B_i from camera i, then the viewing cone VH_i is expressed as generalized cone whose viewpoint is C_i and whose faces F_{ij} is defined by C_i and E_{ij}. Accordingly, a simple algorithm to compute the visual hull is based on computing the intersection of F_{ij} with all the other N-1 cones. Then the PVH is a surface model being made up of multiple polygons. Fig. 2(a) shows the cube and (b) shows the PVH.

The processing of F_{ij} is as follows:

First, project F_{ij} to the image plane I_k and intersect the project result with B_k from camera k to obtain p_k^2.

Second, back-project p_k^2 to F_{ij}, denote the result as p_k^3, then perform the similar intersection with other cones.

Third, compute the intersection FP_{ij} of polygons which are back-projected from image plane to F_{ij}.

Now, we get one face of PVH. The entire process is formulated as equation (2).

$$\begin{aligned} p_k^2 &= \Pi_k(F_{ij}) \cap B_k \\ p_k^3 &= \Pi_k^{-1}(p_k^2) \\ FP_{ij} &= \bigcap_{k=1,k\neq i}^{N} (p_k^3) \\ PVH &= \bigcup_{i=1}^{N} \bigcup_{j=1}^{M} FP_{ij} \end{aligned} \quad . \tag{2}$$

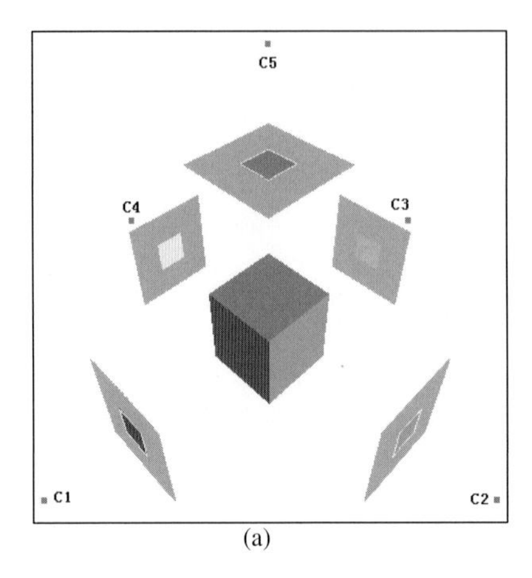

(a)

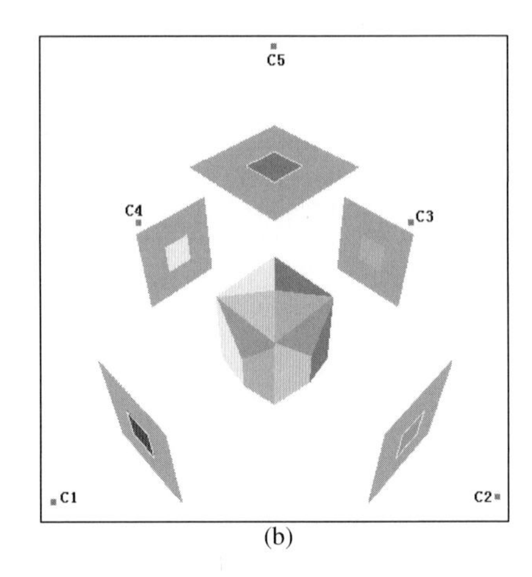

(b)

Fig. 2. (a) shows the cube, (b) shows the PVH

3 Safe Polyhedral Visual Hull

This section presents a novel method to remove phantom volumes in PVH, only using geometric information.

The algorithm contains two main steps: First, according to the information the PVH supplied for image planes, we divide silhouettes into two types of zones, safe zone and unsafe zone, correspondingly, we divide the viewing cone into two parts, one only contains real object, called safe cone, the other may contain phantom volumes. Second, SPVH is constructed, only by reserving the part of PVH lying in safe cone from at least one camera. Fig. 3 shows the processing.

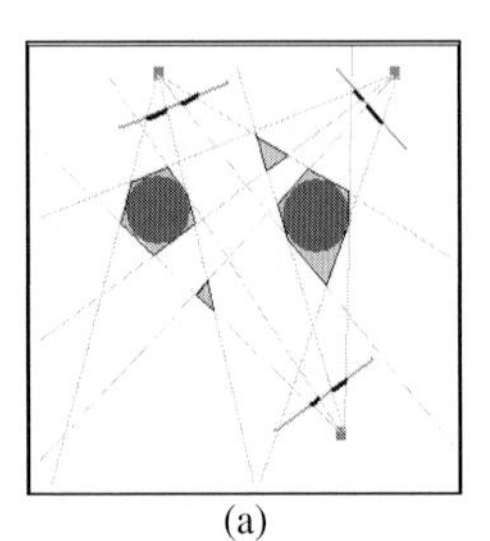
(a)

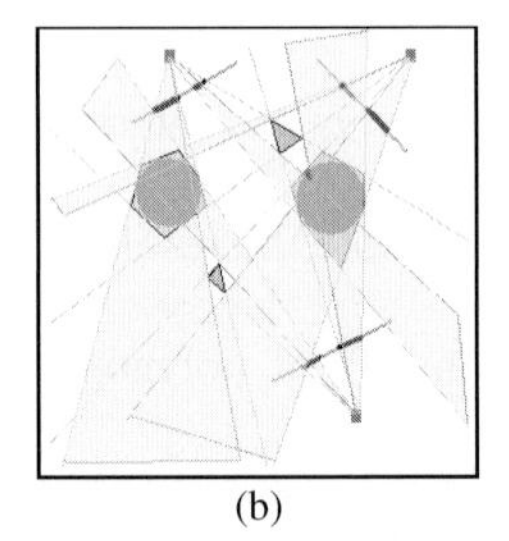
(b)

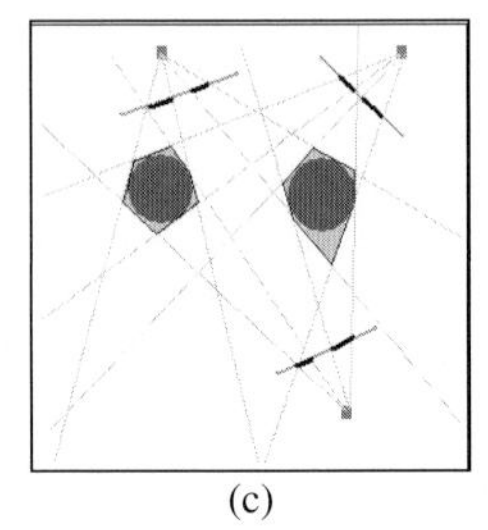
(c)

Fig. 3. There are multiple objects in the scene (the purple circles). Modeling result is shown in (a) with blue shapes, real objects don't contain phantom volumes (the blue regions where do not contain real objects). The partition result is the yellow regions in (b). (c) shows the SPVH reconstruction with phantom volumes removed.

3.1 Divide Silhouettes

To partition the silhouettes into two parts, we project the polygons of the PVH model to the image plane, since PVH is a closed solid, the region in silhouette contains at least two polygons. Thus we define the safe zone $SZone_i$ from camera i as the regions which contain only two polygons, and the remainder regions in silhouette are considered as unsafe zone $NSZone_i$.

In fact, all the regions are made of pixels in silhouette, so we can intersect the ray starting viewpoint and passing through every pixel with every polygon of the PVH model, then calculate the number of corresponding polygons to judge whether it belongs to the safe zone. To simplify the computation, we proposed a novel method to calculate ray-volume intersection with the process of the PVH construction taken into account.

Take the silhouette S_i for example. First, obtain every pixel p_{ij} in S_i; Second, project the ray starting C_i and passing through p_{ij} to the other image plane k and store the edges crossed in O_k; Third, intersect the ray with corresponding faces to the stored edges, calculate the intersection polygons. Therefore, only minorities of polygons are used to intersect with every ray, and the majority was eliminated. Fig. 4 describes the process.

According to plane division, the viewing cone of camera i was divided into safe cone $SCone_i$ and unsafe cone $NSCone_i$. Fig. 5(a) shows two cubes in the scene; and (b) shows the PVH model with phantom volumes (the small blocks); then with the information the PVH provides, we divide the silhouettes into safe zone and unsafe zone respectively, shown in (c), the green region is safe, and the pink is unsafe; relevant division in cone is shown in (d), the blue cones are the unsafe cones, their intersection is the phantom volumes.

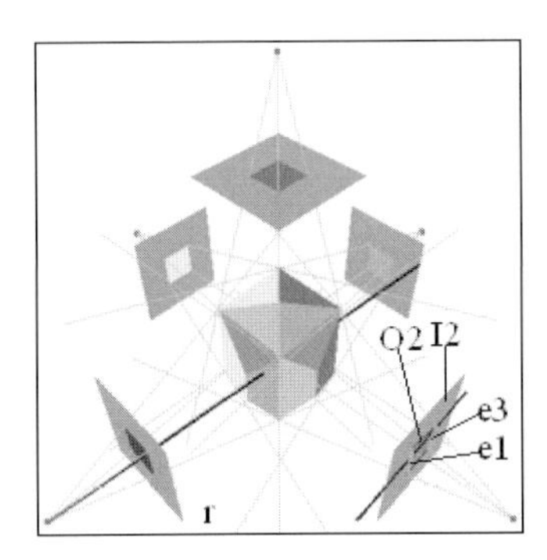

Fig. 4. Fast ray-volume casting. Project r into I2, intersect it with O2, the result is only e1 and e3 (green bold lines), then only test the faces which contains e1 and e3.

3.2 Construct SPVH

After dividing the viewing cone, the PVH model is divided into two parts respectively. One is in the safe cone which only contains real object; the other is the unsafe cone which may contain real object and phantom volumes. Our disposal to the PVH is as follows: if the part lies in the safe cone from at least one camera, reserve it; otherwise, remove it.

In fact, we can not remove the part in one unsafe cone, because it may lie in the safe cone of other cameras, or the real object may be remove by mistake.

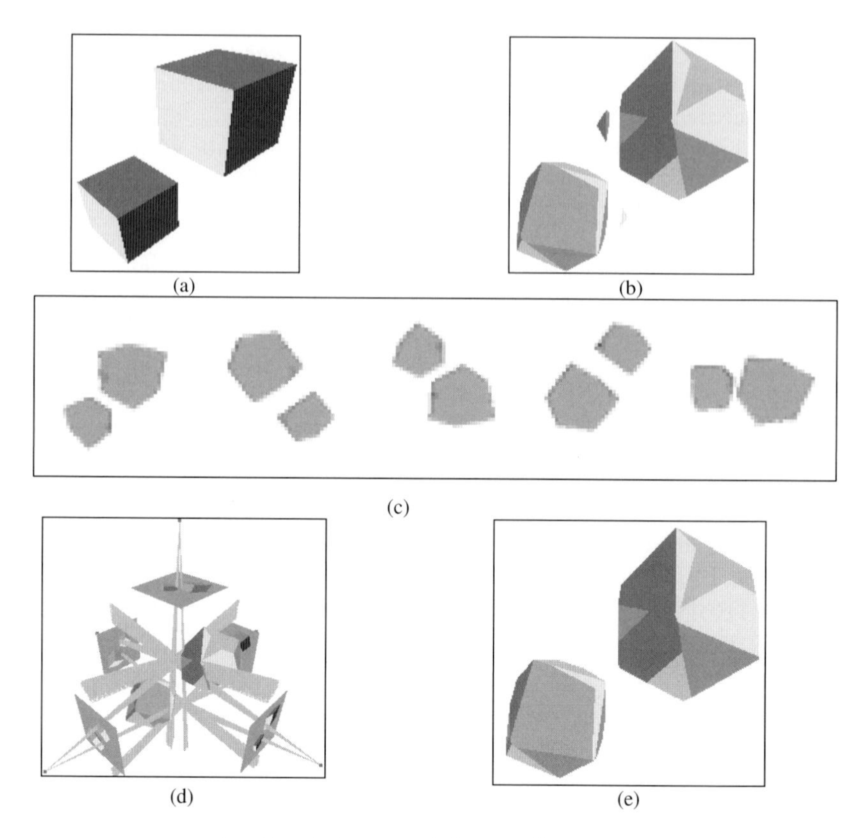

Fig. 5. Safe Polyhedral Visual Hull

Fig. 5(e) shows the SPVH reconstruction with phantom volumes removal.

Polyhedral Division. Suppose the PVH is divided into $SBlock_i$ lying in the safe cone and $NSBlock_i$ lying in the unsafe cone from camera i, described as equation(3).

$$\begin{aligned} SBlock_i &= PVH \cap SCone_i \\ NSBlock_i &= PVH \cap NSCone_i \end{aligned} . \tag{3}$$

The method to calculate 3D volume division is complex, since PVH is made up of multiple polygons; we consider reducing the calculation to 2D conditions. For one polygon *poly* of PVH, take $SCone_i$ for example to describe the process as follows: First, project poly to image plane S_i expressed as equation (4).

$$ppoly = \prod_i (poly) . \tag{4}$$

Second, calculate safe polygon and unsafe polygon which satisfy the expression as equation (5).

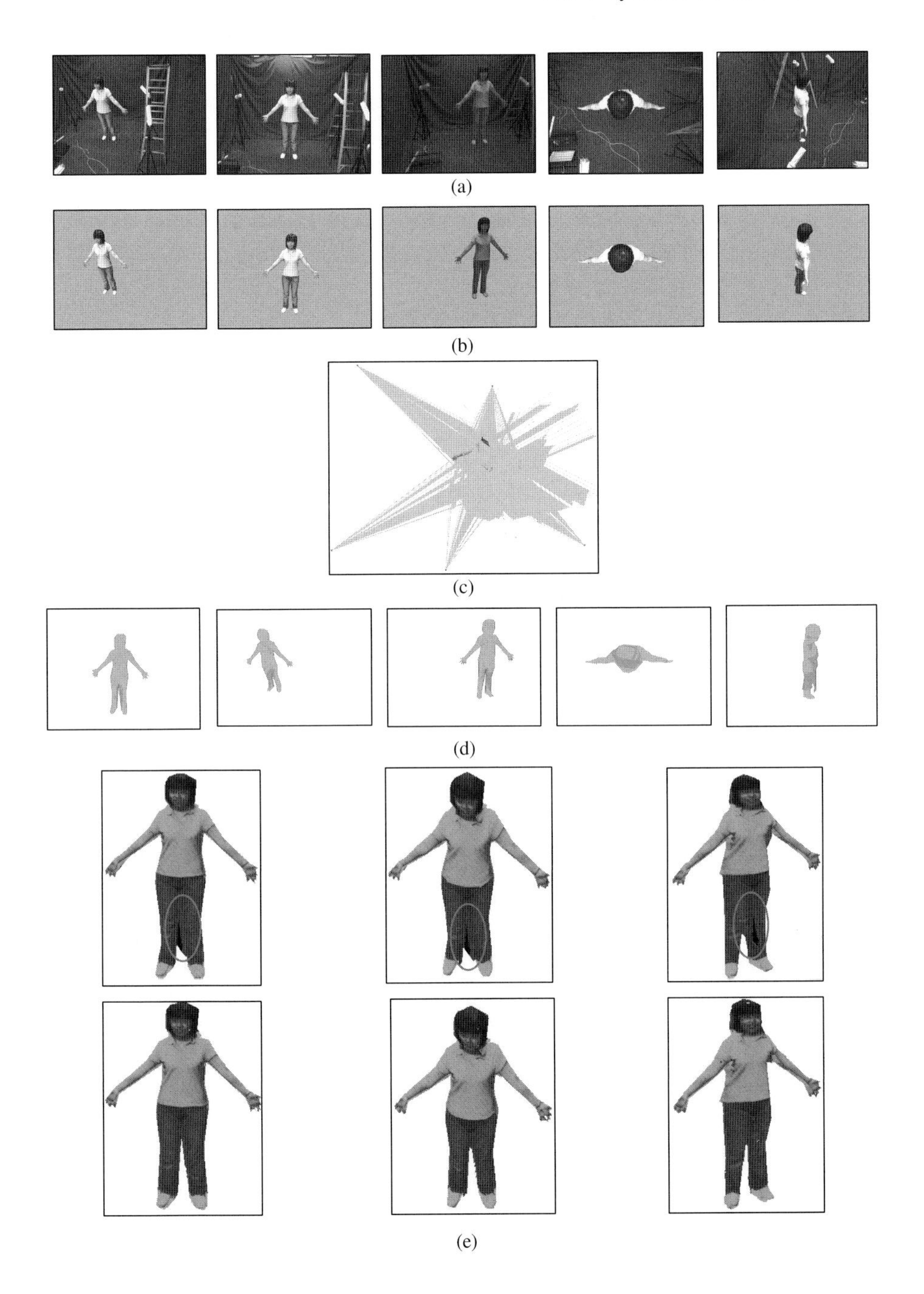

Fig. 6. Views of Safe Polyhedral Visual Hull

$$pspoly_i = ppoly \bigcap SZone$$
$$pnspoly_i = ppoly \bigcap NSZone_i \quad . \qquad (5)$$

Third, back-project $pspoly_i$ and $pnspoly_i$ to $poly$, using the equation (6).

$$wspoly_i = \prod_i^{-1}(pspoly_i)$$
$$wnspoly_i = \prod_i^{-1}(pnspoly_i) \quad . \qquad (6)$$

Then, the $poly$ is the union of $wspoly_i$ and $wnspoly_i$.

We can obtain every division of one polygon from all cameras. Thus, the PVH is divided into different blocks by all cameras.

Polyhedral Division. Given the division of PVH for each camera, the SPVH is illustrated as equation (7).

$$SPVH = \bigcup_{i=1}^{N}(Block_i) \ . \qquad (7)$$

Likewise, we calculate it in 2D space. For $poly$, we have obtained its division with respect to all cameras, thus we merge the safe polygons of $poly$ into $SPoly$, and the equation is as (8).

$$SPoly = \bigcup_{i=1}^{N}(wspoly_i) \ . \qquad (8)$$

Suppose the PVH contains $count$ polygons, the SPVH is expressed as equation (9).

$$SPVH = \bigcup_{t=1}^{count}(SPoly_t) \ . \qquad (9)$$

That is, the union of all the safe polygons.

Obviously, the calculation of the SPVH is similar to that of the PVH. Typically, they both need a great deal of operations in projection, back-projection, and intersection. However, there are two main differences. One is the input data, the PVH only uses the images from cameras, the SPVH, and however, needs the PVH result to be the reference data. The other distinction is that the PVH intersects the back-projected polygons, but the SPVH uses the union.

4 Results and Discussions

This section presents the experiment results of the SPVH reconstruction which improves the quality and accuracy of scene modeling.

The experiment was shown in the reality environment with five cameras which capture the images with 640*480 resolutions at a peak of 15 frames per second. The hardware is an Intel(R) Core(TM)2 Duo CPU E6750 @2.66GHz, ATI Radeon HD 2400 XT with 2GB RAM, and the software platform is Windows XP, Visual C++ 6.0 and OpenGL.

Fig. 6 shows the result of SPVH. (a) shows five human body reference images and (b) shows the five segmented silhouette images with background in pure green color; the PVH with phantom volumes which are apparent between the two legs is shown in (e) on the top row, and on the bottom the three pictures show the SPVH without phantom volumes; the division of safe zones and unsafe zones in silhouettes is shown in (c), with pink zones is unsafe, and the phantom volumes is constructed by unsafe zones in the silhouette images are shown in (d). The results images demonstrate the higher quality of the rendered views using safe polyhedral hull construction rather than polyhedral visual hull.

5 Conclusion and Future Work

This paper presents an efficient geometric method to completely remove phantom volumes from the PVH reconstruction. This method makes full use of the information that the PVH supplied for image plane and takes the construction of the PVH into account to speed up the calculation of the SPVH. Finally, the method produces an intact explicit model, which could be used in many modeling and interaction systems or some rapid acquisition fields.

Nevertheless, there are some limitations, if there are many occlusions in the scene, the number of safe zones will be reduced, then the safe cone doesn't contain all real objects, and the part of the real objects may be removed by mistake. Therefore, this method should guarantee that the whole object is captured at least one camera. The future work is to consider using the hardware for speeding up the calculation to satisfy the need for real-time modeling system.

Acknowledgments. This work is supported by The National Fundamental Research 973 Program of China Grant No.2009CB320805 and the National 863 Project of China Grant No.2009AA01Z333. Thanks are specially given to the anonymous reviewers for their helpful suggestions.

References

1. Baumgart, B.: Geometric Modeling for Computer Vision. PhD thesis, Stanford University (1974)
2. Laurentini, A.: The visual hull concept for silhouette based image understanding. IEEE Trans. Pattern Anal. 16(2), 150–162 (1994)
3. Matusik, W., Buehler, C., Raskar, R., Gortler, S.J., McMillan, L.: Image-based visual hulls. In: Proceedings of the 27th annual conference on Computer graphics and interactive techniques, pp. 369–374 (2000)
4. Miller, G., Hilton, A.: Exact view-dependent visual hulls. In: Proceedings of the 18th International Conference on Pattern Recognition, pp. 107–111. IEEE Computer Society, Los Alamitos (2006)
5. Matusik, W., Buehler, C., McMillan, L.: Polyhedral Visual Hulls for Real-time Rendering. In: Proceedings of the 12th Eurographics Workshop on Rendering, Vienna, pp. 115–125 (2001)

6. Franco, J.-S., Boyer, E.: Exact Polyhedral Visual Hulls. In: Proceedings of the 14th British Machine Vision Conference, Norwich, UK, pp. 329–338 (2003)
7. Buehler, C., Matusik, W., Gortler, S., McMillan, L.: Creating and Rendering Image-based Visual Hulls. MIT Laboratory for Computer Science Technical Report MIT-LCS-TR-780 (1999)
8. Boyer, E., Franco, J.S.: A hybrid approach for computing visual hulls of complex objects. In: Proceedings of the 2003 IEEE Computer Society Conference on Computer Vision and Pattern Recognition, vol. I, pp. 695–701 (2003)
9. Yang, D.B., Gonzalez-Banos, H.H., Guibas, L.J.: Counting people in crowds with a real-time network of simple image sensors. In: Proceedings of the Ninth IEEE International Conference on Computer Vision, Washington, DC, USA, p. 122 (2003)
10. Jarusirisawad, S., Saito, H.: 3DTV view generation using uncalibrated pure rotating and zooming cameras. Image Communication, 17–30 (2009)
11. Miller, G., Hilton, A.: Safe hulls. In: Proceedings of the Conference on Visual Media Production, pp. 1–8 (2007)
12. Bogomjakov, A., Gotsman, C.: Reduced Depth and Visual Hulls of Complex 3D Scenes. Computer Graphics Forum 27(2), 175–182 (2008)

Enhanced Temporal Error Concealment for 1Seg Video Broadcasting

Jun Wang, Yichun Tang, and Satoshi Goto

Graduate School of Information, Production and Systems, Waseda University, Japan

Abstract. Transmission of compressed video over error prone channels may result in packet losses or errors, which can significantly degrade the image quality. Such degradation even becomes worse in 1Seg video broadcasting, which is widely used in Japan and Brazil for mobile phone TV service recently, where errors are drastically increased and lost areas are contiguous. Therefore the errors in earlier concealed MBs (macro blocks) may propagate to the MBs later to be concealed inside the same frame (spatial domain). The error concealment (EC) is used to recover the lost data by the redundancy in videos. Aiming at spatial error propagation (SEP) reduction, this paper proposes a SEP reduction based EC (SEP-EC). In SEP-EC, besides the mismatch distortion in current MB, the potential propagated mismatch distortion in the following to be concealed MBs is also minimized. Compared with previous work, the experiments show SEP-EC achieves much better performance of video recovery in 1Seg broadcasting.

Keywords: Error Concealment, Spatial Error Propagation, 1Seg.

1 Introduction

Transmission of compressed video over error prone channels such as wireless network may result in packet losses or errors in a received video stream. Such errors or losses do not only corrupt the current frame, but also propagate to the subsequent frames [1]. Several error control technologies, such as forward error correction (FEC), automatic retransmission request (ARQ) and error concealment (EC), have been proposed to solve this problem. Compared with FEC and ARQ, EC wins the favor since it doesn't need extra bandwidth and can avoid transmission delays [2].

The EC scheme attempts to recover the lost MBs by utilizing correlation from spatially or temporally adjacent macro blocks (MBs), i.e., spatial error concealment (SEC) or temporal error concealment (TEC) [2]. For TEC, which is focused on by this paper, several related works have been published to estimate the missing motion vector (MV) by using the correctly received MVs around the corrupted MB. In [3], the well-known boundary matching algorithm (BMA) is proposed to recover the MV from the candidate MVs by minimizing a match criterion, the side match distortion (D_{sm}), between the internal and external boundary of the reconstructed MB (see detail in section 3). This algorithm is adopted in H.264 reference software JM [4] and described in detail in [5, 6]. The main improvement of BMA is OBMA (Outer BMA)

S. Boll et al. (Eds.): MMM 2010, LNCS 5916, pp. 45–54, 2010.

[7] also known as decoder motion vector estimation (DMVE) in [8], where the D_{sm} is calculated by the difference of 2 outer boundaries of reconstructed MB instead of internal and external boundary in BMA. Although BMA and OBMA have similar design principle that to minimize mismatch distortion, it is empirically observed that OBMA has better performance in video recovery [7, 8]. 2 variations of OBMA can be found in [7], that multiple boundary layers match, and more candidate MVs search (called refined search). According to evaluations of [7], 1 layer match gives better performance than multiple layers, while refined search gives better performance than original OBMA.

To our knowledge, most of the published works are based on minimization of mismatch for the current lost MB. However, there is no published work considering such observation, that the mismatch error in current lost MB could also propagate to the succeeding MBs in the succeeding EC process. Therefore such kind of propagated mismatch error should also be included in D_{sm} formulation. On the other hand, the lost MBs in 1Seg video broadcasting (see detail in section 2), which is the latest mobile phone TV service widely used in Japan and Brazil, are usually contiguous inside the corrupted frame (spatial domain), therefore in such case the spatial error propagation (SEP) is becoming critical.

Aiming at SEP reduction, this paper proposes a SEP reduction based EC (SEP-EC), where the mismatch not only in current concealed MB but also in the following to be concealed MBs are both minimized.

The rest of this paper is organized as follows. Section 2 presents our work's motivation, which is for 1Seg application. Section 3 gives an overview of well-known BMA and OBMA. Based on OBMA, we present proposed SEP-EC in section 4. Finally section 5 and 6 show our experiments and conclusion.

2 1Seg Application Based Motivation

Our work is targeted to 1seg application [9]. 1Seg is a recently widely used mobile TV service, which is one of services of ISDB-T (Integrated Services of Digital Broadcasting- terrestrial) in Japan and Brazil. Due to errors drastically increased in wireless mobile terminal, EC is critical for 1Seg.

In 1Seg, H.264 format is specified for video compression. Table I shows the specification [9] for video compression and transmission in 1Seg related to this work.

According to Table I, even 1 broken TS packet may cause huge connected area loss (from the start of broken packet to the end of the frame) inside a frame. This is because 1) slice mode and FMO mode [10] are prohibited; 2) for QVGA sized video, each picture/slice usually consists of at least more than 1 TS packet (normally more than 3); and 3) variable length coding (VLC) based entropy coding is included in H.264. Fig. 1 shows a comparison of a typical loss picture (Stefan sequence) in FMO enabled scenario and 1Seg scenario, where green parts are the destroyed area.

Traditional JM [4] can only handle the slice (equivalent to picture in 1Seg) loss problem, whereas can not handle TS packet loss, whose size is much less than slice. In our experiments in section 5, we modified JM to support 1Seg.

Table 1. Specification for Video Part in 1Seg Related to EC

Picture size	320x240(QVGA)
Compression method	H.264, baseline profile, level 1.2
Error resilient tools restriction	No slice mode No FMO (flexible MB ordering) mode
Transmission unit	TS (Transmission Stream) packet (188 Byte)

Fig. 1. A typical broken frame in FMO enable case (left) and 1Seg case (right)

As shown in Fig. 1, the lost MBs are independent in left picture, which means the errors in current concealed MB could not propagate to other corrupted MBs. However, 1Seg case should suffer such situation that the errors in current concealed MB may propagate to other corrupted MBs inside the same picture, this is so called spatial error propagation (SEP). In section 4, we will propose our solution toward this problem.

3 Boundary Matching Algorithm (BMA) and Outer BMA (OBMA) for TEC

For each lost MB, BMA [4, 5, 6] is utilized to recover the lost mv (motion vector) from candidate mvs (mv^{can}s) by the criterion of minimizing the side match distortion (D_{sm}) between the IN-MB and OUT-MB, see Fig. 2. The IN-MB is projected by mv^{can}, which is the candidate mv belongs to one of 4 available neighbors[1] in current frame, either correctly received (OK MB) or concealed MB. To save the space, Fig. 2 only shows left (concealed MB) and top (OK MB) neighbors.

The D_{sm} is determined by the sum of absolute differences between the predicted MB and the neighboring MBs at the boundary, shown in Eq. (1)[2].

---

[1] For simplification, all 4 neighbors are not divided into sub-blocks, meaning each neighbor only has 1 mv.

[2] There is a notation rule needs explanation. In Fig. 2, $Y_i^{IN}(mv^{can})$ is the i-th pixel in the IN-MB, which is projected by mv^{can} in the reference frame. Such rule can be similarly applied in Eq.(3), and (6) in the rest parts.

$$D_{sm} = \frac{1}{B}\left\langle \sum_{i=1}^{N} \left| Y_i^{IN}(mv^{can}) - Y_i^{OUT} \right| \right\rangle \quad (1)$$
$$where, mv^{can} \in \{mv^{top}, mv^{bot}, mv^{lft}, mv^{rt}\}$$

N is the total number of the available boundary pixels, B is the number of available boundaries. In Fig. 2, N=32, B=2. The winning prediction mv mv^{win} is the one which minimizes the side match distortion D_{sm}:

$$mv^{win} = \underset{mv^{can} \in \{mv^{top}, mv^{bot}, mv^{lft}, mv^{rt}\}}{\text{Arg Min}} D_{sm} \quad (2)$$

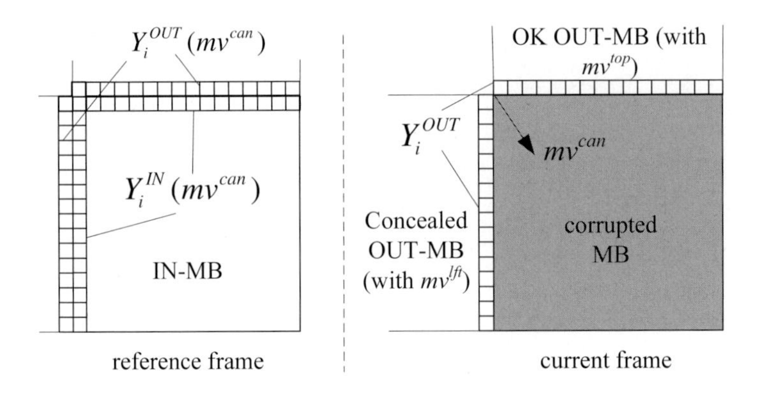

Fig. 2. BMA and OBMA based TEC in H.264

The main improvement of BMA is OBMA, where the main difference is the D_{sm} calculation. The D_{sm} calculation is shown in Eq. (3):

$$D_{sm} = \frac{1}{B}\left\langle \sum_{i=1}^{N} \left| Y_i^{OUT}(mv^{can}) - Y_i^{OUT} \right| \right\rangle \quad (3)$$
$$where, mv^{can} \in \{mv^{top}, mv^{bot}, mv^{lft}, mv^{rt}\}$$

where the D_{sm} is calculated by the difference of 2 outer boundaries ($Y_i^{OUT}(mv^{can})$ and Y_i^{OUT}) of reconstructed MB (see Fig. 2), instead of internal and external boundary in BMA.

According to the evaluation of [7], OBMA provides better image recovery compared with BMA. Also, the reason of it is explained in [7]. Since OBMA has better performance, in this paper, we implement the proposed algorithm with OBMA.

4 Proposed SEP-EC

4.1 Spatial Error Propagation (SEP) Reduction Based Dsm Calculation

It is inevitable that the current concealed MB may cause mismatch distortion compared with the original current MB. As discussed in section 2, in 1Seg, this kind of distortion may propagate to spatial domain, ie., the succeeding MBs inside the corrupted picture, since the current concealed MB might be used to conceal the neighboring corrupted MBs. Therefore, considering the potential propagated distortion, the total side match distortion can be expressed as follows:

$$D_{sm\_total} = (D_{sm\_cur} + D_{sm\_pgt})/2 \tag{4}$$

where, D_{sm_cur} and D_{sm_pgt} are side match distortion for current MB itself, and the potential propagated side match distortion for neighboring corrupted MBs, respectively. Division by 2 is because the distortion should be normalized by boundary.

As for current side match distortion D_{sm_cur} calculation, traditional OBMA can be used without any change. D_{sm_cur} is the minimal D_{sm} for current MB which calculated by mv^{win} . Similar to Eq. (2) and (3), D_{sm_cur} can be easily formulated as follows:

$$D_{sm\_cur} = \underset{mv^{can} \in \{mv^{top}, mv^{bot}, mv^{lft}, mv^{rt}\}}{\text{Min}} D_{sm} \tag{5}$$

where, D_{sm} is calculated with OBMA, see Eq. (3).

The key point is how to formulate $D_{sm\_\_pgt}$. An accurate expression of potential propagated distortion is hard to figure out, yet it has some way to estimate by looking at how current errors propagate to its neighbors. Fig. 3 shows the scenario that the mismatch distortion in current MB propagates to its corrupted neighborhoods.

Let's think about concealment of MB_j which is the j-th corrupted MB next to current concealed MB. Like the EC process of current MB, EC of MB_j should also select a wining mv ($mv_j^{win\_nbr}$) from candidate mv set (mv^{lft} and mv^{top}, in Fig. 3) by the minimal D_{sm}. Note that, mv^{lft} is actually mv^{win_cur} , which is the wining mv of current concealed MB.

Suppose mv^{win_cur} is chosen as winning mv ($mv_j^{win\_nbr}$) for the neighbor MB_j, then the distortion of current MB should propagate to MB_j. Therefore the propagated distortion can be formulated by calculating D_{sm} in MB_j, shown as Eq. (6):

$$D_{sm\_pgt} = \begin{cases} \text{if } mv_j^{win\_nbr} == mv^{win\_cur}, \\ \qquad \frac{1}{M}\sum_{j=1}^{M}(\frac{1}{B}\sum_{i=1}^{N}\left|Y_{i,j}^{OUT\_nbr}(mv_j^{win\_nbr}) - Y_{i,j}^{OUT\_nbr}\right|); \\ else, \\ \qquad 0 \end{cases}$$

$$\text{where, } \quad mv_j^{win\_nbr} = \underset{mv^{can\_nbr} \in \{mv^{top}, mv^{bot}, mv^{lft}, mv^{rt}\}}{\text{Arg Min}} D_{sm\_nbr}^{j} \tag{6}$$

$Y_{i,j}^{OUT\_nbr}(mv_j^{win\_nbr})$ and $Y_{i,j}^{OUT\_nbr}$ are shown in Fig. 3. i is the index of pixel in outer boundary of MB_j, j is the index of neighboring corrupted MB. N is the number of calculated pixels in outer boundary of MB_j, B is the number of available boundaries. For MB_j in Fig. 3, N=32, B=2. M is the number of all neighboring MBs which are not concealed yet. Taking Fig. 3 as an example, M is 2, the number of blue MBs.

Given Eq. (5) and (6), the D_{sm_total} in Eq. (4) can be finally calculated.

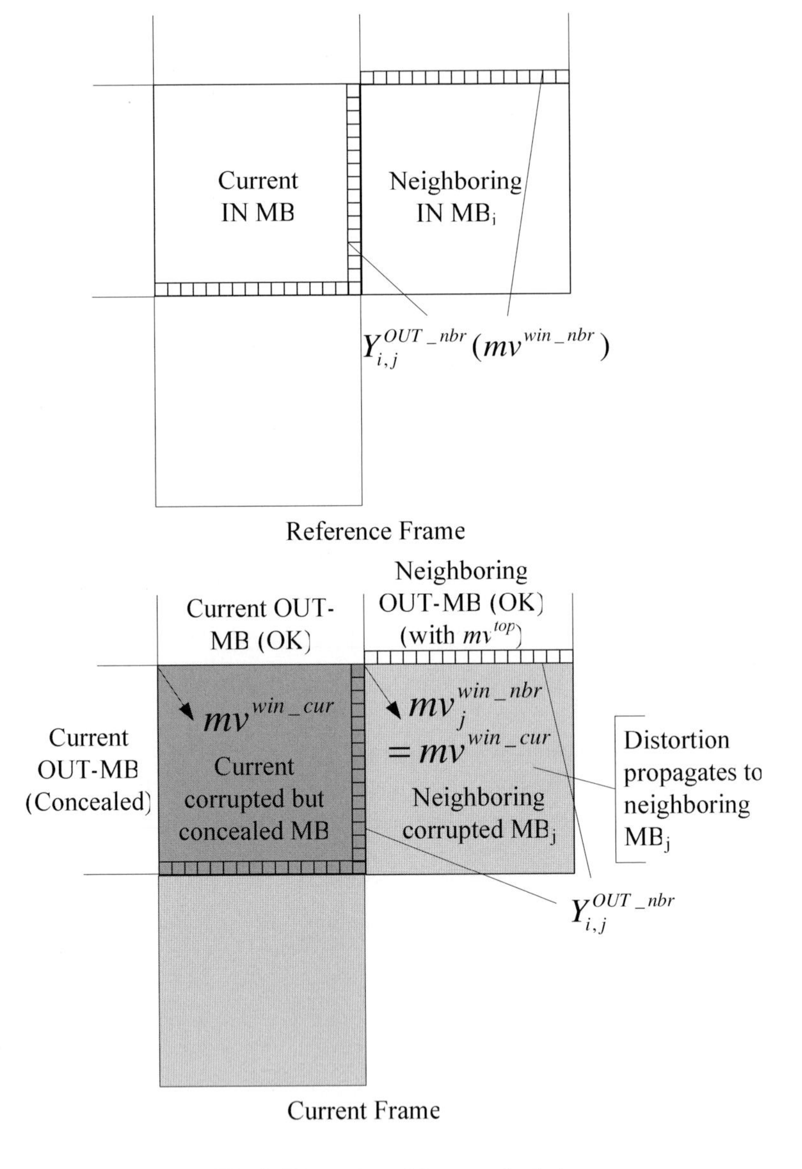

Fig. 3. Illustration of D_{sm_pgt} calculation

As mentioned in section 1, [7] reported that there are 2 variations of OBMA, that multiple boundary layers match, and more candidate MVs search (called selective search (SS)). According to evaluations of [7], 1 layer match gives better performance than multiple layers, while selective search gives better performance than original OBMA. Therefore in this paper, we combined SS with our proposal. In this combination, the mv candidate set is enlarged by SS, while the criterion is our proposed D_{sm_total} in Eq. (4). For more details of SS, please refer to [7].

4.2 Discussion of How Far the Errors May Potentially Propagate

As discussed before, it is difficult to give an accurate expression of potential propagated distortion. Actually the D_{sm_pgt} in Eq. (6) is not an accurate one, since the error may also propagate to a little farer neighbors which are not adjacent to current MB as well as the adjacent neighbors. Fig. 4 shows a scenario that the mismatch distortion propagates to MB_j', which is next to neighboring MB_j. It is easy to find much more such kind of farer neighbors. Therefore, how far the errors in current corrupted MB may potentially propagate deserves discussion. Let L be an integer to show the distance of it. L=1 means errors propagate to the adjacent neighbors, L=2 means errors propagate to the second closest neighbors, etc..

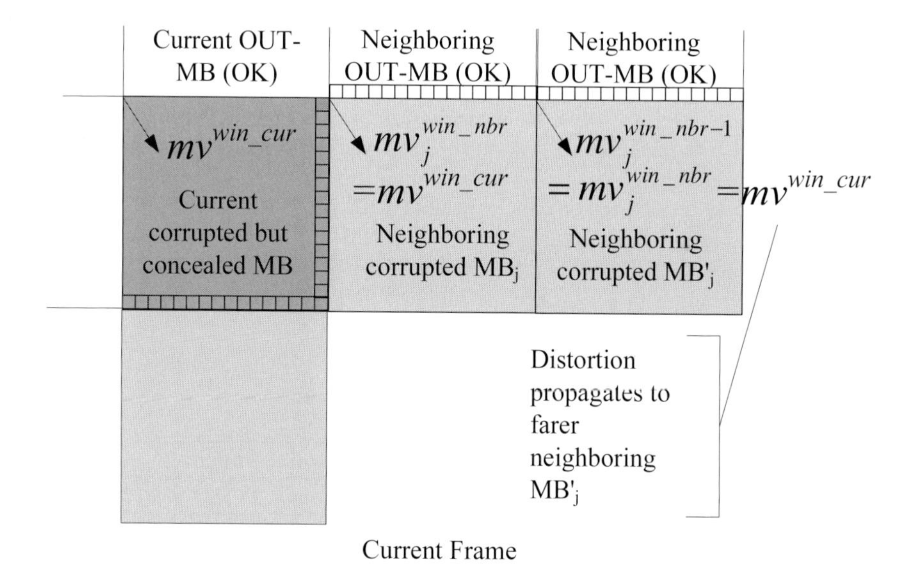

Fig. 4. Illustration of error propagates farer when L=2

We did experiments of SEP-EC to show how L will affect the recovered image quality. In these experiments, we set different Ls in advance, to observe the recovered image quality with PSNR. Fig. 5 shows the result of it. It can be seen L=1 is enough for D_{sm_pgt} formulation, since large L may bring huge computation cost, while almost could not improve image quality any more. Not that all sequences are tested by SEP-EC, without SS.

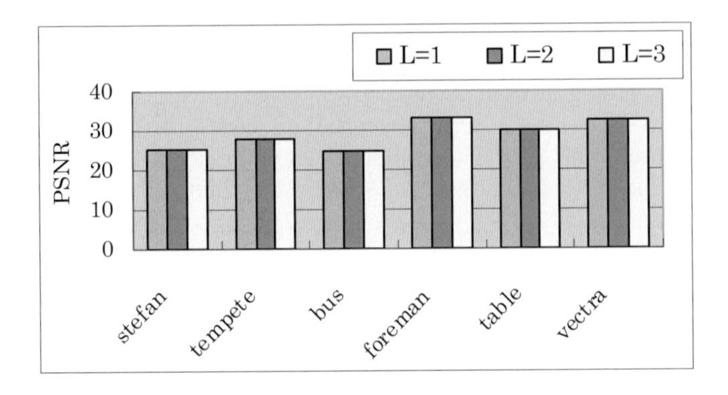

Fig. 5. Discussion of *L* by SEP-EC

5 Experiments

The proposed algorithm is evaluated based on the H.264 codec under the specification of 1Seg application. The JM9.1 reference software is used in the experiment, which is modified to support the packet loss in 1Seg application, while traditional JM can only support slice loss. Note that here a frame normally is divided into several TS packets. 50 frames for stefan, tempete, bus, foreman, table, and vectra sequences in QVGA format are encoded. Slice mode and FMO mode are disabled. "IPP…" GOP pattern is used, and frame rate is set on 30fps. No transmission errors occur in I frames. For each P frame, a number of packets are randomly destroyed to simulate the transmission under the bit error rate of 2%.

As for the objective comparison , the average PSNR of all decoded frames using 5 different methods are presented in Table 2, ie., BMA, OBMA, OBMA with SS (OBMA-SS), proposed SEP-EC, and proposed SEP-EC with SS (SEP-EC-SS). Experiments show our algorithm can provide higher PSNR performance compared with BMA and OBMA.

Table 2. Average PSNR Comparison

	stefan	tempete	bus	foreman	table	vectra
BMA	24.56	27.16	24.22	32.74	29.58	31.51
OBMA	25.15	27.65	24.63	32.95	29.88	32.02
OBMA-SS	25.46	27.78	24.88	33.21	30.16	32.4
SEP-EC	25.61	27.86	24.75	33.13	30.21	32.53
SEP-EC-SS	25.79	28.23	25.15	33.32	30.36	32.8

An example of vision comparison is shown in Fig. 6. Sequence for Stefan is presented. There are some noticeable mismatches in the areas of player, and the advertisement panel, denoted with red circle. However, proposed SEP-EC and SEP-EC-SS can avoid this mismatch efficiently.

(a) damaged frame (b) recovered by BMA

(c) recovered by OBMA (d) recovered by OBMA-SS

(e) recovered by SEP-EC (f) recovered by SEP-EC-SS

Fig. 6. Vision comprarison

6 Conclusion and Discussion

This paper proposed an enhanced EC, where an spatial error propagation reduction is taken as consideration of temporal error concealment in 1Seg broadcasting. In this method, besides the mismatched distortion in current MB, the mismatched distortion in the following MBs, which will be concealed in the succeeding process is also minimized. Compared with BMA and OBMA, the experiments under 1Seg application show our proposal achieves much better performance of video recovery.

It is easy to know that the mismatch errors may not only propagate to following MBs inside the current lost frame, but also to the succeeding frames, this is so called temporal error propagation (TEP). In the future, the D_{sm} calculation should also involve such TEP reduction consideration.

Acknowledgement

This research was supported by Waseda University Global COE Program "International Research and Education Center for Ambient SoC" sponsored by MEXT, Japan, and CREST of JST, Japan, and co-operated with Center for Semiconductor R&D, Toshiba, Japan. It is a part of the outcome of research performed under a Waseda University Grant for Special Research Projects (Project number: 2009A-906).

References

1. Suh, J.-W., et al.: Error concealment techniques for digital TV. IEEE Trans. Broadcasting 48, 209–306 (2002)
2. Wang, Y., et al.: Error control and concealment for video communication: a review. Proc. of IEEE, 947–997 (May 1998)
3. Lam, W.M., et al.: Recovery of lost or erroneously received motion vectors. In: Proceeding of ICASSP, vol. 5, pp. 417–420 (1993)
4. JM, http://iphome.hhi.de/suehring/tml
5. Varsa, V., et al.: Non-normative error concealment algorithm, ITU-T VCEG-N62 (2001)
6. Wang, Y.K., et al.: The error concealment feature in the H.26L test model. In: Proc. ICIP, vol. 2, pp. 729–732 (2002)
7. Thaipanich, T., et al.: low-complexity video error concealment for mobile applications using OBMA. IEEE Trans. Consum. Electron. 54(2), 753–761 (2008)
8. Zhang, J., et al.: A cell-loss concealment techniaue for MPEG-2 coded video. IEEE TCSVT 10(4), 659–665 (2000)
9. Data Coding and Transmission Specification for Digital Broadcasting —ARIB STANDARD, ARIB STD-B24 V.5.1 Vol. 1 (English Translation) (2007)

User-Centered Video Quality Assessment for Scalable Video Coding of H.264/AVC Standard

Wei Song, Dian Tjondronegoro, and Salahuddin Azad

School of Information Technology, Faculty of Science and Technology,
Queensland University of Technology, Brisbane 4000, Australia
wei.song@student.qut.edu.au,
dian@qut.edu.au,
salahuddin.azad@qut.edu.au

Abstract. Scalable video coding of H.264/AVC standard enables adaptive and flexible delivery for multiple devices and various network conditions. Only a few works have addressed the influence of different scalability parameters (frame rate, spatial resolution, and SNR) on the user perceived quality within a limited scope. In this paper, we have conducted an experiment of subjective quality assessment for video sequences encoded with H.264/SVC to gain a better understanding of the correlation between video content and UPQ at all scalable layers and the impact of rate-distortion method and different scalabilities on bitrate and UPQ. Findings from this experiment will contribute to a user-centered design of adaptive delivery of scalable video stream.

Keywords: User perceived quality, scalable video coding, subjective quality evaluation.

1 Introduction

Resource limitation and the diversity of network and user terminals make it hard to adaptively deliver videos in multiple applications. *Scalable Video Coding* Extension of H.264/AVC Standard (H.264/SVC)[1] supports flexible bitstream scaling by using multiple dimension scalable modes, such as temporal, spatial, and quality scalability, thus, providing a number of benefits to heterogeneous delivery applications. These scalabilities of H.264/SVC can be ascribed to a lot of employed advance techniques, such as inter-layer prediction, hierarchical B frames, *coarse-grain scalability* (CGS) or *medium-grain scalability* (MGS), and so on [1]. However, SVC faces one main problem: how to guarantee *user perceived quality* (UPQ). Given that user's satisfaction with video quality is very important for video services, many delivery schemes aim to maximize the UPQ. Previous studies have proved that there is a strong correlation between video content, bitrate and UPQ [2] and [3]. Therefore, it is expected that the correlation can be applied to reduce the bitrate of encoded video without losing the perceptual quality. However, most content-related studies were conducted for single-layer video coding. Wang et al. addressed scalability based on content classification,

S. Boll et al. (Eds.): MMM 2010, LNCS 5916, pp. 55–65, 2010.

but only two scalabilities were considered [4]. Some scalability-related subjective video quality assessment have been done for videos encoded with H.263+ standard [5] and [6]. However, different encoder types will exert different impacts on UPQ [7]. Moreover, most subjective video quality assessments were carried out particularly for small spatial resolutions (e.g., QCIF and CIF) [5],[6], and [7].

Focusing on these limitations, we conducted a subjective video quality assessment for a hierarchical coding standard H.264/SVC at three spatial resolutions (QCIF176×144, CIF 352×288, and 4CIF 704×576). Differing from the recent study on quality assessment of MPEG-4 SVC, which focused on the impact of encoding setting on codec's performance [8], this study aims to find out how video content, bitrate and scalabilities of H.264/SVC influence user perceived quality and what particular factors in a scalable coding structure influence perceptual quality. We obtained some interesting findings in five aspects: content correlations to UPQ, important content features, bitrate saving, scalability strategy, and other user perceived quality loss.

This study will contribute to H.264/SVC-based video streaming, such as video-on-demand (VOD), where it is demanded to guarantee a good user perceived quality under heterogeneous user requirements and network bandwidth. In addition, although the experiment was conducted on a desktop setting, the findings could be valuable for improving video delivery on various platforms, such as mobile TV, whereby bitrate and device limitations need to be tackled for optimal viewing.

The paper is organized as follows: The details of the subjective quality assessment for videos encoded by H.264/SVC are described in Section 2, followed by the experimental results and discussion in Section 3; conclusions and future work are presented in Section 4.

2 Experiments

Our subjective quality evaluation experiment consists of two studies. In Study 1, we collected quality assessment data of all scalable layers to investigate the relationships between: 1) scalable layers and UPQ, and 2) video content, bitrate and UPQ. Based on the results of Study 1, we launched the second study, in which we studied the impact of encoding parameters on bitrate and UPQ for specific video content and compared the perceptual quality loss under different scalabilities. In our experiment, we employed JSVM (Joint Scalable Video Model) v9.16 as the codec tool, which is an open-source software for H.264/SVC made by JVT (Joint Video Team). A detailed description of experimental environment, materials, participants and each of these studies follows.

2.1 Experimental Conditions

The experiment was carried out in a quiet computer room with a comfortable fluorescent light illumination. All test sequences were displayed on a Dell PC with 21inch LCD with a high display resolution of 1600×1200. The background color was set as 50% grey. A total of 26 subjects were enrolled, wherein 20 participated in the first study, and 15 participated in the second study (9 took part in both). The subjects ranged different ages (20~40), gender (12 females and 14 males), education (undergraduate, graduate, and postgraduate), and experience with image processing (3 people claimed they were good at photography; and others are naive). During the test, we allowed participants to freely adjust the viewing distance and angle to a comfortable position because we considered that users would spontaneously adjust the distance between eyes and screens in practical applications.

We adopted the following eight short video clips (each is 10 sec) as the testing materials: "city", "crew", "football", "foreman", "harbour", "mobile", "news", and "soccer", which were used as the standard test set by JVT for testing scalable video coding (ftp://ftp.tnt.uni-hannover.de/pub/svc/testsequences). These clips cover various contents, such as sports, panoramic, news, and people; and have completely different characteristics, shown in Table 1. In our experiment, a demonstration was given at the beginning of each test to acquaint participants with the test procedure, but the test clips were not shown in the demonstration.

Table 1. Content characteristics of test clips

Name	Content Description	Characteristics
City(C)	A panorama scene	Global motion; monotonous colour
Crew(CR)	Many walking astronauts	More details; media motion; large motion area
Football(FB)	An American football match with fast moving players	Large motion area; fast motion with both objects and background
Foreman(FM)	One person's head and a building site	Limited motion; facial details; a scene change
Harbour(H)	A harbour with many masts, several passing ships & flying birds	Local slow motion; monotonous colour
Mobile(M)	A slowly moving toy train, a rolling ball and a scrolling calendar	Medium objects movement; more colourful detail
News(N)	One person, mainly head and shoulder	Limited detail; local fast motion; small motion area
Soccer(S)	Several people playing a football	Small moving regions; slow horizontal background motion

2.2 Study 1: User Perceived Quality on Scalable Video Layers

We adopted *Absolute Category Rating* (ACR) method [9] recommended by ITU-T to evaluate video qualities. Participants were asked to view video sequences and judge the quality using one of the five scales: 1-bad, 2-poor, 3-fair, 4-good, and 5-excellent. The eight video clips were encoded with three spatial layers (QCIF, CIF, and 4CIF of *spatial resolution* (SR)), five temporal layers (we set the size of *group of pictures* (GOP) 16 frames, and thus five temporal layers are gained at 1.875, 3.75, 7.5, 15 and 30 fps of *frame rate* (FR) respectively), and three *medium-grain scalable* (MGS) quality layers (MGS-1, MGS-2, and MGS-3). Other encoding configurations are: intra period of 16 frames; *Quantization Parameters* (QPs) of 36 and 32 for *base layers* (BL) and *enhancement layer* (EL) respectively; MGSVector0=3, MGSVector1=3 and MGSVector2=10 corresponding to the three MGS layers; and Inter-layer Prediction is adaptive. Each clip was extracted and decoded into 36 test sequences at combination points of spatial, temporal and quality layers: {QCIF; CIF; 4CIF}, {7.5fps; 15fps; 30fps}, and {BL, MGS-1; MGS-2; MGS-3}. Since the FR of less 5fps is unacceptable for viewing quality [6]and[10], only three temporal layers were involved into the Study 1. Totally, 288 ($8\times3\times3\times4$) test sequences were used and the evaluation took around 1.3 hour, including 48min of displaying time, 24min of voting time (5sec each sequence) and 5min of break time. To avoid residual impact of the different layers on perception, the displaying order of testing sequences were generated randomly, but the same viewing order was applied to all participants for preventing the order effect. The results of the ACR follow a normal distribution verified by *Kolmogorov-Smirnov* (K-S) Test[1]. In the light of 95% of confidence interval, we discarded 1.1% of scores and then computed *Mean Opinion Scores* (MOSs) of all test sequences as the basis of the follow-up data analysis.

2.3 Study 2: Comparative Evaluation on Scalabilities

In study 2 of the experiment, two comparative evaluation methods, *Degradation Category Rating*(DCR) and *Pair Comparison* (PC)[9], were employed to compare the quality of two sequences with the same content, which were simultaneously displayed on the screen. In the DCR method, subjects were asked to rate the impairment of an impaired sequence in relation to an original sequence using a five-level scale: 5-imperceptible, 4-perceptible but not annoying, 3-slightly annoying, 2-annoying and 1-very annoying. The evaluation is for studying the degree of impact of quantization distortion on UPQ. In the PC method, subjects only needed to choose which of the pair had better quality, aiming to contrast the effect of spatial, temporal and quality scalabilities on UPQ. We refer to the DCR-based evaluation as study 2.1 and the PC-based evaluation as study 2.2, which lasted 5min and 20min respectively.

The study 2.1 only used three test clips,"football","harbour", and "mobile", as they represent the content with high bitrate traffic. Each clip was encoded into three SVC format files with three QP pairs, which are 40 and 36, 36 and 32, 32 and 26 for BL and EL respectively. We compared their impairment quality at CIF and 4CIF layers.

The study 2.2 was conducted for all eight encoded clips. Two sequences in a pair were obtained at the same or approximate bitrate. To compare the effect of the temporal and SNR scalability on UPQ, We extracted one sequence with low FR and high quality (at 15/7.5fps and MGS-3), and the other with high FR and low quality

---

[1] One-sample Kolmogorov-Smirnov test is a test of goodness of fit to compare a sample with a reference probability distribution. Asymp.Sig. (the significance level based on the asymptotic distribution of a test statistic) $> \alpha = 0.05$ means that the samples follow normal distribution.

(at 30fps). When comparing the influence of the spatial scalability with the temporal scalability on UPQ, we scaled one sequence at the layers of QCIF/CIF, 30fps and MGS-3 and the other from the next higher spatial layer (CIF/4CIF) but low FR for maintaining the same bitrate. Before evaluation, we converted the sequence with lower resolution in a pair into the same resolution of the other by normative upsampling (SVC integer-based 6-tap filter) for spatial resolution or frame repeating for temporal resolution. We counted the chosen times of each sequence in a pair, computed their chosen percentages, and then determined the users preference to different scalabilities according to the percentages.

3 Results and Discussions

This section describes the experiment results. Quantitative analysis is given from three aspects: 1) the correlation between video content and UPQ; 2) the effects of bitrate on UPQ; and 3) the effects of three scalabilities of H.264/SVC on UPQ. Additionally, we provide some interview results which are associated with coding performance of H.264/SVC.

3.1 Content Correlation

In single-layer coding such as, MPEG-4 and H.264/AVC the UPQ is closely related to video content [3], [4] and [7]. Similar conclusion was found for multi-layer coding videos (see Fig. 1, which data derived from the study 1 of the experiment). For example, "football" and "mobile" achieved higher MOSs than others under the same encoding configuration. This proves that content characteristics can be used effectively not only in single-layer coding but also in scalable coding with hierarchical layer structure. However, we also observed some unique features. Comparing Fig. 1a with 1b and 1c, it can be seen that at the given frame rate 30fps, when *spatial resolution* (SR) is QCIF the UPQ is greatly influenced by content; on the contrary, when SR is 4CIF and quality is high (MGS-3 layer) the UPQ is hardly affected by content (The MOS only varies in the range of 3.5~4). At 15fps and 7.5fps, similar tendency was observed but not as obvious as that at 30fps. That means that the correlation between UPQ and video content decreases with the increase of the SR, quality and FR. In addition, it can also be seen that participant's sensitivity to the degradation of MGS quality layers decreases as the spatial resolution reduces, and at the QCIF layer, the distinction between quality layers is hardly perceived.

Our experimental results also showed that an obvious inconsistency between subjective video quality and *Peak Signal-to-Noise Ratio*(PSNR) existed at all scalable layers of H.264/SVC (Fig. 1c illustrates a comparison between MOS and PSNR at 4CIF spatial layer). A recent study also demonstrated that the video quality mark of the objective quality metric based on PSNR and framerate was quite different from that of subjective quality assessment [8]. As there are joint impacts of various scalabilities and content on UPQ, an effective object quality metric should consider all these factors and their correlations to video content in order to achieve the best user satisfaction.

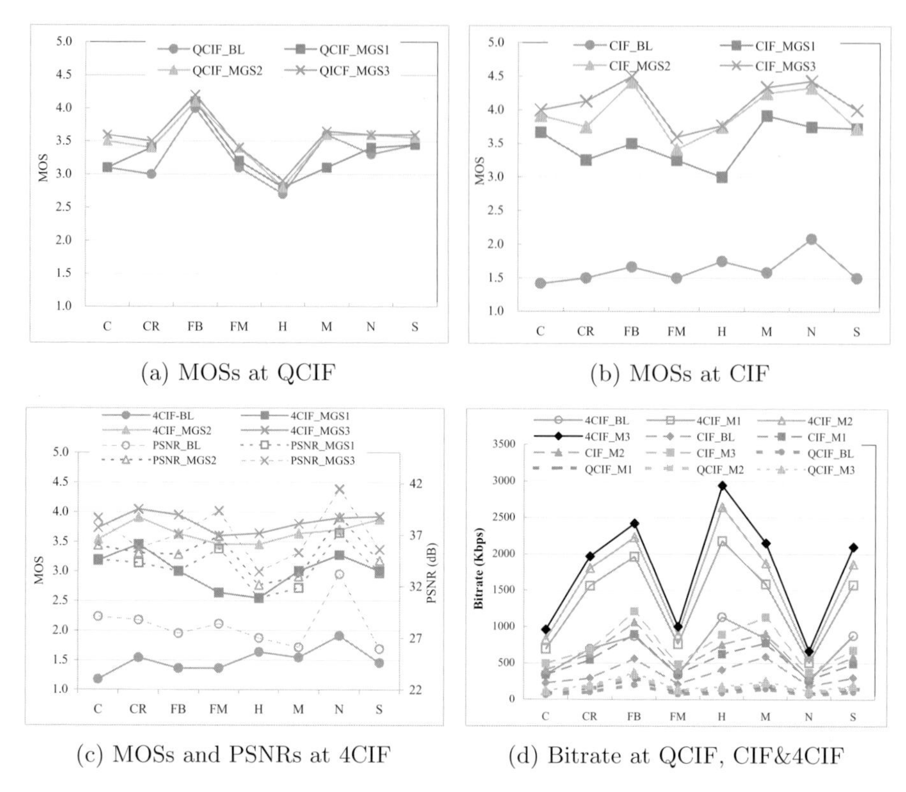

(a) MOSs at QCIF (b) MOSs at CIF

(c) MOSs and PSNRs at 4CIF (d) Bitrate at QCIF, CIF&4CIF

Fig. 1. Correlation between MOS and video content, scalable layers and bitrate at 30fps

3.2 Effects of Bitrate

From the perspective of bitrate, shown in Fig. 1d, it can be observed that reducing spatial resolution can bring a great drop in the bitrate. Comparing bitrate curves with UPQ curves in Fig. 1a- 1c, we can see that for the same video content the difference of bitrate between three quality layers is proportional to the difference of their UPQs; whereas for different video contents bitrate presents significant variations [6]. The high birate can be observed for video sequences with fast motion or large motion region (e.g., "football" and "mobile"), while the low bitrate is for video sequences with slow motion or small motion regions (e.g., "city", "foreman" and "news"). Surprisingly, we noticed that "harbour" with local slow motion reached the highest bitrate at the 4CIF spatial layer but nearly the lowest MOS. Why has "harbour" such a low encoding efficiency at 4CIF layer? Analyzing its content characteristics, we found that slight camera jitter resulted in a small but global and multidirectional motion, and the variation of motion direction was sensitive to coding efficiency in a big spatial resolution.

It is clear that a high bitrate is not conducive to data storage and a great bitrate difference among video contents means a great variability in video traffic, which makes smooth network delivery difficult [11]. In the case of "habour" that a high bitrate does not contribute to a high UPQ, we naturally want to reduce its bitrate. A rate-distortion method is commonly used to adjust bitrate. Therefore, we used three pairs of QPs (32&26, 36&32 and 40&36 for BL&EL) to encode video contents with high bitrate ("football","mobile" and "habour"). Fig. 2a shows bitrate curves of "harbour" under three pairs of QPs for all spatial and temporal layers. An average 47% and 66%of bitrate saving can be observed when QPs are changed from 32&26 to 36&32 and from 32&26 to 40&36. Similar bitrate curves can be obtained for "football" and "mobile". However, does the great saving in bitrate mean a large drop in UPQ? The results of impairment evaluation (study 2.1) are illustrated in Fig. 2b. We noticed that there is no obvious distinction among MOS values under different QP parameters. Thus increasing QP can bring great bitrate saving with a little decrease in UPQ.

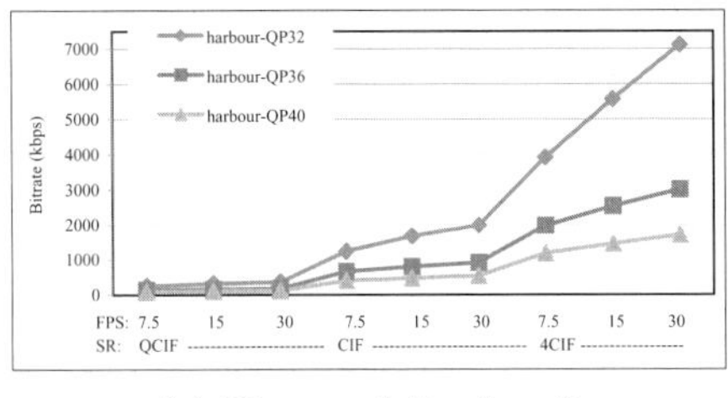

(a) Bitrate of "harbour"

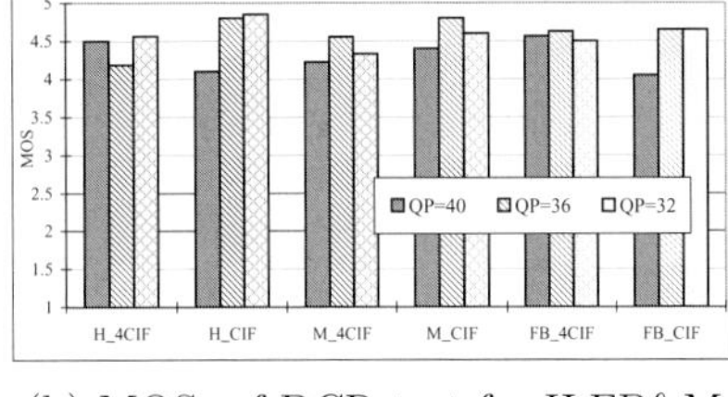

(b) MOSs of DCR test for H,FB&M

Fig. 2. Bitrate saving and impairment evaluation

3.3 Effect of Scalabilities

The influence of frame-rate on UPQ has been studied in many literatures [4], [5], [6], and [7]. It is well known that at small spatial resolutions (QCIF or CIF), the sequences with fast motion (e.g., "football") were more sensitive to the change of the frame-rate [5] and [7]. However, the size of motion region also influences the perceptual quality. The decrease of frame-rate in the sequence with fast motion and large motion region (e.g., "city") is more annoying for viewers than that in the sequence with slow motion and small motion region. At the same time, at the big spatial resolution (4CIF), people's sensitive to frame rate descends but the sensitive to detail distinctly ascends. Next, we will discuss how the scalabilities of H.264/SVC influence user perceived quality.

Temporal scalability vs. quality scalability. In study 2.2, we contrasted the user's preference for temporal scalability and quality scalability under the same bitrate. Fig. 3a shows the results of temporal scaling (at 15fps) versus quality scaling (at 30fps). It can be seen that most choices are given to the sequence with quality loss. This preference is much clearer when the temporal layer is at 7.5fps. That reveals that in this case people prefer high frame rate over high quality. Therefore, considering the effect of scalability on UPQ, under the same bitrate and a given spatial resolution condition the quality scaling should be prior to temporal scaling for streaming videos of H.264/SVC format.

Temporal scalability vs. spatial scalability. While constrasting spatial scalability and temporal scalability (in study 2.2), we found that the bitrate of the layer at QCIF and 30fps is equivalent to that of the layer at CIF and 1.875fps, and the bitrate of the layer at CIF and 30fps approximately equals to that of the layer at 4CIF and 3.75fps. That means that for the H.264/SVC bitstream the bitrate saving achieved by scaling spatial layer is great (in accordance with the conclusion in section 3.2) and equals to the bitrate saving achieved by scaling more than 3 temporal layers. Although as mentioned in section 2.1, a frame rate of below 5fps is inadequate for the perceptual quality, the quality loss caused by four times spatial upsampling is also severe, so we want to know which factor is more negative on UPQ because a large bitrate scaling is sometimes required in an unstable network. The result of subjective evaluation shows that about 90% of participants prefer the blur sequence caused by upsampling rather than the clear but frame-jumping sequence (see Fig. 3b). Thus, to meet the limitation of bandwidth, spatial scalability achieves a better subjective quality than temporal scalability when frame-rate drops below 7.5fps.

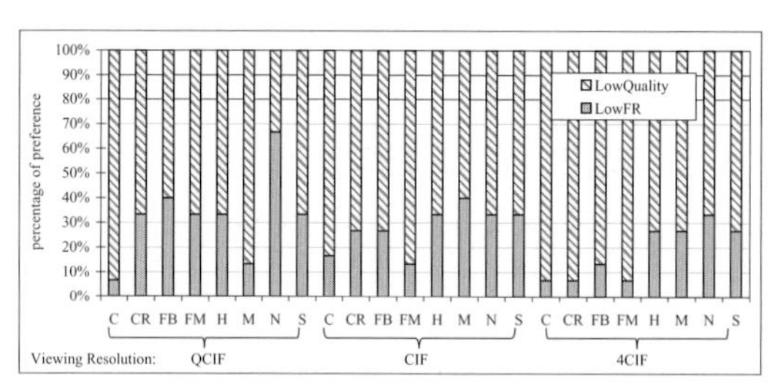

(a) temporal vs. quality scalability

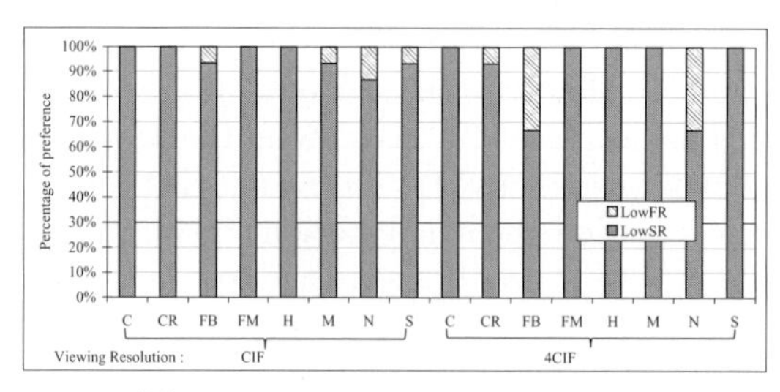

(b) temporal vs. spatial scalability

Fig. 3. Percentages of participants' preference for scalabilities

3.4 Other Effects on UPQ

In order to understand in depth what factors influence people's evaluation behaviors, we designed pre and post questionnaire and we also asked some specific questions during the two experiments. Statistical analysis (ANOVA, Levene's test, and T test)[2] were then carried out to analyse the influencing factors.

Personal preference for content type. There is no evidence that personal preference for content type (e.g., sport, landscape, and figures) is related to the perceptual quality. Even if some respondents believed they gave a higher score to their favorite content, we observed that their scores did not represent any preference. One-way ANOVA analysis on personal preference for "football" shows that there is no different between two groups of "like" and "dislike" (Levene statistic (1,18) = 0.325, p = 0.584; F(1,18) = 0.002, p = 0.969). In other words, the personal preference (like or dislike) does not influence the rating of video quality.

User profile. According to the results of ANOVA analysis, gender has no impact on UPQ ($p>0.4$ for all the eight contents). However, people's experience with image processing to the extent impacts on UPQ. And it is interesting that the impact is dependent on content. The results of independent samples t-test reveal that only for "soccer", "harbour" and "news", there are some evidence that UPQ is influenced by viewer's experience in image ($p<0.1$). Here, we must point out that the above analysis may be affected by the limited number of samples. That is why we did not analyse other factors of user profile.

Double image/shadows & green blocks. Under our encoding parameters, double image/shadows and green blocks were observed at the base layer of CIF and 4CIF spatial resolution (see Fig. 4a and 4b). The phenomena are caused by adaptive inter-layer prediction and missing reference blocks when loopfilter is run at the decoder. Shadows have a badly influence on user perceived quality. Most respondents complained that *double image made them dizzy*. Shadows are more serious in the motion areas and easily perceived under a big spatial resolution. People's complaints about green blocks are far less than shadows, but the green blocks are irritating when they happen frequently (especially in fast motion sequences) or appear in important regions (such as face).

Blur and blockiness. The influence of blur and blockiness on UPQ depends on the degree of blur and blockiness. In our experiment, most participants regarded the two phenomena as just one, that is unclearness, and they considered unclearness was worse than the green blocks.

[2] Analysis of variance (ANOVA) is used to test hypotheses about differences between two or more means. Refer to `http://davidmlane.com/hyperstat/intro_ANOVA.htmlforthedetailsofANOVA/F-test`. Levene's test is for equality of variances, in which $p > 0.05$ tells there is enough evidence that the variances are equal. The Independent Samples T Test is used to compare the means of two independent groups.

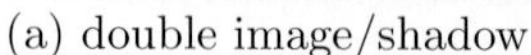

(a) double image/shadow

(b) green blocks

Fig. 4. Coding distortions

4 Conclusion and Future Work

In this paper, we present results from a study on the correlation of video content, scalabilities (spatial, temporal and quality scalability) and bitrate with user perceived quality through two parts of subjective quality assessment. There are five important findings from this study. First, the correlation between video content and user perceived quality exists at all hierarchical layers but is affected by spatial resolution, quantization quality and frame rate. Second, as the most important characteristic of video content, motion exerts a great impact on UPQ. Generally, most research focuses on motion intensity (fast or slow). However, more features of motion should be noticed, such as, the proportion of motion area (used in [3]) and varying frequency of motion direction. Third, content features can contribute to smooth high bitrate and improve the encoding efficiency. For a long video, if different segments and different layers are encoded with proper quantization parameters in terms of content characteristics, bitrate saving and traffic smoothness can be achieved without decreasing perceptual quality greatly. Fourth, during H.264/SVC video delivery, there are some important scalability strategies to adapt bandwidth requirements and at the same time to maintain a good perceptual quality: 1) given a fixed spatial layer, if both quality and temporal scalability can meet the bandwidth limitation, quality scalability should be prioritized; whereas 2) if the frame rate has to be lower than 7.5fps for meeting the bandwidth limitation, spatial scalability should be prior to temporal scalability. Finally, user profile has a slight impact on UPQ, which is relevant to the content; and some quality losses caused by codec negatively affect viewer's perception and should be improved.

Currently, based on this study, content-related bitrate modeling has been developed for scalable videos. Our next work will focus on optimal video adaptive delivery combining the bitrate models with user perceived video quality. Moreover, we are planning to conduct extensive experiments for gaining more user data on different display devices (e.g., mobile devices) and testing various encoding parameters and long video clips which might lead to different influence on UPQ [12]. We would also like to improve coding performance of H.264/SVC.

References

1. Schwarz, H., Marpe, D., Wiegand, T.: Overview of the scalable video coding extension of the H.264/AVC standard. IEEE Transactions on Circuits and Systems for Video Technology 17, 1103–1120 (2007)
2. Ahmad, A.M.A.: Content-based Video Streaming Approaches and Challenges. In: Ibrahim, I.K. (ed.) Handbook of Research on Mobile Multimedia, pp. 357–367. Idea group Reference, London (2006)
3. Ries, M., Crespi, C., Nemethova, O., Rupp, M.: Content based video quality estimation for H.264/AVC video streaming. In: IEEE WCNC 2007, pp. 2668–2673. IEEE Press, Los Alamitos (2007)
4. Wang, Y., van der Schaar, M., Chang, S.-F., Loui, A.C.: Classification-based multidimensional adaptation prediction for scalable video coding using subjective quality evaluation. IEEE Transactions on Circuits and Systems for Video Technology 15, 1270–1279 (2005)
5. Wang, D., Speranza, F., Vincent, A., Martin, T., Blanchfield, P.: Toward optimal rate control: a study of the impact of spatial resolution, frame rate, and quantization on subjective video quality and bit rate. In: Visual Communications and Image Processing 2003, vol. 5150, pp. 198–209. SPIE (2003)
6. McCarthy, J.D., Sasse, M.A., Miras, D.: Sharp or smooth?: Comparing the effects of quantization vs. frame rate for streamed video. In: SIGCHI conference on Human factors in computing systems, pp. 535–542. ACM, New York (2004)
7. Zhai, G., Cai, J., Lin, W., Yang, X., Zhang, W., Etoh, M.: Cross-Dimensional perceptual quality assessment for low Bit-Rate videos. IEEE Transactions on Multimedia 10, 1316–1324 (2008)
8. Niedermeier, F., Niedermeier, M., Kosch, H.: Quality assessment of MPEG-4 scalable video CODEC. In: ICIAP2009. LNCS, vol. 5716, pp. 297–306. Springer, Heidelberg (2009)
9. ITU-T: Subjective video quality assessment methods for multimedia applications (1999)
10. Apteker, R.T., Fisher, J.A., Kisimov, V.S., Neishlos, H.: Video acceptability and frame rate. IEEE Multimedia 2, 32–40 (1995)
11. Van der Auwera, G., David, P.T., Reisslein, M.: Traffic and quality characterization of single-layer video streams encoded with the H.264/MPEG-4 advanced video coding standard and scalable video coding extension. IEEE Transactions on Broadcasting 54, 698–718 (2008)
12. Knoche, H.O., McCarthy, J.D., Sasse, M.A.: How low can you go? The effect of low resolutions on shot types in mobile TV. Multimedia Tools and Applications 36, 145–166 (2008)

Subjective Experiments on Gender and Ethnicity Recognition from Different Face Representations

Yuxiao Hu[1], Yun Fu[2], Usman Tariq[3], and Thomas S. Huang[3]

[1] Microsoft Corporation, 1 Microsoft Way, Redmond, WA 98052 USA
[2] Department of CSE, University at Buffalo (SUNY), NY 14260, USA
[3] Beckman Institute, University of Illinois at Urbana-Champaign, Urbana, IL 61801 USA
Yuxiao.Hu@microsoft.com, raymondyunfu@gmail.com,
{utariq2,huang}@ifp.uiuc.edu

Abstract. The design of image-based soft-biometrics systems highly depends on the human factor analysis. How well can human do in gender/ethnicity recognition by looking at faces in different representations? How does human recognize gender/ethnicity? What factors affect the accuracy of gender/ethnicity recognition? The answers of these questions may inspire our design of computer-based automatic gender/ethnicity recognition algorithms. In this work, several subjective experiments are conducted to test the capability of human in gender/ethnicity recognition on different face representations, including 1D face silhouette, 2D face images and 3D face models. Our experimental results provide baselines and interesting inspirations for designing computer-based face gender/ethnicity recognition algorithms.

Keywords: Face analysis, gender recognition, ethnicity recognition, subjective experiment.

1 Introduction

Humans are quite accurate when deciding the gender of a face even when cues from makeup, hairstyle and facial hair are minimized [1]. Bruce et al. showed that cues from features such as eyebrows and skin texture play an important role in decision making as the subjects were less accurate when asked to judge a face based solely upon its 3D shape representation compared to when 2D face images (with hair concealed and eyes closed) were present. According to [1], the average male face differs from an average female face in 3D shape by having a more protuberant nose/brow and more prominent chin/jaw.

The results from [2] shows that both color and shape are vital for humans in deciding the gender and ethnicity from a 2D face image. Color was more important factor in gender decisions while shape was more important in deciding ethnicity. However, when the both sources of information were combined, the dominant source depended on viewpoint. Shape features proved to be more important in angled views while the color in full-face view.

S. Boll et al. (Eds.): MMM 2010, LNCS 5916, pp. 66–75, 2010.

Males, females and different ethnicities also differ in silhouetted face profile, which can be represented in 1D stimulate. For instance, the ridge of the bone above the eye is more pronounced in males [3]. Also the forehead in case of males is backward sloping while that of the females tends to be more vertical. The female nose tends to be more concave and straighter. The distance between top lip and base of the nose is usually longer in males. The chins of males are much taller than in females.

Overall, humans are able to determine the facial categorization by 3D, 2D and 1D cues or their combination. Different to the case of humans, gender and ethnicity recognition present two of the major challenges of human face analysis in human computer interface researches. Gender recognition is useful in face recognition as it can reduce the problem to matching the face with almost half of the database (if both the genders have equal probability of occurrence in the database). Ethnicity identification will reduce the problem even further. Gender and Ethnicity recognition can also be useful in Human Computer Interface (HCI). The computer shall adapt to the person's gender and ethnic group in terms of speech recognition or offering the person options which is more specific and useful to a particular gender or ethnicity.

The earliest technique for gender classification was based upon neural networks [4]. SEXNET, a fully connected two-layer network, was incorporated in [5] to identify gender from face. A PCA based representation along with radial basis functions was employed by Abdi et al. [6]. A mixed approach based upon collection of neural networks and decision trees was employed by Gutta et al. [7]. In [8], a multi-layered neural network is used to identify gender from multi-resolution face images. Better performance has also been claimed by O' Toole et al. [9] using PCA and neural networks. In [10], Jain et al. used Independent Component Analysis (ICA) to represent each frontal face image as a low-dimensional feature vector for gender identification. They reported superior performance of their approach using support vector machine in ICA space with a 96% accuracy rate. In [11], a skin-color based approach has been employed for gender identification.

2 Setup of Subjective Experiments

2.1 Test Face Samples

The face data used in the subjective experiments were collected by our group with Cyberware 3D laser scanner [12]. We randomly selected 100 3D face data samples with balanced ethnicities. Some examples of the face representations are shown in Fig. 1. The gender and ethnic groups' distribution of the face data are shown in Fig. 2.

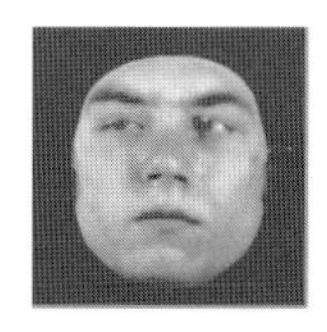

Fig. 1. Different face representations of different faces: silhouette, 2D image, and 3D model

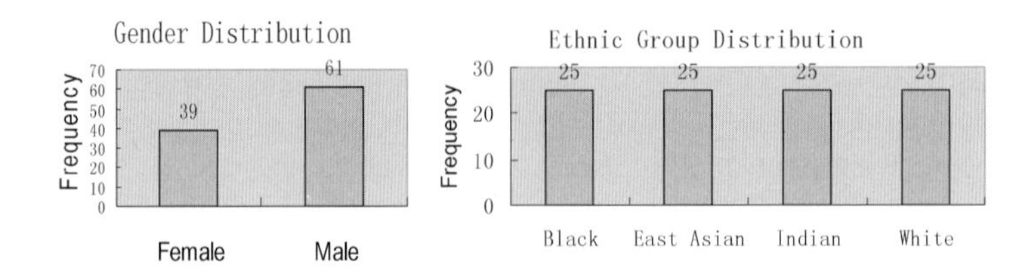

Fig. 2. Gender and ethnics group distribution of the test face

These 100 face data were randomly divided to 3 groups and each group was shown either in silhouette, 2D Face Images or 3D face models representation to each subject. The partition of the data is shown in Table 1.

Table 1. The partition of the test face data

Sample Groups	Number of Samples
SaG#1	33 (#1~#33)
SaG#2	33 (#34~#66)
SaG#3	34 (#67~#100)
Total:	100

2.2 Subjects

Twenty-one graduate students attended the experiments as subjects. They belonged to either "East and South East Asian" or "Indian" or "White" ethnic category. We randomly divided these subjects to 3 groups as shown in Table 2.

Table 2. The partition of the subjects

Subject Groups	Number of Subjects
SuG#1	8
SuG#2	6
SuG#3	7
Total:	21

2.3 Experiments

In the subjective experiments, each subject group was shown three different sample groups in different face representations as shown in Table 3. In this manner, each subject labeled 100 faces, which were grouped into three groups and shown in profile silhouette, 2D Frontal Image or 3D respectively. Each face was shown in three different representations to different subjects, but different representations of a face did not appear in the same test set for a subject. Each face representation was labeled by at least six subjects.

Table 3. The task assignment of different subject groups

Subject Groups	Silhouette	2D texture	3D
SuG#1	SaG#3	SaG#2	SaG#1
SuG#2	SaG#1	SaG#3	SaG#2
SuG#3	SaG#2	SaG#1	SaG#3

Each subject was requested to provide his/her gender and ethnic group information before the experiment. For each face shown on the screen, every subject was asked to judge its gender and ethnicity and specify his/her confidence level of the answer. For gender, the answers could be male or female. For ethnicity, the answers could be one of the following: Black (African or African American), East and Southeast Asian (Chinese, Japanese, Korean, Vietnamese, Filipino, Singaporean ... American Indian), South Asian (Indian, Pakistani ...), White (Caucasian, Hispanic/Latino, Middle Eastern). And the confidence level of above answers could be "Purely guess", "Not sure", "Kind of sure" or "Pretty sure". A screen shot of the experiment program interface is show in Fig. 3, where a profile face is shown.

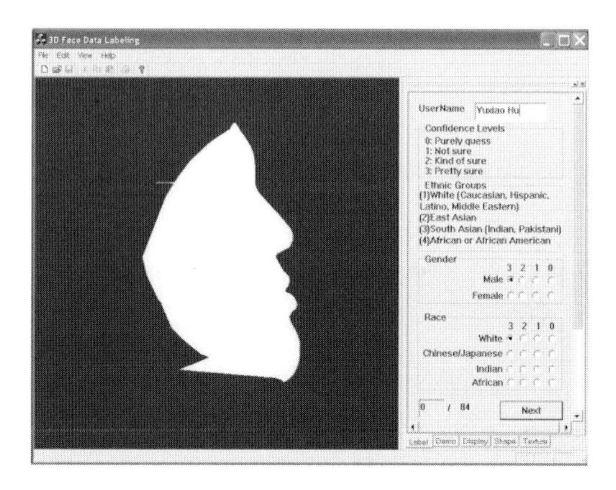

Fig. 3. The experiment program interface

Table 4. Overall accuracy of gender/ethnicity recognition

Accuracy(%)	Gender		Ethnicity	
	Mean	StdDev	Mean	StdDev
Silhouette	57.63%	9.43%	45.08%	18.75%
2D Image	89.41%	6.14%	80.40%	13.52%
3D	92.29%	4.93%	86.00%	12.85%

3 Experimental Results

3.1 Performance Comparison on Different Face Representations

Most of the subjects finished labeling the 100 face samples in 20 minutes. The median time they spent was 896 seconds. Here we didn't use average time because average number is sensitive to outliers. The accuracy of gender and ethnicity recognition was calculated as the ratio between the number of correctly recognized face samples and the total sample number. The overall accuracies of gender recognition on different face representations are shown in Table 4.

It can be seen from these experimental results that gender recognition by human based on silhouette is only a little bit better than chance, which is lower than the accuracy claimed in [13]. On the other hand, ethnicity recognition based on silhouette is significantly above chance. Based upon 2D Image and 3D information, both gender and ethnicity recognitions are pretty accurate. The confusion matrices on gender recognition of different face representation are shown in Table 5. We can see that male faces are better recognized than female faces. As [13] suggested, the reason is that male faces have larger variances on shape and appearances.

The confusion matrices on ethnicity recognition are shown in Table 6. From the confusion matrices on ethnicity, we can see that: Based on silhouette (1D shape), Indians are similar to White and East Asian people; Based on 2D image (texture), Black and Indian are not well separated; Based on 3D information (shape and texture), Black and Indian are better recognized.

Table 5. Confusion matrices on gender recognition based on different face representations

		Recognized As	
Profile Silhouette		Male	Female
GroundTruth	Male	**63.73%**	36.27%
	Female	52.03%	**47.97%**
2D Frontal Face Image		Male	Female
GroundTruth	Male	**94.78%**	5.22%
	Female	18.77%	**81.23%**
3DFace(Shape+Texture)		Male	Female
GroundTruth	Male	**95.33%**	4.67%
	Female	12.55%	**87.45%**

The confusion matrices of the ethnicity recognition accuracy with respect to the subjects from different ethnic group are compared in Table 7. It can be clearly seen that subjects from different ethnic groups recognize the faces of their own ethnic group more accurately.

Table 6. Confusion Matrices on ethnicity recognition based on different face representations

		Recognized As			
Profile Silhouette		White	EastAsian	Indian	Black
Ground Truth	White	64.31%	17.56%	13.47%	4.67%
	EastAsian	25.13%	48.55%	11.71%	14.61%
	Indian	**47.97%**	**26.26%**	**17.67%**	8.09%
	Black	8.75%	14.39%	21.87%	54.99%
2D Frontal Face Image					
Ground Truth	White	83.14%	8.70%	8.17%	0.00%
	EastAsian	5.04%	92.74%	1.68%	0.55%
	Indian	13.24%	5.97%	**60.28%**	**20.51%**
	Black	1.10%	2.72%	**14.05%**	**82.14%**
3D Shape + Texture					
Ground Truth	White	90.58%	7.07%	2.34%	0.00%
	EastAsian	3.45%	95.38%	1.16%	0.00%
	Indian	14.86%	2.28%	**75.43%**	7.43%
	Black	1.13%	2.21%	13.32%	**83.34%**

Table 7. Confusion Matrix of different subject ethnic group based on 3D face representation

		Recognized As			
Indian subjects:		White	EastAsian	Indian	Black
Groud Truth	White	100.00%	0.00%	0.00%	0.00%
	EastAsian	5.57%	94.43%	0.00%	0.00%
	Indian	0.00%	0.00%	**100.00%**	0.00%
	Black	0.00%	0.00%	3.34%	96.66%
Chinese subjects:					
Groud Truth	White	89.31%	8.03%	2.66%	0.00%
	EastAsian	1.74%	**97.41%**	0.85%	0.00%
	Indian	15.33%	2.41%	72.57%	9.70%
	Black	0.85%	2.59%	18.28%	78.27%
Western subjects:					
Groud Truth	White	**90.46%**	7.16%	2.39%	0.00%
	EastAsian	7.33%	90.23%	2.44%	0.00%
	Indian	14.28%	2.04%	81.63%	2.04%
	Black	2.86%	2.86%	5.73%	88.55%

3.2 Performance Comparison of Different Subjects

Among the 21 subjects, the best overall performance of gender and ethnicity recognition is 84%. And the average recognition accuracies of gender and ethnicity recognition are 79.76% and 70.48%. More detailed information about the performance of different subjects is shown in Table 8, Fig. 4, Fig. 5 and Fig. 6; from which we can see that there is no strong correlation between the ethnic group of the subjects and their overall recognition performances.

Table 8. Top ranked subjects on gender and ethnicity recognition accuracy

Rank	Subject No.	Accuracy		
		Gender	Ethnicity	Overall
1	#7, East Asia	85.00%	83.00%	84.00%
2	#9, White	83.00%	78.00%	80.50%
3	#21, Chinese	79.00%	80.00%	79.50%
4	#15, White	83.00%	75.00%	79.00%
5	#6, Indian	86.00%	71.00%	78.50%
Mean Accuracy of All Subjects		79.76%	70.48%	75.12%
Std. Dev. of All Subjects		3.95%	8.45%	4.72%
Max Accuracy of All the 21Subjects		86.00%	83.00%	84.00%
Min Accuracy of All the 21Subjects		73.00%	47.00%	63.00%

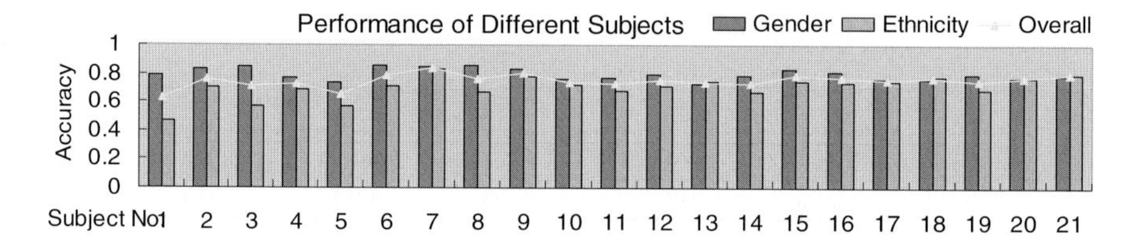

Fig. 4. Performance of gender and ethnicity recognition of different subjects

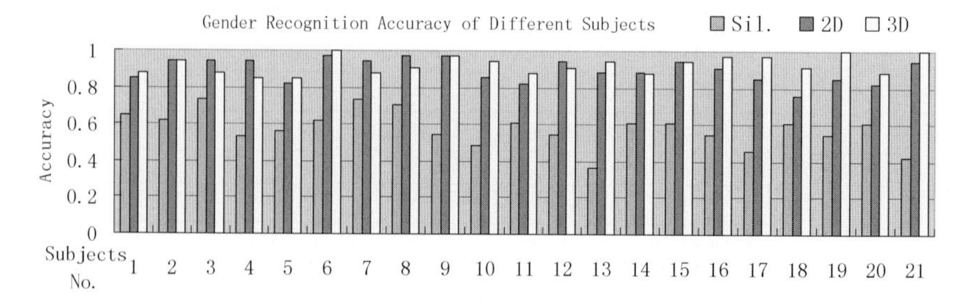

Fig. 5. Gender recognition accuracy of different subjects on different face representations

3.3 Improving the Performance by Combining Multiple Results

In the end, we tried to combine the recognition results of all the subjects by majority vote to see whether this would improve the accuracy of gender and ethnicity recognition. Here each face sample had at least 6 votes in one representation. If there is a tie,

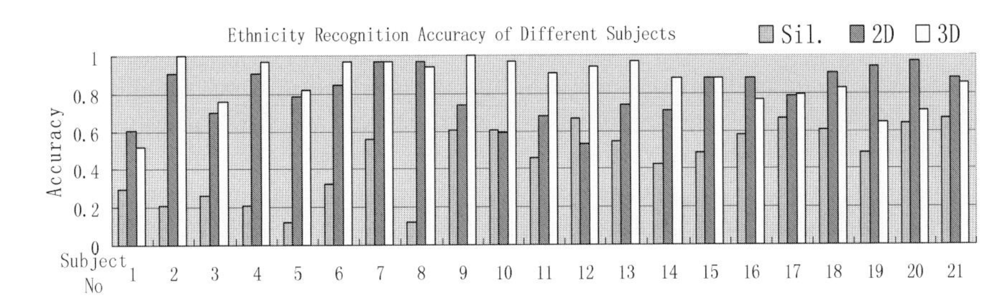

Fig. 6. Ethnicity recognition accuracy of different subjects on different face representations

we would randomly select a result as the decision. We could either vote on the same representation or vote across different representations. The results of voting on same representation are shown in Table 9. The results of voting across different representations are shown in Table 10.

Table 9. Combining the recognition results

Accuracy(%)	Gender		Ethnicity	
	Average	Vote	Average	Vote
Silhouette	57.63%	58%	45.08%	55%
2D Image	89.41%	96%	80.40%	83%
3D	92.29%	97%	86.00%	93%

Table 10. Combining the recognition results across different representations

Accuracy(%)	Gender	Ethnicity
sil+2D+3D	94%	90%
sil+2D	91%	83%
sil+3D	91%	89%
2D+3D	95%	91%

From above results, we can observe that, 1) Voting on gender based on silhouette doesn't help much. 2) Voting on ethnicity based on silhouette increases the accuracy about 10%; 3) Voting on gender/ethnicity based on 2D/3D also improves the accuracy; 4) Combining Silhouette, 2D or 3D does not improve the accuracy. These observations inspire us that if we design multiple classifiers based on different facial feature/representations, their result combination will be better than single classifier on single type of feature.

4 Conclusions

In this work, several subjective experiments were conducted to test the capability of human in gender/ethnicity recognition on different face representations. Based on these results, following conclusions can be made: 1) Gender recognition based on silhouette is slightly better than chance, which indicates the difficulty of silhouette based gender recognition algorithms; 2) Ethnicity recognition based on silhouette is well above chance; 3) Gender/Ethnicity recognition base on 2D/3D information are very accurate. The more information we have, the better the accuracy is; 4) Male faces are better recognized than female faces; 5) Indian faces are better recognized based on 3D information (in balanced ethnic distribution); 6) Subjects from different ethnic groups recognize the faces more accurately from their own ethnic group.

Although above conclusions are drawn from subjective experiments, i.e., the gender/ethnicity recognition is performed by human subjects; computer-based automatic gender/ethnicity recognition algorithms can benefit from these conclusions to infer the effective face representations and discriminant features. That is, computers can "learn from human". Some very encouraging results have already been achieved by our group on silhouetted face profiles. The features used in these experiments were primarily shape context features [14]. In the near future, our plan is to do experiments on other views and come up with an improved classifier for automatic gender and ethnicity identification.

References

1. Bruce, V., Burton, A.M., Hanna, E., Healey, P., Mason, O., Coombes, A., Fright, R., Linney, A.: Sex discrimination: how do we tell the difference between male and female faces? Perception 22, 131–152 (1993)
2. Hill, H., Bruce, V., Akamatsu, S.: Perceiving the sex and race of faces: The role of shape and color. Proc. of the Royal Society - Biological Sciences (Series B) 261, 367–373 (1995)
3. The main differences between male and female faces, `http://www.virtualffs.co.uk/male.femalefacialdifferences.htm`
4. Cottrell, G., Metcalfe, J.: EMPATH: Face, Gender and Emotion Recognition using Holons. In: Proceedings of Advances in Neural Information Processing Systems, pp. 564–571 (1990)
5. Golomb, B.A., Lawrence, D.T., Sejnowski, T.J.: Sexnet: a neural network identifies sex from human faces. In: Lipmann, R.P., Moody, J.E., Touretzky, D.S. (eds.) Proc. of NIPS, vol. 3, pp. 572–577. Morgan Kaufmann, San Mateo (1990)
6. Abdi, H., Valentin, D., Edelman, B., O'Toole, A.: More about the difference between men and women: evidence from linear neural networks and the principal component approach. Perception 24, 539–562 (1995)
7. Gutta, S., Weschler, H., Phillips, P.J.: Gender and Ethnic Classification of Human Faces using Hybrid Classifiers. In: Proc. of IEEE Conf. on AFGR, pp. 194–199 (1998)
8. Tamura, S.H., Kawai, M.H.: Male/Female Identification from 8 x 6 Very Low Resolution Face Images by Neural Network. Pattern Recognition 29, 331–335 (1996)
9. O'Toole, A., Abdi, H., Deffenbacher, K., Valentin, D.: A low-dimensional representation of faces in higher dimensions of space. J. of Optical Society of America 10, 405–411 (1993)
10. Jain, A., Huang, J., Fang, S.: Gender identification using frontal facial images. In: Proc. of IEEE International Conference on Multimedia and Expo., pp. 1082–1085 (2005)

11. Yin, L., Jia, J., Morrissey, J.: Towards race-related face identification: Research on skin color transfer. In: Proc. of IEEE Conf. on AFGR, pp. 362–368 (2004)
12. Hu, Y., Zhang, Z., Xu, X., Fu, Y., Huang, T.S.: Building Large Scale 3D Face Database for Face Analysis. In: Sebe, N., Liu, Y., Zhuang, Y.-t., Huang, T.S. (eds.) MCAM 2007. LNCS, vol. 4577, pp. 343–350. Springer, Heidelberg (2007)
13. Davidenko, N.: Silhouetted face profiles: A new methodology for face perception research. Journal of Vision 7, 1–17 (2007)
14. Belongie, S., Malik, J., Puzicha, J.: Shape matching and object recognition using shape contexts. IEEE Trans. on Pattern Analysis and Machine Intelligence 24, 509–522 (2002)

Facial Parameters and Their Influence on Subjective Impression in the Context of Keyframe Extraction from Home Video Contents

Uwe Kowalik[1], Go Irie[1,2], Yasuhiko Miyazaki[1], and Akira Kojima[1]

[1] NTT Cyber Solutions Laboratories, Nippon Telegraph and Telephone Corporation
[2] Dept. Information and Communication Engineering, University of Tokyo
{kowalik.uwe,irie.go,miyazaki.yasuhiko,kojima.akira}@lab.ntt.co.jp

Abstract. In this paper, we investigate the influence of facial parameters on the subjective impression that is created when looking at photographs containing people in the context of keyframe extraction from home video. Hypotheses about the influence of investigated parameters on the impression are experimentally validated with respect to a given viewing perspective. Based on the findings from a conducted user experiment, we propose a novel human-centric image scoring method based on weighted face parameters. As a novelty to the field of keyframe extraction, the proposed method considers facial expressions besides other parameters. We evaluate its effectiveness in terms of correlation between the image score and a ground truth user impression score. The results show that the consideration of facial expressions in the proposed method improves the correlation compared to image scores that rely on commonly used face parameters such as size and location.

Keywords: Keyframe extraction, home video contents, image ranking, facial expressions.

1 Introduction

Due to the increasing popularity of consumer video equipment the amount of user generated contents (UGC) is growing constantly in recent years. Besides taking photographs also filming at family events or during travel became a common habit for preserving valuable moments of one's life. Whereas taking a good picture of a person with a photo camera requires a decent skill and good timing, a video can easily capture 'good shots' due to its nature. Thus extracting images from video contents is an interesting complement to the task.

In recent years the problem of automatically extracting representative keyframes from video contents attracts the attention of many researchers and various approaches have been proposed to tackle this sub-problem of video abstraction. Existing keyframe extraction techniques can be categorized based on various aspects such as underlying mechanisms, size of keyframe set and representation scope [1]. From the viewpoint of analyzed base units, there exist two

S. Boll et al. (Eds.): MMM 2010, LNCS 5916, pp. 76–86, 2010.

categories: shot based extraction methods and clip based methods. Whereas shot based approaches involve always a shot detection step and thus are limited in their application to structured (i.e. edited) video contents, clip based methods can be applied to unstructured video material such as user generated contents. Despite the fact that there exists a huge variety of proposed keyframe extraction approaches, little attention has been paid to the fact that extracted keyframes should also be visually attractive to the viewer.

In this paper, we address this issue for the home video content which is one specific domain of UGC. Home video contents contain often imagery of people, relatives and friends taken at various occasions and therefore human faces are intuitively important. In addition, the human faces do attract a viewer's attention [2,3]. Provided an application context where the objective is to extract frames from videos with respect to the evoked impression, such as creating a family's photo album and sending picture e-mails, a fully automatic keyframe extraction approach will ideally select images that are considered as 'good shot of the person(s)'. The central problem is how to automatically determine such 'good shots' inside video sequences. An important point is that selected video frames should suffice a certain image quality. This can be achieved by standard techniques such as image de-noising, contrast and color enhancement [4], advanced methods for increasing the image resolution [5], and image de-blurring [6]. In contrast, we focus in this paper on intrinsic image properties that can not be easily changed without altering the image semantics.

More specifically, main contributions of this work are:

- To investigate the influence of facial parameters present in images on the viewer's impression in the context of keyframe extraction from home video contents
- To present results of a conducted subjective user experiment
- To propose an image scoring method based on a weighted combination of extracted face parameters

As a novelty to the field of keyframe extraction, the proposed image score considers facial expressions besides other parameters. We discuss different face parameter combinations and their influence on the performance of a general linear weighting model used for image score estimation. We show results that confirm the effectiveness of included facial expression parameters.

2 Related Work

We focus in this section on the conventional methods that aim on keyframe extraction from unstructured video such as UGC.

In [7] a keyframe extraction approach suitable for short video clips is introduced in the context of a video print system. Keyframes are determined by employing face detection, object tracking and audio event detection. Location of visual objects is taken into account by predefining regions of importance inside the video frame and deriving an appropriate region based frame score. Although the parameters of visual objects (and especially faces) are rather intuitively considered for frame score calculation in the above approach, the example illustrates the awareness of visual object importance in the context of automatic keyframe extraction.

An interesting study focusing on unstructured user generated contents was recently presented in [8]. The authors conducted a psycho-visual experiment in order to derive common criteria for human keyframe selection shared amongst two user groups with different viewing perspectives i.e. first-party users (photographers) and third-party users (external observers). A rule-based keyframe extraction framework is proposed that uses inference of the camera operator's intend from motion cues derived from estimated camera and object motion patterns rather than recognizing the semantic content directly. The authors compare their method with a histogram based and uniform sampling keyframe extraction approach and showed the effectiveness of the algorithm by improved accuracy with respect to the ground truth selected by a human judge. The authors suggest that the accuracy can be further improved by inclusion of semantic information such as facial expressions and gaze at camera. In contrast to the system proposed in [8], this paper focuses on such highly subjective parameters that influence the user's keyframe choice wrt. the *attractiveness* rather than approaching the keyframe extraction problem with the goal to extract interesting images from the video.

In [9], a user-centric weighting model for image objects is proposed that defines an importance measure for images based on users' perception. The authors investigated the relationship between parameters of visual objects such as size, location and the user's perception of 'image aboutness' with respect to one given, object specific query term. It was shown that object size and object location are directly related to the image concept recognized by a user and that their proposed weighting model is efficient for image ranking in the context of concept based image retrieval.

In this paper we view the task of keyframe extraction as a conceptual image ranking problem within a given video sequence. We focus on home video contents since there is a wide range of potential applications for human-centric video indexing technology in this domain. The goal of this study is to provide some insight in how the presence of faces in images influences the viewers' impression in terms of 'a good photograph of a person'.

3 Impression Concept and Investigated Face Parameters

In this work, we define the impression evoked at viewers when looking at a photograph as based on the concept of 'a good picture of a person'. It is assumed that various face parameters contribute to the impression with respect to this conceptual viewpoint. We assess two different face parameter types with regard to their contribution:

1. Image structure related parameters
2. Emotion related parameters

Image structure related face parameter considered in this work are *number of faces*, *face coverage* and *face location*. As a novelty to the task of keyframe extraction, we consider also *emotional* face parameters. We model emotional face parameters by a prototypic facial expression class label assigned to each detected face region according to Ekman's six basic emotions [10]. In particular we focus on two prototypic facial expressions *joy* and *neutral* in this study. Joyful faces are commonly considered to be attractive to viewers. The neutral expression is important, since it reflects the 'normal' state of human faces. In the following, we present our assumptions regarding the relationship between each facial parameter and the impression evoked at viewers where the impression is quantified by means of a user provided impression score.

Face Number N_f: As suggested by related work we assume that images containing more faces are considered to create a better impression at the viewer and thus the *face number* is positively correlated with a user provided impression score.

Face Coverage S: We define face coverage (hereafter: *coverage*) as the ratio between the image area covered by faces and the overall image area as:

$$S_{image} = \frac{1}{A^{image}} \sum_i A_i^{face} \tag{1}$$

where A^{image} denotes the image area and A_i^{face} is the image area covered by i-th face. We assume that larger faces evoke a stronger impression at viewers and thus coverage has a positive correlation with a user provided impression score.

Face Location P_R: We use a region of interest (ROI) approach for describing the face location. P_R is defined with respect to three predefined ROI based on a bary center of a face's rectangular bounding box. Fig. 1 depicts the predefined regions. Our assumption is that there exist preferred ROIs that will lead to a better impression if a face lies inside such a region.

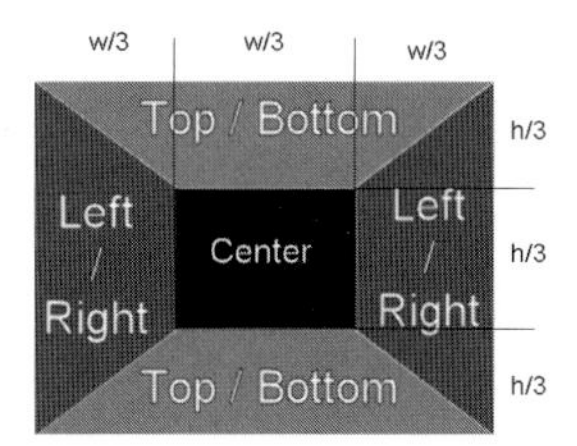

Fig. 1. Three predefined ROI (w: image width, h: image height)

We model the face location P_R by the probabilities that a face lies inside a region R as given in the formula:

$$P_R = \frac{N_f^R}{N_f} \tag{2}$$

where $R \in \{top/bottom, center, left/right\}$ and N_f^R refers to the number of faces inside the region R. We define $P_R = 0$ if $N_f = 0$.

Facial Expressions P_E: Our hypothesis is that facial expressions present in images are important for the overall impression. We assume that the presence of joyful faces will influence the impression positively whereas a present neutral facial expression will have little or no correlation with the user score. We parameterize facial expressions by their occurrence probability P_E in a frame as given in the fomula:

$$P_E = \frac{N_f^E}{N_f} \tag{3}$$

where $E \in \{neutral, joy\}$ and N_f^E equals the number of faces displaying the expression E. This general formulation allows for future extension by adding other prototypic facial expressions. We define $P_E = 0$ if $N_f = 0$.

4 Subjective User Experiment

Goal of the user experiment was to acquire ground truth data in form of a score that quantifies the subjective impression evoked at the participants when looking at the extracted images. We prepared two facial parameter sets based on the images used for the experiment. The first set was manually labeled and provides ground truth data for parameter assessment under the assumption of an ideal system. In order to be able to draw conclusions about the impression-parameter relationship under practical conditions, the second parameter set was generated during the fully automatic keyframe extraction process used for preparing the test images. In the following subsections, we first describe the data preparation, and next provide a description of the experimental conditions.

4.1 Selection of Video Clips

Video clips of three different categories were selected from private home video collections and a video sharing website [12]. We consider creating photo albums from home video as one of the main applications for keyframe extraction techniques and choose therefore two popular categories which roughly relate to the 'home photo' taxonomy proposed in [13], i.e. *travel* and *family Events*. Investigating the thumbnail previews in the family category of [12], we found that nearly 80% of the uploaded video clips are about children and decided therefore to add a special *kids* category to our test set. We selected two typical clips from each of three categories resulting in overall six video clips. The clip properties are listed in Table 1. The number of extracted images per clip used for the experiment is also listed. Fig. 2 illustrates the video contents by providing one example for each category.

4.2 Automatic Keyframe Extraction Approach

Fig. 3 shows the block diagram of the keyframe extraction method. In order to avoid quality degradation by motion blur, the video is first analyzed to detect

Table 1. Video Clip Properties

Category	Title	Length (min)	Num. of Frames	Resolution / FPS	Num. Extracted Frames
Travel	seaside	12:17	11000	320x240 / 15	23
	disney land	02:40	4800	320x240 / 30	7
Family Events	birthday	04:50	8600	320x240 / 30	21
	exhibition	02:40	4800	320x240 / 30	4
Kids	baby	02:00	3700	320x240 / 30	4
	stairs	00:50	1400	320x240 / 30	5

Fig. 2. Example images (left to right: travel, family event, kids)

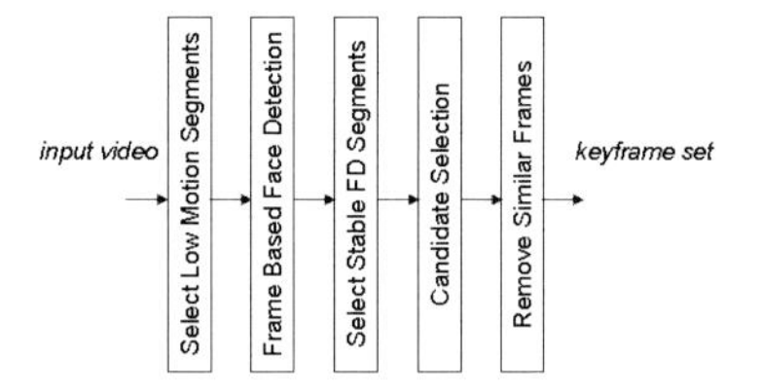

Fig. 3. Keyframe Extraction Process

strong global motion patterns by estimating a segment based motion activity, which is calculated for non-overlapping consecutive groups of N video frames by evaluating the structural similarity between neighboring frames. As for the structural description, we employ a feature vector constructed from Gabor-Jets extracted at equidistant grid positions. We employ the similarity measure introduced in [14] for calculating the similarity between adjacent video frames I_k and I_{k+1}. As for the Jet-configuration, we use five frequencies and eight orientations, the same parameters given in [14].

Video segments that contain high motion activity show usually a low similarity between adjacent frames. We calculate the average similarity for each segment of N video frames and apply a threshold th. Segments holding an average similarity $< th$ are discarded. In our implementation $N = 10$ and $th = 0.9$ lead to good blur suppression results. Only video segments with low motion activity are preselected for further processing and subjected to frame-wise face detection based on the approach introduced in [15]. The number of candidate frames is further reduced by removing video frames with unstable face detection result. We apply a sliding window function of length $K (K = 3)$ and remove frames where the face count in range divided by K is less than 1.0. A redundancy removal step finally removes visually similar frames in sequential order by calculating the inter-frame similarity of adjacent frames as described above and applying a threshold ($th_2 = 0.7$). Frames with a similarity higher than th_2 are removed and the remaining set of keyframes was used for the subjective experiment. Overall 64 images were automatically extracted from the six video clips. Based on the approach given in [11], we have implemented an automatic facial expression detection module for the detection of neutral and joyful facial expressions. It utilizes a neural network for classifying facial expressions based on Gabor-Wavelet feature vectors extracted from detected face image regions. Facial expression detection was applied on the extracted images in order to create the automatically extracted face parameter set mentioned above.

4.3 Experimental Condition

In our experiment, we asked subjects to watch the extracted video frames and provide their opinion with respect to a given question about the images. The experiment was performed at an university amongst students and staff members. Participants were unrelated to the people included in the imagery thus the judgment was given from a 'third-party' viewpoint. The overall number of participants was 22 (19 males, 3 females) in the age between 21 to 49 years. Images were displayed in the center of the computer screen in original size and in random order. Participants were asked to give their feedback with respect to the question

"Do you think this picture is a 'good photograph' of the person?"

The feedback was given directly after watching each single image by selecting a score on a 7-level ordinal scale ranging from -3 for *'not at all'* to +3 for *'I fully agree'*.

5 Experimental Results and Discussion

In this section, we first investigate the relationship between each single face parameter and the subjective user score and draw conclusions regarding the validity of the assumptions made in section 3. This investigation is performed under the assumption of an ideal system, i.e. we use the ground truth face data. Based on the results, we introduce next a general weighting model which uses the validated face parameters for calculating an image score that considers the subjective impression of viewers and is useful for human-centric image ranking with respect to the viewing perspective previously defined. We show the performance of different score predictor combinations in terms of rank-correlation between the predicted image score and the impression score acquired during the user experiment from section 4. Finally we select the best parameter combination and compare the performance of the weighting model for ideally labeled face parameters and the practical case where face parameters are automatically estimated.

5.1 Single Feature Correlation

We use the Spearman rank-correlation measure because the user score acquired during the experiment is ordinal-scaled. In order to remove a possible bias that may have been introduced due to individual differences of the participants, we calculate the normalized average score. Since the original user score is an ordinal scaled value, we decided to calculate the correlation for the median score as well. Table 2 shows the correlation between user score and ground truth face parameters for both score types. In addition the one-side p-value for each feature is calculated for assessing the statistical significance of the correlation value. As can be seen the correlation does not differ much between normalized average score and median score. Thus using either score type is valid. The following discussion refers to the ground truth median user score (hereafter 'user score'). As for the investigated image structure related face parameters only the coverage feature shows statistically significant correlation. The other features, i.e. number of faces and face location show no or little correlation, but more important the correlation is statistically not significant. Thus our hypotheses regarding these face parameters is not validated. A similar result was presented by the authors in [9] and we conclude that our assumption is validated that bigger faces are not only more important to the user, but also contribute to a better overall impression of a photograph, whereas face location and number of faces are not valid features for deriving an image score which relates to the viewers' impression of 'a good photograph of a person'.

Table 2. Correlation of Face Parameters and Median User Score (valid features bold)

Category	Feature	Norm. Avg. Score		Median Score	
		rho	p-one	rho	p-one
Structure	Num Faces	-0.1283	0.156	-0.0876	0.246
	Center	-0.0061	0.480	0.0785	0.269
	Top/Bottom	0.1029	0.211	0.0164	0.449
	Left/Right	-0.1140	0.186	-0.0779	0.269
	Coverage	**0.2410**	**0.027**	**0.2270**	**0.035**
Emotion	**Neutral**	**-0.4183**	**<0.001**	**-0.4497**	**<0.001**
	Joy	**0.4183**	**<0.001**	**0.4497**	**<0.001**

Facial expressions seem to have a quite strong influence on the viewers' opinion. Our assumption that a joyful facial expression contributes positively to the impression was confirmed by the statistically significant positive correlation. Even more interesting we found that neutral facial expressions create a negative impression when looking at a photograph of a person. We conclude that facial expressions are an important factor of the impression created when looking at human photographs. We explain the relatively small correlation values by the influence of other image properties not discussed in this paper. Participants stated after the experiment that they considered also the gaze direction and overall image quality for giving their score. We will investigate these parameters in future work.

5.2 Linear Model for Image Scoring Based on Face Parameter Weighting

We propose a novel image scoring method that takes into account the subjective impression of a viewer evoked by the presence of faces in images by utilizing our findings from previous experiment. The score is calculated as a linear combination of the validated parameters introduced in section 3 based on the general and extendable linear weighting scheme given in the following equation:

$$S(I) = \sum_{i=1}^{N} w_i X_i = w_{coverage} S_{image} + \sum_{E \in \{neutral, joy\}}^{K} w_E P_E \tag{4}$$

where $S(I)$ refers to the image score and $w_i X_i$ are the weighted face parameters. N is the number of face parameters and K refers to the number of facial expressions respectively. We estimate the weights w_i based on the user score by standard multiple linear regression.

Fig. 4 shows the correlation and the related p-values calculated from ground truth face parameters. Results for single feature predictors and the 99% confidence limit are also shown for convenience. The correlation value for the facial expression features is as twice as high as the result for the face coverage feature. Moreover, the correlation between user score and predicted image score is statistically highly significant which confirms our conclusions from 5.1 and we state that facial expressions are an important feature for calculating an impression related image score.

Combining *joy* and *neutral* expression predictors does not improve the correlation. We explain this by the fact that we use only two expression detectors and therefore these two features are 100% negatively correlated. Thus we gain no additional information by including both predictors. This will change, when more facial expressions are added. A calculated cross-correlation between *joy* and *neutral* expression features of $r_{jn} = -1.0$ justifies this statement. A performance improvement is achieved by combining *coverage* and facial expression based predictors. The cross-correlation between *coverage* and either of the facial expression parameters was calculated as $r_{cj} = -0.012$ and $r_{cn} = 0.012$ respectively. Thus we can expect an improvement by combining these predictors in our model.

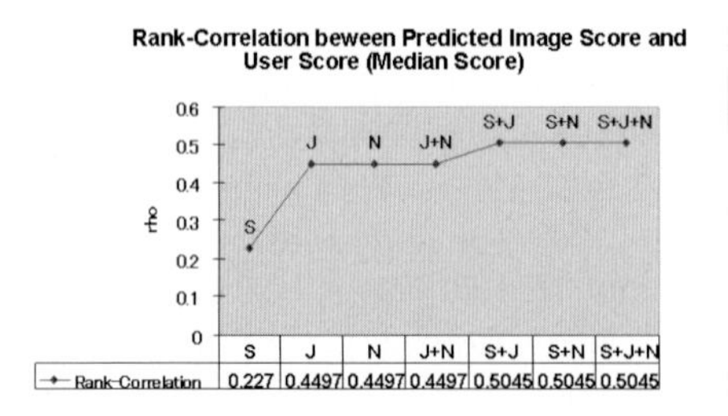

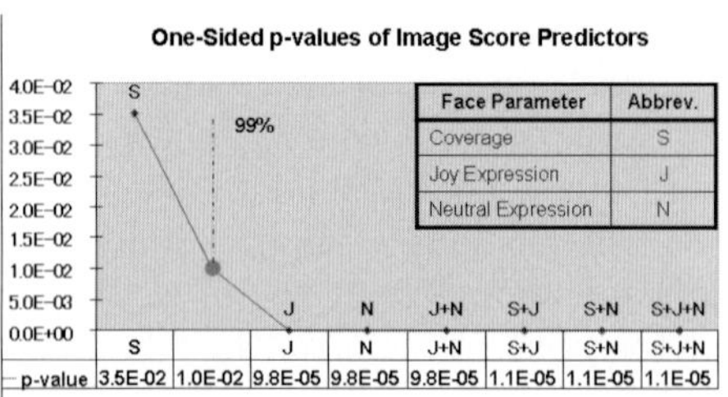

Fig. 4. Rank-Correlation and p-Values for Parameter Combinations

Based on these findings for the ground truth data, we conclude that the combination of the structural image features of *coverage* and the emotional image feature facial expressions leads to the best score prediction result. Therefore, we use this combination in our weighting model for assessing the practical case where all facial parameters are estimated fully automatically. The comparison result is given below:

- Ground Truth : rank-correlation rho=0.5, p=0.00001 (one-sided)
- Automatic : rank-correlation rho=0.34, p=0.003 (one-sided)

The correlation between the image score estimated from automatically estimated face parameters drops by 0.16 compared to the result calculated from ground truth. Analyzing the reason for this, we calculated the cross-correlation r_{xy} (Pearson) between ground-truth labeled and automatically detected face parameters. The results are:

- *coverage* $r_{xy} = 0.75$
- *joy/ neutral* $r_{xy} = 0.65$.

We conclude that the facial expression detection result contributes most to the performance degradation in our automatic system. We will address this issue in our future work.

6 Conclusion and Future Work

We investigated the influence of facial parameters on the subjective impression evoked at viewers when looking at photographs containing people from a 'third-party' viewing perspective. In the present study we focused on images extracted from home video contents in the application context of automatic keyframe extraction. Image structure related parameters such as *face number*, *face coverage* and *face location* were considered. We also investigated the contribution of facial expressions to the viewer's impression. As the results of conducted user experiments, we validated our hypotheses regarding the positive influence of coverage and joyful facial expression at the impression with respect to the predefined viewing concept of 'a good picture of a person'. Moreover, we found that the presence of neutral facial expression influences the impression negatively. The hypotheses about the existence of preferred locations for faces as well as the contribution of the face number were not confirmed, thus we conclude that these parameters do not contribute to the evoked impression. Then, we proposed an extendible linear weighting model that exploits present facial properties for calculating an image score that is correlated to the viewers' impression, and validated its effectiveness for image retrieval tasks.

An open issue to be addressed in the future is the combination of our approach with traditional keyframe extraction methods in order to determine the degree of improvement that can be achieved when emotional cues are taken into account. The relationship with persons depicted in the imagery also influences the keyframe selection and therefore we would like to address the 'first-party' viewing perspective in the future in order to determine how a personal relationship could be possible modeled by using face parameters. Furthermore we are interested in the validation of our model for other facial expressions and image parameters in order to extend our proposed weighting scheme for image score calculation by including these parameters which we expect to increase the correlation between the predicted image score and the subjective impression with respect to the viewing perspective given in this paper.

References

1. Truong, B.T., Venkatesh, S.: Video abstraction: A systematic review and classification. ACM TOMCCAP 3, 1 (2007)
2. Gale, A.: Human response to visual stimuli. In: Hendee, W., Wells, P. (eds.) The Perception of Visual Information, pp. 127–147. Springer, Heidelberg (1997)
3. Senders, J.: Distribution of attention in static and dynamic scenes. SPIE 3016, 186–194 (1997)
4. Russ, J.C.: The Image Processing Handbook. CRC Press, Boca Raton (2006)
5. Park, S.C., Park, M.K., Kang, M.G.: Super-resolution image reconstruction: a technical overview. IEEE Signal Processing Magazine 20(3), 21–36 (2003)
6. Fergus, R., Singh, B., Hertzmann, A., Roweis, S.T., Freeman, W.T.: Removing camera shake from a single photograph. In: ACM SIGGRAPH 2006, pp. 787–794 (2006)
7. Zhang, T.: Intelligent Keyframe Extraction for Video Printing. In: Proc. of SPIE Conference on Internet Multimedia Management Systems V, vol. 5601, pp. 25–35 (2004)
8. Luo, J., Papin, C., Costello, K.: Towards Extracting Semantically Meaningful Key Frames From Personal Video Clips: From Humans to Computers. IEEE Trans. Circuits Syst. Video Techn. 19(2), 289–301 (2009)
9. Martinet, J., Satoh, S., Chiaramella, Y., Mulhem, P.: Media objects for user-centered similarity matching. Multimedia Tools Appl. 39(2), 263–291 (2008)
10. Ekman, P., Keltner, D.: Universal facial expressions of emotion. In: Segerstrale, U., Molnar, P. (eds.) Nonverbal Communication, pp. 27–46. LEA, Mahwah (1997)
11. Kowalik, U., Hidaka, K., Irie, G., Kojima, A.: Creating joyful digests by exploiting smile/laughter facial expressions present in video. In: International Workshop on Advanced Image Technology (2009)
12. ClipLife, `http://cliplife.goo.ne.jp/`
13. Lim, J.H., Tian, Q., Mulhem, P.: Home photo content modeling for personalized event-based retrieval. IEEE Multimedia 10(4), 28–37 (2003)
14. Wiskott, L., Fellous, J.-M., Kruger, N., von der Malsburg, C.: Face Recognition by Elastic Bunch Graph Matching. IEEE Trans. PAMI 19(7), 775–779 (1997)
15. Ando, S., Suzuki, A., Takahashi, Y., Yasuno, T.: A Fast Object Detection and Recognition Algorithm Based on Joint Probabilistic ISC. In: MIRU 2007 (2007) (in Japanese)

Characterizing Virtual Populations in Massively Multiplayer Online Role-Playing Games

Daniel Pittman and Chris GauthierDickey

Department of Computer Science
University of Denver
{dpittman,chrisg}@cs.du.edu

Abstract. Understanding player distributions, sessions, and movements in a Massively Multiplayer Online Role-Playing Game (MMORPG) is essential for research in scalable architectures for these systems. We present the first detailed measurement study and the first models of the virtual populations in two popular MMORPGs, World of Warcraft™ and Warhammer Online™. Our results show that while these two types of MMORPGs are significantly different in play style, the features of their virtual populations can be modeled similarly, allowing future researchers to accurately simulate these types of games.

1 Introduction

Measuring and modeling player distributions, session lengths, and player movements in a virtual world are essential to research in architectures for massively multiplayer online games (MMOs). Accurate models based on empirical evidence significantly strengthens researchers' arguements that one particular architecture is better than another. While prior measurement research has been done on MMOs, it has primarily focused on traffic modeling and characterization. Though traffic measurement is important and useful for research in MMOs, an understanding of how players move around and populate the virtual world allows one to explore architectures that take advantage of this information and analyze those that do not.

Our research addresses this issue. We provide the first set of measurement-based models for population distributions and movements *within* two different massively multiplayer online role-playing games (MMORPGs) that can be used for simulation and analysis of new architectures. MMORPGs are a subset of all possible MMOs, and while other types of MMOs are possible (for example, one based on real-time strategy rules), to date MMORPGs have been the most commercially successful.

The ultimate goal of our research is to design a realistic and empirically-based simulation model from measurements taken from current, commercial MMORPGs. To achieve this, we have measured overall populations, session lengths, player distributions, and player movements over several months on two MMORPGs: World of Warcraft™and Warhammer Online™, both which are classified as MMORPGs but have significantly different play styles. We measured these two games in order to test our hypothesis that regardless of the style of play in the game, players' behavior could be modeled using a unified set of functions. We find our hypothesis to be correct. While

S. Boll et al. (Eds.): MMM 2010, LNCS 5916, pp. 87–97, 2010.

both games have significantly different play styles, the resulting models are quite similar and would allow a researcher or game designer to simply change a few parameters in order to simulate various scenarios. While the models we generated may not translate to all types of MMOs, these two representative games have provided excellent models for future research and simulations.

2 Related Work

Over the last several years, a significant number of measurement studies have been performed on MMOs, though most of these have focused on traffic patterns and network characteristcs. Chambers et al. studied network patterns related to players and the server of small networked multiplayer games [1]. Similar to our data, their measurements also show diurnal patterns of game populations. Kim et al. measured network patterns on Lineage II, a popular MMO in Korea [2]. Their work focused on network packet sizes, RTTs, session times and inter-session arrival times. The data they recorded shows a similar power-law distribution for session times as the times that we have observed. Ye et al. devised a set of performance models for MMORPG servers and networks based on concurrent player population [3]. Chen et al. profiled packet interarrival times, packet load distribution, and bandwidth utilization of *ShenZhou Online*, a popular MMORPG [4]. Svoboda et al. modeled traffic patterns and sessions lengths for players of WoW using both wireless and wired Internet connections [5].

Beyond network traffic measurement, some research has looked at traffic patterns, session lengths, and latency when measured with respect to players and user behaviors. Tarng et al. performed a long term study of WoW in order to see if it was possible to model subscription lengths of players based on how much a user plays an MMO [6]. They showed that while it is possible to predict short term behavior, long term prediction is much more difficult. Claypool and Claypool characterized latency requirements of various online games in terms of the deadline in which a user command must be processed and the precision of the commands the user is issuing [7]. Traffic patterns and session lengths of WoW were profiled with respect to different player action categories by Suznjevic et al. [8]. They hypothesized that mobile devices could be used for some of the less traffic intensive player activities. Fernandes et al. characterized traffic patterns in Second Life during different player activities in the world [9]. Kinicki et al. expanded on the work of Fernandes et al. by considering object and avatar interactions of the player in the virtual world when modeling traffic characteristics [10]. Finally, Szabó et al. provided a model from which you can detect the activity of a user within a MMORPG by correlating the traffic patterns observed through passive monitoring and packet level introspection.

While all of the related work provides important contributions towards modeling MMOs, especially in terms of traffic behavior, our work is the first to provide details of the virtual world, its population distributions, and player movements. Note that our preliminary results were published in [11], but these results only show initial measurements of World of Warcraft over a smaller data set without any modeling. In this paper, we examine a much larger and complete data set of two MMORPGS, we create models for simulation and analysis, and we show the similarity between both MMORPGs even though their play style differs significantly.

3 Methodology

Two methods can be used for measuring virtual populations and behaviors of players in MMORPGs. The first method is to analyze logs generated directly from an MMORPG or from customized clients which log the behavior for you. This method has the advantage of being accurate, though few companies are willing to share logs from their games or allow you to modify their clients and further the logs may not contain the needed information. The second method is to use probing-based measurements to try to infer properties of the system. We use the second method for our research.

In order to measure population information, we designed a set of scripts that run from the game clients using the Lua[1] scripting interface provided by both WoW and WAR[2]. For WoW, we modified the Census+ add-on to collect broad information about all players currently online[3]. WAR's add-on was custom written, but was based on functionality of Census+. We also wrote an additional add-on for both MMORPGs to record continuous detailed information about a randomly selected subset of players.

To measure the games, we performed server queries from the clients using the *who* service, which allows a player to search for another player in the game, and the *friends list*, which is updated by the server whenever the client queries the friends list. Using the who service, we performed back-to-back snapshots of the server population. However, since an entire snapshot takes several minutes, we populated our friends list with a random set of players which were not seen in the previous snapshot. The friends list allows us to track a small subset of players including when they log on and off and where they are at during each query. Using our techniques, we observed over $115,000$ individual players and tracked player movements on over $75,000$ sessions. Note that while all MMOs do not use Lua as a scripting interface, the who service and friends list tends to be universal and therefore similar techniques could be applied to other MMOs.

From both games we measured the total populations over time, the lengths of each session observed, the zones each player visited (including the order visited), and the time spent in each zone. WoW was measured over a 4 month period on the Aerie Peak server while WAR was measured over 2 weeks on the Volkmar server. We examined data from other servers and it was similar to the results presented here, thus these two servers are sufficiently general for both games.

4 Measurements and Models

When analyzing the data we recorded, our goal is two-fold. First, we want to verify previous work regarding arrival rates and session lengths. Prior research has done this through traffic and log analysis; our measurements were taken from within the games themselves. Second, and more importantly, we provide measurements and models for player distributions and movements within the virtual space of both games.

[1] http://www.lua.org
[2] Source code available from http://www.cs.du.edu/~chrisg/measurements
[3] http://www.warcraftrealms.com/censusplus.php

4.1 Daily Populations and Arrival/Departure Rates

In our first set of measurements, we examined population fluctuations over time and arrival and departure rates. Due to space limitations, we only show the measurements on WoW, however we observed similar measurements in WAR. Note that WoW measurements include both factions within the game.

Population Over Time. We measured the total number of players in the game every 15 minutes during our measurement period and averaged the results by hour each day of the week. Our hypothesis was that more players were online during evening and weekend hours, due to weekly obligations such as work and school and therefore architectures would need to address these cycles. Figure 1 shows the 24 hour daily cycle with each line representing one of the days of the week.

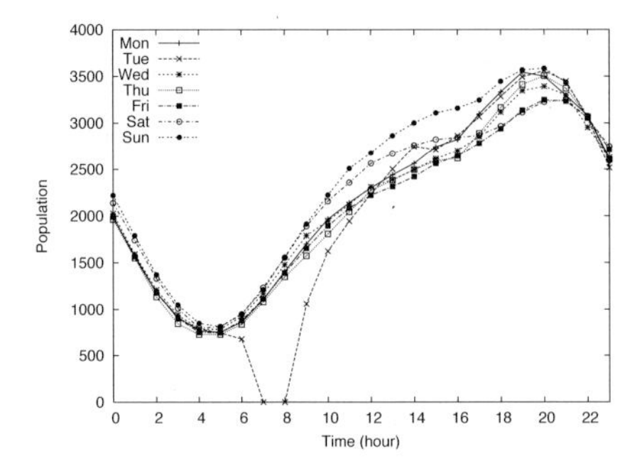

Fig. 1. *Average daily population:* This figure shows the average daily population of the WoW server we measured. Players typically play more in the evenings and both earlier and later on the weekdays. Tuesdays are "patch days", when server maintenance occurs, explaining the empty server at that time.

We note three important aspects of our graphs. First, populations have an average peak at almost 3600 players. Given the imprecision incurred by the measurement method, we estimate that a typical World of Warcraft server will support up to 4000 players. Second, we see that weekend play stands out from weekday play in that the realm experiences a significantly higher average population earlier in the day. This implies that servers must be provisioned for weekend play. Last, we see an almost 5-fold increase in the number of players from the lowest point (at 4AM) to the highest point (7PM) of the realm population. This implies that servers must also be over-provisioned to handle peak loads during the evenings and are only partially loaded during the early mornings.

Arrival and Departure Rates. To further understand the population fluctuations and to help understand the amount of churn that occurs in an MMORPG, we measured the number of arrivals and number of departures per hour and averaged this again on each day of the week. Figures 2 show these results.

In these two figures, we see that the amount of *churn*, or the number of players joining and leaving the game, is high during peak playing times. Figure 2 (a) shows similar

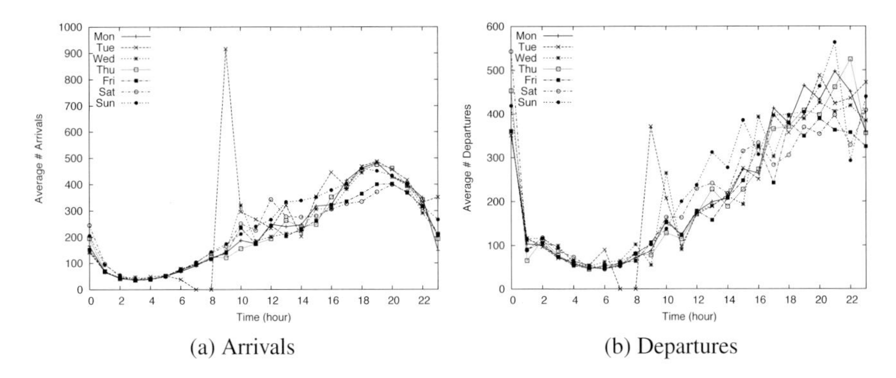

(a) Arrivals (b) Departures

Fig. 2. *Arrival and Departure Rates:* Figure (a) shows the arrival rate in terms of the number of new players seen this hour. Figure (b) shows the departure rate in terms of the number of players seen in the prior hour that are no longer online.

trends of arrivals during the weekdays, but has an increased arrival rate on the weekends during earlier hours of the day. Figure 2 (b) shows that the number of departures increases towards the end of the day. Together, *we see that during peak playing times, over 1,000 players join and leave the game per hour.*

In terms of the magnitude of the difference between minimum and maximum arrival and departure rates, these results show that arrival rates and departure rates differ by a factor of 10 while. In terms of MMO architectures, 1,000 players joining and leaving per hour may not appear to be a huge burden. However, given that WoW claims to have over 10 million subscriptions, a theoretical maximum of 4,000 players per realm indicates that approximately 1 million players log on and off *per hour* of the WoW servers. This is a significant amount of churn that an MMORPG architecture would need to handle.

4.2 Session Lengths

Session lengths were measured by adding a random subset of players seen in the most recent snapshot to the *friends list*, allowing us to track how long a character is played in the game. Our measurements in Figure 3 show that contrary to anecdotal stories, most sessions were short lived.

Figure 3 (a) shows the CDF calculated from all observed session times in WoW. From this figure, we see that only a small percentage of players we observed played for longer than 400 minutes (8 hours), while most players played for less than 200 minutes (3 hours). We calculated the mean session time to be 80 minutes, with a maximum observed session time of 1440 minutes (24 hours) and a minimum session time of 1 second. Note that we did not track players for longer than 24 hours, though for future work we will consider how many players were online for extensive periods of time.

Figure 3 (b) shows the CDF calculated from all sessions observed in WAR. In WAR, almost 90% of all sessions were less than 200 minutes with a mean session time of 89 minutes, a maximum time of 882 minutes (14 hours), and a minimum session time of 1 second.

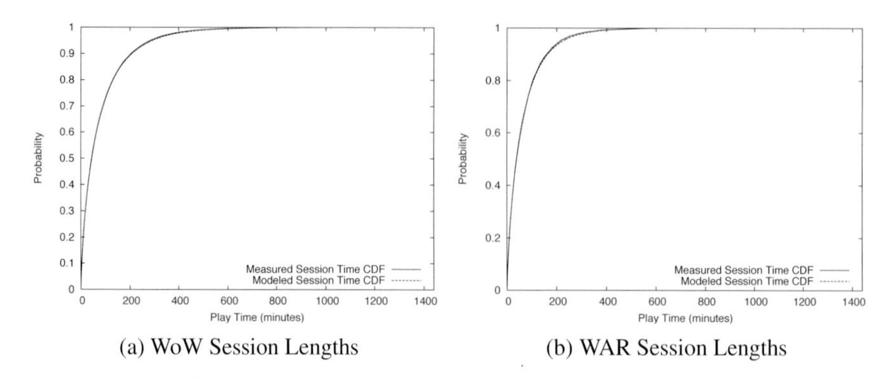

(a) WoW Session Lengths

(b) WAR Session Lengths

Fig. 3. *Session Times Observed:* Figure (a) shows the CDF of all the sessions we recorded in WoW. In addition, the data was fit to a Weibull distribution, and plotted on the same graph. Figure (b) shows the CDF of all the sessions recorded in WAR along with the model created by fitting the data to a Weibull distribution. Both lines in both figures are barely distinguishable due to the close fit of the models.

For both models, we used the least-squares method to find a fit for a model of the measured data. Given the trend of the data, we determined that a Weibull distribution would fit well. The models are plotted in Figure 3 (a) and (b). Next, we validated our models by plotting the residuals between the predicted and measured values (not shown due to space limitations), and found that the residuals for both models had a standard deviation of 0.004, indicating an extremely close fit. Thus, given a uniformly distributed random variable $0 \leq x \leq 1$, the session lengths for WoW and WAR can be modeled as follows:

$$Session_{WoW}[x] = 1 - e^{-(x/69.75)^{0.7522}}, Session_{WAR}[x] = 1 - e^{-(x/59.81)^{0.8322}}$$

This result verifies that MMORPGs experience considerable churn. A large fraction of sessions are short lived while only a small fraction are stable. We believe that what may be happening here is that players may be logging on to check to see if friends or guild members are currently online, checking in-game mail, or checking auctions at the auction house. If this is true, then the implication is that load on an architecture could be reduced by providing an external interface to these services that does not require logging into the game. *Given the predictability of player session time, one may conclude that game developers should target playing experiences for session times that reach the majority of players.* Researchers, on the other hand, can use session times to predict how long players will connect to a given architecture.

4.3 Player Distributions

We next measured the distribution of players in the virtual world of both games. Throughout the world regions are statically divided into *zones*. We measure how many players are in each zone over the measurement period. After examining the data, we

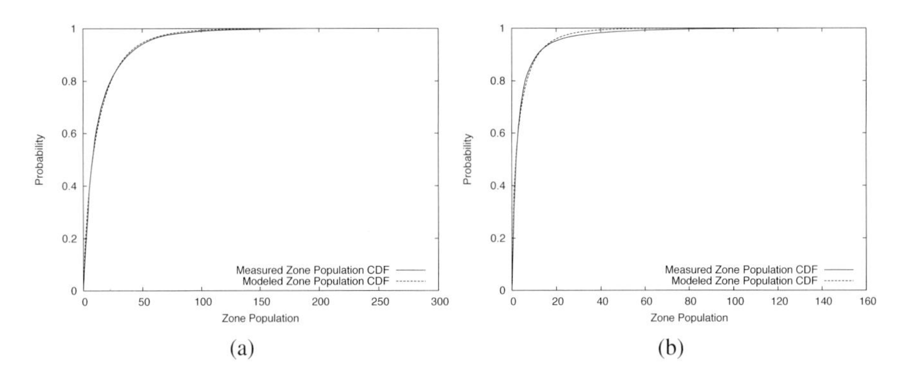

Fig. 4. *Distribution of players per zone:* Figure (a) shows the distribution of players per zone in WoW while Figure(b) shows the distribution of players per zone in WAR. Note that both CDFs do not include the zones without any players: WoW had 36% of the zones empty while WAR had 78% of the zones empty.

realized that a large percentage of the zones had 0 players in them. To model this correctly, we calculated the quantity of 0 population zones we examined (36.44% in WoW and 78.56% in WAR) and removed these from the data set for the purpose of modeling the remaining data. We then stretched the remaining points to cover the probability from 0 to 1. Using the least-squared method, we fit the data using a Weibull distribution.

Figure 4 (a) and (b) show the measured data and the fitted Weibull distributions of the zone populations. Note that in both games, only a few zones have more than 50 players, while the majority of zones have fewer than 10 players. For WoW, we saw an average of 121 players in a zone, with a minimum of 0 and a maximum of 293 players. On WAR, we saw an average of of 74 players in a zone with a maximum of 156 players and a minimum of 0.

As with session times, player distributions were modeled very closely using a Weibull distribution with a standard deviation of 0.007 for WoW and 0.008 for WAR of the residuals from the measured data and models. Thus, given a uniformly distributed random number $0 \leq p \leq 1$, we can model WoW and WAR population distributions as follows:

$$Population_{WoW}[p] = \begin{cases} 0 & \text{if } p \leq .3644 \\ \lfloor 1 - e^{-(((p-.3644)/.6356)/12.744)^{0.7822}} \rfloor & \text{otherwise} \end{cases}$$

$$Population_{WAR}[p] = \begin{cases} 0 & \text{if } p \leq .7856 \\ \lfloor 1 - e^{-(((p-.7856)/.2144)/3.256)^{0.6417}} \rfloor & \text{otherwise} \end{cases}$$

The measurements of player distributions are important because they show that players are *not* uniformly distributed in the virtual world as much of the prior research in scalable game architectures has assumed. Clearly, given a uniform distribution of players, almost any architecture can be reasonably well-balanced so that it scales well. However, a Weibull distribution indicates that players tend to group in large numbers in only a few zones, causing stress on any architecture as it has to handle the increased number of interactions between players. *Therefore, game designers and researchers must consider this Weibull distribution of players in which a few zones contain a large number of players while many zones only have a few (or no) players when characterizing the potential load on a MMO architecture.*

4.4 Player Movements

To model player movements, we measured the number of zones visited in a session, how long they remained in a zone, and what zones they chose to travel to from their current zone.

Number of Zones Visited. We hypothesized that a linear relationship exists between the number of zones visited during a session and the session length. To test this hypothesis, we measured how many zones the players travelled to each session in both WoW and WAR and plotted the results in Figure 5 (a) and (b). The number of zones visited are not unique zones, but the total number of times a player moved from one zone to another. From these figures, we see that our hypothesis held for the 80% of session times in both games, i.e., those 200 minutes and below in WoW and those 100 minutes and below in WAR. On both graphs, the hypothesis no longer seemed to hold for the highest 20% of the sessions. One explanation may be that players who are on for long periods of time behave differently in the game than those on for shorter periods. Note that the game will disconnect players who remain idle for longer than 10 minutes. Thus, even these long sessions consist of active players or bots.

In both cases, we model this behavior using a simple linear equation. For WoW, we found that the equation $y = 0.070x + 0.831$ works well while for WAR we found that the line at $y = 0.014x + 1.20$ works well. For future work, we plan on exploring how the longer sessions can be modeled more accurately.

Time in Zones. We then observed the distribution of time that players spend in any given zone. This information is important because it helps us understand whether players spend an even amount of time in each zone or perhaps spend only a small amount of

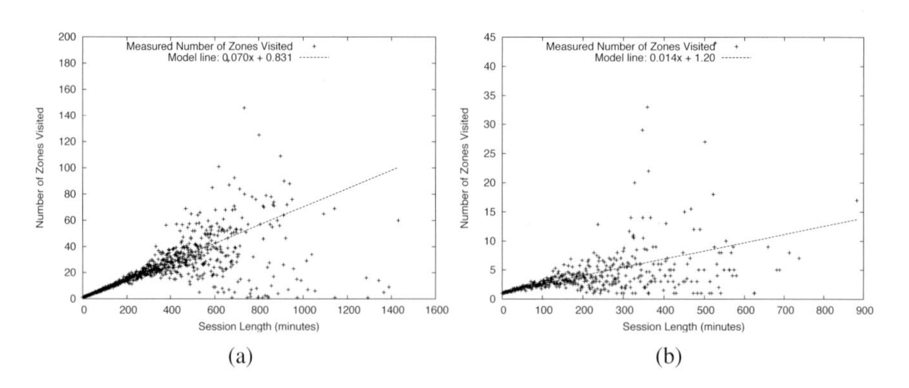

Fig. 5. *Zones Visited vs. Session Lengths:* These figure shows the number of zones visited plotted against the session time in minutes. (a) shows the results from WoW while (b) shows the results from WAR. Simple linear equations are used in both figures to model the data.

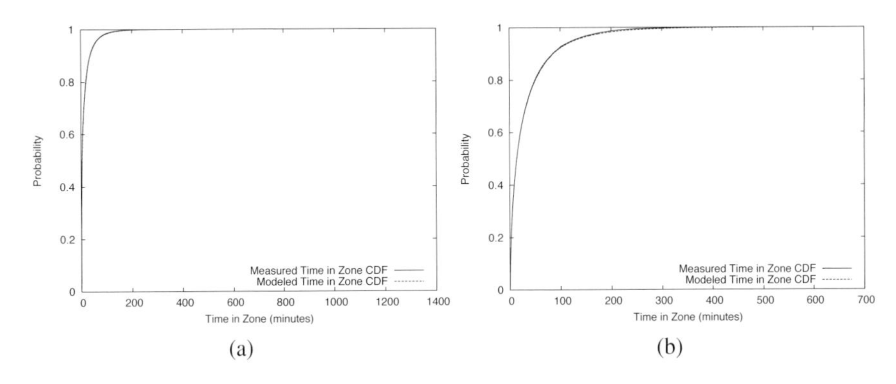

Fig. 6. *Time Spent in a Zone:* Figure (a) shows the measured and modeled CDFs of the time spent in a zone in WoW. Figure(b) shows the measured and modeled CDF time spent per zone in WAR.

time in a majority of zones but a large amount of time in one or two zones. Figure 6 (a) and (b) shows the measured and modeled CDFs of the time players spend in a zone in both WoW and WAR. As with session times, the time in each zone also followed a Weibull distribution, indicating that players did in fact spend most of there time in a few zones, and a small amount of time in the rest of the zones they visited. These distributions are as follows:

$$ZoneTime_{WoW}[p] = 1 - e^{-(x/8.189)^{0.5674}}, ZoneTime_{WAR}[p] = 1 - e^{-(x/24.42)^{0.6669}}$$

As with the previous models, we measured the residuals between the measured and modeled data and found that the standard deviation of the WoW function was 0.009 and for WAR it was 0.002.

Player Movement. The final aspect we measured with regards to player movement was *how* players moved between zones. Our hypothesis was that the random waypoint model of player movement is not accurate for MMORPGs. Our results indicate that instead a log-normal distribution of waypoint choices is more accurate.

To model this type of player movement, we examined the zone locations of all of the players we tracked and recorded where they were from one measurement to the next. We created a Markov chain of zone to zone player movement from this data. A Markov chain is represented by a two dimensional matrix where each row represents the probability of a transition from the row header (or zone in this case) to a given column header (a destination zone). To visualize the matrix, we created a square image where each pixel represents a cell in the matrix and is colored gray according to its probability in the table, with black being a probability of 1 and white a probability of 0. Figure 7 (a) and (b) shows the result of this visualization.

In order to create a player movement model, we also wanted to be able to randomly generate Markov chains that had the same properties as the measured Markov chains. We found that a log-normal CDF worked well in modeling these probabilities. The results are seen in Figure 7 (c) and (d). The modeled CDFs are defined as follows,

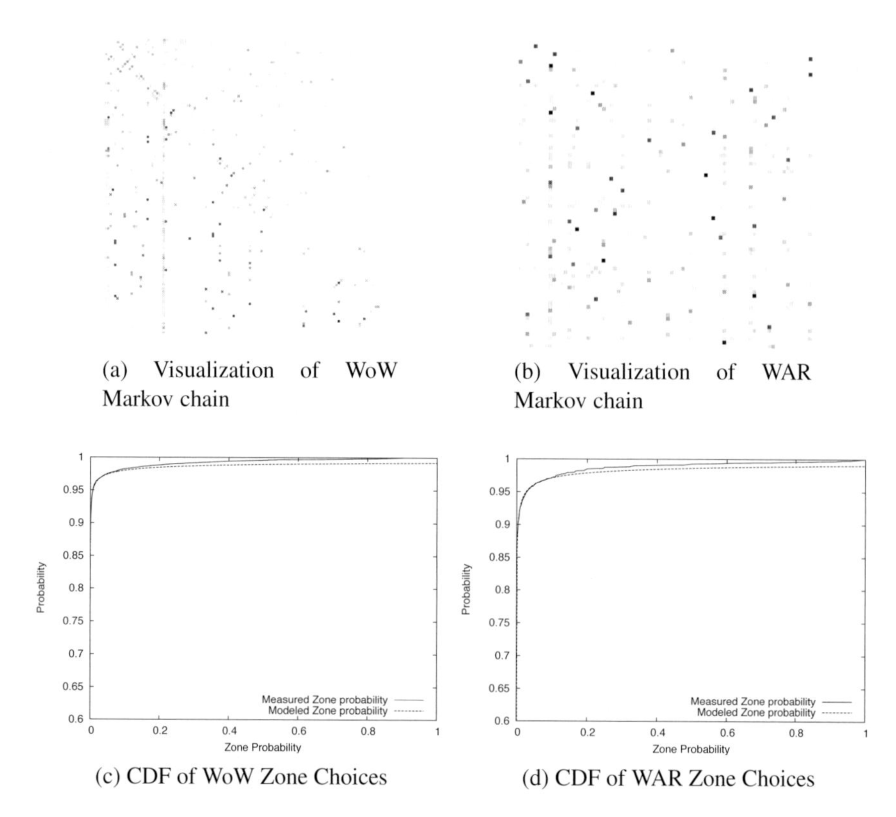

(a) Visualization of WoW Markov chain

(b) Visualization of WAR Markov chain

(c) CDF of WoW Zone Choices

(d) CDF of WAR Zone Choices

Fig. 7. *Visualization and CDF of Markov Chain Probabilities:* Figures (a) and (b) are visualizations of the Markov chains generated from the measured player movements between zones in WoW and WAR. Figures (c) and (d) shows the modeled log-normal CDFs of the probabilities from the measured Markov chains.

where given a uniformly distributed number $0 \leq p \leq 1$ and where *erf* is the Gauss error function:

$$Choice_{WoW}[p] = \frac{1}{2}(1 + \text{erf}(\frac{15.45 + \log(p)}{6.399\sqrt{2}}))$$

$$Choice_{WAR}[p] = \frac{1}{2}(1 + \text{erf}(\frac{12.23 + \log(p)}{5.235\sqrt{2}}))$$

Note that in using this function to create the Markov table would require each line to be adjusted so that it summed to 1, though this should be trivial in practice.

5 Conclusion and Future Work

In this paper we have provided measurement results from and empirically-based models for two popular MMORPGs. Further, by measuring two MMORPGs with significantly different play styles, we show that the models are not unique to a single game, but can be modified slightly to apply to MMORPGs even with different play styles. Given that each distribution used has one or more parameters, these parameters can be altered to create a variety of scenarios to test an MMORPG architecture under. For example, by changing the parameters of the Weibull distributions, one can stretch the function to cover more values or cause its initial rise to cover a larger percentage of the probability space. However, the values provided here are excellent starting points as they model two of the most popular MMORPGs today. As future work, we plan to continue measuring other MMOs to compare the results with our models.

References

1. Chambers, C., Feng, W., Sahu, S., Saha, D.: Measurement-based characterization of a collection of online games. In: Proceedings of the Internet Measurement Conference (2005)
2. Kim, J., Choi, J., Chang, D., Kwon, T., Choi, Y., Yuk, E.: Traffic Characteristics of a Massively Multiplayer Online Role Playing Game. In: Proceedings of ACM NetGames (2005)
3. Ye, M., Cheng, L.: System-performance modeling for massively multiplayer online role-playing games. IBM Systems Journal 45(1), 45–58 (2006)
4. Chen, K.T., Huang, P., Huang, C.Y., Lei, C.L.: Game traffic analysis: an mmorpg perspective. In: NOSSDAV 2005: Proceedings of the international workshop on Network and operating systems support for digital audio and video, pp. 19–24. ACM, New York (2005)
5. Svoboda, P., Karner, W., Rupp, M.: Traffic analysis and modeling for world of warcraft. In: IEEE International Conference on Communications, ICC 2007, June 2007, pp. 1612–1617 (2007)
6. Tarng, P.Y., Chen, K.T., Huang, P.: An analysis of wow players' game hours. In: NetGames 2008: Proceedings of the 7th ACM SIGCOMM Workshop on Network and System Support for Games, pp. 47–52. ACM, New York (2008)
7. Claypool, M., Claypool, K.: Latency and player actions in online games. Communications of the ACM 49(11), 40–45 (2006)
8. Suznjevic, M., Dobrijevic, O., Matijasevic, M.: Mmorpg player actions: Network performance, session patterns and latency requirements analysis. Multimedia Tools and Applications (May 2009)
9. Fernandes, S., Kamienski, C., Sadokn, D., Moreira, J., Antonello, R.: Traffic analysis beyond this world: the case of second life. In: ACM NOSSDAV (June 2007)
10. Kinicki, J., Claypool, M.: Traffic analysis of avatars in second life. In: NOSSDAV 2008: Proceedings of the 18th International Workshop on Network and Operating Systems Support for Digital Audio and Video, pp. 69–74. ACM, New York (2008)
11. Pittman, D., GauthierDickey, C.: A measurement study of virtual populations in massively multiplayer online games. In: Proceedings of ACM NetGames (September 2007)

Browsing Large Personal Multimedia Archives in a Lean-Back Environment

Cathal Gurrin[1,2], Hyowon Lee[1,2], Niamh Caprani[1], ZhenXing Zhang[2,3], Noel O'Connor[1], and Denise Carthy[3]

[1] CLARITY CSET, Dublin City University, Ireland
[2] Centre for Digital Video Processing, Dublin City University, Ireland
[3] Biomedical Diagnostics Inistiture, Dublin City University, Ireland
{cgurrin,hlee,ncaprani,zzhang}@computing.dcu.ie, oconnorn@eeng.dcu.ie, denise.carthy@dcu.ie

Abstract. As personal digital archives of multimedia data become more ubiquitous, the challenge of supporting multimodal access to such archives becomes an important research topic. In this paper we present and positively evaluate a gesture-based interface to a personal media archive which operates on a living room TV using a Wiimote. We illustrate that Wiimote interaction can outperform a point-and-click interaction as reported in a user study. In addition, a set of guidelines is presented for organising and interacting with large personal media archives in the enjoyment oriented (lean-back) environment of the living room.

Keywords: Personal Digital Archives, Browsing Technologies, Lean-back Environment.

1 Introduction

Driven by the increasing penetration of data capture and storage technologies such as digital video cameras and digital photo cameras devices, we note the increasing trend people becoming content creators and not just consumers. At the same time, there can be seen a notable trend recently towards the integration of content management technologies into the lean-back (enjoyment-oriented) environment of the living-room. To take one example, multimedia content analysis technologies are beginning to be integrated into the living room TV, such as the recent incorporation of DVR functionality and internet access into TVs. Such creeping functionality points to the fact that TV manufacturers have identified the living-room TV as an environment in which viewers are relaxed and willing to interact with their own created and downloaded or online media. The next step in this technology convergence process is the integration of personal content organisation facilities into the TV itself, which poses a number of challenges because this needs to be performed by non-expert users, sitting in front of a TV with a remote control in a distractive environment, and not at a desktop computer with use of a keyboard and mouse.

Therefore, in this paper we are concerned with examining how we can support a user in organising and interacting with large personal multimedia archives in the lean-back

S. Boll et al. (Eds.): MMM 2010, LNCS 5916, pp. 98–109, 2010.

environment of the living room TV. We can concerned with large multimedia archives due to the fact that we are all becoming content creators and are gathering much larger digital archives than heretofore. The archive we chose to work with is a (HDM) Human Digital Memory archive, chosen because it represents an extreme challenge in the management of personal archives. To represent the lean-back environment we employ a large-screen TV and utilise the Nintendo Wiimote as an interface tool. Note that we are not focusing on the ideal implementation of a gesture recognition technology, this work is ongoing elsewhere, such as the work of Schlömer et al.[1], rather we focus on an exploration of how useful gesture-based interaction can be for managing personal archives in the lean-back environment.

2 Personal Media Archives and the Lean-Back Environment

Users have been gathering personal digital media archives since the advent of the digital home computer. Whether photos or emails, audio files or video content, organization of, and access to, these personal archives has been the subject of ongoing research. Naaman [2] and O'Hare [3] have shown clearly how it is possible to develop highly effective digital photo search and organization tools as a means of managing ever growing personal digital photo archives. In addition the TRECVid [4] series of workshops has presented many techniques for managing archives of video content, many of the outputs of which can be applied to personal and broadcast content. However, most such techniques are designed for the desktop computer, or in some cases mobile devices, will not necessarily transfer into the lean-back living room environment, for reasons such as user interaction support, ease of querying and even Consumer Electronics (CE) device processor speed. It is our conjecture, however, that since a living-room TV acts as a natural focal point for accessing personal media archives, that taking into account the significant limitations and challenges of developing for such an environment is essential to successfully deploy multimedia content organisation technologies. Indeed initial work in the area by Lee et al. [5] suggests that simplicity of interaction is more crucial for the lean-back environment than in any other digital media domain and that ultimately, this simplicity of interaction determines the success or otherwise of any new applications. The challenge therefore is to marry the competing requirements of supporting complex organization technologies with the simplicity of interaction required for successfully developing content organization technologies for the lean-back environment.

2.1 Human Digital Memory Archives

Human memory is fallible; we find our own limitations every day, for example, the names of people, the dates of events or episodes from our past. Humans have found ways to circumvent these limitations by employing tools and technologies such as diaries, notebooks, digital archives, etc. In recent years, we have noted a new form of extreme personal data capture in the maintenance of HDM (Human Digital Memory archives), which attempt to digitally capture many of a person's life experiences, including continuous image or video capture. The MyLifeBits [6] project at Microsoft Research is perhaps the most famous effort at gathering and organizing life experiences into a HDM. In the MyLifeBits project, Gordon Bell (inspired by Vannevar Bush's MEMEX) is capturing a lifetime of Bell's experiences digitally, everything from books read, photos captured, home movies, emails, and other personal digital sources. Other related research [7] has focused on the contextual gathering and organizing of HDMs with an emphasis on visual capture of user's experiences and the employment of information retrieval techniques to automatically organize the HDM using content and context data. To enable the capture of everyday activities visually, Microsoft Research have developed a device known as the SenseCam, which is a small wearable device that passively captures a person's day-to-day activities as a series of photographs[8]. It is typically worn around the neck and, and so is oriented towards the majority of activities which the user is engaged in. In a typical day, a SenseCam will capture up to 5,000 photos, which are sequential in nature and suitable for summarization to remove inherent duplication. For example photos from a SenseCam, see Figure 1. In a typical year, well over one million SenseCam photos will be gathered, and like conventional digital photos, one of the key automatic organisation methodologies is event segmentation, to segment a continuous stream of visual HDM data into a sequence of meaningful events. Doherty et al. [9], has worked extensively on automatically organising streams of SenseCam photos into events and representing each event with a suitable keyframe. It is a HDM archive that we utilize for this research, as an example of a very large and challenging personal archive.

Fig. 1. Example images captured by a Microsoft SenseCam

2.2 Related Gesture-Based Input Research with the Wiimote

The Wiimote is an input device for the Nintendo Wii games console which incorporates a tri-axial accelerometer to recognize user gestures and utilizes Bluetooth to send gesture data to a host device (typically the games console). There have been a number of uses of the Wiimote as a non-gaming user interaction device, for example work by Gallo and DePietro [11] on using the Wiimote as an interaction mechanism for 3D interaction with medical data. Schlomer et al. [12] have carried out an exploration of how to use the Wiimote to recognise a set of (and arbitrary new) gestures and positively evaluated the average recognition rate of these gestures. Lapping-Carr et al.[13] have utilised the Wiimote as an intuitive robot remote control interface while Shiratori & Hodgins [14] have utilized Wiimotes to dynamically control a simulated animated character. This research suggests that the Wiimote is a capable gesture-based user interaction device with a number of uses beyond gaming and we have chosen it as our input device of choice for this research.

3 Designing for Lean-Back Environments

In this work we employ the lean-back/lean-forward terminology to separate the enjoyment oriented (lean-back) living room environment from the task oriented (lean forward) environment of the office computer or laptop. Lean-back interaction has been the subject of research in the interactive TV community for quite some time. While a number of surveys and ethnographic studies at home have been done to better understand how people interact with TV, the work on translating these understanding of the special characteristics of TV interaction into actionable, prescriptive design guidelines has not been done extensively. With the lack of such transferrable knowledge base and difficulty in leveraging more well-known design guidelines for conventional desktop Graphical User Interface and for the Web, very few interactive TV applications have been commercially successful, and heuristics on some aspects of interactive TV such as social interaction is only appearing now [10].

When developing organization technologies for a certain device, cognissance must be taken of the inherent device limitations. For example, screen size, processor speed, ease of interaction, etc. Studies on interactive TV highlight the special characteristics of lean-back environment and they show design implications and guidelines for a technology operating in such a context. In this section, we summarise the characteristics of lean-back interaction from the perspective of interactive TV literature, serving as a base rationale for our design for Wiimote operated HDM interface on the living room TV.

3.1 Use of Remote Control as an Input Device

The main input device for the TV in the living room is a remote control. Due to the different affordances which a remote control of a TV and keyboard/mouse of a PC exhibit, suitable interaction mechanisms and widget behaviour of the two platforms are inevitably very different. A straightforward menu hierarchy with scroll bar, radio buttons and icons which are all very usable on a PC or Web environment (using a mouse and keyboard) becomes completely unusable when ported to, for example, a TV with remote control. The remote control has very coarse interaction continuity (few buttons for input) therefore the ideal interaction for the remote control should be based on discrete jumping from one area on the TV screen to another, avoiding complex hierarchical navigation but supporting a flat or shallow menu where a few remote control buttons can directly select frequently-used features. The design implication from ethnographic studies (such as [15]) suggests that a small number of frequently-used features should be identified and mapped directly to remote control buttons thus reducing menu navigation burden on the user.

Entering text using a remote control has been a major problem and has been addressed in a number of previous works. Having a virtual keyboard on the TV screen or an SMS text messaging style input have been suggested but currently the research community seems to agree that cumbersome text input with a remote control should be avoided if possible. Allowing each viewer's own mobile device (such as a mobile phone or PDA) as a text input device has been suggested as possible solution [16], [17] but the real utility and experience of such methods is still to be experimented with real users. We envisage that future remote controls will be equipped with motion sensor and operated with a few buttons in conjunction with motion gesture, thus becoming a more similar to Wiimote.

3.2 Viewing Distance

Unlike desktop PC or mobile interaction, lean-back interaction with TV occurs in the user typically sitting 2-2.5m away from the display screen. Due to this distance, the interactive elements on the TV screen need to be large enough to be noticed and read albeit the exact size of these will depend on the size of TV screen itself. Most of the currently available design guidelines for interactive TV suggest a minimum font size of 18pt [18] [19] and the maximum amount of text on the screen of 90 words [18] suggesting a requirement to focus on visual interactive elements, although the ever-increasing consumer TV screen size and resolution today will continue to make relatively smaller font size and more number of words more acceptable over time. Although much more investigation is required to set a standard widget and text sizes for TV interaction, having less details and small amount of comfortably large-size widgets and text is an important implication for designing for lean-back interaction.

3.3 Enjoyment-Oriented Design

The design for lean-back interaction cannot assume a highly-attentive user like traditional usability engineering methods do [20], because a more *enjoyable* interaction is not necessarily a more *efficient* one [21]. There is growing evidence that traditional desktop usability principles do not account for the pleasure of the user experience [22], focusing rather on the task-oriented nature of desktop interaction. Therefore an enjoyment-focused service such as interactive TV requires different designer focus, mindset and priorities from the start. Usability evaluation issues for interactive TV interfaces have been drawn and explored in [23, 24], and a structured evaluation framework for interactive TV has been suggested [9], but these are still based on the theoretical assumptions and past experiences from other media devices and need to mature further. The aesthetic quality of a TV interface is closely related to user enjoyment, therefore priorities for interactive TV design include designing for quick decisions, short attention spans and instant gratification [25].

3.4 Derived Lean-Back Interaction Guidelines

After examining the characteristics of CE devices in lean-back environments, and based on previous research and our own experiences of developing information retrieval systems for lean-back devices [5], we have compiled a set of guidelines for developing interactive multimedia applications for lean-back environments:

- **Minimise user input where possible.** Remove the need for a user to engage with complex query input mechanisms, such as textual querying and rather rely on remote control style browsing interaction. This requires that the system must be able to proactively seek and recommend content to the user or support information seeking via a small number of frequently used interactive features. This will likely require the deployment of hidden back-end technologies that make the user experience better, such as summarizing recorded video content in a DVR [5] or content recommendation technologies.

- **Engage the user with simple, low-overhead and low-learning time interaction methodologies, that are enjoyable to use.** Given the enjoyment oriented scenario of the lean-back environments and the disruptive nature of such environments, it is important that the interaction mechanisms employed must be both intuitive and easy to learn (for example the TV remote control or the Wiimote).
- **Represent complex digital multimedia objects visually.** Complex multimedia objects, such as photo collections, video archives or HDM archives need to be visually represented and manipulatable on screen, with few textual elements so as to maximize user attention in the distractive lean-back environment.

Where the lean-back environment in question includes information presentation on a TV screen, the existing guidelines for conventional or interactive TV [18, 26] can also complement the above suggested assuming the conventional viewing distance. For example, use of a standard iTV font, minimum text size, maximum words/screen, chunking text into small groups, clear menu exit point always visible, correctly sized interactive elements (e.g. thumbnails, icons), etc. should be employed.

4 The HDM Browser, an Experimental Prototype

By taking into account the three guidelines just described, along with the existing conventional interactive TV guidelines, we developed a Wiimote based browsing interface to a HDM archive. While it would be tempting to simply integrate as many interface technologies as possible to organise a HDM (e.g. many axes of search such as location/people/colour, keyframe browsing, textual querying, etc.), this would only serve to complicate the prototype and break our guidleines. In this prototype, the user is presented with a HDM archive segmented temporally into a sequence of days. The daily stream of photos is segmented into a sequence of events using the approach of Doherty et al. [9] and these events are presented in a temporally arranged storyboard at the bottom of the screen, with a keyframe selected for each event [9]. As the user browses a given day, the storyboard moves with the browsing, to give the user context of the temporal surrounding events that took place on that day. No attempt is made to identify the importance of an event on a given day, as this would require a query mechanism to generate a ranked listing of events, which was not the focus of this experiment.

Selecting an event begins playback of that event, which occurs in the large central area of the screen. Playback cycles through images in that event, fast or slow depending on user input. The horizontal arrows support next/previous event switch and the vertical arrows signify next/previous day switch. Jumping to the next day, will begin with the first event of the next day, regardless of what time of day the jump is made. In addition, the small slider control illustrates the speed of playback or rewind and there is minimal textual data on screen (event sequence, date, time and location only). The day-by-day browsing and playback interface as shown in Figure 2.

Fig. 2. The Prototype browsing interface showing playback (paused) from the fourth event of the day, which took place in early afternoon in Dublin, Ireland

Exactly how the three guidelines impacted on the prototype is now illustrated:

- The prototype *minimized user input* by organising the HDM with a calendar as the key access mechanism. A user could select next/previous days (via a simple Wiimote gesture) and then select next/previous event (another simple Wiimote gesture). Upon selecting an event, the event playback began which cycled through the images comprising that event at a fixed speed. The speed of this playback (from pause to fast-forward/fast-rewind) was user controlled by twisting the Wiimote as if one is twisting a dial or a knob.
- The prototype engages the user with *low overhead and low learning time* interaction methodologies that users found enjoyable to use. The Wiimote gestured employed were limited in number and as a result were very intuitive to a user. Simple button presses and gestures controlled all interaction with the HDM archive browser.
- *Represent complex digital multimedia objects visually.* A HDM archive is an enormous repository of data and as such it needs to be summarised and visually easy to browse and interpret. Therefore, the many thousands of images captured daily are treated as a continuous steam (as one would digital video) and undergo an event segmentation process [9], thereby representing a days' images as a set of about thirty individual events that the user can browse through. A keyframe was automatically selected for each event on the basis of its visual significance within that event [9]. These events were then easily played back at varying speed in the interface, allowing the user to quickly and easily view an event.

The HDM browser just described provided one half (gesture-based interface) of our user experiment. In order to do a comparative analysis with a more conventional lean-forward interaction scenario, we provided a baseline system for user evaluation. This baseline system was functionally and visually identical to the gesture-based HDM browser, however it operated via conventional (lean-forward) mouse click interaction, and therefore required lean-forward interaction.

4.1 Experimental Setup

Two weeks of HDM data gathered by one SenseCam wearer was employed for this experiment. This data was chosen from a three year HDM archive and was three years old at the point this experiment was carried out. This data consisted of about 50,000 individual SenseCam photos, each of which was indexed by date/time of capture and textual location (from an accompanying GPS log). For this experiment, we employed seven participants; six users who had no prior experience of HDM archives and one user who was the data owner (i.e. the actual SenseCam wearer from three years before). Since the experiment included the actual sensecam wearer, it was important that the data was not recent, so as to avoid any short-term memory bias from this user. It is our conjecture that including the SenseCam wearer in the experimentation process is important because a HDM is likely to be a private archive (of a person's life experience) and most data access is likely to come from this particular user. We are especially fortunate to have the three-year gap between data capture and this subsequent experiment. We validated that the sensecam wearer did not review the images prior to this experiment.

5 Evaluation and Findings

We focus first on the six novice users. They were each allocated an identical set of six information finding tasks, though organised in such a way as to avoid any bias as a result of learning. Therefore alternate modes of interaction were employed; the Wiimote and the baseline system alternating, which resulted in each topic being evaluated three times by different users on each interface and never at the same point in the task sequence for any user. After completing the experiment (which lasted about 17 minutes), participants completed a post-study system usability questionnaire (adapted from [27]) for both Wiimote and baseline systems. All participants were allowed up to 5 minutes to learn how to use both systems. During these five minutes the participants were encouraged to ask questions. The participants were then given the six search and retrieval tasks to complete. Four of these tasks (task 1, 2, 3, and 6) were single-item (known-to-exist) searches, meaning that the participants were asked to find a described event in the collection as quickly as possible. An example would be, 'find the time the HDM owner was giving his lecture'. The remaining tasks (task 4 and 5) were multi-item searches, meaning that the participants were asked to locate as many events with a particular characteristic as possible in a given time frame (two minutes). For example, 'find as many instances of meal eating as you can'. Each system utilised an identical event segmentation approach [9], therefore offsetting any impact of the event segmentation on the experimental results. No attempt was made to interrupt the user using the gesture-based interface, beyond the interruptions of a research lab environment and the co-ordinator keeping a record of user performance.

Table 1 illustrates the time taken (in seconds) to complete the four single-item retrieval tasks, with the max time allowed. Lower times are considered to be more successful. For the six participant users, there is no significant difference in the findings between the Wiimote interface and the baseline (mouse/desktop) prototypes. This is encouraging because we note that the (less familiar) gesture-based interface did not

Table 1. Participant and HDM Owner performance for single-item retrieval

	Participant Users		HDM Owner	
Task	Wii (mean time)	Baseline (mean time)	Wii (time)	Baseline (time)
1	88 (seconds)	54	-	76
2	33	40	08	-
3	97	*Not Completed*	-	90
6	91	79	03	-

hamper the overall user performance in any way. Indeed, one notable finding is that using the baseline system no user managed to complete the third task at all, while it was completed with the gesture-based interface.

For the HDM owner, the Wiimote prototype significantly outperforms the baseline system, which suggests that when the user has some knowledge of their own archives (e.g. likely time of day of an event taking place), that a gesture based interface may help in locating desired content. For all but one of the four tasks in Table 1, the HDM owner was significantly the fastest user. Note that the HDM owner only evaluated each query on a single interface, hence there is not a score for each query on each system for both tables 1 and 2.

Table 2. Participant and HDM Owner performance for multi-item retrieval

	Participant Users		HDM User	
Task	Wii (mean score)	Baseline (mean score)	Wii (score)	Baseline (score)
4	12	7	19	-
5	8	6	-	17

Table 2 shows the average number of relevant items located by the within the allocated time for multi-item retrieval tasks. Participants performed the score based tasks more effectively using the Wiimote interface for both queries, with users of the baseline system only finding 58% of the items that the users of the Wiimote system found for task 4. The HDM owner significantly outperforms the other six participants, possibly because it would be easier and quicker for the HDM owner to identify relevant images, thereby reducing the requirement to pause or rewind playback; however there was no significant difference between the performance of the Wiimote and baseline systems for the HDM owner. There was no requirement to actually stop and mark the relevant items when located; a simple identification was sufficient and the count of successful identifications was kept by the experiment coordinator.

As expected, the owner of the HDM archive performed overall more successfully on the majority of the tasks compared to the other participants. Immediately post study, a questionnaire (for each of the two systems) was given to the six participant users (but not the HDM owner). This was an 11 item questionnaire using a 7-point likert scale which asks participants to agree or disagree to statements concerning their satisfaction using the both prototype systems. The results of this questionnaire are displayed in Figure 3. It can be seen that participants rated the Wiimote higher for satisfaction, efficiency, productivity, recovery from error and functionality. The baseline (labelled 'mouse' in Figure 3) was rated higher for comfort and ease of learning and there was no difference in the user rating for effectiveness and interface/interaction pleasantness. Overall though, there was no significant difference between the two prototypes in the questionnaire. How much the demanding nature of time-limited topics affected the user questionnaire answers is not known. It would have been anticipated that the Wiimote system would have been rated higher for comfort and pleasantness, which was not the case. Figure 3 shows the mean of the six participant ratings for nine aspects of the interface (two aspects incorporated two merged similar topics).

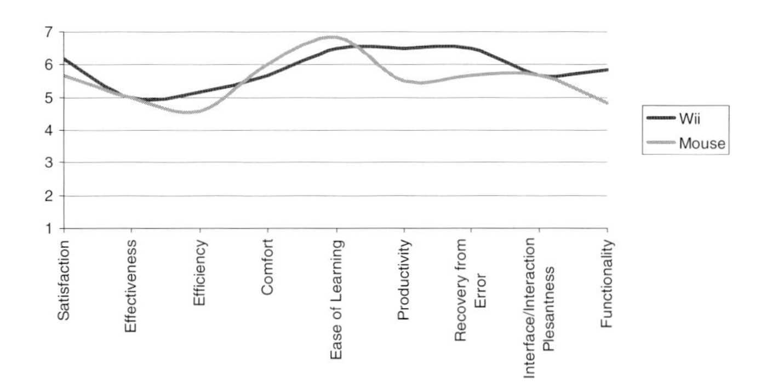

Fig. 3. The averaged user feedback (Likert scale) for the post-experiment questionnaire

Finally, informal feedback from the six participants suggested that although they were all more familiar with a computer mouse, they felt that the Wiimote gesture interface was easy to learn and that it allowed them to complete their tasks more efficiently than the mouse. Overall participants declared themselves to be more satisfied with the Wiimote than baseline, even though some did comment that the gestures were slightly too sensitive as implemented in this prototype.

6 Conclusions

In this paper a prototype HDM browser that implemented a gesture-based interface for a lean-back environment was described and evaluated. The prototype was influenced by a set of guidelines for lean-back environment information systems that were proposed. The findings of a user experiment suggested that the gesture-based prototype was as effective (and sometimes more so) than a functionally and visually similar lean-forward (mouse interaction) desktop prototype. It was found that users were very comfortable with the gesture-based interface and that they found it easy to learn, effective and more satisfying to use than a point-and-click mouse equivalent. It should be noted that to achieve this result, event segmentation technologies and keyframe extraction techniques [9] were employed (behind the interface) to maximize ease to use, which is the key point from all three guidelines presented in section 3.

Future work includes a larger user study, as well as identifying what other axes of organisation are possible in the lean-back environment. For example, implementing a mapping interface (geostamped HDM) would better suit a desktop device than a gesture-based device, but how it could be employed in a gesture-based interface is not known. In addition, it is important to identify how effective is a lean-back, gesture-based interface to other personal archives, such as digital photos or an archive of home movies. Finally, for a HDM owner knowledgeable about their own data, the gesture interface showed significant performance improvements, and needs to be the subject of a larger study, though locating sufficient HDM owners will be a challenge.

Acknowledgments. The authors wish to thank Science Foundation Ireland for the main author's Stokes award, Microsoft Research for providing sustained access to SenseCams and the NDRC (National Digital Research Centre) for providing resources for this work as part of the InsideOut project.

References

1. Schlömer, T., Poppinga, B., Henze, N., Boll, S.: Gesture recognition with a Wii controller. In: Proceedings of the 2nd international Conference on Tangible and Embedded interaction, TEI 2008, Bonn, Germany, February 18-20, ACM, New York (2008)
2. Naaman, M., Harada, S., Wang, Q., Garcia-Molina, H., Paepcke, A.: Context Data in Geo-Referenced Digital Photo Collections. In: proceedings of Twelfth ACM International Conference on Multimedia (ACM MM 2004) (October 2004)
3. O'Hare, N., Lee, H., Cooray, S., Gurrin, C., Jones, G., Malobabic, J., O'Connor, N., Smeaton, A.F., Uscilowski, B.: MediAssist: Using Content-Based Analysis and Context to Manage Personal Photo Collections. In: Sundaram, H., Naphade, M., Smith, J.R., Rui, Y. (eds.) CIVR 2006. LNCS, vol. 4071, pp. 529–532. Springer, Heidelberg (2006)
4. TRECVid, `http://www-nlpir.nist.gov/projects/trecvid/` (Last Visited July 16, 2009)
5. Lee, H., Ferguson, P., Gurrin, C., Smeaton, A.F., O'Connor, N., Park, H.S.: Balancing the Power of Multimedia Information Retrieval and Usability in Designing Interactive TV. In: Proceedings of uxTV 2008 - International Conference on Designing Interactive User Experiences for TV and Video, Mountain View, CA, October 22-24 (2008)
6. Gemmell, F.J., Aris, A., Lueder, R.: Telling Stories with MyLifeBits. In: Proceeding of IEEE International Conference on Multimedia and Expo (ICME 2005), Amsterdam, Netherlands (July 2005)
7. Gurrin, C., Byrne, D., O'Connor, N., Jones, G., Smeaton, A.F.: Architecture and Challenges of Maintaining a Large-scale, Context-aware Human Digital Memory. In: Proceedings of VIE 2008 - The 5th IET Visual Information Engineering 2008 Conference, Xi'An, China, 29 July - 1 August (2008)
8. Hodges, S., Williams, L., Berry, E., Izadi, S., Srinivasan, J., Butler, A., Smyth, G., Kapur, N., Wood, K.: SenseCam: A Retrospective Memory Aid. In: Dourish, P., Friday, A. (eds.) UbiComp 2006. LNCS, vol. 4206, pp. 177–193. Springer, Heidelberg (2006)
9. Doherty, A.R., Smeaton, A.F.: Automatically Segmenting Lifelog Data into Events. In: WIAMIS 2008 - 9th International Workshop on Image Analysis for Multimedia Interactive Services, Klagenfurt, Austria (May 2008)
10. Geerts, D., De Grooff, D.: Supporting the social uses of television: sociability heuristics for social TV. In: Proceedings of ACM CHI 2009, pp. 595–604 (2009)
11. Gallo, L., De Pietro, G., Marra, I.: 3D Interaction with Volumetric Medical Data: experiencing the Wiimote. In: Proceedings of Ambi-sys 2008, Quebec, Canada, February 11-14 (2008)

12. Schlomer, T., Poppinga, B., Henze, N., Boll, S.: Gesture Recognition with a Wii Controller. In: Proceedings of the Second International Conference on Tangible and Embedded Interaction (TEI 2008), Bonn, Germany, February 18-20 (2008)
13. Lapping-Carr, M., Jenkins, O., Grollman, D., Schwertfeger, J., Hinkle, T.: Wiimote interfaces for lifelong robot learning. In: AAAI Symposium on Using AI to Motivate Greater Participation in Computer Science, Palo Alto, CA, USA (March 2008)
14. Shiratori, T., Hodgins, J.K.: Accelerometer-based User Interfaces for the Control of a Physically Simulated Character. ACM Transactions on Graphics (Proc. SIGGRAPH Asia 2008) (December 2008)
15. Darnell, M.: How do people really interact with TV? Naturalistic observations of digital TV and Digital Video Recorder users. Computers in Entertainment 5(2) (2007)
16. Roibas, A.C., Sala, R.: Main HCI issues for the design of interfaces for ubiquitous interactive multimedia broadcast. Interactions 11(2), 51–53 (2004)
17. Park, J., Blythe, M., Monk, A., Grayson, D.: Sharable digital TV: relating ethnography to design through un-useless product suggestions. In: ACM CHI 2006 extended abstracts, New York (2006)
18. Designing for Interactive Television v1.0, BBCi and Interactive TV Programmes. British Broadcasting Corporation (2005)
19. Bonnici, S.: Which channel is that on? A design model for electronic programme guides. In: Proc. 1st European Interactive Television Conference (EuroITV 2003) (April 2003)
20. Chorianopoulos, K.: User interface design and evaluation in interactive TV. The HERMES Newsletter by ELTRUN 32 (May-June 2005)
21. Drucker, S.M., Glatzer, A., Mar, S.D., Wong, C.: SmartSkip: consumer level browsing and skipping of digital video content. In: Proceedings of (CHI 2002) SIGCHI Conference on Human Factors in Computing Systems, New York (2002)
22. Hassenzahl, M., Platz, A., Burmester, M., Lehner, K.: Hedonic and ergonomic quality aspects determine a software's appeal. In: Proceedings of SIGCHI Conference on Human Factors in Computing Systems, New York (2000)
23. Pemberton, L., Griffiths, R.: Usability evaluation techniques for interactive television. In: Proceedings of the Tenth HCI International Conference (June 2003)
24. Chorianopoulos, K., Spilnellis, D.: Affective usability evaluation for an interactive music television channel. Computers in Entertainment 2(3) (2004)
25. Jensen, J.: Interactive television: new genres, new format, new content. In: Proceedings of the Second Australasian Conference on Interactive Entertainment, November 2005, pp. 89–96 (2005)
26. Ahonen, A.: Guidelines for designing easy-to-use interactive television services: experiences from the ArviD. In: Interactive Digital Television - Technologies and Applications. IGI Global (2008)
27. Lewis, J.R.: IBM Computer Usability Satisfaction Questionnaires: Psychometric Evaluation and Instructions for Use. International Journal of Human-Computer Interaction 7(1), 57–78 (1995)

Automatic Image Inpainting by Heuristic Texture and Structure Completion

Xiaowu Chen* and Fang Xu

State Key Laboratory of Virtual Reality Technology and Systems,
School of Computer Science and Engineering, Beihang University, Beijing, P.R. China
chen@buaa.edu.cn, xufang@vrlab.buaa.edu.cn

Abstract. This paper studies an image inpainting solution based on a primal sketch representation model [1], which divides an image into structure (sketchable) and texture (non-sketchable) components. This solution first predicts the missing structures, such as curves and corners, using a tensor voting algorithm [2]. Then the texture parts along structural sketches are synthesized with the patches sampled from the known regions, and the remaining texture parts are defused using a graph cuts algorithm [3]. Compared to the state-of-art image inpainting approaches, the characteristics of this solution include: 1) using the primal sketch representation model to guide completion for visual consistency; 2) achieving fully automatic completion. Finally, the experiments on the public datasets show above characteristics.

1 Introduction

The objective of image inpainting is to fill the missing or unknown components of an input image, which is a well-studied topic in computer vision and graphics research. To achieve visually pleasing results, this paper presents an integrated framework to inpaint images based on a primal sketch representation model [1]. This framework includes automatic structure propagation and texture synthesis for completion process, as illustrated in Fig. 1.

In the literature, Image inpainting was first introduced by Bertalmio et al [4], which uses partial differential equations (PDE) that propagate the information from the boundary of missing region to its interior. These PDE-based methods often result in oversmooth synthesis when the filling pixels are full texture. To solve this problem, the exemplar-based approaches are proposed to propagate missing regions by copy-and-paste from a single image [5] [6], or from a large dataset [7]. Recently, Sun et al. presents an interactive system to achieve the state-of-art results by completing the missing structures with user guidance [8].

This paper studies an image inpainting solution, which automatically completes salient structures and texture of the missing regions. Given an input image,

* Corresponding author. This work was partially supported by National Natural Science Foundation of China (90818003 & 60933006), National High Technology R&D 863 Program of China (2009AA01Z331), National Grand Fundamental Research 973 Program of China (2006CB303007), and Independent Foundation of State Key Lab of VR T&S.

S. Boll et al. (Eds.): MMM 2010, LNCS 5916, pp. 110–119, 2010.

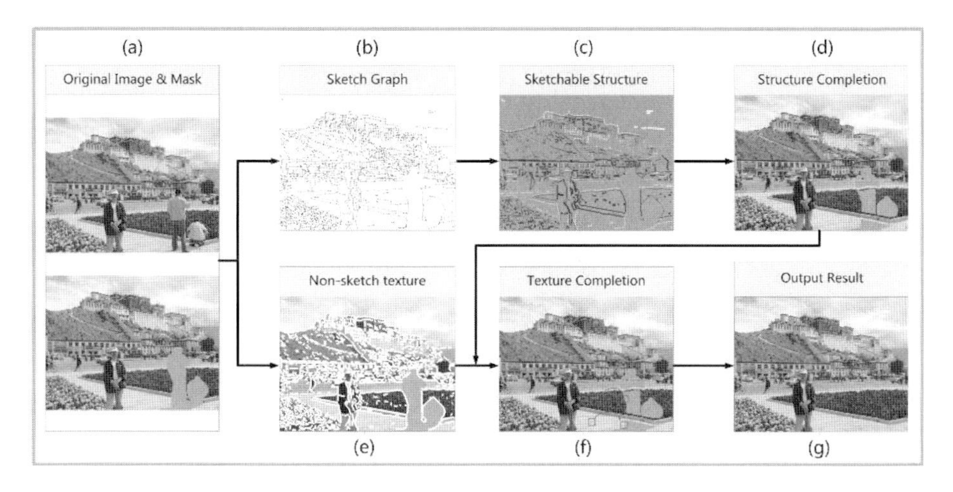

Fig. 1. (a) the input image and marked unknown region. (b) sketch graph computed by primal sketch algorithm. (c) structure part of the input image. (d)structure completion result. (e)texture part of the input image. (f)texture completion process. (g)the final completion result.

its primal sketch graph is firstly computed using the primal sketch representation model [1]. Afterwards, 3-degree-junctions such as T-junctions, Y-junctions around the missing regions are detected by Wu's algorithm [9]. The sketch completion is based on testing the compatibility between any two terminators according to Elastica [10]. The texture along structural sketches are filled by optimal patches set computed by a Belief Propagation algorithm [11]. The filled texture along structural sketches can guide the texture synthesis process of the remaining unknown regions. After structure completion, the remaining missing regions are filled in using exemplar-based texture synthesis. Instead of copying the whole selected patch [5], the proposed solution only pastes an optimal portion of a patch to the unknown region, and the optimal seam is determined by graph cuts algorithm for texture synthesis [3].

2 Problem Formulation of Image Inpainting

We start with a brief review of the primal sketch representation model on which our image inpainting is based. Proposed in Marr's book [12], primal sketch is supposed to be a symbolic and perceptually lossless representation of the input image. A mathematical model for primal sketch was proposed in [1]. As Fig. 1 shows, given an input image I, we compute a sketch graph G whose vertices are image primitives shown in Fig. 1.(b). This sketch graph divides the image lattice into a sketchable parts in Fig. 1.(c) and the nonsketchable parts in Fig. 1.(e). The non-sketchable parts are stochastic texture modeled by Markov random field models. The pixels on the sketchable part are reconstructed by the primitives under some transformations, while the pixels on the non-sketchable part are synthesized from the markov random field models using the sketchable part as boundary condition.

Given an image defined on lattice Λ with unknown region Λ_u,the goal of image inpainting is to infer an idea image I based on the partial observation $I_o|_{\Lambda\backslash\Lambda_u}$.

In the Bayesian framework, image inpainting is defined as maximize the posterior probability of $p(I|I_o,\Lambda)$:

$$p(I|I_o,\Lambda) = \frac{p(I_o|I,\Lambda)p(I_o|\Lambda)}{p(I_o|\Lambda)} \propto p(I_o|I,\Lambda)p(I_o|\Lambda) \tag{1}$$

Essentially, the key to the image inpainting problem is to employ a suitable image prior model $p(I_o|\Lambda)$ or to define a set of energy functions which are the logarithm likelihood functions of the posterior probability $p(I|I_o,\Lambda)$. We adopt the primal sketch model [1] as image prior, which is defined as follows:

$$\begin{aligned} p(I,S) =& \frac{1}{Z}\exp\{-\sum_{i=1}^{n}\sum_{(x,y)\in\Lambda_{str,i}}\frac{1}{2\sigma^2}(I(x,y) \\ &- B_i(x,y|\theta_i))^2 - \gamma_{str}(S_{str}) \\ &-\sum_{j=1}^{m}\sum_{(x,y)\in\Lambda_{tex,j}}\sum_{k=1}^{K}\phi_{j,k}(F_k * I(x,y))) \\ &- \gamma_{tex}(S_{tex})\}. \end{aligned} \tag{2}$$

where S is a set of pixels of discontinuity regions which contain S_{str} and S_{tex}, $B_i(x,y|\theta_i), i = 1,...,n$ are a set of coding functions representing edge and ridge segments in the image,θ_i are geometric and photometric parameters of these coding functions,F_k are filters applied on the pixels in Λ_{tex}.

The computed primal sketch graph divides the image lattice into structure domain S_{str} and texture domain S_{tex} respectively. Image intensities on the structure domain are represented by coding functions with explicit geometric parameters(such as normal directions) and photometric parameters. These parameters provide prior information for structure completion of the image inpainting process. The Markov random field model is assumed to model image intensities on the texture domain S_{tex}, and the texture completion can be achieved by finding similar neighborhoods and synthesizing the missing pixels.

Our solution searches for optimal image patches from known regions to fill the structure and texture components of the unknown region. The optimal sample labels $X = \{x_i\}_{i=1}^{L}$ are obtained by minimizing the energy $E(X)$:

$$E(X) = E_{structure} + E_{texture} \tag{3}$$

$$E_{structure} = E_{elastic} + \sum_{i\in\Lambda_{u,sk}} E_1(x_i) + E_2(x_i) \tag{4}$$

$$E_{elastic} = \int_{\Gamma}(\nu + \alpha\kappa^2)ds \tag{5}$$

where $E_{elastic}$ constrained the curve connection process, ν and α are constant, Γ is the curve, κ is the curvature funciton κ (s),s $\in$ [0,L] and L is the arc-length.

$$E_1(x_i) = d(c_i, c_{x_i}) + d(c_{x_i}, c_i) \quad (6)$$

where $\Lambda_{u,sk}$ is the structure part of unknown region Λ_u, $d(c_i, c_{x_i})$ is the sum of shortest distance between all points in segment c_i and c_{x_i}, c_i is on the connected curves of the unknown region, and c_{x_i} is on the detected curves of the known region.

$E_2(x_i)$ constrained the synthesized patches on the boundary of unknown region to match well with the known pixels. $E_2(x_i)$ is the normalized squared differences of the overlap region between image patches and known region.

$$E_{texture} = \sum_{(i,j)\in\Lambda_{u,nsk}} E_3(x_i + x_j) \quad (7)$$

where $\Lambda_{u,nsk}$ is the texture component of the unknown region Λ_u, $E_3(x_i + x_j)$ encodes the texture consistence constraint between two adjacent pixels i and j.

3 Framework Implementation

3.1 Structure Completion

Structure Sketch Completion. With the result of edge detection using primal sketch algorithm, our algorithm automatically finds long continuous sketches around the unknown regions. 3-degree-junctions such as T-junctions, Y-junctions on the boundary of the unknown regions are detected by Wu's algorithm [9], as illustrated in Fig. 2(a) 3-degree-junctions will be broken into a set of terminators as open bonds, illustrated in Fig. 2(b). These open bonds need to find a match by testing compatibilities on appearance cues and geometric properties. Given tow terminators a_i and a_j,the contour to be completed between them , denoted by Γ^*,is decided by minimizing the Elastica cost function in a contour space Ω_Γ [10].

$$\Gamma^* = \arg\min_{\Gamma\in\Omega_\Gamma} \int_\Gamma [(v_1 + \alpha_1\kappa_1^2) + (v_2 + \alpha_2\kappa_2^2)]ds \quad (8)$$

where κ_1 is the curvature of a_1, κ_2 is the curvature of a_2, v_1 and v_2 are constants, and α_1 and α_2 are scalable coefficients.

First, normal directions of every existing curve points are calculated, these points with normal directions are then represented by 2D stick tensors.

Each point in hole region (blue point in Fig. 3) obtains a curve saliency value $\lambda_1 - \lambda_2$ after receiving and accumulating the collected votes from the existing curve points. For each position x_i, only the point with the highest curve saliency is selected as the curve point P_{x_i}:

$$P_{x_i} = \max\{P_{x_i,y_j}(\lambda_1 - \lambda_2), 1 \leq j \leq S\} \quad (9)$$

where x_i, y_i is the image coordinates, S is the sample density(in pixels).

Structure Texture Completion. To synthesize texture along structural sketches, we employ a Belief propagation algorithm similar to Sun [8], which

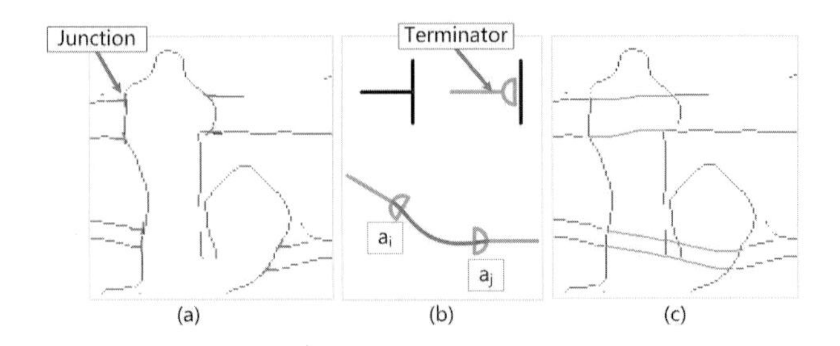

Fig. 2. 3-degree-junctions, terminators and curve connection

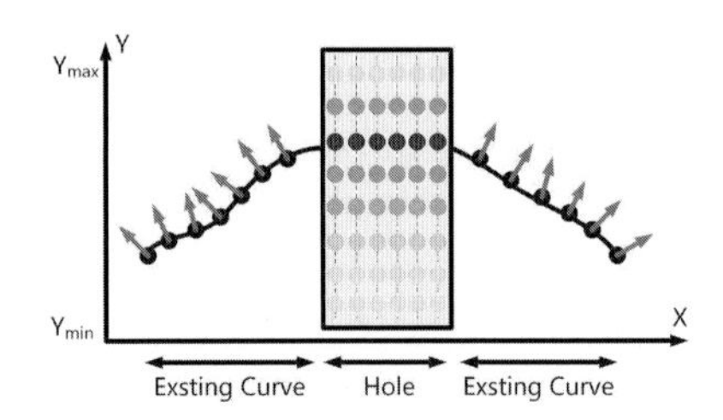

Fig. 3. Curve connection by tensor voting

finds the optimal sample labels for each sample point on the connected curves by minimizing the energy $E_{structure}$(Equation 4).

3.2 Texture Completion

After structure completion, there still exist large unknown regions to be filled. The equivalence of the Julesz ensemble and FRAME models [13] states that texture synthesis can be done without necessarily learning Markov random field models. We adopt this strategy and sample from a subset of the Julesz ensemble by pasting texture patches from the sample texture. To achieve more visual pleasing result, the texture of the remaining unknown regions are filled by the graph cuts texture synthesis algorithm [3]. Before applying the graph cuts based texture synthesis to image inpainting problem we firstly compute a filling order using confidence map, similar to Criminisi [5]. To reduce the time complexity of best patches searching, our approach runs a segmentation algorithm to find adjacent known regions which may have homogenous textures around the unknown regions.

Oversegmentation and Region Adjacent Graph. First we adopt the graph cuts segmentation algorithm of Felzenszwalb [14] with the parameters sigma = 0.5, k = 500, min = 20. This can often group homogenous regions of a image together, and produces reasonable oversegmentations with fewer superpixels (typically less than 100 for an 800x600 image). Unknown regions are always adjacent to some known regions, and these adjacent regions form a region adjacent graph. The proposed solution only need to search in these adjacent regions for best image patches to synthesis the texture of the remaining unknown regions. This search strategy can improve the reliability and efficiency of texture synthesis of the remaining unknown regions.

Completion Priority. The texture completion order is computed by using a confidence map [5]. For a point p on the boundary of unknown region Ω, Ψ_p is the patch centered at point p, and priority $P(p)$ is the product of confidence term $C(p)$ and data term $D(p)$, n_p is a unit vector orthogonal to the boundary at point p:

$$P(p) = C(p)D(p) \tag{10}$$

$$C(p) = \frac{\Sigma_{q \in \Psi_p \cap (I-\Omega)} C(q)}{|\Psi_p|}, D(p) = \frac{|\nabla I_p^{\perp} \cdot n_p|}{\alpha} \tag{11}$$

Graph cuts Texture Synthesis. To synthesis image texture in the remaining unknown regions, we sample optimal patch for each highest priority pixel in known regions. The similarity of tow image patches is measured by the normalized Sum of Squared Differences (SSD). Unlike Criminisi [5], we only copy an optimal portion of a patch to the missing regions instead of copying a whole patch. The portion of the patch to copy is determined by using graph cuts texture synthesis algorithm of Kwatra [3]. Let s and t be two adjacent pixels in the overlap regions between two image patches A and B. Let $A(s)$ and $B(s)$ be the pixel color at the position s in the patch A and patch B. The consistent matching cost M between the two adjacent pixels s and t that copy from patches A and B respectively is:

$$M(s, t, A, B) = (A(s) - B(s))^2 + (A(t) - B(t))^2 \tag{12}$$

We use CIE Lab color space to compute pixel color difference. As Fig. 4 shows, the nodes of the graph are the overlap pixels B and remaining pixels, and the arcs connecting the adjacent pixel nodes are labeled with matching cost $M(s, t, A, B)$. The blue line shows the minimum cut, and this means that pixels $1, 4, 5, 7, 8$ must come from patch A, and pixels $2, 3, 6, 9$ come from patch B. This usually produces more visually pleasing results than directly copying and pasting.

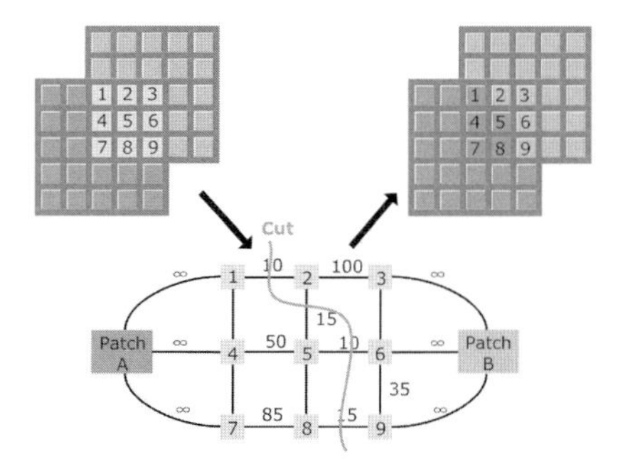

Fig. 4. Graph cuts algorithm for computing optimal patch portion

4 Experiment Results and Conclusion

We apply our algorithm to a number of natural images where the unknown regions are covered by masks. Fig. 5 shows a few representative results by the

Fig. 5. Some typical results of automatic image inpainting

Fig. 6. Comparison of our image completion results and Criminisi's. The first column shows the original images, the second column is the results of Criminisi's algorithm, the third column shows the results of our algorithm.

Fig. 7. The first column shows the original image, the second column is the results of our algorithm

proposed method, due to the space limitation. The first two columns show the input images and marked unknown regions. The third column shows the results of automatic structure completion. The forth column shows the final results.

In addition, we compare our method to Criminisi's algorithm. Results are shown in Fig. 6. The first column shows the original image. The second column shows results using Criminisi's method. The right most column results are ours. These results demonstrate the advantage of this method. Fig. 7 lists more image inpainting results.

This paper has presented a novel image inpainting solution that automatically fills structure and texture components based on primal sketch model. However, this method is still in the preliminary research stage. Currently, we want to extend our method to video inpainting, which promise to impose a new set of challenges.

Acknowledgement

This work was partially supported by National Natural Science Foundation of China (90818003 & 60933006), National High Technology R&D 863 Program of China (2009AA01Z331), National Grand Fundamental Research 973 Program of China (2006CB303007), and Independent Foundation of State Key Lab of VR T&S. And we would like to thank Dr. Liang Lin for helpful suggestions, and Kai Jiang for data processing.

References

1. Guo, C.E., Zhu, S.C., Wu, Y.N.: Primal sketch: Integrating texture and structure. Computer Vision and Image Understanding 106, 5–19 (2006)
2. Medioni, G., Lee, M., Tang, C.: A computational framework for segmentation and grouping. In: USC Computer Vision (2000)
3. Kwatra, V., Schodl, A., Essa, I., Turk, G., Bobick, A.: Graphcut textures: image and video synthesis using graph cuts. ACM Trans. Graph. 22, 277–286 (2003)
4. Bertalmio, M., Sapiro, G.: Image inpainting, pp. 417–424 (2000)
5. Toyama, C.E., Criminisi, A., Prez, P., Toyama, K.: Object removal by exemplar-based inpainting, pp. 721–728 (2003)
6. Efros, A., Leung, T.: Texture synthesis by non-parametric sampling. In: International Conference on Computer Vision, pp. 1033–1038 (1999)
7. Hays, J.H., Efros, A.A.: Scene completion using millions of photographs. ACM Transactions on Graphics (SIGGRAPH 2007) 26(3) (August 2007)
8. Sun, J., Yuan, L., Jia, J., yeung Shum, H.: Image completion with structure propagation. ACM Transactions on Graphics, 861–868 (2005)
9. Wu, T.F., Xia, G.S.: Compositional boosting for computing hierarchical image structures. In: Proc. IEEE. Conf. on Computer Vision and Pattern Recognition (2007)
10. Mumford, D., Shah, J.: Optimal approximations by piecewise smooth functions and associated variational problems. Communications on Pure and Applied Mathematics 42 (1989)

11. Broadway, J.Y., Yedidia, J.S., Freeman, W.T., Weiss, Y.: Generalized belief propagation. In: NIPS, vol. 13, pp. 689–695. MIT Press, Cambridge (2000)
12. Marr, D.: Vision: A Computational Investigation into the Human Representation and Processing of Visual Information. Henry Holt and Co., Inc., New York (1982)
13. Wu, Y.N., Zhu, S.C., Liu, X.: Equivalence of julesz ensembles and frame models. International Journal of Computer Vision 38, 245–261 (2000)
14. Felzenszwalb, P.F., Huttenlocher, D.P.: Efficient graph-based image segmentation (2004)

Multispectral and Panchromatic Images Fusion by Adaptive PCNN

Yong Li[1,2], Ke Wang[1], and Da-ke Chen[1]

[1] College of Communication Engineering, Jilin University, Changchun, 130025 China
liyong8113@sina.com, wangke@jlu.edu.cn, chendake_fm365@yahoo.cn
[2] College of Information Engineering, Jilin Teachers' Institute of engineering&Technology, Changchun, 130052 China

Abstract. As for low resolution of remote sensing images, a novel image fusion algorithm by adaptive PCNN was proposed. The multi-spectral image is firstly converted from RGB to lαβ color space. Then, the input images are adaptively decomposed by simplifying traditional PCNN model and defining image definition as the coupled joint coefficient. The largest entropy ignition time series are finally sent to decision factor to achieve the ultimate fusion image. The experimental results show that the proposed method can not only solve the difficult problem about how to set traditional PCNN parameters adaptively, but also on subjective and objective evaluation, its fusion effect on subjective and objective performance evaluation is better than that of other multi-resolution fusion algorithms such as wavelet transform.

Keywords: Image fusion, pulse-coupled neural network (PCNN), color space conversation, adaptive parameter setting.

1 Introduction

Multispectral and panchromatic image fusion technology has been widely used in environmental observation and surface plants classification [1]. At present, the pixel-level image fusion algorithms can be divided into IHS color space transform method, pyramid decomposition method and wavelet transform method [2]. Among them, IHS method can retain high resolution of panchromatic image, but lead to very serious spectral distortion. Multi-scale analysis methods easily separate the inter-linkages between pixels and bring about high algorithmic complexity [3].

As the single-layer network, pulse coupled neural network (PCNN) is well suited to deal with real-time image fusion, because it can carry out pattern recognition and target classification without training [4]. For example, multi-source image fusion algorithm based on local contrast PCNN model is proposed, which is combined with FPF filters to achieve the best option [5]. The multi-channel PCNN model is used to image fusion, which is combined with Laplace transform based on image block [6].

In the paper, a novel image fusion algorithm based on adaptive PCNN is proposed, which is combined with human visual feature according to PCNN model principle. On one hand, this algorithm not only can achieve adaptive processing by defining

S. Boll et al. (Eds.): MMM 2010, LNCS 5916, pp. 120–129, 2010.

image clarity as the coupling coefficient β, but also avoid the mutual influence of multi-parameter adjustments by simplifying traditional PCNN model. On the other hand, the use of master-slave parallel dual-channel adaptive PCNN structure can solve the problem of high algorithm complexity caused by combination of PCNN model and other multi-resolution algorithms. Experimental results show that compared with other algorithms, it can effectively retain spectral information of multispectral images, and improve spatial resolution.

2 Adaptive PCNN Fusion Principle and Model

2.1 PCNN Model Theory

PCNN is evolved from the mathematical model according to pulse synchronous vibration phenomenon of cat visual cortex, which is put forward by Eckhorn [7]. Every neuron is composed of the reception, modulation and pulse generator. The reception mainly has two functions including feedback input domain and connection input domain, which are respectively connected with the adjacent neuron through weighted synapsis functions M and W . In addition, the external stimulus S is add into the feedback input field. So the mathematical formulae are as follows:

$$F_{ij}[n]=e^{\alpha_F\delta_n}F_{ij}[n-1]+S_{ij}+V_F\sum_{kl}M_{ijkl}Y_{kl}[n-1] \tag{1}$$

$$L_{ij}[n]=e^{\alpha_L\delta_n}L_{ij}[n-1]+V_L\sum_{kl}W_{ijkl}Y_{kl}[n-1] \tag{2}$$

Where F_{ij} is defined as the neuron feedback of point (i, j) , L_{ij} is coupled connective function, Y_{kl} is neuron output of $(n-1)$ time iteration, V_F and V_L are respectively the inherent potential of F_{ij} and L_{ij} . Weighted connective matrix M and W as information transfer degree between adjacent neuron and central neuron require iterative computing according to exponential decay rule.

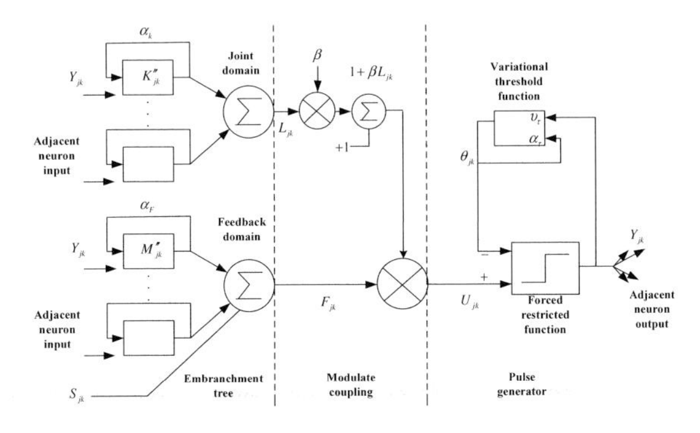

Fig. 1. PCNN neuron model

Internal neuron activity is formed of the two above-mentioned functions by non-linear multiplication, where β is defined as connective coefficient among synapsis. When the neuron is stimulated by external inspiration and influenced by the adjacent feedback input domain and coupling connection domain, internal neuron activity isn't gradually increasing until it is greater than dynamic threshold Θ . Then neurons stimulate excitement and bring timing pulse timing sequence. At the same time, Θ is suddenly increasing and then gradually reduced by means of exponential rule. So this process can be described as:

$$U_{ij}[n] = F_{ij}[n](1 + \beta L_{ij}[n]) \tag{3}$$

$$Y_{ij}[n] = \begin{cases} 1 & U_{ij}[n] > \Theta_{ij}[n] \\ 0 & others \end{cases} \tag{4}$$

$$\Theta_{ij}[n] = e^{\alpha_\Theta \delta_n} \Theta_{ij}[n-1] + V_\Theta Y_{ij}[n] \tag{5}$$

From the above analysis, we can see that PCNN is a multi-parameter neural network model and its application depends largely on parameters setting, so there is a problem of finding the optimal parameters. But so far the connection factor, threshold amplification factor and the number of iterations have been set through repeated experiments. Thus, this situation isn't suited for PCNN further application.

2.2 Improved PCNN Model

In order to solve the problem of poor adaption in image fusion, the traditional PCNN model is improved and a novel adaptive PCNN expansion model in master-slave parallel mode is proposed in this paper.

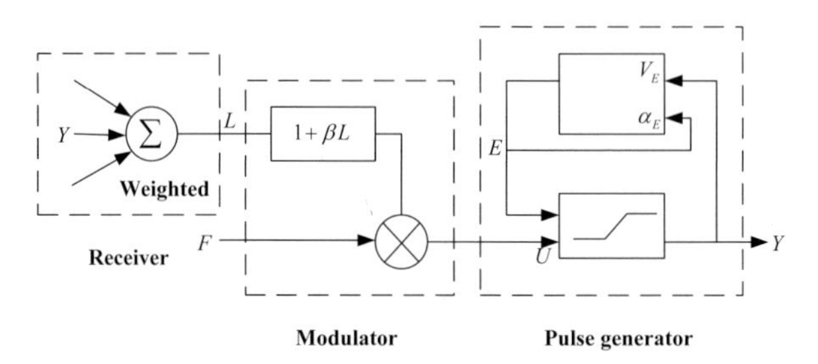

Fig. 2. Simplified PCNN neuron structure

From the above figure, we can see that the simplified PCNN model not only reduces parameters setting, but also maintains several important features of the original model to some extent. Firstly, it retains connective characteristics and the neurons in similar situation have the synchronize output pulse. Secondly, internal activities are formed by the output and connection domain according to the non-linear way. Thirdly, the dynamic threshold is still declined by means of exponential rule. When the neurons output the pulse, the threshold will be reset as V_E. The mathematical expressions of whole process are as

$$H_{ij}^k[n] = f^k(Y[n-1]) + S_{ij}^k \tag{6}$$

$$U_{ij}^k[n] = I_{ij}^k(1 + \beta_{ij}^k H_{ij}^k[n]) + \sigma \tag{7}$$

$$Y_{ij}[n] = \begin{cases} 1 & U_{ij}[n] > E_{ij}[n-1] \\ 0 & others \end{cases} \tag{8}$$

$$E_{ij}[n] = \exp(-\alpha_E)E_{ij}[n-1] + V_E Y_{ij}[n] \tag{9}$$

Where I_{ij}^k is defined as external incentive input of k $(k=1,2)$ channel, where point (i, j) is pixel gray value of input image. Function $f(\bullet)$ is the impact of adjacent on neurons on its own. σ is the internal balance factor of neurons. U_{ij}, Y_{ij} and E_{ij} are respectively the internal activity, pulse output and dynamic threshold of each neuron.

Connective coefficient is directly related to the weighted value of fusion image shared by input source image. In this model, each neuron (i, j) has its own connection coefficient β_{ij}^k, which reflects different coupling differences among neurons and adaptively adjusts according to different stimulus. Therefore, can enter the incentive to adjust the different adaptive. It is assumed that I_1 and I_2 are defined as the source input images, β_{ij}^1 and β_{ij}^2 are connection coefficients of the corresponding channels, so the mathematical expressions are as follows:

$$\beta_{ij}^1 = \frac{1}{1+e^{-\eta D(i,j)}}, \beta_{ij}^2 = \frac{1}{1+e^{\eta D(i,j)}} \tag{10}$$

In above formulae, $D(i, j)$ is defined as mean filter for input $d(i, j)$ to remove the wrong definition value, namely $D(i, j) = \sum_{m=-r/2}^{r/2} \sum_{n=-r/2}^{r/2} d(i+m, j+n)$. The input $d(i, j)$ is represented as $d(i, j) = g(I_1(i, j)) - g(I_2(i, j))$ and $g(I(i, j))$ is neighborhood clarity of image $I(i, j)$ by means of gradient method.

$$g(I(i,j)) = \sqrt{\begin{matrix}[I(i,j) - I(i+1,j)^2 + I(i,j) - I(i-1,j)^2 + \\ I(i,j) - I(i,j+1)^2 + I(i,j) - I(i,j-1)^2]\end{matrix}} \tag{11}$$

From above formula, we come to conclusion that β_{ij}^1 is a increasing function, while β_{ij}^2 is a declined function. It means that the point is more clearly, thus the connective coefficient of corresponding neuron is greater, which lead to higher weighted value in fusion image. So the adaptive setting of β_{ij}^1 and β_{ij}^2 is completed.

3 The Process of Fusion Algorithm

3.1 Algorithm Steps

There is the processing of image fusion algorithm based on adaptive PCNN model in figure 3. The whole process is described in detail as follows:

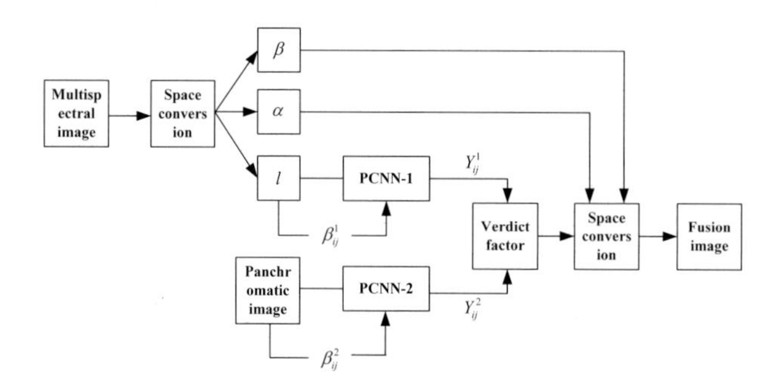

Fig. 3. Adaptive PCNN image fusion

(1) Color space conversion
Registered multi-spectral image is transformed from RGB to $l\alpha\beta$ color space, which is proposed by Welsh [8]. In this color space, l is defined as the achromatic channel, α is defined as the yellow - blue channel and β is defined as the red-green) channel. These three channels are orthogonal relationship.

$$\begin{bmatrix} L \\ M \\ S \end{bmatrix} = \begin{bmatrix} 0.3811 & 0.5783 & 0.0402 \\ 0.1967 & 0.7244 & 0.0782 \\ 0.0241 & 0.1288 & 0.8444 \end{bmatrix} \begin{bmatrix} R \\ G \\ B \end{bmatrix} \tag{12}$$

$$\begin{bmatrix} l \\ \alpha \\ \beta \end{bmatrix} = \begin{bmatrix} 1/\sqrt{3} & 0 & 0 \\ 0 & 1/\sqrt{6} & 0 \\ 0 & 0 & 1/\sqrt{2} \end{bmatrix} \begin{bmatrix} 1 & 1 & 1 \\ 1 & 1 & -2 \\ 1 & -1 & 0 \end{bmatrix} \lg \begin{bmatrix} L \\ M \\ S \end{bmatrix} \tag{13}$$

(2) PCNN parameters setting
In this algorithm, the gray pixel of l channel is selected as the input of main PCNN neurons, and each neuron is linked with other neurons in its neighboring 3×3 domain. The connective coefficient β_{ij}^1 is calculated according to formula

(10)-(11).The output of each neuron has only two situations: firing or not firing. The main parameter settings are as follows: the balance factor $\sigma = -0.1$, the decay time of threshold $\alpha_E = 0.5$, the threshold magnification factor $V_E = 220$, the internal activity U, pulse output Y and coupling output H are set as zero.

The gray pixel of panchromatic image is selected as the input of slaved PCNN neuron. The connective coefficient β_{ij}^2 is calculated according to formula (10)-(11), too. The difference between main and slaved PCNN channel is that the external stimulus S is removed in slaved PCNN. The main parameters are as follows: threshold decay time and magnification factor of slaved PCNN are less than that of main PCNN, and other parameter settings are the same.

(3) Adaptive fusion processing

According to the above parameters setting, the pixel value of l channel and panchromatic image is input into PCNN, then the received signals of each neuron are integrated based on formula (6) - (9). In this algorithm, the connective weighting function of adjacent neurons $f(\bullet)$ in main and slaved PCNN is the reciprocal of Euclidean distance square [9], namely weighted matrix of neuron ij and kl is as follows:

$$M_{ijkl} = W_{ijkl} = \frac{1}{(i-k)^2 + (j-l)^2} \tag{14}$$

The acts of main PCNN neurons are as follows: neurons with larger pixel value are first ignited and carried out pulse, which neurons with similar pixel and adjacent location are ignited one after the other based on pulse coupling characteristics. So the neurons clusters of synchronous pulse are formed and image decomposition will be completed. The acts of slaved PCNN neurons are as follows: all neurons are naturally ignited, and then the connection between neurons depends on their own coupling intensity. Each region has ignition phenomenon by controlling the number of iterations, and then feature points are selected. On determining the number of iterations, the maximum entropy principle [10] is used and then ignition time maps Y_{ij}^1 and Y_{ij}^2 with the largest entropy are obtained after dual-channel PCNN decomposition. Then they are input into verdict factor according to ignition situation of neurons to determine whether the target is in l channel image or in panchromatic image.

$$F(i,j) = \begin{cases} I_A(i,j) & Y_{ij}^1(i,j) > Y_{ij}^2(i,j) \\ I_B(i,j) & Y_{ij}^1(i,j) < Y_{ij}^2(i,j) \\ (I_A(i,j) + I_B(i,j))/2 & Y_{ij}^1(i,j) = Y_{ij}^2(i,j) \end{cases} \tag{15}$$

(4) Color space inversed conversion

The l channel image that is obtained by $l\alpha\beta$ space conversion on original multispectral image is replaced by gray fusion image $F(i, j)$ that is input from verdict factor. Then the final fusion image of RGB space is obtained by means of $l\alpha\beta$ inversed conversion on the new component l, α and β.

$$\begin{bmatrix} \Gamma \\ \Omega \\ \Psi \end{bmatrix} = \begin{bmatrix} 1 & 1 & 1 \\ 1 & 1 & -1 \\ 1 & -2 & 0 \end{bmatrix} \begin{bmatrix} 1/\sqrt{3} & 0 & 0 \\ 0 & 1/\sqrt{6} & 0 \\ 0 & 0 & 1/\sqrt{2} \end{bmatrix} \begin{bmatrix} l \\ \alpha \\ \beta \end{bmatrix} \tag{16}$$

$$\begin{bmatrix} R \\ G \\ B \end{bmatrix} = \begin{bmatrix} 4.4679 & -3.5873 & 0.1193 \\ -1.2186 & 2.3809 & -0.1624 \\ 0.0497 & -0.2439 & 1.2045 \end{bmatrix} \begin{bmatrix} 10^{\Gamma} \\ 10^{\Omega} \\ 10^{\Psi} \end{bmatrix} \tag{17}$$

4 Simulation and Analysis

To verify the algorithm performance, the simulation makes use of two groups of remote sensing images to simulate and contrast with the common methods such as IHS, discrete wavelet transform(DWT), Contourlet algorithm [11]. The multispectral and panchromatic image photographed by IKONOS satellite in Fredericton of Canada is adopted in the first experiment, while the Nanking zone images photographed by Quick Bird satellite is adopted in the second experiment. In the first experiment,

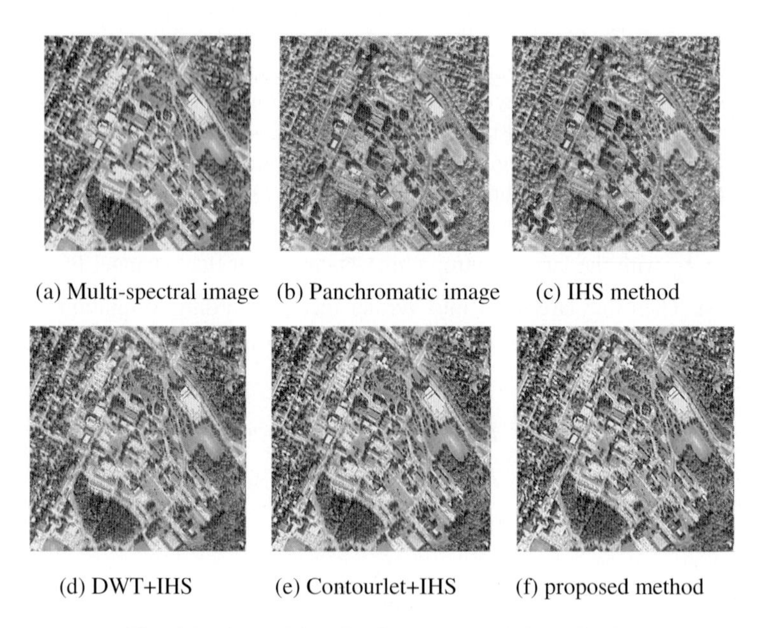

(a) Multi-spectral image (b) Panchromatic image (c) IHS method

(d) DWT+IHS (e) Contourlet+IHS (f) proposed method

Fig. 4. Fusion results of different methods in IKONOS

DWT algorithm uses the biorthogonal wavelet "bior4.4" and decomposition is 4 levels, while the decomposition of Contourlet algorithm is 5 levels. The above fusion methods use the low-frequent coefficients to substitute the intensity component and use the maximum regional energy as the high-frequent coefficients in order to validate the impact of different multi-scale decomposition tools on fusion performance. In the second experiment, the parameters settings of above algorithms is unchanged, but the pictures are appended into Gaussian white noise, which the mean is 0 and variance is 0.05 in order to validate the effects of noise interference on different algorithms. In the proposed algorithm, the initial value of parameters is set according to Step (2). The simulation results are shown in Figure 4 and 5.

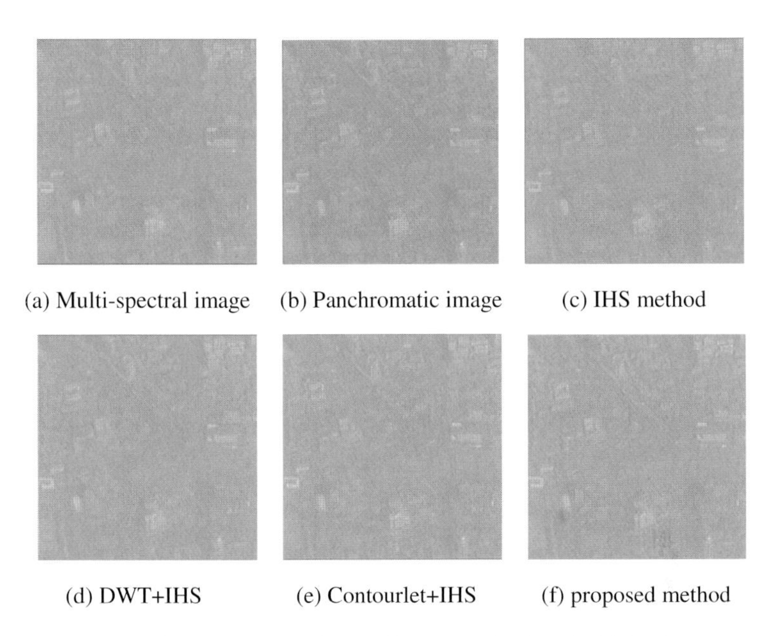

(a) Multi-spectral image (b) Panchromatic image (c) IHS method

(d) DWT+IHS (e) Contourlet+IHS (f) proposed method

Fig. 5. Fusion results of different methods in Quick Bird

The results of the objective evaluation on the above fusion methods are shown in Table 1 and 2. The edge maintainability and spectral distortion [12] are compared in the average of three-band R, G, and B.

Table 1. Fusion Performance of Different Methods in IKONOS

Fusion Method	Edge Maintainability	Spectral Distortion	SNR
IHS method	0.6560	46.9975	12.1286
DWT +IHS method	0.7148	28.7321	18.5569
Contourlet+IHS method	0.7256	23.4805	18.5102
The proposed method	0.7260	21.8591	19.4394

Table 2. Fusion Performance of Different Methods in Quick Bird

Fusion Method	Edge Maintainability	Spectral Distortion	SNR
IHS method	0.3783	2.9020	30.2798
DWT +IHS method	0.3929	2.4462	33.8287
Contourlet+IHS method	0.4135	2.1515	34.4391
The proposed method	0.4303	2.0417	34.9347

From Fig. 4, 5 and Table 1, 2, the conclusions are as below:

(1) From the view of edge maintainability, the effect of IHS algorithm is the worse owing to no directional information extraction. Although DWT+IHS and Contourlet+IHS methods can achieve detailed integration in each frequent band, "virtual shadow" is appeared in fusion image owing to the sub-sampled processing of multi-resolution decomposition. The effect of the proposed method is the best owing to the scale and displacement invariance of improved model.

(2) From the view of spectral distortions, the relationship among each channel of IHS space conversion isn't orthogonal so the color distortion fusion image is obvious. Although IHS space conversion is combined with DWT and Contourlet transform, the effect isn't satisfied owing to its own shortcoming. In the proposed method, orthogonal $l\alpha\beta$ space conversion is used to reduce the cross distortion of each color channel.

(3) From the view of SNR, IHS algorithm has the worst anti-noise ability; DWT + IHS and Contourlet + IHS methods carry out direction decomposition on image pixels with noise, which easily leads to signal-to-noise-aliasing. The proposed method has the optimal anti-noise ability, which removes majority of high frequent noise by neuron coupling.

(4) From the view of calculation complexity, for the image with $N \times N$ sizes the number of the proposed method are less than that of DWT+IHS method about $\frac{4}{3}KN^2$ times, and less than that of Contourlet+IHS method about $\frac{2}{3}(K+2)N^2$, which K is the length of filter.

5 Conclusion

With the problem of multi-parameter setting of traditional PCNN model, a novel image fusion algorithm based on adaptive PCNN model and $l\alpha\beta$ color space conversion is proposed in this paper. It can achieve self-adaptive processing by means of simplifying traditional PCNN model and defining image definition as the coupled joint coefficient. The experimental results show that the proposed method can adequately take the correlation between pixels and noise influence into account the correlation between pixels and noise impact, and then the fusion effect is better than that of other multi-resolution decomposition algorithms both in subjective visual analysis and objective evaluation standards.

References

1. Park, J.H., Kim, K.K., Yang, Y.K.: Image Fusion Using Multiresolution Analysis. In: J. Geoscience and Remote Sensing Symposium, pp. 709–711 (2001)
2. Garper, W.J., Lillesand, T.M., Kiefer, R.W.: The Use of Intensity-Hue-Saturation Transformations for Merging SPOT Panchromatic and Multispectral Image Data. J. Photogrammetric Engineering and Remote Sensing, 459–467 (1990)
3. Zhou, J., Civco, D.L., Silander, J.A.: A Wavelet Transform Method to Merge Landsat TM and SPOT Panchromatic Data. International Journal of Remote Sensing, 743–757 (1998)
4. Broussard, R.P., Rogers, S.K., Oxley, M.E.: Physiologically Motivated Image Fusion for Object Detection using Pulse-Coupled Neural Network. J. IEEE Transactions on Neural networks, 554–563 (1999)
5. Miao, G.-q., Wang, B.-s.: A Novel Image Fusion Algorithm based on Local Contrast and Adaptive PCNN. J. Chinese Journal of Computers, 875–880 (2008)
6. Huang, W., Jing, Z.: Multi-Focus Image Fusion using Pulse Couple Neural Network. J. Pattern Recognition Letters, 1123–1132 (2007)
7. Eckhorn, R.: Neural Mechanisms of Scene Segmentation: Recordings from the Visual Cortex Suggest Basic Circuits or Linking Field Models. J. IEEE Transactions on Neural Network, 464–479 (1999)
8. Welsh, T., Ashikhmin, M., Mueller, K.: Transferring Color to Grayscale Images. J. ACM Transactions on Graphics, 277–280 (2002)
9. Kuntimad, G., Ranganath, H.S.: Perfect Image Segmentation using Pulse Coupled Neural Networks. J. IEEE Transactions on Neural Networks, 591–598 (1999)
10. Liu, Q., Ma, Y.-d., Qian, Z.-b.: Automated Image Segmentation using Improved PCNN Model based on Cross-Entropy. J. Journal of Image and Graphics, 579–584 (2005)
11. Li, G.-x., Wang, K.: Color Image Fusion Algorithm using Contourlet Transform. J. Acta Electronica Sinica, 112–117 (2007)
12. Wang, Q., Shen, Y., Zhang, Y., Zhang, J.-q.: A Quantitative Method for Evaluating the Performances of Hyperspectral Image Fusion. J. IEEE Transactions on Instrumentation and Measurement, 1041–1047 (2003)

A Dual Binary Image Watermarking Based on Wavelet Domain and Pixel Distribution Features

Wei Xia, Hongwei Lu*, and Yizhu Zhao

College of Computer Science and Technology,
Huazhong University of Science and Technology,
Wuhan, China
xw7932@126.com, luhw@hust.edu.cn, missbamboofirst@163.com

Abstract. Considering that the binary images are featured little capability in data hiding, difficulty in watermarking embedding and two values, in order to improve the robustness and invisibility of watermarkings embedded into the binary images, a novel algorithm is presented which is based on DWT (Discrete Wavelet Transformation). The original watermarking signal is embedded into the lowest frequency sub-band of the wavelet domain. Combined with the technique of encryption, the extracted invariant pixel distribution features of the binary image establish another layer virtual watermarking and a mapping relationship between the two layer watermarkings. The mapping enables the self-restoration of the original watermarking with virtual watermarking when the binary image is attacked. The watermarking can be well extracted without the original binary image. The invisibility and robustness of the proposed algorithm are demonstrated by the Simulations such as Gaussian noise, JPEG compression and some geometric attacks.

Keywords: DWT, pixel distribution features, virtual watermarking, watermarking restoration.

1 Introduction

As an effective method of multimedia copyright protection and information security maintenance, digital watermarking technique is becoming a hot topic in the area of information processing. At present, a large amount of references mainly involve gray image, color image and the research of the video and audio watermarking. The binary image is bi-level and it is difficult to embed watermarking into it, so the binary image watermarking is seldom considered. With the global progress of the information digitalization, a lot of important binary text data, such as personal files, medical records, academic certificates, patent certificates, handwriting signatures, design patterns, library books, confidential documents have turned into digital documents by scan. In

* Corresponding author. This work is partially supported by the special funds for university basic research work of China (No.M2009020).

S. Boll et al. (Eds.): MMM 2010, LNCS 5916, pp. 130–140, 2010.

addition, with the increasingly prevalent e-commerce and e-government, the copy tracing and the integrity authentication of the electronic binary text documents are urgently required. We can deem most of these documents as binary text images. In some fields, the binary images are much more valuable than common gray and color images or video and audio files. So the copyright protection and information security maintenance of the binary images are especially important.

Currently, few domestic papers about binary image watermarking are found. The familiar binary image watermarking methods are line space encoding and character space encoding. Line space encoding is proposed by Brassil etc [1] in bell lab and implements watermarking embedding by means of changing the line space. Character space encoding is proposed by Huang etc [2] and implements watermarking embedding by means of left-right shifts of a word.

In general, we can classify image watermarking into two categories depending on the embedding domain: the spatial domain techniques and the frequency domain techniques. In spatial domain, the watermarking is embedded in the original image by modifying the pixel values or the least significant bits (LSB). All of the above bi-level image watermarking schemes are based on spatial domain. Compared with the schemes based on transform domain, these schemes are featured little capability in data hiding, poor robustness and complicated watermarking extraction process. Moreover, Chinese characters have no character space and baseline in the sense of English, so the effectiveness of Chinese character watermarking embedding is not always ideal by above methods. Recently, some works has focused on the frequency domain watermarking. Among the transform domains, the wavelet, due to its similarity to human visual system (HVS), is a proper domain for watermarking embedding. By using this transformation, modification is imposed on those regions that are less sensitive to human eyes. Therefore it causes effective achievement in the fidelity and robustness requirements. In addition, both newly JPEG2000 image and MPEG-4 video compression standard adopt this transformation, so the watermarking scheme is in well compatible with the new standards.

Dual digital watermarking technique embeds two layer watermarkings in one image and effectively increases information quantity of the embedded watermarkings. Traditional dual watermarking mainly emphasizes on realizing the functionally expansion of the watermarking to meet various practical requirements [3, 4]. Generally speaking, it can actuate two or more functions simultaneously and obviously has advantage in information quantity, but it is also far inferior to others in invisibility [5]. The algorithm features such as higher complex computation restrict the self-development of the dual watermarking.

The paper presents a new dual digital watermarking scheme which is based on wavelet domain and invariant distribution features of pixels in binary images. Therefore, a blind binary image watermarking algorithm with excellent robustness, invisibility and low computational complexity is realized.

The rest of the paper is structured as follows. As the backgrounds of the watermarking scheme, the pixel distribution features are presented in section 2. The entire watermarking scheme, including the two layer watermarking embedding process and the watermarking restoration, is introduced in detail in section 3. In section 4, the performance of the proposed watermarking method is evaluated by applying some familiar attacks to the watermarked binary images. Finally in section 5 we conclude our method.

2 Pixel Distribution Features of Binary Images

As for a binary image $f(x, y)$, target pixels (If the coordinate of the target pixel is (a, b), then $f(a,b)=1$) are parts of the global image region, so gravity center $(\bar{x}, \bar{y})$ can be selected as the centroid of the image and it meets the below equations.

$$\bar{x}=(\sum_i\sum_j i\times f(i,j))/(\sum_i\sum_j f(i,j)) . \tag{1}$$

$$\bar{y}=(\sum_i\sum_j j\times f(i,j))/(\sum_i\sum_j f(i,j)) . \tag{2}$$

$$f(i,j)=\begin{cases}1,,(i,j)\in R\\0,(i,j)\notin R\end{cases} . \tag{3}$$

" R " represents the target pixel region of the binary image. Computing the Euclidean distance between the centroid and every target pixel and finding out the maximum distance (denoted as $D_{\max}$) must be executed after getting the centroid. Subsequently, we obtain a circumcircle of the target pixel region. The circle center of the circumcircle is the centroid and the radius of it is $D_{\max}$.

Assuming that there are K bits of watermarkings to be embedded into the original image and the total number of target pixels is X, we can divide the circumcircle of the target pixel region into one circle sub-region and $\lfloor K/(\lfloor \log_2 X \rfloor+1)\rfloor-1$ concentric ring sub-regions and all sub-regions have the same circle center (the centroid). These sub-regions are separately denoted as $R_1, R_2, \cdots, R_M$ from inside to outside ($M=\lfloor K/(\lfloor \log_2 X \rfloor+1)\rfloor$). There are two sub-region division methods namely equidistant division method and equal-area division method [6].

Here we use the equal-area division method. Every sub-region interval has the same area with the assumption that $1\le i\le M$. It is described in equation (4).

$$R_i=\{(x,y)\,|\,(i-1)\times D_{\max}{}^2/M<(x-\bar{x})^2+(y-\bar{y})^2\le i\times D_{\max}{}^2/M\} \tag{4}$$

Fig. 1 shows the results under the condition of $M=4$. Certainly, in the real application, M is much larger than 4.

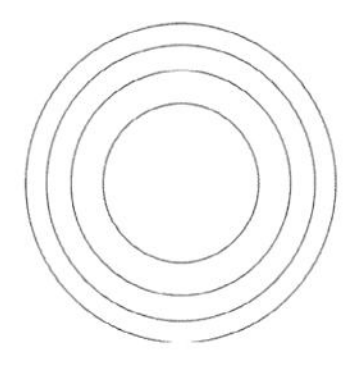

Fig. 1. One circle and three concentric ring sub-regions divided by equal-area method under the condition of $M = 4$

The target pixel number of every sub-region can be represented to $S_i (i = 1,2,\cdots,M)$. With the maximum value of S_i ($S_{\max} = \max_{i=1,2,\cdots,M}(S_i)$), we can calculate the probability density of the target pixels of every sub-region.

$$r_i = S_i / S_{\max} (i = 1,2,\cdots,M) . \tag{5}$$

In theory, $r_i (i = 1,2,\cdots,M)$ is comparatively invariant under all kinds of familiar geometric attacks such as rotation, translation and scaling, so we can encode every element of $r_i (i = 1,2,\cdots,M)$ with $\lfloor \log_2 X \rfloor + 1$ binary bits and then get $M \times (\lfloor \log_2 X \rfloor + 1)$ binary bits. If $M \times (\lfloor \log_2 X \rfloor + 1)$ is less than K, in order to make the length of the obtained binary code be equal to K, we fill it with the lowest $K - M \times (\lfloor \log_2 X \rfloor + 1)$ bits of $S_{\max}$. In the end, the obtained binary code with the length of K bits are denoted as B. We can define the binary code as the pixel distribution features of the binary image.

3 The Watermarking Scheme

Without affecting the results, we suppose that the size of the original host image F and the watermarking image W_1 are $b \times b$ and $a \times a$ respectively, as shown below.

$$W_1 = \{ w(i,j) \mid w(i,j) = 0,1; 1 \le i \le a, 1 \le j \le a \} . \tag{6}$$

$$F = \{ f(i,j) \mid f(i,j) = 0,1; 1 \le i \le b, 1 \le j \le b \} . \tag{7}$$

The watermarking embedding steps are described as follows.

3.1 The First Layer Watermarking Scheme

As shown in Fig. 2, after One-level wavelet transformation, the original image is decomposed into four sub-bands: LL, HL, LH, HH. As for binary image, the higher frequency sub-bands include ample texture features, poor energies and are liable to the affection of interference, so the coefficients in LL (lowest frequency sub-band) sub-band are selected to implement watermarking scheme.

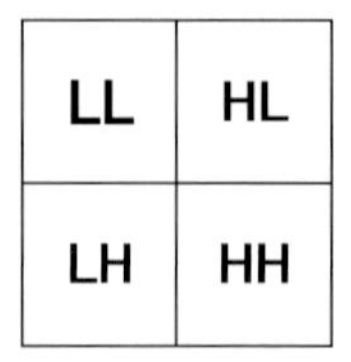

Fig. 2. Four sub-bands after One-level wavelet decomposition

Using L-level wavelet transformation, the original image is decomposed into its sub-bands. The coefficient matrix (denoted by FA_L) in the lowest frequency sub-band of the last level is selected and L can be obtained by the following formula.

$$L \le \lfloor \log_2(b/a) \rfloor . \tag{8}$$

The robustness is better and the decomposition and reconstruction time becomes longer when L is increasing.

Finally we can embed watermarking into FA_L by means of modifying the coefficients in it and a binary logical table in relation to the original watermarking is produced. It is actually a superposition method. Some related formulae are listed here.

$$FA_L^{'}(i, j) = FA_L(i, j) + \alpha W_1(i, j) . \tag{9}$$

$$FA_L^{*}(i, j) = round(FA_L^{'}(i, j)/t) . \tag{10}$$

$$FA_L^{''}(i, j) = FA_L^{*}(i, j) \times t . \tag{11}$$

$$kk(i, j) = FA_L^{*}(i, j) \bmod 2 . \tag{12}$$

$$key(i, j) = kk(i, j) \oplus W_1(i, j) . \tag{13}$$

"α" represents the watermarking embedding strength. Aiming at the robustness and invisibility of the watermarking scheme, the value range of the quantization parameter t is 0~0.5 according to the coefficients in FA_L. "round" in equation (**10**) is the integral function by rounding rule. After watermarking embedding and quantization, the selected coefficient matrix is transformed to $FA_L^{''}$. HASH function is used to assure the reliability of the binary logical table (denoted as $key(i, j)$). The binary logical table is necessary for original watermarking extraction. As the secret key, it can be used to apply to the third party for copyright protection.

Watermarking extraction is the reverse process of the watermarking embedding. L-level wavelet decomposition of the watermarked image is carried out to extract the coefficient matrix in the lowest frequency sub-band of the last level. The matrix is denoted as $T\_FA_L(i,j)$. We can easily extract watermarking (denoted as $W^*(i,j)$) according to the matrix and the produced binary logical table. The original image is not required in this method. It is described as below equations.

$$T\_FA_L^*(i,j) = round(T\_FA_L(i,j)/t) \ . \tag{14}$$

$$kk^*(i,j) = T\_FA_L^*(i,j) \bmod 2 \ . \tag{15}$$

$$W^*(i,j) = kk^*(i,j) \oplus key(i,j) \ . \tag{16}$$

3.2 The Second Layer Watermarking Scheme

Using the method discussed in section 2, a pixel distribution feature matrix (denoted as B) of the watermarked image can be obtained and the size of the matrix is $b \times b$. We get the second layer watermarking W_2 using the following formula.

$$W_2(i,j) = E(B(i,j) \oplus W_1(i,j)) \ . \tag{17}$$

E is the encryption function for the purpose of increasing the security of the watermarking. A variety of encryption methods can be employed. Here we use disorder processing with M and S_{max} being the secret keys. $\oplus$ is XOR operator($1 \oplus 1 = 0, 0 \oplus 0 = 0, 1 \oplus 0 = 1, 0 \oplus 1 = 1$). W_2 is virtual because in fact, it is not embedded into the host image. On the contrary, together with M and S_{max}, it is stored in an encrypted XML document [7]. When the watermarked image suffer some serious attacks, the virtual watermarking may be useful to restore the first layer original watermarking which is in fact embedded into the host binary image. We can see it in section 3.3 and well realize it later from experiments. Furthermore, the equation (17) establishes a mapping relationship between the first and the second layer watermarking.

3.3 Watermarking Restoration

It has been widely recognized that the robustness of the watermarking is an important issue in applications of the watermarking algorithm. It can be defined as "ability to detect the watermarking after common signal processing operations". Watermarkings could be removed intentionally or unintentionally after some distortions such as compression, noise adding, shear transformation. For image watermarking, the robustness of the watermarking under geometric transformations, among other possible distortions, has to be addressed first and foremost, because they can dramatically affect the correct detection of the watermarking.

In the sequel we propose a low complexity watermarking restoration scheme which is computationally simple. It is based on the stability of the pixel probability density which is formerly put forward. The restorarion scheme is especially effective under attacks such as rotation, scaling and translation. Even under other attacks, the pixel probability density of an image is comparatively stable only if the watermarked image is not damaged a lot under these attacks. So we can also compute the pixel probability density of every target pixel sub-region of the watermarked image under attacks.

According to the discussion in section 3.2, with the extraction and using of M and S_{max} that are stored in the XML document, after computing the pixel probability density of every target pixel sub-region of the watermarked image under attacks, we can get a matrix T. The size of the matrix is equal to that of the second layer virtual watermarking. The formula of watermarking restoration process is a simple XOR operation.

$$W_1^{'}(i,j) = T(i,j) \oplus \underline{E}(W_2(i,j)) . \quad (18)$$

$\underline{E}$ is the decryption function which is the reverse process of E in equation (17). The matrix $W_1^{'}$ is the final result of the watermarking restoration process.

4 Experimental Results

In order to evaluate the proposed scheme, we choose an original binary image of size 384×384 and a watermarking image of size 96×96 and perform experiments as following.

Fig. 3 shows the process of embedding and extraction of the watermarking without any attacks. The result of comparing the original image with the watermarked image shows that the algorithm has good imperceptibility and the result of comparing the original watermarking and the extracted watermarking shows the effectiveness of the extraction.

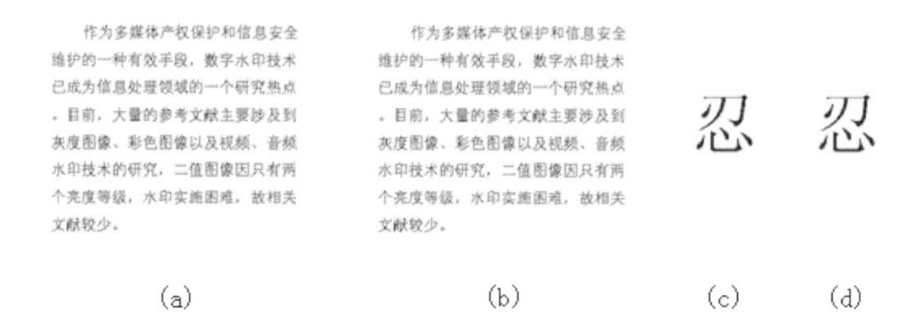

Fig. 3. (a) Original image, (b) Watermarked image, (c) Original watermarking, (d) Extracted watermarking

Subsequently, let us see the robustness of the dual watermarking scheme under some familiar attacks.

Fig. 4 shows the robustness under JPEG compression attacks with compression rate of 10%, 50% and 90% respectively. According to the extracted watermarkings

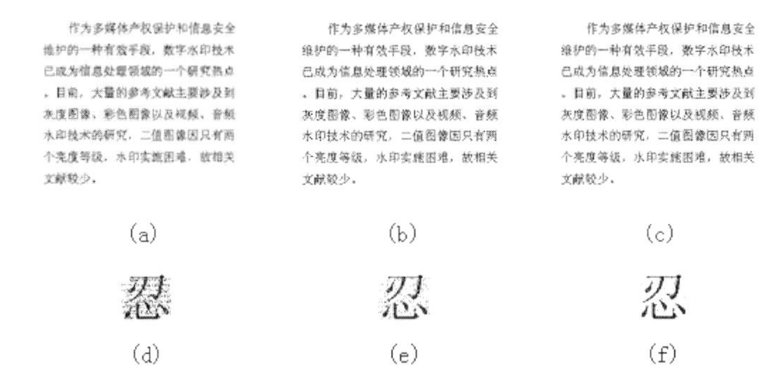

Fig. 4. (a) 10% compression, (b) 50% compression, (c) 90% compression, (d) Watermarking from (a), (e) Watermarking from (b), (f) Watermarking from (c)

from the three watermarked images, even the compression rate is 10%, the watermarking is very easy to be seen.

Fig. 5 shows the robustness under 20% shearing attack. The result shows that the extracted watermarking is slightly fuzzy in a small part of the center region using the usual single DWT watermarking scheme. Using the restoration scheme described in section 3.3, we can see that the restored watermarking image can be clearly identified. By the way, the issue "restored watermarking" in this paper is always obtained by the restoration scheme described in section 3.3.

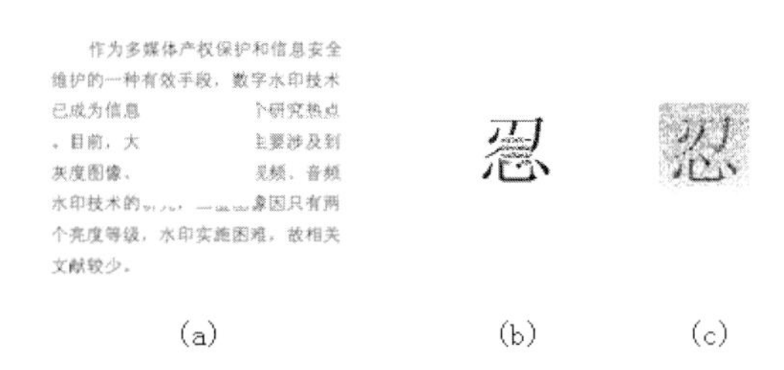

Fig. 5. (a) 20% sheared image, (b) Extracted watermarking using usual DWT scheme, (c) Restored watermarking using dual watermarking scheme

Fig. 6 shows the good robustness under Gaussian noise attacks with intensity of 0.005, 0.010, 0.015 and 0.020 respectively. From the extracted watermarking images, it is obvious that the scheme has good ability of resisting Gaussian noise attacks. Comparing the extracted watermarkings using the usual single DWT watermarking scheme and the restored watermarkings, it implies that at the same intensity level of the noise, the quality of the restored watermarking image is higher than that of the extracted watermarking image using the usual single DWT watermarking scheme.

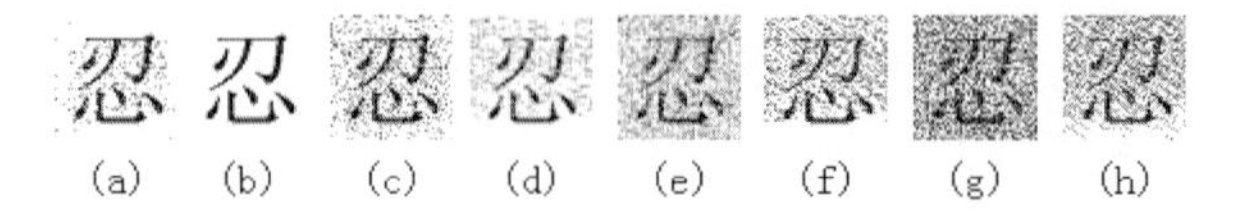

Fig. 6. (a) Extracted watermarking under 0.005 Gaussian noise using usual DWT scheme, (b) Restored watermarking under 0.005 Gaussian noise using dual watermarking scheme, (c) Extracted watermarking under 0.010 Gaussian noise using usual DWT scheme, (d) Restored watermarking under 0.010 Gaussian noise using dual watermarking scheme, (e) Extracted watermarking under 0.015 Gaussian noise using usual DWT scheme, (f) Restored watermarking under 0.015 Gaussian noise using dual watermarking scheme, (g) Extracted watermarking under 0.020 Gaussian noise using usual DWT scheme, (h) Restored watermarking under 0.020 Gaussian noise using dual watermarking scheme

Fig. 7 shows the robustness under translation attack. After translating the watermarked image down and to the right by 25 pixels, the extracted image is very fuzzy using the usual single DWT watermarking scheme, but as we see, the restored watermarking image is almost the same as the original watermarking image.

Fig. 7. (a) Extracted watermarking under translation attack using usual DWT scheme, (b) Restored watermarking under translation attack using dual watermarking scheme

Fig. 8 shows the robustness under 0.75-time scaling attack. The quality of the watermarked image under the scaling attack declines obviously. After scaling to the original size of the watermarked image, we obtain a satisfied result of the extracted watermarking using the usual single DWT watermarking scheme. But sometimes, not knowing the original size of the watermarked image, we are still satisfied with the quality of the restored watermarking.

Fig. 8. (a) Watermarked image after 0.75-time scaling attack, (b) Extracted watermarking after scaling to the original size using usual DWT scheme, (c) Restored watermarking using dual watermarking scheme

Fig. 9 shows the powerful restoration scheme again. The extracted watermarkings are too fuzzy to be identified using the usual single DWT watermarking scheme. The result shows that the quality of the restored watermarkings are all fairly good by rotating the watermarked image by 20, 40 and 60 degrees respectively.

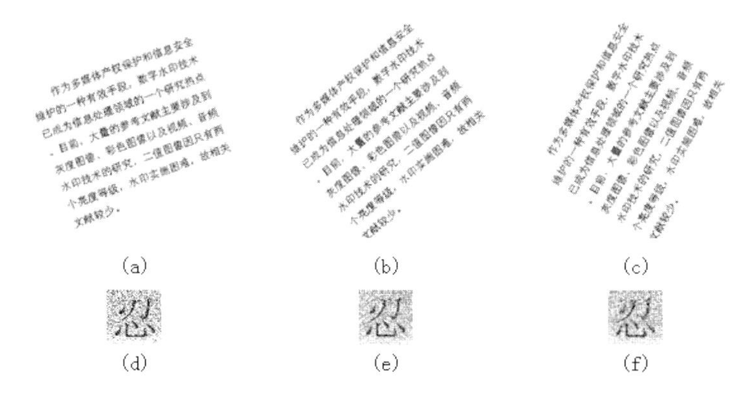

Fig. 9. (a) Watermarked image after rotated by 20 degrees, (b) Watermarked image after rotated by 40 degrees, (c) Watermarked image after rotated by 60 degrees, (d) Restored watermarking from (a), (e) Restored watermarking from (b), (f) Restored watermarking from (c)

5 Conclusion

In this paper, a dual DWT and pixel distribution features based blind binary image watermarking scheme has been proposed. The comparatively invariant pixel distribution features against a lot of attacks are used to form the second layer virtual watermarking and establish a mapping relationship between the two layer watermarkings. So the watermarking scheme is robust against all kinds of familiar attacks. Using DWT decomposition, the lowest frequency sub-band of the last level is selected to embed the first layer original watermarking because it is not liable to the interference. Therefore the scheme has excellent invisibility. The invariant features described in this paper can be easily obtained. In other words, the watermarking algorithm has low complexity and it is easy to manipulate. In addition, compared with other schemes, less virtual watermarkings are required to be stored in the encrypted XML document. These arguments are strongly supported by the experiment results.

References

1. Brassil, J., Low, S., Maxemchuk, N.F.: Copyright Protection for the Electronic Distribution of Text Documents. Proceedings of IEEE 87(7), 1181–1196 (1999)
2. Huang, D., Yan, H.: Interword Distance Changes Represented by Sine Waves for Watermarking Text Images. IEEE Trans. on Syst. Video Technology 11(12), 1237–1245 (2001)
3. Mohanty, S.P., Ramakrishnan, K.R., Kankanhall, M.: A Dual Watermarking Technique for Images. In: Proceedings of the 7th ACM International Multimedia Conference, pp. 49–51. ACM Press, New York (1999)
4. Lie, W.N., Hsu, T.L., Lin, G.S.: Verification of Image Content Integrity by Using Dual Watermarking On Wavelets Domain. In: International Conference on Image Processing, Barcelona, vol. 3, pp. 487–490 (2003)

5. Barni, M., Bartolini, F.: Data Hiding for Fighting Piracy. IEEE Signal Processing Magazine 21(2), 28–39 (2004)
6. Li, G., Xing-hua, S., Yuan-yuan, H., et al.: Distance Distribution Histogram and Its Application in Trademark Image Retrieval. Journal of Image and Graphics 7A(10), 1027–1031 (2002)
7. Frattolillo, F., D'Onofrio, S.: An Image Watermarking Procedure Based on XML Documents. In: The 11th International Conference on Distributed Multimedia Systems, Banff, pp. 22–27 (2005)

PSF-Constraints Based Iterative Blind Deconvolution Method for Image Deblurring

Xuan Mo[1], Jun Jiao[2], and Chen Shen[3]

[1] State Key Laboratory of Intelligent Technology and Systems
Tsinghua National Laboratory for Information Science and Technology
Department of Automation, Tsinghua University, Beijing, China
blademamama@gmail.com

[2] Dept. of Computer Science and Technology. Nanjing University, Nanjing, China
failedjj@gmail.com

[3] School of Software and Electronics, Peking University, Beijing, China
scv119@gmail.com

Abstract. In recent years, Image Deblurring techniques have played an essential role in the field of Image Processing. In image deblurring, there are several kinds of blurred image such as motion blur, defocused blur and gaussian blur. Many methods to address this problem have been proposed by researchers in previous research, among which the iterative blind deconvolution (IBD) method is the most popular method to solve this problem. However, the convergence of this method is not ensured, and there is no effective method to choose a proper initial estimate image and PSF(point spread function). In this paper, we improve the iterative blind deconvolution method by adding several constraints, which could be the type or parameters range of the PSF, on the PSF in each iteration. Experiment results validate that, with the help of these newly added constraints, our method are more likely to converge and has better deblurring performance than the IBD.

1 Introduction

Blind image deblurring is the process of estimating both the true image and the blur from the blurred image characteristics, using partial information about the imaging system[1]. In classical linear image restoration, the blurring function is given, and the deblurring process is inverted using one of the many already known deblurring algorithms. The various approaches that have appeared in the literature depend upon the particular blurring and image models[2][16].

In many practical image processing applications, the following linear model can accurately represent the blur of the true image.

$$g(x, y) = f(x, y) * h(x, y) + n(x, y) \tag{1}$$

where g(x,y) is the blurred image,f(x,y) is the true image, h(x,y) is the PSF, and n(x,y) represents the additive noise. The noise term is important because additive noise cannot be neglected in practical imaging situations[17].

S. Boll et al. (Eds.): MMM 2010, LNCS 5916, pp. 141–151, 2010.

There are two main approaches to blind deconvolution of images:

1) Identifying the PSF separately from the true image, in order to use it later with one of the known classical image deblurring methods. Estimating the PSF and the true image are disjoint procedures. This approach leads to computationally simple algorithms.

2) Incorporating the identification procedure with the deblurring algorithm. This mergence involves simultaneously estimating the PSF and the true image, which leads to the development of more complex algorithms.

The first approach uses some simple algorithms to estimate the PSF, which requires lots of prior knowledge that may be wrong in some situation. In addition, the result of estimated image is far from perfect. Therefore, in this paper, we only focus on the second approach.

The iterative blind deconvolution (IBD) method[5] proposed by Ayers and Dainty is the most popular method in the second class of approaches. This method is popular for its low computational complexity, but lack of reliability. The uniqueness and convergence properties are, as yet, uncertain. In addition, the restoration is sensitive to the initial image estimate, and the algorithm also exhibits instability[3].

In this paper, we improve the iterative blind decovolution method by adding several constraints on the PSF in each iteration. These constraints are based on prior knowledge of the blurred image. For example, if we know that the type of blurred image is Gaussian blur, then we can add the Gaussian constraint on the PSF to make sure that the PSF is Gaussian type, and set the gaussian parameters to approximate the PSF best, which is calculated after each iteration.

The rest of this paper is organized as follows. We briefly review the related work in Section2. Section 3 introduces our PSF-constrains based iterative blind deconvolution method. In section 4, we report the results of our method. Finally, Section 5 concludes and indicates several issues for future work.

2 Related Work

This paper presents a novel PSF-constraints based iterative blind deconvolution method to restore blurred image. There exists a substantial amount of researches related to restore the blurred image. The following subsections will first discuss the types and models of PSF, then we will discuss the traditional iterative blind deconvolution.

2.1 PSF Models

In the Equation 1, we can observe that in order to restore the blurred image, the estimation of PSF is the most important, if we can estimate the PSF precisely, we will get a clear and perfect image easily. So, we first discuss the model of three different types of PSF[14].

Motion Blur PSF. Motion blurred image can be reconstructed, provided that the motion is shift-invariant, or at least locally, and that the blur function that caused the blur is known. Thus, the associated parameter value(s) can be uniquely determined. As the blur function is not usually known, one often need to estimate the blur function from the image itself, or from the motion model analysis of the images[12].

Generally speaking, the first one is the typical vertical camera motion which can be completely identified as follows:

Vertical camera motion blur of length L

$$h(x,y) = \begin{cases} 0 \text{ if } x \neq 0, \ -\infty \leq y \leq \infty \\ \frac{1}{L} \text{ if } x = 0, \ -L \leq y \leq L \end{cases} \tag{2}$$

The second one, in general, can approximate the pendulous motion as linear motion with a small angle: pendulous motion blur of length L and angle ϕ

$$h(x,y) = \begin{cases} \frac{1}{L} \text{ if } \sqrt{x^2+y^2} \leq L \\ 0 \text{ otherwise} \end{cases} \tag{3}$$

$$tan(\phi) = \frac{x}{y} \tag{4}$$

Particularly, the first case can be seen as the special case of the second one with no swing angle ($\phi = 0$)[4].

Defocus Blur PSF. Defocused image has only one parameter:

$$h(x,y) = \begin{cases} 0 \quad \text{if } \sqrt{x^2+y^2} > r \\ \frac{1}{\pi r^2} \text{ if } \sqrt{x^2+y^2} \leq r \end{cases} \tag{5}$$

where r is the radius of the defocus PSF model.

Gaussian Blur PSF. Gaussian blur describes blurring an image by a Gaussian function:

$$h(x,y) = \frac{1}{2\pi\sigma^2} e^{-\frac{x^2+y^2}{2\sigma^2}} \tag{6}$$

The visual effect of this blurring technique is a smooth blur resembling that of viewing the image through a translucent screen.

2.2 Traditional Iterative Blind Deconvolution

The iterative blind deconvolution (IBD) method proposed by Ayers and Dainty is the most popular method. The method requires that the PSF be nonnegative with known finite support[13].

The general method makes use of the fast-Fourier transform (FFT) algorithm. The basic structure of the algorithm is presented in Figure 1. The image estimate is denoted by $\hat{f}$, the PSF estimate by $\hat{h}$, and the blurred image by $g(x,y)$. The capital letters represent fast-Fourier transformed versions of the corresponding signals. Subscripts denote the iteration number of the algorithm[10].

After a random initial guess is made for the true image, the algorithm alternates between the image and Fourier domains, enforcing known constraints in

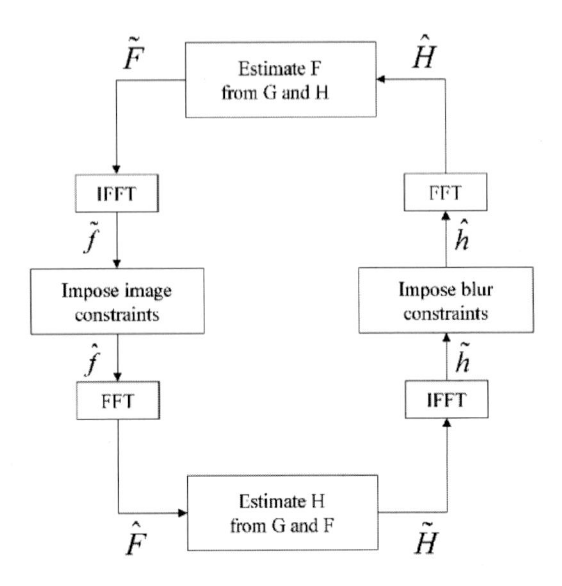

Fig. 1. iterative blind deconvolution method[5]

each. The constraints are based upon information available about the image and PSF. The image domain constraints can be imposed by replacing negative-valued pixels within the region of support with zero and nonzero pixels outside the region of support with the background pixel value. The Fourier domain constraint involves estimating the PSF (image) using the FFT of the degraded image and image (PSF) estimate. That is, at the kth iteration[3],

$$\tilde{H}_k(u,v) = \frac{G(u,v)\hat{F}^*_{k-1}(u,v)}{|\hat{F}_{k-1}(u,v)|^2 + \alpha/|\tilde{H}_{k-1}(u,v)|^2} \tag{7}$$

$$\tilde{F}_k(u,v) = \frac{G(u,v)\hat{G}^*_{k-1}(u,v)}{|\hat{H}_{k-1}(u,v)|^2 + \alpha/|\tilde{F}_{k-1}(u,v)|^2} \tag{8}$$

where $(.)^*$ denotes the complex conjugate of (.). The real constant, α, represents the energy of the additive noise and is determined by prior knowledge of the noise contamination level, if available. The value of α must be chosen carefully for reliable restoration. The algorithm is run for a specified number of iterations, or until the estimates begin to converge[5].

The IBD method is popular for its low computational complexity. Many different implementations of this basic algorithm have been suggested. They differ in their assumptions about the true image and PSF, and how these assumptions are imposed in the image and Fourier domains [5][6][7]. Extensions have been proposed for situations in which several degraded versions of the same image are available [8][9]. Another advantage of this technique is its robustness to noise because of the Wiener-like filters (Equation 7 and 8) used in the Fourier domain.

Robustness to noise refers to the ability of the algorithm to suppress noise amplification that results from the ill-posed nature of the restoration problem[11].

The major drawback of the IBD method is its lack of reliability. The uniqueness and convergence properties are, as yet, uncertain. In addition, the restoration is sensitive to the initial image estimate, and the algorithm can exhibit instability[15].

3 Iterative Blind Deconvolution Method Based on PSF-Constraints

3.1 Constraints on PSF

By using the traditional iterative blind deconvolution, after every iteration, estimated PSF will be changed into any possible shape which may have no meaning. After iterating again and again, the estimated image may be unable to converge. The estimated image could be worse than the blurred image.

Firstly, We can identify which of the three common types that the blur belongs to. Base on prior knowledge, we can determine which of the three common types that the exact one is: defocus blur, motion blur and Gaussian blur. Once the blur type is confirmed, It is advisable to set the PSF to the corresponding type, so that we can not only ensure the convergence of iteration, but also achieve a better result from deconvolution.

Secondly, in order to set the PSF to the corresponding type, we have to find a PSF $h_c(x, y)$ which belongs to the certain common type:

$$h_c(x,y) \xrightarrow{indefinitely \quad approach} h(x,y) \tag{9}$$

In addition, least square error is used to measure the difference between certain type of PSF and the PSF from iteration, and we need to seek a $h_c(x, y)$ to minimize this difference:

$$\min \quad L(h(x,y), h_c(x,y)) = \sum_{x,y} (h(x,y) - h_c(x,y))^2 \tag{10}$$

In order to solve this optimization problem, parameters of the according PSF need to be restrained in a certain range. Take Gaussian blur for example, it has two parameters: the radius r and the variance σ. We can set the upper limit and the lower limit to make sure:

$$r \in (r_l, r_u), \sigma \in (\sigma_l, \sigma_u) \tag{11}$$

Since we confined the range of parameters, simple enumeration method is adopted here to solve this optimization problem. A step length is manually set for the enumeration. For example, in the case of Gaussian Blur, our method solves the following optimization problem by enumerating the parameters in certain step length:

Algorithm 1. PSF-constraints based Iterative Blind Deconvolution Method

Parameter:
N: When number of iteration is equal to N, the algorithm will terminates.
Input:
$H_0(x, y)$: the blurred image.
Output:
$\tilde{f}_N(x, y)$: the deblurred image.
Process:

1. Step 1: Choose a certain type PSF and select the initial parameters:$h_i(x, y)$, use traditional deconvolution to get the initial image estimate:$\tilde{f}_0(x, y)$.
2. Step 2: This function is Fourier transformed to frequency yield:$\tilde{F}_0(u, v)$.
3. Step 3: $\tilde{F}_0(u, v)$ is then inverted to form an wiener filter and multiplied by $G(u, v)$ to form a first estimate of the PSF's spectrum $H_0(u, v)$.
4. Step 4: The $H_0(u, v)$ is inverse transformed to give $h_0(x, y)$.
5. *Step 5: Use the equation 12, select the range and step length, solve the optimization problem to get* $h_{0c}(x, y)$.
6. *Step 6: Use the equation 13, choose an* α *to get modified PSF:* $h_{0m}(x, y)$.
7. Step 7: $h_{0m}(x, y)$ is consequently formed that is Fourier transformed to give the spectrum:$\tilde{H_{0m}}(u, v)$.
8. Step 8: $\tilde{H_{0m}}(u, v)$ is inverted to form another wiener filter and multiplied by $G(u, v)$ to give the next spectrum estimate $F_1(u, v)$.
9. Step 9: $F_1(u, v)$ add a Hann Window, and then be completed by inverse Fourier transforming $F_1(u, v)$ to give $f_1(x, y)$.
10. Step 10: Constrain the $f_1(u, v)$ to be nonnegative to give $\tilde{f}_1(x, y)$.
11. Step 11: Repeat the Step 2 to Step 10, until the iteration becomes convergent or we can satisfy with result of deblurred image.

$$\begin{aligned} \min \quad & L(h(x, y), h_c(x, y)) = \sum_{x,y} (h(x, y) - h_c(x, y))^2 \\ S.T. \quad & r \in (r_l, r_u) \\ & \sigma \in (\sigma_l, \sigma_u) \end{aligned} \tag{12}$$

3.2 Modified PSF

After obtaining the $h_c(x, y)$, we use the linear combination of $h_c(x, y)$ and $h(x, y)$ to replace the original PSF in the traditional method:

Finally, we use the modified PSF: $h_m(x, y)$ to replace the original PSF $h(x, y)$:

$$h_m(x, y) = \alpha * h_c(x, y) + (1 - \alpha) * h(x, y) \tag{13}$$

From the Equation 13 we can see that if $\alpha = 0$, $h_m(x, y) = h(x, y)$, it is the traditional iterative blind deconvolution method: if $\alpha = 1$, $h_m(x, y) = h_c(x, y)$, $h_c(x, y)$ will completely dominate the iteration. In different application, α should be adjusted to different values to achieve the best performance.

3.3 Hann Window

Hann Window is a famous window function which can be formulated as follow:

$$\omega(n) = 0.5(1 - \cos(\frac{2\pi n}{N-1})) \tag{14}$$

We can use Hann Window at the end of each iteration to make the edge of image more smooth.

The whole procdure of our proposed method is listed in Algorithm 1.

4 Experimental Result and Analysis

4.1 Result of Motion Blur

In order to get a motion blurred image, we choose the parameters:$L = 15, \phi = 45^0$. Original clear image and the motion blurred image are presented in Figure 2.

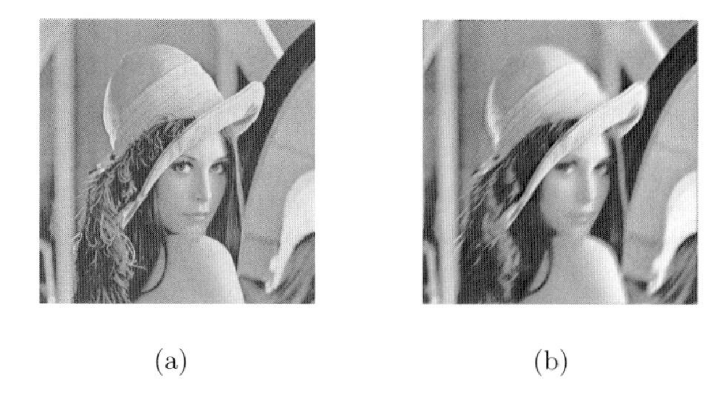

(a) (b)

Fig. 2. (*a*): Original clear image, (*b*): Motion blurred image

We can see from Figure 2 that, with the previously given parameters, the motion blurred image is obviously blurred at the direction which has an approximate 45-degree angle with horizontal.

We set the range of parameters:$L \in (1, 20)$,$\phi \in (30, 60)$, the step length:$\Delta L = 0.1, \Delta\phi = 0.1$ to search. By solving the optimization problem we can get a $h_c(x, y)$.

We choose the parameter: $\alpha = 0.5$. According to Equation 13, after each iteration we can get a new PSF:$h_m(x, y)$, and we would use this new PSF:$h_m(x, y)$ to substitute the previous PSF:$h_c(x, y)$ in the next iteration. Repeating the process mentioned above and after 5 iterations, we will get the deblurred image in Figure 3(b).

The result of traditional iterative blind deconvolution method is shown in Figure 3(a). By comparison, the PSF-constraints IBD method achieves better performance than the traditional one. In addition, if we increase the number of iteration, the traditional method is going to become divergent, while our method would still keep convergent.

(a) (b)

Fig. 3. (*a*): Deblurring result of traditional IBD method for motion blur (*b*): Deblurring result of PSF-constraints IBD method for motion blur

4.2 Result of Defocus Blur

We choose the parameters:$r = 8$. Then, we get the original clear image and the defocus blurred image which are presented in Figure 4.

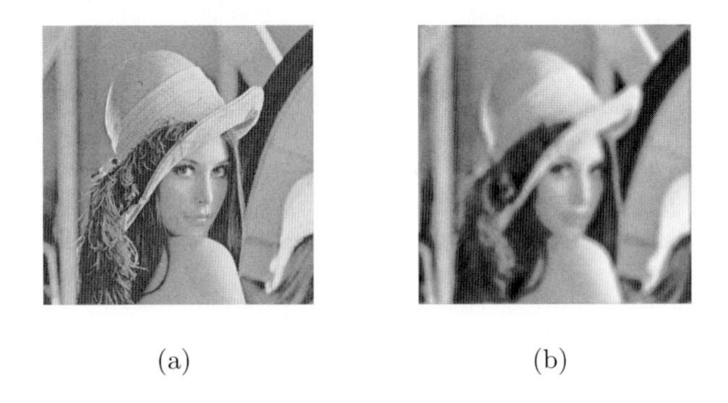

(a) (b)

Fig. 4. (*a*): Original clear image, (*b*): Defocus blurred image

From Figure 4, we can observe that the defocus blurred image has a obvious blur. Every time we get a $h(x, y)$ from the iteration:

We set the range of parameters:$r \in (1, 20)$, the step length:$\Delta r = 0.1$ to search. By solving the optimization problem we can get a $h_c(x, y)$.

We choose the parameter: $\alpha = 0.5$. By using the Equation 13, we can get a modified PSF:$h_m(x, y)$. Then put the modified PSF into the iteration. Follow the Process which is mentioned above and after 5 iterations, we will get the deblurred image in Figure 5(b).

As same as the first experiment, the PSF-constraints IBD method performs better than the traditional one.

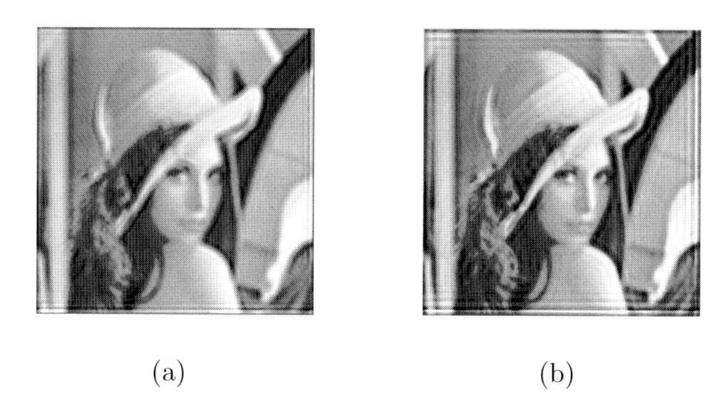

(a) (b)

Fig. 5. (*a*): Deblurring result of traditional IBD method for defocus blur (*b*): Deblurring result of PSF-constraints IBD method for defocus blur

4.3 Result of Gaussian Blur

We select the parameters:$r = 8, \sigma = 5$ to get a Gaussian blurred image. Then, we show the original clear image and the gaussian blurred image in Figure 6.

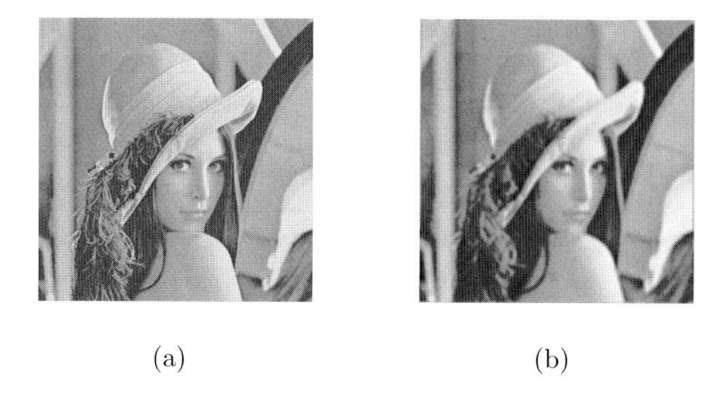

(a) (b)

Fig. 6. (*a*): Original clear image, (*b*): Gaussian blurred image

As is shown in the figure that the Gaussian blurred image has a blur which is a smooth blur resembling that of viewing the image through a translucent screen. Every time we get a $h(x, y)$ from the iteration:

We set the range of parameters:$r \in (1, 30)$,$\sigma \in (1, 20)$, the step length:$\Delta r = 0.1, \Delta\sigma = 0.1$ to search. By solving the optimization problem we can get a $h_c(x, y)$.

We choose the parameter: $\alpha = 0.5$. By using the Equation 13, we can get a modified PSF:$h_m(x, y)$. Then put the modified PSF into the iteration. Follow the Process which is mentioned above and after 5 iterations, we will get the deblurred image in Figure 7(b).

By comparison, the PSF-constraints IBD method is also better than the traditional one. The PSF-constraints IBD method is more reliable. The uniqueness and convergence properties are certain.

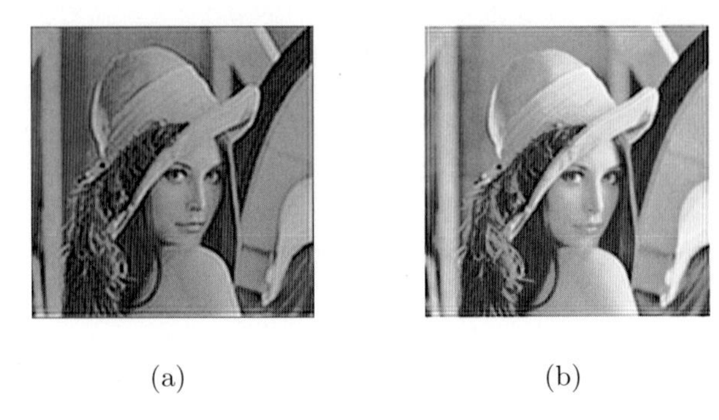

(a) (b)

Fig. 7. (*a*): Deblurring result of traditional IBD method for gaussian blur (*b*): Deblurring result of PSF-constraints IBD method for gaussian blur

5 Conclusion and Future Work

In this paper, we proposed a PSF-constraints iterative blind deconvolution method. The PSF-constraints IBD method performs better by adding the PSF-constraints on PSF calculated from iteration to make sure that the modified PSF will tend to be a certain type. In this way, iteration will be more reliable, and exhibit better characteristic of uniqueness and convergence. We can see from the results that PSF-constraints IBD method is better than the traditional one, besides the Hann Window makes the edge of image smooth.

It is true that the PSF-constraints IBD method is better than the traditional one. However, there are still some future work to do. First of all, we do not verify the convergence and uniqueness properties of the PSF-constraints IBD method in mathematical field. What's more, it is better if we can find a proper number of iteration which would produce the most wonderful result. Finally, we have to figure out how to deal with blurred image with more than one type of blur.

References

1. Katsaggelos, A.K.: Digital Image Restoration. Springer Series in Information Science. Springer, Berlin (1991)
2. Andrews, H.C., Hunt, B.R.: Digital Image Restoration. Prentice-Hall Signal Processing Series. Prentice-Hall, Englewood Cliffs (1977)
3. Kundur, D., Hatzinakos, D.: Blind image deconvolution revisited. IEEE Signal Processing Mag. 13, 43–64 (1996)
4. Fu, S.Y., Zhang, Y.C., Cheng, L., Liang, Z.Z., Hou, Z.G., Tan, M.: Motion Based Image Deblur Using Recurrent Neural Network for Power Transmission Line Inspection Robot. In: International Joint Conference on Neural Networks, IJCNN 2006, pp. 3854–3859 (2006)
5. Ayers, G.R., Dainty, J.C.: Iterative blind deconvolution method and its applications. Optics Letters 13, 547–549 (1988)
6. Davey, B.L.K., Lane, R.G., Bates, R.H.T.: Blind deconvolution of a noisy complex-valued image. Optics Communications 69, 353–356 (1989)

7. Miura, N., Baba, N.: Extended-object reconstruction with sequential use of the iterative blind deconvolution method. Optics Communications 89, 375–379 (1992)
8. Tsumuraya, F., Miura, N., Baba, N.: Iterative blind deconvolution method using Lucys algorithm. Astron Astrophys. 282, 699–708 (1994)
9. Miura, N., Ohsawa, K., Baba, N.: Single -frame blind deconvolution by means of frame segmentation. Optics Letters 19, 695–697 (1994)
10. Thibaut, E., Conan, J.: Strict a priori constraints for maximum-likelihood blind deconvolution. Journal of the Optical Society of America A (1995)
11. Biemond, J., Lagendijk, R.L., Mersereau, R.M.: Iterative methods for image deblurring. Proc IEEE 78, 856–883 (1990)
12. Money, J.H., Kang, S.H.: Total variation minimizing blind deconvolution with shock filter reference. Image and Vision Comput. 26, 302–314 (2008)
13. Prasath, V.B.S., Singh, A.: Ringing artifact reduction in blind image deblurring and denoising problems by regularization methods. In: Seventh International Conference on Advances in Pattern Recognition (2009)
14. Carasso, A.: Direct blind deconvolution. SIAM Journal on Applied Mathematics 61, 1980–2007 (2001)
15. Liang, X.B., Wang, J.: A recurrent neural network for nonlinear optimization with a continuously differentiable objective function and bound constraints. IEEE Transactions on Neural Networks 11, 1251–1262 (2000)
16. Rekleitis, I.M.: Optical Flow Recognition From The Power Spectrum of A Single Blurred Image. In: Image Processing, 1996. Proceedings, International Conference, vol. 3, pp. 791–794 (1996)
17. Yang, Y., Galatsanos, N.P., Stark, H.: Projection-based blind deconvolution. Journal of the Optical Society of America 11, 2401–2409 (1994)

Face Image Retrieval across Age Variation Using Relevance Feedback

Naoko Nitta, Atsushi Usui, and Noboru Babaguchi

Graduate School of Engineering, Osaka University
2-1 Yamada-oka, Suita, Osaka, 565-0871 Japan
{naoko,usui,babaguchi}@nanase.comm.eng.osaka-u.ac.jp

Abstract. Given a single face image of a specific person as a query, it is very difficult to retrieve all of his/her images from a personal image collection stored for a long term due to age-related changes in facial appearances. This paper proposes to apply relevance feedback to enhance the performance of image retrieval from the image collections with age variation. Specifically, we propose two types of update schemes: i) query expansion and ii) weight updating and show the effects of each scheme by experiments with two actual image collections. For an image collection, the recall rate improved from 40.8% to 72.5% after five iterations of relevance feedback.

Keywords: Face-based Image Retrieval, Relevance Feedback, Aging Effects.

1 Introduction

Digital cameras are now widely used by ordinary people to take images of everyday life. The growth in the number of such images has increased the demand for a technique to easily find specific images, for example, all images of a specific person. Images of a specific person are usually searched based on his/her face; therefore, such image retrieval techniques are deeply related to the techniques of face recognition. Many face recognition techniques have been developed considering variations in various factors such as lighting conditions, facial directions, and facial expressions. Additionally, in order to retrieve images of a specific person from the image collection stored for a long term, it is crucial to consider aging effects in the facial appearances.

There are mainly two approaches to handle aging effects in face recognition: 1) to predict how faces are going to age and 2) to use facial features invariant to aging. Especially, the former has been the major approach, and it is known that aging can be simulated by changing the facial parameters of facial shapes and textures[2][3]. Although the synthesized faces usually look plausible, the differences between the synthesized and real faces make it difficult to recognize faces with high accuracy. On the other hand, the latter approach has used features extracted from ears, which usually do not change over the years[4], or features invariant to wrinkles[5]. However, since these features can only be useful for

S. Boll et al. (Eds.): MMM 2010, LNCS 5916, pp. 152–162, 2010.

sideways face images or face images of grown-ups, they are inadequate for image retrieval from personal image collections. Based on the discussions above, our purpose is to realize the face image retrieval with facial features invariant to aging targeted on personal image collections from babies to elderly.

For content-based image retrieval, relevance feedback[6][7] has been widely used as an effective technique to bridge the gap between the low-level features and the high-level semantics of images. Since we also need to bridge the gap between the facial features and the personal identity, relevance feedback can be very effective. Therefore, this paper proposes a system which adapts relevance feedback in retrieving images of the person of a query image from an image collection with age variation. Multiple features are extracted from faces and the similarity between the query image and each image in the collection is calculated based on the extracted features. After a user gives feedback on the relevance of the M most similar images to the query, the system applies two update schemes: i) query expansion, which adds queries, and ii) weight updating, which modifies the weight of each feature, to obtain the better retrieval results. In this paper, experiments are conducted on two types of image collection with age variation to evaluate the effectiveness of relevance feedback and to examine the effect of each update scheme.

2 Face Image Retrieval with Relevance Feedback

Figure 1 shows the outline of the proposed system. Given a face image of a specific person as a query image, our system retrieves his/her images from an image collection of various people of various ages. The system firstly retrieves the M most similar images to the query Q based on the similarities of their low-level features. The user gives his/her feedback whether the person of the query image is in each retrieved image, and the system adds queries and modifies the weights of features to give better retrieval results. Here, the query Q is composed of the initial query Q_0 in the first iteration and the additional queries $Q_q (q = 1, 2, \cdots)$ are added after each iteration to construct $Q = \{Q_q | q = 0, 1, \cdots\}$.

The proposed system is composed of the following three steps.

Step1) Feature Extraction: Features are extracted from Q_0 and each image $I_n (n = 1, \cdots, N)$ in the image collection, where N is the total number of images.

Step2) Similarity Calculation: Similarity between Q and I_n is calculated based on the extracted features and the images $I_n (n = 1, \cdots, N)$ are ranked in the descending order of the similarity. The top M images are presented to the user.

Step3) Update by Relevance Feedback: The user selects the images of the person of Q from the presented images. The selected images are called relevant images and other presented images are called irrelevant images. After applying the following two update schemes based on these images, the system repeats **Step2** and **Step3**.

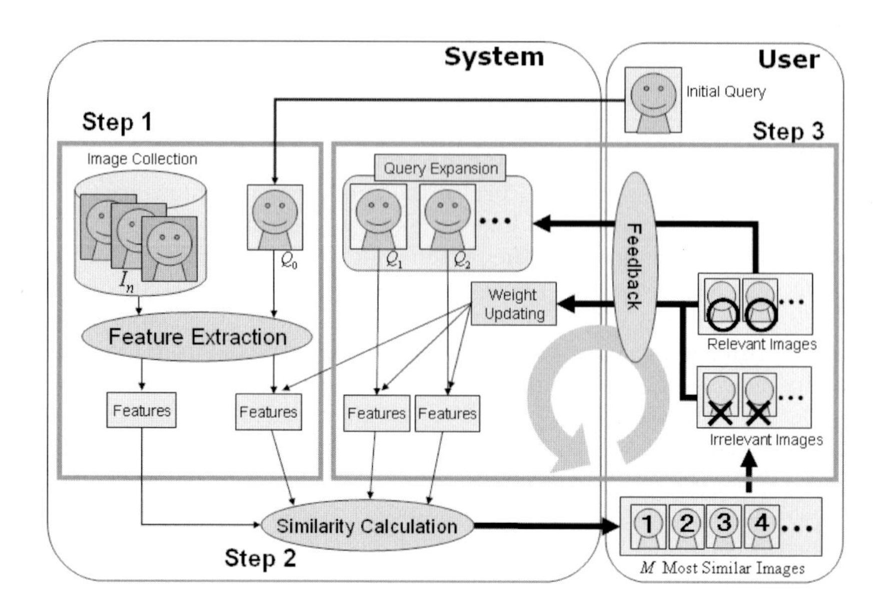

Fig. 1. Outline of Proposed System

- Query Expansion: The relevant images are added to Q as queries $Q_q(q = 1, 2, \cdots)$.
- Weight Updating: The weights of features are modified to increase and decrease the similarities of the relevant and irrelevant images to $Q_q(q = 0, 1, \cdots)$ respectively.

2.1 Feature Extraction

Our target is the personal image collections. When people have their photos taken, they often look toward the camera and smile. Therefore, their face directions and facial expressions are relatively consistent, while the lighting conditions differ largely due to the variations in the time and place of the photo shooting. Moreover, the facial features should be robust to the changes in the local facial appearance by glasses, hair style, beard, etc.

Considering above, we use *jets*, F^g, the set of convolution coefficients for Gabor kernels of 8 different orientations and 5 different frequencies around a single point, which are robust to moderate lighting changes and small deformations[1]. This feature is also robust to the changes in the local facial appearance because they are extracted from local feature points. Here, F^g are extracted from J facial points shown in Fig. 2 and represented as

$$F^g = (F_1^g, \cdots, F_j^g \cdots, F_J^g) \tag{1}$$

$$F_j^g = (f_{j,1}^g, \cdots, f_{j,k}^g, \cdots, f_{j,40}^g), \tag{2}$$

where F_j^g is the 40-dimensional coefficients for the jth facial point (x_j, y_j).

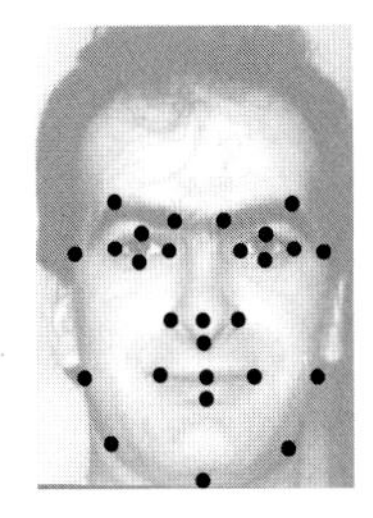

Fig. 2. Facial Points

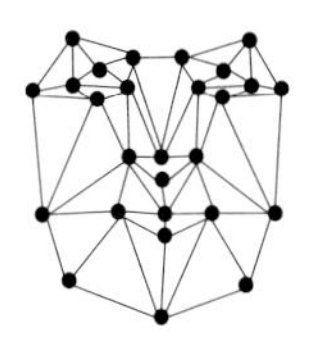

Fig. 3. Face Graph

Moreover, geometric features F^d, which represent the physical relationships among the facial points, are combined. Here, the facial points shown in Fig. 2 are connected by L edges as shown in Fig. 3, generating a face graph. When f_l^d represents the length of the lth edge, F^d is represented as

$$F^d = (f_1^d, \cdots, f_l^d, \cdots, f_L^d). \tag{3}$$

Thus, the facial feature $\mathcal{F}$ can be represented as $\mathcal{F} = (F^g, F^d)$ and has a hierarchical structure as shown in Fig. 4.

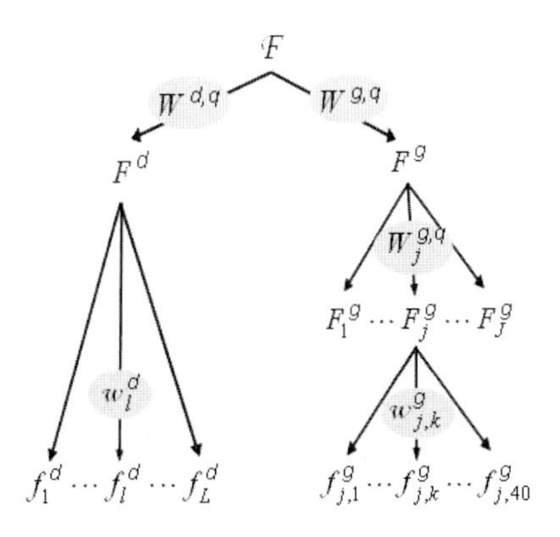

Fig. 4. Hierarchical Structure of Facial Features

2.2 Similarity Calculation

$S(Q_q, I_n)$, the similarity between the query image Q_q and the face image I_n in the image collection, is calculated as follows.

$$S(Q_q, I_n) = W^{g,q} S^g(Q_q, I_n) + W^{d,q} S^d(Q_q, I_n) \tag{4}$$

$$S^g(Q_q, I_n) = \sum_j W_j^{g,q} S_j^g(Q_q, I_n) \tag{5}$$

$$S^d(Q_q, I_n) = \frac{\sum_l w_l^d f_l^{d,Q_q} f_l^{d,I_n}}{\sqrt{\sum_l \left(w_l^d f_l^{d,Q_q}\right)^2 \sum_l \left(f_l^{d,I_n}\right)^2}} \tag{6}$$

$$S_j^g(Q_q, I_n) = \frac{\sum_k w_{j,k}^g f_{j,k}^{g,Q_q} f_{j,k}^{g,I_n}}{\sqrt{\sum_k \left(w_{j,k}^g f_{j,k}^{g,Q_q}\right)^2 \sum_k \left(f_{j,k}^{g,I_n}\right)^2}} \tag{7}$$

where $S^g(Q_q, I_n)$, $S^d(Q_q, I_n)$, and $S_j^g(Q_q, I_n)$ represent the similarity between Q_q and I_n based on F^g, F^d, and F_j^g, $W^{g,q}$, $W^{d,q}$, and $W_j^{g,q}$ represent the weights of F^g, F^d, and F_j^g for Q_q, and w_l^d and $w_{j,k}^g$ are the weights of f_l^d and $f_{j,k}^d$ for Q. The suffixes Q_q and I_n for the features represent whether the features are extracted from Q_q or I_n. The weights are initialized to no-bias weights: $W^{g,0} = W^{d,0} = \frac{1}{2}$, $W_j^{g,0} = \frac{1}{J}$, and $w_{j,k}^g = w_l^d = 1$.

2.3 Update by Feedback

Relevance feedback is generally used to establish the link between high-level semantics and low-level features[6]. Here, we use relevance feedback to link the facial features and the personal identity. In order to improve the performance of face image retrieval across age variance, we need to i) expand the age range of the retrieved images and ii) improve the capability of discrimination between relevant and irrelevant images. Therefore, we use two schemes to update the retrieval results for each purpose: i) query expansion and ii) weight updating. The details of each scheme are as follows.

[Query Expansion]

Since, the larger the age gaps, the greater the changes in the facial appearance usually are, a single query is expected to retrieve only his/her images of similar ages. Therefore, after the user feedback, we use the relevant images as the additional queries to expand the age range of the retrieval results.

Additionally, while faces usually look very similar during infancy, they start to bear distinct features for each person as people get older. This infers that the face images are distributed differently in the feature space according to their ages. Figure 5 shows an example where Q_A and Q_B are two query images of a person, I_C is an image of the same person in the image collection, and Xs represent the images of other people. Let us assume that the images of similar ages are distributed around each query image and the images are distributed more densely around Q_A than around Q_B since the person in Q_A is too young to bear distinct personal features. In this case, since placing equal importance on Q_A and Q_B would retrieve only the images around Q_A and fail to retrieve I_C, more importance should be placed on Q_B than on Q_A. Thus, we also set the weight $\mathcal{W}_q$ for the query Q_q as follows so that the importance of each query is determined based on the distribution of images around the query:

$$\mathcal{W}_q = \frac{1}{\mu_q}, \tag{8}$$

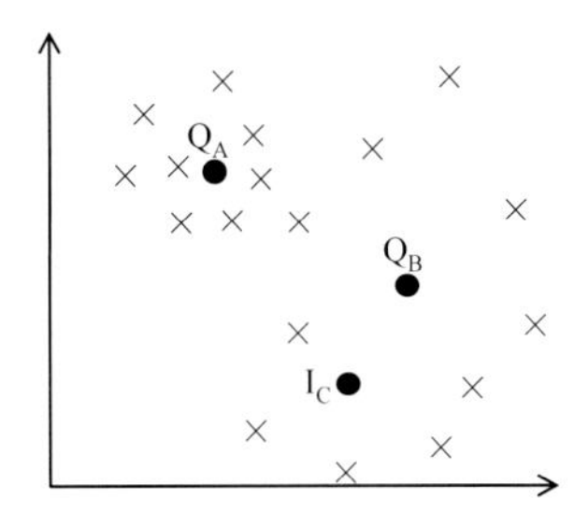

Fig. 5. An Example of How Images are Distributed in the Feature Space

where μ_q is the average similarity of the M most similar images to the query Q_q. Then, $S(Q, I_n)$, the similarity between the set of queries Q and the image I_n is determined as

$$S(Q, I_n) = \max_q W_q S(Q_q, I_n) \quad (q = 0, 1, 2, \cdots). \tag{9}$$

[Weight Updating]

Both the relevant and irrelevant images have high similarities to the query before user feedback. Thus, the weights of features should be updated to lower the similarities between the query and irrelevant images while keeping the similarities between the query and the relevant images high. Based on this idea, we further refine the weight updating scheme proposed by Rui et al.[6] by considering the similarities between the relevant/irrelevant images and query images as follows.

Firstly, the intra-weights[6], $w^g_{j,k}$ and w^d_l, are updated so that the weights of features which contribute to the distinction between the relevant and irrelevant images are increased, and vice versa. For simplicity, we consider how to update the weight w_k for f_k, the kth component of a feature vector. Figure 6 shows how images can be distributed in the f_k-axis, where Q, P, and N represent the query, relevant, and irrelevant images. The weight w_k is increased when the relevant images are distributed closely to the query image while the irrelevant images are distantly distributed from the query image as shown in Fig. 6(a). On the other hand, w_k is decreased when relevant and irrelevant images are both evenly distributed in the f_k-axis as shown in Fig. 6(b), since f_k is not a reliable feature component to distinguish between relevant and irrelevant images. Thus, w_k is determined as

$$w_k = \frac{\sigma^{PN}_k}{\sigma^P_k}, \tag{10}$$

where σ^P_k represents the standard deviation of f_k of the query and relevant images and σ^{PN}_k represents the standard deviation of f_k of the query, relevant, and irrelevant images.

The inter-weights[6], $W^{g,q}_j$, $W^{d,q}$, and $W^{g,q}$, are updated in a similar way. Let us assume W^q is the weight for a vector F. After user feedback, all images

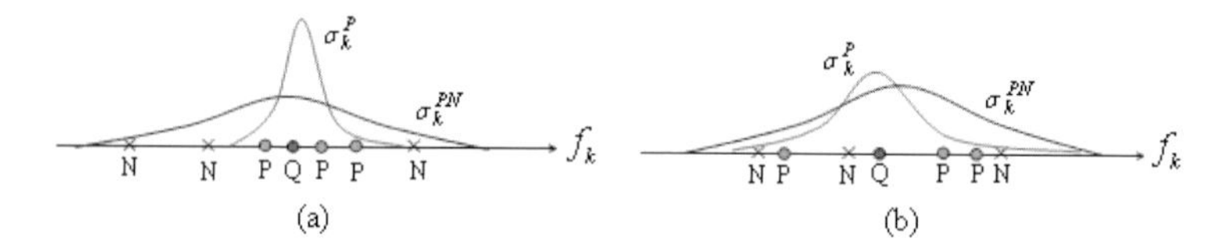

Fig. 6. Weight Updating for Intra-Weight

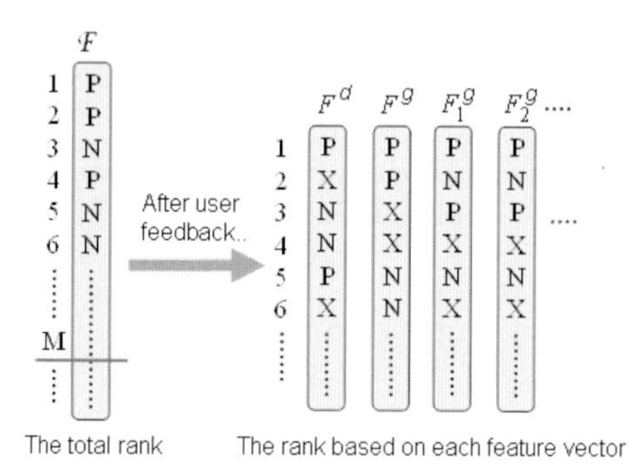

Fig. 7. Weight Updating for Inter-Weight

I_n in the image collection are ranked according to the similarity between I_n and Q_q based only on the feature vector F as shown in Fig. 7, where P, N, and X represent the relevant, irrelevant, and other images which were not presented to the user. The weight W^q is increased when the relevant images are ranked higher and the irrelevant images are ranked lower. Thus, W'^q is determined as

$$W'^q = \begin{cases} W^q + \frac{\mu^N \nu^P}{\mu^P \nu^N} & (\mu^P < \mu^N) \\ W^q - \frac{\mu^P}{\mu^N} & (\mu^P > \mu^N) \end{cases}, \tag{11}$$

where μ^P and μ^N represent the average ranks of relevant and irrelevant images respectively, and ν^P and ν^N represent the Mean Average Precision (MAP) of relevant and irrelevant images respectively. Here, MAP is defined as below and indicates how high the relevant/irrelevant images are ranked.

$$\nu = \frac{1}{\sum_{i=1}^{N} u_i} \sum_{i=1}^{N} \left(\frac{u_i}{i} \sum_{j=1}^{i} u_j \right), \tag{12}$$

where $u_i = 1$ when the ith image is relevant/irrelevant and $u_i = 0$ otherwise. Eq. 11 indicates that the weight is increased when the average rank of relevant images is higher than that of irrelevant images, and vice versa. Moreover, the weight is increased more when the relevant images are concentrated in the high ranks and the irrelevant images are concentrated in the low ranks. Finally, $W^{q'}$ is normalized to 0-1 scale and assigned to W_j^q.

Table 1. Image Collection Specification

	# of images	# of people	age range
DB1	1002	82	0 ‘ 69
DB2	120	11	0 ‘ 54

3 Experiments

The experiments were conducted with FG-NET Aging Database [8] (DB1) and the collection of images of our laboratory members (DB2). 27 feature points shown in Fig. 2 are given for DB1 and are manually obtained for DB2. Table 1 shows the specification of each image collection. Figure 8 shows the example images of a person in each collection.

The effects of relevance feedback are evaluated by using every image in the collection as a query with the three types of update scheme: I) weight updating, II) query expansion, and III) weight updating and query expansion. The feedback is given 5 times by setting $M = 15$. The images of the person of the query image are called *correct images* and the M most similar images to the query are called *retrieved images*. The results are evaluated after each iteration using MAP and the following two types of measurement.

Average Age Gap G_{ave}: G_{ave} evaluates the capability of retrieving the correct images with large age gaps.

$$G_{ave} = \frac{\sum_b |a^{I_b} - a^{Q_0}|}{N_b}, \quad (13)$$

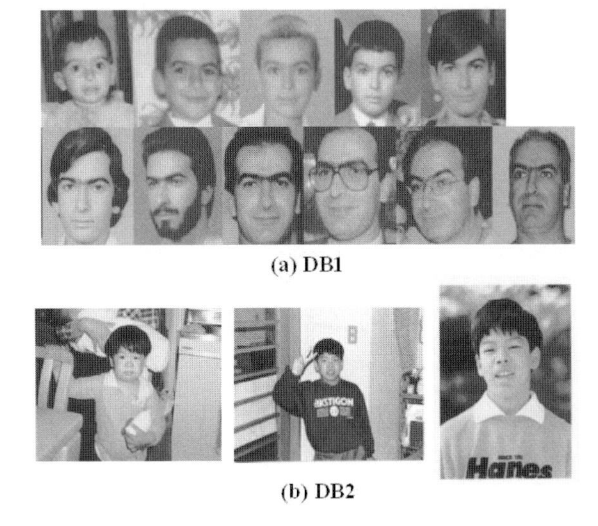

Fig. 8. Example images in each collection

where $I_b(b = 1, \cdots, N_b)$ are the correct images retrieved for the first time in the current iteration, N_b is the number of such images, and a^{I_b} and a^{Q_0} are the ages of the queried person in I_b and Q_0.

Max Age Gap G_{max}: G_{max} evaluates the capability of expanding the age range of retrieved correct images.

$$G_{max} = \max_c \left| a^{I_c} - a^{Q_0} \right|, \tag{14}$$

where $I_c(c = 1, 2, \cdots)$ are the retrieved correct images and a^{I_c} and a^{Q_0} are the ages of the queried person in I_c and Q_0. MAP, G_{ave}, and G_{max} are averaged over all trials, each of which is the retrieval by a single image in the image collection, and Figs. 9(a)-(c) and (d)-(f) show the results for DB1 and DB2 respectively. Firstly, Figs. 9(a) and (d) show how the correct images moved to the higher ranks after each iteration. Especially, using both update schemes resulted in the best retrieval performance compared to using each scheme.

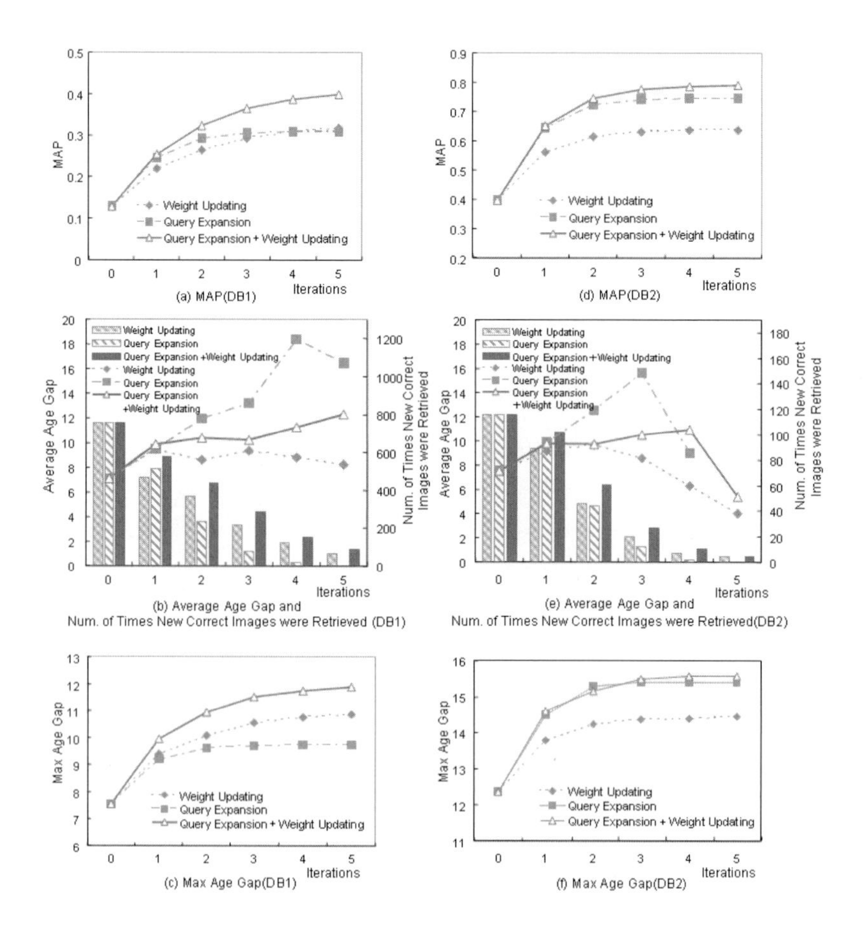

Fig. 9. Experimental Results

In Figs. 9(b) and (e), the line plot represents G_{ave} after each iteration and the bar line represents the number of trials $N_b > 0$ correct images were retrieved in each iteration. According to these figures, weight updating is effective in retrieving more correct images, while query expansion is effective in retrieving correct images of larger age gaps.

Figures 9(c) and (f) show G_{max} after each iteration. For both image collections, combining both update schemes expanded the age gaps among the retrieved correct images more than using each update scheme. Especially for DB2, query expansion retrieved the correct images of larger age gaps than weight updating, supporting the previous assertion about the effects of each update scheme. On the contrary, for DB1, weight updating retrieved the correct images of larger age gaps than query expansion. This is because although query expansion did retrieve the correct images of larger age gaps than weight updating, weight updating retrieved much more correct images than query expansion, resulting in the larger G_{max} by taking average of all trials.

Finally, the results are evaluated with the recall rate, which is the ratio of the number of the retrieved correct images to the number of the correct images. After 5 iterations, the recall rate improved from 15.7% to 38.0% for DB1 and from 40.8% to 72.5% for DB2. These results show the effectiveness of relevance feedback in retrieving images of a specific person across age variance. Figure 10 shows an example of retrieved correct images with her ages. Here, Fig. 10(a) is the query image which shows the 21-year old female, Fig. 10(b) shows her images at the age of 16 and 20, which were retrieved before feedback, and Fig. 10(c) shows her images aged between 5 and 36, which were retrieved after 5 feedback iterations. As can be seen, the relevance feedback was effective in retrieving images with age-related facial changes and in expanding their age range. Other

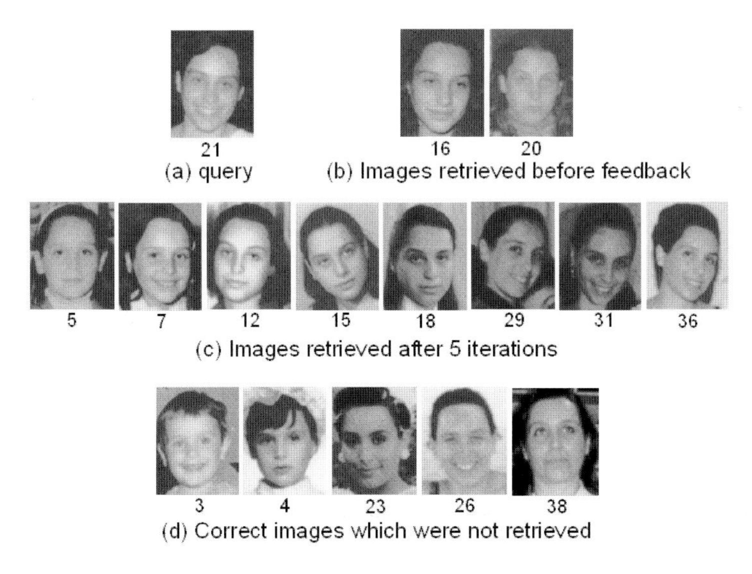

Fig. 10. Example of Retrieved Correct Images

correct images were not retrieved due to the variance in the face directions and expressions as shown in Fig. 10(d).

4 Conclusion

This paper proposed a system which adapts relevance feedback in face image retrieval across age variation and evaluated the effects of two update schemes: query expansion and weight updating. The experiments with two actual image collections show that query expansion is effective in retrieving correct images of larger age gaps, while weight updating is effective in retrieving more correct images. By combining both update schemes, the recall rate improved from 40.8% to 72.5% after 5 iterations for one of the image collections. The future work includes more experiments with a larger image collections and the development of more practical system by incorporating the facial feature point extraction techniques.

References

1. Wiskott, L., Fellous, J.M., Kruger, N., von der Malsburg, C.: Face Recognition by Elastic Bunch Graph Matching. IEEE Trans. on PAMI 19(7), 775–779 (1997)
2. Suo, J., Min, F., Zhu, S., Shan, S., Chen, X.: A Multi-Resolution Dynamic Model for Face Aging Simulation. In: Proc. CVPR, pp. 1–8 (2007)
3. Wang, J., Shang, Y., Su, G., Lin, X.: Age Simulation for Face Recognition. In: Proc. ICPR, pp. 913–916 (2006)
4. Sato, K., He, N., Takahashi, Y.: Human Face Extraction and Recognition Using Radial Basis Function Network. IEICE Trans. E86-D(5), 956–963 (2003)
5. Ling, H., Soatto, S., Ramanathan, N., Jacobs, D.W.: A Study of Face Recognition as People Age. In: Proc. ICCV, pp. 1–8 (2007)
6. Rui, Y., Huang, T.S., Ortega, M., Mehrotra, S.: Relevance Feedback: A Power Tool for Interactive Content-Based Image Retrieval. IEEE Trans. on Circuits and Systems for Video Technology 8(5), 644–655 (1998)
7. Yuen, P.C., Man, C.H.: Human Face Image Searching System Using Sketches. IEEE Trans. on Systems, Man and Cybernetics, Part A 37(4), 493–504 (2007)
8. FG-Net Aging Database, `http://www.fgnet.rsunit.com/`

Visual Reranking with Local Learning Consistency

Xinmei Tian[1,⋆], Linjun Yang[2], Xiuqing Wu[1], and Xian-Sheng Hua[2]

[1] University of Science and Technology of China, Hefei, Anhui, 230027, China
[2] Microsoft Research Asia, 100190, Beijing, China
xinmei@mail.ustc.edu.cn, linjuny@microsoft.com,
xqwu@ustc.edu.cn, xshua@microsoft.com

Abstract. The graph-based reranking methods have been proven effective in image and video search. The basic assumption behind them is the ranking score consistency, *i.e.*, neighboring nodes (visually similar images or video shots) in a graph having close ranking scores, which is modeled through a regularizer term. The existing reranking methods utilise pair-wise regularizers, *e.g.*, the Laplacian regularizer and the normalized Laplacian regularizer, to estimate the consistency over the graph from the pair-wise perspective by requiring the scores to be close for pairs of samples. However, since the consistency is a term defined over the whole set of neighboring samples, it is characterized by the local structure of the neighboring samples, *i.e.*, the multiple-wise relations among the neighbors. The pair-wise regularizers fail to capture the desired property of consistency since they treat the neighboring samples independently. To tackle this problem, in this paper, we propose to use local learning regularizer to model the multiple-wise consistency, by formulating the consistent score estimation over a local area into a learning problem. Experiments on the TRECVID benchmark dataset and a real Web image dataset demonstrate the superiority of the local learning regularizer in visual reranking.

Keywords: Image/video search, visual reranking, regularizer, local learning.

1 Introduction

Most of the frequently-used image/video search engines build on text-based search by using the associated textual information, such as URL, surrounding text from the Web pages, speech transcript, *etc.* However, it is commonly observed that the text-based search results are not satisfying enough. Irrelevant or little-relevant images are often returned as top results, which is caused by the mismatching between images and its surrounding texts, the low ability to discriminate images' relevance levels with only textual information and so on.

⋆ This work was performed when Xinmei Tian was visiting Microsoft Research Asia as a research intern.

S. Boll et al. (Eds.): MMM 2010, LNCS 5916, pp. 163–173, 2010.

To tackle these difficulties, visual reranking is proposed to improve the text-based search result by incorporating visual information. Up to now, many reranking methods have been proposed, *e.g.*, classification-based [8,13], clustering-based [3] and graph-based [4,6,7,10]. Among them, the graph-based methods have drawn increasing attention recently and have already shown promising results on image and video search [6,10]. In such methods, a graph is constructed with the samples, *i.e.*, images or video shots, as the nodes and the edges between them being weighted by their visual similarity. In [4,6,7], reranking is formulated as random walk over the graph and the ranking scores are propagated through the edges. The stationary probability of the random walk process is used as the reranked score directly.

The basic assumption behind the graph-based reranking methods is that: neighboring nodes in the graph should have similar ranking scores, *i.e.*, the ranking score consistency (or smoothness) over the graph topology. By modeling this assumption explicitly, a general reranking framework, *i.e.*, Bayesian reranking, is proposed in [10]. Bayesian reranking maximizes the ranking score consistency while minimizes the ranking distance, which represents the disagreement between the reranked result and the initial text-based ranking result. As illustrated in [10], the other graph-based reranking methods [4,6,7] can also be unified into this framework.

The ranking score consistency over the graph is represented through a regularizer term, which plays a crucial role in the graph-based reranking methods. In [10], the widely used regularizers in graph-based classification and video annotation, *i.e.*, Laplacian regularizer and the normalized Laplacian regularizer, are directly adopted. In [4,6,7], a variant of normalized Laplacian regularizer is utilised, as discussed in [10].

However, all the regularizers mentioned above model the ranking score consistency by approximating from the pair-wise perspective. Specifically, for each sample, a set of pairs is formed between it and each of its neighbors, and then, the overall consistency is measured by aggregating the individual consistency over each pair. In fact, a sample's consistency on a local area of the graph is multiple-wise instead of pair-wise, since the consistency is a term defined over the whole neighboring samples instead of over each pair of them. In other words, the consistency is characterized by the structure of the neighboring set of data. The pair-wise regularizers treats each neighbors independently with the multiple-wise relationship among them unconsidered. From this point of view, the consistency approximated with pair-wise regularizers is not satisfactory enough.

In this paper, we propose to use a local learning regularizer to model the desired multiple-wise consistency. Specifically, for each sample, instead of calculating the consistency with each of its neighboring samples individually, the local learning based regularizer considers the consistency with all of its neighboring samples simultaneously. In this regularizer, a local model is firstly trained for each sample with its neighbors and then this model is used to predict this sample's consistent ranking score. Finally, by minimizing the difference between the objective ranking score and this locally predicted one, the desired multiple-wise consistency over the graph is guaranteed.

The rest of this paper is organized as follows. In Section 2, we will briefly introduce the graph-based visual reranking method. The proposed local learning regularized reranking method and its solution are detailed in Section 3. Experimental results are presented and analysed in Section 4. The parameter sensitivity analyses are presented in Section 5, followed by the conclusion in Section 6.

2 Graph-Based Visual Reranking

Given N samples (images or video shots) $\{\mathbf{x}_1, \mathbf{x}_2, \cdots, \mathbf{x}_N\}$ and their initial text search result represented by a ranking score vector $\bar{\mathbf{r}} = [\bar{r}_1, \bar{r}_2, \cdots, \bar{r}_N]^T$ where $\bar{r}_i$ is the ranking score of $\mathbf{x}_i$. A larger ranking score indicates the sample is more relevant and thus should be ranked higher.

Bayesian reranking [10], a general graph-based visual reranking framework is proposed recently. In this framework, a graph $\mathcal{G}$ is constructed with nodes being the samples while weights on edges being the similarity between the corresponding samples. Specifically, the weight w_{ij} on the edge between nodes $\mathbf{x}_i$ and $\mathbf{x}_j$ is computed using Gaussian kernel $w_{ij} = \exp(-\frac{\|\mathbf{x}_i - \mathbf{x}_j\|^2}{2\sigma^2})$, where σ is the scaling parameter. Then, the reranking is formulated into an optimization problem and the optimal reranking score vector $\mathbf{r} = [r_1, r_2, \cdots, r_N]^T$ is derived by minimizing the following energy function,

$$E(\mathbf{r}) = Reg(\mathcal{G}, \mathbf{r}) + c \times Dist(\mathbf{r}, \bar{\mathbf{r}}). \quad (1)$$

Here, the first term, termed regularizer term, penalizes the ranking score inconsistency over the graph topology while the second term, termed ranking distance term, penalizes the derivation of the reranked result from the initial ranking. c is a trade-off parameter which balances the influence of the two terms. We will briefly introduce the two terms respectively in the following.

2.1 Ranking Distance

The ranking distance term is measured by the disagreement between the two ranking score vectors $\mathbf{r}$ and $\bar{\mathbf{r}}$. The preference strength distance is proposed in [10]:

$$Dist(\mathbf{r}, \bar{\mathbf{r}}) = \frac{1}{2}\sum\nolimits_{i,j} dist_{ij} = \frac{1}{2}\sum\nolimits_{i,j,i\neq j} (1 - \frac{r_i - r_j}{\bar{r}_i - \bar{r}_j})^2, \quad (2)$$

where $dist_{ij}$ denotes pair $(\mathbf{x}_i, \mathbf{x}_j)$'s distance which is measured by the two samples' preference strength difference before and after reranking. The preference strength which means the score difference of two samples, *i.e.* $r_i - r_j$ for pair $(\mathbf{x}_i, \mathbf{x}_j)$, reflects the degree of $\mathbf{x}_i$ being ranked before $\mathbf{x}_j$. The preference strength distance has been proven to be effective [10], and will be directly utilised in this paper.

2.2 Regularizer

The regularizer term aims to model the ranking score consistency over the graph topology. The widely used regularizers in semi-supervised classification and video annotation, *i.e.*, Laplacian regularizer [15] and normalized Laplacian regularizer [14], are directly adopted in [10].

When Laplacian regularizer is adopted,

$$Reg_{Lap}(\mathcal{G}, \mathbf{r}) = \mathbf{r}^T \mathbf{L} \mathbf{r} = \frac{1}{2} \sum\nolimits_{i,j} w_{ij} (r_i - r_j)^2, \tag{3}$$

where $\mathbf{L} = \mathbf{D} - \mathbf{W}$ is the Laplacian matrix. $\mathbf{D} = Diag(\mathbf{d})$ is a degree matrix with $\mathbf{d} = [d_1, d_2, \cdots, d_N]^T$ and $d_i = \sum_j w_{ij}$.

When normalized Laplacian regularizer is adopted:

$$Reg_{NLap}(\mathcal{G}, \mathbf{r}) = \mathbf{r}^T \mathbf{L}_n \mathbf{r} = \frac{1}{2} \sum\nolimits_{i,j} w_{ij} \left(\frac{r_i}{\sqrt{d_i}} - \frac{r_j}{\sqrt{d_j}}\right)^2, \tag{4}$$

where $\mathbf{L}_n = \mathbf{I} - \mathbf{D}^{-\frac{1}{2}} \mathbf{W} \mathbf{D}^{-\frac{1}{2}}$. $\mathbf{I}$ is the unit matrix, $\mathbf{W}$ and $\mathbf{D}$ are the same as in Laplacian matrix.

However, both regularizers model the ranking score consistency pair-wisely and have less ability to capture the multiple-wise property of the ranking score consistency. As will be discussed in the next section, local learning based regularizer models the multiple-wise consistency by formulating the score estimation into a learning problem without any heuristic assumptions.

3 Visual Reranking with Local Learning Regularizer

3.1 Local Learning Regularizer

With the ranking score consistency assumption, the desired property of $\mathbf{r}$ is that: for each sample $\mathbf{x}_i$, its ranking score r_i should be consistent with its neighboring samples' scores simultaneously. In Eq. (3) and Eq. (4), this multiple-wise consistency is approximated by accumulating $\mathbf{x}_i$'s consistency with each of its neighboring samples individually.

To reveal the intrinsic multiple-wise consistency, we tackle this problem from the local learning perspective. If a sample's ranking score can be estimated from its neighbors, this sample's multiple-wise consistency with its neighbors is guaranteed. From this point of view, we can model the ranking score consistency from the machine learning perspective. Specifically, for sample $\mathbf{x}_i$, we first learn the desirably consistent score $\hat{r}_i$ from its neighboring samples. By requiring the objective r_i be close to this predicted value $\hat{r}_i$, the multiple-wise consistency is guaranteed. The details will be discussed in the following.

For each sample $\mathbf{x}_i$, a local model $o_i(\cdot)$ is trained locally with the data $\{(\mathbf{x}_j, r_j)\}_{\mathbf{x}_j \in \mathcal{N}(\mathbf{x}_i)}$, where $\mathcal{N}(\mathbf{x}_i)$ denotes the set of $\mathbf{x}_i$'s neighboring samples. $\mathbf{x}_i$'s ranking score can be predicted by this learned model. Then the regularizer term is derived by aggregating the local model's prediction loss on each sample:

$$Reg_{Local}(\mathcal{G}, \mathbf{r}) = \sum\nolimits_i (r_i - o_i(\mathbf{x}_i))^2. \tag{5}$$

The task of the local model $o_i(\cdot)$ is to predict sample $\mathbf{x}_i$'s ranking score r_i from its neighboring samples accurately. Many approaches can be used as the local model. A linear one is adopted in [12]. However, due to the complexity of the real-world images, it is hard to predict the scores accurately by using simple linear model. To handle this difficulty, we propose to use a local kernel model. Since it is apparently a regression problem, the kernel ridge regression statistical model [2], which is well-known and simple to implement, is adopted in this paper.

In kernel ridge regression, by using a kernel mapping $\phi(\cdot)$ operating from input space $\mathcal{X}$ to a kernel space $\mathcal{F}$, $\phi : \mathbf{x} \in \mathcal{X} \mapsto \Phi(\mathbf{x}) \in \mathcal{F}$, the dependencies between $\mathcal{N}(\mathbf{x}_i)$ and its score vector $\mathbf{r}_i = [r_j]^T_{\mathbf{x}_j \in \mathcal{N}(\mathbf{x}_i)}$ is modeled as:

$$o_i(\mathbf{x}) = \mathbf{w}^T \phi(\mathbf{x}). \tag{6}$$

Its cost function is:

$$\sum\nolimits_{j, \mathbf{x}_j \in \mathcal{N}(\mathbf{x}_i)} (r_j - \mathbf{w}^T \phi(\mathbf{x}_j))^2 + \lambda \left\| \mathbf{w} \right\|^2, \tag{7}$$

λ is a coefficient to balance the capacity and complexity of this model.

Differentiating Eq. (7) w.r.t. $\mathbf{w}$ and then equating it to zero, we obtain:

$$\mathbf{w} = \Phi_i (\Phi_i^T \Phi_i + \lambda \mathbf{I})^{-1} \mathbf{r}_i,$$

where Φ_i denotes matrix $[\phi(\mathbf{x}_j)]^T$ for $\mathbf{x}_j \in \mathcal{N}(\mathbf{x}_i)$. Then, for $\mathbf{x}_i$, the score predicted by its local model $o_i(\cdot)$ is:

$$o_i(\mathbf{x}_i) = \mathbf{w}^T \phi(\mathbf{x}_i) = \mathbf{k}^T (\lambda \mathbf{I} + \mathbf{K})^{-1} \mathbf{r}_i = \beta_i^T \mathbf{r}_i, \tag{8}$$

where $\mathbf{k}$ is a vector with $k_j = \phi(\mathbf{x}_i)^T \phi(\mathbf{x}_{t_j}) = k(\mathbf{x}_i, \mathbf{x}_{t_j})$, and $\mathbf{K}$ is a matrix with $k_{mn} = \phi(\mathbf{x}_{t_m})^T \phi(\mathbf{x}_{t_n}) = k(\mathbf{x}_{t_m}, \mathbf{x}_{t_n})$, $j, m, n = 1, 2, \cdots, |\mathcal{N}(\mathbf{x}_i)|$ and $\mathbf{x}_{t_j}, \mathbf{x}_{t_m}, \mathbf{x}_{t_n} \in \mathcal{N}(\mathbf{x}_i)$ with t_p is the subscript of the pth sample in $\mathcal{N}(\mathbf{x}_i)$. As for kernel based methods, we only need to define the kernel function k without defining $\phi(\cdot)$ explicitly. The Gaussian kernel is adopted as the kernel function in this paper.

The local learning regularizer is formulated as:

$$\begin{aligned} Reg_{Local}(\mathcal{G}, \mathbf{r}) &= \sum\nolimits_i (r_i - o_i(\mathbf{x}_i))^2 \\ &= \sum\nolimits_i (r_i - \beta_i^T \mathbf{r}_i)^2 = \mathbf{r}^T \mathbf{R}_{Local} \mathbf{r} \end{aligned} \tag{9}$$

$\beta_i^T = \mathbf{k}^T (\lambda \mathbf{I} + \mathbf{K})^{-1}$. $\mathbf{R}_{Local} = (\mathbf{I} - \mathbf{B})^T (\mathbf{I} - \mathbf{B})$ is the local learning regularization matrix and $\mathbf{B} = [b_{ij}]_{N \times N}$ where b_{ij} equals the corresponding element of β_i if $\mathbf{x}_j \in \mathcal{N}(\mathbf{x}_i)$, otherwise $b_{ij} = 0$.

3.2 Solution

With the local learning regularizer in Eq. (9) as the regularizer term and the preference strength distance in Eq. (2) as the distance term respectively, the objective function of the local learning regularized Bayesian reranking is formulated as:

$$E(\mathbf{r}) = Reg_{Local}(\mathcal{G}, \mathbf{r}) + c \times Dist(\mathbf{r}, \bar{\mathbf{r}})$$
$$= \mathbf{r}^T \mathbf{R}_{Local} \mathbf{r} + c \times \frac{1}{2} \sum_{i,j,i \neq j} (1 - \frac{r_i - r_j}{\bar{r}_i - \bar{r}_j})^2. \quad (10)$$

The optimal solution $\mathbf{r}^*$ is obtained by minimizing Eq. (10). Denote $\alpha_{ij} = 1/(\bar{r}_i - \bar{r}_j)$, then we can get

$$\mathbf{r}^* = \arg\min_{\mathbf{r}} \mathbf{r}^T \mathbf{R}_{Local} \mathbf{r} + c \times \frac{1}{2} \sum_{i,j,i \neq j} (1 - \alpha_{ij}(r_i - r_j))^2$$
$$= \arg\min_{\mathbf{r}} \mathbf{r}^T \mathbf{R}_{Local} \mathbf{r} + c \times \frac{1}{2} \sum_{i,j,i \neq j} \alpha_{ij}^2 (r_i - r_j)^2$$
$$- c \sum_{i,j,i \neq j} \alpha_{ij}(r_i - r_j)$$
$$= \arg\min_{\mathbf{r}} \mathbf{r}^T \mathbf{R}_{Local} \mathbf{r} + c \times \mathbf{r}^T \mathbf{L}_A \mathbf{r} - c \sum_{i,j,i \neq j} \alpha_{ij}(r_i - r_j)$$
$$= \arg\min_{\mathbf{r}} \mathbf{r}^T (\mathbf{R}_{Local} + c\mathbf{L}_A)\mathbf{r} - \tilde{\mathbf{c}}^T \mathbf{r} \quad (11)$$

where $\tilde{\mathbf{c}} = 2c(\mathbf{A}\mathbf{e})$ with $\mathbf{e}$ is a vector with all elements equal 1 and $\mathbf{A} = [\alpha_{ij}]_{N \times N}$ is an anti-symmetric matrix. $\mathbf{L}_A$ is a Laplacian regularization matrix defined over the graph $\mathcal{G}_A$. $\mathcal{G}_A$ has the same structure with $\mathcal{G}$ but the weight between nodes $\mathbf{x}_i$ and $\mathbf{x}_j$ is $|\alpha_{ij}|$ instead of w_{ij}.

Differentiating (11) w.r.t $\mathbf{r}$ and then equating it to zero, it gives:

$$2(\mathbf{R}_{Local} + c\mathbf{L}_A)\mathbf{r}^* = \tilde{\mathbf{c}}$$
$$\mathbf{r}^* = \frac{1}{2}(\mathbf{R}_{Local} + c\mathbf{L}_A)^{-1}\tilde{\mathbf{c}} \quad (12)$$

Then a closed-form solution for local learning regularized reranking is obtained. The solutions for Bayesian reranking with Laplacian regularizer and normalized Laplacian regularizer have the same form with Eq. (12) only by replacing the $\mathbf{R}_{Local}$ with $\mathbf{L}$ and $\mathbf{L}_n$ respectively.

4 Experiments

4.1 Experimental Setting

To evaluate the effectiveness of the proposed local learning regularizer for reranking, we conducted experiments on two datasets. One is the widely used video search benchmark, *i.e.*, TRECVID 2005-2007 test set [11], which consists of 508 videos and 143,392 shots totally. Its text search baseline is obtained based on the Okapi BM-25 formula [9] using ASR/MT transcripts at shot level. For each of the 72 queries, 24 for each year, the top 1400 shots returned by the search system are used as the initial text search result.

The other dataset is a real-world Web image dataset collected from Google[1]. We selected 29 queries from a commercial image search query log as well as

[1] http://images.google.com/

popular tags of Flickr[2]. For each query, at most top 1000 images returned by Google image search are collected. Thus, there are 24,676 images in total. The initial search result given by Google is regarded as the text search baseline.

For both datasets, the visual feature used is 225-dimentional block-wise colour moments extracted over 5*5 fixed grid partitions. The K-Nearest Neighbor is adopted to find the neighboring samples.

For TRECVID 2005-2007, the video shot's relevance is provided by NIST [11] on two levels, *i.e.*, "Relevant" or "Irrelevant". The most often used performance measure for this dataset is the non-interpolated Average Precision (AP) [1], which is also adopted in this paper. We average the APs over all the 24 queries in each year to get the Mean Average Precision (MAP) to measure the overall performance.

For the Web image dataset, each image's relevance degree is judged by three participants on four levels, *i.e.*, "Excellent", "Good", "Fair" and "Irrelevant", and its final relevance degree is set as the middle one among the scores given by the three judges. The performance on this dataset is measured by the Normalized Discounted Cumulated Gain(NDCG) [5], which is a measure commonly used in information retrieval when there are more than two relevance levels.

4.2 Experimental Results and Analysis

We compare the local learning regularizer with the other two popular ones, *i.e.*, Laplacian regularizer and normalized Laplacian regularizer, in the Bayesian reranking framework. Bayesian reranking with the three different regularizers are denoted as LocalReg, LapReg and NLapReg respectively. We also compare LocalReg with another well-known graph-based method - VisualRank [6], which also measures the consistency pair-wisely, as discussed in [10]. The parameters are globally set for each method respectively.

The overall performance on the two datasets are summarized in Table 1 and Fig. 1 respectively. We can see that LocalReg outperforms LapReg, NLapReg as well as VisualRank over the two datasets consistently and stably. This result

Table 1. MAP comparison for different reranking methods on TRECVID 2005-2007

Method	TRECVID 2005		TRECVID 2006		TRECVID 2007	
	MAP	Gain	MAP	Gain	MAP	Gain
Baseline	0.0441	-	0.0381	-	0.0306	-
VisualRank	0.0506	14.74%	0.0401	5.25%	0.0333	8.82%
LapReg	0.0487	10.43%	0.0461	21.00%	0.0465	51.96%
NLapReg	0.0534	21.09%	0.0434	13.91%	0.0471	53.92%
LocalReg	**0.0583**	**32.20%**	**0.0497**	**30.45%**	**0.0485**	**58.50%**

[2] http://www.flickr.com/

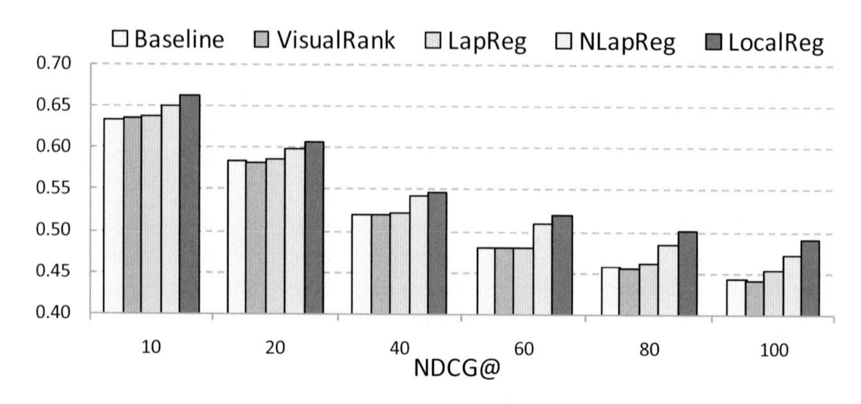

Fig. 1. NDCG comparison for different reranking methods on Web image dataset

demonstrates that pair-wise approximation of the consistency over the graph is not precise enough. The local learning based regularizer captures the multiple-wise consistency and thus provides more satisfactory results.

Besides the overall performance, we also investigate the effectiveness of LocalReg over each query, as shown in Fig. 2. Here, due to the space limitation, we only take TRECVID 2005 for illustration. From Fig. 2, we can see that most of the queries benefit from LocalReg after reranking and some queries show significant gains, such as query 152[3], 156[4], 169[5] and 171[6]. By further comparing LocalReg with other three reranking methods, we find that although the other ones achieve high improvements on certain queries, *e.g.*, query 154[7] for

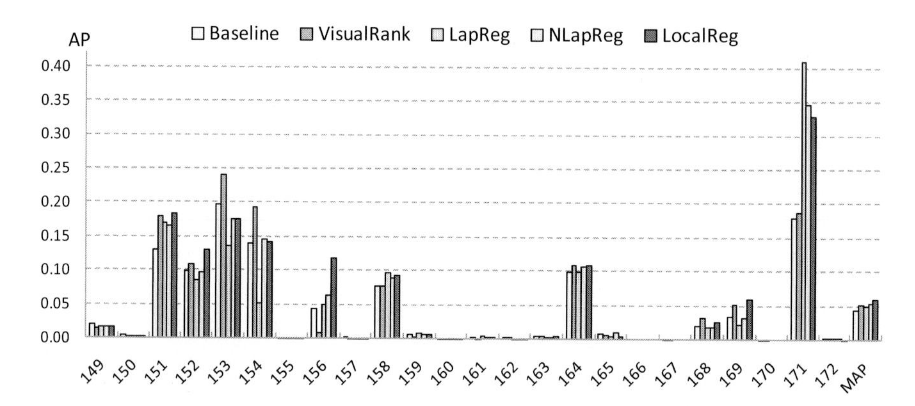

Fig. 2. Performance comparison over each query on TRECVID 2005

[3] Find shots of Hu Jintao.

[4] Find shots of tennis players on the court.

[5] Find shots of George Bush entering or leaving a vehicle.

[6] Find shots of a goal being made in a soccer match.

[7] Find shots of Mahmoud Abbas.

VisualRank, query 171 for LapReg and NLapReg, they show dramatic performance decrease on other queries, *e.g.*, query156 for VisualRank and query 154 for LapReg. In contrast, LocalReg presents stable performance improvements on most of queries with slight performance decrease on few ones. This phenomenon further demonstrates the superiority of the multiple-wise consistency derived from local learning regularizer.

5 Parameter Sensitivity

In the local learning regularizer, there are two important parameters, *i.e.*, the number of nearest neighbors K for local model learning and the trade-off parameter λ in Eq. (7) of kernel ridge regression. In this section, we will analyse the sensitivity of LocalReg to them, by taking experiments conducted on the TRECVID dataset as illustration.

5.1 The Number of Nearest Neighbors K

The number of nearest neighbors K plays an important role in local learning regularizer. With too few neighbors, over-fitting is likely to occur and thus the learned model gives poor prediction for the centroid sample. A larger K can ensure more samples are involved for local model training. However, the samples which are far away from the centroid sample have less to do with it and thus gives little contribution to the local model. Besides, more noises will be introduced at the same time when enlarging the training size. A proper K is required to achieve a good performance.

The MAP-K curves are presented in Fig. 3. As illustrated, larger Ks are preferred on both TRECVID 2005 and 2007, while on TRECVID 2007 a small K gives better performance. As analysed from the data, the average numbers of relevant samples across queries are 41, 55 and 24 for TRECVID 2005 - 2007 respectively. We can observe that the optimal K is roughly proportional to the average number of relevant samples. TRECVID 2007 has the fewest relevant samples, thus, a smaller training set is preferred to prevent too many noises are involved. This observation can provide a rough guideline in setting K in the future practical application.

5.2 The Trade-Off Parameter λ

We also investigated the influence of the trade-off parameter λ in kernel ridge regression. Figure 4 shows the performance of LocalReg with different λ in terms of MAP on TRECVID 2005-2007.

From the figures, we can see that, compared with the text search baseline, the performance is improved with variant λs consistently. For further observation, we can find that moderate, larger and smaller λs are preferred on the three years respectively. The reason is that, on TRECVID 2007, the local learning problem is the hardest (with the lowest baseline and the fewest relevant samples) thus a

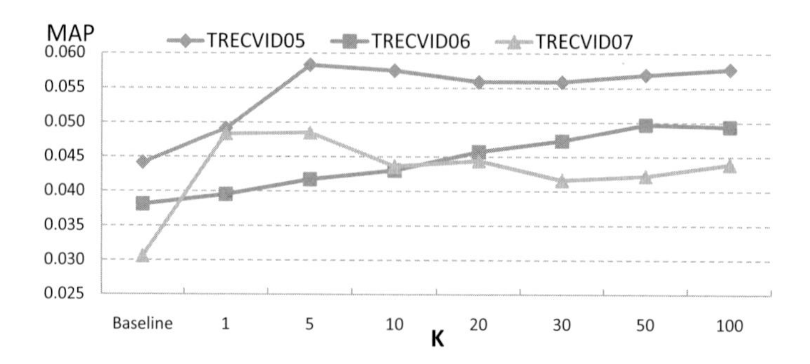

Fig. 3. The performance of LocalReg with different K on TRECVID dataset

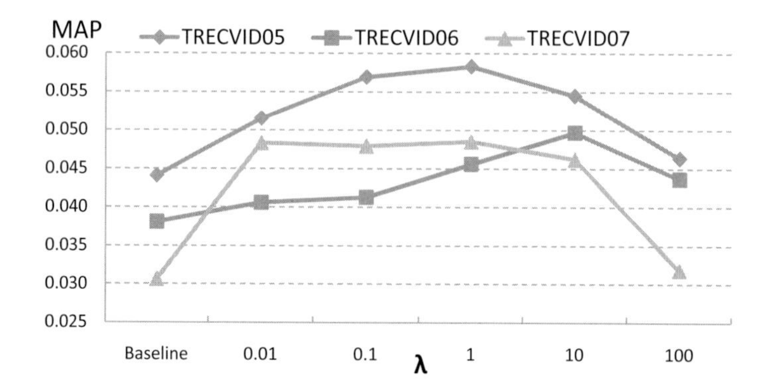

Fig. 4. The performance of LocalReg with different λ on TRECVID dataset

complex model is required with small λ. For TRECVID 2006, it has the most relevant samples and the local learning problem is the easiest. A simple model is effective enough and there for a larger λ is preferred. On TRECVID 2005, although it has the highest baseline, its relevant samples are fewer than that of TRECVID 2006. As a consequence, a moderate λ is optimal for it.

6 Conclusion

By investigating the effect of the regularizer term in graph-based visual reranking, an effective regularization approach, *i.e.*, local learning regularizer, is adopted in this paper from the local learning perspective. The local learning regularizer models the multiple-wise consistency by formulating the score estimation into a learning problem without any heuristic assumptions. Extensive experiments conducted on both TRECVID dataset and real Web image dataset demonstrate the superiority of the local learning regularizer.

References

1. Trec-10 proceddings appendix on common evaluation measures, http://trec.nist.gov/pubs/trec10/appendices/measures.pdf
2. Cristianini, N., Shawe-Taylor, J.: An introduction to support vector machines and other kernel-based learning methods. Cambridge University, Cambridge (2000)
3. Hsu, W.H., Kennedy, L.S., Chang, S.-F.: Video search reranking via information bottleneck principle. ACM Multimedia, 35–44 (2006)
4. Hsu, W.H., Kennedy, L.S., Chang, S.-F.: Video search reranking through random walk over document-level context graph. ACM Multimedia, 971–980 (2007)
5. Järvelin, K., Kekäläinen, J.: Ir evaluation methods for retrieving highly relevant documents. In: SIGIR, pp. 41–48 (2000)
6. Jing, Y., Baluja, S.: Visualrank: Applying pagerank to large-scale image search. IEEE Transactions on PAMI 30, 1877–1890 (2008)
7. Liu, J., Lai, W., Hua, X.-S., Huang, Y., Li, S.: Video search re-ranking via multi-graph propagation. ACM Multimeida, 208–217 (2007)
8. Liu, Y., Mei, T., Hua, X.-S., Tang, J., Wu, X., Li, S.: Learning to video search rerank via pseudo preference feedback. In: ICME, pp. 297–300 (2008)
9. Robertson, S.E., Walker, S., Hancock-Beaulieu, M., Gatford, M., Payne, A.: Simple, proven approaches to text retrieval. Cambridge University Computer Laboratory Technical Report TR356 (1997)
10. Tian, X., Yang, L., Wang, J., Yang, Y., Wu, X., Hua, X.-S.: Bayesian video search reranking. ACM Multimedia, 131–140 (2008)
11. TRECVID. TREC Video Retrieval Evaluation, http://www-nlpir.nist.gov/projects/trecvid/
12. Wu, M., Schölkopf, B.: Transductive classification via local learning regularization. In: 11th International Conference on AISTATS, pp. 624–631 (2007)
13. Yan, R., Hauptmann, A.G., Jin, R.: Multimedia search with pseudo-relevance feedback. In: Bakker, E.M., Lew, M., Huang, T.S., Sebe, N., Zhou, X.S. (eds.) CIVR 2003. LNCS, vol. 2728, pp. 238–247. Springer, Heidelberg (2003)
14. Zhou, D., Bousquet, O., Lal, T.N., Weston, J., Schölkopf, B.: Learning with local and global consistency. In: NIPS (2003)
15. Zhu, X., Ghahramani, Z., Lafferty, J.D.: Semi-supervised learning using gaussian fields and harmonic functions. In: ICML, pp. 912–919 (2003)

Social Image Search with Diverse Relevance Ranking

Kuiyuan Yang[1,⋆], Meng Wang[2], Xian-Sheng Hua[2], and Hong-Jiang Zhang[3]

[1] University of Science and Technology of China,
Hefei, Anhui, 230027 China
ustc_yky@hotmail.com
[2] Microsoft Research Asia,
49 Zhichun Road Beijing, 100080 China
{mengwang,xshua}@microsoft.com
[3] Microsoft Adv. Tech. Center,
49 Zhichun Road Beijing, 100080 China
hjzhang@microsoft.com

Abstract. Recent years have witnessed the success of many online social media websites, which allow users to create and share media information as well as describe the media content with tags. However, the existing ranking approaches for tag-based image search frequently return results that are irrelevant or lack of diversity. This paper proposes a diverse relevance ranking scheme which is able to simultaneously take relevance and diversity into account. It takes advantage of both the content of images and their associated tags. First, it estimates the relevance scores of images with respect to the query term based on both the visual information of images and the semantic information of associated tags. Then we mine the semantic similarities of social images based on their tags. With the relevance scores and the similarities, the ranking list is generated by a greedy ordering algorithm which optimizes Average Diverse Precision (ADP), a novel measure that is extended from the conventional Average Precision (AP). Comprehensive experiments and user studies demonstrate the effectiveness of the approach.

Keywords: Image search, diversity.

1 Introduction

There is an explosion of community-contributed multimedia content available online, such as Youtube, Flickr and Zooomr. Such media repositories promote users to collaboratively create, evaluate and distribute media information. They also allow users to annotate their uploaded media data with descriptive keywords called tags, which can greatly facilitate the organization of the social media. However, performing search on large-scale social media data is not an easy task.

⋆ This work was performed when Kuiyuan Yang was visiting Microsoft Research Asia as research intern.

S. Boll et al. (Eds.): MMM 2010, LNCS 5916, pp. 174–184, 2010.

Commonly-used tag-based image search is a straightforward approach, which returns the images annotated with a specific query tag. Currently, Flickr provides two ranking options for tag-based image search. One is "most recent", which orders images based on their uploading time, and the other is "most interesting", which ranks the images by "interestingness", a measure that integrates the information of click-through, comments, etc. These two search options both rank images according to other measures (interestingness or time) instead of relevance levels and thus may bring many irrelevant images in the returned ranking lists. In addition to relevance, lack-of-diversity is also a problem. Many images on social media websites are actually close to each other. For example, several users used to upload continuously captured images in batch, and many of them will be visually and semantically close. When these images appear simultaneously in the top results, users will get only limited information. Therefore, a ranking scheme that can simultaneously generate relevant and diverse results is highly desired.

In this work we propose a Diverse Relevance Ranking (DRR) scheme for social image search. It is able to rank the images based on their relevance levels with respect to query tag while simultaneously considering the diversity of the ranking list. The scheme works as follows. First, we estimate the relevance score of each image with respect to the query term as well as the semantic similarity of each image pair. The relevance estimation incorporates both the visual information of images and the semantic information of their associated tags into an optimizations framework, and the semantic similarity is mined based on the associated tags of images. With the estimated relevance scores and similarities, we then implement the DRR algorithm, which can be viewed as a greedy ordering algorithm that optimizes Average Diverse Precision (ADP), a novel measure that is extended from conventional Average Precision (AP). Different from AP that only considers relevance, ADP further takes diversity into account, and thus the derived DRR algorithm can generate results that are both relevant and diverse. The main contribution of this paper can be summarized as follows:

(1) Propose a diverse relevance ranking scheme for social image search, which is complementary to the existing ranking approaches.
(2) Propose a method to estimate the relevance scores of images with respect to a query tag. It leverages both the visual information of images and the semantic information of tags.
(3) Extend the conventional AP measure to ADP to take diversity into account, and then derive a greedy ordering algorithm accordingly that compromises relevance and diversity.

The organization of the rest of this paper is as follows. In Section 2, we provide a short review on the related work. In Section 3, we introduce the diverse relevance ranking approach. In Section 4, we detail the relevance and semantic similarity estimation of social images. Empirical study is presented in Section 5. Finally, we conclude the paper in Section 6.

2 Related Work

The last decade has witnessed a great advance of image search technology [10,5,17]. Different from general web images, social images are usually associated with a set of user-provided descriptors called tags, and thus tag-based search can be easily accomplished by using the descriptors as index terms. But user-provided tags are usually very noisy [8,11,14], and this fact usually makes search results unsatisfactory. In comparison with the extensive studies on how to help users better perform tagging or leveraging tags as an information source in other applications, the literature regarding tag-based image search is still very sparse. Li et al. have proposed a tag relevance learning method which is able to assign each tag a relevance score, and they have shown its application in tag-based image search [11]. But they simply adopt a visual neighborhood voting method, and the semantic information of tags is not utilized. Their method also cannot deal with the aforementioned lack-of-diversity problem.

It has been long acknowledged that diversity plays an important role in information retrieval. In 1964, Goffman have recognized that the relevance of a document must be determined with respect to the documents appearing before it [4]. Carbonell et al. propose a ranking method named Maximal Marginal Relevance (MMR), which attempts to maximize relevance while minimizing similarity to higher ranked documents [2]. Zhai et al. propose a subtopic search method, which aims to return the results that cover more subtopics [19]. Many related works can be found in [19] and the references therein.

The diversity problem is actually more challenging in image search, as it involves not only the semantic ambiguity of queries but also the visual similarity of search results [9,16]. Currently there are two typical approaches to enhancing the diversity in image search: search results clustering [1,7,12,9] and duplicate removal [6,15,13]. The clustering and duplicate elimination techniques are both useful, but they have their limitations due to the involved heuristics. These two methods actually both accomplish diversification by removing several images in the ranking list (near-duplicates or images that are not the representatives of clusters), and this thus introduces a dilemma. If we adopt too many clusters or a small threshold for near-duplicate detection, then the diversity of search results cannot be guaranteed, and contrarily if clusters are too few or we set a large threshold for near-duplicate detection, many informative images will be removed. The diverse relevance ranking scheme proposed in this work adopts a different approach. We just rank all images and keep the diversity of top results. Therefore, users will not miss information since we do not remove any image, and the relevance and diversity of top results can still both be kept.

3 Diverse Relevance Ranking

We introduce the DRR approach in this section. Here we present it as a general ranking algorithm algorithm and leave the two flexible components, i.e., relevance score and similarity estimation of images, to the next section. We first prove that ranking by relevance scores can be viewed as optimizing the mathematical expectation of the conventional Average Precision (AP) measure. Then we analyze the limitation of AP and generalize it to an Average Diverse Precision (ADP) measure to take diversity into account. The DRR algorithm is then derived by greedily optimizing the mathematical expectation of ADP measurement.

3.1 Average Precision

AP is a widely-applied performance evaluation measure in information retrieval. Given a collection of images $\mathcal{D} = \{x_1, x_2, \ldots, x_n\}$, denote by $y(x_i)$ the binary relevance label of x_i with respect to the given query, i.e., $y(x_i) = 1$ if x_i is relevant and otherwise $y(x_i) = 0$. Denote by $\boldsymbol{\tau}$ an ordering of the images, and let $\tau(i)$ be the image at the position of rank i (lower number means higher ranked image). Let R be the number of true relevant images in the set $\mathcal{D}$. Then the non-interpolated AP is defined as

$$AP(\boldsymbol{\tau}, \mathcal{D}) = \frac{1}{R} \sum_{j=1}^{n} y(\tau(j)) \frac{\sum_{k=1}^{j} y(\tau(k))}{j} \quad (1)$$

Obviously, ranking images with their relevance scores in decreasing order is the most intuitive approach if we do not consider other factors. Now we prove that the ranking list generated in this way actually maximizes the mathematical expectation of AP measurement.

Denote by $r(x_i)$ the relevance score of x_i (how to estimate it will be introduced in the next section), and it is reasonable for us to assume that $r(x_i) = P(y(x_i) = 1)$, i.e., we regard the relevance score $r(x_i)$ as the probability that x_i is relevant. Since R can be regarded as a constant, we do not take it into account in the expectation estimation. We also assume that the relevance of an image is independent with other images, and hence the expected value of $AP(\boldsymbol{\tau}, \mathcal{D})$ can be computed as follows

$$\begin{aligned} E\{AP(\boldsymbol{\tau}, \mathcal{D})\} &= \frac{1}{R} \sum_{j=1}^{n} \sum_{k=1}^{j} \frac{E\{y(\tau(k))y(\tau(j))\}}{j} \\ &= \frac{1}{R} \sum_{j=1}^{n} \frac{1}{j} \left(r(\tau(j)) + \sum_{k=1}^{j-1} r(\tau(k)) r(\tau(j)) \right) \end{aligned} \quad (2)$$

Then we have the following theorem:

Theorem 1. *Ranking the images in $\mathcal{D}$ with relevance scores $r(x_i)$ in non increasing order maximizes $E\{AP(\boldsymbol{\tau}, \mathcal{D})\}$.*

Proof. Denote by $\boldsymbol{\tau}^*$ the ranking of images in $\mathcal{D}$ with relevance scores in non increasing order, i.e., $r(\tau^*(i)) \geq r(\tau^*(i+1))$. Then we only need to prove $E\{AP(\boldsymbol{\tau}^*, \mathcal{D})\} \geq E\{AP(\boldsymbol{\tau}, \mathcal{D})\}$ for every possible $\boldsymbol{\tau}$.

Without loss of generality, we consider an ordering $\boldsymbol{\tau}'$ that has exchange the documents at the positions of rank i and $i+1$ in $\boldsymbol{\tau}^*$, i.e., $\tau'(i) = \tau^*(i+1)$ and

$\tau'(i+1) = \tau^*(i)$. Actually it is not difficult to find that any change on the τ^* can be decomposed into a series of such adjacent exchanges. So, our task is simplified to prove $E\{AP(\boldsymbol{\tau}^*, \mathcal{D})\} \geq E\{AP(\boldsymbol{\tau}', \mathcal{D})\}$.

For simplicity, we denote $r_i = r(\tau^*(i))$ and $r'_i = r(\tau'(i))$. Since $r'_i = r_{i+1}, r'_{i+1} = r_i$, and $r'_k = r_k$ when $k \neq i$ and $i+1$, we have

$$\begin{aligned}
\triangle &= E\{AP(\boldsymbol{\tau}^*, \mathcal{D}\} - E\{AP(\boldsymbol{\tau}', \mathcal{D})\} \\
&= \frac{1}{R}\left(\sum_{1 \leq j \leq n, j \neq i, j \neq i+1} \frac{r_j + \sum_{k=1}^{j-1} r_k r_j}{j} + \frac{r_i + \sum_{k=1}^{i-1} r_k r_i}{i} + \frac{r_{i+1} + \sum_{k=1}^{i} r_k r_{i+1}}{i+1}\right) \\
&- \frac{1}{R}\left(\sum_{1 \leq j \leq n, j \neq i, j \neq i+1} \frac{r'_j + \sum_{k=1}^{j-1} r'_k r'_j}{j} + \frac{r'_i + \sum_{k=1}^{i-1} r'_k r'_i}{i} + \frac{r'_{i+1} + \sum_{k=1}^{i} r'_k r'_{i+1}}{i+1}\right) \\
&= \frac{r_i - r_{i+1} + \sum_{k=1}^{i-1} r_k (r_i - r_{i+1})}{i} - \frac{r_i - r_{i+1} + \sum_{k=1}^{i-1} r_k (r_i - r_{i+1})}{i+1} \\
&= (1 + \textstyle\sum_{k=1}^{i-1} r_k)(r_i - r_{i+1})(\frac{1}{i} - \frac{1}{i+1})
\end{aligned} \tag{3}$$

Since $r_i \geq r_{i+1}$, we have $\triangle \geq 0$, i.e., $E\{AP(\boldsymbol{\tau}^*, \mathcal{D})\} \geq E\{AP(\boldsymbol{\tau}', \mathcal{D})\}$, which completes the proof.

This proof demonstrates that adopting the AP performance evaluation measure will prioritize images with high relevance. However, the measure may not be consistent with users' experience due to the neglect of diversity. Therefore, the AP measure can be enhanced by considering diversity.

3.2 Average Diverse Precision

Here we generalize the existing AP measure to Average Diverse Precision (ADP) to take diversity into account, which is defined as

$$ADP(\boldsymbol{\tau}, \mathcal{D}) \triangleq \frac{1}{R} \sum_{j=1}^{n} y(\tau(j)) Div(\tau(j)) \left(\frac{\sum_{k=1}^{j} y(\tau(k)) Div(\tau(k))}{j} \right) \tag{4}$$

where $Div(\tau(k))$ indicates the diversity score of $\tau(k)$. We define $Div(\tau(k))$ as its minimal difference with the images appearing before it, i.e.,

$$Div(\tau(k)) = \min_{1 \leq t < k} (1 - s(\tau(t), \tau(k))) \tag{5}$$

where $s(.,.)$ is a similarity measure between two images. It is worth noting that it needs not to be visual similarity. Actually in the next section we will introduce a semantic similarity for social images. Comparing the definition of AP and ADP (see Eq. 1 and Eq. 4), we can see that the only difference is that we have changed $y(\tau(k))$ to $y(\tau(k))Div(\tau(k))$. For an image in the ranking list, its contribution to the ADP measure is not only determined by its relevance with respect to the query but also its difference with the images appearing before it. If an image is identical to one of the previously appeared images, it will contribute zero to the ADP measurement. Thus the ADP measure takes both relevance and diversity into account. Denote by $\boldsymbol{\tau}^*$ the optimal ranking list under the ADP performance evaluation measure, i.e., the list that achieves the highest ADP measurement, we can prove that $y(\tau(i))Div(\tau(i)) \geq y(\tau(j))Div(\tau(j))$ for any $i < j$. This indicates that the top images will tend to be more relevant and diverse. Here we omit its proof since it is analogous to Theorem 1.

3.3 Diverse Relevance Ranking

The DRR algorithm is actually a greedy approach to optimizing the expected value of the ADP measurement. Analogous to AP, we can estimate the expected value of ADP as

$$\begin{aligned} E\{ADP(\boldsymbol{\tau}, \mathcal{D})\} &= \frac{1}{R}\sum_{j=1}^{n}\sum_{k=1}^{j}\frac{E\{y(\tau(k))y(\tau(j))Div(\tau(k))Div(\tau(j))\}}{j} \\ &= \frac{1}{R}\sum_{j=1}^{n} r(\tau(j))Div(\tau(j))\left(\frac{Div(\tau(j)) + \sum_{k=1}^{j-1} r(\tau(k))Div(\tau(k))}{j}\right) \end{aligned} \tag{6}$$

The direct optimization of $E\{ADP(\boldsymbol{\tau}, \mathcal{D})\}$ is a permutation problem and the solution space scales is $O(n!)$. Thus here we propose a greedy method to solve it. Considering the top $i-1$ documents have been established, based on Eq. 6 we can derive that the i-th image should be decided as follows

$$\tau(i) = arg \max_{x \in \mathcal{D} - \mathcal{S}_i} \frac{r(x)}{i} Div(x)(C + Div(x)) \tag{7}$$

where

$$\mathcal{S}_i = \{\tau(1), \tau(2), \ldots, \tau(i-1)\} \tag{8}$$

$$C = \sum_{k=1}^{i-1} r(\tau(k))Div(\tau(k)) \tag{9}$$

4 Relevance and Similarity of Social Images

In this section, we introduce the estimation of relevance scores and similarities of social images, which are the two necessary components of the DRR algorithm. The following notations will be used. Given a query tag t_q, denote by $\mathcal{D} = \{x_1, x_2, \ldots, x_n\}$ the collection of images that are associated with the tag. For image x_i, denote by $\mathcal{T}_i = \{t_1^i, t_2^i, \ldots, t_{|\mathcal{T}_i|}^i\}$ the set of its associated tags. The relevance scores of all images in $\mathcal{D}$ are represented in a vector $\mathbf{r} = [r(x_1), r(x_2), \ldots, r(x_n)]^T$, whose element $r(x_i) > 0$ denotes the relevance score of image x_i with respect to query tag t_q. Denote by $\mathbf{W}$ a similarity matrix whose element W_{ij} indicates the visual similarity between images x_i and x_j.

4.1 Relevance Estimation

Our relevance estimation approach is accomplished by leveraging both the visual information of images and the semantic information of tags. Our first assumption is that the relevance of an image should depend on the "closeness" of its tags to the query tag. Thus we first have to define the similarity of tags. Different from images that can be represented as sets of low-level features, tags are textual words and their similarity exists only in semantics. Recently, there are several works aim to address this issue [3,18]. Here we adopt an approach that is analogous to Google distance [3], in which the similarity between tag t_i and t_j is defined as

$$sim(t_i, t_j) = \exp(-\frac{\max(\log c(t_i), \log c(t_j)) - \log c(t_i, t_j)}{\log M - \min(\log c(t_i), \log c(t_j))}) \quad (10)$$

where $c(t_i)$ and $c(t_j)$ are the numbers of images associated with t_i and t_j on Flickr respectively, $c(t_i, t_j)$ is the number of images associated with both t_i and t_j simultaneously, and M is the total number of images on Flickr. Therefore, the similarity of the query tag t_q and the tag set of image x_i can be computed as

$$sim(t_q, \mathcal{T}_i) = \frac{1}{|\mathcal{T}_i|} \sum_{t \in \mathcal{T}_i} sim(t_q, t) \quad (11)$$

Our second assumption is that the relevance scores of visually similar images should be close. The visual similarity between two images can be directly computed based on Gaussian kernel function with a radius parameter σ, i.e.,

$$W_{ij} = \exp(-\frac{\| x_i - x_j \|^2}{\sigma^2}) \quad (12)$$

Note that this assumption may not hold for several images, but it is still reasonable in most cases. Based on the two assumptions, we formulate a regularization framework as follows

$$Q(\mathbf{r}) = \sum_{i,j=1}^{n} W_{ij} \left(\frac{r(x_i)}{\sqrt{D_{ii}}} - \frac{r(x_j)}{\sqrt{D_{jj}}} \right)^2 + \lambda \sum_{i=1}^{n} (r(x_i) - sim(t_q, \mathcal{T}_i))^2$$
$$\mathbf{r}^* = arg \min Q(\mathbf{r}) \quad (13)$$

where $r(x_i)$ is the relevance score of x_i, and $D_{ii} = \sum_{j=1}^{n} W_{ij}$. The above equation can be written in matrix form as

$$Q(\mathbf{r}) = \mathbf{r}^T(\mathbf{I} - \mathbf{D}^{-1/2}\mathbf{W}\mathbf{D}^{-1/2})\mathbf{r} + \lambda \| \mathbf{r} - \mathbf{v} \|^2 \quad (14)$$

where $\mathbf{D} = Diag(D_{11}, D_{22}, \ldots, D_{nn})$ and $\mathbf{v} = [sim(t_q, \mathcal{T}_1), sim(t_q, \mathcal{T}_2), \ldots, sim(t_q, \mathcal{T}_n)]^T$. $\mathbf{r}^*$ can be obtained in an iterative way:

(1) Construct the image affinity matrix $\mathbf{W}$ by Eq. 12 if $i \neq j$ and otherwise $W_{ii} = 0$.
(2) Initialize $\mathbf{r}^{(0)}$. The initial values will not influence the final results.
(3) Iterate $\mathbf{r}^{(t+1)} = \frac{1}{1+\lambda}\mathbf{D}^{-1/2}\mathbf{W}\mathbf{D}^{-1/2}\mathbf{r}^{(t)} + \frac{\lambda}{1+\lambda}\mathbf{v}$ until convergence.

The method can be viewed as a random walk process, and it will converge to a fixed point.

4.2 Semantic Similarity Estimation

We define a semantic similarity for social images, which is estimated based on their associated tag sets. Note that we have obtained the similarity of tag pair in Eq. 10. Consequently, we estimate the semantic similarity of images x_i and x_j as

$$s(x_i, x_j) = \frac{1}{2|\mathcal{T}_i|} \sum_{k=1}^{|\mathcal{T}_i|} \max_{t \in \mathcal{T}_j} sim(t_k^i, t) + \frac{1}{2|\mathcal{T}_j|} \sum_{k=1}^{|\mathcal{T}_j|} \max_{t \in \mathcal{T}_i} sim(t_k^j, t) \quad (15)$$

We can see that the above definition satisfies the following properties:

(1) $s(x_i, x_j) = s(x_j, x_i)$, i.e., the semantic similarity is symmetry.
(2) $s(x_i, x_j) = 1$ if $\mathcal{T}_i = \mathcal{T}_j$, i.e., the semantic similarity of two images is 1 if their tag sets are identical.
(3) $s(x_i, x_j) = 0$ if and only if $sim(t', t'') = 0$ for every $t' \in \mathcal{T}_i$ and $t'' \in \mathcal{T}_j$, i.e., the semantic similarity is 0 if and only if every pair formed by the two tag sets has zero similarity.

5 Empirical Study

5.1 Experimental Settings

We evaluate our approach on a set of social images that are collected from Flickr. We first select a diverse set of popular queries, including *airshow*, *apple*, *beach*, *bird*, *car*, *cow*, *dolphin*, *eagle*, *flower*, *fruit*, *jaguar*, *jellyfish*, *lion*, *owl*, *panda*, *starfish*, *triumphal*, *turtle*, *watch*, *waterfall*, *wolf*, *chopper*, *fighter*, *flame*, *hairstyle*, *horse*, *motorcycle*, *rabbit*, *shark*, *snowman*, *sport*, *wildlife*, *aquarium*, *basin*, *bmw*, *chicken*, *decoration*, *forest*, *furniture*, *glacier*, *hockey*, *matrix*, *Olympics*, *palace*, *rainbow*, *rice*, *sailboat*, *seagull*, *spider*, *swimmer*, *telephone*, and *weapon*. For simplicity, we use 1 to 52 to denote the IDs of these queries, respectively. We then perform tag-based image search with "ranking by most recent" option, and the top 2,000 returned images for each query are collected together with their associated information, including tags, uploading time, user identifier, etc. In this way, we obtain a social image collection consisting of 104,000 images and 83,999 unique tags. But many of the raw tags are misspelling and meaningless. Hence, we adopt a pre-filtering process on these tags. Specifically, we match each tag with the entries in a Wikipedia thesaurus and only the tags that have a coordinate in Wikipedia are kept. In this way, 12,921 unique tags are kept for our experiment, and there are 7.74 tags associated with an image in average.

For each image, we extract 428-dimensional features, including 225-dimenaional block-wise color moment features generated from 5-by-5 fixed partition of the image, 128-dimensional wavelet texture features, and 75-dimensional edge distribution histogram features. The ground truth of the relevance of each image is voted by three human labelers. The radius parameter σ in Eq. 12 is empirically set to the median value of all the pair-wise Euclidean distances between images.

5.2 Experimental Results

We first compare the following three ranking methods:

(1) Time-based ranking, i.e., order the images according to their uploading time.

(2) Relevance-based ranking, i.e., order the images according to their estimated relevance scores $r(x_i)$.

(3) Diverse Relevance Ranking (DRR), i.e., the method proposed in this work.

Fig. 1 and 2 illustrate the AP and ADP measurements obtained by different methods, respectively. We also illustrate the mean AP (MAP) and mean ADP (MADP) measurements that are averaged over all queries. First, we observe Fig. 1 and it can found that the time-based ranking performs the worst in terms of relevance. This is understandable since the ranking list is generated merely based on time information. The relevance-based ranking performs much better, and this demonstrates the effectiveness of our relevance estimation method. The AP measurements of DRR degrade slightly in comparison with the relevance-based ranking. The MAP measurements of relevance-based ranking and DRR are 0.684 and 0.663, respectively. However, from Fig. 2 we can see that the ADP measurements of DRR are much higher. The MADP measurements of time-based ranking, relevance-based

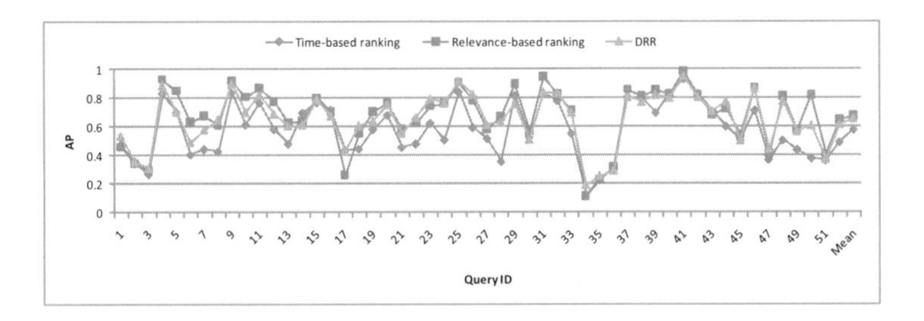

Fig. 1. The comparison of AP of different ranking schemes. The MAP measurements of time-based ranking, relevance-based ranking and DRR are 0.576, 0.676 and 0.655, respectively.

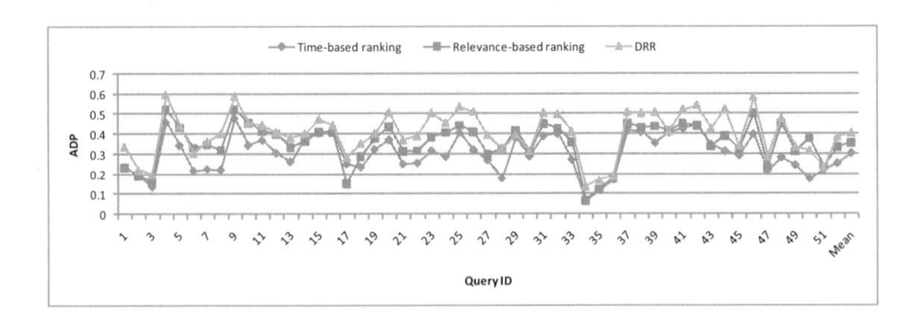

Fig. 2. The comparison of ADP of different ranking schemes. The MADP measurements of time-based ranking, relevance-based ranking and DRR are 0.304, 0.356 and 0.401, respectively.

Table 1. The average rating scores and variances converted from users study results

DRR vs. Time-based ranking		DRR vs. Relevance-based ranking	
DRR	Time-based	DRR	Relevance-based
2.40±0.386	1.03±0.033	**2.40±0.455**	1.133±0.189

Table 2. The ANOVA test results on comparing DRR and time-based ranking

The factor of ranking schemes		The factor of users	
F-statistic	p-value	F-statistic	p-value
108.57	2.58×10^{-11}	0.656	0.894

Table 3. The ANOVA test results on comparing DRR and relevance-based ranking

The factor of ranking schemes		The factor of users	
F-statistic	p-value	F-statistic	p-value
46.74	2.5×10^{-11}	0.25	0.999

ranking and DRR are 0.308, 0.361 and 0.412, respectively. This shows that DRR achieves a good trade-off between relevance and diversity.

We then conduct a user study to compare the three ranking schemes. To avoid bias, a third-party data management company is involved. The company invites 30 anonymous participants, who declare they are regular users of Internet and familiar with image search and media sharing websites. We ask them to freely choose queries and compare DRR with each of the other two ranking approaches. The users are asked to give the comparison results using ">", "≫" and "=", which mean "better", "much better" and "comparable". To quantitate the results, we convert the results into ratings. We assign score 1 to the worse scheme and the other scheme is assigned a score 2, 3 and 1 if it is better, much better and comparable than this one, respectively. The average rating scores and the variances are illustrated in Table 1. From the results we can clearly see the preference of users towards the DRR scheme. We also perform an ANOVA test, and the results are illustrated in Table 2 and Table 3. The results demonstrate that the superiority of the DRR is statistically significant.

6 Conclusion

This paper proposes a diverse relevance ranking scheme for social image search, which is able to simultaneously take relevance and diversity into account. It leverages both visual information of images and the semantic information of tags. The ranking algorithm optimizes an Average Diverse Precision (ADP) measure, which is generalized from the conventional AP measure by taking diversity into account. Experimental results have demonstrated the effectiveness of the approach. In addition, we have also shown the application of the DRR scheme in web image search diversification. In the future, we will test our scheme with more queries as well as comprehensively investigate the dependence of users' search experience with relevance and diversity.

References

1. Cai, D., He, X., Li, Z., Ma, W.Y., Wen, J.R.: Hierarchical clustering of www image search results using visual, textual and link information. In: Proceedings of ACM Multimedia (2004)
2. Carbonell, J., Goldstein, J.: The use of MMR, diversity-based reranking for reordering documents and producing summaries. In: Proceedings of SIGIR (1998)
3. Cilibrasi, R., Vitanyi, P.M.B.: The google similarity distance. IEEE Transactions on Knowledge and Data Engineering (2007)
4. Goffman, W.: A searching procedure for information retrieval. Information Storage and Retrieval 2 (1964)
5. Hsu, W.H., Kennedy, L.S., Chang, S.F.: Video search reranking via information bottleneck principle. In: Proceedings of ACM Multimedia (2006)
6. Jaimes, A., Chang, S.F., Loui, A.C.: Detection of non-identical duplicate consumer photographs. In: Proceedings of ACM Multimedia (2003)
7. Jing, F., Wang, C., Yao, Y., Deng, K., Zhang, L., Ma, W.Y.: Igroup: web image search results clustering. In: Proceedings of ACM Multimedia (2006)
8. Kennedy, L.S., Chang, S.F., Kozintsev, I.V.: To search or to label? predicting the performance of search-based automatic image classifiers. In: Proceedings of the 8th ACM international workshop on Multimedia information retrieval (2006)
9. Leuken, R.H.V., Garcia, L., Olivares, X., Zwol, R.: Visual diversification of image search results. In: Proceedings of WWW (2009)
10. Li, J., Wang, J.: Real-time computerized annotation of pictures. IEEE Transactions on Pattern Analysis and Machine Intelligence 30(6) (2008)
11. Li, X.R., Snoek, C.G.M., Worring, M.: Learning tag relevance by neighbor voting for social image retrieval. In: Proceeding of the ACM International Conference on Multimedia Information Retrieval (2008)
12. Song, K., Tian, Y., Huang, T., Gao, W.: Diversifying the image retrieval results. In: Proceedings of ACM Multimedia (2006)
13. Srinivasan, S.H., Sawant, N.: Finding near-duplicate images on the web using fingerprints. In: Proceedings of ACM Multimedia (2008)
14. Tang, J., Yan, S., Hong, R., Qi, G.-J., Chua, T.-S.: Inferring semantic concepts from community-contributed images and noisy tags. In: ACM Multimedia (2009)
15. Wang, B., Li, Z., Li, M., Ma, W.Y.: Large-scale duplicate detection for web image search. In: Proceedings of ICME (2006)
16. Wang, M., Hua, X.-S., Tang, J., Hong, R.: Beyond distance measurement: Constructing neighborhood similarity for video annotation. IEEE Transactions on Multimedia 11(3) (2009)
17. Wang, M., Yang, K., Hua, X.-S., Zhang, H.-J.: Visual tag dictionary: Interpreting tags with visual words. In: ACM Workshop on Web-Scale Multimedia Corpus, in association with ACM MM (2009)
18. Wu, L., Hua, X.S., Yu, N., Ma, W.Y., Li, S.: Flickr distance. In: Proceedings of ACM Multimedia (2008)
19. Zhai, C., Cohen, W.W., Lafferty, J.: Beyond independent relevance: methods and evaluation metrics for subtopic retrieval. In: Information Processing and Management (2006)

View Context: A 3D Model Feature for Retrieval

Bo Li and Henry Johan

School of Computer Engineering, Nanyang Technological University
Block N4, Nanyang Avenue, 639798, Singapore
{libo0002,henryjohan}@ntu.edu.sg

Abstract. 3D model feature extraction is an important issue for 3D object retrieval. We propose a novel 3D model feature named view context and based on this we construct a view context shape descriptor for 3D model retrieval. The view context of a particular view captures the distribution of visual information differences between this view and a set of arranged views. We select a set of feature views, compute their view contexts and use them as the shape descriptor of a 3D model. In order to enhance the retrieval accuracy, we perform an approximate symmetric axis-based 3D model alignment and propose a combined shape distance between two models, by incorporating the dissimilarity between two view context shape descriptors and the Zernike moments feature difference between the feature views of two models. Experiment results show that the view context shape descriptor is comparable with the state-of-the-art descriptors in retrieval performance.

Keywords: 3D model feature, view context, 3D model retrieval, shape deviation, visual information.

1 Introduction

With the increase in the number of available 3D models, the ability to accurately and efficiently search for 3D models is crucial in many applications. As a result, 3D model retrieval has become an important research area. In recent years, several algorithms that extract different 3D model features such as shape histogram [1], shape distribution [13], moment [5], Light Field [3], 3D harmonics [10], have been proposed for 3D model retrieval. Funkhouser et al. [6] present a multi-modal web-based search engine which supports queries based on 3D sketches, 2D sketches, 3D models and/or text keywords. Even though there is much work done on 3D model retrieval, it is still a challenging problem to find a good 3D model feature for retrieval.

In this paper, we propose a novel 3D model feature named "view context". When we look at a 3D model from a view V (i.e. the viewer is located at V), the visible features of the model are the important visual information of the model from this view V. We assume a 3D model is represented as a triangle mesh. We present two definitions of the visual information of a 3D model from a view V: (1) A hybrid point feature set composed of contours, suggestive contours [4], boundaries as well as silhouettes of the model as seen from the view V;

S. Boll et al. (Eds.): MMM 2010, LNCS 5916, pp. 185–195, 2010.

(2) Zernike moment features of the silhouette image rendered from the view V. The view context of a view V is then defined as the differences of the visual information between V and a set of arranged views.

We apply the view context for 3D model retrieval. We concentrate on 3D model retrieval using 3D models as queries. In order to apply view context in retrieval, we propose a view context shape descriptor of a 3D model. The proposed shape descriptor consists of the view contexts computed at several sampling views. To improve the retrieval accuracy, we first align the 3D model using a proposed simple approximate symmetric axis-based approach before computing the shape descriptor for a 3D model. To further improve the retrieval accuracy, we also add the difference between the feature views sets of two models. Our main contributions are as follows:

(1) We devise a new 3D model feature named view context for identifying models. Similar models in general have similar view contexts and the view contexts of models from different classes are often distinctively different.

(2) We propose a retrieval algorithm based on view context and through experiments verify that it can often achieve comparable retrieval performance with the state-of-the-art descriptors.

The remaining of this paper is organized as follows. In Section 2, we review the related work in 3D model retrieval. Section 3 describes in detail the idea of view context. In Section 4, we propose the view context shape descriptor. In Section 5, an algorithm for 3D model retrieval using the view context shape descriptor is explained. The results of retrieval experiments are demonstrated in Section 6. Section 7 contains the conclusions and lists some topics for future work.

2 Related Work

The existing 3D model retrieval techniques can be classified into two categories: geometry-based techniques and visual information-based techniques. Geometry-based techniques use the distribution of geometric elements, such as vertices and faces, or some intrinsic topological structures to characterize the features of 3D models while visual information-based techniques extract features based on the rendered view images.

Most of the previous work in 3D model retrieval belong to the geometry-based techniques. The key idea of shape distribution [13] is to represent the shape signature as a probability distribution sampled from a shape function measuring the global geometric properties of a 3D model. Shape histogram [1] is an extension of the shape matching techniques from 2D to 3D. It encodes the distance distribution of the surface points as a function of the distance from the center of mass and spherical angle. It includes three descriptors: SHELL (only use distance), SECTOR (only use spherical angle), and SHELL and SECTOR (use both). 3D harmonics [10] uses spherical harmonics to compute the similarity between 3D models without the need to align their orientations. It decomposes a spherical function into orthogonal components while preserving the norms.

In the visual information-based techniques, the visual similarities between view images of different models are compared with each other to measure the difference of the models. Multiple view descriptor [7], Light Field descriptor [3] and our view context shape descriptor belong to this category. Multiple view descriptor classifies models by comparing the views rendered from the primary, secondary and tertiary viewing directions of principle axes after an alignment with PCA. Chen et al. [3] proposed the Light Field descriptor to define the distance of two models as the minimum distance between their 10 corresponding silhouette views, rendered from the vertices of a dodecahedron using the orthographic projection. The features in each image are encoded using the Zernike moments and Fourier descriptor. Their Light Field descriptor compares the views of different models directly, while our view context shape descriptor compares and encodes the difference of views of the same model first and then we compare the view context features of different models to measure their difference.

3 View Context

3.1 Definition

The view context of a particular view of a 3D model encodes the visual information differences between this view and a set of arranged views. It captures the shape appearance deviation of a 3D model. In our work, a 3D model is represented as a triangle mesh.

Assume that the 3D model is centered in the origin of a 3D coordinate system. Its view context from a view V_0 is defined as follows. First, we rotate the 3D model such that view V_0 coincides with the z-axis of the coordinate system. Then, we orderly sample a set of views $\{(\varphi, \theta)\}$ based on the current initial pose. For example, view (φ, θ) can be generated by first rotating the model φ degrees about the x axis and then θ degrees about the y axis. The view context of a view V_0 is composed of a set of feature vectors:

$$\{(\varphi, \theta, d)|(\varphi, \theta) \in \mathbf{V}\} \ , \tag{1}$$

where d is the view appearance distance between view (φ, θ) and view V_0. $\mathbf{V}$ is a sequence of sampling views. The methods of view sequence sampling and view distance computation will be presented in Sections 3.2 and 3.3, respectively.

View context represents the relative appearance deviations with other views. Figure 1 gives an example of the view contexts of the initial poses of six models (Figures 1(a)~(f)) in Princeton Shape Benchmark Database [14]. Figures 1(g)~(l) show the matrix representation of the six view contexts. In these examples, the view sequence $\mathbf{V}$ consists of 12 views based on the uniform step sampling (Section 3.2) and it is organized into a 3×4 matrix. The top-left element represents the view V_0 and other elements indicate other relative views with respect to V_0. The gray scale values represent the view appearance distances and darker means smaller values. Figure 1(m) shows the plot of the six view context. The graphs are obtained by ordering the elements of the matrix based on the following rule: from top to bottom

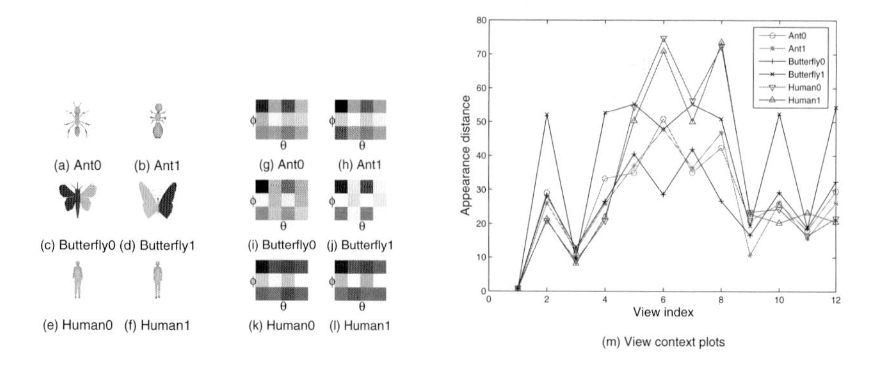

Fig. 1. View context of six models

and from left to right. We can see that similar models will have similar view contexts and the view contexts of different models are often distinctively different.

3.2 View Sequence Sampling

To decide the view sequence **V**, we need to determine the values of φ and θ in Equation 1. The ranges of their values are $\varphi \in [0, 180]$ and $\theta \in [0, 360)$. To consider different types of features extracted, we adopt two sampling methods: uniform step-based sampling and cube-based sampling. The former is used for contour-based features, and the latter is for region-based features.

Uniform Step-Based Sampling. Using sampling steps $\Delta\varphi$ and $\Delta\theta$, the (φ, θ) for a view sequence **V** can be defined as follows.

$$\{(i * \Delta\varphi, j * \Delta\theta) | 0 \leq i \leq \left\lfloor \frac{180}{\Delta\varphi} \right\rfloor, 0 \leq j < \left\lfloor \frac{360}{\Delta\theta} \right\rfloor\} \ . \qquad (2)$$

The view sequence **V** is written as $\mathbf{V} = \{V_0, V_1, \cdots, V_{n-1}\}$, where $n = (\left\lfloor \frac{180}{\Delta\varphi} \right\rfloor + 1) * \left\lfloor \frac{360}{\Delta\theta} \right\rfloor$. In our experiments, we set $\Delta\varphi = 90$ and $\Delta\theta = 90$. Then, the view sequence **V** consists of the following 12 views.

$$\{(\varphi, \theta) | \varphi \in \{0, 90, 180\}, \theta \in \{0, 90, 180, 270\}\} \ . \qquad (3)$$

Cube-Based Sampling. By adopting region-based image descriptors, for every view we only need to generate a sequence of silhouette images rendered from a set of cameras using orthogonal projection. Considering the symmetry property of the projection, we render 13 views by setting the cameras on the surface of a cube. The camera locations are (1,1,1), (-1,1,1), (-1,-1,1), (1,-1,1), (1,0,0), (0,1,0), (0,0,1), (1,0,-1), (0,1,-1), (1,1,0), (0,1,1), (1,0,1), (1,-1,0). They comprise 4 top corner views, 3 adjacent face center views and 6 middle edge views. Based on these camera locations, we then compute the view sequence $\{(\varphi, \theta)\}$.

3.3 View Appearance Distance

To compute the view appearance distance d between two view images V_0 and V_i, we use hybrid feature matching cost if extracting contour-based features and use Zernike moment distance if extracting region-based features.

Hybrid Feature Distance. We first extract a hybrid feature point set of a view and then use a 2D shape matching algorithm to compute the view distance.

(1) **Feature extraction.** We define a hybrid feature set of a view image (Figure 2) by integrating contours, suggestive contours [4], boundaries as well as silhouettes. Contours are the point sets whose normal vectors are perpendicular to the view vector. Suggestive contours are contours in nearby views. A view image that includes the suggestive contours already integrates the contour features in the nearby views. Therefore, it facilitates a sparse view sampling and thus improves the efficiency of 3D model retrieval. Beside contours and suggestive contours, boundaries and silhouettes are also essential to accurately represent the features of 3D models. Boundary comprises edges that only belong to a single triangle face. Silhouette is an outline of the model from a certain view. Figure 2 shows one example of these features.

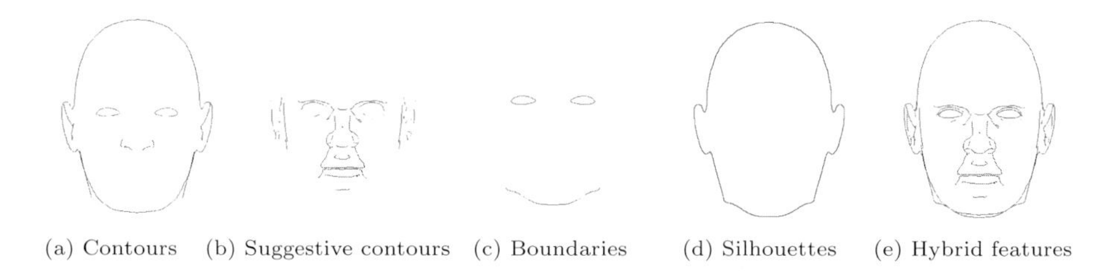

(a) Contours (b) Suggestive contours (c) Boundaries (d) Silhouettes (e) Hybrid features

Fig. 2. One example of the features of a view image

(2) **Feature sampling.** To reduce the time for feature matching, we need to first resample the feature points. In our experiments, we sample 100 points for every view image by adopting the following feature points sampling steps: curve extraction, cubic B-Spline curve interpolation and uniform sampling. During the uniform sampling, we set the number of sampling points for each curve to be proportional to its original length (in pixel).

(3) **View distance computation.** After sampling two feature point sets from the view images V_0 and V_i, we use the shape context matching algorithm [2] to compute the view distance of these two views.

Zernike Moments Distance. We use moments to extract the feature of the silhouette views V_0 and V_i. To achieve rotation invariance when measuring the difference of two images, we adopt the Zernike moments [11] and compute up to 10th order moments (36 moments in total) for the feature of every view image. To compare the difference between two moments feature vectors, we adopt the Canberra distance [12]: $\sum_{i=1}^{36} |x_i - y_i|/|x_i + y_i|$, where $(x_1, x_2, \ldots, x_{36})$ and $(y_1, y_2, \ldots, y_{36})$ are the Zernike moment feature vectors of two views V_0 and V_i.

4 View Context Shape Descriptor

4.1 Definition

To build the view context shape descriptor for a 3D model, we can first align the model (Section 4.2) or use the original database directly, then select a set of feature views and compute the view contexts of these views, and finally concatenate them into a matrix as the view context shape descriptor.

For the feature views set selection, generally, the front, right, back and left views are the standard views for viewing a 3D model. Since we can perform an approximate alignment (Section 4.2) on the 3D model beforehand, we can only select these 4 views. However, if we do not perform the alignment, we select the 13 cube-based sampling views (Section 3.2) as our feature views.

4.2 Symmetric Axis-Based Alignment

Symmetry is an important factor of human perception to align similar models in similar poses. Considering this, we propose a simple method to align a 3D model based on the approximate symmetric-axis. The algorithm is as follows:

(1) **Align the 3D model using the Principal Component Analysis method [8].**

(2) **Find the best approximate symmetric axis.** We only consider the x, y and z axes after the alignment in step (1). We assume the vertices of the model are $\{v_i =< v_{i,x}, v_{i,y}, v_{i,z} > |i = 1, \cdots, t\}$, where t is the number of vertices. First, we find the best symmetric axis A by summing up x, y and z coordinates of all the vertices respectively and then find the axis that has the minimum absolute value,

$$A = argmin\{|\sum_{i=1}^{t} v_{i,x}|, |\sum_{i=1}^{t} v_{i,y}|, |\sum_{i=1}^{t} v_{i,z}|\} \ . \tag{4}$$

If A=$x/y/z$ (/ means or), it means that the symmetric plane is roughly yoz /zox /xoy. Then, we need to find the axis (y/z, z/x, x/y) that has bigger variance (larger eigenvalue) and regard it as the best symmetric axis.

(3) **Rotate the model such that the best symmetric axis coincide with the y axis.** If the three eigenvalues are all negative, we flip the model about the x axis.

We tested this simple and approximate algorithm on the 3D models in the Princeton Shape Benchmark Database [14] and found that in most cases it can align similar models in similar way except for some reflection errors about x axis. These errors may be due to the ambiguity of the principle axes, which is an intrinsic shortcoming of PCA. Some alignment results are shown in Figure 3.

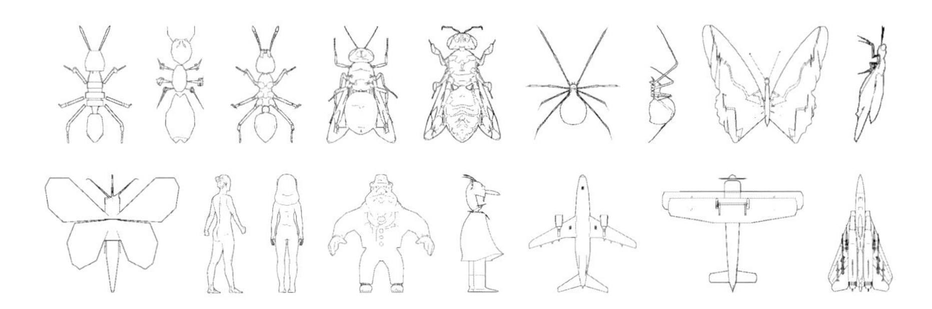

Fig. 3. Some alignment results

5 3D Model Retrieval with View Context

5.1 Retrieval Algorithm

We focus on retrieval using 3D models as queries. Given a query model, we propose a 3D model retrieval algorithm as follows.

(1) **Coordinates normalization.** To achieve translation and scale invariance, we translate the centers of all the models to the origin of the coordinate system and then scale them such that their bounding spheres have the same radius.

(2) **Pose normalization.** To enhance the retrieval accuracy, we perform the approximate symmetric axis-based 3D model alignment (Section 4.2).

(3) **Compute view context shape descriptor.** First, we select a set of feature views (e.g. 4 or 13), then compute the view context for each feature view and finally construct a view context shape descriptor by concatenating them.

(4) **Compute the shape distance matrix and ranking.** We design two shape distance metrics to measure the difference between two types of view context shape descriptors. We also propose a combined shape distance by combining the dissimilarity between two view context shape descriptors and the Zernike moments feature difference between two models' feature views sets. We describe these three distances in Section 5.2.

5.2 Shape Distance Metrics

Two candidate metrics that can be used to measure the distance between two view contexts are correlation and χ^2 distance. After comparing their differentiation capabilities through experiments, we found that correlation performs better. Therefore, we use correlation to measure the difference of two view contexts. As depicted in Section 4.1, we may choose 4 standard views (front/right/back/left) or 13 cube-based sampling views as feature views set. Accordingly, we design one shape distance metric for each, described as follows.

Ordered Correlation Distance for 4 Feature Views. The initial pose of a model can be in any view in the feature views set (front/right/back/left).

Therefore, to achieve rotation invariance among these 4 views, we need to design a metric that can encompass these differences. We assume that the feature views of two models are $\mathbf{V}=\{V_0, V_1, V_2, V_3\}$ and $\widetilde{\mathbf{V}}=\{\widetilde{V}_0, \widetilde{V}_1, \widetilde{V}_2, \widetilde{V}_3\}$. We design an ordered correlation distance d_v based on the best correspondence between the feature views of two models,

$$d_v = \min_i\{\sum_{j=0}^{3} ViewCost[j][(i+j) \bmod 4]\}, \;\; i \in \{0,1,2,3\} \; . \tag{5}$$

$ViewCost[i][j] = 1 - C(V_i, \widetilde{V}_j)$ and $C(V_i, \widetilde{V}_j)$ is the correlation between the view contexts of V_i and $\widetilde{V}_j$.

LAP Correlation Distance for 13 Feature Views. For 13 feature views, we get a 13 × 13 view cost matrix $ViewCost$ and we use the Jonker's Linear Assignment Problem (LAP) algorithm [9] to correspond these two sets of feature views and use the minimal matching cost as the distance between them.

Combined Shape Distance. To improve the retrieval performance, we can also consider the Zernike moments feature difference between two feature views sets. We propose a combined shape distance which combines the dissimilarity between two view context shape descriptors, depicted by d_v, and the Zernike moments feature difference between two sets of feature views, depict by d_m. d_m is computed in the same way as view context shape descriptor except that we use the Canberra distance [12] rather than the correlation distance. We combine the two distances based on an automatic weighting method according to the differentiation ability of each type of distance. First, we normalize d_v and d_m into $\widetilde{d}_v$ and $\widetilde{d}_m$ by their respective maximum distances. Then, we compute the weights ω_v and ω_m for the view context feature distance and moment feature distance,

$$\omega_v = \frac{s_1}{s_1 + s_2}, \;\; \omega_m = \frac{s_2}{s_1 + s_2} \; . \tag{6}$$

s_1 and s_2 are the standard deviations of d_v and d_m over all the models in the database. Finally, we combine these two normalized features by their corresponding weights,

$$d = \omega_v * \widetilde{d}_v + \omega_m * \widetilde{d}_m \; . \tag{7}$$

6 Experiments and Discussion

To investigate the retrieval performance and the characteristics of our view context shape descriptor, we tested our view context descriptor on the Princeton Shape Benchmark Database (PSB) [14] and the National Taiwan University Benchmark (NTU) [3] and compared it with other state-of-the-art or related descriptors. We mainly use the precision-recall plots [14] to measure the retrieval performance. Recall means how much percentage of a class has been retrieved among the top K list while precision indicates how much percentage of the top K models belongs to the same class as the query model.

6.1 Princeton Shape Benchmark Database (PSB)

We used the test dataset of the PSB database [14]. It contains 907 models which are classified into 131 classes. Figure 4 gives the precision-recall plots of several variations of our proposed view context based retrieval algorithm as well as other three shape descriptors. For the precision-recall plots of D2 [13] and SHELL [1], we referred to the experiment results in [14]. For Light Field descriptor [3], we generated the plots using their provided execution file. Figure 4(a) shows the results on the approximately aligned PSB database, while Figure 4(b) shows the results on the original PSB database. We can see for this case, compared with the Light Field descriptor, view context shape descriptor can achieve better or comparable performance when retrieving a certain percentage of models (for example around 20 percent). Compared to the shape distribution (D2) and SHELL descriptor (one type of shape histogram descriptor), view context descriptor performs apparently better. We also found that for the same method (e.g. View Context-M-13), the approximate alignment process can improve the performance.

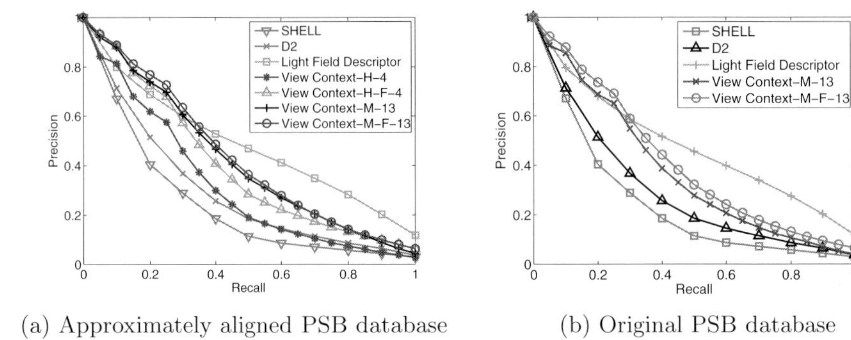

(a) Approximately aligned PSB database (b) Original PSB database

Fig. 4. Precision-recall plots of our view context and other three descriptors: (a) Approximately aligned PSB database; (b) Original PSB database. "H" means the hybrid feature based view context and "M" means the Zernike moments based view context. "4": use the uniform step-based sampling and the 4 standard views as feature views set. "13": use the 13 cube-based sampling views as feature views set. "F": combined shape distance by integrating the differences in the view context shape descriptors and the feature views' Zernike moments.

6.2 National Taiwan University Database (NTU)

This database [3] contains 1833 3D models and only 549 3D models are clustered into 47 classes and the rest 1284 models are classified as the "miscellaneous". Therefore, we do the approximate alignment only for the 549 models. Since there are no PCA information in the database, we complete the approximate alignment process manually.

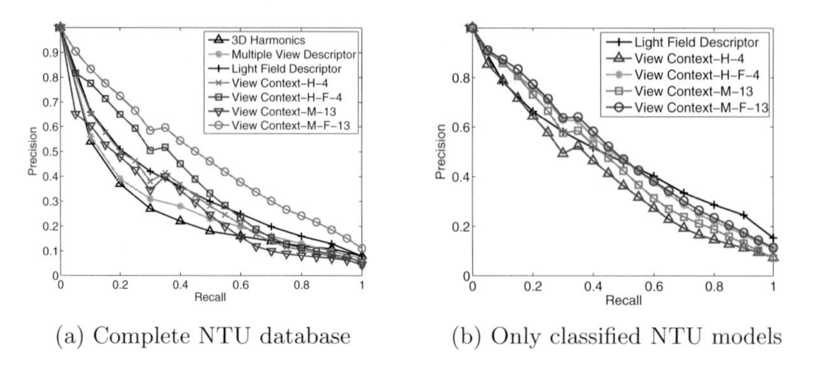

(a) Complete NTU database (b) Only classified NTU models

Fig. 5. Precision-recall plots of our view context and other three descriptors: (a) 1833 approximately aligned NTU database; (b) 549 approximately aligned NTU database (only classified models). The denotations of "H" ,"M" ,"F", "4" and "13" are described in Figure 4.

For the full approximately aligned database (1833 models), we compared our view context with the three descriptors mentioned in [3]: Light Field [3], 3D harmonics [10], and multiple view descriptor [7]. Figure 5(a) shows the comparison of the performance. For Light Field descriptor, we generated the plots using their provided execution file, while for 3D harmonics and multiple view descriptor, we referred to the experiment results in [3]. We can see that our view context shape descriptor can achieve a similar and comparable performance as the Light Field descriptor. It outperforms the 3D harmonics and the multiple view descriptor on average.

Figure 5(b) gives the performance comparison between our four methods and Light Field descriptor over the approximately aligned 549 classified models database. We can see that adding more feature views or integrating the Zernike moments difference of feature views sets can apparently improve the accuracy. Overall, we can achieve better results than the Light Field.

7 Conclusions and Future Work

We have presented a new 3D model feature: view context, which captures the shape deviation distribution feature of a 3D model. It can differentiate models effectively because similar models have similar view contexts and different models in general have apparently different view contexts. To improve the retrieval accuracy, we perform an approximate alignment and propose a combined shape distance. Experiment results show that the view context shape descriptor is comparable with the state-of-the-art descriptors in retrieval performance.

There are still many facets about the view context to be explored. For instance, we can adopt a different view sampling method by setting the cameras on the 20 vertices of a regular dodecahedron. We can also organize a view context scale-space by dividing the view space uniformly at a series of scales, arranged from coarse to fine. For example, we set the uniform sampling steps $\Delta\varphi = \Delta\theta = 60°, 30°, 15°$ and then define three view context features for a model at different scales, which are 3×6, 6×12 and 12×24 matrices.

References

1. Ankerst, M., Kastenmuller, G., Kriegel, H.P., Seidl, T.: 3D shape histograms for similarity search and classification. In: Güting, R.H., Papadias, D., Lochovsky, F. (eds.) SSD 1999. LNCS, vol. 1651, pp. 207–226. Springer, Heidelberg (1999)
2. Belongie, S., Malik, J., Puzicha, J.: Shape matching and object recognition using shape contexts. IEEE Transactions on Pattern Analysis and Machine Intelligence 24(4), 509–522 (2002)
3. Chen, D.Y., Tian, X.P., Shen, Y.T., Ming, O.: On visual similarity based 3D model retrieval. In: Eurographics, Computer Graphics Forum, pp. 223–232 (2003)
4. Doug, D., Adam, F., Szymon, R., Anthony, S.: Suggestive contours for conveying shape. ACM Transactions on Graphics 22(3), 848–855 (2003)
5. Elad, M., Tal, A., Ar, S.: Content based retrieval of VRML objects - an iterative and interactive approach. In: Proc. of 6th Eurographics Workshop on Multimedia, pp. 97–108 (2001)
6. Funkhouser, T., Min, P., Kazhdan, M., Chen, J., Halderman, A., Dobkin, D., Jacobs, D.: A search engine for 3D models. ACM Transactions on Graphics 22(1), 83–105 (2003)
7. Jeannin, S., Cieplinski, L., Ohm, J.R., Kim, M.: MPEG-7 Visual part of eXperimentation Model Version 7.0. ISO/IEC JTC1/SC29/WG11/N3521, Beijing (2000)
8. Jolliffe, I.: Principal Component Analysis, 2nd edn. Springer, Heidelberg (2002)
9. Jonkerand, R., Volgenant, A.: A shortest augmenting path algorithm for dense and sparse linear assignment problems. Computing 38(4), 325–340 (1987)
10. Kazhdan, M., Funkhouser, T., Rusinkiewicz, S.: Rotation invariant spherical harmonic representation of 3D shape descriptors. In: Proc. of the Symposium on Geometry Processing 2003, pp. 156–164 (2003)
11. Khotanzad, A., Hong, Y.: Invariant image recognition by Zernike moments. IEEE Transactions on Pattern Analysis and Machine Intelligence 12(5), 489–497 (1990)
12. Laga, H., Nakajima, M.: Supervised learning of similarity measures for content-based 3D model retrieval. In: Tokunaga, T., Ortega, A. (eds.) LKR 2008. LNCS (LNAI), vol. 4938, pp. 210–225. Springer, Heidelberg (2008)
13. Osada, R., Funkhouser, T., Chazelle, B., Dobkin, D.: Matching 3D models with shape distributions. In: Proc. of Shape Modeling and Applications 2001, pp. 154–166 (2001)
14. Shilane, P., Min, P., Kazhdan, M., Funkhouser, T.: The princeton shape benchmark. In: Proc. of Shape Modeling and Applications 2004, pp. 167–178 (2004)

Scene Location Guide by Image-Based Retrieval

I-Hong Jhuo[1,2], Tsuhan Chen[3], and D.T. Lee[1,2]

[1] Department of Computer Science and Information Engineering, National Taiwan University, Taipei, Taiwan
[2] Institute of Information Science, Academia Sinica, Taipei, Taiwan
[3] School of Electrical and Computer Engineering, Cornell University, Ithaca, NY 14853, USA

Abstract. In this paper, we propose a new image-based algorithm to identify where a tourist is when visiting unfamiliar places. When the tourist takes a photo of an unfamiliar place, our algorithm can recognize where the tourist is by retrieving similar images from an image database, where location information is associated with each image. Our method is not only fusing global and local information but using a coarse-to-fine three-stage search process. We first extract image descriptors from the image taken by the tourist and retrieve a number of most relevant images from the database. Then, we re-rank these relevant images based on geometric consistency. Finally, our method determines where the tourist is by using an image-to-class distance measure. Promising performance of the proposed algorithm is demonstrated by the experiments.

Keywords: Image Retrieval, Image-to-Class, Information Fusion.

1 Introduction

"Where are we?" It is a frequently asked question when tourists visit unfamiliar places. Since camera-equipped mobile devices are now almost ubiquitous, it seems applicable to utilize these devices to identify where tourists are with an image-based positioning system.

With such devices, tourists can take a photo of a prominent building or a scene spot. This image can be used as a query to find similar images from a database consisting of pre-collected images using content-based image retrieval (CBIR) techniques [18,20,22]. CBIR techniques analyze an image in its context, where shapes, colors, textures, or any other features may be used, to retrieve similar images from a collection on the basis of syntactical image features. Recently, CBIR methods extract salient features as multi-dimensional descriptors from images and then cluster these descriptors into vocabularies of bag-of-word (BoW) [11]. In fact, such BoW feature has become a dominant representation of images for both object categorization and scene classification [11,5,16,10,2]. The assumption of bag-of-word is that a different scene can generally maintain the co-occurrence of a number of visual components which play as the role of 'visual textures'. Therefore, an image is composed of a collection of local features which are computed on interest points or on points in a densely sampled grid. The

S. Boll et al. (Eds.): MMM 2010, LNCS 5916, pp. 196–206, 2010.

orderless model has a successful application to scene classification and achieves promising performance. However, such models still focus on local features and ignore global information about spatial information. It may limit its descriptive abilities.

There are other retrieval methods for location identification in the literature. A hybrid keyword-and-image system benefiting from both modalities is proposed in [21]. Based on a data-driven scene matching approach, a simple algorithm is proposed to estimate a distribution over geographic locations from a single image [8]. Multi-view geometric feature-based matching approaches have also been applied to location recognition [19,13,17]. Specially, the holistic image analysis provides another view to recognize [14,15]. For scene perception, Oliva and Torralba proposed a "shape of a scene" concept which regards a scene as an object with a unitary shape. They found that the shape of a scene, which is spatial information, can be inferred from its spatial layout and plays an important role in scene understanding. The concept of *spatial envelope* achieved promising performance on the scene classification. In other words, they all provide different viewpoints for visual processing.

In order to consider both the local and global properties, we try to fuse salient features that include not only global spatial information but also local feature properties for improving retrieval abilities in our experiment. Furthermore, we also present a hierarchical approach, which is a coarse-to-fine image-based method for gradually filtering out the irrelevant images in building identification. There are three stages in our method. In the first stage, we compute and organize each image's contextual information for coarse irrelevant image filtering. In the second stage, we apply the RANSAC algorithm [6] to refine correspondence matching and re-rank the relevant images retrieved in the first stage. In the final stage, the location of the query image is determined by an *image-to-class* [1] distance measure, where each distinct location of the relevant images is represented as a distinct class.

This paper is organized as follows: in Section 2, we will describe how to fuse global and local information for roughly filtering out irrelevant images. In Section 3, we will exploit geometric consistency for refining correspondence matching and re-ranking the remaining images. In Section 4, we will apply an *image-to-class* method to determine the location of the query image. And in Section 5, we will show our experimental results and compare our approach to existing ones. Finally, the conclusion will be depicted in Section 6.

2 Stage I: Coarse Irrelevance Filtering

The goal of this stage is to filter out from the database as many images irrelevant to the query as possible. With a restricted computational cost at this coarse stage, we only require a low *false negative* rate of the excluded images since the remaining images will be further verified in the successive stages. The task of the stage can be accomplished by thresholding the similarities to the query based on some off-the-shelf image descriptors. Motivated by the fact that it is generally difficult to find a descriptor which gives good performances for all images in a large

(a) (b)

Fig. 1. (a) Two frames showing the same building from different camera view points, with feature points extracted by Harris corner detection, and their correspondences, and (b) final geometrically consistent points as evaluated by the RANSAC algorithm

dataset, we adopt several image descriptors for feature extraction, each of which captures distinctive visual cues, such as shape, color, and some perceptually sensitive properties. In the following, we first give a brief introduction to the three adopted descriptors, and then provide an effective way for their combination.

2.1 Pyramid HOG

Shape-based features provide a strong evidence for image categorization, and this phenomenon has been reported in the literature of object recognition. To utilize the discriminant power of shape information, we adopt the *PHOG* (pyramid histogram of oriented gradients) descriptor [3] for shape feature extraction. We set the number of bins of the histograms to 8 and the number of levels in the pyramid to 3, and use χ^2 distance to measure the dissimilarity between two images.

2.2 Color Histogram

Inspired by image retrieval literature, we also consider the use of color information for image similarity measure. It makes sense in the application since many buildings have their distinctive distributions over colors. To this end, we implement color histograms in CIE *Lab* color space, and set the numbers of bins to 21, 40, and 40 for channels L, a, and b respectively. For images under the representation, χ^2 distance is also used as the dissimilarity measure.

2.3 Gist

We adopt the *gist* descriptor [15] as the third feature for its compactness and high performance. The gist descriptor performs Fourier transform analysis to each individual sub-region of an image, and the image is then summarized by a set of perceptual properties. The usefulness of gist has been demonstrated in a broad range of applications, such as scene categorization [14] and image retrieval. Euclidean distance is used to estimate the distance between a pair images under gist descriptor.

Fig. 2. (a) The query. (b) The responsiveness map of the three images (c $\sim$ e) to the query. It illustrates that none of the three images individually explains the query well, but they do jointly. (c) $\sim$ (e) Three images of the building that pass the first two stages.

2.4 Descriptor Fusion

The three image descriptors capture diverse image properties and complement each other. However, the relative importance among them mostly depends on the dataset under consideration. To decide the relative importance for descriptor fusion, we define the distance between the query $\mathbf{q}$ and some images $\mathbf{x}$ in the database as

$$d(\mathbf{q}, \mathbf{x}) = d_{PHOG}(\mathbf{q}, \mathbf{x}) + \alpha \cdot d_{Lab}(\mathbf{q}, \mathbf{x}) + \beta \cdot d_{gist}(\mathbf{q}, \mathbf{x}), \quad (1)$$

where constants α and β determine the weights for the corresponding descriptors. Their optimal values can be decided by using the five-fold cross validation method. By thresholding the distances of images to the query, we can exclude a large portion of irrelevant images while keeping relevant ones.

3 Stage II: Geometric Consistency Checking

Since the distance function in the first stage does not reflect the geometric consistency, we would like to check the geometric consistency by robust feature matching in this stage. We first extract local feature points from the query and the top k relevant images and then perform robust correspondence matching between the query image and each relevant image. We re-rank the top k relevant images according to the number of matched feature points.

Specifically, we detect spatial Harris features [7] from the images. For correspondence matching, we adopt the RANSAC algorithm in [6], where a fundamental matrix is built to remove outliers. The algorithm randomly samples matched points and iteratively estimates the parameters until the fundamental matrix is found.

Figure 1(a) shows the corresponding points without removing outliers, while Figure 1(b) shows corresponding points by the robust matching algorithm. To verify the effectiveness of the re-ranking method in this stage, we use the Normalized Discounted Cumulative Gain (NDCG) [9] measure to evaluate the quality of the top k relevant images before and after the re-ranking. The NDCG score is defined as

$$S_k = \sum_{j=1}^{k} \frac{\delta(c_j, c_q)}{log_2(1+j)} \tag{2}$$

where j is the the index of the jth relevant image, c_j and c_q are the classes of the jth relevant image and the query image, respectively. $\delta(c_j, c_q)$ is 1 if $c_j = c_q$, 0 otherwise.

4 Stage III: Decision Making

Unlike prior stages in which *image-to-image* similarities are computed, the concept of *image-to-class* similarities [1] is employed in the stage. That is, the query is interpreted not by a single image *individually* but instead by the set of survival images *jointly*.

We argue that the computation of image-to-class distances makes sense in our application, and this point is illustrated in Figure 2. The query and the three survivals of the building images, i.e., those that pass the first two stages, are shown in Figure 2a, 2c $\sim$ 2e respectively. According to the *responsiveness map* in Figure 2b, it can be seen that although none of the three survivals individually explain the query well, they do *show* a good match jointly.

To implement the idea, we apply the *DoG* detector [12] to the query $\mathbf{q}$, and use the *SIFT* descriptor [12] to depict each detected point. That is $\mathbf{q} = \{\mathbf{d}_1, \mathbf{d}_2, ..., \mathbf{d}_n\}$, where n is the number of detected points and $\mathbf{d}_i$ is the feature vector of point i. Except for the query, we pre-compute the same representation for each image in the database. Now we are ready to compute the image-to-class distance from query $\mathbf{q}$ to some class (building) c. Let $S = \{\mathbf{x}_1, \mathbf{x}_2, ..., \mathbf{x}_m\}$ denote the collection of the feature vectors of detected points in the survival images that belong to class c. Then the image-to-class distance $d_{I2C}(\mathbf{q}, c)$ is defined as follows:

$$d_{I2C}(\mathbf{q}, c) = \sum_{i=1}^{n} \|\mathbf{d}_i - NN_S(\mathbf{d}_i)\|^2, \text{ where} \tag{3}$$

$$NN_S(\mathbf{d}_i) = \arg\min_{\mathbf{x}} \{\|\mathbf{x} - \mathbf{d}_i\|^2 | \mathbf{x} \in S\}. \tag{4}$$

Fig. 3. Sample images in the dataset, each row representing images taken from different views around the same building

We compute the distances from the query to all the classes in which there are still images left, and complete the prediction by finding the shortest one.

5 Experimental Result

The performance of the proposed approach is evaluated in the section. In the following sections, we describe the used dataset, experiment settings, and quantitative results.

5.1 Experimental Dataset

The motivation of our task is to help a tourist identify the location when visiting unfamiliar places. Therefore, we collect an image database that consists of 13 distinct buildings located in our campus for performance evaluation. The number of images of each building ranges between 75 and 85, and these images have intraclass variations caused by serval different factors, such as photographying viewpoints, scales, and camera settings. We consider images from the same building as relevant, and irrelevant otherwise. The rule serves as the groundtruth. Some examples are given in Figure 3.

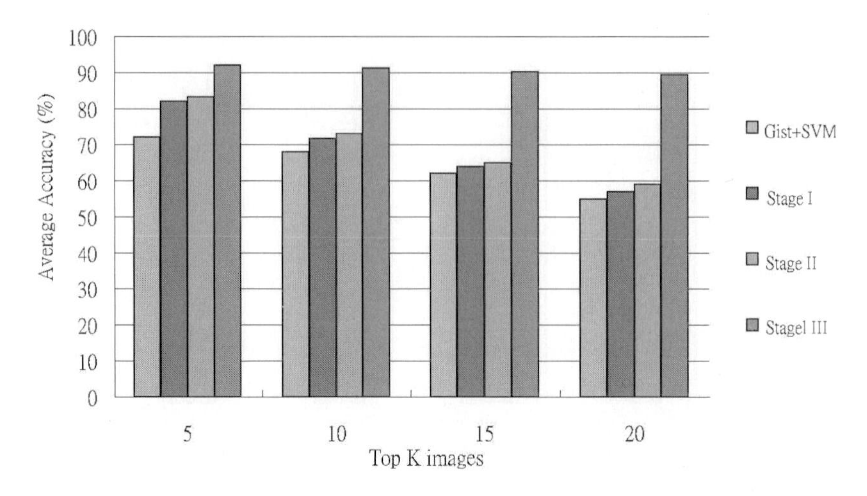

Fig. 4. Accuracy rates broken down into top five, ten, fifteen,and twenty images, respectively. As we can see, image-to-class method present an effective performance at each first k images.

5.2 Results

In this section we present experiments for building recognition. We show that our proposed approach gives promising performance over the state of the art method in this task. We concurrently implement global feature representation, Gist with SVM [4] (Gist+SVM) and two kinds of local information, Color with SVM (Color+SVM) and bag-of-features [11] as our baseline in this experiment. In addition, for each image, we build joint histograms of color in CIE L*a*b color space.

To analyze the effectiveness of the three-stage retrieval process, we report the performance stage by stage. All the results in the following are measured by using five-fold cross validation. In the first stage, the query image is compared to those in the dataset, and the most similar k images are selected. Then these k images are re-ranked by confirming their geometric consistency to the query in the second stage. By setting the value of k to 5, 10, 15, and 20 respectively, the average NDCG scores, defined in (1), are reported in Table 1. According to the scores, many irrelevant images are filtered out in the first stage, and geometric consistency checking in the second stage is helpful for score improvement.

In Figure 4, we report the accuracy rates of both "Gist+SVM" approach and our proposed method in the three stages respectively. Obviously, the usage of image-to-class distances in the third stage significantly improves the effectiveness of our approach. Although the computational cost of image-to-class distances is relatively high, it doesn't limit the applicabilities of our method since image-to-class distances are computed only for images that pass the first two stages, instead of for the images in the whole dataset. We compare our strategy with the state of the art model for this task. For the "Gist+SVM", we use four-fifth

Table 1. NDCG Score

Top k	5	10	15	20	25	30
Stage I	2.323	3.241	3.811	4.305	4.845	5.245
Stage II	2.354	3.303	3.883	4.415	4.932	5.313

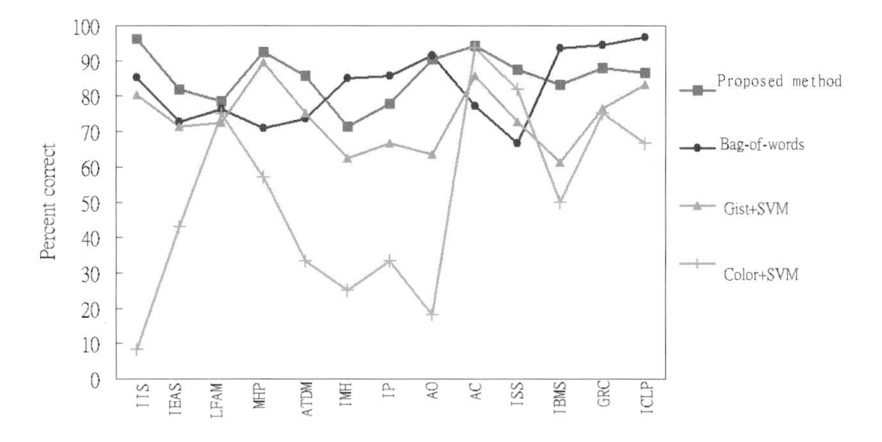

Fig. 5. Comparison with "Gist+SVM", "Color+SVM" and bag-of-words with our dataset. Average performance for the different methods are: our proposed approach: 86.26%, bag-of-words: 82.29%, "Gist+SVM": 73.94%, and "Color+SVM":50.79%.

images of each class/building for training and one-fifth images for testing. As we can see from Figure 4, our approach fusing local and global information shows promising performance.

The confusion matrix in Table 2 depicts the classification performance of our proposed approach. The average classification performance is 86.26%. We compare our result with those of the state of the art approaches including bag-of-words [11,5,16,10,2], "Gist+SVM" [14,15] and "Color+SVM" approaches in Figure 5. The classification results by "Color+SVM", "Gist+SVM" and bag-of-words are 50.79%, 73.94% and 82.29%, respectively. As we can see, our approach achieves better performance over other methods.

Figure 6 shows an illustration of our proposed approach. The query image is given in the first column and the first row shows the top 5 images selected by employing both local and global information described in Section2. The second row shows the top 5 images after executing RANSAC algorithm and re-ranking. The third row shows the result via the image-to-class method.

In summary, we have demonstrated the effectiveness by fusing both global and local information for scene building recognition. The decision making in our third stage improves the accuracy and outperforms the state of the art approaches in this task.

Table 2. Confusion Matrix of the proposed approach on the thirteen buildings. The average classification rates of each building are shown in the diagonal and the average accuracy rate of retrieval is 86.26%.

	IIS	IEAS	LFAM	MHP	ATDM	IMH	IP	AO	AC	ISS	IBMS	GRC	ICLP
IIS	96.1	0	10.67	0	0	0	0	3.33	0	0	7.33	6.33	0
IEAS	0	81.81	0	0	0	14.52	15.63	0	0	0	3.47	0	0
LFAM	0	0	83.74	0	0	0	0	0	0	0	2.41	0	0
MHP	0	0	0	92.30	0	0	0	0	3.06	0	0	0	0
ATDM	0	0	0	0	83.71	0	0	0	0	0	0	0	0
IMH	0	1.24	0	0	0	76.3	0	6.19	0	0	0	0	0
IP	0	11.11	0	0	0	0	77.78	0	0	0	0	0	0
AO	0	0	0	0	0	8.05	5.86	90.33	0	0	0	3.18	0
AC	0	0	0	0	0	0	0	0	94.25	0	0	0	0
ISS	0	0	0	0	9.42	0	0	0	0	87.5	2.47	2.54	3.12
IBMS	3.13	5.34	4.31	5.53	0	0	0	0	0	12.11	83.33	0	9.33
GRC	0	0	0	0	0	0	0	0	0	0	0	87.58	0
ICLP	0	0	0	2.07	6.17	0	0	0	2.63	0	0	0	86.67

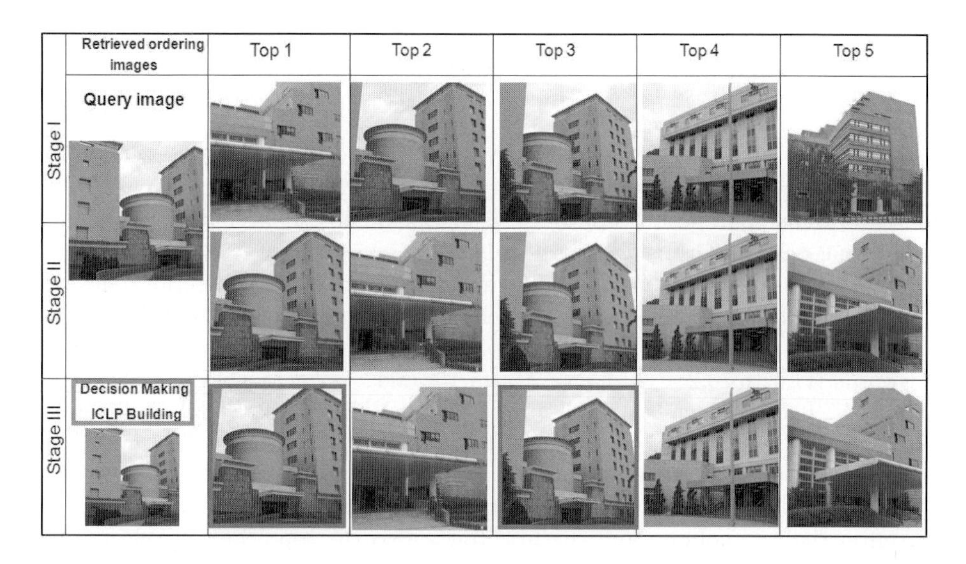

Fig. 6. Illustration of retrieved images by our three stage approach, respectively. The first row shows the top 5 images via stage I. The second row represents the result after re-ranking images of stage I. Final row shows the result that the query image belongs to 'ICLP' building, since the class shown in red-lined box has the shortest distance (d_{I2C}) among the three classes of buildings.

6 Conclusion

In this paper, we have presented an image-based approach for recognizing building based on fusion of local and global information. Our coarse-to-fine approach first filters out irrelevant images in first two stages. In decision making stage, image-to-class method is employed based on Euclidean metric for evaluating distances and determines the result of the query image from the top k survival images. We compare our approach with the Color+SVM, bag-of-word approaches and the Gist+SVM method which use local and global information, respectively. The proposed approach shows promising results with respect to the state of the art methods. In the future, we would like to conduct different experiments, such as a larger number of categories and achieve better performance.

References

1. Boiman, O., Shechtman, E., Irani, M.: In defense of nearest-neighbor based image classification. In: IEEE Computer Society Conference Vision and Pattern Recognition (2008)
2. Bosch, A., Zisserman, A., Muñoz, X.: Scene classification via pLSA. In: Leonardis, A., Bischof, H., Pinz, A. (eds.) ECCV 2006. LNCS, vol. 3954, pp. 517–530. Springer, Heidelberg (2006)
3. Bosch, A., Zisserman, A., Munoz, X.: Image classification using random forests and ferns. In: IEEE International Conference on Computer Vision (2007)
4. Chang, C., Lin, C.: Libsvm: a library for support vector machines (2005)
5. Dance, C., Willamowski, J., Fan, L., Bray, C., Csurka, G.: Visual categorization with bags of keypoints. In: European Conference on Computer Vision International Workshop on Statistical Learning in Computer Vision. LNCS. Springer, Heidelberg (2004)
6. Fishler, M., Bolles, R.: Random sample consensus: a paradigm for model fitting with application to image analysis and automated cartography. Communications of the ACM 24, 381–395 (1981)
7. Harris, C., Stephens, M.: A combined corner and edge detector. In: Proceedings of the Alvey Vision Conference, pp. 147–152 (1988)
8. Hays, J., Efros, A.A.: IM2GPS: estimating geographic information from a single image. In: IEEE Computer Society Conference Vision and Pattern Recognition (2008)
9. Jarvelin, K., Kekalainen, J.: Cumulated gain-based evaluation of IR techniques. ACM Transactions on Information Systems (TOIS) 20, 422–446 (2002)
10. Lazebnik, S., Schmid, C., Ponce, J.: Beyond bags of features: Spatial pyramid matching for recognizing natural scene categories. In: IEEE Computer Society Conference Vision and Pattern Recognition, pp. 2169–2178 (2006)
11. Li, F.F., Perona, P.: A bayesian hierarchical model for learning natural scene categories. In: IEEE Computer Society Conference Vision and Pattern Recognition (2005)
12. Lowe, D.G.: Distinctive image features from scale-invariant keypoints. International Journal of Computer Vision 2, 91–110 (2004)
13. Luo, Z., Li, H., Tang, J., Hong, R., Chua, T.S.: ViewFoucus: Explore places of interests on google maps using photos with view direction filtering. In: Proceeding of Multimedia. ACM, New York (2009)

14. Oliva, A., Torralba, A.: Modeling the shape of the scene: A holistic representation of the spatial envelope. International Journal of Computer Vision 42, 145–175 (2001)
15. Oliva, A., Torralba, A.: Building the gist of a scene: The role of global image features in recognition. Progress in Brain Research 155, 23–26 (2006)
16. Philbin, J., Chum, O., Isard, M., Sivic, J., Zisserman, A.: Object retrieval with large vocabularies and fast spatial matching. In: IEEE Computer Society Conference Vision and Pattern Recognition (2007)
17. Schaffalitzky, F., Zisserman, A.: Multi-view matching for unordered image sets, or how do I organize my holiday snaps? In: Heyden, A., Sparr, G., Nielsen, M., Johansen, P. (eds.) ECCV 2002. LNCS, vol. 2350, pp. 414–431. Springer, Heidelberg (2002)
18. Smeulders, A.W.M., Worring, M., Santini, S., Gupta, A., Jain, R.: Content-based image retrieval at the end of the early years. IEEE Transactions Pattern Analysis and Machine Intelligence 22, 1349–1380 (2000)
19. Szeliski, R.: Where am I?: ICCV 2005 computer vision contest, `http://research.microsoft.com/iccv2005/contest/`
20. Vogel, J., Schiele, B.: Semantic modeling of natural scenes for content-based image retrieval. International Journal of Computer Vision 72, 133–157 (2007)
21. Yeh, T., Tollmar, K., Darrell, T.: Searching the web with mobile images for location recognition. In: IEEE Computer Society Conference Vision and Pattern Recognition (2004)
22. Zhang, H., Low, C., Smoliar, S., Wu, J.: Video parsing, retrieval and browsing: an integrated and content-based solution. In: Proceeding of Multimedia, pp. 15–24. ACM, New York (1995)

Learning Landmarks by Exploiting Social Media

Chia-Kai Liang, Yu-Ting Hsieh, Tien-Jung Chuang,
Yin Wang, Ming-Fang Weng, and Yung-Yu Chuang*

National Taiwan University

Abstract. This paper introduces methods for automatic annotation of landmark photographs via learning textual tags and visual features of landmarks from landmark photographs that are appropriately location-tagged from social media. By analyzing spatial distributions of text tags from Flickr's geotagged photos, we identify thousands of tags that likely refer to landmarks. Further verification by utilizing Wikipedia articles filters out non-landmark tags. Association analysis is used to find the containment relationship between landmark tags and other geographic names, thus forming a geographic hierarchy. Photographs relevant to each landmark tag were retrieved from Flickr and distinctive visual features were extracted from them. The results form ontology for landmarks, including their names, equivalent names, geographic hierarchy, and visual features. We also propose an efficient indexing method for content-based landmark search. The resultant ontology could be used in tag suggestion and content-relevant re-ranking.

1 Introduction

As digital cameras and storage get cheaper, many of us have thousands of photographs in our own albums. Their effective management becomes increasingly important but more difficult nevertheless. *Image annotations* have been shown effective to facilitate organization and retrieval of photograph collections. However, automatic image annotation algorithms for generic semantic are still far from being applicable. A good news is that automatically collected metadata, such as time and location, and their derived information have been proved helpful in management of photo collections [1]. Almost all digital cameras record *time stamps* when pictures were taken. Some *location-aware cameras* can augment location information about where pictures were taken by using GPS, cellular or Wi-Fi Networks. Unfortunately, although automatically-collected location context is useful, location-aware cameras do not grow at a rapid rate because of cost, power consumption and image quality. Thus, most images still lack of geographic metadata for effective organization and retrieval.

Even though it is useful to automatically add geographic tags to photographs taken by non-location-aware cameras, limited by available content analysis technology, it is still hopeless to automatically annotate general geographic names for photographs in the near future. Thus, this paper focuses on *landmark photographs*, pictures of a specific but useful category. Figure 1 shows the overview of our system. There are two phases in our system, the *pre-processing phase* and the *application phase*. In the pre-processing phase, we downloaded from Flickr a total of 11,028,186 *geotagged* photos which were uploaded by 140,948 users during 2005/01/01 and 2008/01/01. Geotags

* This work was supported by grants NSC97-2622-E-002-010-CC2 and NTU-98R0062-04.

S. Boll et al. (Eds.): MMM 2010, LNCS 5916, pp. 207–217, 2010.

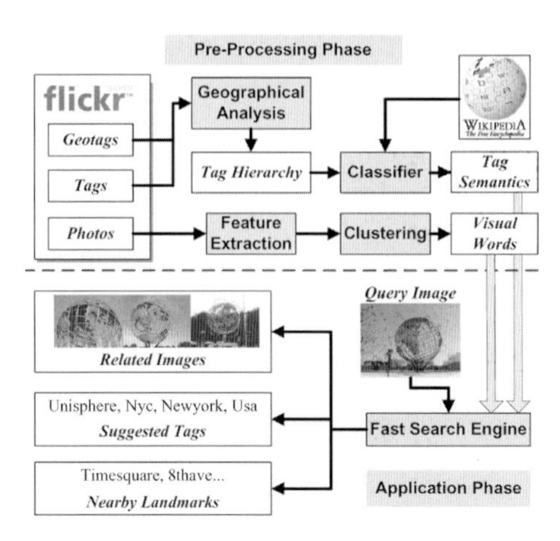

Fig. 1. Overview of the proposed system

record latitude and longtitude coordinates where the pictures were taken. They are either recorded by cameras or labeled by users, and may contain errors. Figure 2(a) shows the spatial distributions of all these retrieved photographs over the world. We perform statistical analysis on geotags to find textual tags with strong spatial coherence. These tags more likely refer to geographic terms. Landmark tags are further classified from these geographic tags using corresponding Wikipedia articles. In addition, association analysis is used to build a *tag hierarchy*, from which we can find the containment and equivalence relationships among geographic terms. Section 3 describes our methods for extracting the above information. In Section 4, for each landmark tag, Flickr is queried to retrieve a collection of relevant images. From each set of images, we extract a set of distinctive features associated to a specific landmark tag. Hence, we have a database containing a set of landmark tags and each tag has its own set of *visual words*. For efficient indexing, these visual words are clustered into a hierarchical tree.

With the learnt landmark ontology (landmark names, synonyms, hierarchy, visual words and so on), many interesting applications become possible. The application phase utilizes the ample and precise data extracted from the pre-processing phase for various applications. For example, after the user uploads a new photo for query, our system can immediately identify the landmark in the photo. It can then suggest the name of the landmark, return the related or representative images of the landmark, recommend other proper tags for this image, and even show the user the nearby landmarks within the same city. Although similar to Kennedy and Naaman's work [2], our paper can be taken as a further step by providing the following improvements:

– It puts more focus on landmarks and proposes better methods to identify landmarks' textual tags while they simply borrowed the idea from Ahern *et al.*'s paper [3]. Along this line, this paper introduces tag hierarchy construction by association analysis and landmark classification using corresponding Wikipedia articles. Thus, we can extract more structured information, such as synonyms of landmarks and their containment relationships, to form a more complete landmark ontology.

(a) (b)

Fig. 2. Spatial distributions of the retrieved photos and landmark tags. (a) The spatial distribution for geotags. There are totally 11,028,186 geotagged photos in our database mirrored from Flickr. A warmer color represents locations with more photos and a colder color means the ones with less photos. (b) The spatial distribution of landmark tags. We have identified a total of 3,821 landmark tags over the world. These tags are shown in red. Green pixels represents the coverage of all geotags.

- Their work mainly focus on finding representative images, but this paper proposes methods for more efficient landmark search by visual contents.
- They only demonstrated their method within San Francisco area. On the contrary, this paper explores the world's landmarks. In addition, this paper demonstrates a set of interesting applications enabled by the discovered landmark ontology.

2 Related Work

In recent years, there are quite a few work on exploring the usage of geographic metadata. Toyama *et al.* described WWMX, a system for browsing geo-referenced photo collections [4] and various issues related to alike systems. Mor Naaman and his colleagues have done lots of exciting work on various topics related to geo-referenced tags and photographs, including tag visualization [5,3], extracting the event and place semantics [6,7], and ranking representative images [8,2]. The main differences distinguishing our work from theirs are that most of them assume existence of geographic metadata and confine the usage to the geotagged images. On the contrary, we perform visual analysis to tag images and use the hierarchical visual words to avoid the expensive pair-wise similarity measurement, making the system more scalable and robust.

University of Oxford conducted a series of researches on applying the text search techniques to the image search problem [9,10,11]. The key idea is to treat the local distinct features, or *visual words*, in an image as the words in a document. However, in their approach, each image in the dataset are considered as an unique entity and therefore a large index storage is required. On the contrary, in our approach, hundreds or thousands of images from a single landmark are automatically grouped together. Their visual word distributions are aggregated into a single one. Therefore, our system can be easily scaled to deal with millions of on-line images. Also our system provides novel applications beyond simply finding similar images, such as automatic tag suggestion. Finally, Hays and Efros use global image features to match an image to the geotagged images [12]. However, this method is not accurate enough for serious applications.

3 Textual Tags for Landmarks

The geotags reveal the geographic distribution of a tag, which consequently can be used as the signal for landmarks. For example, the distributions of "birthday" and "beach" are wide and sparse on Earth, but that of "Statue of Liberty" is local and clustered. We exploit this property to identify geographically related tags. Furthermore, we build a tag hierarchy of those tags according to their co-occurrence and geographical distributions. However, pure geographical analysis is not sufficient to identify landmarks precisely. Tag distributions of either cities or villages are sometimes similar to that of landmarks. One of the reasons is that photos are not uniformly distributed in those areas, and might be clustered at some specific places. Therefore, differences between landmark and non-landmark tags are sometimes not distinguishable from spatial distributions. In order to tell whether a tag is a landmark, we use knowledge stored in Wikipedia to build a classification model.

3.1 Geographical Analysis

A photo P_i has the following attributes relevant to our application: (1) geotags $L_i = (x_i, y_i)$, (2) photographer u_i and (3) tag set $T_i = \{t_{i_1}, ..., t_{i_{n_i}}\}$ where n_i is the number of tags associated with P_i. To identify which tags are geographic terms, photographs' geographic locations are grouped by tag names to form clusters. Specifically, the geographic cluster of a tag t is $C_t = \{L_j | t \in T_j\}$. If a geographic cluster of a tag is localized in a small region, this tag is likely to refer to a place. One issue is that photos with geographic tags are not always at the right place due to labeling errors or other reasons. These photos are considered as *noises*, which are handled with RANSAC. For each tag t, we first create its geographic cluster C_t with size $|C_t|$. We randomly pick up one point $\mathbf{x}$ from C_t and use it as the center of a Gaussian. The deviation σ is determined by taking the $(68\% \times |C_t|)$-th closest point (68% is confidence level of Gaussian model in 1σ). The fitness of the hypothetic Gaussian is evaluated as $\sum_i G(L_i; \mathbf{x}, \sigma)$, where $L_i \in C_t$. The process is repeated several times and the Gaussian $G_t(\mathbf{x}, \sigma)$ with the best fitness is chosen to describe the spatial distribution of the geographic cluster.

After deciding $G_t(\mathbf{x}, \sigma)$, we collect photos located within 3σ as inliers. The area $A(t)$ of the tag t is defined as area of the convex hull of all inliers. Because non-geographic tags tend to be distributed wildly over the world, we keep the tags whose areas are smaller than a threshold and send them to the tag hierarchy construction stage in the next section. Among 60,449 tags that go through geographic analysis, 13,854 of them are identified as geographic terms.

3.2 Tag Hierarchy

A photo can have multiple tags, which are usually semantically or geographically related. Because we have eliminated non-geographic tags in the previous step, the remaining co-occurred tags are very likely geographically related. For example, photos labelled with "Statue of Liberty" are often labelled with "New York" as well, but "Golden Gate Bridge" is not likely to appear with "Africa". We use association analysis to formulate their closeness. Assuming $P(b|a)$ denotes the probability that photos labelled with tag b given that tag a is labelled; $N(a, b)$ denotes the number of

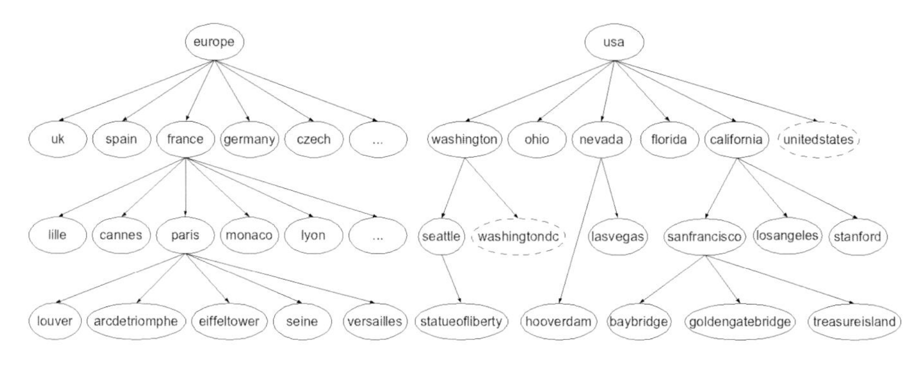

Fig. 3. Examples from the tag hierarchy. The synonyms are shown in dotted circles.

photos with both tags a and b; and $N(a)$ the number of photos with tag a, we have $P(a|b) = \frac{P(a \cap b)}{P(b)} = \frac{N(a,b)}{N(a)}$. The most related tag to tag a, $M(a)$, is defined as $M(a) = \arg\max_b P(b|a)$. Given a tag a, if we iteratively evaluate $M(a), M(M(a)), \cdots$, eventually it will reach a tag like "USA," "Europe," "Asia," "Africa," etc. A sequence of tags generated in this way is called a *trace*. An example trace beginning from `spaceneedle` is shown as follows:

$$M(\texttt{spaceneedle}) = \texttt{seattle} \text{ with } P = 0.959924$$
$$M(\texttt{seattle}) = \texttt{washington} \text{ with } P = 0.294886$$
$$M(\texttt{washington}) = \texttt{usa} \text{ with } P = 0.0914492$$

Some interesting properties can be observed in the traces. First, the synonyms are usually the most related tag to each other; i.e., $M(M(a)) = a$. Second, a tag which is the ancestor of other tags is less likely to be a landmark tag. It usually corresponds to a district or an even larger area. We create the trace of each geographic tag independently and then merge them into a tag hierarchy. Two examples in Figure 3 show the subsets of the tag hierarchy. We can clearly see the hierarchical relationships between the tags, from continents of root nodes to individual landmarks of leaf nodes.

3.3 Wikipedia Knowledge

The leaf nodes of the tag hierarchy may still not correspond to a landmark. It could be a local event or an unattractive static object. We show that the exact semantics of the tag can be inferred from the corresponding article in Wikipedia, and thus the accuracy of the landmark identification can be further improved. For each tag, we find the corresponding article on Wikipedia. Note that The synonyms for a landmark are already merged in the tag hierarchy. Thus, for a group of synonyms, we only use the one with the highest count. Among 13,854 tags which pass geographic analysis, less than 10 tags did not have a Wikipedia article. In these cases, we classify them as the class of "other".

Inspired by the spam detection algorithms, we formulate our problem as a classification problem. Each article should belong to exactly one of three classes *landmark*, *city*, and *others*. The *city* class contains not only the city-scale tags, but also all the areas that are larger than a specific attractive, natural or man-made structure, such as districts, towns, and beaches. The *other* class contains all other things, including local events or even non-geographical elements. We find that this three-class formulation can significantly improve the accuracy. We add the class of *city* because the articles in the *city* class contain several unique descriptions (population, etc) and therefore they should not be mixed with the ones in the *other* class. In addition, identifying the names of those large areas would potentially provide novel applications.

We use the occurrence of the tokens (words of the "wiki text") as the features of the article to perform classification. Because the articles on Wikipedia are relatively terser and more precise than general documents, the naïve-Bayesian model can provide fairly accurate results. Let $P(W|C)$ denote probability that the word W appears in the documents of class C and $P(C|A)$ denote the probability that document A belongs to the class C. Using Bayesian rule, we have

$$P(C|A) = P(C|T_1, ..., T_n) = \frac{p(C) * p(T_1, ..., T_n|C)}{p(T_1, ..., T_n)}, \quad (1)$$

where T_i is the i-th token in the document. The most likely class of A would be $c^*_A = \arg\max_c P(C = c|A)$. In addition to the individual words, using n-gram as tokens can significantly improve the accuracy. This is because that the Wiki article uses many deterministic sentences and formal keywords when describing cities and landmarks.

To verify the performance of our classifier, we manually label 634 tags as the ground truth for validation. For building the classifier, we randomly choose 50 tags for each class as the training data. The accuracy of three-class classification using single words is 78.9% and improved to 86.1% when 2-gram and 3-gram are included as tokens.

3.4 Results and Discussions

To summarize, there are a total of 2,068,833 distinct textual tags we retrieved from 11,028,186 geotagged photos of Flickr. Among them, 60,449 tags were used by more than 15 users. After geographic analysis, 13,854 of them are considered geography-related. Among them, 4,633 tags are classified as landmarks by the wiki-article classifier (with an accuracy around 85%). Considering the tag hierarchy, 3,821 of them appear at the leaf and are considered landmark tags. Figure 2(b) shows the spatial distribution of these landmark tags. Note that the distribution of landmarks are biased to Flickr users' patterns. As an example, below are some landmarks we identified within London area.

```
greenwich bigben londoneye waterloo docklands battersea kew (kewgardens) canarywharf
tate (tatemodern) westminsterabbey towerbridge riverthames londres brixton sciencemuseum
housesofparliament heathrow batterseapowerstation trafalgar (trafalgarsquare) leicester
nationalgallery harrods cuttysark clapham gherkin britishmuseum crystalpalace
```

Tags in parenthesis are synonyms. The off-the-shelf databases could give landmark name as well. However, our approach has the following advantages. First, most of those databases are more interested in administrative hierarchy. Landmarks are often not the main focus. Thus, landmarks are not necessarily listed. Second, even if they are, landmarks could have multiple names, but not all are listed in the off-the-shelf databases. Finally, off-the-shelf database can be outdated, but information extracted from social media keeps updated and reflects how landmarks are really tagged in social media.

Our approach shares a similar goal and part of the methodology as Ahern *et al.*'s world explorer [3]. However, our system has the following features: more emphasis on landmarks and the incorporation of tag hierarchy and Wikipedia-classification. These give better results. Using London as an example, here are the tags that are at the leaf under London but classified as "others" by our wiki-article classifier, `guesswherelondon`, `londonbus`, `londonist` and `londonunderground`. It means that all four have a landmark-scale cluster in the geographic analysis. With only geographic analysis like Section 3.1 and Ahern *et al.* did, they can't be distinguished from real landmarks. In addition, the tf-idf measure used by Ahern *et al.* can find a better tag to represent a group of tags in one area, but it does not change the number of clusters. On the contrary, we use the tag hierarchy to merge the synonyms. Also, two landmarks in a small area are not mixed together in our method. Finally, Ahern *et al.* segmented the earth into many regions in a multi-level pyramid. On the contrary, we perform the analysis globally. This can remove some non-geographical tags such as `baseball` and `soccer`.

4 Visual Features for Landmarks

This section shows how to exploit the visual information of the landmark photos (i.e., images with the tags classified as landmarks) for content-based image retrieval. The system must be *robust* and *fast*, returning the results immediately after given the query image. Additionally, system should be *scalable* to handle millions of photos.

4.1 Hierarchical Visual Words Construction

Since many landmarks are made of similar materials and shot under similar illumination conditions, traditional global image features such as color histogram can hardly be used to distinguish one landmark from another. A landmark is recognizable due to its unique structures and thus it is better to use locally distinct features. Here we apply SIFT [13] to detect the interest points in the photo. There are usually hundreds to thousands of SIFT features in a single image and therefore it is impractical to store all features in the database and perform pairwise feature matching to all of them in the query phase. Here we adopt the concept of *visual word* [14]. All features are coarsely quantized into many clusters using k-means and each image can be considered as an article written using those clusters. In this way, many techniques in text retrieval can be readily applied [9,11]. To recognize thousands of landmarks, we still have to use a large number of clusters to preserve the distinctness. This could significantly slow the matching process. Here we quantize the features in a hierarchical fashion [15]. In the beginning all feature are clustered into k clusters and the features in one cluster are further clustered into k sub-clusters. This process is perform recursively until a specific storage limit is reached. All the leaf nodes are the final visual words.

4.2 Efficient Indexing and Search

In the search phase, features in the query image are detected and each is assigned to the nearest visual word. This can be done very efficiently by traversing the tree using best-first search if the approximate nearest visual word is sufficient. We also use a modified

(a) the 15 landmarks used in the evaluation (b) a subset of images for `agbar`

Fig. 4. (a) The 15 landmarks used in the experiments. (b) A subset of images retrieved from Flickr using `agbar` as the keyword. It shows a great deal of visual diversity and contains a few "noises".

n-best search method [16] to improve the accuracy. Instead of only traversing the best path, we traverse the first n best paths in parallel.

In previous methods, each visual word is attached with a backward index to the images containing that word, and the ranking of the retrieved images depends on the characteristics of the indices [10]. However, this approach requires huge storage when there are millions of images to be indexed (in [10], only 5k positive images were indexed). Also, retrieving the very similar images may not be useful in many applications since they give no more information than the original query image. To resolve these problems, we propose to index the landmarks instead of photos. For each visual word, we record the backward index to landmarks containing that word. This method is more useful than the per-image indexing for many reasons. First, the number of landmarks in much fewer than the number of photos, and increase at a much slower rate. Second, together with our tag semantics and geographical analysis, identifying the landmark is enough for many applications. Third, this indexing method is more robust to the noisy tag inputs. Few irrelevant images in the training data would not affect the search results. In terms of the text retrieval, our method attempts to categorize the input article, not to retrieve the similar ones from the database. After this step, other related but not similar articles in that category can be retrieved using other existing techniques. Specifically, at each leaf node v of the hierarchical tree, we store the number of the landmarks that containing the visual words N_v and the number of occurrences in the i-th landmark $C_v(i)$. When a feature is assigned to a the visual word v, we increase the score to the i-th landmark by a modified tf-idf function $C_v(i)log(N/N_v)$, where N is the total number of images in the database.

4.3 Evaluation

The construction of ground truth for the landmark image query is labor-intensive, so we only choose 15 landmarks for evaluation (Figure 4(a)). For each landmark, we manually examined and selected at least 500 Flickr photos that indeed capture the landmark structure. These images naturally covers many different illuminations and view positions. For training, we randomly pick 2,700 images (180 for each landmark) from the ground truth to construct the hierarchical tree. There are totally 1,942,243 raw feature

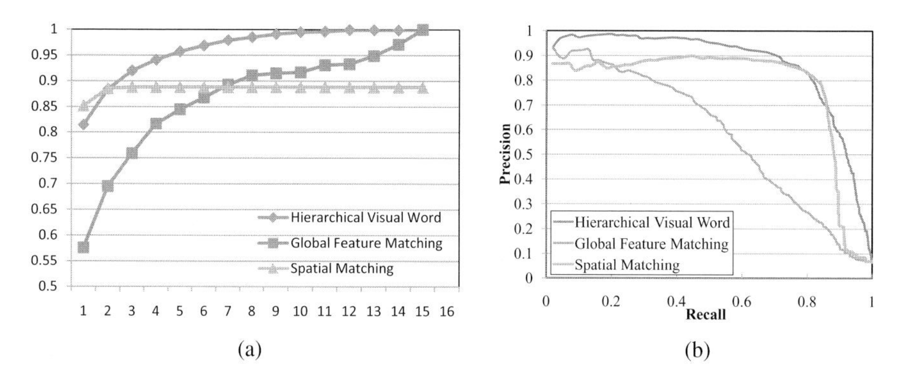

Fig. 5. (a) The average accuracy using three different methods. (b) Their average PR curves.

vectors. The degree of the clustering at each level is 8, and the maximal tree depth is 11. The final hierarchical tree contains 778,011 visual words. The tree consumes 528MBs and the backward index consumes 17.5MBs.

Two methods are compared. The first one uses the global feature to measure the per-image similarity. (We used six features including histogram, Gabor texture and others; SVM is used for classification.) The second one uses the spatial matching to measure the similarity [10]. (Count the number of feature matches between the query image and each of the images in the database. The feature matches are verified by spatial constraints.) For the first method, the best accuracy of the top return is only 57.6% although it is slightly faster than our algorithm (0.211 seconds). The spatial matching method has a higher accuracy than our method in the first return, but its performance soon saturates after the first 3 returns. It is because many tested images can never find a match image in the database when it is not big enough. Also the spatial matching is much slower than our method. On average, each query takes 54.843 seconds. Figure 5(a) shows the overall performance for three methods. Figure 5(b) shows the PR curves. These show that our method is compared favorably to the other two methods.

For testing robustness, we replace the training data with photos queried from Flickr by tags, which are more convenient to obtain but also noisier. For example, Figure 4(b) shows part of the retrieved images from Flickr for *Torre Agbar*, a 21st-century skyscraper in Spain. It contains many photos without the landmark. Although many of those photos are not visually related to the landmarks, the performance is only decreased by 2%. The degradation can be easily compensated by increasing the number of visual words. This shows that the hierarchical tree combined with the per-landmark indexing is very robust to noise in training data.

Finally, for testing in a larger scale, we increase the number of landmarks to 150. For dataset of this size, we can only use Flickr's returned images as both data for training and ground truth for evaluation. In this case, the average accuracy of the top return is 30%, which is much higher than that of the random guess (0.6%), and again can be increased by increasing the number of visual words. The real performance should be better since the "ground truth" here are actually retrieved using Flickr's search engine containing much noise (Figure 4(b)). This shows that our method is highly scalable and robust. Categorizing more landmarks does not require much labeling effort to

(a) automatic tag suggestion (b) image re-ranking

Fig. 6. Applications using the discovered landmark ontology. (a) Automatic tag suggestion. Once the landmark in the image is classified, the tags in the same trace of the tag hierarchy could be added. The synonyms are listed in parenthesis. (b) The result of the image re-ranking. The top shows the top 24 images returned by Flickr when using `Ferrybuilding` as the keyword. The red-framed images are the obvious outliers. The bottom are the results after visual re-ranking. We can see that, after re-ranking, they have the lowest scores.

build the training data, and the storage only increases linearly with the number of the landmarks.

5 Applications

This section presents a set of applications which uses the built landmark ontology. Other potential applications include *attraction map construction* and *album management*.

Landmark identification from images. As shown in the previous section, our system can identify the presence of a landmark in an image efficiently and accurately. Once we identify this, the landmark tag and its derived tags can be automatically added or more photographs related to this landmark could be displayed depending on the application.

Automatic tag suggestion. Our landmark ontology eases landmark image annotation by borrowing tags learnt from those who tagged photos of the same landmark. Once a landmark is detected, the ontology suggests a set of potential tags. Figure 6(a) gives some results. For example, our system recognizes that the top-left photo of Figure 6(a) contains the *Agbar Tower*. A set of tags are then suggested, `agbar`, `barcelona`, `spain` and `europe`. In addition, `torreagbar` is suggested since the system recognizes that `agbar` and `torr agbar` are both dialects referring to the Agbar Tower by Flickr users.

Visual relevance re-ranking. The ontology can also be used to re-rank results for landmark image search by considering not only textual relevance but also visual content. Figure 6(b) shows an example using the keyword `ferrybuilding`. On the top, we see the top 24 images returned by Flicker. The bottom show the results after re-ranking. We can see that all the irrelevant images now have lower scores. The overall processing time is 4.53 seconds in this example.

6 Conclusion

This paper proposes methods to automatically transfer tags to unlabeled photographs from annotated landmark photographs of a photo-sharing website. We use geographic analysis, tag hierarchy construction and wiki-article classification to identify landmarks' textual keywords. These also tell us their synonyms and geographic hierarchy. The ability to assign structure to tags makes tagging systems more useful. In addition, we propose an efficient indexing method for content-based landmark search. With all these, we demonstrate a set of interesting applications related to landmarks. In the future, we plan to develop more interesting applications using the discovered landmark ontology and make the visual search for landmarks more efficient.

References

1. Naaman, M., Harada, S., Wang, Q., Garcia-Molina, H., Paepcke, A.: Context data in geo-referenced digital photo collections. In: Proceedings of ACM Multimedia, pp. 196–203 (2004)
2. Kennedy, L.S., Naaman, M.: Generating diverse and representative image search results for landmarks. In: Proceedings of WWW, pp. 297–306 (2008)
3. Ahern, S., Naaman, M., Nair, R., Yang, J.: World explorer: Visualizing aggregate data from unstructured text in geo-referenced collections. In: Proceedings of ACM/IEEE JCDL (2007)
4. Toyama, K., Logan, R., Roseway, A., Anandan, P.: Geographic location tags on digital images. In: Proceedings of ACM Multimedia, pp. 156–166 (2003)
5. Jaffe, A., Naaman, M., Tassa, T., Davis, M.: Generating summaries and visualization for large collections of geo-referenced photographs. In: Proceedings of MIR, pp. 89–98 (2006)
6. Rattenbury, T., Good, N., Naaman, M.: Towards extracting Flickr tag semantics. In: Proceedings of WWW, pp. 1287–1288 (2007)
7. Rattenbury, T., Good, N., Naaman, M.: Towards automatic extraction of event and place semantics from Flickr tags. In: Proceedings of ACM SIGIR, pp. 103–110 (2007)
8. Kennedy, L., Naaman, M., Ahern, S., Nair, R., Rattenbury, T.: How Flickr helps us make sense of the world: Context and content in community-contributed media collections. In: Proceedings of ACM Multimedia, pp. 631–640 (2007)
9. Chum, O., Philbin, J., Sivic, J., Isard, M., Zisserman, A.: Total recall: Automatic query expansion with a generative feature model for object retrieval. In: Proceedings of IEEE ICCV (2007)
10. Philbin, J., Chum, O., Isard, M., Sivic, J., Zisserman, A.: Object retrieval with large vocabularies and fast spatial matching. In: Proceedings of IEEE CVPR (2007)
11. Philbin, J., Chum, O., Isard, M., Sivic, J., Zisserman, A.: Lost in quantization: Improving particular object retrieval in large scale image databases. In: Proceedings of CVPR (2008)
12. Hays, J., Efros, A.: IM2GPS: estimating geographic information from a single image. In: Proceedings of IEEE CVPR (2008)
13. Lowe, D.G.: Distinctive image features from scale-invariant keypoints. Internatioanl Journal of Computer Vision 60(2), 91–110 (2004)
14. Sivic, J., Zisserman, A.: Video Google: A text retrieval approach to object matching in videos. In: Proceedings of IEEE ICCV, vol. 2, pp. 1470–1477 (2003)
15. Nistér, D., Stewénius, H.: Scalable recognition with a vocabulary tree. In: Proceedings of IEEE CVPR, vol. 2, pp. 2161–2168 (2006)
16. Schindler, G., Brown, M., Szeliski, R.: City-scale location recognition. In: Proceedings of IEEE CVPR (2007)

Discovering Class-Specific Informative Patches and Its Application in Landmark Charaterization

Shenghua Gao, Xiangang Cheng, and Liang-Tien Chia

CeMNet, School of Computer Engineering
Nanyang Technological University, Singapore
{gaos0004,ch0061ng,asltchia}@ntu.edu.sg

Abstract. Discovering class-specific informative regions for a given concept with a few images is an interesting but very challenging task, due to occlusion, scale changes of objects, as well as different views under varying lighting conditions. This paper proposes a new perspective to discover the informative regions by using several images. To achieve this, we introduce a new representation of image: Ordered-BoW Image (BoWI), whose elements summarizes information of the patch centered at the element in original image. Because of its "structured pixels", BoWI is robust and informative enough for an object class representation. Histogram-based Multi-Ranking Amalgamation Strategy (MRAS) is adopted to explore the most informative patches for an object in BoWI. Experiments on Landmark-National Icon data set that our approach is robust to occlusion, scale and illumination, and achieves promising performance in discovering class-specific informative regions.

Keywords: BoW Image, Multi-Ranking Amalgamation Strategy, Informative Patch.

1 Introduction

Discovering class-specific informative patches for a given concept is a very interesting and meaningful question because its potential application in many research and real application areas, for example, content based image resizing and object recognition. However, the challenging problems in computer vision, especially in the presence of pose and view point changes, intra-class variation, occlusion and background clutter, make this problem very difficult. For example, we have lots of pictures taken for a given landmark on hand. However, these pictures will vary a lot due to a wide range of views, resolution, illumination and distance to object. 'understanding' the informative regions in these images and automatically discover them will be quite meaningful for us. In this paper, we will focus on this problem.

"Bag-of-Word" (BoW) model[2] and part-based model are two widely used image representation methods in image recognition and classification. By assigning descriptors of local invariant regions to "visual words", each image is represented as a histogram of word frequency. To some extent, this model is robust and insensitive to scale and illumination change, but it suffers from the lack of any

S. Boll et al. (Eds.): MMM 2010, LNCS 5916, pp. 218–228, 2010.

Fig. 1. Examples of National Icons varying in illumination, scale, occlusion and view angle

available spatial information. Lazebnik et al. [6] extended the BoW model with Pyramid Matching Kernel (PMK) by partitioning the images into increasingly fine sub-regions. But SPM implicitly assumes the corresponding sub-regions are scale or position invariant, which applies in scene classification. Current BoW based methods regard images as a unit, where objects and background are not distinguished. So they are sensitive to background clutter.

Part-based model represents object as a spatial layout of multiple parts, where the deformable configuration is characterized by spring-like connections between them[15]. The disadvantage of existing part-based model is that it heavily depends on the representations of each part. Both interesting points based parts detection [14] and correlation-of-patches based parts [8] introduce many local ambiguities and limited stable parts.

Discovering informative patches is very useful to resolve the problems above, both for removing the background and for learning the discriminative object parts. Here, a patch is regarded as informative if it has highly similar patches which can be frequently found in the other instances of the same class. A robust and discriminative image representation is critical for such work.

In this paper, we propose a novel image representation, referred as "BoW Image(BoWI)", which has varying region size in a uniformly-spaced sampled grid, thereby characterizing the structure of the local region in its element. The BoWI will be shown to be robust to illumination and scale variation. Based on BoWI, patches with different sizes are represented by local histograms, and the similarities between them can be computed by certain distance metric. To handle the problem of parts occlusion, a Multi-Ranking Amalgamation Strategy (MRAS) was implemented to select the most informative parts within the objects.

The following parts of this paper are organized as follows. Section 2 describes the BoWI image representation. In section 3, the Multi-Ranking Amalgamation Strategy is explained. Experimental results are reported in section 4. Finally, we conclude by listing the contributions of this paper in Section 5.

2 BoW Image Representation

In recent years, BoW model has become one of the widely used image representation method for object recognition and classification. The earlier work in this area uses detectors and descriptors to detect and represent interesting keypoints in the image and a clustering of the interesting keypoints into "bags of visual Words". In 2005, Feifei and Perona [12] was the first to proposed the use of an uniformly-spaced sampled grid for scene classification. Lazebnik *et al* [6] improved on the scene classification results by using the spatial pyramid matching method on a uniformly-spaced sampled grid. It is generally agreeable that uniformly spaced sampled grid, with a significantly larger number of patches generated will do well for the matching of scene categories as images in the same scene categories are likely to show similar geometric structure.

For object classification and detection, both Bosch *et al* [11] and Harzallah *et al* [13] has recently applied uniformly spaced sampled grid (with various patch sizes and different grid spacing) for object classification.

The above results further strengthen our view that extracting local regions with uniformly-spaced sampled grid is a good approach. Methods with various patch sizes and different grid spacing has been proposed for those methods, but what is a good size patch and how many different grid size should be considered. By increasing the number of patches for each image, there is a danger that the discriminative nature of patches may be lost and a large number of false matches may occur during the matching phase.

2.1 Detecting a Suitable Region Size

As discussed, our intention is to work with a uniformly-spaced sampled grid but not with a fixed size patch or region. For the first step, there is a need to determine the right size of a region at every sampled point in the dense grid. By selecting the right region size, we expect the proposed method to be robust for objects of different sizes appearing in different images and increase discriminative power when comparing multiple objects with relatively different physical size.

Mikolajczyk [5] has suggested that DoG[1] is an efficient detector that can estimate a suitable scale for each stable region and explore the distribution of gradient related features. For a start, potential interest points that are invariant to scale changes are efficiently identified by using a difference-of-Gaussian(DoG) function.

$$G = \frac{1}{2\pi\sigma^2} e^{-(x^2+y^2)/2\sigma^2} \tag{1}$$

From experiment, we have observed that neighboring interest points are likely to be detected at similar or the same scale, therefore adopting the same scale for spatially close points is acceptable. For each uniformly-spaced sampled grid, the scale of its spatially nearest detected keypoint will be highly related to the region size l. And each region will be represented as an picture element in our new image representation.

$$l = k\bar{\sigma} \tag{2}$$

In our experiments, we set $k = 8$. However, for those smooth regions, such as sky, no keypoint is detected, we will use an estimated scale as its scale, which is the average scale of all detected keypoints. From experimental observation, our

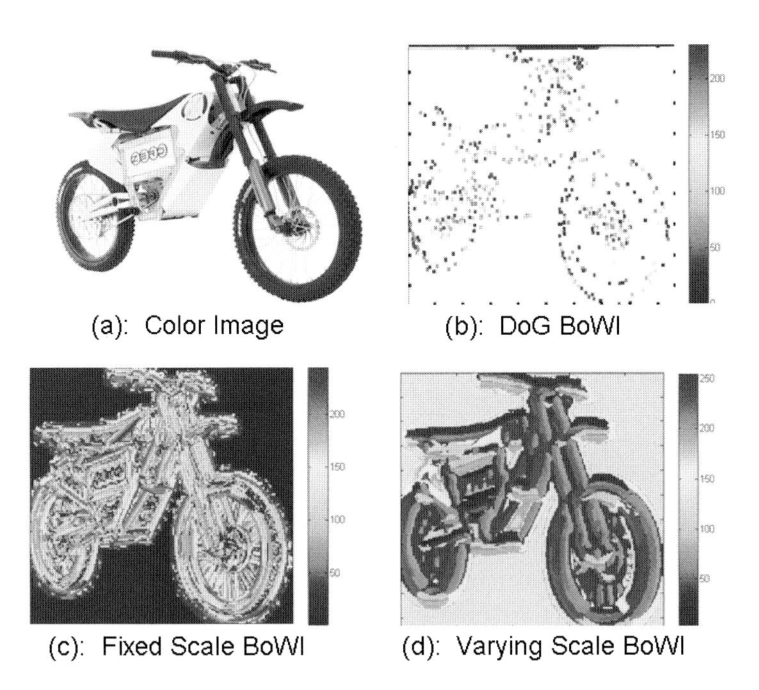

Fig. 2. BoWI images for varying/fixed scale grid sampling and DOG detector

conjecture is that the scales of both foreground objects and background scene will change or remain stable within their respective scale range. Once the proper region size is determined, we will adopt the widely accepted SIFT descriptor[1] to describe the local region.

Fig 2 shows an example of BoWI of the same image with three different methods of local regions extraction. *fig 2b* has keypoints extracted with the DoG detector, *fig 2c* is an image with uniformly-spaced sampled grid and finally *fig 2d* is an image with our proposed uniformly space sampled grid with varying patch size. From this, we can see that our adaptive scale selecting strategy can outperform both the two traditional image representation methods.

2.2 Labeling the Image Region for BoWI

It is very complex and time-consuming if we use the high-dimensional descriptors directly. Varying in cardinality and lacking meaningful ordering of descriptors result in difficulty of finding an acceptable BoWI model to represent the image. To address these problems, we proposed using a fixed-number label set $\mathcal{L}$ by clustering techniques and assigning descriptors to their respective labels (or Visual Words), where similar descriptors are assigned with the same label. We adopt the index of this label to represent the picture element of our BoWI.

As we know, a generic continuous-tone grayscale image has pixels representing intensity values, and neighboring pixels exhibits highly correlated intensity

Fig. 3. Both image and BoWI with different viewangle and background are shown for the same bike object. Similar structure of the wheels and parts of the bike are not lost in the BoWI representation.

values. Can we create a meaningful image, and retain the spatial relationship of neighboring pixels, from the collection of Visual Words? Such an image, which we will name as an Ordered-BoW Image (BoWI), is expected to have similar properties that the neighboring descriptor element has a high probability of falling under the same visual word. To visually observe such a property, we need the clusters and its associated labels to be ordered.

We therefore use the hierarchical K-means clustering method, which groups data simultaneously over a variety of levels and builds a hierarchical relationship between different clusters. The picture element of the BoWI adopts the labels found on the leaves of Hierarchical K-means. Suppose the hierarchical cluster tree has L $levels(L = 0, 1, ...)$ and the branch number for each node is K, then we will get a full K-nodes tree. The total number of leaf nodes is $N = K^L$.

The new representation of BoWI has many advantages: (1): The picture element in BoWI represents the local descriptor, which highlights the important features of the local region. (2): Compared to high dimensional descriptors, BoWI reduces the representation of a descriptor to a one-dimension label by clustering. The distribution of local patches in BoWI can easily be computed to give a robust summarization of important local features. (3): Spatial information can be explored in BoWI. *Figure3* give an example our BoWI with different viewangle and background, and it shows the robustness of our BoWI.

3 Multi-ranking Amalgamation Strategy

In this paper, we will utilize BoWI to help in identifying similar and informative patches across images in the same concept class. To discover the informative patches for the concept, which can be at different scale, we generate patches of different sizes from the BoWI. This flexibility to form different sizes of patches will be exploited in the ranking of similarity between patches.

3.1 Representing Patches with Local Histogram

As shown in [7], spatial information will be useful for object representation. In this work, different patches sizes with half size overlapped are used. Denote the histogram for the patch centered at (x, y) as $H(x, y)$. To improve the discriminative power, we also consider the neighboring eight patches to each central patch. Let w is the weight assigned to the central and its nearby patches. Therefore, each patch is finally represented by itself and its neighbors as:

$$\hat{H} = w. * H \tag{3}$$

where $H = H_{ij}, (i, j \in [-1, 0, 1])$ and $w = \omega_{ij}, (i, j \in [-1, 0, 1])$ is the weighting window, which satisfies $\sum_{ij}(\omega_{ij}) = 1$.

The local histogram $\hat{H}$s cover the whole image as well as different scales, so that we can mine the most discriminative patterns at any location and scale among images in the same class.

3.2 Similarity of Patches

The χ^2 distance will be used to measure the similarity between the local histograms of two patches. Let $X = \{x_1, x_2, \ldots, x_N\}$ be a set of learning instances in one concept class. Each instance, x_i, has a collection of M_i patches and they are represented by its local histograms as: $H_i = \{H_{i1}, H_{i2}, \ldots, H_{iM_i}\}$, in BoWI. For each patch H_{ik}, the closest distance to the patch in x_j is: $S^j_{ik} = \min_{l \in M_j} S(H_{ik}, H_{jl})$. Therefor a distance set: $S^j_i = [S^j_{i1}, S^j_{i2}, \ldots, S^j_{iM_i}]^T$ can be generated by comparing to all patches in instance x_j. Extending the comparison of patches from x_i to all other instances in the concept set, x_j $(x_j \in x_N \ \& \ j \neq i)$, will result in a similarity matrix: $S_i = [S^1_i, \ldots, S^j_i, \ldots, S^N_i]$, where $(j \neq i)$.

3.3 Combination of Multi-ranking

Images from the same concept class may have a large variation in their illumination conditions and/or pose positions. For images with such a large variation, the similarity distance of all the patches from one to all other instances may be very large. If a ranking method is used, the relatively large similarity distance will not effect the degree of relevance for patches in x_i to image x_j. Therefore the ranking set will be $R^j_i = [R^j_{i1}, R^j_{i2}, \ldots, R^j_{iM_i}]^T$. Correspondingly, the $M_i \times (N-1)$ ranking matrix is: $R_i = [R^1_i, \ldots, R^j_i, \ldots, R^N_i]$, where $(j \neq i)$.

In the same class, the ranking positions in an instance may be affected by missing patches when comparing the occluded object instance with any complete object instances. To handle instances which contain occluded object in a concept class. Only the first W ranks is used to calculate the ranking set for each patch. That is, for patch k from x_i, we re-rank R^j_{ik} with all j and keep the significant W ranks $R_{ik} = [R_{ik}(1), R_{ik}(2), ...R_{ik}(W)]$, for which, $R_{ik}(m) < R_{ik}(n), \ \ and \ \ m < n < (N-1)$.

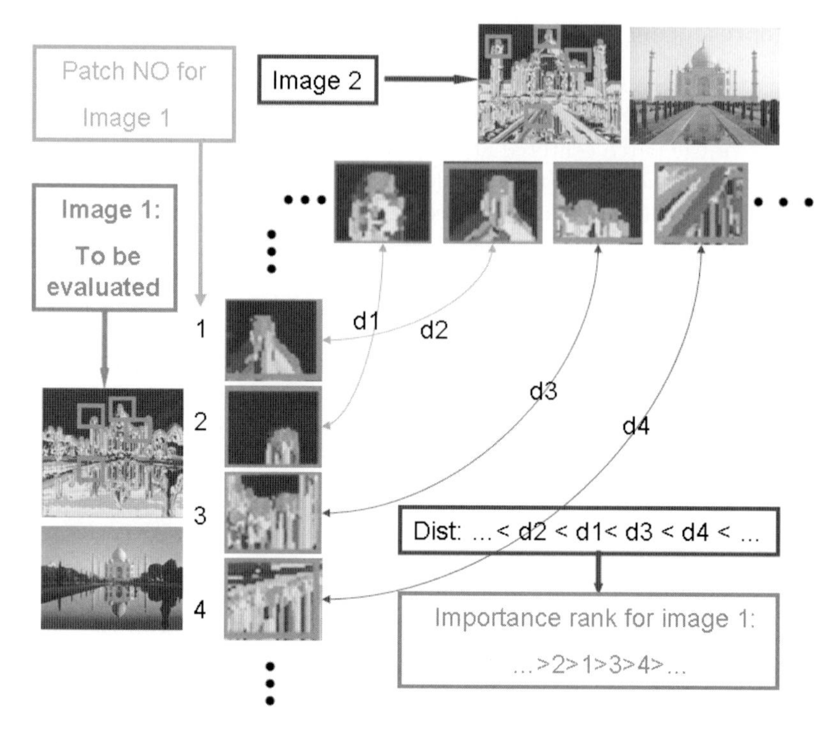

Fig. 4. An Illustration of the Multi-Ranking Amalgamation Strategy (MRAS). Image 1 and Image 2 are two BoWIs. For each patch i in Image1, the most similar patch in image2 can be found with a distance of d_i. We rank these distances and get the importance ranking for the patches in Image1 in this pairwise comparison.

To evaluate the importance for each patch, we assign each patch with a rank-based discount factor[10] according to their ranking position: the more significant the ranking position is, the more discriminative the patch is, therefore significant ranking positions should contribute a larger portion to the final value. The *inverse of log function* (discount function) is used to progressively reduce the voting weight. Therefore, the final ranking score is computed as follows:

$$\overline{R_{ik}} = \frac{1}{W} \sum_{w=1}^{W} \frac{1}{log(1 + R_{ik}(w))} \tag{4}$$

Based on $\overline{R_{ik}}$, the discriminative patches for each concept class can be selected.

4 Experiment

The implementation details and evaluation method will be discussed in this section. Available and widely accepted dataset are not appropriate for assessing the performance of our work on discovering class-specific informative patches.

Table 1. Comparison of the AP for the discriminative object parts detection

	BB	BG	CT	ET	HB	KT	Merlion	RRS	Redeemer
NC(%)	20.48	44.04	7.27	9.66	16.35	12.13	27.25	20.70	13.00
HOG(%)	35.30	59.57	16.30	17.79	31.66	30.27	35.91	33.83	32.33
SIFT(%)	32.29	65.15	15.85	14.56	26.10	32.07	22.67	26.92	16.10
Ours(%)	**57.55**	**62.98**	**21.48**	**41.64**	**39.86**	**57.79**	**48.03**	**49.54**	**49.50**

	SL	SOASM	SOH	TM	LM	TB	VSO	$\overline{MAP}$	
NC(%)	11.89	28.93	24.35	22.85	17.02	29.91	44.97	21.93	
HOG(%)	24.30	39.88	41.36	33.09	40.99	46.93	71.70	36.95	
SIFT(%)	29.73	30.72	36.35	27.71	23.92	39.86	50.72	30.67	
Ours(%)	**43.63**	**60.15**	**71.38**	**34.80**	**54.28**	**57.68**	**80.43**	**51.92**	

BB: Big Ben, BG: Brandenburg Gate, CT: CN Tower, ET: Eiffel Tower, HB: Harbour Bridge, KT: Kuwait Tower, RS: Red Square, SL: Statue of Liberty, SOASM: Sultan Omar Ali Saifuddin Mosque, SOH: Sydney Opera House, TM: Taj Mahal, LM: The Little Mermaid, TB: Tower Bridge, VSO: Vienna State Opera

A challenging dataset can be created for different landmark concepts where each concept contains many real images from different users and with wide variant in viewangle, color, illumination, scale of object and partial occlusion. We collected a Landmarks (National Icons) dataset according to the list of National Icons in Wikipedia[1]. For each national icon category, we selected at least 7 web images from the images returned by the DBpedia-Flickr wrapper (Flickr website)[2]. The dataset currently contains 16 categories in all(images in the dataset are shown in Fig 1). The Ground truth masks of the landmark for every image in every are manually segmented. concept is manually created.

4.1 Experiment Setup

Rectangles with different sizes: 60×60, 40×40, 60×40, 60×40 are used to handle objects with different aspect ratio. To reduce the effect of occlusion, we set $W = 5$ in the multi-rank amalgamation strategy. As for Hierarchical K-means clustering, we set $K = 2, L = 8$. In experiment, the weighting factor for the central patch is 0.5 and 0.0625 for its 8 neighbor patches. Spatial information is important for an object, especially for Landmarks, whose spatial layout are fixed.

We evaluate the performance using the Precision/Recall (PR) curve. B_p and G_t are the regions enclosed by the informative patches and the segmentation masks of ground truth respectively. The recall $r = \frac{|B_p \cap G_t|}{|G_t|}$ shows the proportion of object instances in the image set that has been detected. Whereas the precision $p = \frac{|B_p \cap G_t|}{|B_p|}$ measures the proportion of the detections corresponding to correct object instances. We use the average precision (AP) for the final evaluation.

[1] http://en.wikipedia.org/wiki/National_icons
[2] http://www4.wiwiss.fu-berlin.de/flickrwrappr/

4.2 Performance for Informative Patches Detection

To examine the performance of the proposed MRAS-BoWI approach for informative patches detection, we compare our method with three other algorithms based on normalized correlation(NC), SIFT descriptor and HOG[9] descriptor respectively. More specifically, Each patch of the image is represented by SIFT/HOG descriptor, or the original gray-scale value of every pixel, then euclidian distance is used to calculate the distance for each pair represented by SIFT/HOG descriptor, and Normalized correlation is used to calculate the similarity of the two patches represented by the gray-scale value of each pixel. All the patches are ranked according to this distance/similarity. MRAS is also applied to calculate the final ranking for each patch. As for other parameters, including the patch size, we follow the same settings with MRAS-BoWI.

We list the APs of all the object classes in Table 1. The gradient based, HOG and SIFT methods outperform the Normalized Correlation(NC) method. The proposed MRAS-BoWI method has the best performance for all classes. Fig 5 shows the top nine informative patches detected for each object instance in the two classes. From the results, we can see that that the top informative patches

Fig. 5. Top 9 important patches for each image. Red rectangles with solid line indicate the top 3 patches. Green rectangles with dashed line indicate the top 4-6 patches. Blue rectangles with dash-dot line indicate the top 7-9 patches.

detected are mostly correct and our method of discovering informative patches is very efficient in characterization of landmarks images.

5 Conclusion

In this paper, We present a novel approach to detect discriminative parts for a given class with only a few instances. The contributions are two-folder. Firstly, our proposed BoW Image (BoWI), which uses the characterization of local region as its element, is robust to various object variations. Secondly, the Multi-Ranking Amalgamation Strategy (MRAS) based on BoWI can well detect the most discriminative patches for an object instance. Experiments illustrate that our method can detect those discriminative parts correctly. These informative patches offer a valuable preparation for the object recognition and object detection.

References

1. Lowe, D.G.: Distinctive Image Features from Scale-Invariant Keypoints. International Journal of Computer Vision (November 2004)
2. Sivic, J., Zisserman, A.: Video Google: A Text Retrieval Approach to Object Matching in Videos. In: IEEE Proceedings on International Conference Computer Vision (October 2003)
3. Lazebnik, S., Schmid, C., Ponce, J.: Beyond Bags of Features: Spatial Pyramid Matching for Recognizing Natural Scene Categories. In: Proceedings of IEEE Computer Society Conference on CVPR (June 2006)
4. Fergus, R., Perona, P., Zisserman, A.: A sparse object category model for efficient learning and exhaustive recognition. In: IEEE Computer Society Conference on Computer Vision and Pattern Recognition (2005)
5. Mikolajczyk, K., Tuytelaars, T., Schmid, C., Zisserman, A., Matas, J., Schaffalitzky, F., Kadir, T., Gool, L.: A Comparison of Affine Region Detectors. International Journal of Computer Vision 65(1), 43–72 (2005)
6. Lazebnik, S., Schmid, C., Ponce, J.: Beyond Bags of Features: Spatial Pyramid Matching for Recognizing Natural Scene Categories. In: Proceedings of IEEE Computer Society Conference on CVPR (June 2006)
7. Quack, T., Ferrari, V., Leibe, B., Gool, L.V.: Efficient Mining of Frequent and Distinctive Feature Configurations. In: Proceedings of IEEE International Conference on Computer Vision (October 2007)
8. Felzenszwalb, P., Huttenlocher, D.: Pictorial Structures for Object Recognition. International Journal of Computer Vision 61(1), 55–79 (2005)
9. Dalal, N., Triggs, B.: Histograms of oriented gradients for human detection. In: IEEE Computer Society Conference on Computer Vision and Pattern Recognition, vol. 1 (2005)
10. Järvelin, K., Kekäläinen, J.: IR evaluation methods for retrieving highly relevant documents. In: Proceedings of the 23rd annual international ACM SIGIR conference on Research and development in information retrieval, pp. 41–48. ACM, New York (2000)
11. Bosch, A., Zisserman, A., Munoz, X.: Representing shape with a spatial pyramid kernel. In: Proceedings of the International Conference on Image and Video Retrieval (2007)

12. Fei-Fei, L., Perona, P.: A Bayesian Hierarchical Model for Learning Natural Scene Categories. In: IEEE Computer Society Conference on Computer Vision and Pattern Recognition (2005)
13. Harzallah, H., Schmid, C., Jurie, F., Gaidon, A.: Classification aided two stage localization. In: PASCAL Visual Object Classes Challenge Workshop, in conjunction with ECCV (October 2008)
14. Fergus, R., Perona, P., Zisserman, A.: A sparse object category model for efficient learning and exhaustive recognition. In: IEEE Computer Society Conference on Computer Vision and Pattern Recognition (2005)
15. Fischler, M., Elschlager, R.: The representation and matching of pictorial structures. IEEE Transactions on Computers (1973)

Mid-Level Concept Learning with Visual Contextual Ontologies and Probabilistic Inference for Image Annotation

Yuee Liu[1,2], Jinglan Zhang[1], Dian Tjondronegoro[1], Shlomo Geva[1], and Zhengrong Li[1]

[1] Faculty of Science and Technology, Queensland Univerisity of Technology, Brisbane, Australia
[2] Northwest Agriculture and Forestry University, Yangling, China
{yuee.liu,jinglan.zhang,dian,shlomo,zhengrong.li}@qut.edu.au

Abstract. To date, automatic recognition of semantic information such as salient objects and mid-level concepts from images is a challenging task. Since real-world objects tend to exist in a context within their environment, the computer vision researchers have increasingly incorporated contextual information for improving object recognition. In this paper, we present a method to build a visual contextual ontology from salient objects descriptions for image annotation. The ontologies include not only partOf/kindOf relations, but also spatial and co-occurrence relations. A two-step image annotation algorithm is also proposed based on ontology relations and probabilistic inference. Different from most of the existing work, we exploit how to combine representation of ontology, contextual knowledge and probabilistic inference. The experiments in the LabelMe dataset show that image annotation results are improved using contextual knowledge.

Keywords: Image Annotation, Salient Objects, Visual Context, Ontology, Probabilistic Inference, multi-level concept.

1 Introduction

Object-based image analysis is hard to achieve in real situations since accurate object segmentation is still hard to achieve using current technologies [1]. Salient objects are proposed as a more practical mid-level representation of image content [2]. Salient objects do not exactly correspond to the real objects, but they could capture most common visual properties of object classes. It is defined as visually distinguishable image compounds that can characterize visual properties of corresponding object classes. For example, a salient object "sky" could be described as a set of connected image regions with dominant image components that can be detected semantically by human as "sky".

In order to recognize the semantics of salient objects (mid-level concept), two types of information can be used: 1) the visual appearance of each object; 2) the knowledge about the contextual information. Since real-world object tend to exist in

S. Boll et al. (Eds.): MMM 2010, LNCS 5916, pp. 229–239, 2010.

a context, incorporating contextual information for object recognition has been increasingly realized in the computer vision field [3-5]. *Visual context* corresponds to the likelihood of an object which is expected to be found in some scenes but not others. Most recent methods incorporate co-occurrence of objects to model the contextual information. Wolf and Bileschi used "semantic layers" which indicate the presence of a particular object in the training images, to describe the semantic context [6]. Rabinovich et al. point out that the presence of a certain object class in an image probabilistically influences the presence of a second class [4]. Rabinovich et al. derived semantic context from external knowledge obtained from Google Sets web application which generates a list of possibly related items from a few examples. However, such models have not yet incorporated the explicit spatial context between objects in different scenes. Recently, the work by Galleguillos et al. proposed a new method of object categorization which incorporates both of the co-occurrence context and the spatial context using a conditional random field (CRF) formulation in order to maximize contextual constraints over the object labels of context into a unified framework [7]. However, in these methods context is considered in a much more local way (e.g. pair-pixel or pair regions), and does not model the whole scene. Moreover, contextual information (i.e. spatial or semantic context) is more implicitly adopted and does not make sense to users. In this paper, we learn contextual knowledge and represent it explicitly in ontology to guide image interpretation. We propose visual-contextual ontologies with mid-level concepts for semantics interpretation. Ontologies are not only enriched with visual information but also contextual knowledge (e.g. spatial and co-occurrence relations). Moreover, the enriched ontologies can support the reasoning process to recognize the abstract concepts.

The remainder of this paper is organized as follows: Section 2 presents how the visual-contextual ontologies are built including extracting relations and populating ontologies. In section 3 the ontologies are employed to do the annotation work. Section 4 demonstrates and discusses the evaluation results on the LabelMe image dataset. Section 5 provides the conclusion and future work.

2 Visual-Contextual Ontologies

The most important advantage of ontology is that it could provide a formal framework for supporting explicit and machine-processable semantic definition and enables deriving new knowledge through inference. Accordingly, this section presents our ontologies in detail. They contain not only high-level concepts but also contextual information. They are derived from annotations of a dataset because these annotations provide a snapshot of visual contents of image collections and discover relationships between visual contents that were not immediately obvious.

The LabelMe image database is the most comprehensive public database, and all images are annotated online by multiple volunteers [8]. These annotations contain textual information of objects as well as the contour locations of those objects. Therefore, our ontologies are built based on LabelMe images and their annotations.

2.1 Concept Ontologies Construction

In this paper, subjective concepts (e.g. emotion) are not considered. This paper focuses on the scenes which can be described using components (e.g. field scene, beach scene). Concepts in ontologies are divided into two classes: abstract and concrete concepts. An abstract concept describes the scene of an image, whereas the concrete concept corresponds to the specific objects in this image. A simple observation of the real world reveals that an abstract concept is composed of several concrete objects. For example, an image shows a "beach" scene (i.e. beach is the abstract concept), and it contains several perceivable objects as sand, sea and sky (i.e. these objects are concrete).

The LabelMe images are stored in separate folders whose names strongly indicate the scene of the containing images, whereas each image in the folder is labeled online by multi-users. Therefore, the concepts and the hierarchy among these concepts are extracted based on the folder name in which images are stored and their annotations [9].

The abstract concept is extracted by analyzing the folder name using the standard text processing technologies. Firstly, the most common stop words are removed, and then the meaning words are separated by the concept. The relationships between concepts are traced using the holonymy/hypernymy relationships in the WordNet. The concrete concepts are from the users' manual annotations, and the relationships between the concrete concepts are also obtained based on WordNet.

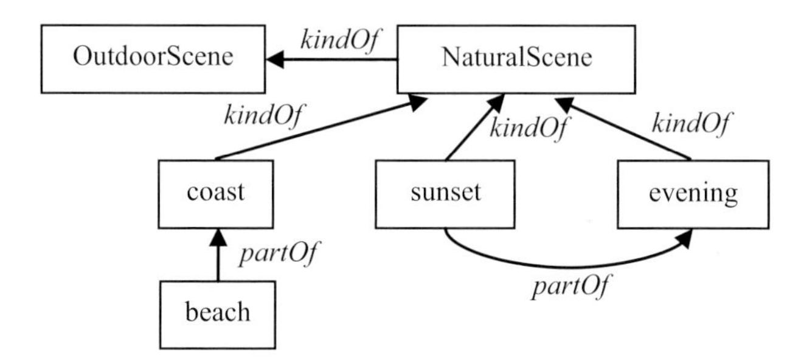

Fig. 1. The abstract concept ontology for image annotation

As the images are labeled by multiple users, the same object could be labeled with slightly different or related phrases. For example, the "sea" region is annotated using "water sea", "sea water", "water", "ocean" etc. These words are slightly different, but have similar meanings. Hence they are converted to a uniform word "sea". Additionally, some concepts are annotated using verbal nouns like "person man standing", "boat cropped". These nouns represent the major meaning so all adjunct words are removed (e.g. "person man standing" to "person", "boat cropped" to "boat"). There are also some concepts which appear less frequently in the database, such as "rainbow", "person", etc. We remove all these concepts because they are not dominant concepts and not sufficient as the training set from the classification perspective or they are less visible and hard to be extracted from the segmentation perspective). Figure 1 and Figure 2 shows the hierarchy among concepts.

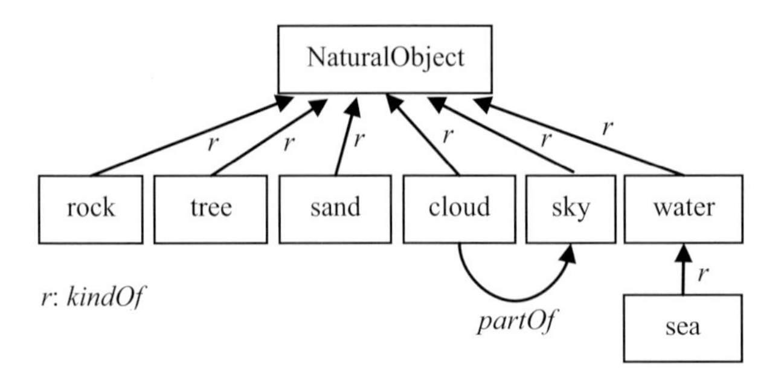

Fig. 2. The concrete concept ontology for image annotation

2.2 Relationships among Concepts

In the real world, objects are naturally related to other objects. For example, the "sea" region is similar to the "sky" region. Ontology, as a popular representation method of the real world, could contain any kind of relations (semantic, spatial, temporal, spatiotemporal) among which semantic and spatial relations are the most suitable for description of image content [10]. Therefore, we only use semantic and spatial realtions. Additionally, one concept naturally coexist with other concepts. For example, "sea" often appear with "sand". However, this natural property is usually neglected. In this paper, co-occurrence relations are incorporated to provide a view of contextual knowledge together with other relations.

Co-occurrence Relations between Concepts. These relations R^{co} are used for evaluating which concepts are more inclined to appear together. They are defined statistically on the training dataset. It is defined as

$$R^{co} = \{co(c_i, c_j)\}, \quad c_i, c_j \in C$$

where $co(c_i, c_j) = \dfrac{\sum |c_i \in I^p \wedge c_j \in I^p|}{\sum |c_i \in I^p \vee c_j \in I^p|}$, C is a set of concepts c_i.

It could be represented by a co-occurrence matrix between any concept pair as below. It is a symmetric matrix, and the horizontal and vertical axes represent every concept in the dataset. The white blocks indicate the selected concept pair with significant correlations. The brighter the block is, the higher of possibility the concept pair occurs together. For example, the concept pair <*sea*, *sky*> occurs together more frequently than the concept pair <*sea*, *sand*> does in the database. The matrix also could reflect the high-level concepts. For example, if concept *sky*, *sea* and *sand* appear more frequently, it should be a beach scene with high possibility than the sunset scene. It is also noticed that the co-occurrence of the "*sky*" and "*cloud*" is low. The reason is that in the testing dataset, there are not as many "*cloud*" regions as the "*sky*" regions, and in most cases it is hard to distinguish between "*cloud*" and "*sky*". In this paper, we only consider sky detection.

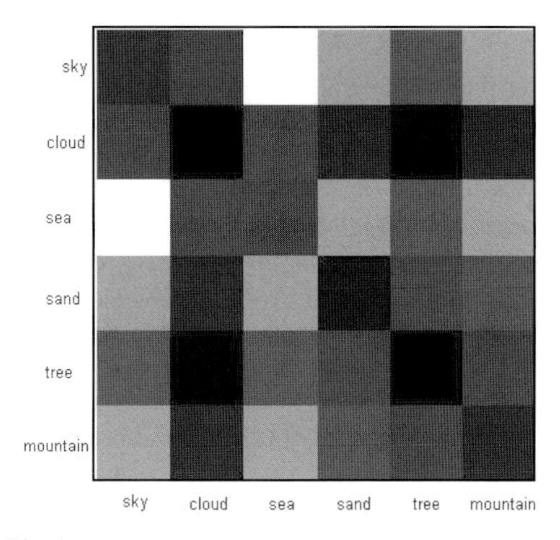

Fig. 3. Mutual information among the concept pairs

Semantic Relations between Concepts. The MPEG-7 standard provides a set of semantic relations R^{sem} that have proven to be useful for image analysis. It is defined as

$$R^{sem} = \{sem(c_i, c_j)\} \quad sem = \{similarity, partOf, specification\} \quad i, j = 1, ..., N$$

The similarity relation measures the *similarity* among any pair of concepts (e.g. sky/sea), the *partOf* relation represent the belonging relation between the concept pair (e.g. cloud/sky) and specification relation is the *kindOf* relation (e.g. sea/water). The *partOf* and *kindOf* relations are used to create the hierarchy of abstract and concrete concepts.

Spatial Relations between Concepts. It is a more straightforward relation. It could be used to refine the concept interpretation. For example, "sky" is often above the "sea", and "grass" is usually below the "sky". In this paper, four spatial relations are used as defined below equations. These 4 directions are defined based on the bounding box of the segmented objects as shown in Figure 4.

$$R^{spat} = \{spat(c_i, c_j)\} \quad spat = \{above, below, left, right\} \quad i, j = 1, ..., N$$

In training dataset, each image is segmented and labeled with a concrete concept. The relations of different segments are calculated for future region annotation optimization.

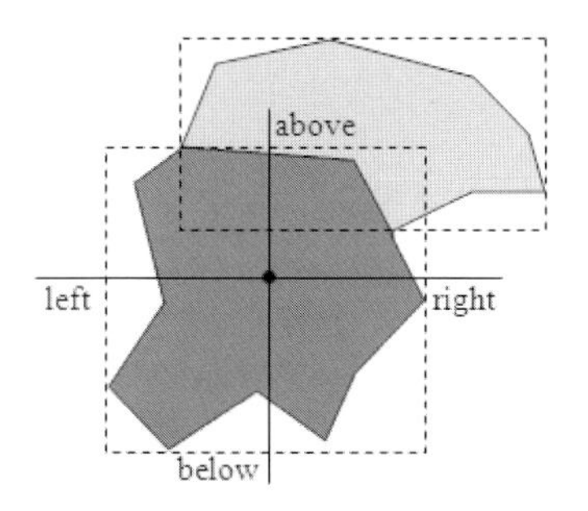

Fig. 4. Spatial relationship between two objects

All above defined relations $R = \{R^{co}, R^{sem}, R^{spat}\}$ work together to comprise the context model for concepts. A scene description could be represented based on these relations in Table 1. For example, a beach scene contains sand, sea, sky and (probably) rock. The beach scene is also a coast scene through tracing the *kindOf* relation.

2.3 Ontology Population with Region Types

When the backbone (concepts and relation hierarchy among the concepts) of ontologies are built, the concepts in the ontology are populated with instances. Usually the concepts are populated with images directly. However, concept may have several visually-different instances. The concept "sky" may have many instances differentiating in the color features. It could be blue when in the day, and could be red when in the evening. Therefore, we select region type as the representative instance of a set of regions for concepts [11]. The creation process of region types is performed by extracting visual features and performing an unsupervised clustering method. After clustering, the regions with similar features are grouped into a cluster and represent this concept in the specific scene. The region types are automatically obtained using the centers of clusters.

3 Image Annotation with Visual-Contextual Ontology

The process of creating visual-contextual ontology is described. It could be used to perform multi-level image annotation. We propose a two-step annotation algorithm. The salient object are labeled first using concrete concepts, and refined using ontological contextual knowledge, and then high-level concepts are calculated based on concrete concepts and probabilistic inference. This idea is derived from a simple observation of the real world which reveals that an abstract concept is composed of several concrete objects. For example, the abstract concept "beach" could be described via several concrete concepts. The *partOf* relationships between concrete and abstract concepts are shown in Figure 5.

This process of image annotation is achieved in two steps. Initially, candidate concepts are obtained by comparing the dissimilarity between the submitted image and the region types in ontologies. If a similar region type is found, this image is annotated with the higher level concepts, resulting in finding more comprehensive annotations. Salient objects with similar features (e.g. color features) could belong to totally different concept categories (e.g. *sky* regions could be mixed with *sea* regions based on their blue color), but they do not have different spatial relations (e.g. *sky* regions are usually above *sea* regions), so annotations are refined based on ontology relations defined in section 2.2. The algorithm of salient object labeling is as below.

program C_p = SalientObjectLabelling(p)
  C_p : return annotations of image p;
  var p : an input image;
     sf_i^p : segmentation region i of the input image p;

```
        rt : region types of concept c
        d : visual distance between two regions;
        C_i^p : annotations of segment i in the image p;
        C_p : annotations of the image p;
        τ : threshold;
begin
  segment p into sf_i^p, and extract visual features;
  // initial annotation of concrete concepts
  for each region sf_i^p
    calculate the distance d = dist(rt, sf_i^p);
    if d < τ;
      annotate sf_i^p with concept c to which rt is related
        C_i^p = C_i^p + {c};
    end
  end
  // Optimization using ontology relations
  for region sf_i^p with its all candidate concepts C_i^p
    check c_i^p (c_i^p ∈ C_i^p) based on ontology relations R^spat and
    locations of sf_i^p;
   if location of sf_i^p meets location of c_i^p
     annotate sf_i^p with c_i^p  C_p = C_p + c_i^p;
   end;
 end;
```

After detection of concrete concepts, abstract concepts are calculated based on the co-occurrence and semantic relations in section 2.2. In order to exploit the conditional dependence among concrete and abstract concepts, the generative probabilistic graphical model – Bayesian Network (BN) is used to enable probabilistic concept reasoning. Thus, annotations could be detected with certain probability.

The structure of that BN is based on Figure 5. At the training stage, parameters of BN are learned via maximum likelihood estimate and expectation maximization algorithm. At the testing stage, given an unknown image, we segment the image into regions and get the annotations for each region from the concrete concept ontology. After that, posterior probabilities of concrete concepts of the given image at the given leaf concept in the abstract concept ontology is calculated based on Bayes' Rule. For a given high-level image concepts AC_k from the abstract concept ontology, its posterior probability P is calculated from the children nodes ($CC_1,\ldots,CC_N$) (i.e. the leaf nodes from concrete concept ontology). We introduce binary random variables $O_1,\ldots,O_N \in [0,1]$ to represent the existence of each concrete concept respectively. $O_i=1$ shows that the ith concept appears in the scene, otherwise $O_i=0$. Mathematically, P is calculated as following:

$$P(AC|O_1,\ldots,O_N) = P(O_1,\ldots,O_N|AC)\cdot P(AC) \quad (1)$$

The image annotations could be attached with the most related abstract concept with the maximum value of posterior probability via Maximum A Posterior criterion as below

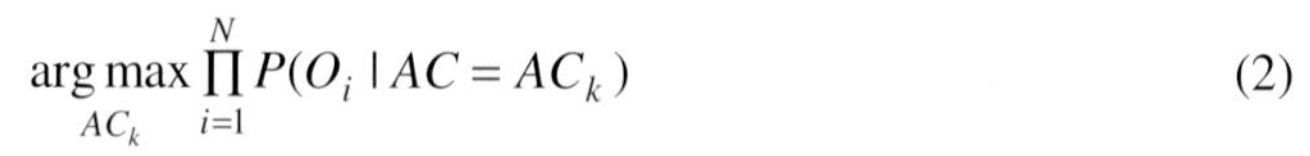

$$\arg\max_{AC_k} \prod_{i=1}^{N} P(O_i \mid AC = AC_k) \quad (2)$$

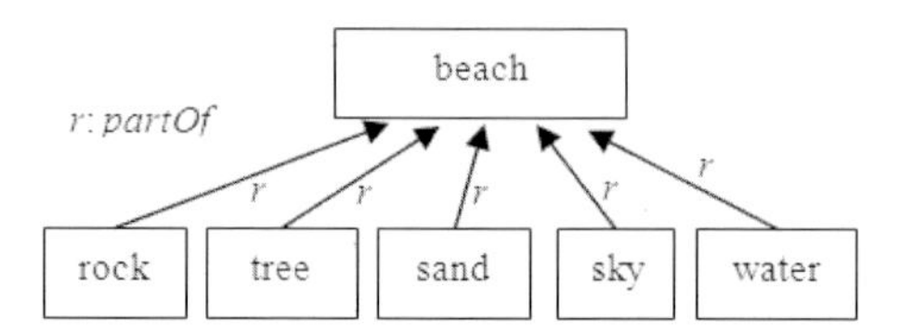

Fig. 5. The relationships between abstract and concrete concepts for ontology inference

4 Experiments

In order to evaluate the proposed image annotation algorithm, the LabelMe database is adopted because this dataset is fully annotated by multiple users. The following 6 concepts, which represent the meaningful objects in the real world, are obtained by using WordNet and eliminating the less-frequent concepts (e.g. building) as described in Section 2: *Sand*, *Tree*, *Sky*, *Cloud*, *Rock*, and *Sea*. These concepts cover a majority of the concepts found in the dataset, and are included in the concrete ontology.

As the proposed annotation algorithm is based on the salient objects, each image is segmented using an improved JSEG segmentation algorithm [12], and we develop a human-computer interaction tool to facilitate the association of segment fragment with the corresponding concept. Color features (*color moments in HSV color space)* are extracted from each segment. K-means are employed to calculate region types through running on different concept categories respectively. In order to avoid bias of the testing set, images are listed in the order of their names, and then every 5 images are divided among which the first 3 images are assigned as training images and the last 2 images as testing images.

The results are summarized in Table 1 and Figure 6. From Figure 6, after employing ontology relations, the annotation results are improved. While tracing the hierarchy among the concepts, images also could be labeled with abstract concepts, thus leading to multi-level annotations. Table 1 shows that regions are labeled with more precision concepts after using the ontology relations. Concept *cloud* has less precision as we observe that *cloud* and *sky* are hard to distinguish from each other, and sometimes they are mixed together as shown in Figure 6(b) and 6(d). They are so similar that they are annotated with the same concept *sky*.

As this paper exploits the relationship between concrete concepts and abstract concepts which could be described by these concrete concepts, images in the database are segmented using an improved JSEG segmentation algorithm [12] and regions are labeled first (i.e. detecting concrete concept). After that Bayesian Network are learned to bridge the concrete and abstract concepts.

The precision p and recall r is used to measure the performance of this high-level concept detecting algorithm using Bayesian inference.

$$p = a/(a+b) \qquad r = a/(a+c)$$

where a is the number of images that are correctly annotated; b is the number of images which are not correctly annotated; c is the number of annotations which are missed.

The experimental results showed that the abstract "beach" scene could be recognized by combination of several concrete concepts (e.g. tree, sand, sky, water). The precision and recall of the experimental results are 72.1% and 59.7 respectively.

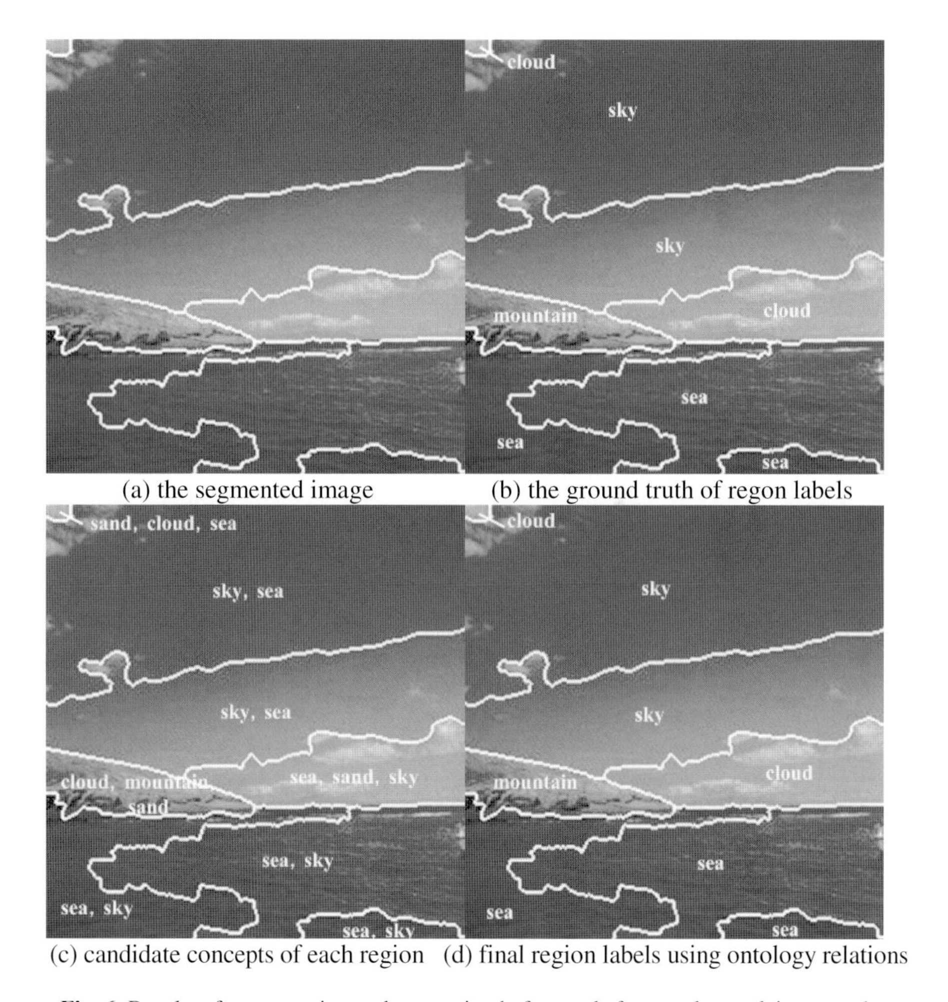

(a) the segmented image (b) the ground truth of regon labels

(c) candidate concepts of each region (d) final region labels using ontology relations

Fig. 6. Results of segmentation and annotation before and after ontology relaionts used

Table 1. Precision of concept detection before and after employing ontologies

concept	before	after
sky	69%	75%
sea	68%	76%
cloud	55%	60%
sand	65%	70%
mountain/rock	60%	69%
tree	60%	65%

5 Conclusion and Future Work

In order to recognize the semantics of salient objects (mid-level concept), this paper presents visual-contextual ontologies which contain not only the MEPG-7 semantic relations but also include the co-occurrence and spatial relations. Ontologies are populated with region types which are created using the clustering techniques. Initially each region is labeled by several concepts from the concrete ontology by calculating the similarity between this region and the region types. The regions are finally annotated by employing the relations in ontologies. High-level concepts are obtained through the probabilistic inference based on semantic and co-occurrence relations. The evaluation shows that employing the relations in ontologies could improve the annotation results. In the future, more image features will be used for salient object labeling. The proposed visual-contextual ontology will be applied to a large set of images which contain different scenes.

References

1. Ge, F., Wang, S., Liu, T.: Image Segmentation Evaluation From the Perspective of Salient Object Extraction. In: IEEE International Conference on Computer Vision and Pattern Recognition (CVPR), New York, USA (2006)
2. Fan, J., Gao, Y., Luo, H., Xu, G.: Salient Objects: Semantic Building Blocks for Image Concept Interpretation. In: 3rd ACM International Conference on Image and Video Retrieval (2004)
3. Heitz, G., Koller, D.: Learning Spatial Context: Using Stuff to Find Things. In: Forsyth, D., Torr, P., Zisserman, A. (eds.) ECCV 2008, Part I. LNCS, vol. 5302, pp. 30–43. Springer, Heidelberg (2008)
4. Rabinovich, A., Vedaldi, A., Galleguillos, C., Wiewiora, E., et al.: Objects in Context. In: IEEE International Conference on Computer Vision (ICCV), Janeiro, Brazil (2007)
5. Tu, Z.: Auto-context and its application to high-level vision tasks. In: IEEE International Conference on Computer Vision and Pattern Recognition (CVPR), Anchorage, Alaska (2008)
6. Wolf, L., Bileschi, S.: A Critical View of Context. International Journal of Computer Vision 69(2), 251–261 (2006)
7. Galleguillos, C., Rabinovich, A., Belongie, S.: Object Categorization using Co-Occurrence, Location and Appearance. In: IEEE International Conference on Computer Vision and Pattern Recognition (CVPR), Anchorage, Alaska (2008)

8. Russell, B.C., Torralba, A., Murphy, K.P., Freeman, W.T.: LabelMe: a Database and Web-based Tool for Image Annotation. International Journal of Computer Vision 77(1-3), 157–173 (2008)
9. Gao, Y., Fan, J.: Incorporating concept ontology to enable probabilistic concept reasoning for multi-level image annotation. In: 8th ACM SIGMM International Workshop on Multimedia Information Retrieval, pp. 79–88 (2006)
10. Spyrou, E., Mylonas, P., Avrithis, Y.: Using region semantics and visual context for scene classification. In: Proceedings of 15th International Conference on Image Processing (ICIP 2008), pp. 53–56 (2008)
11. Mylonas, P., Spyrou, E., Avrithis, Y.: Enriching a context ontology with mid-level features for semantic multimedia analysis. In: 1st Workshop on Multimedia Annotation and Retrieval enabled by Shared Ontologies, co-located with SAMT 2007 (2007)
12. Liu, Y., Zhang, J., Tjondronegor, D., Geva, S., et al.: An Improved Image Segmentation Algorithm for Salient Object Detection. In: The 23rd International Conference on Image and Vision Computing New Zealand (IVCNZ 2008), pp. 1–6 (2008)

A Color Saliency Model for Salient Objects Detection in Natural Scenes

Minghui Tian, Shouhong Wan, and Lihua Yue

Computer Science Department, University of Science and Technology of China, Hefei, Anhui, 230027, P.R. China
mhtian@mail.ustc.edu.cn, {wansh,llyue}@ustc.edu.cn

Abstract. Detection of salient objects is very useful for object recognition, content-based image/video retrieval, scene analysis and image/video compression. In this paper, we propose a color saliency model for salient objects detection in natural scenes. In our color saliency model, different color features are extracted and analyzed. For different color features, two efficient saliency measurements are proposed to compute different saliency maps. And a feature combination strategy is presented to combine multiple saliency maps into one integrated saliency map. After that, a segmentation method is employed to locate salient objects' regions in scenes. Finally, a psychological ranking measurement is proposed for salient objects competition. In this way, we can obtain both salient objects and their rankings in one natural scene to simulate location shift in human visual attention. The experimental results indicate that our model is effective, robust and fast for salient object detection in natural scenes, also simple to implement.

Keywords: Color saliency, natural scenes, object detection, visual attention.

1 Introduction

Salient objects detection plays a very important role in many research areas which are related to computer vision. It helps object recognition, scene analysis and content-based image/video retrieval. The ability of the human visual system to detect salient objects is extraordinarily fast and reliable. So in recent years there have been many researches to use this visual attention mechanism to solve some problems in detecting salient objects.

The human brain and visual system pay more attention to some parts of an image. This is called visual attention [1]. The visual attention mechanism has been studied by researchers in physiology, psychology, neural systems, and computer vision for a long time. This mechanism can basically help to detect regions of interest and to allocate computation and memory resource rationally. However, computational modeling of this basic intelligent behavior still remains a big challenge.

How is the visual attention mechanism achieved in the human vision system? According to literature [2], it is believed that two stages of visual processing are involved: first, the parallel, fast, but simple pre-attentive process; and then, the serial, slow, but complex attention process. The former refers to the sensory attention driven

S. Boll et al. (Eds.): MMM 2010, LNCS 5916, pp. 240–250, 2010.

by environmental events, commonly called bottom-up or stimulus-driven. The latter is the voluntary attention that refers to both external and internal stimuli, commonly called top-down or goal-driven.

Previous methods for salient object detection can be grouped into three classes. The first class of methods aims to directly group or segment all the image pixels into some disjoint regions which are expected to coincide with the underlying salient objects, and tries to locate the salient objects in them based on region competition. Recently we have witnessed some methods, such as region based method [3], [4] and [5]. Those methods usually are segmentation-dependent, and have difficulties in incorporating perceptual rules and keeping object-integrality. The second class of methods is designed based on some spectral or subspace transformations, such as spectral residual approach [6] and subspace analysis [7]. However, many of those methods usually have difficulties in finding globally optimal boundaries of salient objects. The third class of methods aims to compute multiple feature saliencies and combine different saliencies into one integrated saliency map to locate the salient objects based on Treisman's Feature Integration Theory [8], such as Itti's neural model [9, 10]. The challenge of those methods is how to measure different feature saliencies and combine them dynamically to keep object-integrality.

In this paper we focus on the fast pre-attention process and propose a fast robust bottom-up visual saliency model based on FIT [8] and the plausible architecture proposed by Koch and Ullman [2]. In our bottom-up model, different color features are extracted from the original image, and two efficient saliency measurements for color features are introduced. After that, different color saliency maps are combined into a single integrated saliency map. Salient regions, which are regarded as salient object candidates, can "pop up" automatically in this integrated saliency map. Then, we employ a saliency segmentation method and a region filter to obtain the locations and contours of salient objects in scenes. Finally, a fast psychological measurement for salient object competition is given to compute the rankings of salient objects. Comparing with precious works above, our contribution is the global measurement for saliency computation and the computational method for simulation of location shift of visual attention. These benefit our model to have the chance to keep the object-integrity and more suitable for further object recognition/classification tasks.

The rest of the paper is organized as the following. In next section, our visual saliency model is introduced in detail, including feature extraction, saliency map computation and feature combination strategy. In section 3, the salient object detection method based on our model is presented, including salient object extraction and salient objects competition. Section 4 presents the experimental results and evaluations for our model. Conclusions and future works will be drawn in Section 5.

2 Visual Saliency Model

In our model, different color features are extracted to describe different feature saliency and combined into a single topographical saliency map. And the purpose of the saliency map is to represent the local conspicuity at every location in the visual field by a scalar quantity and to guide the selection of visual salient objects based on the spatial distribution of saliency. Different spatial locations compete for saliency within each feature map, so that only locations that locally stand out from their surround can persist, which means only salient objects will be analyzed for further.

The framework of our color saliency model is shown as Fig. 1. First, for natural scene images, we choose HSI color space instead of RGB color space as color features to describe visual channels, because HSI color space is more similar to human's visual sense. For different factors in color theory, two efficient global saliency measurements are proposed to compute saliency maps. These saliency maps describe how different each location of the input image is from the average saliency value in different color features. After that, all these color saliency maps are normalized by an exponential amplification method to enhance salient objects, and combined into one integrated saliency map in the third stage. Comparing with Itti's model [9], the global saliency measurements which we propose benefit our model more efficient to keep the object-integrity (including the contours and the shapes) during the extraction of salient objects. And also the computational complexity of our model is much lower than Itti's.

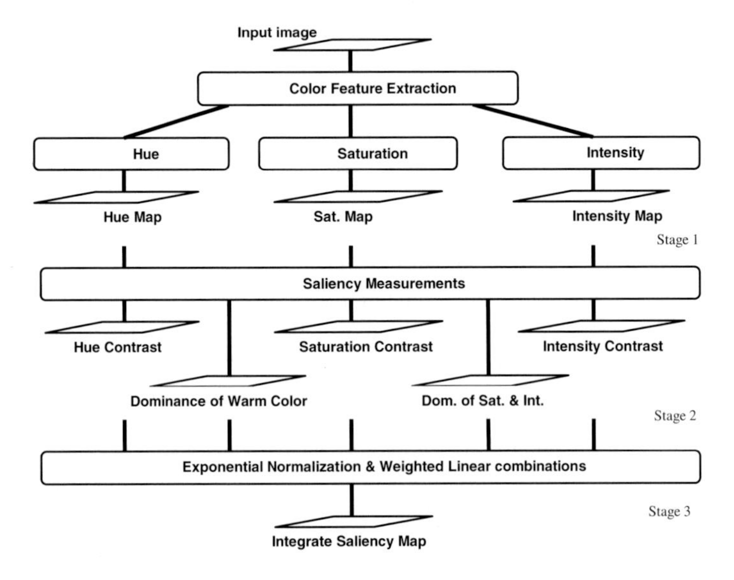

Fig. 1. The Framework of our saliency model. This framework is based on Treisman's Feature Integration Theory and Koch's architecture. It supports multiple parallel computations for different feature channels.

2.1 Saliency Maps Computation

In early vision system [11], we simply choose color features to describe a static natural colored image. For one colored image, we transform the image into a perceptually uniform HSI color space. And we mark each color channel with a unified token F_i (i=1, 2, 3) which denote 3 color feature channels (H, S, I). Then we mainly consider about the contrast theory of color [12] and dominance in color [3, 13].

1. ***Contrast of Hue***: The difference of hue angle on the color wheel contributes to creation of contrast. High difference will obviously cause more effective contrast. Due to circular nature of hue, the largest difference between two hue values can be 180°.
2. ***Contrast of Saturation***: A contrast is produced by low and highly saturated colors. The value of contrast is directly proportional to the magnitude of the saturation difference. Highly saturated colors tend to attract attention in such situations unless a low saturated region is surrounded by highly saturated one.
3. ***Contrast of Intensity***: A contrast will be visible when dark and bright colors co-exist. The greater is the difference in intensity the more is the effect of contrast. Bright colors catch the eye in this situation unless the dark one is totally surrounded by the bright one.
4. ***Dominance of Warm Color***: The warm colors dominate their surrounding whether or not there exists a contrast in the environment.
5. ***Dominance of Brightness and Saturation***: Highly bright and saturated colors are considered as active regardless of their hue value. Such colors have more chances of attracting attention.

To describe the saliency of color, two different global saliency measurements are proposed here for the contrast-based features (1, 2, 3) and dominance-based features (4, 5) respectively. Now we use a uniform token for the three color channels (H, S, I) as F_i (i=1,2,3). And the corresponding feature saliency map which we talked above is marked as S_i (i=1,2,3,4,5). So all the five saliency maps can be computed as:

For ***Contrast-Based Features***:

$$DF_i(x,y) = \left| F_i(x,y) - \overline{F_i} \right| \quad i = 1,2,3 \tag{1}$$

$$S_i(x,y) = 1 - \exp\left\{ -\frac{DF_i(x,y)}{\overline{DF_i}} \right\} \tag{2}$$

where DF_i is the diff-image of feature map F_i (H, S or I); S_i is the saliency map of F_i.

For ***Dominance-Based Features***:

$$S_4(x,y) = \begin{cases} Int(x,y) \cdot Sat(x,y) \cdot \cos(\theta) \cdot \delta & \text{if } \frac{\sqrt{2}}{2} \leq \cos(\theta) \leq 1 \\ 0 & \text{otherwise} \end{cases} \tag{3}$$

$$S_5(x,y) = Int(x,y) \cdot Sat(x,y) \cdot \delta \tag{4}$$

where $Int(x, y)$ and $Sat(x, y)$ denote the intensity and the saturation respectively; θ is the Hue for pixel (x, y) and normalized to [0, 2 π]; δ is a constant value.

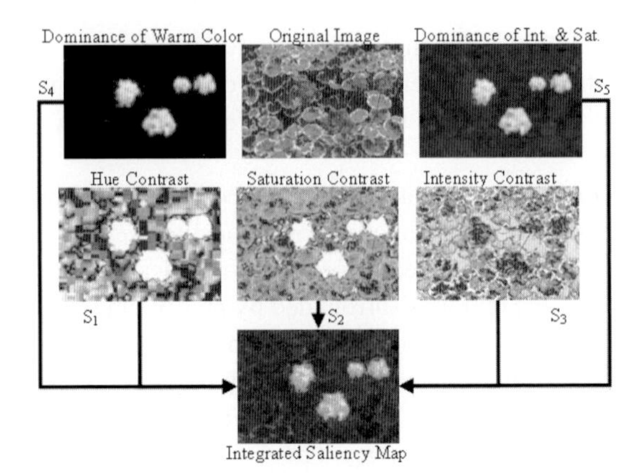

Fig. 2. An Example of Color Saliency Maps. Different color features capture different saliencies for the same scene. And in the integrated saliency map, we can see that the red flowers and some edges of leaves survive from the feature maps combination.

2.2 Feature Combination Strategy

Before we combine multiple saliency maps, we normalize and enhance each saliency map by improving the normalization operator N(.) in [9]. According to literature [14], it would be better to use an exponential coefficient to enhance the integrated saliency map. So all color saliency maps S_i (i=1, 2, 3, 4, 5) are enhanced, normalized by our exponential normalization operator $N_{\exp}(.)$, and finally combined into one integrated saliency map S as bellow:

$$N_{\exp}(S_i) = (M - \overline{m})^2 \cdot (S_i)^{\gamma} \tag{5}$$

$$S = \sum_{i=1}^{FNum} w_i \cdot N_{\exp}(S_i) \tag{6}$$

where M and m are the global maximum value and the local maximum value in S_i; $FNum$ denotes the number of feature categories (in this paper, it is 5); w_i is the weight of Feature F_i ; γ is a constant value and is set to 3 in our experiment. An example for this section is given in Fig. 2.

3 Salient Objects Detection

One point we have to emphasize is that in this paper the "object" here refers to a perceptual object rather than a real object or a natural object, such as a person, a cat, or a house. What we are studying here is low-level visual process and specifically data-driven or bottom-up visual attention. Without high-level knowledge or top-down information, it is impossible to put those regions that are perceptually heterogeneous together to compose a real object. For example, at early stage, it is infeasible to group the regions of a red roof and a white wall into a whole region representing a house. It is more likely that we focus on the red region and the white region separately. After using high-level knowledge to find the relationship between them, we can group them together to form the concept of a house. However, a perceptual object may be or part of a real object. In other words, a real object may consist of one or multiple perceptual objects.

Although it is hard to give a rigorous definition of a perceptual object, we can still get some clues from our common sense. A perceptual object is spatially connected and homogenous in color or intensity, and has high contrast compared with its surrounding. Hence a perceptual object should have the following defining characteristics:

- Being contrasted relatively to the background.
- Having a bounded spatial extension and one or several closed contours.
- Usually being a main or important part of one natural scene, and has a measurable size.

3.1 Salient Object Extraction

The integrated saliency map describes the local saliency at every location in the visual field. So in this map, only salient locations that locally stand out from their surround can persist. In other words, most locations represent low saliency value and several salient regions represent high value. Hence it's not very difficult to segment this integrated saliency map. The classic Minimum-Error segmentation method [15] is employed to segment the integrated saliency map and extract the salient objects from their background.

According to the characteristics of perceptual objects which are defined above, one salient object region in a natural scene usually cannot be some very small region. So in order to clear some noises and blurs we collect the salient regions by their sizes in our region filter. And if a region is too small, it will be removed from the final saliency map. Only those salient regions whose sizes are in top N or above a threshold T (T>0) can persist in the ROI Map. Some morphology methods, such as dilation and erosion, are also employed to fix holes inside the salient objects. An example of this process is shown as Fig. 3.

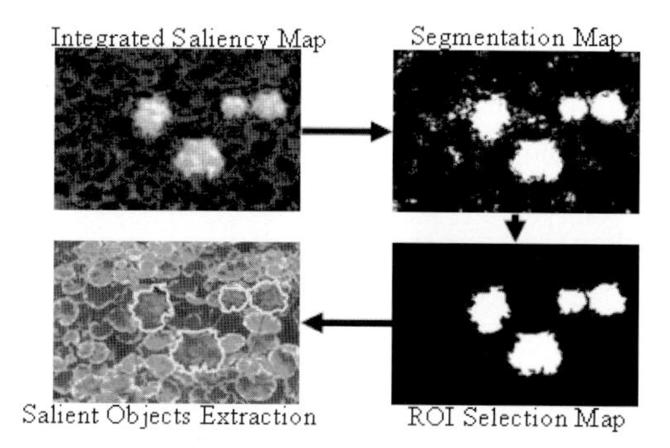

Fig. 3. An Example of Salient Objects Extraction. In this figure, we can see that the edges of green leaves cannot compose a salient object and are ignored, so we only get the red flowers as salient objects for this example scene.

3.2 Salient Objects Competition

We mainly consider several factors below and effects in psychology for location shift of focus of attention in human vision system according to [16, 20].

1. ***Area factor*:** It is obvious that larger objects will have larger effect to others. This is represented as area factor which is simply the ratio of the area of the object region to the area of the whole image and is computed as:

$$\varepsilon_1(A_i) = \frac{A_i}{AreaOf\,\mathrm{Im}\,age} \tag{7}$$

where A_i is the area size of Salient Object i.

2. ***Global effect*:** In human vision system, the attractiveness of a unit is affected by the nearer neighbors much more than by the father ones. So, this property could be represented as a distance factor, which is an exponential function of the spatial distance between regions. The factor is regarded as $\varepsilon_2(DS_{i,j})$, which is calculated:

$$\varepsilon_2(DS_{i,j}) = 1 - \exp(-DS_{i,j} / 2\sigma^2) \tag{8}$$

where $DS_{i,j}$ is the relative spatial distance between Object i and Object j, normalized to [0, 1].

3. ***Central effect*:** It is referred as central effect that while watching an image observers have a general tendency to stare at the central locations. Here a position factor is used to evaluate the central effect, which is represented as an exponential function:

$$\varepsilon_3(P_i) = 1 - \exp\left(-\frac{P_i^2}{2\sigma^2}\right) \tag{9}$$

where P_i is the relative distance of the region away from the center of the image and is normalized to [0, 1]; σ determines saliency of marginal regions.

Considering all the psychological factors above, in a natural scene the comprehensive competitive power of one salient object can be obtained as:

$$CP_i = w_I \cdot \overline{VS_i} \cdot \varepsilon_1(A_i) \cdot \varepsilon_3(P_i) + w_O \sum_{j=1}^{N-1} \left(\overline{VS_j} \cdot \varepsilon_1(A_j) \cdot \varepsilon_2(DS_{i,j})\right) \tag{10}$$

where w_I and w_O are the weights of inner-factors and outer-factors for salient objects respectively, and both are set to 0.5 in our experiment; VS_i and VS_j are the average saliencies of the corresponding Salient Object i and j respectively in the final integrated saliency map S. An example of this subsection is shown as Fig. 4.

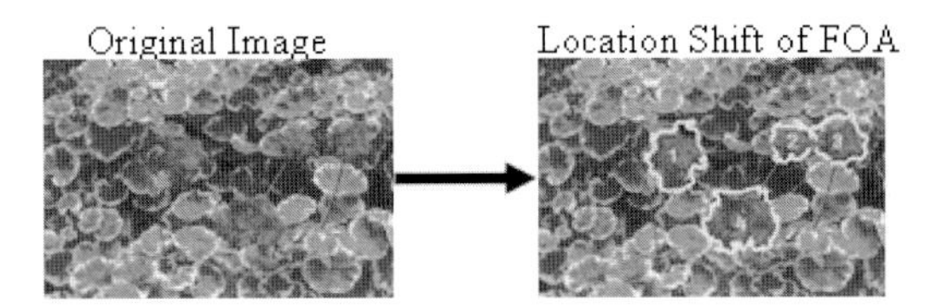

Fig. 4. Location Shift of Focus of Attention. This figure shows the results of salient objects competition to simulate location shift of focus of attention in human vision system. From the figure, we can see the simulation for location of FOA shifts is just similar to human's eyes.

4 Experiment and Evaluation

Our experiments are performed on a PC with AMD Athlon™ XP 2600+ (1.91GHz) processor and 1G memory. The operating system is Microsoft Windows XP Professional SP2, and the software environment is Matlab-7.0.4.

In our experiment, we provide 200 natural scene images with naïve subjects. These images are taken partly from Internet and partly from [7], [17], [18] and [19]. Each subject is instructed to "select regions where objects are presented". All these are resized into the resolution of 800*600 for preprocess. From the results shown in Fig. 5 and Fig. 6, we can see that salient objects are extracted from their background effectively. And we can also obtain their rankings to simulate the location shift of focus of attention between multiple salient objects in human vision system, such as Fig. 6. In Fig. 6, we also compared the results between our model and Itti's model. We can see that our results are much better than Itti's on the point of keeping the integrity of objects, and more suitable for further processing tasks.

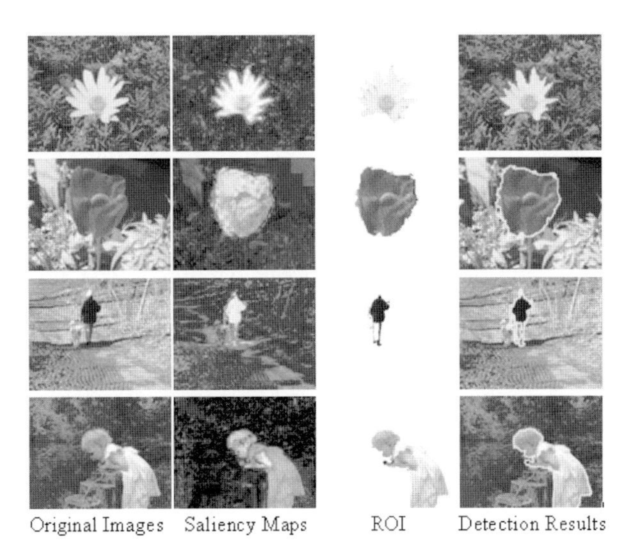

Fig. 5. Examples of Single Salient Object Detection. This figure shows some results of our model for single salient object detection. It is obvious that our model captured the most salient object in these natural scenes.

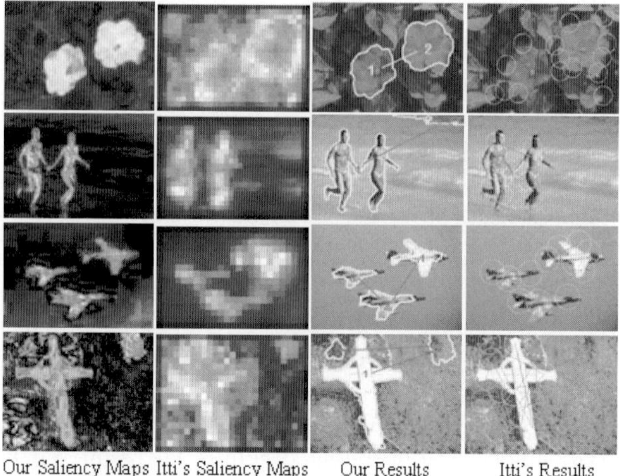

Fig. 6. Examples of Location Shift between Multiple Salient Objects in Comparison with Itti's Model. From this figure, we can see that our saliency map is better than Itti's, and this location shift of our results is better in keeping the physical object-integrity and shapes.

For further discussion, we modified a simple and efficient evaluation method in [6] to evaluate the effectiveness and robustness of our model. For each input $I(x,y)$, the binary image obtained from hand-labeler is denoted as $O(x,y)$, in which 1 denotes for target objects, 0 for background. Given the generated saliency map $S(x,y)$ which is normalized to [0, 1], the Hit Rate (*HR*) and the False Alarm Rate (*FAR*) can be obtained as:

$$HR = E\left(O(x,y) \cdot S(x,y)\right) \tag{11}$$

$$FAR = E\left(\left(1 - O(x,y)\right) \cdot S(x,y)\right) \tag{12}$$

$$O(x,y) = \begin{cases} 1 & \text{if it is target object} \\ 0 & \text{if background} \end{cases} \tag{13}$$

And we compared the *HR* and *FAR* between our model and Itti's model. From Table 1, it is obvious that the *HR* of our model is much higher than that of Itti's model, and our *FAR* is lower. The average time cost of our model is less than half of Itti's model, with all the images in resolution of 800*600.

Table 1. Performance of Our Model and Itti's

	Ours	**Itti's Model**
HR	0.8022	0.4953
FAR	0.1571	0.3051
Average Time	6.315s	14.731s

According to the results above, we can see that our approach is effective for salient object detection in natural scenes. Comparing with Itti's model, our model separates salient objects from their backgrounds more accurately and more effectively. In addition, the speed of our model is even faster than that of Itti's model. In this point, our approach is more similar to human's visual attention process.

5 Conclusion

In this paper we proposed a color saliency model for salient objects detection in natural scenes. In our model, different color features are extracted, analyzed, and finally fused into a single saliency map. Two efficient saliency measurements are given for different color factors. In addition, we presented an approach for salient objects competition to simulate the location shift of focus of attention in human vision system. The experimental results indicate that our model is effective and robust for salient objects detection in natural scenes. However, the deficiency of our model is that detection results are perceptual objects, not real objects. For future work, we plan to experiment with novel invariant features to improve the hit rate of our model. We also would like to analyze the relationship between different salient objects to conduct semantic model for scene understanding and integrate top-down techniques for physical objects recognition.

Acknowledgement

This research work was funded by Natural Science Funds of China (No. 60833005). And this paper was finally completed when the first author was visiting Vision Research Lab, Electrical and Computer Engineering Department, University of California, Santa Barbara as a visiting scholar. We also thank *IEEE Fellow, Prof. B. S. Manjunath* for his valuable suggestions.

References

1. James, W.: The Principles of Psychology. Harvard University Press, Oxford (1890)
2. Koch, C., Ulman, S.: Shifts in Selection in Visual Attention: Toward the Underlying Neural Circuitry. Human Neurobiology 4(4), 219–227 (1985)
3. Aziz, M.Z., Mertsching, B.: Fast and Robust Generation of Feature Maps for Region-Based Visual Attention. IEEE Transaction on Image Processing 17(5), 633–644 (2008)
4. Liu, F., Gleicher, M.: Region Enhanced Scale-Invariant Saliency Detection. In: Proceeding of 2006 IEEE International Conference on Multimedia & Expo., pp. 1477–1480 (2006)
5. Liu, H., Jiang, S., Huang, Q., Xu, C., Gao, W.: Region-Based Visual Attention Analysis with Its Application in Image Browsing on Small Displays. In: Proceeding of 2007 ACM International Conference on Multimedia, pp. 305–308 (2007)
6. Hou, X., Zhang, L.: Saliency Detection: A Spectral Residual Approach. In: Proceeding of 2007 IEEE Conference on Computer Vision and Pattern Recognition, pp. 1–8 (2007)
7. Hu, Y., Rajan, D., Chia, L.-T.: Robust Subspace Analysis for Detecting Visual Attention Regions in Images. In: Proceeding of the 13th annual ACM International Conference on Multimedia, pp. 716–724 (2005)

8. Treisman, A.M., Gelade, G.: A Feature-Integration Theory of Attention. Cognitive Psychology 12(1), 97–136 (1980)
9. Itti, L., Koch, C., Niebur, E.: A model of saliency-based visual attention for rapid scene analysis. IEEE Transaction on Pattern Analysis and Machine Intelligence 20(11), 1254–1259 (1998)
10. Itti, L., Koch, C.: Computational Modelling of Visual Attention. Nature Reviews Neuroscience 2(3), 194–203 (2001)
11. Treisman, A., Gormican, S.: Feature analysis in early vision: evidence from search asymmetries. Psychology Review 95, 15–48 (1988)
12. Itten, J.: The Elements of Color. John Wiley & Sons Inc., New York (1961)
13. Mahnke, F.: Color, Environment, and Human Response. Van Nostrand Reinhold, Detroit (1996)
14. Ouerhani, N., Bur, A., Hügli, H.: Linear vs. Nonlinear Feature Combination for Saliency Computation: A Comparison with Human Vision. In: Franke, K., Müller, K.-R., Nickolay, B., Schäfer, R. (eds.) DAGM 2006. LNCS, vol. 4174, pp. 314–323. Springer, Heidelberg (2006)
15. Kitterler, J., Illingworth, J.: Minimum Error Thresholding. Pattern Recognition 19(1), 41–47 (1986)
16. Zabrodsky, H., Peleg, S.: Attentive transmission. Visual Communication and Image Representation 1(2), 189–198 (1990)
17. Liu, T., Sun, J., Zheng, N.-N., Tang, X., Shum, H.-Y.: Learning to Detect A Salient Object. In: Proceeding of the 2007 IEEE Conference on Computer Vision and Pattern Recognition, pp. 1–8 (2007)
18. Zhou, Q., Ma, L., Celenk, M., Chelberg, D.: Content-Based Image Retrieval Based on ROI Detection and Relevance Feedback. Multimedia Tools and Applications 27, 251–281 (2005)
19. Walther, D., Itti, L., Riesenhuber, M., Poggio, T., Koch, C.: Attentional Selection for Object Recognition - a Gentle Way. In: Bülthoff, H.H., Lee, S.-W., Poggio, T.A., Wallraven, C. (eds.) BMCV 2002. LNCS, vol. 2525, pp. 472–479. Springer, Heidelberg (2002)
20. Luo, J., Singhal, A., Etz, S.P., Gray, R.T.: A computational approach to determination of main subject regions in photographic images. Image Vision Computing 22(3), 227–241 (2004)

Generating Visual Concept Network from Large-Scale Weakly-Tagged Images

Chunlei Yang[1], Hangzai Luo[2], and Jianping Fan[1]

[1] Department of Computer Science, UNC-Charlotte, Charlotte NC 28223, USA
[2] Software Engineering Institute, East China Normal University, Shanghai, China

Abstract. When large-scale online images come into view, it is very attractive to incorporate visual concept network for image summarization, organization and exploration. In this paper, we have developed an automatic algorithm for visual concept network generation by determining the diverse visual similarity contexts between the image concepts. To learn more reliable inter-concept visual similarity contexts, the images with diverse visual properties are crawled from multiple sources and multiple kernels are combined to characterize the diverse visual similarity contexts between the images and handle the issue of sparse image distribution more effectively in the high-dimensional multi-modal feature space. Kernel canonical correlation analysis (KCCA) is used to characterize the diverse inter-concept visual similarity contexts more accurately, so that our visual concept network can have better coherence with human perception. A similarity-preserving visual concept network visualization technique is developed to assist users on assessing the coherence between their perceptions and the inter-concept visual similarity contexts determined by our algorithm. Our experimental results on large-scale image collections have observed very good results.[1]

1 Introduction

With the exponential availability of high-quality digital images, there is an urgent need to develop new frameworks for image summarization and interactive image navigation and exploration [1-2]. The project of Large-Scale Concept Ontology for Multimedia (LSCOM) is the first one of such kind of efforts to facilitate more effective end-user access of large-scale image/video collections in a large semantic space [3-4]. By exploiting large amounts of image/video concepts and their inter-concept similarity relationships for image/video knowledge representation and summarization, concept ontology can be used to navigate and explore large-scale image/video collections at the concept level according to the hierarchical inter-concept relationships such as "IS-A" and "part-of" [4].

Concept ontology may also play an important role in learning more reliable classifiers for bridging the semantic gap [14-19]. By exploiting only the hierarchical inter-concept or inter-object similarity contexts, some pioneer work have

[1] This work is supported by Shanghai Pujiang Program under 08PJ1404600 and NSF-China under 60803077.

S. Boll et al. (Eds.): MMM 2010, LNCS 5916, pp. 251–261, 2010.

been done recently to integrate the concept ontology and multi-task learning for improving image classifier training, and the concept ontology can be used to determine the inter-related learning tasks more precisely [12].

Because of the following issues, most existing techniques for concept ontology construction may not be able to support effective navigation and exploration of large-scale image collections: (a) Only the hierarchical inter-concept relationships are exploited for concept ontology construction [22-23]. When large-scale online image collections come into view, the inter-concept similarity relationships could be more complex than the hierarchical ones (i.e., concept network) [21]. (b) Only the inter-concept semantic relationships are exploited for concept ontology construction [22-23], thus the concept ontology cannot allow users to navigate large-scale online image collections according to their visual similarity contexts at the semantic level. It is well-accepted that the visual properties of the images are very important for users to search for images [1-4, 21]. Thus it is very attractive to develop new algorithm for visual concept network generation, which is able to exploit more precise inter-concept visual similarity contexts for image summarization and exploration.

Based on these observations, this paper will focus on: (a) integrating multiple kernels to achieve more precise characterization of the diverse visual similarity contexts between the images in the high-dimensional multi-modal feature space; (b) incorporating kernel canonical correlation analysis (KCCA) to enable more accurate characterization of inter-concept visual similarity contexts and generate more precise visual concept network; and (c) supporting similarity-preserving visual concept network visualization and exploration for assisting users on perceptual coherence assessment.developing new techniques to exploit the inter-concept visual similarity contexts for visual concept network generation.

The remainder of this paper is organized as follows. Section 2 introduces our approach for image content representation and similarity characterization. Section 3 introduces our work on inter-concept visual similarity determination and automatic visual concept network generation. We describe our visual concept network visualization algorithm in section 4. More discussions on our experimental observations are given in section 5. We conclude this paper at section 6.

2 Data Collection Feature Extraction and Image Similarity Characterization

The images used in our benchmark experiment are partly from Caltech-256 [8] and LabelMe [9] and are partly crawled from the Internet. To determine the meaningful text terms for crawling images from the internet like Google or Flickr, many people use the keywords which are sampled from WordNet. Unfortunately, most of the keywords on WordNet may not be meaningful for image concept interpretation. Based on this understanding, we have developed a taxonomy for nature objects and scenes interpretation. Thus we follow this pre-defined taxonomy to determine the meaningful keywords for image crawling as shown in Fig. 3. Because there is no explicit correspondence between the image semantics

Fig. 1. Two image content representation and feature extraction frameworks: (a) image-based; (b) grid-based

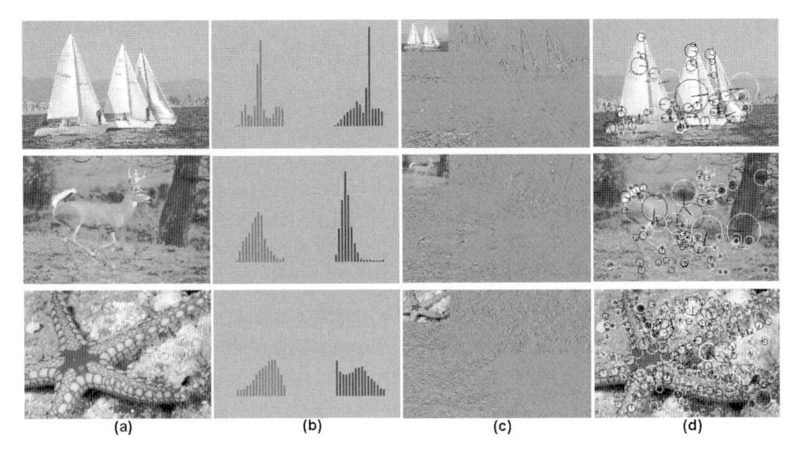

Fig. 2. Image feature extraction for similarity characterization: (a) original images; (b) RGB color histograms; (c) wavelet transformation; (d) interesting points and SIFT features

and the keywords extracted from the associated text documents, images returned are sometimes junk images or weakly-related images. We apply the algorithms introduced in [13] for cleansing the images which are crawled from the Internet (i.e., filtering out the junk images and removing the weakly-related images).

For image retrieval application, the underlying framework for image content representation and feature extraction should be able to: (a) characterize the image contents effectively and efficiently; (b) reduce the computational cost for feature extraction and image similarity characterization significantly. Based on these observations, we have incorporated two frameworks for image content representation and feature extraction as shown in Fig. 1: (1) image-based; and (2) grid-based. In the image-based approach as shown in Fig. 1(a), we have extracted both the global visual features and the local visual features from whole images [12]. In the grid-based approach as shown in Fig. 1(b), we have extracted the grid-based local visual features from a set of image grids [6].

The global visual features such as color histogram can provide the global image statistics and the perceptual properties of entire images, but they may not be able to capture the object information within the images [5, 7]. On the other hand, the local visual features such as SIFT (scale invariant feature transform) features and the grid-based visual features can allow object recognition against the cluttered backgrounds [5, 7]. In our current implementations, the global visual features consist of 36-bin RGB color histograms and 48-dimensional texture features from Gabor filter banks. The local visual features consist of a number of interest points and their SIFT features and a location-preserving union of grid-based visual features. As shown in Fig. 2, one can observe that our feature extraction operators can effectively characterize the principal visual properties for the images.

By using high-dimensional multi-modal visual features (color histogram, wavelet textures, SIFT, and location-preserving union of grid-based visual features) for image content representation, it is able for us to characterize the diverse visual properties of the images more sufficiently. On the other hand, the statistical properties of the images in such the high-dimensional multi-modal feature space may be heterogeneous because different feature subsets are used to characterize different visual properties of the images, thus the statistical properties of the images in the high-dimensional multi-modal feature space may be heterogeneous and sparse. Therefore, it is impossible for us to use only one single type of kernel to characterize the diverse visual similarity relationships between the images precisely.

Based on these observations, the high-dimensional multi-modal visual features are first partitioned into multiple feature subsets and each feature subset is used to characterize one certain type of visual properties of the images, thus the underlying visual similarity relationships between the images are more homogeneous and can be approximated more precisely by using one particular type of kernel.

In this paper, the high-dimensional multi-modal visual features are partitioned into five feature subsets: (a) color histograms; (b) wavelet textural features; (c) SIFT features; (d) location-preserving union of grid-based color histograms; and (e) location-preserving union of grid-based SIFT features. We have also studied the statistical property of the images under each feature subset. The gained knowledge for the statistical property of the images under each feature subset has been used to design the basic image kernel for each feature subset. Because different basic image kernels may play different roles on characterizing the diverse visual similarity relationships between the images, and the optimal kernel for diverse image similarity characterization can be approximated more accurately by using a linear combination of these basic image kernels with different importance.

For a given image concept C_j, the diverse visual similarity contexts between its images can be characterized more precisely by using a mixture of these basic image kernels (i.e., mixture-of-kernels) [10-13].

$$\kappa(u,v) = \sum_{i=1}^{5} \alpha_i \kappa_i(u_i, v_i), \qquad \sum_{i=1}^{5} \alpha_i = 1 \tag{1}$$

where u and v are the visual features for two images in the given image concept C_j, u_i and v_i are their ith feature subset, $\alpha_i \geq 0$ is the importance factor for the ith basic image kernel $\kappa_i(u_i, v_i)$.

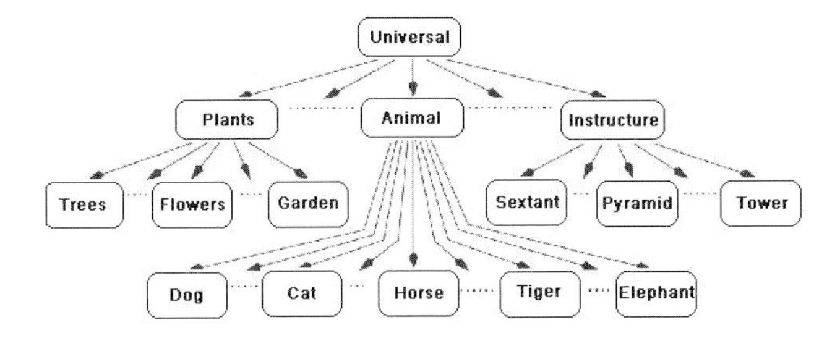

Fig. 3. The taxonomy for text term determination for image crawling

Obviously, combining different kernels can allow us to achieve more precise characterization of the diverse visual similarity contexts between the images in the high-dimensional multi-modal feature space. On the other hand, the weights for all these five kernels may be different for different image concepts. Ideally, for different image concepts, we should be able to identify different sets of these weights for kernel combination. Given a set of images, the weights for these five basic image kernels are determined automatically by searching from a given set of all the potential weights and their combinations.

3 Inter-concept Visual Similarity Determination

After the image concepts and their most relevant images are available, we can use these images to determine the inter-concept visual similarity contexts for automatic visual concept network generation as shown in Fig. 4. The inter-concept visual similarity context $\gamma(C_i, C_j)$ between the image concepts C_i and C_j can be determined by performing kernel canonical correlation analysis (KCCA) [20] on their image sets S_i and S_j:

$$\gamma(C_i, C_j) = \max_{\theta, \vartheta} \frac{\theta^T \kappa(S_i)\kappa(S_j)\vartheta}{\sqrt{\theta^T \kappa^2(S_i)\theta \cdot \vartheta^T \kappa^2(S_j)\vartheta}} \tag{7}$$

where θ and ϑ are the parameters for determining the optimal projection directions to maximize the correlations between two image sets S_i and S_j for the image concepts C_i and C_j, $\kappa(S_i)$ and $\kappa(S_j)$ are the cumulative kernel functions for characterizing the visual correlations between the images in the same image sets S_i and S_j.

$$\kappa(S_i) = \sum_{x_l, x_m \in S_i} \kappa(x_l, x_m), \quad \kappa(S_j) = \sum_{x_h, x_k \in S_j} \kappa(x_h, x_k) \tag{8}$$

where the visual correlation between the images is defined as their kernel-based visual similarity $\kappa(\cdot, \cdot)$ in Eq.(1).

The parameters θ and ϑ for determining the optimal projection directions are obtained automatically by solving the following eigenvalue equations:

$$\kappa(S_i)\kappa(S_i)\theta - \lambda_\theta^2 \kappa(S_i)\kappa(S_i)\theta = 0$$

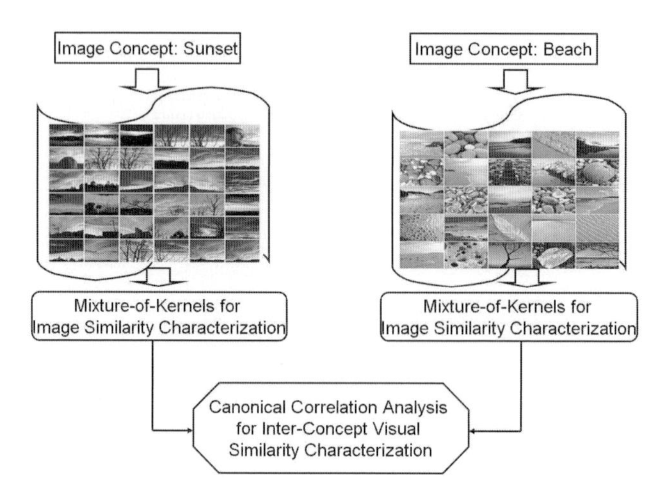

Fig. 4. Major components for inter-concept visual similarity determination

$$\kappa(S_j)\kappa(S_j)\vartheta - \lambda_\vartheta^2\kappa(S_j)\kappa(S_j)\vartheta = 0 \tag{9}$$

where the eigenvalues λ_θ and λ_ϑ follow the additional constraint $\lambda_\theta = \lambda_\vartheta$.

When large numbers of image concepts and their inter-concepts visual similarity contexts are available, they are used to construct a visual concept network. However, the strength of the inter-concept visual similarity contexts between some image concepts may be very weak, thus it is not necessary for each image concept to be linked with all the other image concepts on the visual concept network. Eliminating the weak inter-concept links can increase the visibility of the image concepts of interest dramatically, but also allow our visual concept network to concentrate on the most significant inter-concept visual similarity contexts. Based on this understanding, each image concept is automatically linked with the most relevant image concepts with larger values of the inter-concept visual similarity contexts $\gamma(\cdot,\cdot)$ (i.e., their values of $\gamma(\cdot,\cdot)$ are above a threshold $\delta = 0.65$ in a scale from 0 to 1).

Compared with Flickr distance [21], our algorithm for inter-concept visual similarity context determination have several advantages: (a) It can deal with the sparse distribution problem more effectively by using a mixture-of-kernels to achieve more precise characterization of diverse image similarity contexts in the high-dimensional multi-modal feature space; (b) By projecting the image sets for the image concepts into the same kernel space, our KCCA technique can achieve more precise characterization of the inter-concept visual similarity contexts.

4 Concept Network Visualization

To allow users to assess the coherence between the visual similarity contexts determined by our algorithm and their perceptions, it is very important to enable graphical representation and visualization of the visual concept network, so

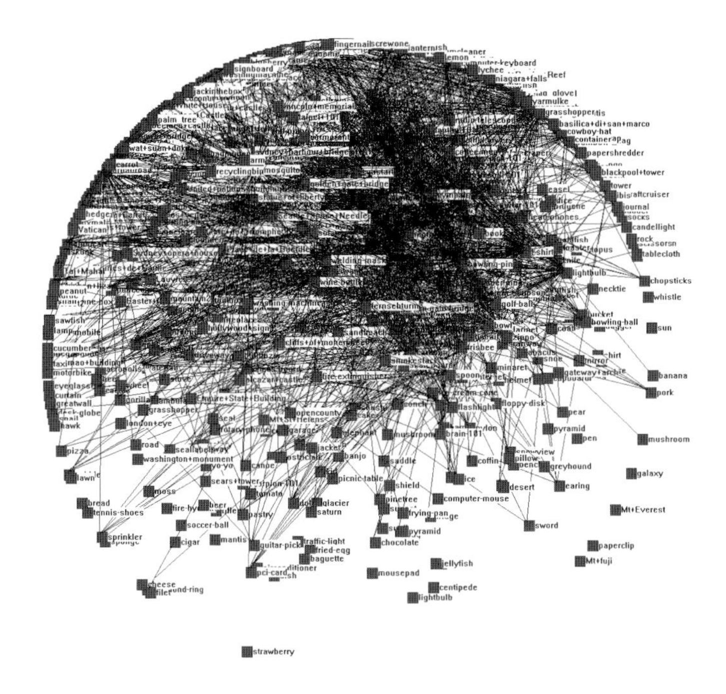

Fig. 5. Visual concept network for our 600 image concepts and objects

that users can obtain a good global overview of the visual similarity contexts between the image concepts at the first glance. It is also very attractive to enable interactive visual concept network navigation and exploration according to the inherent inter-concept visual similarity contexts, so that users can easily assess the coherence with their perceptions.

Based on these observations, our approach for visual concept network visualization exploited hyperbolic geometry [24]. The hyperbolic geometry is particularly well suited for achieving graph-based layout of the visual concept network and supporting interactive exploration. The essence of our approach is to project the visual concept network onto a hyperbolic plane according to the inter-concept visual similarity contexts, and layout the visual concept network by mapping the relevant image concept nodes onto a circular display region. Thus our visual concept network visualization scheme takes the following steps: (a) The image concept nodes on the visual concept network are projected onto a hyperbolic plane according to their inter-concept visual similarity contexts by performing multi-dimensional scaling (MDS) [25] (b) After such similarity-preserving projection of the image concept nodes is obtained, Poincare disk model [24] is used to map the image concept nodes on the hyperbolic plane onto a 2D display coordinate. Poincare disk model maps the entire hyperbolic space onto an open unit circle, and produces a non-uniform mapping of the image concept nodes to the 2D display coordinate.

The visualization results of our visual concept network are shown in Fig. 5, where each image concept is linked with multiple relevant image concepts with larger values of $\gamma(\cdot,\cdot)$. By visualizing large numbers of image concepts according to their inter-concept visual similarity contexts, our visual concept network can allow users to navigate large amounts of image concepts interactively according to their visual similarity contexts.

5 Algorithm Evaluation

For algorithm evaluation, we focus on assessing whether our visual similarity characterization techniques (i.e., mixture-of-kernels and KCCA) have good coherence with human perception. We have conducted both subjective and objective evaluations. For subjective evaluation, we have conducted a user study to evaluate the coherence between the inter-concept visual similarity contexts and their perceptions. For objective evaluation, we have integrated our visual concept network for exploring large-scale image collections and evaluating the benefits on using the visual concept network.

For subjective evaluation, users are involved to explore our visual concept network and assess the visual similarity contexts between the concept pairs. In such an interactive visual concept network exploration procedure, users can score the coherence between the inter-topic visual similarity contexts provided by our visual concept network and their perceptions. For the user study listed in Table 2, 21 sample concept pairs are selected equidistantly from the indexed sequence of concept pairs. The first one is sampled from the top and the following samples are derived every 20,000th in a sequence of 179,700 concept pairs. By averaging the scores from all these users, we get the final scores as shown in Table 2, one can observe that our visual concept network has a good coherence with human perception on the underlying inter-concept visual similarity contexts.

By clicking the node for each image concept, our hyperbolic concept network visualization technique can change the view into a star-schema view, which can allow users to easily assess the coherence between their perceptions and the inter-concept visual similarity contexts determined by our algorithm.

We incorporate our inter-concept visual similarity contexts for concept clustering to reduce the size of the image knowledge. Because the image concepts and their inter-concept similarity contexts are indexed coherently by the visual concept network, a constraint-driven clustering algorithm is developed to achieve more accurate concept clustering. For two image concepts C_i and C_j on the visual concept network, their constrained inter-concept similarity context $\varphi(C_i, C_j)$ depends on two issues: (1) inter-concept similarity context $\gamma(C_i, C_j)$ (e.g., similar image concepts should have larger values of $\gamma(\cdot,\cdot)$); and (2) constraint and linkage relatedness on the visual concept network (e.g., similar image concepts should be closer on the visual concept network). The constrained inter-concept similarity context $\varphi(C_i, C_j)$ between two image concept C_i and C_j is defined as:

$$\varphi(C_i, C_j) = \gamma(C_i, C_j) \times \begin{cases} e^{-\frac{l^2(C_i, C_j)}{\sigma^2}}, if \;\; l(C_i, C_j) \leq \Delta \\ 0, \qquad otherwise \end{cases} \tag{11}$$

Table 1. Image concept clustering results

group 1	group 2	group 3	group 4
urban-road	knife	electric	bus
street-view	humming	-guitar	earing
touring-bike	-bird	suv-car	t-shirt
school-bus	cruiser	fresco	school-bus
city-building	spaghetti	crocodile	screwdriver
fire-engine	sushi	horse	hammock
moped	grapes	billboard	abacus
brandenberg	escalator	waterfall	light-bulb
-gate	chimpanzee	golf-cart	mosquito
buildings			

Table 2. Evaluation results of perception coherence for inter-concept visual similarity context determination: KCCA and Flickr distances

concept pair	user score	γ	Flickr Distance
urbanroad-streetview	0.76	0.99	0.0
cat-dog	0.78	0.81	1.0
frisbee-pizza	0.56	0.80	0.26
moped-bus	0.50	0.75	0.37
dolphin-cruiser	0.34	0.73	0.47
habor-outview	0.42	0.71	0.09
monkey-humanface	0.52	0.71	0.32
guitar-violin	0.72	0.71	0.54
lightbulb-firework	0.48	0.69	0.14
mango-broccoli	0.48	0.69	0.34
porcupine-lion	0.58	0.68	0.22
statue-building	0.72	0.68	0.32
sailboat-cruiser	0.70	0.66	0.23
doorway-street	0.54	0.65	0.58
windmill-bigben	0.40	0.63	0.85
helicopter-city	0.30	0.63	0.34
pylon-highway	0.34	0.61	0.06
tombstone-crab	0.22	0.42	0.40
stick-cupboard	0.28	0.29	0.51
fridge-vest	0.20	0.29	0.43
journal-grape	0.22	0.19	0.02

where the first part $\gamma(C_i, C_j)$ denotes the inter-topic visual similarity context between C_i and C_j, the second part indicates the constraint and linkage relatedness between C_i and C_j on the visual concept network, $l(C_i, C_j)$ is the distance between the physical locations for the image concepts C_i and C_j on the visual concept network, σ is the variance of their physical location distances, and Δ is a pre-defined threshold which largely depends on the size of the nearest neighbors to be considered. In this paper, the first-order nearest neighbors is considered, $\Delta = 1$.

Our concept clustering results are given in Table 1. Because our KCCA-based measurement can characterize the inter-concept visual similarity contexts more precisely, our constraint-driven concept clustering algorithm can effectively generate the concept clusters, which may significantly reduce the cognitive load for human coherence assessment on the underlying inter-concept visual similarity contexts. By clustering the similar image concepts into the same concept cluster, it is able for us to deal with the issue of synonymous concepts effectively, e.g., multiple image concepts may share the same meaning for object and scene interpretation. Because only the inter-concept visual similarity contexts are used for concept clustering, one can observe that some of them may not semantic to human beings, thus it is very attractive to integrate both the inter-concept visual similarity contexts and their inter-concept semantic similarity contexts for concept clustering.

As shown in Table 2, we have also compared our KCCA-based approach with Flickr distance approach [21] on inter-concept visual similarity context determination. The normalized distance to human perception is 0,92 and 1.42 respectively in terms of Euclidean distance, which means KCCA-base approach performs 54% better than Flickr distance on the random selected sample data.

6 Conclusions

To incorporate the visual concept network for summarizing and exploring large-scale image collections, we have developed a novel algorithm for determining the diverse visual similarity contexts between large amounts of image concepts. Multiple kernels and kernel canonical correlation analysis are combined to characterize the diverse inter-concept visual similarity relationships more precisely in a high-dimensional multi-modal feature space. Our experimental results on large-scale image collections have observed very good results.

References

1. Smeulders, A.W.M., Worring, M., Santini, S., Gupta, A., Jain, R.: Content-based image retrieval at the end of the early years. IEEE Trans. PAMI 22(12), 1349–1380 (2000)
2. Hauptmann, A., Yan, R., Lin, W.-H., Christel, M., Wactlar, H.: Can high-level concepts fill the semantic gap in video retrieval? A case study with broadcast news. IEEE Trans. on Multimedia 9(5), 958–966 (2007)
3. Benitez, A.B., Smith, J.R., Chang, S.-F.: MediaNet: A multimedia information network for knowledge representation. In: Proc. SPIE, vol. 4210 (2000)
4. Naphade, M., Smith, J.R., Tesic, J., Chang, S.-F., Hsu, W., Kennedy, L., Hauptmann, A., Curtis, J.: Large-scale concept ontology for multimedia. IEEE Multimedia (2006)
5. Lowe, D.: Distinctive image features from scale invariant keypoints. Intl. Journal of Computer Vision 60, 91–110 (2004)
6. Wu, L., Li, M., Li, Z., Ma, W.-Y., Yu, N.: Visual lanuage modeling for image classification. In: ACM MIR (2007)

7. Bay, H., Ess, A., Tuytelaars, T., Van Gool, L.: Speeded-up robust features (surf). Comput. Vis. Image Underst. 110(3), 346–359 (2008)
8. Grin, G., Holub, A., Perona, P.: Caltech-256 object category dataset. Technical Report 7694, California Institute of Technology (2007)
9. Russell, B.C., Torralba, A., Murphy, K.P., Freeman, W.T.: Labelme: A database and web-based tool for image annotation. Int. J. Comput. Vision 77(1-3), 157–173 (2008)
10. Sonnenburg, S., Ratsch, G., Schafer, C., Scholkopf, B.: Large scale multiple kernel learning. Journal of Machine Learning Research 7, 1531–1565 (2006)
11. Zhang, J., Marszalek, M., Lazebnik, S., Schmid, C.: Local features and kernels for classification of texture and object categories: A comprehensive study. Intl. Journal of Computer Vision 73(2), 213–238 (2007)
12. Torralba, A., Murphy, K.P., Freeman, W.T.: Sharing features: efficient boosting procedures for multiclass object detection. In: IEEE CVPR (2004)
13. Gao, Y., Peng, J., Luo, H., Keim, D., Fan, J.: An Interactive Approach for Filtering out Junk Images from Keyword-Based Google Search Results. IEEE Trans. on Circuits and Systems for Video Technology 19(10) (2009)
14. Huang, J., Kumar, S.R., Zabih, R.: An automatic hierarchical image classification scheme. In: ACM Multimedia, Bristol, UK (1998)
15. Vasconcelos, N.: Image indexing with mixture hierarchies. In: IEEE CVPR (2001)
16. Li, J., Wang, J.Z.: Automatic linguistic indexing of pictures by a statistical modeling approach. IEEE Trans. on PAMI 25(9), 1075–1088 (2003)
17. Fei-Fei, L., Perona, P.: A Bayesian hierarchical model for learning natural scene categories. In: IEEE CVPR (2005)
18. Barnard, K., Forsyth, D.: Learning the semantics of words and pictures. In: IEEE ICCV, pp. 408–415 (2001)
19. Naphade, M., Huang, T.S.: A probabilistic framework for semantic video indexing, filterig and retrieval. IEEE Trans. on Multimedia 3(1), 141–151 (2001)
20. Hardoon, D.R., Szedmak, S., Shawe-Taylor, J.: Canonical correlation analysis: An overview with application to learning methods, Technical Report, CSD-TR-03-02, University of London (2003)
21. Wu, L., Hua, X.-S., Yu, N., Ma, W.-Y., Li, S.: Flickr distance. ACM Multimedia (2008)
22. Cilibrasi, R., Vitanyi, P.: The Google similarity distance. IEEE Trans. Knowledge and Data Engineering 19 (2007)
23. Fellbaum, C.: WordNet: An Electronic Lexical Database. MIT Press, Boston (1998)
24. Lamping, J., Rao, R.: The hyperbolic browser: A focus+content technique for visualizing large hierarchies. Journal of Visual Languages and Computing 7, 33–55 (1996)
25. Cox, T., Cox, M.: Multidimensional Scaling. Chapman and Hall, Boca Raton (2001)

Automatic Image Annotation with Cooperation of Concept-Specific and Universal Visual Vocabularies

Yanjie Wang, Xiabi Liu, and Yunde Jia

Beijing Laboratory of Intelligent Information Technology,
School of Computer Science, Beijing Institute of Technology, Beijing, P.R. China
{wangyanjie,liuxiabi,jiayunde}@bit.edu.cn

Abstract. This paper proposes an automatic image annotation method based on concept-specific image representation and discriminative learning. Firstly, the concept-specific visual vocabularies are generated by assuming that localized features from the images with a specific concept are of the distribution of Gaussian Mixture Model (GMM). Each component in the GMM is taken as a visual token of the concept. The visual tokens of all the concepts are clustered to obtain a universal token set. Secondly, the image is represented as a concept-specific feature vector by computing the average posterior probabilities of being each universal visual token for all the localized features and assigning it to corresponding concept-specific visual tokens. Thus the feature vector for an image varies with different concepts. Finally, we implement image annotation and retrieval under a discriminative learning framework of Bayesian classifiers, Max-Min posterior Pseudo-probabilities (MMP). The proposed method were evaluated on the popular Corel-5K database. The experimental results with comparisons to state-of-the-art show that our method is promising.

Keywords: Image annotation, Image retrieval, Visual vocabulary, Bag-of-features, Max-Min posterior Pseudo-probabilities (MMP).

1 Introduction

Many users of image retrieval systems prefer keyword query to visual query. Therefore, with the rapid growth of the number of available images, it becomes more and more important to automatically annotate images with keywords. An increasing interest in solving this problem of automatic image annotation (AIA) has been shown in recent literature. The AIA problem can be described as the one of associating the image descriptor with its concepts. The direct descriptor of an image is the visual features from it. Recently, the researchers also try to explore indirect textural cues, such as the text around the image in the web page, to supplement the visual features [17].

This paper investigates the problem of determining the concepts of an image according to localized visual features from it. A corresponding AIA method is proposed based on a novel concept-specific image representation schema and

S. Boll et al. (Eds.): MMM 2010, LNCS 5916, pp. 262–272, 2010.

a discriminative learning framework of Bayesian classifier, Max-Min posterior Pseudo-probabilities (MMP) [12]. Firstly, a concept-specific visual vocabulary is generated for each concept by assuming that localized features from the images with this concept are of the distribution of Gaussian Mixture Model (GMM). The GMM for each concept is learned from the training images with this concept by using the Expectation-Maximization (EM) algorithm with the Minimum Description Length criterion (MDL). Each component in the GMM is taken as a visual token of this concept. Secondly, the Concept-specific Visual Tokens (CVTs) of all the concepts are clustered to obtain a universal visual vocabulary. The correspondences between concept-specific and universal visual tokens are recorded. Thirdly, an image is represented through the cooperation between concept-specific and universal visual vocabularies. In fact, the posterior probabilities of being Universal Visual Tokens (UVTs) for localized features in the image are computed. Then the average posterior probability is calculated for each UVT and assigned to the corresponding CVT. This probability can be seen as the possibility of a CVT occurring in the image. Therefore, the average posterior probabilities for all the CVTs of each concept are arranged orderly to represent the image. It means that the feature vector for an image varies with different concepts. Finally, feature vectors extracted from images with a specific concept are also assumed to be of the distribution of GMM. By embedding this GMM into the discriminative learning framework of MMP, we get our image annotation and retrieval algorithm and evaluate it on the Corel-5K database. In the experiments, the localized feature of the image is obtained by partitioning an image into fix-sited rectangular blocks and extract Discrete Cosine Transform (DCT) features in YBR color space from each block. The performance comparison between our method and other state-of-the-art counterparts shows that the proposed method is promising. The main contributions of this work are:

(1) A novel image representation is realized for AIA through the cooperation between concept-specific and universal visual vocabularies. This image representation brings the advantages of more tolerance to background clutter, free of dimensionality reduction, and flexibility to the change of concept set. The last advantage seems useful for real applications where the number of concepts is not easy to be preset beforehand.

(2) A new discriminative learning framework of Bayesian classifier, Max-Min posterior Pseudo-probabilities (MMP), is tailored to tackle the AIA problem. Because the dimensionality of the feature vectors of an image varies with different concepts, the corresponding statistical models of the concepts learned by traditional generative learning methods such as EM algorithm are not appropriate for image classification. As a discriminative learning approach, MMP is a suitable solution to this problem.

The rest of this paper is organized as follows. Section 2 reviews the related work of AIA based on statistical modeling and learning of concepts. Section 3 presents the generation method of concept-specific and universal visual vocabularies. The cooperative image representation strategy and subsequent image annotation by

MMP is put forward in Section 4. Section 5 discusses the experimental results on Corel-5K database. We conclude the paper in Section 6.

2 Related Work

Many AIA algorithms have been developed based on statistical modeling and learning of concepts. Bayes theory is the basis of these algorithms, where the posterior probabilities of being keywords for images are used as annotation confidence but mostly reflected by joint probabilities of keywords and images or class-conditional probabilities of images given keywords. An image is usually a combination of several concepts. The regional features of an image should be expressed to realize multi-label annotation for images with multiple concepts. This task can be implemented by segmenting an image into regions or partitioning it into blocks (also called grids). Based on the localized features from segmented regions or partitioned blocks, we can directly associate the regions with concepts, or holistically classify the image according to combined regional features. Many statistical models have been explored in AIA algorithms with segmentation or partition strategy, including the translation model [5], Cross-Media Relevance Model (CMRM) [8], Continuous-space Relevance Model (CRM) [10], Coherent Language Model (CLM) [9], CORRespondence Latent Dirichlet Dl-location (CORR-LDA) [1], Hidden Concept Model [19], GMM [2], probabilistic latent semantic analysis [13], Bayes method [2], HMM [11], Bayes Point Machines (BPM) [3], etc.

Visual tokens based image representation was used behind some of AIA methods mentioned above, such as translation model, CMRM, CLM, etc. In these methods, image regions are clustered to obtain the basic elements for representing images. These basic elements are called blob-token, visual term, visual word, or visual token. In the field of object categorization which is closely related to AIA, visual tokens based image representation has also recently become a hot topic. However, most of visual token based image representation methods construct visual vocabulary by unsupervised manner, without taking class information into account. Recently, statistical modeling of visual tokens has been advised to improve its effectiveness for image classification [6,18,15]. The relation between local features and visual tokens can be described more accurately and reliably through statistical modeling of visual tokens. Furthermore, a local feature is allowed to be softly mapped to multiple visual tokens in this way, so the aliasing effects can be reduced.

3 Visual Vocabularies Generation

This section describes our method of generating concept-specific visual tokens (CVTs) and universal visual tokens (UVTs) from localized features. As illustrated in Fig. 1, we firstly generate CVTs of each concept according to localized features which are extracted from images with this concept, and then perform

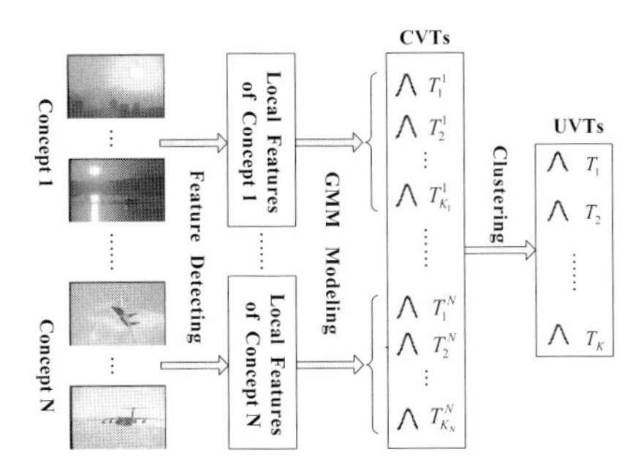

Fig. 1. The flowchart of concept-specific and universal visual vocabularies generation

the clustering on CVTs from all the concepts to obtain UVTs. The correspondences between CVTs and UVTs are recorded for subsequent use in the image representation.

3.1 Concept-Specific Visual Vocabularies Generation

In order to generate CVTs, it is assumed that the distribution of localized features extracted from images with a specific concept is a GMM. Let $\mathbf{x}$ be a localized feature, K_i be the number of components in the GMM for the i-th concept, Θ be the parameter set of the GMM for the i-th concept, which includes the weights $w_k|_{k=1}^{K_i}$, the means $\boldsymbol{\mu}_k|_{k=1}^{K_i}$, and the covariance matrices $\boldsymbol{\Sigma}_k|_{k=1}^{K_i}$. Then we have

$$p(\mathbf{x}|\Theta) = \sum_{k=1}^{K_i} w_k N(\mathbf{x}|\boldsymbol{\mu}_k, \boldsymbol{\Sigma}_k), \tag{1}$$

where

$$\begin{aligned} & N(\mathbf{x}|\boldsymbol{\mu}_k, \boldsymbol{\Sigma}_k) \\ &= (2\pi)^{-\frac{D}{2}} |\boldsymbol{\Sigma}_k|^{-\frac{1}{2}} \exp\left(-\frac{1}{2}(\mathbf{x}-\boldsymbol{\mu}_k)'\boldsymbol{\Sigma}_k^{-1}(\mathbf{x}-\boldsymbol{\mu}_k)\right). \end{aligned} \tag{2}$$

The covariance matrix $\boldsymbol{\Sigma}_k$ is considered as a diagonal matrix for simplicity.

The parameter set Θ of the GMM is estimated by Expectation-Maximization algorithm [4] with maximum likelihood setting, which is implemented using Torch machine learning library [1] in this paper. The component number of the GMM K_i is determined by the Minimum Description Length (MDL) principle [7]. After the GMM for a concept is learned from the data by MDL-EM algorithm, each Gaussian component in the GMM is regarded as a CVT. All the Gaussian components constitute a set of CVTs for this semantic concept. Let T_j^i be the j-th CVT of the i-th CVT, then the set of CVTs for the concept is denoted as $\{T_1^i, T_2^i, \cdots, T_{K_i}^i\}$.

[1] Torch Machine Learning Library. Available: http://www.torch.ch

3.2 Universal Visual Vocabulary Generation

After the CVTs of all the concepts are generated, the k-means clustering algorithm is performed on mean vectors of the GMMs for all the CVTs to combine similar CVTs. Each resultant cluster corresponds to a universal visual token (UVT), which is composed of one or several CVTs. We get local features corresponding with CVTs in each cluster and then compute the mean vector and covariance matrix of these local features to form a UVT. Let T_i be the i-th UVT, K be the number of UVTs, then the set of UVTs is $\{T_1, T_2, \cdots, T_K\}$.

The correspondences between UVTs and CVTs are recorded, which will be used in subsequent image representation procedure. Fig. 2 illustrates the correspondences between UVTs and CVTs.

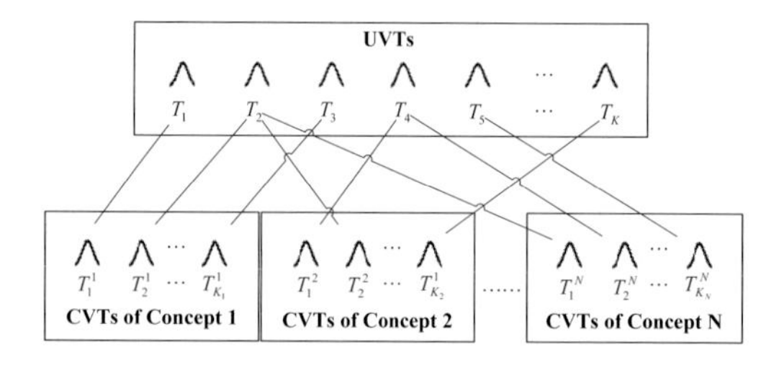

Fig. 2. Illustration of correspondences between CVTs and UVTs

4 Image Representation, Annotation and Retrieval

This section explains how to represent images through the cooperation between class-specific and universal visual vocabularies. Then the image annotation method by applying the proposed image representation scheme under GMM-MMP classification framework is described.

4.1 Cooperative Image Representation

An image is represented as a soft histogram over CVTs of each concept. The value in each bin of the histogram represents the possibility of a CVT occurring in the image, but it is measured according to the set of UVTs. Actually, the posterior probabilities of being each UVT for localized features in the image are computed using Bayes formula. Then the average posterior probability is calculated for each UVT as the measure of its occurrence possibility and assigned to the corresponding CVTs which are classified into the cluster represented by this UVT. Finally, the average probabilities for all the CVTs of a concept is arranged orderly to obtain concept-specific feature vector of the image. It means the feature vector of the image varies with different concepts. The flowchart of our image representation strategy described above is shown in Fig. 3, where the average posterior probability for a UVT and its corresponding CVTs are displayed in a same color.

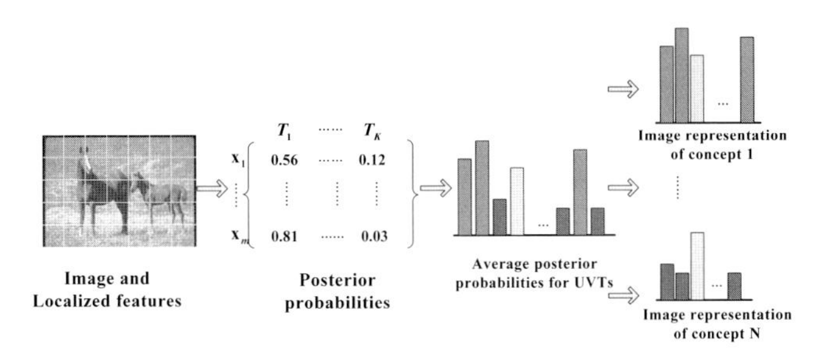

Fig. 3. Flowchart of image representation for different concepts

The details of our image representation scheme are given as follows. Firstly, since the distribution of a UVT T_i is a Gaussian model, the class-conditional probability of a localized feature $\mathbf{x}$ give T_i is computed as

$$p(\mathbf{x}|T_i) = N(\mathbf{x}|\boldsymbol{\mu}_i, \boldsymbol{\Sigma}_i), \tag{3}$$

According to Bayes formula and the assumption of the same $P(T_i)$ for all the UVTs, the posterior probability of being T_i for $\mathbf{x}$ is obtained as

$$P(T_i|\mathbf{x}) = \frac{P(\mathbf{x}|T_i)}{\sum_{k=1}^{K} P(\mathbf{x}|T_k)}. \tag{4}$$

The assignment of average posterior probabilities for UVTs to corresponding CVTs is illustrated in Fig. 3, where the red and blue columns represent the average posterior probabilities for CVTs of these two concepts and assigned from different UVTs, respectively. While the yellow columns represent the average posterior probabilities for CVTs of these two concepts but assigned from a common UVT.

The concept-specific image representation described above is more compact and more discriminative than traditional histograms over universal visual vocabulary. In traditional histograms over universal visual vocabulary, localized features of an image are mapped to thousands of the UVTs. It leads to a sparse representation. Oppositely, we obtain a compact representation by only considering the CVTs related to a specific concept. The dimensionality of feature vectors is reduced greatly. Furthermore, the occurrence possibilities of CVTs are expected to be measured as high values for images with the corresponding concept and measured as low values for images without corresponding concept. It means that our concept-specific image representation could have better discriminability than traditional histograms over universal visual vocabulary.

4.2 GMM-MMP Based Image Annotation and Retrieval

MMP is a new kind of discriminative learning approach for Bayesian classifiers. In the following, we briefly introduce the MMP algorithm designed for image annotation and retrieval. The reader is referred to our paper for more details of MMP [12].

4.2.1 Image Annotation and Retrieval by Posterior Pseudo-probabilities

Let $\mathbf{X}^C$ be a concept-specific feature vector of a concept C, which is extracted from an arbitrary image. Let $p(\mathbf{X}^C|C)$ be the class-conditional probability density function. Then the posterior pseudo-probability of being C for $\mathbf{X}^C$ is computed as

$$f(p(\mathbf{X}^C|C)) = 1 - \exp(-\lambda p^t(\mathbf{X}^C|C)), \tag{5}$$

where λ, t are positive numbers. Consequently, $f(p(\mathbf{X}^C|C))$ is a smooth, monotonically increasing function of $p(\mathbf{X}^C|C)$, and $f(0) = 0$ and $f(+\infty) = 1$. The form of class-conditional probability density function $p(\mathbf{X}^C|C)$ in Eq. 5 should be provided for using posterior pseudo-probabilities based classifers, which is also assumed to be the GMM with diagonal covariance matrix in this paper.

Given an input image, we compute the posterior pseudo-probability for each concept according to Eq. 5. Then the image is annotated through ranking concepts in descending order of their posterior pseudo-probabilities. The corresponding semantic retrieval is realized by ranking images for query concept. Given a query concept, we retrieval the images by ranking the images in descending order of the posterior pseudo-probabilities for this concept and each image in the database, which have been computed in the image annotation stage.

4.2.2 MMP Training

There are unknown parameters in Eq. 5, including λ, t, and those in $p(\mathbf{X}^C|C)$. A method called Max-Min posterior Pseudo-probabilities (MMP) is used to learn these parameters. The main idea behind MMP learning is to optimize the classifier performance through maximizing posterior pseudo-probabilities towards 1 for each class and its positive samples, while minimizing those towards 0 for each class and its negative samples. More formally, let $f(\mathbf{X};\tilde{\mathbf{\Lambda}})$ be the posterior pseudo-probability measure function of a class, where $\tilde{\mathbf{\Lambda}}$ denote the set of unknown parameters in it. Let $\hat{\mathbf{X}}_i$ be the feature vector of arbitrary positive sample of the concept, $\bar{\mathbf{X}}_i$ the feature vector of arbitrary negative sample of the concept, m and n be the number of positive and negative samples of the concept, respectively. According to the idea above of the MMP learning, the objective function for estimating parameters is designed as

$$F(\tilde{\mathbf{\Lambda}}) = \frac{1}{m}\sum_{i=1}^{m}[f(\hat{\mathbf{X}}_i;\tilde{\mathbf{\Lambda}}) - 1]^2 + \frac{1}{n}\sum_{i=1}^{n}[f(\bar{\mathbf{X}}_i;\tilde{\mathbf{\Lambda}})]^2. \tag{6}$$

$F(\tilde{\mathbf{\Lambda}}) = 0$ means the perfect classification performance on the training data. Consequently, we can obtain the optimum parameter set $\tilde{\mathbf{\Lambda}}^*$ of the posterior pseudo-probability measure function by minimizing $F(\tilde{\mathbf{\Lambda}})$:

$$\tilde{\mathbf{\Lambda}}^* = \arg\min_{\tilde{\mathbf{\Lambda}}} F(\tilde{\mathbf{\Lambda}}). \tag{7}$$

The gradient descent algorithm is employed to optimize the parameter set $\tilde{\mathbf{\Lambda}}^*$.

5 Experiments

The image annotation and retrieval experiments were conducted on the popular Corel-5K database [5]. There are 5000 images from 50 Stock Photo CDs in this database, and each CD contains 100 digital photos of the same topic. One to five keywords are provided for each of these images. Following the commonly used evaluation scheme on Corel-5K database [5], we used 4500 images as the training set and the remaining 500 images as the test set. The total 371 semantic concepts are involved in the database, but only 260 concepts coexist in both of the training set and the test set. These 260 concepts are considered in the experiments.

We obtain localized features by partitioning an image into fix-sized rectangular blocks and extract Discrete Cosine Transform (DCT) features in YBR color space from each block. This localized feature extractor is similar with that used by Carneiro et al. [2]. The size of rectangular blocks is set to be 8×8 through experiments.

In the MMP training, the positive samples of each concept are images with the concept, and other images are its negative samples. Before using MMP training algorithm, we obtain the initial parameters by using the MDL-EM algorithm described in Section 3 on positive samples to get the parameters in the GMM, and set λ and t through experiments. The initial parameters are then revised by performing MMP training on all the samples including positive samples and negative samples. For the MDL based model selection of the GMM, we evaluate the component numbers from 20 to 300 at intervals of 10 for concept-specific visual vocabularies, and the component numbers from 1 to 20 for concepts. The resultant component numbers for concept-specific visual vocabularies vary from 20 to 280, while those for concepts are 6 to 15. As for the number of universal visual tokens, we set it to 3000 through careful experiments.

In order to compare our method with other related work on Corel-5K database, we annotate the images with top-five concepts. Then the image annotation performance is evaluated by the mean recall rate and precision rate, as well as the number of concepts with nonzero recall rate. And the subsequent image retrieval performance is evaluated by the mean average precision (MAP) [2]. Fig. 4 and Fig. 5 show the effectiveness of our image retrieval and annotation algorithm through some example. In Fig. 4, the concepts automatically annotated by the proposed method are compared with the ground-truth of human annotation for the test set. In Fig. 5, top-5 images retrieved for some query concepts are displayed from left to right.

The performance of our image annotation algorithm is compared with those recently reported on the Corel-5K database, including the co-occurrence model [14], the translation model [5], the continuous-space relevance model [10,16], the multiple-Bernoulli relevance model (MBRM) [16], and supervised multiclass labeling model (SML)[2]. We further compare our retrieval results with those from SML and MBRM. The comparison result of image annotation and retrieval is listed in Table 1-2, respectively. As shown in Table 1, our proposed algorithm achieves 26.8% recall rate and 23.5% precision rate, which are comparable to

Images				
Truth	jet, plane, sky, smoke	bengal, cat, forest, tiger	frost, frozen, ice	filed, foals, horses, mare
Our Results	smoke, plane, jet, formation, sand	tiger, cat, bengal, forest, river	crystals, frost, frozen, ice, sculpture	horses, mare, porcupine, foals, field
Images				
Truth	people, pool, swimmers, water	locomotive, railroad, smoke, train	bear, polar, snow	cars, formula, tracks, wall
Our Results	swimmers, pool, people, water, boats	train, railroad,smoke, locomotive,monastery	bear, polar, snow, herd, bush	formula, car, tracks, wall, street

Fig. 4. Comparison of the annotations by the proposed algorithm with the ground-truth

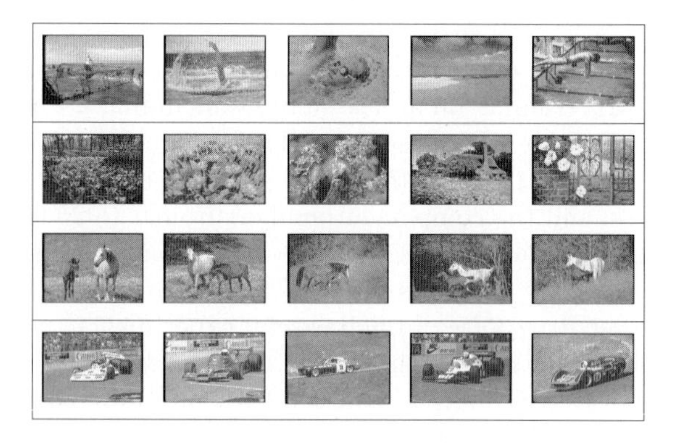

Fig. 5. Each row shows the top five retrieved images for a semantic concept. From top to bottom: water, flower, horses, and cars.

the previous best results achieved by SML and MBRM, and outperform others in both recall rate and precision rate. The number of recalled concepts is 137, which is the same as the previous best one reported by SML. As shown in Table 2, our mean average precision over the total 260 concepts is a little worse than those from SML and MBRM, while the result over concepts with nonzero recall rate is in the middle of those from SML and MBRM. It should be noted that the dense sampling features are adopted in SML. Compared overlapping blocks used there, the number of non-overlapping blocks in this work is much smaller. We expect to further show the advantages of our method by analyzing its efficiency and effectiveness according to more experiments on localized features from dense sampling in the future.

Table 1. Comparison of Automatic Annotation on Corel-5K

Methods	Co-occurrence	Translation	CRM	MBRM	SML	Our
#concepts with recall > 0	19	49	107	122	137	137
Results on all 260 words						
Mean Per-concept Recall Rate	0.02	0.04	0.19	0.25	0.29	0.268
Mean Per-concept Precision Rate	0.03	0.06	0.16	0.24	0.23	0.235

Table 2. Comparison of Semantic Retrieval on Corel-5K

Mean Average Precision for Corel-5K		
Methods	All 260 Concepts	Concepts with Recall>0
Ours	0.286	0.475
MBRM	0.30	0.35
SML	0.31	0.49

6 Conclusions

In this paper, a novel image annotation method has been proposed through representing images based on the cooperation between concept-specific and universal visual vocabulary. The main feature of the proposed method is that the image representation is defined on concept level, instead of on universal level. And the feature vector of the image varies with different concepts. For each concept, the posterior probabilities for concept-specific visual tokens and localized features in the image are measured according to the universal visual vocabulary. They are arranged orderly to represent the image. The advantages of this representation strategy are summarized as follows: 1) The image is represented with only concept-specific information to obtain more robustness to background clutter; 2) The dimensionality of feature vector of the image is small. So it is unnecessary to perform dimensionality reduction which risks the loss of discriminative information; 3) Clustering is performed on the concept-specific visual tokens, instead of on the huge set of localized features. Thus the visual token generation is more flexible to the change of concept sets. This feature seems useful for real applications where the number of concepts is not easy to be preset beforehand.

By embedding our image representation scheme into a new discriminative learning framework of Bayesian classifiers, Max-Min posterior Pseudo-probabilities (MMP), we get the corresponding image annotation and retrieval algorithm which achieved the comparable performance to the previous best methods on the Corel-5K database.

Our future work includes: 1) More sophisticated localized features such as dense sampling features will be considered to improve the effectiveness of the proposed method; 2) The proposed method will be evaluated on other more complicated databases, such as Corel-30K, PSU, etc.

Acknowledgements

This work was partially supported by National Natural Science Foundation of China (Grant No. 60973059 and Grant No. 90920009).

References

1. Blei, D.M., Jordan, M.I.: Modeling annotated data. In: ACM SIGIR Conference (2003)
2. Carneiro, G., Chan, A., Moreno, P., Vasconcelos, N.: Supervised learning of semantic classes for image annotation and retrieval. IEEE Transaction on Pattern Analysis and Machine intelligence 29(3), 394–410 (2007)
3. Chang, E., Goh, K., Sychay, G., Wu, G.: CBSA: Content-based soft annotation for multimodal image retrieval using bayes point machines. IEEE Transactions on Circuits and Systems for Video Technology 13(1), 26–38 (2003)
4. Dempster, A., Laird, N., Rubin, D.: Maximum likelihood from incomplete data via the EM algorithm. Journal of the Royal Statistical Society 39(1), 1–38 (1977)
5. Duygulu, P., Barnard, K., de Freitas, J.F.G., Forsyth, D.: Object recognition as machine translation: Learning a lexicon for a fixed image vocabulary. In: Heyden, A., Sparr, G., Nielsen, M., Johansen, P. (eds.) ECCV 2002. LNCS, vol. 2353, pp. 97–112. Springer, Heidelberg (2002)
6. Farquhar, J., Szedmak, S., Meng, H., Shawe-Taylor, J.: Improving "bag-of-keypoints" image categorisation: Generative models and pdf-kernels. Technical report, University of Southampton (2005)
7. Hansen, M.H., Yu, B.: Model selection and the principle of minimum description length. Journal of American Statistical Association 96(454), 746–774 (2001)
8. Jeon, J., Lavrenko, V., Manmatha, R.: Automatic image annotation and retrieval using cross-media relevance models. In: ACM SIGIR Conference (2003)
9. Jin, R., Chai, J.Y., Si, L.: Effective automatic image annotation via a coherent language model and active learning. In: ACM Multimedia Conference (2004)
10. Lavrenko, V., Manmatha, R., Jeon, J.: A model for learning the semantics of pictures. In: Neural Information Processing Systems (2003)
11. Li, J., Wang, J.Z.: Automatic linguistic indexing of pictures by a statistical modeling approach. IEEE Transactions on Pattern Analysis and Machine Intelligence 25(9), 1075–1078 (2003)
12. Liu, X., Jia, Y., Chen, X., Deng, Y., Fu, H.: Image classification using the max-min posterior pseudo-probabilities method. Technical Report BIT-CS-20080001, Beijing Institute of Technology (2008), http://www.mcislab.org.cn/member/~xiabi/papers/2008_1.PDF
13. Monay, F., Gatica-Perez, D.: On image auto-annotation with latent space models. In: ACM Multimedia Conference (2003)
14. Mori, Y., Takahashi, H., Oka, R.: Image-to-word transformation based on dividing and vector quantizing images with words. In: Workshop Multimedia Intelligent Storage and Retrieval Management (1999)
15. Perronnin, F.: Univeral and adapted vocabularies for generic visual categorization. IEEE Transactions on Pattern Analysis and Machine Intelligence 30(7), 1243–1256 (2008)
16. Feng, R.M.S., Freitas, D.: Multiple bernoulli relevance models for image and video annotation. In: IEEE Conference on Computer Vision and Pattern Recognition (2004)
17. Wang, X., Zhang, L., Li, X., Ma, W.: Annotating images by mining image search results. IEEE Transactions on Pattern Analysis and Machine Intelligence 30(11), 1919–1932 (2008)
18. Winn, J., Criminisi, A., Minka, T.: Object categorization by learned universal visual dictionary. In: IEEE International Conference on Computer Vision (2005)
19. Zhang, R., Zhang, Z.: Effective image retrieval based on hidden concept discovery in image database. IEEE Transactions on Image Processing 16(2), 562–572 (2007)

Weak Metric Learning for Feature Fusion towards Perception-Inspired Object Recognition

Xiong Li[1], Xu Zhao[1], Yun Fu[2], and Yuncai Liu[1]

[1] Institute of Image Processing & Pattern Recognition
Shanghai Jiao Tong University, Shanghai 200240, China
{lixiong,zhaoxu,whomliu}@sjtu.edu.cn
[2] Department of CSE, University at Buffalo (SUNY), NY 14260, USA
raymondyunfu@gmail.com

Abstract. With extracted local features of a given image, computing its global feature under perceptual framework has shown promising performance in object recognition. However, under some tough applications with large intra-class variance, using only one kind of local feature is inadequate to build a robust classification system. To integrate the discriminability of complementary local features, in this paper, we extend the efficacy of perceptual framework to adapt to heterogeneous features. Given multiple raw global features, we propose a fusion strategy through metric learning, which is called weak metric learning in this work, for fusing high dimensional features. The fusion model is solved with the maximal kernel canonical correlation formulation with the multiple global features as outputs. Experimental results show that our method achieves significant improvements about 5% to 11% than the benchmark perceptual framework system, HMAX, on several difficult categories of object recognition with much less training samples and feature elements.

Keywords: Object recognition, feature fusion, weak metric learning, perceptual distance.

1 Introduction

Object recognition has seen rapid progress in recent years, motivated by innovative studies in relative fields such as statistical learning and cognition science. However, it is still in a long arduous travel for machine to approach human being's vision capability which can distinguish about 30,000 categories with very few training samples [1]. As a highlight of current researches, some human perception-inspired models [2,3] reach state-of-the-art performance.

Studies of human perception construct a basic framework for object recognition. Rosch [4] argued that categories are not defined by lists of features but by similarity to prototypes. Similarities defined on prototype examples, or equivalently perceptual distances rather than feature spaces attract the focus of researches. In this framework, scaling to a large number of categories just requires enough prototypes instead of adding new features. It is also possible to train the

S. Boll et al. (Eds.): MMM 2010, LNCS 5916, pp. 273–283, 2010.

   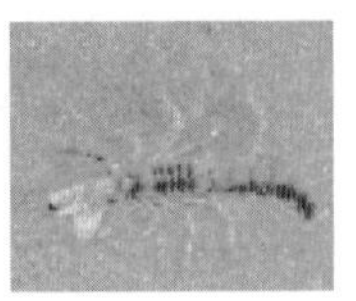

Fig. 1. Four images from scorpion category of the Caltech 101 dataset [1], which show large appearance variations. The main variations include mixed scorpion species with different biological morphology, texture blur, pose change, and cluttered background.

model with very few samples because the invariance to certain transformations or intra-class variation can be built into the perceptual distance function.

Serre and Poggio [5,3] modeled the ventral stream of primate visual cortex as a hierarchical structure (HMAX) for object recognition. The model is composed of S_1, C_1, S_2 and C_2 layers, of which C_1 produces local feature invariant to scaling and rotation and C_2 computes global features by defined perceptual distance (or similarity function). Corresponding to visual cortex, S layers improve invariance while C layers improve selectivity. The tradeoff between invariance and selectivity is achieved through alternate procedures. Frome [2] chose to learn a perceptual distance function for each example with metric learning algorithms [6], which determine weights for elements of all global features. In nature, these algorithms learn a transformation for the entire sample space.

These models and most perceptual inspired models follow the insight of Rosch [4] and share a basic outline: (1) For a test or training image, select a set of interest regions and extract patches from them. (2) Compute a local feature for each patch, which gives a set of local features for each given image. (3) For image pairs, return a value of distance by defining a distance function on their feature sets [7,2]. Or, given a local feature set from a image and the learned prototypes, return a set of distances as the global feature by defining a distance function between image and prototypes [3]. (4) Assign a category label to the image using the distance function or global feature. In step (3), both distance functions, known as perceptual distance, are defined on the local feature space.

However, local features developed for tasks like image registration can lead to a problem that they tend to fail under extreme lighting and pose conditions (for instance, SIFT [8] will be failure on binary images), and therefore could not provide enough discriminative information to classify complex objects, where even images from the same category show large intra-class variance. See Fig. 1.

Recently, some multiple local feature representations [9,10] were proposed to attack the above problem under "distance function learning" framework [2](learn distance functions instead of computing global feature with local features). Motivated by the capability of perceptual inspired models, we explore to integrate multiple local features with global feature computation models, such as HMAX.

In this paper, we propose an integrated solution that extends feature computation model and fusion strategy of global features, as illustrated in Fig. 2. It extends the HMAX model to adapt multiple local features, however, in general

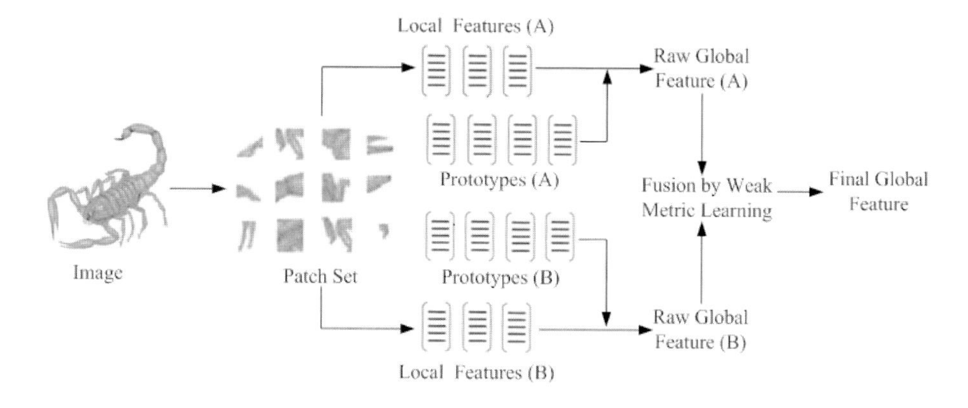

Fig. 2. Illustration of our framework

the extension to other models [7,2] is straightforward. The model could adapt to multiple kinds of local features. For each kind of feature, corresponding raw global feature is computed by measuring its distances from the pre-computed prototypes with the same feature representation[1]. Then an algorithm, namely *weak metric learning*, is developed to fuse these raw global features for object recognition. In nature, it aligns features at the metric level. For the feature fusion task, a criterion, maximal kernel canonical correlation [11,12], is used to solve the weak metric learning model. In sum, our main contributions are two folds.

1. Extend the global feature computation model, HMAX, to adapt to multiple complementary local features. It greatly improves the capability of the system to recognize ambiguous categories.
2. Introduce the kernel canonical correlation to learn the metrics to fuse different global features. Each set of metric weights is derived from the same template function with few free parameters so that the fusion model could be solved with few training samples.

For the improved features, experiments performed on Caltech 101 [1] show consistently significant improvement about $5 \sim 11\%$ than the benchmark model HMAX [3] with robust performance.

2 Model Extensions

The original HMAX model is composed of three steps: (1) compute C_1 response for the given image, (2) learn prototypes from C_1 responses of images, (3) for each prototype, compute the maximal response between the prototype and the C_1 response, which produces an element of the global feature C_2. To extend it for multiple local features, the point is to represent an image with a set of

[1] The prototypes are extracted from a set of randomly selected images, such as natural images.

patches and define the C_1 response on these image patches instead of the whole image (the patch based C_1 response is called C_1 descriptor in the paper). Then the following steps are updated accordingly and other local features could be introduced into the model by replacing C_1 descriptor.

In the extended model, the final global feature is computed according to the following four steps.

1. Extract patches from experimental images and arbitrary natural images. The natural images are used for learning prototypes.
2. Consider a type of local feature, compute a set of such features for the extracted patches.
3. Learn prototypes from the local feature set of natural images, and compute the raw global feature for an image with its local feature set and the learnt prototypes.
4. Multiple kinds of raw global features could be computed by replacing the feature type at step 2. Fuse the raw global features as the final global feature.

In this section, we describe how to compute the raw global features from local features. The fusion scheme will be introduced in Section 3.

Given image I and its local patch set $\mathcal{P}(I) = \{\mathbf{p}_i\}_{i=1}^n$ with varying sizes at detected interest regions, the local feature $\mathbf{c}$ is computed for each patch. Let $\mathcal{C}(I) = \{\mathbf{c}_i\}_{i=1}^n$ represent the local feature set of image I. At learning step, prototypes are extracted from local feature set $\cup_k \mathcal{C}(I_k)$ of natural images randomly and the learnt prototypes set is represented as $\mathcal{T} = \{\mathbf{c}_i^*\}_{i=1}^m$.

For candidate image I with its local feature set $\{\mathbf{c}_i\}_{i=1}^n$ and learnt prototype set $\{\mathbf{c}_i^*\}_{i=1}^m$, the element of raw global feature $\mathbf{x}(I) \in \mathbb{R}^m$ which corresponds to the local feature set, namely, the perceptual distance, is defined as

$$x_i \stackrel{def}{=} \min_{\mathbf{c} \in \mathcal{C}(I)} d(\mathbf{c}_i^*, \mathbf{c}), \tag{1}$$

where function d is a distance measurement of local features $\mathbf{c}_i^*$ and $\mathbf{c}$. We employ Euclidean distance and normalized inner product to measure C_1 and SIFT based perceptual distance respectively in this work. Further, the minimum distance could be regarded as an implementation of the maximal neural response. For C_1 feature, the simulated neural response corresponds to the shape tuning process of visual cortex. On the other hand, x_i could be interpreted as the baseline representation of patch $\mathbf{c}$, based on the prototype set $\mathcal{T}$.

Some descriptors such as SIFT [8], shape context [13] and geometric blur [14], can be used in the extended model. Most of them follow the scale space theory [15] and are invariant to rotation, scaling, or affine translation. In our solution, two complementary descriptors, C_1 and SIFT, are introduced into the extended model because C_1 encodes rich contour and shape information and the complementary SIFT encodes rich gradient information. In the following sections, the C_1 based raw global feature and the SIFT based raw global feature are called as C_2 and SIFT$_2$ respectively. What should be mentioned here is that only two descriptors from patches with the same size could be used for computing the perceptual distance.

3 Weak Metric Learning for Feature Fusion

With previous steps, two raw global features are computed. However, global features derived from different local features have different metrics even though they share the same perceptual distance function. Common schemes suggest to learn two weights for them. Moreover, recent work in [9] shows that multiple features fusion could benefit from subspace learning. However, it is hard to merge the metric difference of features in this task with these methods. In this section, a novel fusion scheme towards eliminating metric difference through metric learning is proposed. We also develop a novel metric learning method called as *weak metric learning* to deal with high dimensional feature. A criterion, maximal canonical correlation is used to solve the metric weights.

3.1 Formulation

Metric learning [2,6] is originally proposed to learn distance or similarity function by weighting each feature dimension. In [16], a correlation metric for feature extraction and similarity measurement is proposed. The technique can eliminate metric difference between feature dimensions implicitly. However, the metric learning scheme has to determine large number of independent weights therefore the scheme tend to fail for high dimensional feature and relative few training samples. In the weak metric learning scheme, a set of nonlinearly dependent weights are assigned to feature dimensions, and only few function parameters, instead of large numbers of weights, have to be determined.

For a given image, suppose similar feature elements correspond to the similar prototypes therefore they have similar metrics with similar weights. The continuous function $h \in H$ is used for assigning weights $w_i = h(x_i)/x_i$ to the global feature $\mathbf{x}(I) \in \mathbb{R}^m$. Then the weighed feature $\mathbf{x}'(I)$ could be formulated as

$$\begin{aligned} \mathbf{x}'(I) &= diag(w_1, \ldots, w_m)\mathbf{x}(I) \\ &= (h(x_1), \ldots, h(x_m))^T \\ &= h \circ \mathbf{x}(I). \end{aligned} \tag{2}$$

It suggests that weighting feature with template derived weights equals to applying a nonlinear transformation on the feature. Because weights for a raw global feature are derived from the same template function h, the task of determining weight set $\{w_i\}_{i=1}^m$ is converted to determine the parameter set of the template function h. The weights are nonlinearly dependent because the number of free parameters of $\{w_i\}_{i=1}^m$ (equals to the parameter number of h) is much smaller than m. It leads to a weak learning scheme. However, with capacity increasing of the template function h, the weak metric learning scheme will approach the general metric learning.

Similar to [2,6], the scheme can also be used to learn the distance function and solved with maximal margin formulation on the triplets training set. For fusion tasks, however, we develop a different model and solving scheme.

3.2 Metric Solving and Feature Fusion

For raw global features introduced in Section 2, we try to fuse them by weighting their elements. We next focus on fusing two features and the way to fuse multiple features is similar. The weight set for a raw global feature in Eq. (2) is derived from the same template function. As to fuse two features, two independent weight sets, equally two template functions have to be determined.

In [17], the canonical correlations of within-class sets and between-class sets for discriminative learning is explored. [18] uses canonical correlation analysis for feature fusion by determining pairs of projective matrices, given two candidate features. Different from above works, we employ canonical correlation to determine template functions (or the derived weight set equally) instead of projective matrices, though a weight set could be regarded as a special projective matrix.

The kernel version of canonical correlation [11,12] is used in the process of feature fusion, in our work, because it increases the flexibility of the feature selection through kernel trick. For the training image set $\mathcal{I}_N$, raw global features $X_{N\times p} = (\mathbf{x}_1, \ldots, \mathbf{x}_N)^T$, $Y_{N\times q} = (\mathbf{y}_1, \ldots, \mathbf{y}_N)^T$ are computed from two different kinds of local features respectively with Eq. (1). Given nonlinear transformations $g, h \in H$, the kernel canonical correlation of two weighted global features is defined as

$$\phi(g, h, \alpha, \beta) = \mathrm{corr}_{\mathrm{ker}}(\alpha^T (g \circ X), \beta^T (h \circ Y)), \tag{3}$$

where $g \circ X$ represents applying transformation g on feature matrix X, as Eq. (2) formulated, and vectors $\alpha, \beta \in \mathbb{R}^N$ represent the combination coefficients of canonical correlation. We choose optimum nonlinear transformations by maximizing Eq. (3) stepwise

$$(g^*, h^*) = \arg \max_{g,h\in H} \widehat{\max_{\alpha,\beta\in\mathbb{R}^N}} \phi(g, h, \alpha, \beta), \tag{4}$$

where $\widehat{\max}_{\alpha,\beta}$ is a constrained maximizing process. We maximize Eq. (4) by enumerating g, h in function space H firstly. After g and h are given, we then further maximize $\phi(h, g, \alpha, \beta)$ in space $\mathbb{R}^N$. That is to maximize kernel canonical correlation

$$\widehat{\max_{\alpha,\beta\in\mathbb{R}^N}} \mathrm{corr}_{\mathrm{ker}}(\alpha^T (g \circ X), \beta^T (h \circ Y))$$
$$\texttt{s.t.}: \mathrm{var}(\alpha^T (g \circ X)) = \mathrm{var}(\beta^T (h \circ Y)) = 1.$$

It can be solved using Lagrange method which leads to an eigenvalue decomposition problem. Then $\phi_{\max}(g, h)$ could be substituted into Eq. (4) to continue maximizing in function space. It is time consuming to enumerate function space H. A specific yet effective solving procedure is to solve the optimization problem in the parameter space of a certain function instead of in the function space. Specially, let H be a function family parameterized by $\theta \in \mathbb{R}^S$. Eq. (4) can be formulated as

$$(\theta_g^*, \theta_h^*) = \arg \max_{\theta_g,\theta_h\in R^S} \widehat{\max_{\alpha,\beta\in R^N}} \phi(\theta_g, \theta_h, \alpha, \beta). \tag{5}$$

According to our experiments, enumerating θ_g and θ_h on an experiential range can satisfy this problem. To ensure optimization, for each candidate image, two local feature sets should be derived from the same patch set.

After parameter sets θ_g^* and θ_h^* are determined, two weighted global features could be given by Eq. (2), leading to the final global feature $(\mathbf{x}'(I)^T, \mathbf{y}'(I)^T)^T$. In the metric learning based fusion scheme, weighting on each feature element can be regarded as the adjusting process with feedback signals in visual cortex.

4 Experiments

Object recognition experiments with the fused global feature are performed on Caltech 8 to (1) show the advantage of the extended model and the fusion scheme; (2) examine the stability of fused features under varying number of samples and feature elements. The HMAX is chosen as the benchmark system because it provides the basic framework for our method. The SVM is used as classifier.

4.1 Dataset and Experimental Setup

A subset of Caltech 101 is chosen in the experiments. Although some categories in Caltech 101 are relative easy to classify, many categories with images taken under extreme lighting and large variations on view and pose are hard to be recognized. The same difficult may also come up in several sub categories with large intra-class variance. To validate the efficacy of our model, we deliberately select 8 difficult categories and the background category with the size of samples ranging from 80 to 800 for test. To speed up feature computation, all the images are normalized to gray images with 140 pixels high and a fixed aspect ratio.

We extract patches from interest regions. Several interest region detectors such as MSER [19], Harris-Affine, and Hessian-Affine [20] can be embedded into our framework. According to the comparison studies in [21], we select the Hessian-Affine as the detector of the interest regions. For a candidate image, patches with the sizes of 4×4, 8×8, 12×12, and 16×16 are extracted from all the interest regions respectively. The C_1 descriptors are constructed for each patch while SIFT descriptors are constructed for 12×12 and 16×16 patches. For prototype learning, patch extraction and descriptor construction are similar to the candidate images, except that for 500 patches per size are randomly extracted from interest points. Although descriptors could be constructed for all size of patches, descriptors from 12×12 and 16×16 patches work well. Then two prototype sets are learnt from natural images for C_1 and SIFT respectively.

At feature fusion step, two independent Gaussian functions are chosen for the weak metric learning procedure, with scale factor ranging from 6.5 to 10, variance ranging from 0.4 to 1.1 and mean ranging from 1.6 to 2.8. Larger range might improve the performance with more expensive time cost. Given the set sizes, the training set and testing set are sampled randomly from corresponding categories, as the same scheme for raw global features. For each setting, we sample data set and raw features about 20 to 40 rounds respectively. Then the average performance and its variance are reported in the final results.

Table 1. Performance comparison of three global feature settings: the feature with 2000 C_2 elements, the combinational feature of 1600 C_2 and 400 SIFT_2 elements, and the fused feature of 1600 C_2 and 400 SIFT_2 elements using our proposed method. Experiments are conducted under a configuration that the number of positive training samples, negative training samples, positive testing samples and negative testing samples are 30, 50, 50, and 50 respectively.

Data set	C_2	Combination of C_2 and SIFT_2	Fusion of C_2 and SIFT_2
Butterfly	0.8092	0.8515	0.8879
Brain	0.8112	0.8458	0.8833
Bonsai	0.7969	0.8130	0.8681
Chandelier	0.7783	0.7891	0.8281
Car-side	0.9737	0.9791	0.9929
Airplanes	0.9674	0.9735	0.9800
Buddha	0.7947	0.8349	0.8729
Scorpion	0.7754	0.8058	0.8438

4.2 Results

To test the performance under different configurations, the size of the positive training set and the length of the global feature are varying in our experiments. The experimental setting is as follows: the sizes of negative training set, positive testing set and negative testing set are taken as 50 respectively.

We run a series of experiments using 30 positive training images per category and 2000 elements (corresponding to 2000 prototypes) per global feature on the 9 categories dataset, with 30 random training sets (also 30 testing sets) and 20 random subsets of global feature (30×20 rounds overall). To evaluate performance, three global feature settings which share the same feature dimension but different element configurations, the feature with 2000 C_2 elements, the combinational feature of 1600 C_2 elements and 400 SIFT_2 elements, and the fused features of 1600 C_2 elements and 400 SIFT_2 elements, are compared.

As shown in Table 1, C_2 feature achieves high performance about 96.7% to 97.4% on Car-side and Airplanes categories and relative low performance about 77.5% to 81.1% on other categories. Similar situation appeared in other two settings. This is because that images of Car-side or Airplanes have small variance or similar appearance even though they are taken from different lighting and pose conditions. For all 8 categories, the combinational feature of C_2 and SIFT_2 elements outperforms C_2 feature about 0.6% to 4.2%. On the other hand, experiments in [3] indicate that increasing the number of feature elements is hard to improve the performance when the number is more than 1000. These evidences suggest that only appending other complementary descriptors based feature elements may be helpful. Compared with C_2 feature, our fusion scheme reaches an improvement about 5.0% to 7.9% on categories except Car-side and Airplanes (improvement about 1.3% to 1.9%).

To validate the fusion scheme on scalable positive training images, varying number of positive training images are tested. Fig. 3 shows these results for 5

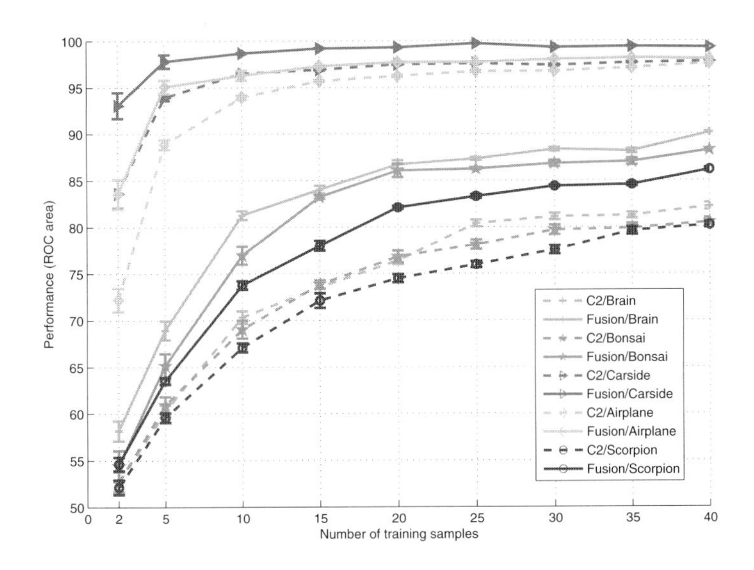

Fig. 3. Comparison between the feature of 2000 C_2 elements and the fused feature of 1600 C_2 and 400 SIFT_2 elements on Caltech 8 for varying number of training examples

categories in Caltech 101. Our fusion scheme outperforms C_2 on all tested categories. For easy categories Car-side and Airplanes, fusion scheme with 5 positive training images achieve satisfying performance about 97.8% and 95.1% with improvements about 3.9% and 9.2%. For other three categories, fusion scheme with 20 positive training images reach significant performance more than 83.1% while the performance of C_2 feature is under 77.2%, and it also outperforms C_2 about 5.1% to 7.9% when the number of positive training images is more than 20.

We also perform a series of experiments for varying number of feature elements from 2 to 2500 to test the fusion scheme. As shown in Fig. 4, the fusion scheme outperforms C_2 feature under all settings. For easy categories Car-side and Airplanes, the fused feature with 50 elements reach a satisfying performance about 97.5% and 95.5% with improvements about 4.0% and 4.5%. For other tested categories, the fused feature with 100 elements reaches significant performance exceeding 80% when the performance of C_2 feature is no more than 72.5%. Under settings of 50 or more feature elements, the fusion scheme outperforms C_2 feature at least 5.9%, especially 11.8% for Butterfly category.

In the fusion scheme, the optimization process of Eq. (5) consumes more time than other steps. Using Gaussian function as the template function to solve Eq. (5) by enumerating 448 parameter points on a normal computer takes about 110 seconds per 80 training images. When the size of dataset grows, the computation complexity mainly depends on the maximization in $\mathbb{R}^N$ and linearly depends on the enumeration number on $\mathbb{R}^S$. In our solution, the patch set that represents the candidate image is extracted from interest regions instead of overlapping regions [3] so that Eq. (1) takes only 1/70 time of it to compute global features.

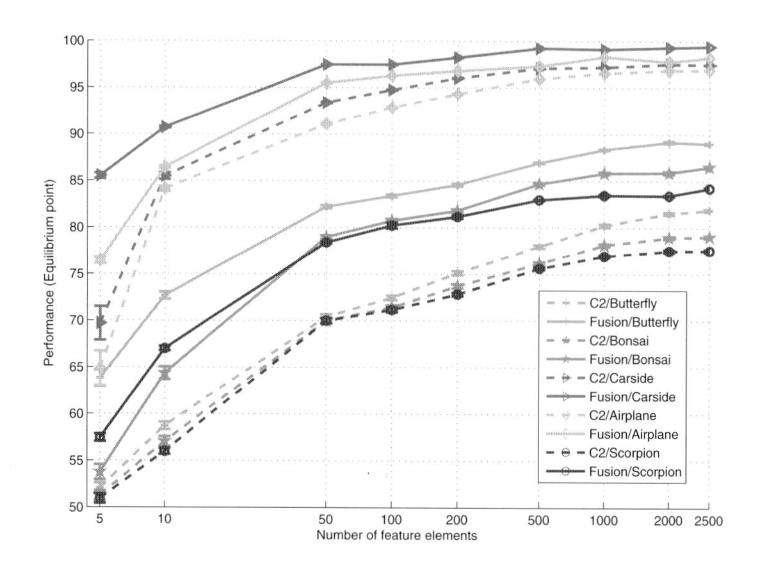

Fig. 4. Comparison between the feature with pure C_2 elements and the fused feature of 75% C_2 and 25% SIFT_2 elements on Caltech 8 for varying number of feature elements

5 Conclusions

In this paper, the perception inspired framework, HMAX, is extended to adapt multiple local features, producing multiple raw global features. A weak metric learning algorithm is developed for high dimensional features towards constructing the feature fusion model. The metric learning based model is solved through maximal canonical correlation formulation, giving the final global feature for object recognition towards difficult categories. Experiments on Caltech 8 show significant improvements under settings of varying number of training images and feature elements, which also confirms the validity and stability of our scheme. The fusion scheme, however, reaches the performance at the cost of much computing time, which will be the further research topic of this model.

Acknowledgment

Thanks to China National 973 Program 2006CB303103, China NSFC Key Program 60833009 and National 863 program 2009AA01Z330 for funding.

References

1. Fei-Fei, L., Fergus, R., Perona., P.: Learning generative visual models from few training examples: an incremental Bayesian approach tested on 101 object categories. In: IEEE CVPR, Workshop on Generative-Model Based Vision (2004)

2. Frome, A., Singer, Y., Malik, J.: Image retrieval and classification using local distance functions. In: NIPS (2007)
3. Serre, T., Wolf, L., Bileschi, S., Riesenhuber, M., Poggio, T.: Robust object recognition with cortex-like mechanisms. IEEE TPAMI 29(3), 411–426 (2007)
4. Rosch, E.: Natural Categories. Cognitive Psychology 4(3), 328–350 (1973)
5. Riesenhuber, M., Poggio, T.: Hierarchical models of object recognition in cortex. Nature Neuroscience 2, 1019–1025 (1999)
6. Schultz, M., Joachims, T.: Learning a distance metric from relative comparisons. In: NIPS (2004)
7. Zhang, H., Berg, A., Maire, M., Malik, J.: SVM-KNN: Discriminative nearest neighbor classification for visual category recognition. In: IEEE CVPR (2006)
8. Lowe, D.: Object recognition from local scale-invariant features. In: IEEE ICCV (1999)
9. Fu, Y., Cao, L., Guo, G., Huang, T.S.: Multiple feature fusion by subspace learning. In: ACM CIVR, pp. 127–134 (2008)
10. Lin, Y., Liu, T., Fuh, C.: Dimensionality Reduction for Data in Multiple Feature Representations. In: NIPS (2008)
11. Lai, P., Fyfe, C.: Kernel and nonlinear canonical correlation analysis. International Journal of Neural Systems 10(5), 365–378 (2000)
12. Hardoon, D., Szedmak, S., Shawe-Taylor, J.: Canonical correlation analysis: an overview with application to learning methods. Neural Computation 16(12), 2639–2664 (2004)
13. Belongie, S., Malik, J., Puzicha, J.: Shape matching and object recognition using shape contexts. IEEE TPAMI, 509–522 (2002)
14. Berg, A., Malik, J.: Geometric blur and template matching. In: IEEE CVPR (2001)
15. Lindeberg, T.: Scale-space: A framework for handling image structures at multiple scales. European Organization for Nuclear Research-Reports-CERN, 27–38 (1996)
16. Fu, Y., Yan, S., Huang, T.: Correlation metric for generalized feature extraction. IEEE TPAMI, 2229–2235 (2008)
17. Kim, T., Kittler, J., Cipolla, R.: Discriminative learning and recognition of image set classes using canonical correlations. IEEE TPAMI 29(6), 1005–1018 (2007)
18. Sun, Q., Zeng, S., Liu, Y., Heng, P., Xia, D.: A new method of feature fusion and its application in image recognition. Pattern Recognition 38(12), 2437–2448 (2005)
19. Matas, J., Chum, O., Urban, M., Pajdla, T.: Robust wide-baseline stereo from maximally stable extremal regions. Image and Vision Computing 22(10), 761–767 (2004)
20. Mikolajczyk, K., Schmid, C.: An affine invariant interest point detector. In: Heyden, A., Sparr, G., Nielsen, M., Johansen, P. (eds.) ECCV 2002. LNCS, vol. 2350, pp. 128–142. Springer, Heidelberg (2002)
21. Mikolajczyk, K., Tuytelaars, T., Schmid, C., Zisserman, A., Matas, J., Schaffalitzky, F., Kadir, T., Gool, L.: A comparison of affine region detectors. International Journal of Computer Vision 65(1), 43–72 (2005)

The Persian Linguistic Based Audio-Visual Data Corpus, AVA II, Considering Coarticulation

Azam Bastanfard[1], Maryam Fazel[2], Alireza Abdi Kelishami[3], and Mohammad Aghaahmadi[3]

[1] Information Technology Research Group , Department of Engineering, Islamic Azad University Karaj branch, Iran
bastanfard@kiau.ac.ir
[2] Islamic Republic of Iran Broadcast University, Tehran, Iran
mfazzel@yahoo.com
[3] Department of Electrical, Computer and IT Engineering, Qazvin Islamic Azad University, Qazvin, Iran
alireza.abdi@yahoo.com, aghaahmadi@ymail.com

Abstract. Collecting an audio visual data corpus based on the linguistic rules is an unquestionable, must-take step in order to conduct major research in multimedia fields as AVSR, lip synchronization and visual speech synthesis. Building up a reliable data corpus where it covers all phonemes in all phonemic combinations of a language is a difficult and time consuming task. To partially deal with this problem, in this research, vc, cv and vcv combinations, instead of the entire possible phonemic combinations were used, where they carry the most language information. This paper gives an indication on the new data corpus, capturing 14 respondents. To better perceive coarticulation effect in speech, continuous speech was considered other than isolated and continuous digits. This makes the collection process a more time and cost-saving one, maintaining the efficiency high.

Keywords: Audio visual database design, linguistic approach, coarticulation, Persian data corpus, multimedia modeling, Farsi audio visual data corpus, AVA II.

1 Introduction

Along with the great advances of computer science in current years in multimedia fields and the noticeable improvement in human and machine interaction, there's an increasing consideration on linguistics and its applications in language-based computer science fields. Many defects in uni-modal data can be remedied by multi-modal data [4]. Building audio visual data corpora is among the initial prerequisites for establishing this interaction and improving its quality. It goes with no doubt that depending on the target language and application, various data corpora are needed. For researches as audio-visual speech recognition (AVSR), lip reading, or identity recognition to be conducted, there is a certain need for a data corpus to be built first, considering the aimed application's requirements. AVA [35], the first comprehensive

S. Boll et al. (Eds.): MMM 2010, LNCS 5916, pp. 284–294, 2010.

Persian database, has been built aiming for speech therapy and viseme extraction. The extracted visemes in AVA are not applicable to some other uses; such as audio visual speech recognition or lip reading, for there are two respondents captured, covering all possible phonemic combinations in it; although it has well met its aiming application's requirements. In this project, AVA II, there has been a study on Persian language, after which more effort is put into optimizing the spoken material such that not only it covers most effective phonemic combinations in Persian language, but also does it consider coarticuation effect on visual information.

In the following, section 2 prepares an overview of current AV datasets. Section 3 offers the pre-recording analysis undertakings, including speaker population and the spoken materials. Section 4 describes the recording phase comprising the studio settings and the recording approach. Data labelling and addressing comes in the 5th section, where section 6 concludes the paper and offers future work in the area.

2 Related Work

A data corpus is developed in a language according to its matchless phonemic, visemic characteristics. Putting a step forward, many have argued that not only phonemic and visemic information vastly changes among languages, but also this information changes depending on the phoneme's position in a phonemic combination in a single language. This variability in an articulator's pose caused by the assimilation of a speech unit to its precedent unit is called the coarticulation effect. A comprehensive analysis former to corpus design, could be the key to developing a corpus which could at the first place be a further useable one.

A number of corpora are designed for recognizing digits such as CUAVE containing 36 speakers [22,23] and M2VTS [25], and its later extended version, XM2VTS [19]; where some of these as [2] have their respondents rotate their head or capture the profile view. Recording the profile view of respondents enables the corpus to be used for 3-D featured applications and makes corpora more robust for face recognition purposes. Some of the corpora as Manssa L.K [18] is developed to enable speech recognition. AVOZES [10,11], Mandrain Chinese [17], Czech [5,26] and Dutch [27] are of this kind. Some have captured their subjects saying rather unforeseen phonemic combinations, and are mostly covering all probable phonemic combinations, but do not consider coarticulation effect [1, 6]. [7] has video recordings of 1000 sentences spoken by each of its 34 speakers. M3 [14] aims to support research in multi-biometric technologies for pervasive computing using mobile devices. The IBM AV consists of 304 speakers and has identification and verification experiments based on its database [32]. The DAVID [33] has challenging visual conditions including illumination changes and variable scene background complexity.

Few have narrowed their target application range [34, 16, 21] and were filmed in office or car environments or are built for covering emotional speaking faces [12]. A number of them are built to be used as computer assistive technology [24]. Some are developed aimed at talking head generation [8,29]. The AMP/CMU [30] includes 78 isolated words commonly used for scheduling applications. A survey on audio visual databases is prepared in [3] listing many corpora, most of which cover mono-modal or have limited spoken material.

AVA, the first corpus collected in Persian, which is aimed at providing data for developing a speech therapy application had 2 subjects saying 5500 utterances, where it falls short of appropriately meeting the new requirements of some applications as AVSR, where more subjects are needed to provide an acceptable population in number.

Although many AV corpora have been developed in many languages, none is a thoroughly reusable one. They are designed so that they meet their former target applications' requirements. Hence, for a latter unforeseen project, they usually fall short of meeting the new requirements. By choosing the framework described in the next sections, there is an effort on building up a corpus which could be widely used by all researchers in the area to access the needed data and develop their desirable applications using this data which is the first of its kind in Persian language.

3 Pre-recording Phase

To enhance the database usability, an analysis was involved at the beginning. It is now considered that data corpora are designed based on a comprehensive analysis over their target application. Some works in this area have led to proposing new frameworks such as [20]. An analysis was performed, besides considering an appropriate lighting and acoustic environment in a professional studio for recording. In the following, the spoken material, the prompts and the speaker population are to be discussed.

3.1 Spoken Material and Prompts

Persian language has 23 consonants and 6 vowels. Figure 1 compares the Persian vowels with IPA [31] standard [28,15], where the vertical axis is the mouth opening size; and the horizontal axis shows place of articulation. Persian vowels are in dark italic font. Firstly, to later detect some parameters including minimum and maximum mouth openness, interior and posterior lip movement range, the subject is asked to utter sounds like /ɒ/, /i/, /ɒ/, /i/, /ɒ/, /i/ and /i/, /u/, /i/, /u/, /i/, /u/.

Secondly, subjects are asked to pronounce phonemic combinations in single words. In AVA, due to its target application, speech therapy, a huge set of utterances had to be captured covering all possible phonemic combinations, which itself was a time consuming task for a subject to utter that huge set of utterances even including many utterances which are not used in the language's vocabulary. Nonetheless in this version, AVA II, all consonants and vowels in cv, vc, vcv combinations are uttered, making the set considerably narrowed in size, and it has lessened the needed filming time per each speaker where at the same place it carries most of phoneme articulation and coarticulation information. To further describe the phonemic combinations, table 1 presents all possible and captured forms of consonant /b/ in the three stated combinations.

Next, to have continuous speech captured as well, subjects are asked to utter 20 common sentences in contemporary Persian [9]. In this set, phoneme proportion is balanced. Table 2 includes two sentence examples. The place and manner of articulation of the consonants in Persian language are indicated in table 3 [13].

Then, numbers are pronounced either in isolated or continuous format. For continuous digits, it was first decided to have the speakers count from 0 to 20, but since it was experimentally realized that the last digit is unconsciously uttered incompletely by the speakers, subjects count till 21 instead. Moreover, other digits as 30, 40, 100, 200, 1000, million and billion were uttered in addition. At the end, to have continuous digits, the subjects are asked to continuously read digits 367, 549 and 821. In Persian continuous digit counting system, /o/ sound is used to attach digits when reading them continuously; hence, the pronunciation pattern of continuous digits would be accessible through these three.

Some corpora have used headphones for subjects to hear the spoken material and repeat it. But in this project, prompts are shown on a screen instead, for the speaker to read. The following two experimental reasons justify using prompts on a screen, instead of dictating them on headphones:

- The voice which is played on the headphone for the subject unconsciously dictates its manner of articulation on that of the respondents, which is not what corpora designers will to happen.
- Reading utterances from a text causes speakers pronounce the utterances in formal Persian accent, Tehrani and unconsciously avoid accent variations that exist in many forms in Persian.

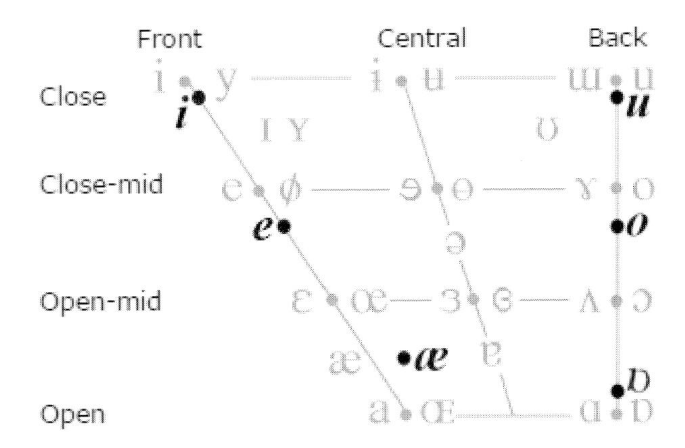

Fig. 1. Vowels in IPA and Persian language

Table 1. Phonemic combinations of consonant /b/

c=b	/u/	/o/	/ɒ/	/æ/	/e/	/i/
cv	bu /bud/	bo /boro/	bɒ /bɒbɒ/	bæ /bæle/	be /bede/	bi /bitɒ/
vc	ub /xub/	ob /sob/	ɒb /xɒb/	æb/ædæb/	eb /teb/	ib /sib/
vcv	vbv	vbv	vbv	vbv	vbv	vbv

Table 2. Example of two common sentences

c=b	/u/
IPA form	/dæstet dærd nækone/
Persian form	هنکن درد تتسد
IPA form	/pam dærd mikone /
Persian form	پام درد می کنه

Table 3. Consonants in Persian language. Here, PoA and MoA stand for place of articulation and manner of articulation, respectively.

PoA → /MoA↓	Labial	Alveolar	Post-alveolar	palatal	Velar	Uvular	Glottal
Nasal	m	n			[ŋ]		
Plosive	p b	t d			K g	ɢ	[ʔ]
Affricate			tʃ dʒ				
Fricative	f v	s z	ʃ ʒ	x ɣ			h
Tap	[ɾ]						
Trill	r						
Approxi-mant	l		j				

The spoken materials are prepared in power point slides. Prompts are shown on a laptop, having a minimum noise, next to the front camera. The studio layout is mentioned in the following sections. To avoid unwanted phonemic and visemic coarticulation, each slide is passed in 5 seconds of interval.

3.2 Speaker Population

In this project, 7 female and 7 male are chosen for recording. Respondents are between 18 and 30 years of age, and have the Tehrani accent. The subjects have no articulatory disorders. A speech and language pathologist in the group checks every step. Respondents speak in their normal speech rate and make no alternation in their prosody state. Before filming each of the respondents, they were asked to sign a written permission to let AVA group use their audio-video data in future research.

4 Recording Phase

The TV studio of IRIBU, Iran's broadcasting university, was chosen to film the data in an appropriate professional environment. Three digital cameras are utilized, consisting of one N9000 Panasonic, named cam1 for quick referencing, and two Canon XL2 cameras, named cam2 and cam3. Cameras 2 and 3 record the profile and front view respectively, where cam 1 captures the lip area with a slight angular difference with cam2 and is placed beneath it to record a better view of the tongue and teeth image. Figure 2 illustrates the studio setup in different perspectives. Prior to data recording, cameras' lens focus, white-balance and brightness were automatically calibrated. Here, in AVA II, another high quality camera is selected to film the lip area, and is placed beneath the camera shooting the frontal view. Experiments demonstrate that with teeth and tongue being visible and having a high quality lip view video captured by a high quality camera, better analysis and computations results can be produced. The studio has a perfect lighting system and an acoustic environment. To remove shadows from the speaker's face, several hanging light projectors other than 4 portable light projectors were used. The background of the cameras is a blue curtain. Figure 3 depicts the position of the cameras and the portable light projectors.

Here, two microphones are used, one an AKG color microphone, placed at the top of the speaker's head to minimize the noise commonly made because of the friction against the scarf and clothes. The other microphone is a Road NT 1000 one which records the voice in studio's internal environmental noise which is placed besides the cameras. Moreover, all cameras automatically record voice on their own microphone. Recording voice via microphones in addition to the automatically recorded voice of the cameras brings a higher quality to the captured database for either of the future expected or unanticipated projects; and enhances future reusability, for there could be more robust signal processing. The microphones' output is plugged into a mixer, and from the mixer to the MOTU external sound card in the control room. The MOTU output is then recorded via the Adobe Audition software. In this data base, video is in AVI format, 25 fps, 720x576 pixels; and the audio is in WAV, 16 bps and 48 KHz.

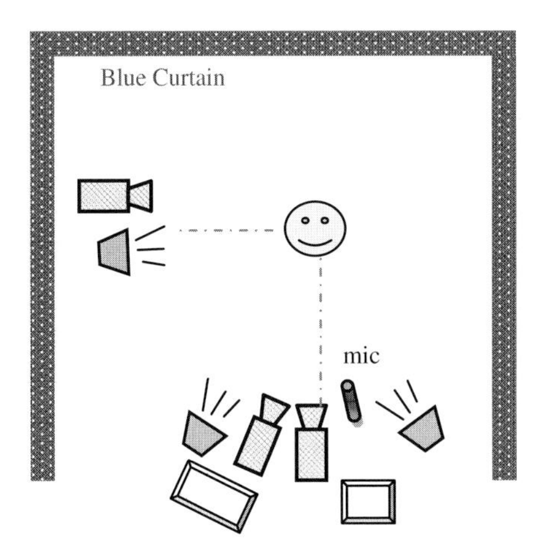

Fig. 2. Studio layout

Every speaker has 90 minutes filmed, if not paused in the middle. The process may be paused for various reasons, e.g. when the speaker gets tired or for any other problems that might arise.

The speaker is asked to clap hands at the beginning of each local start, which could be more than one, according to the stated reasons above. This makes it possible to better and easier synchronize the separated recorded audio and video by simple signal processing of a sound file and observing the sudden oscillation, or the video when clapping hands. Neglecting this may cause a corpus designer encounter problems and more difficulties with the synchronization matter, for voice and video normally do not start at the same time, if recorded separately.

Prompts are shown on a laptop located close to the front camera. Two additional monitors were used, to control the filming process. The first monitor, which is a feedback monitor here, is the output of cam1. It is employed as a hint for the speaker to keep her face in frame by looking at this monitor frequently, and exclude the distraction and noise caused by another person even inaudibly telling him how to keep her face in frame. The other monitor is positioned in the control room. This one is used to control the video of the speaker and to correct the probable problems in the recording process. Figure 3 illustrates the studio and speaker's position.

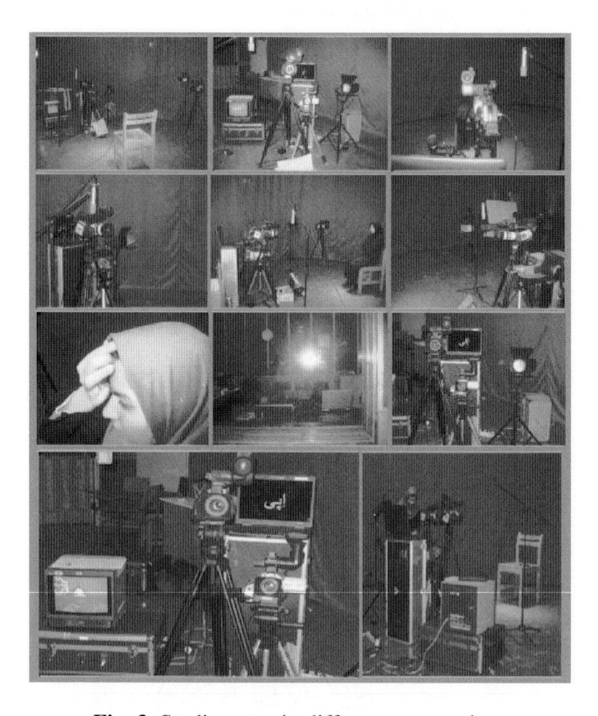

Fig. 3. Studio setup in different perspectives

Formerly in test section, respondents were told to wear dark clothes. Then it experimentally proved that white clothes could prevent dark light reflections of dark clothes; for which respondents wore white clothes. The entire recording phase took place in 5 sessions, during two months.

5 Data Capturing and Arrangement

The videos are captured on a hard drive after filming each respondent's audio-video data. Then, for a faster data access and reference, the large amount of data is captured as follows:

- Every respondent is given an ID code and the data corresponding to the speaker is captured in its own folder.
- For each speaker, the data corresponding to the frontal, the profile and the lip area views are stored in different folders separately.
- Each of the stated folders consists of three subfolders: continuous speech, serial numbers and phoneme combinations.

Figure 4 illustrates a sample of stated data categorizing approach. Phonemic combinations' folders are each divided based on the corresponding consonant within the combination.

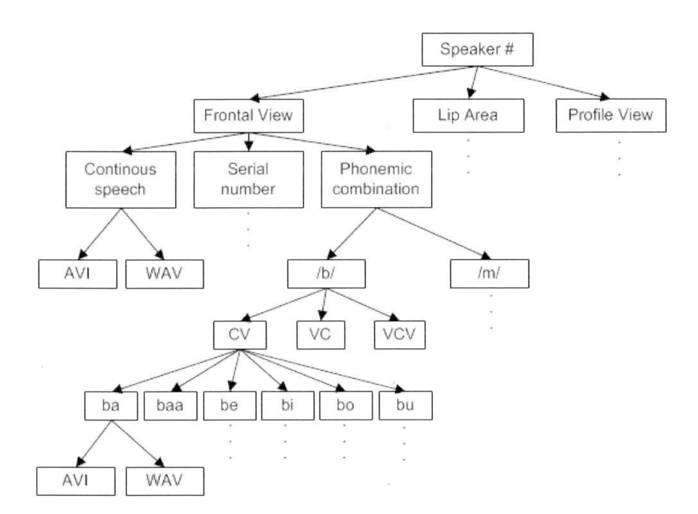

Fig. 4. Categorizing approach of audio-video data for each speaker and a sample of phoneme combination arrangement

6 Conclusion and Future Works

This paper presented the design and collection process of AVA II, the Persian audio-visual data corpus. AVA II is collected considering linguistic principles and phonetics. It has an optimized set of spoken material and was captured in an appropriate, noiseless studio with great care on the light and sound conditions, beside the effort made to record both video and audio in good quality. Respondents speak the shared utterances including all cv, vc and vcv combinations where they carry the most information on phoneme articulation and coarticulation. Moreover, to better identify coarticulation effect, continuous speech, isolated and continuous digits are captured. Video sequence includes 3 different views of frontal, profile and lip area. Figure 5 shows images of two of the respondents while uttering /u/ sound. This data corpus is employed in AVSR, lip synchronization and some other state of the art technological fields.

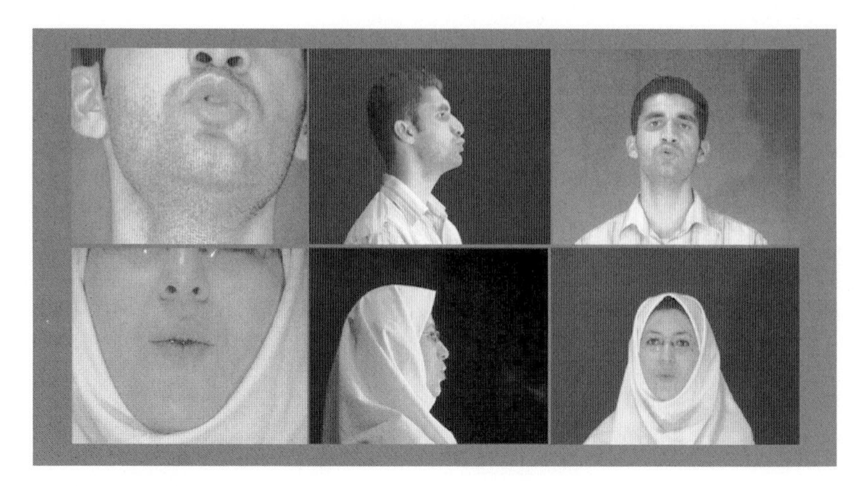

Fig. 5. Categorizing approach of audio-video data for each speaker and a sample of phoneme combination arrangement

To extend the AVA II data corpus in future, various emotional states during speech will be combined aimed at recognizing changes of lip posture in different emotional states.

References

1. Bailly-Bailliere, E., Bengio, S., Bimbot, F., Hamouz, M., Kittler, J., Mariethoz, J., Matas, J., Messer, K., Popovici, V., Poree, F., Ruiz, B., Thiran, J.P.: The BANCA Database and Evaluation Protocol. In: Kittler, J., Nixon, M.S. (eds.) AVBPA 2003. LNCS, vol. 2688, pp. 625–638. Springer, Heidelberg (2003)
2. Movellan, J.R.: Visual speech recognition with stochastic networks. In: Tesauro, G., Toruetzky, D., Leen, T. (eds.) Advances in Neutral Information Processing Systems, vol. 7. MIT Press, Cambridge (1995/2000); Neti, C., Pontamianos, G., Luettin, J., Matthews, I.
3. Chibelushi, C.C., Deravi, F., Mason, J.S.D.: Survey of audio visual speech databases. Tech. Rep., Department of Electrical and Electronic Engineering, University of Wales, Swansea, UK (1996)
4. ChiŃu, A.G., Rothkrantz, L.J.M.: Building a Data Corpus for Audio-Visual Speech Recognition. In: AGC, April 2007, pp. 88–92 (2007)
5. Cisar, P., Zelezny, M., Krnoul, Z., Kanis, J., Zelinka, J., Müller, L.: Design and recording of Czech speech corpus for audio-visual continuous speech recognition. In: Proceedings of the Auditory-Visual Speech Processing International Conference, AVSP 2005, Vancouver Island, pp. 1–4 (2005)
6. Sanderson, C., Paliwal, K.K.: Identity verification using speech and face information. Digital Signal Processing 14(5), 449–480 (2004)
7. Cooke, M., Barker, J., Cunningham, S., Shao, X.: An audio-visual corpus for speech perception and automatic speech recognition. Journal of the Acoustical Society of America 120(5), 2421–2424 (2006)
8. Ezzat, T., Poggio, T.: Visual Speech Synthesis by Morphing Visemes. International Journal of Computer Vision 38, 45–57 (2000)

9. Movalleli, G.: Sara Lip-Reading Test: Construction, Evaluation and operating on a group of people with hearing disorder. MSc Thesis, Department of Rehabilitation in Tehran University of medical sciences (2002) (in Persian)
10. Goecke, R., Millar, J.B., Zelinsky, A., Robert-Ribes, J.: A detailed description of the AVOZES data corpus. In: Proceedings of IEEE International Conference on Acoustics, Speech, and Signal Processing (ICASSP 2001), Salt Lake City, Utah, USA, May 2001, pp. 486–491 (2001)
11. Goecke, R., Millar, J.B.: The Audio-Video Australian English Speech Data Corpus AVOZES. In: Proceedings of the 8th International Conference on Spoken Language Processing, ICSLP 2004, vol. III, pp. 2525–2528 (2004)
12. Grimm, M., Narayanan, S.: The Vera am Mittag German audio-visual emotional speech database. In: ICME 2008, pp. 865–868. IEEE, Los Alamitos (2008)
13. Jahani, C.: The Glottal Plosive: A Phoneme in Spoken Modern Persian or Not? In: Csató, É.Á., Isaksson, B., Jahani, C. (eds.) Linguistic Convergence and Areal Diffusion: Case studies from Iranian, Semitic and Turkic, pp. 79–96. RoutledgeCurzon, London (2005)
14. Meng, H.M., Ching, P.C., Lee, T., Mak, M.-W., Mak, B., Moon, Y.S., Siu, M.-H., Tang, X., Hui, H.P.S., Lee, A., Lo, W.-K., Ma, B., Sio, E.K.T.: The Multi-Biometric, Multi-Device and Multilingual (M3) Corpus. In: Proceedings of The Second Workshop on Multimodal User Authentication (MMUA), Toulouse, France, May 11-12 (2006)
15. Ladefoged, P.: Vowels and Consonants, 2nd edn. Blackwell Publishers Pub., Malden (2004)
16. Lee, B., Hasegawa-Johnson, M., Goudeseune, C., Kamdar, S., Borys, S., Liu, M., Huang, T.: AVICAR: Audio- Visual Speech Corpus in a Car Environment. In: Proceedings of International Conference on Spoken Language Processing – INTERSPEECH, Jeju Island, Korea, October 4-8 (2004)
17. Liangi, L., Luo, Y., Huang, F., Nefian, A.V.: A multi-stream audio-video large-vocabulary mandarin Chinese speech database. In: IEEE International Conference on Multimedia and Expo., vol. 3, pp. 1787–1790 (2004)
18. Marassa, L.K., Lansing, C.R.: Visual Word Recognition in Two Facial Motion Conditions: full-face versus Lip plus Mandible. Journal of speech and hearing Research 38(6), 1387–1394 (1995)
19. Messer, K., Matas, J., Kittler, J., Luettin, J.: XM2VTSDB: the extended M2VTS database. In: Proceedings of the 2nd International Conference on Audio-and Video-Based Biometric Person Authentication (AVBPA 1999), Washington, DC, USA, March 1999, pp. 72–77 (1999)
20. Millar, J.B., Wagner, M., Goecke, R.: Aspects of Speaking-Face Data Corpus Design Methodology. In: Proc. 8th Int. Conf. Spoken Language Processing, ICSLP, Jeju, Korea, vol. II, pp. 1157–1160 (2004)
21. Mostefa, D., Moreau, N., Choukri, K., Potamianos, G., Chu, S.M., Tyagi, A., Casas, J.R., Turmo, J., Christoforetti, L., Tobia, F., Pnevmatikakis, A., Mylonakis, V., Talantzis, F., Burger, S., Stiefelhagen, R., Bernardin, K., Rochet, C.: The CHIL audiovisual corpus for lecture and meeting analysis inside smart rooms. Journal of Language Resources and Evaluation 41, 389–407 (2008)
22. Patterson, E., Gurbuz, S., Tufekci, Z., Gowdy, J.N.: Moving-talker, speaker-independent feature study, and baseline results using the CUAVE multimodal speech corpus. EURASIP Journal on Applied Signal Processing 2002, 1189–1201 (2002)
23. Patterson, E., Gurbuz, S., Tufekci, Z., Gowdy, J.N.: CUAVE: a new audio-visual database for multimodal human computer-interface research. In: Proceedings of IEEE International Conference on Acoustics, Speech, and Signal Processing (ICASSP 2002), Orlando, Fla, USA, May 2002, vol. 2, pp. 2017–2020 (2002)

24. Pera, V., Moura, A., Freitas, D.: LPFAV2: a New Multi-Modal Database for Developing Speech Recognition Systems for an Assistive Technology Application. In: SPECOM 2004: 9th Conference Speech and Computer, St. Petersburg, Russia, September 20-22 (2004)
25. Pigeon, S., Vandendorpe, L.: The m2vts multimodal face database (release 1.00). In: Bigün, J., Borgefors, G., Chollet, G. (eds.) AVBPA 1997. LNCS, vol. 1206, pp. 403–409. Springer, Heidelberg (1997)
26. Trojanová, J., Hrúz, M., Campr, P., Železný, M.: Design and Recording of Czech Audio-Visual Database with Impaired Conditions for Continuous Speech Recognition. In: Proceedings of the Sixth International Conference on Language Resources and Evaluation, LREC 2008 (2008)
27. Wojdeł, J.C., Wiggers, P., Rothkrantz, L.J.M.: An audiovisual corpus for multimodal speech recognition in Dutch language. In: Proceedings of the International Conference on Spoken Language Processing, ICSLP 2002, Denver CO, USA, September 2002, pp. 1917–1920 (2002)
28. Samareh, Y.: Persian phonetics. Markaze Nashre Daneshgahi Pub., Tehran (1998) (in Persian)
29. Yotsukura, T., Nakamura, S., Morishima, S.: Construction of audio-visual speech corpus using motion-capture system and corpus based facial animation. The IEICE Transaction on Information and System E 88-D(11), 2377–2483 (2005)
30. Chen, T.: Audiovisual speech processing. IEEE Signal Processing Mag. 18, 9–21 (2001)
31. International Phonetic Association. In: Handbook of the International Phonetic Association: A guide to the use of the International Phonetic Alphabet, pp. 124–125. Cambridge University Press, Cambridge (1999) ISBN 978-0521637510
32. Chaudhari, U.V., Ramaswamy, G.N.: Information fusion and decision cascading for audiovisual speaker recognition based on time-varying stream reliability prediction. Paper presented at the Int. Conf. Multimedia Expo. (1999)
33. Chibelushi, C.C., Gandon, S., Mason, J.S.D.: Design issue for a digital audio-visual integrated database. Integrated Audio-visual Processing for Recognition, Synthesis and Communication (1996)
34. Fox, N.A., O'Mullane, B., Reilly, R.B.: The realistic multi-modal VALID database and visual speaker identification comparison experiments. Paper presented at the 5th International Conference on Audio- and Video-Based Biometric Person Authentication (2005)
35. Bastanfard, A., Fezel, M., Kelishami, A.A., Aghaahmadi, M.: A comprehensive audio-visual corpus for teaching sound Persian phoneme articulation. In: IEEE International Conference on Systems, Man, and Cybernetics (accepted, 2009)

Variational Color Image Segmentation via Chromaticity-Brightness Decomposition

Zheng Bao[1,2], Yajun Liu[2,3], Yaxin Peng[4], and Guixu Zhang[5,*]

[1] United Data Information Technology Co. Ltd, Shanghai 200081, China
[2] Department of Mathematics, East China Normal University, Shanghai 200241, China
[3] Department of Mathematics, Xinjiang Normal University, Urumqi 830054, China
[4] Department of Mathematics, Shanghai University, Shanghai 200444, China
[5] Department of Computer Science, East China Normal University, Shanghai 200241, China

Abstract. A region-based variational model for color image segmentation is proposed using the chromaticity-brightness decomposition. By this decomposition, we extend the Wasserstein distance based method to color images. The chromaticity term of the proposed functional follows the data term of the color Chan-Vese model with constraint on unit sphere, and the brightness term is formulated by the Wasserstein distance between the computed probability density function in the local windows (e.g. 3 by 3 or 5 by 5 window) and its estimated counterparts in classified regions. Experimental results on synthetic and real color images show that the proposed method performs well for the segmentation of different image regions.

Keywords: Variational model, chromaticity-brightness decomposition, color image segmentation.

1 Introduction

Image segmentation is an essential problem in image processing. One of the most successful models for image segmentation is the Chan-Vese model [1], a well known region-based active contour model. Chan et al. [2] also proposed a generalized Chan-Vese model for vector-valued or color images. Recently, Ni et al. proposed a new segmentation model [3] based on their previous work [4], which obtains the data term by comparing the histogram of each region with the local histogram. The distance between the two histograms is measured by the Wasserstein distance. However, it is difficult to extend this work directly to color images, since there is no closed form for the Wasserstein distance in three dimensional case so far.

It is well known that a color image can be presented in various ways [5], such as the RGB (red, green and blue) model, the YCbCr (Luminance, blue-difference and red-difference Chroma) model, and the HSV (hue, saturation and value) model. As far as we know, most of color image segmentation methods use the RGB model, since this model can be mathematically viewed as 3 dimensional vectorial functions. However, the RGB model is a linear or channel-by-channel model, and directly applying

* Corresponding author.

S. Boll et al. (Eds.): MMM 2010, LNCS 5916, pp. 295–302, 2010.

the gray level method to each channel does not get satisfactory results. A better color model, chromaticity and brightness (CB) model, which has been studied well for image colorization or denoising of color images in [6, 7], can also be used in color image segmentation.

It has been shown in [8] that the essential geometrical information is contained in the gray level of color images, and as suggested in [9], color and texture are separated phenomena that can be treated individually. Besides, from the viewpoint of human perception, we can find edges or segment color images by two ways: one is the luminance or brightness and the other is the chrominance or color information. Generally speaking, treating the brightness separately from the chromaticity can provide more flexible and satisfactory results in denoising, colorization or segmentation problems.

In this paper, we tackle the color image segmentation problem by chromaticity and brightness decomposition, and it is a feasible way to extend the Wasserstein distance to color images. The chromaticity term of the proposed functional follows the data term of the color Chan-Vese model with constraint on unit sphere. The brightness term is formulated by the Wasserstein distance between the computed probability distributions in the local windows (e.g. 3 by 3 or 5 by 5 window) and its estimated counterparts in classified regions. Their contributions to the energy functional depend on their own weights. The proposed model is solved by a recently developed fast global minimization method.

The outline of the paper is given as follows: in section 2, the segmentation model for color images based on chromaticity and brightness decomposition is described along with the corresponding Euler-Lagrange (EL) equations. Section 3 presents the fast global minimization of the functional by introducing an intermediate variable. We perform the numerical experiments in section 4. Segmentation results show the effectiveness of our algorithm. In section 5, we conclude the paper and give some perspectives of our work.

2 Proposed Model

In this section, we first present our variational model via chromaticity-brightness decomposition and give a fast global minimization algorithm. We relax the original energy functional, separate it into two minimization problems by introducing an immediate variable, and solve the minimization problems by the dual method for Total Variation (TV) norm. Then we describe the choice of data term for chromaticity and brightness components and show the optimal conditions to solve them respectively.

Let $\Omega \in \mathbb{R}^2$ be an open and bounded image domain; $\mathbf{I} = (I_1, I_2, I_3) : \Omega \to \mathbb{R}^3$ be the observed color image with RGB representation. In CB model, image $\mathbf{I}$ is separated into the brightness component $B=|\mathbf{I}|:\Omega \to [0, L]$, where L is the maximum of component B, and the chromaticity component $\mathbf{C} = (C_1, C_2, C_3) = \mathbf{I}/|\mathbf{I}|:\Omega \to \mathbb{S}^2$. Assuming that Γ separates Ω into two regions Ω_i with $i = 1,2$; $|\Gamma|$ is the length of Γ; P_i is the probability density function (PDF) in Ω_i of brightness component B; P_x is the PDF or histogram of the local window centered at x in Ω, could be estimated directly from B; F_i and F_x are the corresponding probability distributions.

Let W_x be a local window centered at x. The local probability distribution or cumulative histogram is given by [3]:

$$F_x(l) = \frac{|y \in W_x \cap \Omega : B(y) \le l|}{|W_x \cap \Omega|} \quad \text{for } 0 \le l \le L.$$

We define a new energy functional with three parts: the first regularization part corresponds to the arc length term, the second fidelity part is the brightness term based on the Wasserstein distance of probability distribution, and the last is the chromaticity term using the vectorial Chan-Vese model and the unit sphere constraint. The proposed segmentation model is:

$$\min_{\Gamma, P_i, \boldsymbol{\mu}_i} \{E_1(\Gamma, P_i, \boldsymbol{\mu}_i \mid \mathbf{I}) = |\Gamma| + \sum_i \alpha \int_{\Omega_i} W_1(P_i, P_x)dx + \beta \int_{\Omega_i} (\mathbf{C}(x) - \boldsymbol{\mu}_i)^2 dx\} \tag{1}$$

where α and β are parameters of chromaticity term and brightness term, and $W_1(P_i, P_x)$ is the Wasserstein distance with exponent 1 defined as:

$$W_1(P_1, P_2) = \int_0^L |F_1(l) - F_2(l)| \, dl .$$

The Wasserstein distance is a natural way to compare histograms, and does not need a user-selected parameter to controlling the degree of smooth, like the Parzen window method for non-parametric approaches. We refer to [3, 4, 10] for details about the Wasserstein distance between two random variables.

Since $\boldsymbol{\mu}_i$ is the mean value of chromaticity in different partitions, it should also take value on $\mathbb{S}^2$. Thus, for problem (1), we add a penalty term for $\boldsymbol{\mu}_i$ with weight γ similar to [7]:

$$\begin{aligned} \min_{\Gamma, P_i, \boldsymbol{\mu}_i} \{E_1(\Gamma, P_i, \boldsymbol{\mu}_i \mid \mathbf{I}) = |\Gamma| + \sum_i \alpha \int_{\Omega_i} W_1(P_i, P_x)dx \\ + \beta \int_{\Omega_i} (\mathbf{C}(x) - \boldsymbol{\mu}_i)^2 dx + \gamma(|\boldsymbol{\mu}_i| - 1)^2\} \end{aligned} \tag{2}$$

Minimization problem (2) means that we wish to find an optimal segmentation combining the chromaticity and brightness components by two weights α and β. (2) can be rewritten as the functional with respect to the characteristic function χ_i of classified region Ω_i. For our two-phase case, $\chi_1 = \chi$ and $\chi_2 = 1 - \chi$. Let

$$R_i(x) = \alpha W_1(P_i, P_x) + \beta(\mathbf{C}(x) - \boldsymbol{\mu}_i)^2 = \alpha \int_0^L |F_i(l) - F_x(l)| \, dl + \beta(\mathbf{C}(x) - \boldsymbol{\mu}_i)^2,$$

then (2) changes to

$$\begin{aligned} \min_{\chi, F_1, F_2} \{\tilde{E}_1(.,.,. \mid I) = \int_\Omega |\nabla \chi(x)| dx + \int_\Omega \chi(x) R_1(x) dx \\ + \int_\Omega (1 - \chi(x)) R_2(x) dx + \gamma \sum_i (|\boldsymbol{\mu}_i| - 1)^2\} \end{aligned} \tag{3}$$

Since functional (3) is non-convex with respect to χ, based on [11, 12], we relax (3) by choosing a suitable function u in $BV_{[0,1]}(\Omega)$:

$$\min_{u \in BV_{[0,1]}, F_1, F_2} \{E_2(u, F_1, F_2 \mid I) = \int_\Omega |\nabla u(x)| dx + \int_\Omega u(x) R_1(x) dx \\ + \int_\Omega (1-u(x)) R_2(x) dx + \gamma \sum_i (|\boldsymbol{\mu}_i| - 1)^2 \} \tag{4}$$

If F_i and $\boldsymbol{\mu}_i$ are fixed, it is easy to prove the existence of solutions for u by classical variational arguments.

Since the problem (4) is a convex problem with respect to u, any minimizer of (4) is a global minimizer. Similar to [11 - 13], if F_i and $\boldsymbol{\mu}_i$ are fixed, and $u(x)$ is any minimizer of E_2, then the characteristic function $1_{\{x:u(x)>\mu\}}$ is a global minimizer of $\tilde{E}_1$ for almost every $\mu \in [0,1]$.

3 Minimization Algorithm

In the following we show how to solve the minimization problems. First, we fix u and minimize with F_i and $\boldsymbol{\mu}_i$, respectively. For all $l \in [0, L]$, variations with respect to F_1 and F_2 yield the optimality conditions:

$$\int_\Omega u(x) \frac{F_1(l) - F_x(l)}{|F_1(l) - F_x(l)|} dx = 0 \tag{5}$$

and

$$\int_\Omega (1-u(x)) \frac{F_2(l) - F_x(l)}{|F_2(l) - F_x(l)|} dx = 0 \tag{6}$$

If u is the characteristic function χ_1 of Ω_1, (5) means the total number of $F_1(l) > F_x(l)$ is equal to the total number of $F_1(l) < F_x(l)$ in Ω_1, so F_1 is the median of F_x in Ω_1. Similarly F_2 is the median of F_x in Ω_2 by (6). Now, since u is no longer the characteristic function, F_1 and F_2 are viewed as the weighted medians over their own partitions [3].

As for $\boldsymbol{\mu}_1$ and $\boldsymbol{\mu}_2$, since we add an unit sphere constraint, they can be updated by:

$$\boldsymbol{\mu}_1^{n+1} = \frac{\gamma \frac{\boldsymbol{\mu}_1^n}{|\boldsymbol{\mu}_1^n|} + \beta \int_\Omega u(x) \mathbf{C}_1(x) dx}{\gamma + \beta \int_\Omega u(x) dx}$$

and

$$\boldsymbol{\mu}_2^{n+1} = \frac{\gamma \frac{\boldsymbol{\mu}_2^n}{|\boldsymbol{\mu}_2^n|} + \beta \int_\Omega (1-u(x)) \mathbf{C}_2(x) dx}{\gamma + \beta \int_\Omega (1-u(x)) dx}.$$

Next, we fix F_i and $\boldsymbol{\mu}_i$, solve u. Of course, we can use the gradient descent method to minimize u as in [13]. However, there exists a more efficient way [11]. By introducing an intermediate variable v, the problem (4) can be regularized as:

$$\min_{u,0\le v\le 1} \int_\Omega |\nabla u| dx + \frac{1}{2\theta}\int_\Omega (u-v)^2 dx + \int_\Omega (R_1(x) - R_2(x)) v dx \tag{7}$$

where the parameter θ is chosen to be small enough to guarantee that u and v are close to each other under the L^2-norm. Then the convex functional (7) can be separated into two sub-problems with respect to u and v respectively:

$$\min_u \int_\Omega |\nabla u| dx + \frac{1}{2\theta}\int_\Omega (u-v)^2 dx \tag{8}$$

and

$$\min_{0\le v\le 1} \frac{1}{2\theta}\int_\Omega (u-v)^2 dx + \int_\Omega (R_1(x) - R_2(x)) v dx \tag{9}$$

The first sub-problem (8) is easily solved by Chambolle's dual projection method of the TV-norm, we refer to [14] for more details about the algorithm and its convergence. The solution is

$$u = v - \theta \operatorname{div} q$$

where $q = (q^1, q^2)$ can be obtained by the fixed point method: let q^0=0, then iterating

$$q^{n+1} = \frac{q^n + \delta t \nabla(\operatorname{div} q^n - v/\theta)}{1 + \delta t |\nabla(\operatorname{div} q^n - v/\theta)|}.$$

Since the second sub-problem (9) is a constraint problem, it is equivalent to the following unconstraint problem [11, 13]:

$$\min_v \frac{1}{2\theta}\int_\Omega (u-v)^2 dx + \alpha \int_\Omega w(v) dx + \int_\Omega (R_1(x) - R_2(x)) v dx \tag{10}$$

where $w(.)$ is a function penalizing v if v is outside [0, 1].

The sub-problem (10) with respect to v can be solved approximately by cutting v off as follows:

$$v = \min\left\{\max\left\{u(x) - \theta(R_1(x) - R_2(x)), 0\right\}, 1\right\}.$$

4 Experimental Results

We present several experiments on synthetic and natural color images in this section. In our experiments, the minimization of TV-norm realized with the classical gradient descent method takes 3 to 5 minutes, but our algorithm takes less than 15 seconds.

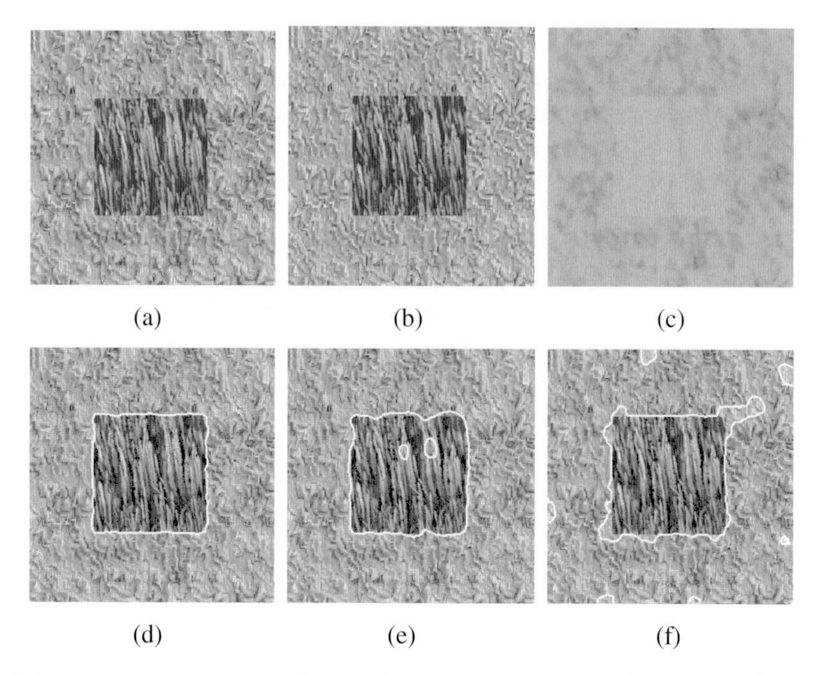

Fig. 1. The segmentation results of our method on a synthetic color image consisting of two different textures.

Fig. 2. The segmentation results of our method on natural color images

Figure 1 presents the segmentation results of a synthetic image with two different textures. 1(a) is the original image. 1(b) is the image of the brightness component. 1(c) is the chromaticity component. 1(d) is the segmentation result with weights $\alpha = 0.01$ and $\beta = 20$; 1(e) is the segmentation result with weights $\alpha = 0.01$ and $\beta = 0$, which means the chromaticity term does not take effect, so it can be viewed as Wasserstein distance method for gray level image; 1(f) is the segmentation result with weights $\alpha = 0$ and $\beta = 20$, so the brightness term is not used in the segmentation, and this is similar to the vectorial Chan-Vese model for color image. Comparing with the last three results verifies that combining the color information and the brightness information gives better segmentation.

Figure 2 is the segmentation results on natural color images with weights $\alpha = 0.01$ and $\beta = 20$. Those results show that the expected goal is achieved, i.e., two different regions are successfully segmented.

5 Conclusion

We have proposed a region-based active contour model for color images using chromaticity and brightness decomposition. Our model is an improvement to the segmentation of gray images based on the Wasserstein distance. Our variational energy can reach the global minimum because the characteristic function of the region is used. By the Chambolle's scheme, we significantly improve the speed of calculation. Experimental results show that the algorithm is effective.

Our future work will include the generalization of this model to the case of multiphase.

Acknowledgments. The research is supported by National Science Foundation of China (Nos. 10971066 and 60773119), Shanghai Rising-Star Program (No. 07QH14005) and Program for New Century Excellent Talents in University (No. NCET-08-0193).

References

1. Chan, T.F., Vese, L.A.: Active contours without edges. IEEE Trans. Image Processing 10(2), 266–277 (2001)
2. Chan, T.F., Sandberg, B.Y., Vese, L.A.: Active contours without edges for vector-valued images. Jour. Visual Comm. Image Represent. 11(2), 130–141 (2000)
3. Ni, K., Bresson, X., Chan, T.F., Esedoglu, S.: Local histogram based segmentation using Wasserstein distance. UCLA CAM reports 08-47, USA (2008)
4. Chan, T.F., Esedoglu, S., Ni, K.: Histogram Based Segmentation Using Wasserstein Distances. In: Sgallari, F., Murli, A., Paragios, N. (eds.) SSVM 2007. LNCS, vol. 4485, pp. 697–708. Springer, Heidelberg (2007)
5. Gonzalez, R., Wood, R.: Digital image processing, 2nd edn. Publishing House of Electronics Industry, Beijing (2002)
6. Chan, T.F., Kang, S.H., Shen, J.: Total variation denoising and enhancement of color images based on the CB and HSV color models. Jour. Visual Comm. Image Represent. 12(4), 422–435 (2001)

7. Kang, S.H., March, R.: Variational models for image colorization via chromaticity and brightness decomposition. IEEE Trans. Image Processing 16(9), 2251–2261 (2007)
8. Caselles, V., Lions, P., Morel, J., Coll, T.: Geometry and color in natural images. Jour. Math. Imag. Vis. 16(2), 89–105 (2002)
9. Maenpaa, T., Pietikainen, M.: Classification with color and texture: jointly or separately. Pattern Recogn. 37(8), 1629–1640 (2004)
10. Villani, C.: Topics in optimal transportation. Graduate students in Mathematics, vol. 58. American Mathematical Society, Providence (2003)
11. Bresson, X., Esedoglu, S., Vandergheynst, P., Thiran, J., Osher, S.: Fast global minimization of the active contour/snake model. Jour. Math. Imag. Vis. 28(2), 151–167 (2007)
12. Mory, B., Ardon, R.: Fuzzy region competition: A convex two-phase segmentation framework. In: Sgallari, F., Murli, A., Paragios, N. (eds.) SSVM 2007. LNCS, vol. 4485, pp. 214–226. Springer, Heidelberg (2007)
13. Nikolova, M., Esedoglu, S., Chan, T.F.: Algorithms for finding global minimizers of image segmentation and denoising models. SIAM Jour. Appl. Math. 66(5), 1632–1648 (2006)
14. Chambolle, A.: An algorithm for total variation minimization and applications. Jour. Math. Imag. Vis. 20(1-2), 89–97 (2004)

Image Matching Based on Representative Local Descriptors

Jian Hou[1], Naiming Qi[1], and Jianxin Kang[1,2]

[1] School of Astronautics, Harbin Institute of Technology, Harbin, China, 150001
[2] School of Engineering, Northeast Agriculture University, Harbin, China, 150030
dr.houjian@gmail.com, qinm@hit.edu.cn, kjx.kjx@gmail.com

Abstract. While straightforward image matching with keypoint based local descriptors produces a high matching accuracy, it is usually accompanied by enormous computation load. In this paper we present a representative local descriptors (RLDs) based approach to improve image matching efficiency without sacrificing matching accuracy. Firstly, local descriptors in one image are clustered with a similarity based method where descriptors are clustered into one group if they are similar enough to their mean. Then only the RLD in each group is used in matching and the number of matched RLDs is used to evaluate the similarity of two images. Experiments indicate that the RLDs approach produces better matching accuracy than both straightforward matching with original descriptors and visual words matching.

Keywords: Representative Local Descriptors, Straightforward Matching, Visual Words.

1 Introduction

In many image processing applications such as image retrieval and object recognition, we need to do image matching to evaluate the similarity of two images. Feature based matching is one of the most commonly used image matching methods. Currently available features can be roughly classified into local and global features. Global features usually extract the distribution of one or several kinds of primitives in one image and express it in a feature vector. While the relatively compact descriptors of global features lend them to fast matching with a large image database, they often contain less information than needed to guarantee a high matching accuracy. On the other hand, local features detect keypoints and compute a descriptor for each keypoint. All these descriptors are then used in image matching. Usually the extraction and matching of local features is computationally more expensive than that of global features. However, local descriptors are usually designed to be invariant to some extent to viewpoints, illumination, scale and rotation, etc. Furthermore, the matching strategy of local descriptors make them robust to occlusion. These factors often enable local features to obtain better matching accuracy than global ones. As in this paper we intend to reduce computation load in local descriptor matching, we briefly review some important research on local features.

S. Boll et al. (Eds.): MMM 2010, LNCS 5916, pp. 303–313, 2010.

In currently available local features, the scale invariant feature transform (SIFT) proposed by Lowe [1] has been shown to be very successful. SIFT descriptor is designed to be invariant to scale and rotation and also partially invariant to illumination, viewpoint and affine transform. In evaluations [2,3] with several image databases SIFT and SIFT based descriptors such as PCA-SIFT [4] and GLOH [3] were shown to be the most distinctive descriptors among over 20 state-of-the-art global and local features. Unlike SIFT based features using blob like areas as keypoints, in [5] the authors proposed to extract multi-scale oriented patches (MOPS) which concentrate on edges and corners. Regions have also been exploited to tackle the image matching problem under various transformations [6,7,8]. In [6] a Maximally Stable Extremal Region (MSER) is defined as a connected component of an appropriately thresholded image. MSER is preserved under a broad class of geometric and photometric changes and outperforms several other affine region detectors in evaluation [9]. Keypoint based matching method may not work well when only very few or even no keypoints can be reliably detected. On the other hand, edge points can be easily extracted with various edge detectors. Using sampled edge points as keypoints, [10] proposed to describe shapes with shape context that allows for measuring shape similarity and recovering of point correspondences. In [3] only SIFT based features perform slightly better than shape context. Some other commonly used local features include moment invariants [11], steerable filters [12] and spin images [13], etc. Since the 128-d SIFT descriptor brings about large computation load in matching, some descriptors of smaller size, such as PCA-SIFT and shape context, have been proposed. While these descriptors may produce better performance than SIFT in some domains, SIFT still outperforms them in general applications [3]. This means that with descriptors of smaller size, the reduction in computation load is at the cost of reduced accuracy.

The straightforward local feature matching method matches keypoints in two images and uses the number of matched keypoints to evaluate the similarity of the two images [2,14]. This method produces superior matching performance at the cost of enormous computation load. Another popular matching method of local features is the so-called visual words method. Visual words representation of an image, also known as codebook, is a global feature used to represent the distribution of different local image patterns [15,16,17,18]. Keypoints are extracted from all database images and then clustered into a number of groups. By treating each group as a visual word, we get a visual word vocabulary describing all kinds of local image patterns. An image can then be represented as a vector containing the count of each visual word in the image. This vector serves as the feature vector in image matching. Instead of using all visual words, Li et al. [19] presented an approach to learn optimal compact vocabulary by selecting a subset of discriminative visual words from a large vocabulary. The proposed approach produces superior results compared to k-means method with the same size of vocabulary. In [20] the authors proposed to map unordered feature sets into multi-resolution histograms and the pyramid matching method is shown to improve matching efficiency dramatically. [21] further presented a spatial pyramid matching method that produces superior scene categorization performance on challenging image databases. To improve matching efficiency with a large database, [22] proposed to build a vocabulary tree by hierarchical k-means clustering. The method is designed for efficient lookup of visual words with a large vocabulary and a large image database. There is still no theoretical guidelines on how to select a suitable vocabulary size. However, empirical studies [16,18] show that a size of at least several thousands seems necessary for the visual words vector to be effective and that a large size tends to improve matching performance. As a global feature, visual words vector produces good performance in object recognition [2]. However, the matching accuracy of visual words method is usually lower than that of straightforward matching. In other words, visual words matching reduces computation load at the cost of reduced accuracy too.

In this paper we present a representative local descriptors (RLDs) based matching approach. While RLDs method is also clustering based, it has some important differences with visual words method. The latter method does clustering among local descriptors in ALL database images, whereas our method clusters local descriptors in ONE image and use only the RLDs in each group in matching. Besides, the clustering in our method is based on a similarity threshold, whereas visual words method usually use k-means method where k needs to be given beforehand. Our approach builds on existing local descriptors and reduces matching computation load to a large extent. At the same time the matching accuracy is not reduced but increased. The validity of the approach using both distance ratio and similarity criterion is verified by recognition experiments with challenging image databases.

This paper is organized as follows. In Section 2 we present the details of RLDs based matching method. Section 3 and 4 compare the performance of RLDs based method with straightforward matching and visual words matching with experiments respectively. In Section 5 we conclude the paper with some discussions.

2 Representative Local Descriptors

With distance ratio criterion [1], a pair of keypoints are regarded as a match if the distance ratio between the closest match and the second closest match is below a threshold

$$\frac{d(f, f_{1st})}{d(f, f_{2nd})} < th_d \tag{1}$$

where f is the descriptor to be matched and f_{1st} and f_{2nd} are the closest and the second closest descriptors from the model image, with $d(.,.)$ denoting the Euclidean distance between two descriptors. Distance ratio criterion helps to reject many false matches. However, in the context of buildings or other artificial objects, it may reject many possible correct matches since there are many repetitive keypoints in these images. In order to solve this problem, we propose to use the following representative local descriptors based approach.

Firstly we cluster the local descriptors in one image. The clustering is similarity based, that is, a number of descriptors are clustered into one group if their similarity with their mean is above a threshold. After clustering each group are represented by a representative local descriptor d (mean descriptor in our implementation) and a number n that is the count of descriptors in the group. The set of representative local descriptors (RLDs) in the groups no longer contain repetitive descriptors and can be used for matching with distance ratio criterion. While one original local descriptor represents one keypoint, one RLD represents one type of keypoints. As in general the number of keypoint types is smaller than that of keypoints, it is obvious that matching with RLDs saves computation. Although a RLD involves a clustering step, experiments indicate that time consumption in this step is limited compared to the time it saves in matching.

In matching with RLDs, let's assume that one RLD d_i with a number n_i in one image is matched to one RLD d_j with a number n_j in another image. If we still use the number of matched keypoints as similarity metric, the number of matched keypoints provided by this match is $min\{n_i, n_j\}$. However, it was found in experiments that if we discard the number n_i and n_j and use the number of matched RLDs as metric, we will get a higher matching accuracy, which is even higher than using original descriptors. This is interesting since it means that with RLDs, we reduce computation load without sacrificing matching accuracy. Instead the matching accuracy is increased.

Though both RLDs method and feature selection technique [19,23] involve reducing the number of features in matching, they are not of the same kind. Feature selection aims to select some "important" features and discard less important ones by learning from training images, whereas in RLDs method no feature is discarded actually. Instead, all the features in one image are represented by a relatively small number of RLDs. As each original feature is represented by a similar RLD, this method does not involve a learning process for a particular object.

Unlike k-means clustering method where k need to be set beforehand, in similarity based clustering k is totally determined by a clustering threshold th_c. For two local descriptors f and g, the similarity score is defined as

$$S(f,g) = \frac{f^T g}{\|f\|_2 \|g\|_2} \tag{2}$$

where $\|f\|_2$ represents the L2-norm of the vector. The similarity based clustering for all descriptors in one image can be performed in the following steps:

1. Label all descriptors as ungrouped, and set $i = 0$.
2. Label the first ungrouped descriptor as in group i and as the base descriptor.
3. Find the most similar descriptor to base descriptor from all ungrouped ones. If the similarity is larger than th_c, go to next step. Otherwise, let i be increased by 1 and go to step 2, until all descriptors are grouped.
4. Label the most similar descriptor as in group i and compute the mean descriptor as the new base descriptor.

5. Compare each ungrouped descriptor with base descriptor. If the similarity is larger than th_c, label it as in group i and compute the new mean descriptor as base one.
6. Let i be increased by 1 and go to step 2, until all descriptors are grouped.

Obviously it's possible to use other procedures to do this similarity based clustering and it's not our intention in this paper to propose a new clustering method. The essence of this kind of methods is that after clustering we know to which extent the descriptors in one group are similar to their RLD. Only in this way can we know to which extent one RLD represents the descriptors in its group. In the following sections we will see that this similarity extent impacts on matching performance.

For a given set of local descriptors, the "optimal" number of clusters is actually an intrinsic characteristic of the dataset and decided by both the number of descriptors and the relative similarity distribution of descriptors. Hence with only the number of descriptors it is hard to select a suitable k for k-means clustering unless empirically. On the other hand, similarity based method uses the intrinsic properties of dataset, i.e., the similarity distribution of descriptors, to determine the number of groups. Hence if we find a suitable similarity threshold th_c, it is possible to apply the method to a new set of descriptors. This's the reason why we choose similarity based clustering method over k-means method.

3 Comparison with Straightforward Matching

We now use object recognition experiments to demonstrate the advantage of RLDs over original descriptors in straightforward matching. Two publicly available datasets ZuBud [24] and UKY [22] are adopted in experiments. In ZuBud we use 115 query and 1005 database images. The UKY dataset consists of 10200 images of 2550 objects. In the 4 images of each object the first image is used as query and the other three as database. Due to the enormous computation load and our limited computation power, we use for testing only the first 700 queries at present. The two datasets contain images of various types of objects with rotation and large variances of illumination, scale and viewpoints. We think these images are rather representative in image matching tasks and the performance shown in experiments with these datasets is convincing.

We have tested the method using both SIFT and MOPS as local descriptors. Due to limited space, in this paper we only show the results with SIFT and the conclusion with MOPS is the same. In experiments we use both number of matched keypoints and number of matched RLDs as similarity metric of two images to compare their performance. Besides distance ratio criterion, we also use similarity criterion to judge the match of a pair of local descriptors. With similarity criterion, two descriptors f and g are regarded as a match if $S(f, g) > th_s$ where th_s is a similarity threshold. In order to obtain a reliable conclusion with thorough experiments, we use 9 values from 0.1 to 0.9 as th_d for distance ratio criterion and as th_s for similarity criterion. For the clustering threshold th_c we only adopt three values 0.7, 0.8 and 0.9 as lower values will obviously cluster dissimilar descriptors into one group. For the convenience of comparison, the original local descriptors are regarded as RLDs with $th_c = 1$. As with original descriptors number of matched RLDs equals to number of matched keypoints, the recognition rates for both metrics are the same in this case. The experimental results are shown in Table 1 to 4.

We now analyze the results in the following aspects:

(1) similarity metric

From the recognition results we find that with distance criterion, in most cases number of matched RLDs metric performs better than number of matched keypoints. With similarity criterion, when th_s is within the reasonable range (0.8 and above), number of matched RLDs metric performs better than num-

Table 1. Recognition rate of ZuBud with distance ratio criterion

th_c	Similarity metric	0.1	0.2	0.3	0.4	0.5	0.6	0.7	0.8	0.9
0.7	Number of matched RLDs	27.0	60.0	79.1	89.6	90.4	89.6	89.6	87.0	55.7
	Number of matched keypoints	27.0	59.1	61.7	42.6	31.3	20.0	29.6	24.3	16.5
0.8	Number of matched RLDs	30.4	76.5	89.6	94.8	93.9	94.8	98.3	97.4	93.0
	Number of matched keypoints	30.4	70.4	72.2	69.6	75.7	76.5	81.7	86.1	83.5
0.9	Number of matched RLDs	28.7	80.0	89.6	93.0	96.5	97.4	98.3	98.3	96.5
	Number of matched keypoints	28.7	76.5	80.9	87.8	92.2	93.9	95.7	97.4	96.5
1.0	Number of matched RLDs Number of matched keypoints	28.7	76.5	86.1	90.4	93.9	94.8	96.5	96.5	96.5

Table 2. Recognition rate of ZuBud with similarity criterion

th_c	Similarity metric	0.1	0.2	0.3	0.4	0.5	0.6	0.7	0.8	0.9
0.7	Number of matched RLDs	3.5	3.5	3.5	3.5	3.5	5.2	36.5	87.8	88.7
	Number of matched keypoints	7.0	7.0	7.0	7.0	7.0	7.0	11.3	27.8	29.6
0.8	Number of matched RLDs	3.5	3.5	3.5	3.5	3.5	3.5	24.3	97.4	96.5
	Number of matched keypoints	17.4	17.4	17.4	17.4	17.4	18.3	40.0	82.6	77.4
0.9	Number of matched RLDs	3.5	3.5	3.5	3.5	3.5	3.5	30.4	95.7	98.3
	Number of matched keypoints	23.5	23.5	23.5	23.5	23.5	25.2	62.6	96.5	95.7
1.0	Number of matched RLDs Number of matched keypoints	3.5	3.5	3.5	3.5	3.5	3.5	27.8	95.7	98.3

Table 3. Recognition rate of UKY with distance ratio criterion

th_c	Similarity metric	0.1	0.2	0.3	0.4	0.5	0.6	0.7	0.8	0.9
0.7	Number of matched RLDs	2.7	27.9	47.3	59.1	68.0	73.3	73.4	56.3	12.6
	Number of matched keypoints	2.7	20.7	21.3	17.1	14.4	13.7	10.4	9.3	4.6
0.8	Number of matched RLDs	4.0	37.7	61.3	76.0	85.7	91.9	90.1	78.9	38.6
	Number of matched keypoints	4.0	27.0	30.1	36.6	42.0	49.4	56.7	60.3	44.1
0.9	Number of matched RLDs	4.3	36.9	62.0	75.9	85.6	88.3	85.9	75.6	47.3
	Number of matched keypoints	4.3	27.9	42.3	53.6	65.4	73.3	77.1	72.9	49.6
1.0	Number of matched RLDs Number of matched keypoints	4.0	33.4	52.9	62.3	68.7	72.9	73.9	69.3	49.9

Table 4. Recognition rate of UKY with similarity criterion

th_c	Similarity metric	0.1	0.2	0.3	0.4	0.5	0.6	0.7	0.8	0.9
0.7	Number of matched RLDs	0.1	0.1	0.1	0.1	0.1	0.3	10.1	69.4	73.6
	Number of matched keypoints	0.3	0.3	0.3	0.3	0.3	0.6	2.7	7.9	15.4
0.8	Number of matched RLDs	0.1	0.1	0.1	0.1	0.1	0.1	2.9	80.0	90.7
	Number of matched keypoints	2.3	2.3	2.3	2.4	2.4	3.4	10.3	59.3	54.1
0.9	Number of matched RLDs	0.1	0.1	0.1	0.1	0.1	0.1	3.4	73.3	90.3
	Number of matched keypoints	1.9	1.9	2.0	2.4	2.9	3.4	16.1	72.4	80.4
1.0	Number of matched RLDs Number of matched keypoints	0.1	0.1	0.1	0.1	0.1	0.1	3.1	74.0	87.6

ber of matched keypoints. In all cases the best performance are obtained with number of matched RLDs metric. This means that in matching with RLDs, we no longer need to account for the number of keypoints represented by one RLD. Instead, we use the number of matched RLDs to evaluate the similarity of two images. In the following analysis we will use only the results with number of matched RLDs as the similarity metric.

(2) matching accuracy

From Table 1 to 4 we also find that the best recognition rate are always obtained in matching with RLDs but not original descriptors. This observation holds with different object types, descriptors and matching criterions.

(3) related parameters

We have observed that the best matching accuracy is obtained in matching with RLDs but not original descriptors. However, with so many parameter values in RLDs matching (e.g. 0.1 to 0.9 in distance ratio thresholds, 0.7 to 0.9 in clustering thresholds), if we can't find a fixed set of parameters (th_c and th_d/th_s) that produces better performance with RLDs than with original descriptors for different object types, we can not apply the approach to other datasets. For a new matching task, it is impossible to test all these parameters just to obtain a moderate increase in matching accuracy.

Fortunately, we find that for each combination of descriptor and matching criterion, a better result with RLDs than with original descriptors can be located at a fixed set of parameters which are listed as follows:

SIFT with distance ratio criterion: $th_c = 0.8$, $th_d = 0.7$

SIFT with similarity criterion: $th_c = 0.9$, $th_s = 0.9$

(4) time consumption

Besides improving matching accuracy, one major advantage of RLDs over original local descriptors is that matching with RLDs reduces computation load to a large extent. The average time consumption of each query in recognition experiments and in clustering step are presented in Table 5.

It can be seen from Table 5 that for both criterions, using RLDs in matching reduces the time consumption in matching. Although one RLD involves a

Table 5. Time spent on recognition per query (seconds)

Dataset	th_c=0.8	th_c=0.9	Original descriptors	Clustering
ZuBud	542.6	855.2	1136.2	1.0
UKY	2835.8	4128.5	5025.7	0.9

clustering step, the time spent in this step is negligible compared to that saved in matching.

The reduction in time consumption using RLDs can be explained by the fact that the number of RLDs is small compared with that of original descriptors. In fact, the ratio of number of RLDs to number of original descriptors is 0.581 with th_c=0.8 and 0.864 with th_c=0.9 for ZuBud. The two parameters for UKY is 0.515 and 0.859 respectively.

We now analyze the reason why in RLDs matching number of matched RLDs metric performs better than number of matched keypoints, and why matching with RLDs performs better than matching with original local descriptors. One RLD, whatever the number of descriptors it represents, expresses only one type of image pattern. Hence as a similarity metric, number of matched RLDs conveys the similarity of two images more globally and number of matched keypoints expresses the similarity more locally. It's natural to think that global similarity has priority over local similarity. Besides, the number of matched RLDs is more robust to occlusion. As there usually exist repetitive keypoints in many types of object images, one RLD often represents more than one keypoints. When some of these keypoints are occluded the number of matched RLDs is more likely to remain unchanged. Matching with original local descriptors can be regarded as matching with RLDs using number of matched keypoints metric. Therefore it is reasonable for RLDs to perform better than original descriptors. Based on these analysis we have reason to expect RLDs to be applicable to other keypoint based local descriptors, though we have only tested with SIFT and MOPS by far.

In appearance there exist some similarities between RLDs and visual words matching: both methods cluster local descriptors and use clusters in matching. Before we compare the performance of RLDs matching with that of visual words matching, we'd like to highlight the difference between two methods. Firstly, RLDs method clusters local descriptors in ONE image while visual words method does clustering among local descriptors in ALL database images. Secondly, the number of visual words is usually decided empirically as there is still no theory to guide the selection of this parameter. We know the descriptors in one group are similar to each other, but we do not know to which extent they are similar to each other. On the other hand, in RLDs method local descriptors are clustered into one group only when their similarity with their centers is above a clustering threshold. As our experiments have indicated that the similarity extent of descriptors in one group has an impact on matching accuracy and a particular clustering threshold produces the best performance for different image databases, it is possible for us to do clustering based on this clustering threshold for other image datasets. Thirdly, in current applications visual words method often evaluates the similarity of two images based on the similarity of the distribution of local descriptors in two images. Whereas RLDs method is a special kind of straightforward matching as it uses the number of matched RLDs to evaluate the similarity of two images. In other words, the advantage of RLDs over original local descriptors actually indicate number of matched RLDs as a more distinctive similarity metric between images than number of matched keypoints.

4 Comparison with Visual Words Matching

In this section we compare the matching performance of RLDs method with that of visual words method. In order to achieve high matching efficiency, we select vocabulary tree structure to do visual words matching. A vocabulary tree is a tree built by hierarchical k-means clustering [22]. Firstly, the training descriptors are partitioned into k groups based on k-means clustering. In the next level, a k-means clustering is run on each group and we obtain a total of $k \times k$ groups. The same process is recursively applied to each group until the expected level is reached. With the clustering parameter k and level parameter l we get a tree with k^l leaf nodes. In matching each descriptor is propagated down the tree by comparing it to the k candidate group centers at each level. This process only involves a total of $k \times l$ dot product operations and results in an efficient way to compute the visual words vector of an image. In order to improve performance, usually a weighting scheme is applied before the visual words vector is used in matching.

In experiments we use the ZuBud dataset and the full version of UKY dataset. The vocabulary size was selected to be 10^6 empirically. We build the tree with $k = 10$ and $l = 6$ and adopted inverse document frequency as the weighting scheme. The performance comparison of two methods with distance ratio criterion and similarity criterion is shown in Table 6. It is obvious from the Table that RLDs matching performs much better than visual words matching.

Table 6. Recognition rate comparison of RLDs matching with visual words matching

Dataset	Distance ratio	Similarity	Visual words
ZuBud	98.3	98.3	86.9
UKY	91.0	91.2	48.2

5 Conclusion

In keypoint based local feature matching, straightforward matching method produces good matching performance at the cost of enormous computation load. On the other hand, visual words method represents one image as the distribution of different types of local descriptors in the image. This global descriptor leads to very efficient matching at the cost of reduced matching accuracy.

In this paper we propose to cluster local descriptors in one image with a similarity based method and use in matching only the representative local descriptor (RLD) in each group. We show with extensive experiments that instead of number of matched keypoints, using number of matched RLDs as metric reduces computation load to a large extent and improves matching accuracy moderately. RLDs method also shows advantage over visual words method in matching accuracy. This conclusion holds with different local descriptors, object types and matching criterions. As one RLD represents one type of local descriptors, the experimental results indicate number of matched keypoint types as a more distinctive similarity metric between images. We attribute this result to that number of matched RLDs conveys the global similarity of two images and is more robust to occlusion than number of matched keypoints. This property may and should be explored for better performance in image matching.

References

1. Lowe, D.G.: Distinctive Image Features from Scale-Invariant Keypoints. Int. J. Comput. Vis. 60(2), 91–110 (2004)
2. Deselaers, T., Keysers, D., Ney, H.: Features for Image Retrieval: an Experimental Comparison. Inf. Retr. 11(2), 77–107 (2008)
3. Mikolajczyk, K., Schmid, C.: A Performance Evaluation of Local Descriptors. IEEE Trans. Pattern Anal. Machine Intell. 27(10), 1615–1630 (2005)
4. Ke, Y., Sukthankar, R.: PCA-SIFT: a More Distinctive Representation for Local Image Descriptors. In: IEEE Computer Society Conference on Computer Vision and Pattern Recognition, vol. 2, pp. 511–517. IEEE Press, New York (2004)
5. Brown, M., Szeliski, R., Winder, S.: Multi-Image Matching Using Multi-Scale Oriented Patches. In: IEEE Computer Society Conference on Computer Vision and Pattern Recognition, vol. 1, pp. 510–517. IEEE Press, New York (2005)
6. Matas, J., Chum, O., Urban, M., Pajdla, T.: Robust Wide-Baseline Stereo from Maximally Stable Extremal Regions. In: 13th British Machine Vision Conference, vol. 1, pp. 384–393. British Machine Vision Association, London (2002)
7. Tuytelaars, T., Gool, L.V.: Wide Baseline Stereo Matching Based on Local, Affinely Invariant Regions. In: 11th British Machine Vision Conference, pp. 412–425. British Machine Vision Association, London (2000)
8. Kadir, T., Zisserman, A., Brady, M.: An Affine Invariant Salient Region Detector. In: Pajdla, T., Matas, J.G. (eds.) ECCV 2004. LNCS, vol. 3021, pp. 228–241. Springer, Heidelberg (2004)
9. Mikolajczyk, K., Tuytelaars, T., Schmid, C., Zisserman, A., Matas, J., Schaffalitzky, F., Kadir, T., Gool, L.V.: A Comparison of Affine Region Detectors. Int. J. Comput. Vis. 65(1-2), 43–72 (2006)
10. Belongie, S., Malik, J., Puzicha, J.: Shape Matching and Object Recognition Using Shape Contexts. IEEE Trans. Pattern Anal. Machine Intell. 24(4), 509–522 (2002)
11. Gool, L.V., Moons, T., Ungureanu, D.: Affine/Photometric Invariants for Planar Intensity Patterns. In: Buxton, B.F., Cipolla, R. (eds.) ECCV 1996. LNCS, vol. 1064, pp. 642–651. Springer, Heidelberg (1996)
12. Freeman, W.T., Adelson, E.H.: The Design and Use of Steerable Filters. IEEE Trans. Pattern Anal. Machine Intell. 13(9), 891–906 (1991)

13. Lazebnik, S., Schmid, C., Ponce, J.: Sparse Texture Representation Using Affine-Invariant Neighborhoods. In: IEEE Computer Society Conference on Computer Vision and Pattern Recognition, vol. 2, pp. 319–324. IEEE Press, New York (2003)
14. Zhang, W., Kosecka, J.: Hierarchical Building Recognition. Image Vis. Comput. 26(5), 704–716 (2007)
15. Sivic, J., Zisserman, A.: Video Google: a Text Retrieval Approach to Object Matching in Videos. In: 9th IEEE International Conference on Computer Vision, pp. 1470–1477. IEEE Press, New York (2003)
16. Deselaers, T., Keysers, D., Ney, H.: Discriminative Training for Object Recognition Using Image Patches. In: IEEE Computer Society Conference on Computer Vision and Pattern Recognition, vol. 2, pp. 157–162. IEEE Press, New York (2005)
17. Mikolajczyk, K., Leibe, B., Schiele, B.: Multiple Object Class Detection with a Generative Model. In: IEEE Computer Society Conference on Computer Vision and Pattern Recognition, vol. 1, pp. 26–36. IEEE Press, New York (2006)
18. Yang, J., Jiang, Y., Hauptmann, A., Ngo, C.W.: Evaluating Bag-of-Visual-Words Representations in Scene Classification. In: 9th ACM SIGMM International workshop on Multimedia Information Retrieval, pp. 197–206. ACM Press, New York (2007)
19. Li, T., Mei, T., Kweon, I.S.: Learning Optimal Compact Codebook for Efficient Object Categorization. In: IEEE 2008 Workshop on Applications of Computer Vision, pp. 1–6. IEEE Press, New York (2008)
20. Grauman, K., Darrell, T.: The Pyramid Match Kernel: Discriminative Classification with Sets of Image Features. In: 10th IEEE International Conference on Computer Vision, vol. 2, pp. 1458–1465. IEEE Press, New York (2005)
21. Lazebnik, S., Schmid, C., Ponce, J.: Beyond Bags of Features: Spatial Pyramid Matching for Recognizing Natural Scene Categories. In: IEEE Computer Society Conference on Computer Vision and Pattern Recognition, vol. 2, pp. 2169–2178. IEEE Press, New York (2006)
22. Nister, D., Stewenius, H.: Scalable Recognition with a Vocabulary Tree. In: IEEE Computer Society Conference on Computer Vision and Pattern Recognition, vol. 1, pp. 2161–2168. IEEE Press, New York (2006)
23. Dorko, G., Schmid, C.: Selection of Scale-Invariant Parts for Object Class Recognition. In: 9th IEEE International Conference on Computer Vision, vol. 1, pp. 634–639. IEEE Press, New York (2003)
24. Shao, H., Svoboda, T., Gool, L.V.: ZUBUD-Zurich Building Database for Image Based Recognition. Technical report No. 260, Swiss Federal Institute of Technology (2003)

Stereoscopic Visual Attention Model for 3D Video

Yun Zhang[1,2,3], Gangyi Jiang[1,2], Mei Yu[1], and Ken Chen[1]

[1] Faculty of Information Sciences and Engineering, Ningbo University, Ningbo, 315211, China
[2] Institute of Computing Technology, Chinese Academic of Sciences, Beijing, 100080, China
[3] Graduate School of the Chinese Academic of Sciences, Beijing, 100080, China
zhangyun_8851@163.com, {jianggangyi,yumei,chenken}@nbu.edu.cn

Abstract. Compared with traditional mono-view video, three-dimensional video (3DV) provides user interactive functionalities and stereoscopic perception, which makes people more interested in pop-out regions or the regions with small depth value. Thus, traditional visual attention model for mono-view video can hardly be directly applied to stereoscopic visual attention (SVA) analysis for 3DV. In this paper, we propose a bottom-up SVA model to simulate human visual system with stereoscopic vision more accurately. The proposed model is based on multiple perceptual stimuli including depth information, luminance, color, orientation and motion contrast. Then, a depth based dynamic fusion is proposed to integrate these features. The experimental results on multi-view video test sequences show that the proposed model maintains high robustness and is able to efficiently simulate SVA of human eyes.

Keywords: Three dimensional video, visual attention, depth perception.

1 Introduction

Multi-view video is capable of providing users with three-dimensional (3D) depth impression and allows the users to freely choose a view of a visual scene [1]. Multi-view video plus depth (MVD) [2] [3] supports high image quality and low complexity of rendering a continuum of output views. It has been the main representation of 3D video (3DV) content and been used in various multimedia applications, including free view-point video, 3D television and immersive teleconference. High compression efficiency requirement and content-related functionalities of these multimedia applications call for efficient tools that can guide or extract interesting objects in 3DV. Human visual attention is being studied as a key technology to region-of-interest detection and intelligent content-based system [4] [5].

Attention is a neurobiological conception. It implies the concentration of mental powers upon an object by closer or more careful observing or listening. Attention area in a picture is the area where it tends to catch more human visual attention. Visual attention cues of human visual system (HVS) are generally classified into two categories: top-down and bottom-up. Bottom-up (stimuli-driven) visual attention is driven by external low-level stimuli including luminance, color, orientation and motion contrast etc; it is automatic and has a transient time course. On the other hand, top-down

S. Boll et al. (Eds.): MMM 2010, LNCS 5916, pp. 314–324, 2010.

(task-driven) visual attention involves pattern, shape, and other cognitive processing related features. It is an effortful and voluntary process based on the cognitive knowledge of human brain.

Many efforts have been devoted to visual attention detection [5-11]. Itti *et al* developed bottom-up visual attention model for still images based on Treisman's stimulus integration theory [6] [7]. It generates saliency map with the integration of perceptual stimuli from intensity contrast, color contrast and orientation contrast. Then, winner-take-all networks and inhibit of return process are implemented to obtain attended visual attentions. Zhai *et al* used the low-level features as well as cognitive features, such as skin color and captions, in their visual attention model [8]. Motion is another important cue for visual attention detection in video, thus, a bottom-up spatio-temporal visual attention model is proposed for video sequences [9]. Wang *et al* proposed segment based video attention detection method [10]. Lu *et al* combine face detection with other low-level stimuli for visual attention modelling [11]. Ma *et al* proposed a bottom-up and top-down combined visual attention model by integrating multiple features, including contrast in image, motion, face detection, audition and text etc. [5].

However, these existing visual attention models for mono-view video can hardly be directly applied to visual attention analysis for 3DV because they have not taken the stereoscopic perception into account. In this paper, we have proposed a bottom-up stereoscopic visual attention (SVA) model to simulate HVS more accurately by integrating depth information with other low-level features, including motion, intensity, color and orientation contrast.

2 Stereoscopic Visual Attention Model

2.1 The Proposed SVA Model

Attention area in a picture or video is the area where it tends to catch more human visual attention. Many computational attention systems for image mainly focus on three features: intensity, orientation, and color. A special case of color computation is the separate computation of skin color. In video, motion information is an important feature in human perception and added for video visual attention detection. But 3DV provides the most effective depth perception (also called stereoscopic perception) that obtained by viewing a scene from slightly different viewing positions. The depth perception that makes people feel more real in 3DV and it is another important factor to affect human visual attention while comparing with motion and texture contrasts in traditional two dimensional (2D) video. For example, people are often interested in the regions popping out from video screen and interest ratio is of attention regions decrease as they are getting far away. Because the stereoscopic perception can also be represented by the 2D video and depth which indicates the relative distance between video object and camera system, each SVA object is modeled by three attributes with low-level features, including depth, image saliency and motion saliency, in our work. Therefore, the SVA model is defined as

$$S_{SVA} = \{D, S_s, S_m\}, \tag{1}$$

where S_{SVA} is SVA saliency map, D is the intensity of depth maps, S_s and S_m are image saliency and motion saliency, respectively.

Fig.1 shows architecture of the proposed SVA model. For image saliency detection, color, intensity and orientation contrasts are employed to compute the saliency maps within image. For motion saliency detection, motion information is obtained from temporal successive frames in each view. Additionally, depth maps can be generated from the disparities between neighboring views of each time instant in 3DV. Then, a novel dynamic model fusion method is used to integrate the obtained pixel-wise image saliency map, motion saliency and depth map. Finally, post-processing operation is implemented for extract attention area based on SVA saliency map. The proposed model is limited to the bottom-up control of attention, i.e. to control selective attention by the properties of the visual stimuli. The proposed model does not incorporate any top-down, volitional component because it relies on the cognitive knowledge which differs from person to person.

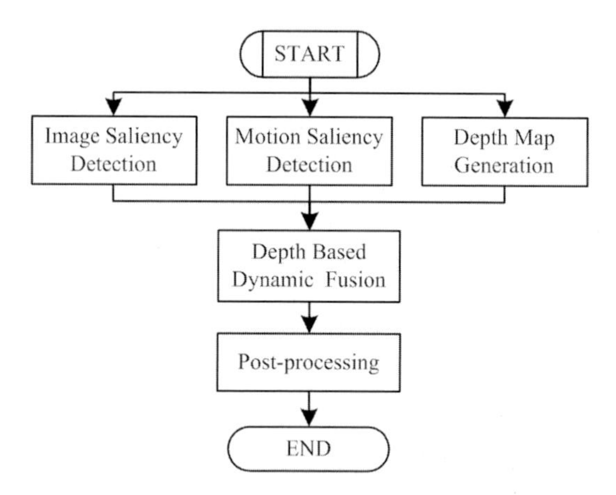

Fig. 1. Flowchart of the proposed SVA model

2.2 Spatial Attention Model

We adopts Itti's bottom-up attention model [6] to implement our spatial visual attention model. For static scene, retinal input is processed in parallel by seven multi-scale low-level feature maps, which detect local spatial discontinuities using simulated centre-surround neurons. The seven neuronal features implemented are sensitive to color contrast (red/green and blue/yellow), intensity contrast and four orientations (0°, 45°, 90° and 135°) for static images [6]. Centre and surround scales are obtained using dyadic Gaussian pyramids with nine levels, from level 0 the original image, to level 8, reduced by a factor 256 horizontally and vertically. Centre-surround differences are then computed as point-wise differences across pyramid levels, for combinations of three centre scales (c = {2,3,4}) and two centre-surround scale differences (s-c={3,4}); and then, six feature maps are computed for each of the seven features, yielding a total of 42 feature maps. Finally, all feature maps are integrated into the unique scalar image saliency, S_s with normalized summation strategy proposed in [7].

2.3 Temporal Attention Model

Motion is one of the major stimuli on visual attention. Block based optical flow algorithm is utilized to estimate motion of image objects between consecutive frames. Frame group consists of $2n+1$ temporal consecutive frames in view v, $\mathbf{F}(v,t)=\{f_{v,t} \mid t = t + k, -w \le k \le w, k \in Z\}$, are employed to extract robust motion magnitude. The horizontal and vertical motion channels of frame $f_{v,t}$ are determined by frame group $\mathbf{F}(v,t)$, respectively; then they are combined together as the motion

$$M_{v,t}^{k} = \Theta_{m,n}\left[\left|\Phi_{m,n}^{h}\left(f_{v,t}, f_{v,t+k}\right)\right| + \left|\Phi_{m,n}^{v}\left(f_{v,t}, f_{v,t+k}\right)\right|\right], \tag{2}$$

where $\Phi_{m,n}^{h}$ and $\Phi_{m,n}^{v}$ denotes horizontal and vertical optical flow operator with $m \times n$ block size. '$|\cdot|$' is the magnitude of motion velocity. $\Theta_{m,n}$ performs $m \times n$ times up-sampling operation with Gaussian low-pass filter. Here, 4×4 block, i.e. $m = n = 4$, provides a robust performance. Therefore, $M_{v,t}^{k}$ is with the same resolution as $f_{v,t}$. Forward and backward motion is intersected so as to eliminate the background exposure phenomena.

$$M_{v,t}(k) = \begin{cases} \left(M_{v,t}^{k} + M_{v,t}^{-k}\right)/2 & \text{if } \min\left(M_{v,t}^{k}, M_{v,t}^{-k}\right) > 0 \\ 0 & \text{else} \end{cases}. \tag{3}$$

And then, $M_{v,t}(k)$ are weighted combined to from a robust motion map, M, and it is calculated as

$$M = \sum_{k=1}^{w} \varsigma_k \cdot M_{v,t}(k), \tag{4}$$

where ς_k are weighted coefficients satisfying $\sum_{k=1}^{w} \varsigma_k = 1$. Usually, motion attention level increases with relative motion, which is the object motion against the background [11]. For static camera arrangement, absolute motion is relative motion and motion object is attention object. But for the video with camera motion, object with motion contrast relative to background motion is attentive object. Motion map are decomposed with dyadic Gaussian pyramids with nine levels and center-surround difference model is adopted to separate the attentive object motion from background motion. So final motion saliency map is

$$S_m = \Psi\left(\bigoplus_{c=2}^{4} \bigoplus_{s=c+3}^{c+4} \left(\mathcal{N}\left(\left|M(c) \ominus M(s)\right|\right)\right)\right), \tag{5}$$

where $\ominus$ is across-level denotes the across-scale difference between two maps at the center (c) and the surround (s) levels of the respective feature pyramids; $\oplus$ is across-level addition; $\mathcal{N}(\cdot)$ is an normalization operator; $\Psi(\cdot)$ is a linear normalize function which adjusts the saliency value from 0~255, where value 255 indicates the most salient.

2.4 Depth's Impacts on SVA

3DV can provide users with this unique depth perception and interactive view switching functionalities, which make users feel more personally on the scene. The stereoscopic perception can also be represented by 2D video plus depth map which indicates the relative distance between video object and camera system. Hence, we use depth map to analyze the differences between 3D video and traditional 2D video. Compared with traditional 2D video, the depth perception effects on human SVA in four aspects listed as follow:

1) While watching 3D video, people are usually more interested in the regions getting out of the screen, i.e. pop-out regions, which are with small depth values or large disparities, than other regions.
2) Interested ratio of video object usually decreases as depth increases.
3) The objects with the depth out of depth of field (DOF) of the camera system are usually not the attention areas. For example, defocusing blur for background object or foreground object.
4) Depth discontinuous regions or depth contrast regions are usually the attention areas while view angles or view positions are changed.

Depth map is an 8-bit gray image that can be captured by depth camera or generated from multi-view video by using algorithms. In this paper, we firstly estimate disparity for each pixel in multi-view video by using stereo matching method and graph cuts algorithm [12]. Then, the disparity is converted into perceive depth. Finally, intensity of each pixel in depth map is calculated with irregular spaces conversion.

HVS perceive depth, Z, is shown as

$$Z = B \cdot f / d_c \,, \tag{6}$$

where f is the focal length of the cameras, B is the baseline between the neighboring cameras, d_c is the physical disparity (measured by centimeter) between the corresponding points of the neighboring views. However, disparity estimated by stereo matching is measured by pixel. So we use a centimeter-to-pixel ratio, λ, i.e. a ratio of CCD size to image resolution, to convert physical disparity to pixel disparity.

$$d_p = d_c / \lambda \,. \tag{7}$$

Because near object is usually more important than far object, the depth value Z which corresponds to the pixel (x, y) is transformed into the 8-bit intensity D with irregular spaces [13].

$$D = \left\lfloor 255 \cdot \frac{z^n}{Z} \cdot \frac{z^f - Z}{z^f - z^n} + 0.5 \right\rfloor , \tag{8}$$

where '$\lfloor \cdot \rfloor$' is floor operation, z^f and z^n indicate the farthest and nearest depth, $z^f = \frac{Bf}{\min\{d_p\}} \cdot \frac{1}{\lambda}$, $z^n = \frac{Bf}{\max\{d_p\}} \cdot \frac{1}{\lambda}$. The space between z^f and z^n is divided into narrow spaces around the z^n plane and is divided into wide spaces around the z^f plane.

2.5 Depth Based Dynamic Fusion for SVA Model

Psychological studies reveal that, HVS is more sensitive to motion contrast compared to color, intensity and orientation contrast in single view video. 3DV provides the most effective depth perception sensation obtained by viewing a scene from slightly different viewing positions, i.e. depth. Depth perception is another key factor in stereoscopic vision. As mentioned in subsection 2.4, people are more likely attracted by pop-up regions and the objects with small depth value in stereoscopic vision. Moreover, the attractiveness of the stereoscopic attention objects decrease as they are getting further while within DOF. Therefore, the SVA saliency map of an image in the 3DV is constructed as

$$S_{SVA} = \Psi\left[D \cdot \left(k_s S_s + k_m S_m + k_D D - \sum_{uv \in \{sm, sD, mD\}} e_{uv} C_{uv}\right)\right], \tag{9}$$

where k_D, k_s and k_m are the weighted coefficients for depth, motion saliency and static image saliency, respectively. C_{uv} is correlation between channels u and v, e_{uv} is weighted coefficients for the correlation channel C_{uv}, $C_{sm} = \min(S_s, S_m)$, $C_{sD} = \min(S_s, D)$ and $C_{mD} = \min(S_m, D)$. If strong motion contrast is present in the sequence, temporal attention model should be more dominant over the spatial attention model. However, if the motion contrast is low in the sequence, the spatial attention model is dominant. That is the relative importance of spatial attention and motion attention is changed according to the strength of motion contrast. Therefore, we set the weighted coefficients of motion and static images saliency as

$$\begin{cases} k_m = (1 - k_D)\dfrac{p}{p+C} \\ k_s = (1 - k_D)\dfrac{C}{p+C} \end{cases}, \tag{10}$$

where C is a constant and parameter p is a parameter which is direct proportional to motion contrast.

3 Experimental Results and Analyses

To evaluate the performance of our SVA model, we perform attention detection experiments with the multi-view video sequences provided by Heinrich Hertz Institute (HHI), Microsoft Research (MSR), Nagoya University and Mitsubishi Electric Research Laboratories (MERL) [14-16]. Table 1 shows the parameters of the test multi-view video sequences, in which the depth maps of Breakdaners and Ballet, marked as 'A' in last column, are available [16]. The depth maps of the rest videos, marked as 'N/A', is generated by the algorithm described in subsection 2.4. Usually, image and motion saliency is more important than depth perceptual saliency except for the 3DV with very strong depth sensation. Thus, k_D is relative smaller than k_s and k_m, and it is set as 0.2. In the experiment the dynamic fusion coefficients, e_{sm} is set as $\min(k_s, k_m)$. There is almost no implicit correlation between depth and image saliency, depth and motion saliency, thus e_{mD} and e_{sD} are set as 0. Additionally, C and p are set as, 0.3 and (max (M) – median (M)) / max (M), respectively.

Table 1. Parameters of the Test Multi-view videos

MVV	Provided by	Image Size	Frame rate	Camera array	Views	Depth
Door flower	HHI	1024×768	16.7fps	6.5cm /1D	16	N/A
Alt Moabit	HHI	1024×768	16.7fps	6.5cm /1D	16	N/A
Akko&kayo	Nagoya Univ.	640×480	25fps	5cm /2D	5×5	N/A
Ballroom	MERL	640× 480	25fps	20cm /1D	8	N/A
Ballet	MSR	1024×768	15fps	20cm/1D arc	8	A
Breakdancers	MSR	1024×768	15fps	20cm/1D arc	8	A

Fig. 2 shows the experimental results. Fig.2 (a) is the original video. Fig.2 (b) shows the saliency maps of static images detected by using Itti's algorithm in [6]. The spatial attention model can simulate HVS well for the sequences with simple background, such as Door flower, Ballet and Akko&kayo. However, for the sequences with complex background, i.e. Alt Moabit, Ballroom and Breakdancers, the spatial attention model is not accurate enough to detect the stereoscopic attention for 3D video because complex image is full of color, orientation and intensity contrast. Fig.2(c) shows motion saliency maps. Large motion contrast areas are the potential attention areas. However, it is not always true. For example, for Ballet and Breakdancers sequence, the shadow of the dancing girl/man is with high motion contrast, but it is not an attentive area. Fig. 2(d) illustrates the depth maps generated by our depth generation algorithm. Obviously, we can not expect to detect visual attention area simply according to depth maps because the areas close to cameras are not always the visual attention regions, such as the floor in Ballet and Breakdancers sequences. Therefore, these features should be integrated to improve attention model and overcome their short-coming in each channel.

Fig. 2(e) shows SVA saliency maps of the 3DV. In Fig. 2(f), the high attention area is highlighted according to the SVA saliency. Taking Ballet sequence as an example, the proposed model can not only depress the noise in spatial saliency map (black region on the wall in color image), but also depress the noise in motion saliency map (shadow of the dancing girl) and noise in depth (the foreground floor). Then, a favorable saliency map is created. For Door Flower sequence, multiple attention cues including motion (two men and the door), static image attention (clock, painting and chair) and depth (the sculpture) are integrated together very well by the proposed model. For Akko&kayo sequence, only one girl is detected as attention area in spatial saliency and large holes are in motion saliency. In contrast, our SVA model is more biologically plausible as we can see from SVA saliency. Similar results can be found for other multi-view video sequences. Therefore, the proposed model detects the SVA accurately and simulates HVS well by combining depth information, static image saliency and motion saliency. Additionally, though the generated depth map is not accurate or some obvious error region exists in the generated depth map, such as the left most area of Ballroom and Akko&kayo sequences, the proposed model can generate satisfactory SVA saliency maps by jointly using depth, motion and texture information. Thus, the proposed model is with high error reliance and maintains high robustness.

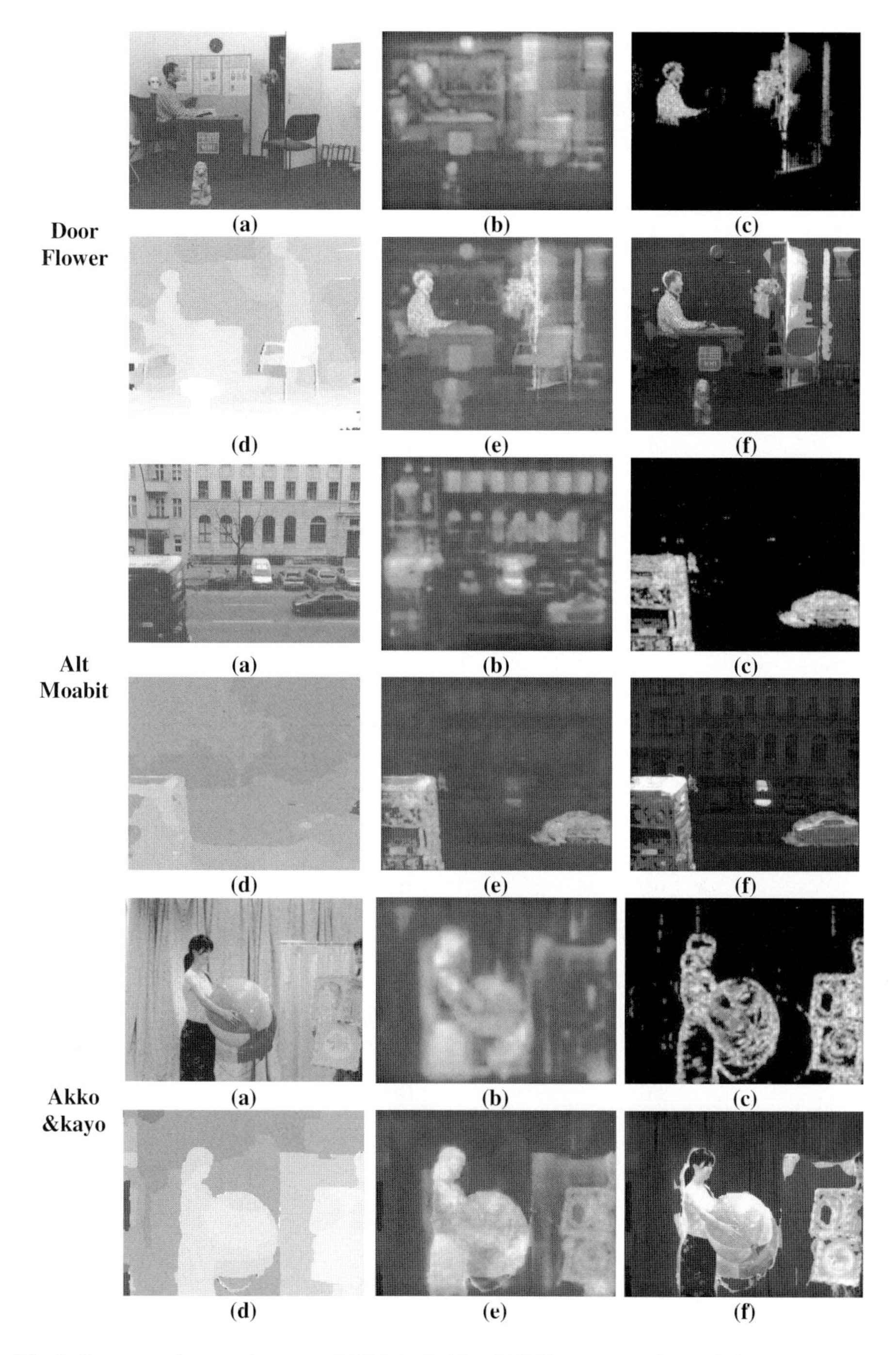

Fig. 2. Stereoscopic attention maps (a)Original video (b)Saliency maps for static images (Itti), (c) Motion saliency maps, (d) Depth maps, (e)SVA saliency maps (the proposed), (f)High stereoscopic attention areas

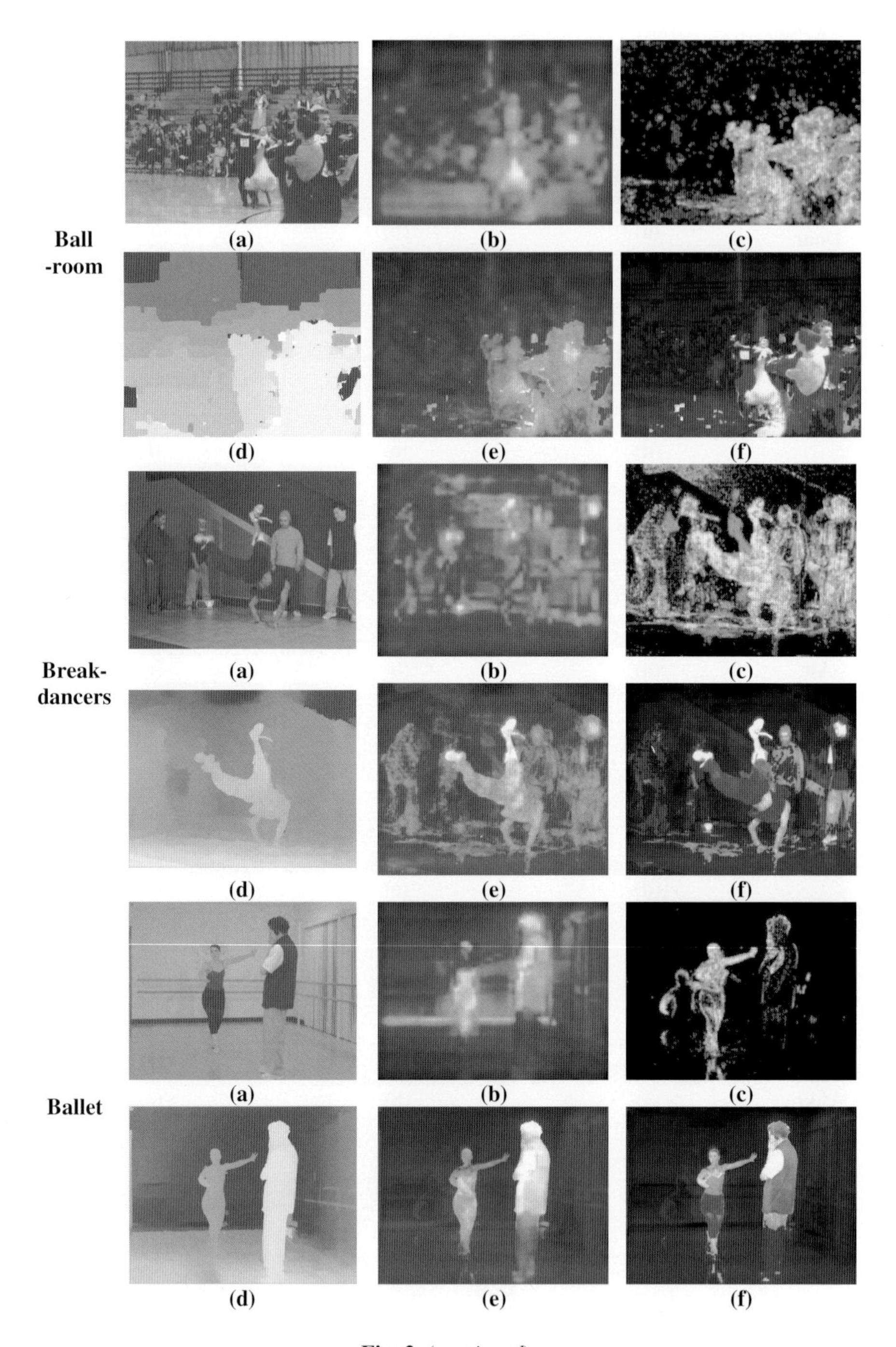

Fig. 2. (*continued*)

4 Conclusions

In this paper, we proposed a bottom-up stereoscopic video attention model to simulate human vision with stereoscopic perception by using multiple perceptual stimuli. And a depth based dynamic fusion is proposed to integrate depth with other low-level features, including motion, intensity, color, orientation contrast. The proposed model is not only able to efficiently simulate stereoscopic visual attention of human eyes, but also can depress noise and maintain high robustness. The stereoscopic visual attention will play an important role in the research fields of content oriented three-dimensional video processing, video retrieval and computer vision.

Acknowledgments. This work was supported by the Natural Science Foundation of China (60672073, 60832003) and the Innovation Fund Project for Graduate Student of Zhejiang province (YK2008044). Thanks for the HHI, MSR, Nagoya University and MERL providing us multi-view video sequences.

References

[1] Tanimoto, M.: Overview of Free Viewpoint Television. Signal Proc.: Image Commun. 21(6), 454–461 (2006)
[2] Muller, K., Merkle, P., Wiegand, T.: Compressing time- varying visual content. IEEE Signal Processing Magazine 24(6), 58–65 (2007)
[3] Smolic, A., Mueller, K., Merkle, P., et al.: Multi-view video plus depth (MVD) format for advanced 3D video systems. MPEG and ITU-T SG16 Q.6, JVT-W100, San Jose, USA (April 2007)
[4] Han, J., Ngan, K.N., Li, M., Zhang, H.: Unsupervised extraction of visual attention objects in color images. IEEE Trans. CSVT 16(1), 141–145 (2006)
[5] Ma, Y.F., Hua, X.S., Lu, L., et al.: A generic framework of user attention model and its application in video summarization. IEEE Trans. Multimedia 7(5), 907–919 (2005)
[6] Itti, L., Koch, C.: Computational Modeling of Visual Attention. Nature Reviews Neuroscience 2(3), 194–203 (2001)
[7] Itti, L., Koch, C.: Feature combination strategies for saliency-based visual attention system. J. Electron. Imaging 10, 161–169 (2001)
[8] Zhai, G.T., Chen, Q., Yang, X.K., Zhang, W.J.: Scalable visual sensitivity profile estimation. In: ICASSP, Las Vegas, Nevada, USA, April 2008, pp. 876–879 (2008)
[9] Zhai, Y., Shah, M.: Visual attention detection in video sequences using spatiotemporal cues. In: Proceedings of the 14th ACM Multimedia, Santa Barbara, CA, USA, pp. 815–824 (2006)
[10] Wang, P.P., Zhang, W., Li, J., Zhang, Y.: Real-time detection of salient moving object: a multi-core solution. In: ICASSP, Las Vegas, Nevada, USA, April 2008, pp. 1481–1484 (2008)
[11] Lu, Z., Lin, W., Yang, X., Ong, E.P., Yao, S.: Modeling Visual Attention's Modulatory Aftereffects on Visual Sensitivity and Quality Evaluation. IEEE Trans Image Proc. 14(11), 1928–1942 (2005)
[12] Kolmogorov, V., Zabih, R.: Multi-camera scene reconstruction via graph cuts. In: Heyden, A., Sparr, G., Nielsen, M., Johansen, P. (eds.) ECCV 2002. LNCS, vol. 2352, pp. 82–96. Springer, Heidelberg (2002)

[13] Tanimoto, M., Fujii, T., Suzuki, K.: Improvement of Depth Map Estimation and View Synthesis, ISO/IEC JTC1/SC29/WG11 M15090, Antalya, Turkey (January 2008)
[14] Vetro, A., McGuire, M., Matusik, W., et al.: Multiview Video Test Sequences from MERL, ISO/IEC JTC1/SC29/WG11, MPEG05/m12077, Busan, Korea (April 2005)
[15] Feldmann, I., Mueller, M., Zilly, F., et al.: HHI Test Material for 3D Video, ISO/IEC JTC1/SC29/WG11, M15413, Archamps, France (April 2008)
[16] Zitnick, C.L., Kang, S.B., Uyttendaele, M., et al.: High-quality video view interpolation using a layered representation. In: ACM SIGGRAPH and ACM Trans. on Graphics, Los Angeles, CA, August 2004, pp. 600–608 (2004)

Non-intrusive Speech Quality Assessment with Support Vector Regression

Manish Narwaria, Weisi Lin, Ian Vince McLoughlin, Sabu Emmanuel, and Chia Liang Tien

School of Computer Engineering, Nanyang Technological University
Singapore, 639798
{mani0018,wslin,mcloughlin,asemmanuel,asltchia}@ntu.edu.sg

Abstract. We propose a new non-intrusive speech quality assessment algorithm based on Support Vector Regression (SVR) and Mel Frequency Cepstral Coefficients (MFCCs). The basic idea is to map the MFCCs into the desired quality score using SVR. The sensitivity of the MFCCs to external noise is exploited to gauge the changes in the speech signal to evaluate its perceptual quality. The use of SVR exploits the advantages of machine learning with the ability to learn complex data patterns for an effective and generalized mapping of features into a perceptual score, in contrast with the oft-utilized feature pooling process in the existing speech quality estimators. Experimental results indicate that the proposed approach outperforms the standard P.563 algorithm for non-intrusive assessment of speech quality with a total of 1792 speech files and the associated subjective scores.

Keywords: Speech quality assessment, Support Vector Regression, Mel frequency cepstral coefficients (MFCC).

1 Introduction

Evaluating speech quality is important in telephone networks, voice over Internet, mobile communications and numerous other telecommunication, as well as multimedia applications. Besides this, speech quality evaluation is a valuable assessment tool for the development of speech coding and enhancement techniques. To overcome the limitations of subjective quality evaluation (such as costs, unsuitability for many in-process, in-service and real-time applications), objective speech quality assessment has attracted substantial research over the past years [1-7].

The aim of objective speech quality assessment is to predict and replace human judgment of perceived speech quality by machine evaluation. Objective measures, which assess speech quality by using the extracted physical parameters and computational models, are less expensive to administer, save time, and give more consistent results. The majority of the existing objective methods are based on input/output comparisons, i.e., intrusive methods, which estimate speech quality by measuring the "distortion" between the input and output signals, and mapping the distortion values to a predicted quality. However, in many practical situations like wireless communications, voice over IP and other in-service networks requiring speech quality monitoring, an

S. Boll et al. (Eds.): MMM 2010, LNCS 5916, pp. 325–335, 2010.

intrusive approach is not applicable because the input speech signal is unavailable. In such cases a non-intrusive measurement which depends only on the altered speech signal is desirable.

Non-intrusive evaluation, which is also termed as no-reference, single-ended or output-based evaluation, is a challenging problem since the measurement of speech quality has to be performed with only the output speech signal of the system under test, without using the original signal as a reference. An early attempt towards nonintrusive speech quality measure based on spectrogram of the perceived signal is presented in [1]. The method described in [2] uses Gaussian Mixture Models (GMMs) to create an artificial reference model to compare the degraded speech; while in [3] speech quality is predicted by the Bayesian inference and minimum mean square error (MMSE) estimation based on a trained set of GMMs. A perceptually motivated speech quality assessment algorithm based on temporal envelope representation of speech is presented in [4]. In [5] a low complexity, non-intrusive speech quality assessment scheme has been proposed based on features computed from commonly used speech coding parameters (e.g. spectral dynamics). Recently in [6], a comprehensive study has been reported which assesses the correlation of existing objective measures with the quality of noise-suppressed/enhanced speech. The ITU-T has released P.563 as its non-intrusive objective quality measurement standard algorithm [12]. The P.563 resulted from a collaboration of Psytechnics' NiQA algorithm [14], SwissQual's NiNA [15], and Opticom's P3SQM, and represents the most relevant state of the art non-intrusive speech quality evaluation scheme.

Different speech features have been detected for quality evaluation. In [5], spectral flatness, spectral dynamics, spectral centroid, speech variance, pitch period and excitation variance have been used as features for speech quality estimation. Perceptual linear prediction (PLP) cepstral coefficients have been used in [7] as speech features for quality assessment. Speech quality estimation utilizing the temporal envelope representation of speech has been reported in [4].

Appropriate speech features are essential for effective speech processing and quality estimation. Recent studies have found speech features combined with perceptual models of the human auditory system (HAS) to be more effective for speech quality evaluation [13]. The MFCCs approximate the HAS response more closely and therefore provide a good basis for non-intrusive speech quality assessment. They are the most commonly used acoustic features in currently available speech recognition systems. The MFCCs have been used for speaker verification and recognition applications [16]. The MFCCs are also increasingly being used in music information retrieval applications such as genre classification and audio similarity measures [18]. It is also well known that the performance of MFCC based speech recognition systems deteriorates substantially in noisy environments. This is because the MFCCs are sensitive to external noise [8, 18], and by using them for speech quality assessment our aim is to exploit their high sensitivity to noise for quality prediction.

Another important issue of speech quality evaluation is the mapping of detected features into a final score to represent the perceived quality. Literature survey shows that scant research effort has been directed towards optimal pooling of features into a single number to represent perceived speech quality. In [4] and [7], the overall speech quality has been obtained by a linear combination of the features.

The ITU-T standard P.862 [17] for intrusive speech quality assessment also uses a linear combination of the average disturbance value and the average asymmetrical disturbance value to obtain the overall quality score. In general, the relationship between the features and quality score needs not be linear and is difficult to be determined apriori. Thus, a machine learning approach will be more reasonable and convincing to establish mapping between detected features and the quality score.

In this paper, we propose a Mel Frequency Cepstral Coefficient (MFCC) and machine learning based non-intrusive speech quality predictor. The proposed method formulates speech quality prediction as a regression problem, and uses SVR to find a mapping between speech features and quality score. Machine learning is useful to discover the underlying complex relationship between a set of acoustic features and the perceptual quality score. Figure 1 shows a block diagram of the proposed system. During the training phase, switch 1 will be at position A and switch 2 will be on. The model parameters will be updated resulting in a trained model. For the system test, switch 1 is placed in position B while switch 2 will be turned off so that the system outputs the predicted quality score based on the trained model and the test speech features.

The rest of the paper is organized as follows. Section 2 describes the proposed MFCC-based feature method. Section 3 presents a brief description of the SVR algorithm, and experimental results are presented in Section 4, while Section 5 gives the concluding remarks.

2 MFCC for Speech Quality Assessment

As aforementioned, MFCCs are sensitive to external distortions [8, 18], and thus can be used to estimate the changes in the speech signal due to the added perturbations. Ref. [19] demonstrates the use of MFCCs of noisy speech (lower SNR) signal for the reconstruction of clean speech (higher SNR) signal i.e. MFCCs have been shown to be effective in perceptual quality restoration. Furthermore, in a recent and related work [20], the inversion of MFCCs has been used for speech enhancement. The basic idea here was to estimate the MFCCs of the underlying clean speech signal by direct inversion of the MFCCs of noisy speech signal and significant improvements in terms of the perceptual quality were achieved. Thus, MFCCs convey meaningful information about speech quality and are an effective representation of perceptual quality variations of speech signal. This motivates us to use MFCCs for perceptual speech quality assessment which to the best of our knowledge, have not been exploited so far for non-intrusive speech quality assessment. An additional advantage in using MFCC is that, it has decorrelating effect on the spectral data and maximizes the variance of the coefficients. This is a desirable property since it makes machine learning more effective and meaningful.

The input speech signal, sampled at 8 kHz, is segmented into frames of length 30 ms with a frame rate of 100 Hz. Each individual frame f(n) is windowed using a Hamming window to minimize signal discontinuities at the beginning and end of each frame.

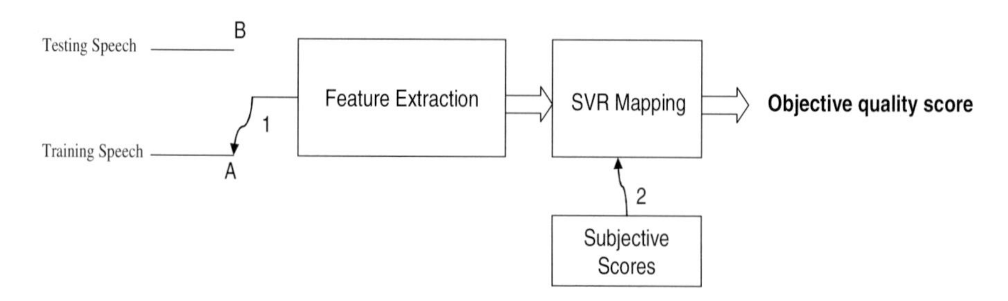

Fig. 1. Block Diagram of the proposed scheme

Let $w(n)$ denote the window (n=0 to N-1; N being the number of samples in each frame), the windowed frame $f_t(n)$ is then given by

$$f_t(n) = f(n)\, w(n) \tag{1}$$

The Hamming window used is of the form

$$w(n) = 0.54 - 0.46\cos\left(\frac{2\pi n}{N-1}\right), \quad 0 \le n \le N-1$$

The magnitude spectrum is then fitted to a mel-scaled filter bank. The Mel-filter banks are used to emulate the basilar membrane of the human ear. In the human ear basilar membrane, there are more receptors for frequencies between 0 to 1 KHz and their number decreases rapidly thereafter. Therefore, thirteen linearly spaced and twenty seven log spaced triangular filters are applied in grouping the FFT bins. The lowest frequency is chosen to be 133.33 Hz, a linear spacing of 66.66 Hz and log spacing of 1.07 are used. The MFCCs are the coefficients obtained by the Discrete Cosine Transform (DCT) of the log Mel spectrum. The DCT is performed to decorrelate the log Mel spectrum coefficients and facilitate the subsequent dimension reduction.

Assume that there are a total of K frames in a speech signal, then the MFCC vector $\boldsymbol{x}_j$ for the jth frame is represented as

$$\boldsymbol{x}_j = (m_{j1}, m_{j2}, \ldots m_{j13})^T$$

where each m_{ji} (i =1 to 13 and j =1 to K) denotes a Mel frequency cepstral coefficient. In order to obtain a global feature vector for the entire speech signal, we compute the mean MFCC vector $\boldsymbol{x}$ which is represented as

$$\boldsymbol{x} = (M_1, M_2 \ldots M_{13})^T$$

where each M_i ($= \frac{1}{K}\sum_{j=1}^{K} m_{ji}$) denotes the ith mean Mel cepstral frequency coefficient.

3 Support Vector Regression (SVR)

The HAS is a highly complex and non-linear system and it's a challenge to determine the exact relationship between the changes in a speech signal due to the noise and the corresponding change in the perceived speech quality. In other words, it is not easy to combine the features detected from speech signal into a single number that represents the quality score. The task becomes even more difficult when the original signal is unavailable. The rationale behind using machine learning techniques like SVR [9] for speech quality prediction is that in general, the extracted speech features may have a complex, possibly non-linear relationship to the final quality score and in practice it is very difficult to exactly determine the underlying relationship. The changes in the MFCC of a speech signal due to noise leads to quality degradation. However, the relation between changes in MFCC and the corresponding degradation of quality cannot be known accurately apriori. By using the SVR algorithm our aim is to estimate the underlying complex relationship between MFCC and the quality score of the speech signal. Although other choices of machine learning techniques are possible, in this paper we use the SVR for feature fusion because of its higher generalization ability.

The SVR formulation introduces the concept of a loss function that ignores error that is within a distance of the true value. This type of function is referred to as ε insensitive loss function and can control a parameter that is equivalent to the margin parameter for separating hyper planes. More specifically, the SVR is to find a function, which approximates mapping from an input domain to the real numbers, based on a training sample. Suppose that $\boldsymbol{x}_i$ is the feature vector of the pth speech sample in the training image set (i=1, 2… p_m; p_m is the number of training speech samples). In ε-SV regression [9], the goal is to find a function $f(\boldsymbol{x}_i)$ that has the deviation of ε at most from the actually obtained targets y_i (being the corresponding subjective quality score) for all the training data, and at the same time is as flat as possible. The function to be learned is $f(\boldsymbol{x}) = \mathbf{w}^T \varphi(\boldsymbol{x}) + b$; where $\varphi(\boldsymbol{x})$ is a non-linear function of feature vector $\boldsymbol{x}$, $\mathbf{w}$ is the weight vector and b is the bias term. The aim is to find the unknowns $\mathbf{w}$ and b from the training data such that the error

$$|y_i - f(\boldsymbol{x}_i)| \leq \varepsilon \tag{2}$$

for the ith training sample $\{\boldsymbol{x}_i, y_i\}$. It has been shown [9] that

$$\mathbf{w} = \sum_{i=1}^{n_{sv}} (\eta_i^* - \eta_i)\ \varphi(\boldsymbol{x}_i) \tag{3}$$

where η_i^* and η_i ($0 \leq \eta_i^*, \eta_i \leq C$) are the Lagrange multipliers used in the optimization of the Lagrange function, n_{sv} is the number of support vectors and C is the trade off error parameter. For data points for which Inequality (2) is satisfied, i.e. the points which lie within the ε tube, the corresponding η_i^* and η_i will be zero such that the Karush Kuhn Tucker (KKT) conditions are satisfied [9]. Therefore we have a sparse expansion of $\mathbf{w}$ in terms of $\boldsymbol{x}_i$ (i.e. we do not need all $\boldsymbol{x}_i$ to describe $\mathbf{w}$). Support vectors are the samples that come with nonvanishing coefficients (i.e. non zero η_i^* and η_i). The function to be learned then becomes

$$f(\boldsymbol{x}) = \mathbf{w}^T \varphi(\boldsymbol{x}) + b = \sum_{i=1}^{n_{sv}} (\eta_i^* - \eta_i)\ \varphi(\boldsymbol{x}_i)^T \varphi(\boldsymbol{x}) + b$$

$$= \sum_{i=1}^{n_{sv}} (\eta_i^* - \eta_i)\ K(\boldsymbol{x}_i, \boldsymbol{x}) + b \tag{4}$$

where $K(\boldsymbol{x}_i, \boldsymbol{x}) = \varphi(\boldsymbol{x}_i)^T \varphi(\boldsymbol{x})$, being the kernel function.

In the training phase, the SVR system is presented with the training set { $\boldsymbol{x}_i$, y_i }, and the unknowns **w** and γ are estimated to obtain the desired function given by (4). During the test phase, the trained system is presented with the feature test vector $\boldsymbol{x}_j$ of the jth test speech sample and it predicts the estimated objective score y_j ($j = 1$ to q_m; q_m is the number of test speech samples). We used the Radial Basis Function (RBF) as the kernel function $K(\mathbf{x_i}, \mathbf{x}) = \exp(-\rho \, \|\mathbf{x_i} - \mathbf{x}\|^2)$ where ρ is a positive parameter controlling the radius. The parameters C, ρ and ε were determined by using a validation set.

4 Experiments and Performance

For the experiments in this work, we use a third-party database, NOIZEUS [6], which comprises speech corrupted by four types of noise (babble, car, street, and train) at two SNR levels (5 and 10 dB) and processed by 13 different noise suppression algorithms; a total of 1792 speech files are available. The noise suppression algorithms fall under four different classes: spectral subtractive, subspace, statistical-model based, and Wiener algorithms. A complete description of the algorithms can be found in [6, 10]. The subjective evaluation of the NOIZEUS database was performed according to ITU-T Recommendation P.835 [11].

4.1 MFCC Components

As aforementioned, MFCC features are sensitive to distortions. For example, adding noise to the speech signal will affect the speech power spectrum at all frequencies. Other types of noise disturb the speech in different ways. To illustrate this point further, we utilize a clean speech signal from the database [6] used in the experiments. The clean speech signal has been corrupted by car, street, train and babble noise (all at 5 dB). Further, these corrupted speech signals have been processed by a speech enhancement scheme. In Figure 2, we show the MFCCs of the clean speech signal, degraded speech and the enhanced speech. As can be seen, with the added noise, each MFCC component of the speech signal is changed from its clean version, and the speech enhancement brings about meaningful restoration in the MFCCs (making them closer to the original ones). We can see from Figure 2 that the MFCCs are reasonable representation of the effect of noise injection and signal restoration. Thus, they are expected to provide an effective basis for non-intrusive speech quality assessment.

4.2 Speech Quality Evaluation

Although there are 1792 speech files available in the NOIZEUS database, for comparison between objective and subjective score a usual way is to compare the per-condition MOS with the per-condition average objective score [11]. The aforementioned 13 different speech enhancement algorithms were used for processing noisy speech files, and by including the unprocessed noisy speech files also, we get a total of 14 algorithms (i.e., conditions). This analysis involved the use of mean objective scores and subjective ratings computed across a total of 112 conditions (= 14 algorithms x 2 SNR levels x 4 noise types). In order to test the robustness of the proposed

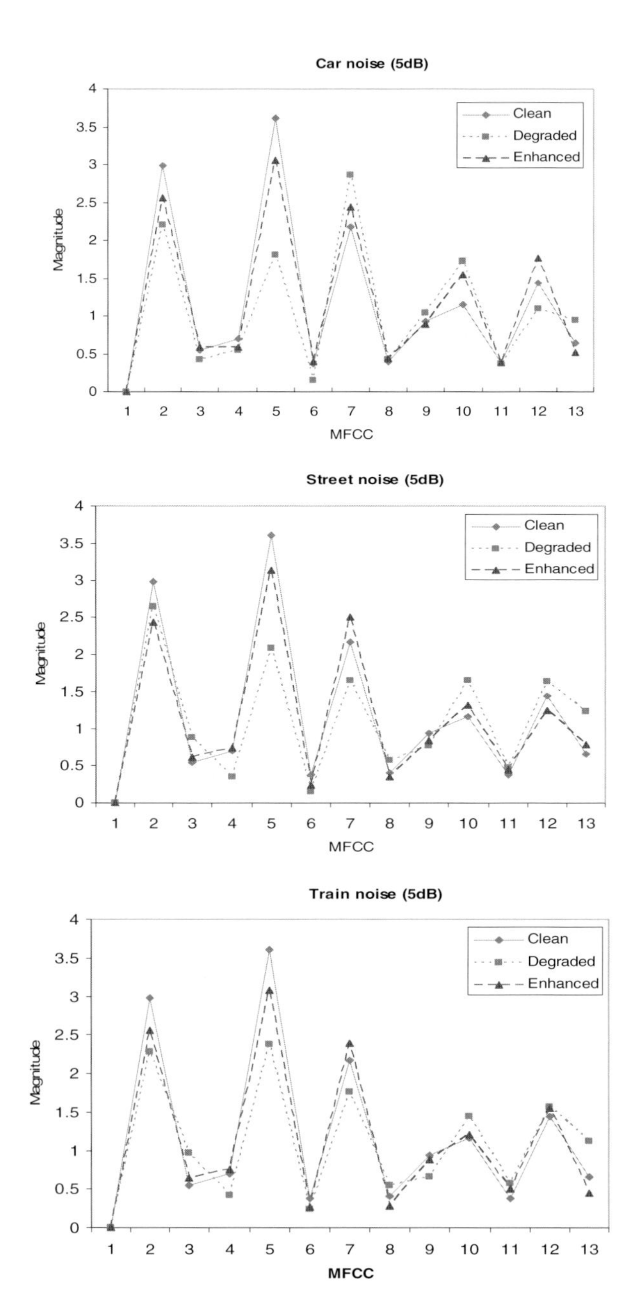

Fig. 2. MFCC vector components for clean, degraded and enhanced speech (shown on an exponential scale for visual clarity in plotting)

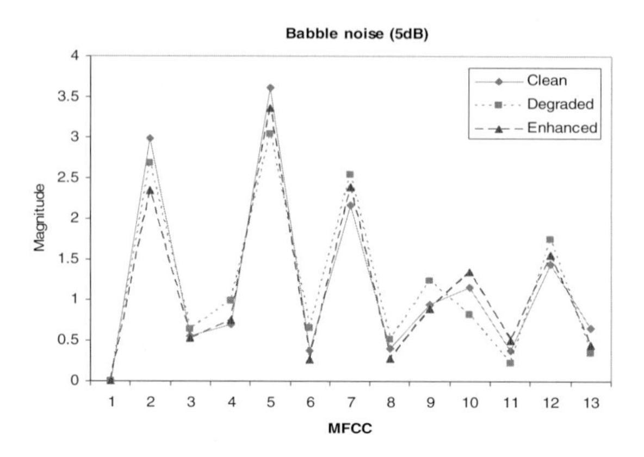

Fig. 2. (*continued*)

system we used 10 fold cross validation, for which the data is split into 10 chunks, one chunk is used for testing and the remaining 9 chunks are used for training. The experiment is repeated with each of the 10 chunks used for testing. The average of the accuracy of the tests over the 10 chunks is taken as the performance measure. In [12], it is suggested that offsets and non-linearities between the scales of objective and subjective MOSs be eliminated by applying a 3rd order monotonic function to map the objective scores onto the subjective scale. Following this, we use a 3rd order polynomial to map the objective scores and subjective MOSs. We use three common criteria [6, 7] for performance evaluation: Pearson linear correlation coefficient C_P (for prediction accuracy), Spearman rank order correlation coefficient C_S (for prediction monotonicity) and Root Mean Squared Error (RMSE), between the subjective rating and the objective prediction. For an ideal match between the objective and subjective scores, $C_P = C_S$ and RMSE = 0.

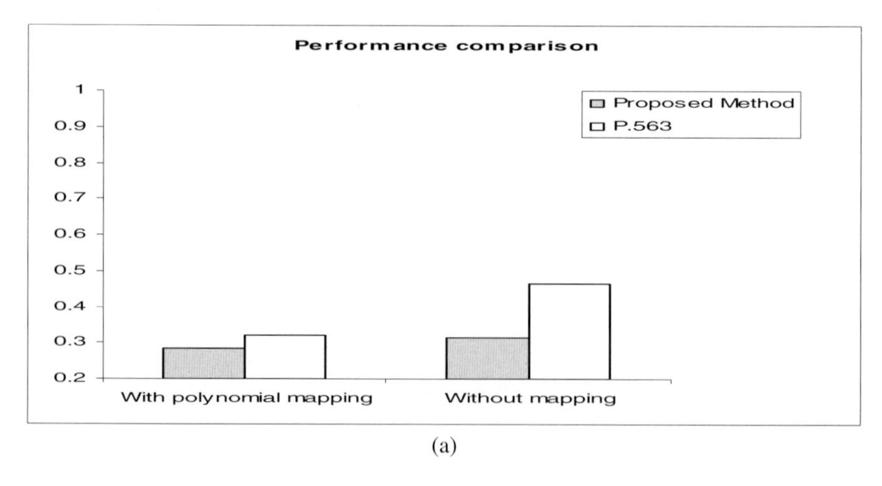

(a)

Fig. 3. Comparison of (a) Root Mean Square Error (RMSE), (b) Pearson correlation coefficient and (c) Spearman correlation coefficient. The 95% confidence intervals are indicated in (b) and (c).

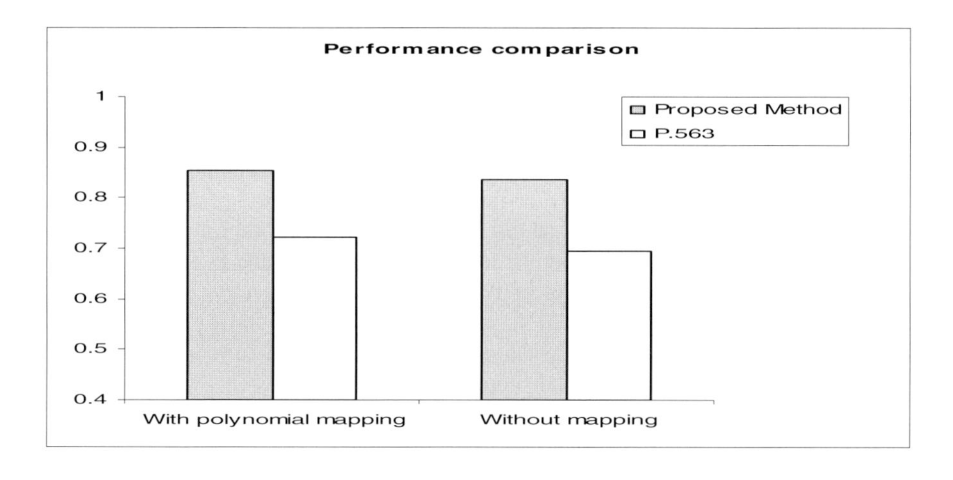

(b)

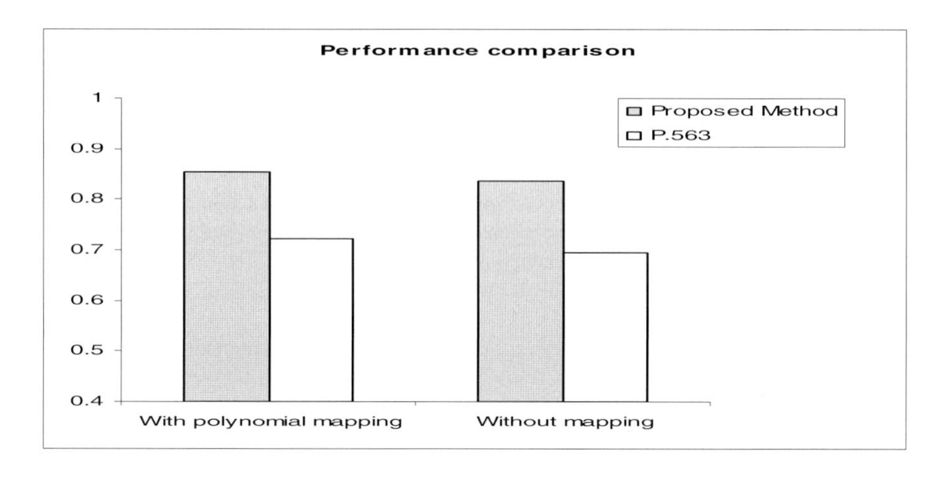

(c)

Fig. 3. (*continued*)

Figure 3 shows the comparison with P.563 algorithm [12], with 95% confidence intervals indicated. We find that the polynomial mapping has very little effect on the performance of the proposed system which indicates that the predicted objective scores are nearly linearly correlated with the subjective scores. In either case (with or without mapping) the proposed system outperforms the P.563 algorithm by a significant margin. As can be seen from the experimental results, both C_p and C_s are improved by nearly 20% while RMSE reduces by about 10% with respect to P.563 algorithm.

5 Conclusions

We have proposed a new non-intrusive speech quality prediction system based on Support Vector Regression (SVR). It uses Mel Frequency Cepstral Coefficients (MFCCs) as the acoustic features and SVR to find an optimal mapping between the MFCCs and the quality score, by formulating speech quality prediction as a re gression problem. The advantages of the proposed system are that it can effectively predict speech quality without reference to the original signal and is effective in predicting quality of speech in terms of high correlation with subjective scores. The experimental results have demonstrated its improvement over the most relevant existing metric, with a third-party database containing 1792 speech files (a total of 112 conditions) of different noise additions and suppression schemes.

Acknowledgment. The authors would like to thank Prof. Philipos C. Loizou (University of Texas at Dallas, USA) for providing the **NOIZEUS** database along with subjective scores and his advice during the experiments.

References

1. Au, O., Lam, K.: A novel output-based objective speech quality measure for wireless communication. In: Proc. 4th Int. Conf. Signal Process., vol. 1, pp. 666–669 (1998)
2. Falk, T., Xu, Q., Chan, W.Y.: Non-intrusive GMM-based speech quality measurement. In: Proc. IEEE Int. Conf. Acoust., Speech, Signal Process., vol. 1, pp. 125–128 (2005)
3. Chen, G., Parsa, V.: Bayesian model based non-intrusive speech quality evaluation. In: Proc. IEEE Int. Conf. Acoust., Speech, Signal Process., vol. 1, pp. 385–388 (2005)
4. Kim, D.: ANIQUE: An auditory model for single-ended speech quality estimation. IEEE Trans. Speech Audio Process. 13(5), 821–831 (2005)
5. Grancharov, V., David, Y., Jonas, L., Bastiaan, W.: Low Complexity Non Intrusive Speech Quality Assessment. IEEE Trans. Speech Audio Process. 14(6), 1948–1956 (2006)
6. Hu, Y., Loizou, P.C.: Evaluation of Objective Quality Measures for Speech Enhancement. IEEE Trans. Speech Audio Process. 16(1), 229–230 (2008)
7. Falk, T., Chan, W.Y.: Single-ended speech quality measurement using machine learning methods. IEEE Trans. Audio, Speech, Lang. Process. 14(6), 1935–1947 (2006)
8. Zhu, Q., Alwan, A.: The Effect of Additive Noise on Speech Amplitude Spectra: A Quantitative Analysis. IEEE Signal Processing Letters 9(9), 275–277 (2002)
9. Scholkopf, Smola, A.J.: Learning with kernels. MIT Press, Cambridge (2002)
10. Hu, Y., Loizou, P.C.: Subjective comparison and evaluation of speech enhancement algorithms. Speech Commun. 49, 588–601 (2007)
11. ITU-T.: Subjective test methodology for evaluating speech communication systems that include noise suppression algorithms. In: ITU-T Rec. P.835, Geneva, Switzerland (2003)
12. ITU-T.: Single-ended method for objective speech quality assessment in narrow-band telephony applications. In: ITU-T P.563Geneva, Switzerland (2004)
13. Rix, A.: Perceptual speech quality assessment - A Review. In: Proc. IEEE Int. Conf. Acoust., Speech, Signal Process., May 2004, vol. 3, pp. 1056–1059 (2004)
14. Psytechnics Limited: NiQA - Product Description. Tech. Rep. (January 2003), http://www.psytechnics.com/pages/products/niqa.php

15. SwissQual Inc.: NiNA - SwissQual's non-intrusive algorithm for estimating the subjective quality of live speech. Tech. Rep. (June 2001), http://www.swissqual.com/HTML/ninapage.htm
16. Murty, K., Yegnanarayana, B.: Combining evidence from residual phase and MFCC features for speaker recognition. IEEE Signal Processing Letters 13(1), 52–55 (2006)
17. ITU-T.: Perceptual evaluation of speech quality. In: ITU-TP.862 Recommendation (Febrauary 2001)
18. Müller, M.: Information Retrieval for Music and Motion. Springer, New York (2007)
19. Shao, X., Milner, B.: Clean speech reconstruction from noisy Mel frequency cepstral coefficients using a sinusoidal model. In: Proc. IEEE Int. Conf. Acoust., Speech, Signal Process., vol. I, pp. 704–707 (2003)
20. Boucheron, L., Philip, L.: On the inversion of Mel frequency cepstral coefficients for Speech Enhancement Applications. In: Proc. ICSES, pp. 485–488 (2008)

Semantic User Modelling for Personal News Video Retrieval

Frank Hopfgartner and Joemon M. Jose

Dept. of Computing Science, University of Glasgow, Glasgow, G12 8RZ, UK
{hopfgarf,jj}@dcs.gla.ac.uk

Abstract. There is a need for personalised news video retrieval due to the explosion of news materials available through broadcast and other channels. In this work we introduce a semantic based user modeling technique to capture the users' evolving information needs. Our approach exploits the Linked Open Data Cloud to capture and organise users' interests. The organised interests are used to retrieve and recommend news stories to users. The system monitors user interaction with its interface and uses this information for capturing their evolving interests in the news. New relevant materials are fetched and presented to the user based on their interests. A user-centred evaluation was conducted and the results show the promise of our approach.

1 Introduction

A challenging problem in the user profiling domain is to create profiles of multimedia retrieval system users. Due to the Semantic Gap, it is not trivial to understand the content of multimedia documents and to find other documents that the users might be interested in. A promising approach to ease this problem is to set multimedia documents into their semantic contexts. For instance, a video about Barack Obama's speech in Ghana can be put into different contexts. First of all, it shows an event which happened in Accra, the capital of Ghana. Moreover, it is a visit by an American politician, the current president. Retrieving a video about Obama's visit to Ghana might indicate that someone is interested in either Barack Obama, Ghana, or in both.

Another challenge in user profiling research is the identification of users' interests in various events. Multiple interests lead to a sparse data representation and approaches need to be studied to tackle this sparsity.

In this paper, we introduce a semantic user profiling approach for news video retrieval, which exploits a generic ontology to put news stories into a context. In order to identify a user's interest in specific topics, we exploit his/her relevance feedback which is provided implicitly while interacting with the system. The remainder of this paper is structured as followed: In Section 2, we introduce various research domains which are relevant within our study and discuss research challenges in Section 3. In Section 4, we introduce our system from capturing daily news to presenting them to the users and evaluate it in Section 5. Section 6 provides a conclusion and a discussion of our findings.

S. Boll et al. (Eds.): MMM 2010, LNCS 5916, pp. 336–346, 2010.

2 Related Work

In this section, we introduce state-of-the-art methodologies to address research challenges that our work builds upon.

User profiling is the process of learning a user's interests over a long period of time. Several approaches have been studied to capture users' news interests in a profile. Chen and Sycara [5] analyse internet users during their information seeking task and explicitly ask them to judge the relevance of the webpages they visit. Exploiting the created user profile of interest, they generate a personalised newspaper containing daily news. However, providing explicit relevance feedback is a demanding task and users tend not to provide much feedback [8]. Bharat et al. [2] created a personalised online newspaper by unobtrusively observing the user's web-browsing behaviour. Although their system is a promising approach to release the user from providing feedback, their main research focus is on developing user interface aspects, ignoring the sophisticated retrieval issues. The web-based interface of their system provides a facility to retrieve news stories and recommends stories to the user based on his/her interest.

An interesting approach for news personalisation is to map relationships between concepts in the user profile by using ontologies. Fernández et al. [7] argue that ontologies can be exploited to structure news items and to annotate them with additional information. In the news video domain, Bürger et al. [3] have shown that such structured data can be used to assist the user in accessing a large news video corpus. Dudev et al. [6] propose the creation of user profiles by creating knowledge graphs that model the relationship between different concepts in the Linked Open Data Cloud. This collection of ontologies unites information about many freely available different concepts. The backbone of the cloud is DBpedia, an information extraction framework which interlinks Wikipedia content with other databases on the Web such as Geonames or WordNet. In this paper, we exploit this data cloud to link automatically segmented story videos. Challenges and open research questions are introduced in the next section.

3 Research Challenges

Various challenges arise when aiming at creating semantic user profiles in the multimedia domain.

State-of-the-art user profiling approaches exploit the textual content of relevant documents to identify user's interests. Considering the short length of news video stories, creating useful profiles is rather problematic. The development of Semantic Web technologies promise a solution for this problem. If a story contains various concepts, additional information about these concepts might help to model user interests more accurate. Järvelin et al. [10] already showed that a concept-based query expansion is helpful to improve retrieval perfomance. Adapting their approach, we hypothesise that ontologies can be exploited to organise user interests and also the pre-fetched relevant documents.

The next problem is how the user's evolving interests can be captured in a long-term user profile. What a user finds interesting on one day might be completely irrelevant on the next day. In order to model this behaviour, we incorporate the Ostensive Model of developing Information Need [4]. In this model, providing feedback on a document is considered as ostensive evidence that this document is relevant for the user's current interest. As argued before, however, users tend not to provide constant feedback on what they are interested in. Thus, one condition we set is that a user profile should be automatically created by capturing users' *implicit* interactions with the retrieval interface. Our next hypothesis is hence that implicit relevance feedback techniques can efficiently be employed to create efficient long-term user profiles.

Another problem in the context of user profiling is the users' multiple interests in various topics. For example, users may be interested in Sports and Politics or in Business news. Further, they can even be interested in sub categories such as Football, Baseball or Hockey. A specification for a long-term user profile should therefore be to automatically identify these multiple aspects. We hypothesise that separating user profiles based on broader news categories can lead to a structured representation of the users' interests. This will lead to a more indicative presentation of materials. Moreover, we hypothesise that a hierarchical agglomerative clustering of the content of these category-based profiles can be used to effectively identify sub categories.

Summarising, we address the following hypotheses in this work:

1. Implicit relevance feedback techniques can be exploited to create efficient long-term user profiles.
2. Separating user profiles based on broader news categories can lead to a structured representation of the users' interests.
3. Hierarchical agglomerative clustering of the content of these category-based profiles can be used to effectively identify sub categories.
4. Ontologies can be exploited to organise user interests and also the pre-fetched documents.

In order to study these hypotheses, we introduce a novel news video retrieval system which automatically captures users' interests. The system and its components will be introduced in the next section.

4 System Description

The architecture of the introduced news video retrieval system can be segmented into three conceptual parts: A data processing phase, the graphical user interface and the profiling module. Since we want to provide an up-to-date news video collection, the data processing phase is called twice a day, starting with the actual capturing of the broadcast and the decoding of the teletext transmission. In this study, we focus on the daily BBC One O'Clock News and the ITV Evening News, the UK's largest news programmes. Each bulletin with a running time of thirty minutes is enriched with a teletext signal. Following Hopfgartner et al. [9], we segment these news videos into coherent news stories. In the remainder of this section, we introduce the steps from annotating these news stories using external sources and indexing them. Moreover, we introduce the system interface and discuss our user profiling approach.

4.1 Semantic Annotation

Usually, news content providers classify their news in accordance to the IPTC standard, a news categorisation thesaurus developed by the International Press Telecommunications Council. We use OpenCalais[1], a Web Service provided by Thomson Reuters, to classify each story into one or more of the following IPTC categories: Business & Finance, Entertainment & Culture, Health, Medical & Pharma, Politics, Sports, Technology & Internet and Other.

In a next step, we aim to identify concepts that appear in the stories. Once these concepts have been positively identified, the Linked Open Data Cloud can be exploited to further annotate the stories with related concepts. Three problems arise when conducting this procedure.

First of all, how can we determine concepts in the story which are strong representatives of the story content? In the text retrieval domain, named entities are considered to be strong indicators of the story content, since they carry the highest content load among all terms in a document. Therefore, we extract persons, places and organisations from each story transcript using OpenCalais.

The second question is, how can these named entities be positively matched with a conceptual representation in the Linked Open Data Cloud. For resolving the identity of an entity instance, we again rely on the OpenCalais Web Service, which compares the actual entity string with an up-to-date database of entities and their spelling variations. Once entities have been disambiguated, OpenCalais maps these entities with a uniform resource identifier (URI) and their representation in DBpedia.

Since the link between the story and the Linked Open Data Cloud has been established, the next problem is how can the structured knowledge represented in the Linked Open Data Cloud be exploited to augment the story. A long-term user profile which is created using implicit evidence will contain many entries, which makes a weighted semantic network approach as suggested by Dudev et al. [6] infeasible. Therefore, we consider only direct links from the identified concept to other concepts in the Semantic Web. We therefore augment the stories with all URIs that are directly associated with these entities in the Cloud.

4.2 User Interface

Figure 1 shows a screenshot of the news video retrieval interface. It can be split into three main areas: Search queries can be entered in the search panel on top, results are listed on the right side and a navigation panel is placed on the left side of the interface. When logging in, the latest news will be listed in the results panel. Search results are listed based on their relevance to the query. Since we are using a news corpus, however, users can re-sort the results in chronological

[1] http://www.opencalais.com/

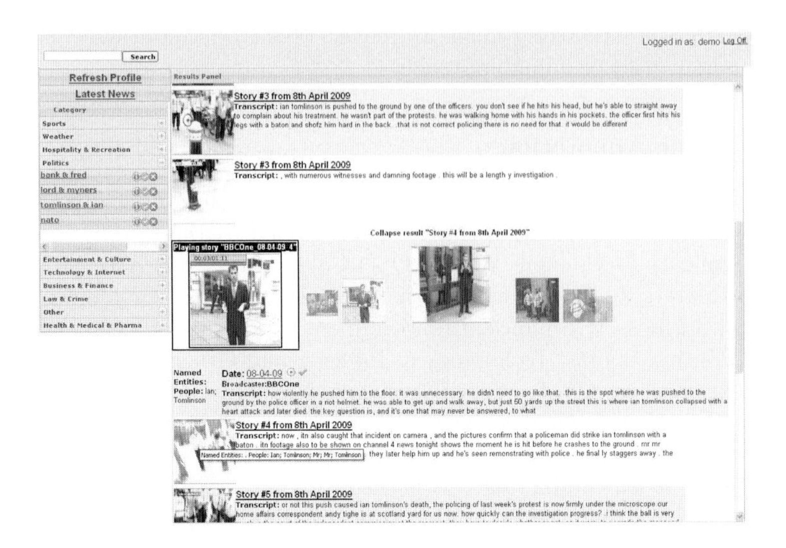

Fig. 1. Graphical User Interface of the System

order with latest news listed first. Each entry in the result list is visualised by an example keyframe and a text snippet of the story's transcript. Keywords from the search query are highlighted to ease the access to the results. Moving the mouse over one of the keyframes shows a tooltip providing additional information about the story. A user can get additional information about the result by clicking on either the text or the keyframe. This will expand the result and present additional information including the full text transcript, broadcasting date, time and channel and a list of extracted named entities. In the example screenshot, the third search result has been expanded. The shots forming the news story are represented by animated keyframes of each shot. Users can browse through these animations either by clicking on the keyframe or by using the mouse wheel. This action will center the selected keyframe and surround it by its neighboured keyframes. The keyframes are displayed in a cover-flow view, meaning that the size of the keyframe grows larger the closer it is to the focused keyframe. In the expanded display, a user can also select to play a video, which opens the story video in a new panel.

The user's interactions with the interface are expoited to identify multiple topics of interests (see Section 4.3). On the left hand side of the interface, these interests are presented by different categories. Clicking on any of these categories in the navigation panel will reveal up to four sub categories for the according category. The profiling approach will be introduced in the following section.

4.3 User Profiling

When a user interacts with a result, he leaves a "semantic fingerprint" that he is interested in the content of this item to a certain degree. In this work, we employ a *weighted story vector* approach to capture this implicit fingerprint in a profile. The weighting of the story will be updated when the system submits a new weighted story to the profile starting a new iteration j. Hence, we represent the interaction I of a user i at iteration j as a vector of weights

$$\boldsymbol{I}_{ij} = \{W_{ij1} ... W_{ijs}\}$$

where s indexes the story in the whole collection. The weighting W of each story expresses the evidence that the content of this story matches the user's interest. The higher the value of W, the closer this match is. In this work, we define a static value for each possible implicit feedback feature:

$$W = \begin{cases} 0.1, & \text{when a user browses through the keyframes} \\ 0.2, & \text{when a user uses the highlighting feature} \\ 0.3, & \text{when a user expands a result} \\ 0.5, & \text{when a user starts playing a video} \end{cases}$$

Note that some of these features are independent, while others depend on a previous action (e.g. a video cannot be played without being clicked on).

As explained before, each news story has been classified as belonging to one or more broad news categories C. Since we want to model the user's multiple interests, we use this classification as a splitting criteria. Thus, we represent user i's interest in C in a category profile vector $\boldsymbol{P}_i(C)$, containing the story weight $SW(C)$ of each story s of the collection:

$$\boldsymbol{P}_i(C) = \{SW(C)_{i1} ... SW(C)_{is}\}$$

In the user interface, each category profile is represented by an item in the navigation panel.

In our category profile, the story weight for each user i is the combination of the weighted stories s over different iterations j: $SW(C)_{is} = \sum_j a_j W_{ijs}$. Following Campbell and van Rijsbergen [4], we include the ostensive evidence

$$a_j = \frac{1 - C^{-j+1}}{\sum_{k=2}^{j_{max}} 1 - C^{-k+1}} \tag{1}$$

to introduce an inverse exponential weighting which will give a higher weighting to stories which have been added more recently to the profile, compared to stories which were added in an earlier stage.

The above introduced methodology results in a category-based representation of the user's interests. Each category profile consists of a list of weighted stories, with the most important stories having the highest weighting. A challenge is here to identify different contextual aspects in each profile. We approach this problem by performing a hierarchical agglomerative clustering of stories with the highest story weight at the current iteration. Following Bagga and Baldwin [1], we treat the transcripts extracted from these stories as term vectors and compare them by cosine. Unlike their approach, however, we use the whole transcript rather than sentences linked by coreferences and use the square root of raw counts as our term frequencies rather than the raw counts. We use complete-link clustering since this approach results in more compact clusters. Moreover, we do not use inverse-document frequency normalisation since this value can be important for discremination. For tokenisation, we use standard filters (conversion to lower case, stop word removal and stemming). The numbers of clusters k is a parameter. Since each cluster should contain stories associated with an aspect of the user's interest, k should be equal to the number of different interests that a user has. In this study, we have set $k = 4$. In the interface, the clusters represent the four sub categories under each category in the navigation panel. The two most frequent named entities in each cluster are used as a label for each sub category.

The content of the users' profiles is displayed on the navigation panel of the left hand side of the interface. Since the idea of such navigation panel is to assist the users in finding other stories that match their interests, the next challenge is to identify more stories in the data corpus that might be of the users' interests. Assuming that each of the sub categories contains stories that cover one or more (similar) aspects of a user's interest, the content of each sub category can be exploited to recommend more documents belonging to that cluster. The simplest method is to create a search query based on the content of each cluster and to retrieve stories using this query. A promising source to create such queries is the use of most frequent named entities within each cluster. Due to the rather short length of the story transcript, we identify additional named entities by performing an additional pseudo relevance feedback step.

For the initial search, we first extract all URIs from the cluster and retrieve stories containing these URIs. Then, we extract all named entities from the stories in the result list and finally use the most frequent entities as a search query.

5 Evaluation

In order to evaluate the hypotheses which have been introduced in Section 3, we performed a user study which will be described in the remainder of this section.

5.1 Experimental Design

Since the proposed profiling approach includes the capturing of long-term user interests, we had to study the effectiveness of our system over several days. We therefore captured six month of news video broadcasts and paid participants to use the system as additional source of information in their daily news consumption routine. Their interactions with the system were logged to evaluate the approach. They were asked to use the system for up to ten minutes each working day for up to seven days to search for any topic that they were interested in. In addition, we also created a simulated search task situation. Our expectation was twofold: First of all, we wanted to guarantee that every user had at least one topic to search for. Moreover, we wanted the participants to actually explore the data corpus. Therefore, we chose a scenario which had been a major news story over the last few months:

> "Dazzled by high profit expectations, you invested a large share of your savings in rather dodgy securities, stocks and bonds. Unfortunately, due to the credit crunch, you lost about 20 percent of your investment. Wondering how to react next and what else there is to come, you follow every report about the financial crisis, including reports about the decline of the house's market, bailout strategies and worldwide protests."

Each participant started with an individual introductory session, where they were asked to fill in an entry questionnaire and could familiarise themselves with the interface. Every day, they were asked to fill in an online report where they were encouraged to comment on the system as they used it. At the end of the experiment, everyone was asked to fill in an exit questionnaire to provide feedback on their experience during the study.

5.2 Participants

16 users with an average age of 30.4 years participated in our experiment. Their favourite sources for gathering information on the latest news stories are news media web portals, word-of-mouth and the television. The typical news consumption habit they described was to check the latest news online in the morning and late at night after dinner. We hence conclude that the participants represent the main target group for the introduced retrieval system.

5.3 Results

By asking for daily reports, our goal was to evaluate the users' opinion about the system at various stages of the experiment. The first question was to find out what the participants actually used the system for. The majority of participants used it to retrieve the latest news, followed by identifying news stories they were not aware of before.

One of our main research interests was to determine whether the system provides satisfactory access to the data collection. Therefore, we asked the participants to judge various statements on a Five-Point-Likert scale from 1 (Agree) to 5 (Disagree). The order of the agreements varied over the questionnaire to reduce bias. Figure 2 shows the average judgement of all users over all seven days for two statements that were aimed at determing the general usability of the system. The first statement posed was "The interface structure helped me to explore the news collection", denoted "explore collection" in the figure. The second statement was "The interface helped me to explore various topics of interest", denoted "explore topics". As can be seen, the average user found the interface useful and was satisfied with its usability.

With the aim of evaluating our first hypothesis that implicit relevance feedback can be used to create long-term user profiles, we asked the participants

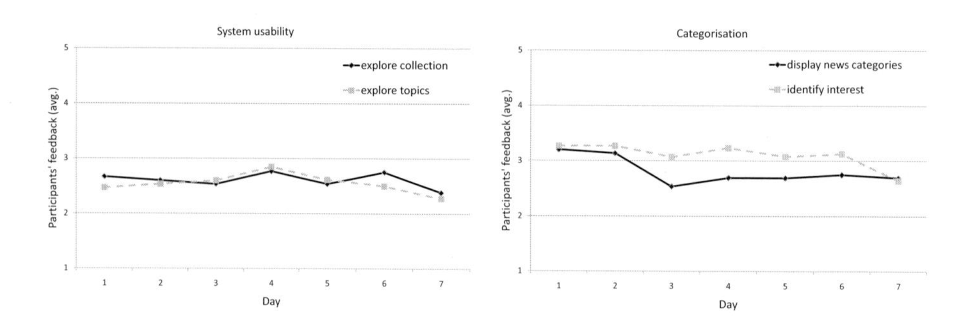

Fig. 2. Feedback on system usability (Lower is better)

Fig. 3. Feedback on interest capturing (Lower is better)

to judge if the system was effective in automatically identifying their interests. Another statement was "the system sucessfully identified and displayed news categories I was interested in". As Figure 3 illustrates, the participants did neither agree or disagree to these statements, which corralates with the observation that the use of implicit features for short-term user modelling provides weaker evidence of relevance than explicit relevance feedback [11]. Nevertheless, a tendency towards a positive rating for the use of implicit indicators is visible, in particular towards the end of the user study. This suggests that implicit relevance feedback can be used to create long-term user profiling. However, further research is necessary to differentiate positive and negative indicators of relevance, which is beyond the scope of this work.

With the goal of evaluating the news categorisation, we asked the users to judge the following two statements: "The displayed sub categories represent my interests in various topics" and "the displayed results for each sub category were related to each other". Figure 4 shows the average answer over the whole time of the experiment. The first question, denoted "relevant sub categories", aimed to evaluate whether separating user profiles based on broader categories leads to

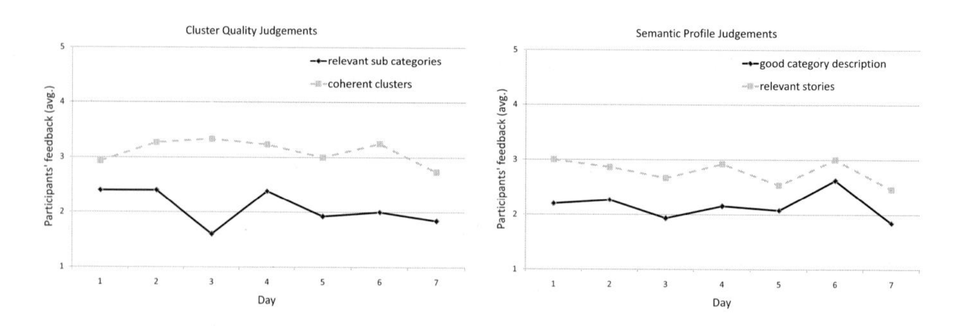

Fig. 4. Feedback on sub categories (Lower is better)

Fig. 5. Feedback on ontology-based recommendations (Lower is better)

a structured representation of the users' interests, our second hypothesis. The second question aimed to evaluate the coherence of each category-based profile, targeting the third hypothesis. As can be seen, the participants had a positive perception of the relevance of the sub categories. This could indicate that our clustering approach was successful in identifying diverse aspects of the same news category, supporting our hypothesis that categorising the user interests into broader categories provides a structured representation to the users' interests. Concerning the coherence of each cluster, the participants tended towards a neutral perception.

The last set of statements aimed to evaluate the fourth hypothesis that ontologies can be exploited to organise user interests and also the pre-fetched relevant documents. Thus, participants judged the following differentials: "The displayed results for each category matched with the category description" and "the displayed results for each category contained relevant stories I did not retrieve otherwise". Figure 5 shows the relevant responses. Again, a tendency towards a positive perception of the results which are determined using semantics is visible. In order to explore this hypothesis further, we analysed user transaction patterns which were captured in the log files. Our analysis revealed that the participants used the provided subcategories extensively. Roughly 40% of all search queries were triggered by clicking on one of these categories. This high percentage suggests that the participants found the results given by these categories to be useful, which would suggest that our semantics based user profiling is effective in recommending relevant results.

6 Conclusion

In this paper, we introduced a semantic based user modeling technique which automatically captures the users' evolving information needs and represents this interest in dynamic user profiles. Therefore, we introduce a novel news video retrieval system which automatically captures daily broadcasting news and segments the bulletins into coherent news stories. The Linked Open Data Cloud is exploited to set these stories into context. This semantic augmentation of the news stories is used as the backbone of our user profiling methodology. The profiles can be used to identify the users' multiple interests in diverse aspects of news over a longer period of time. The semantic augmentations of the stories in the user profile are used to fetch new relevant materials.

Our preliminary study is based on four hypotheses, which are evaluated using a user-centred evaluation scheme. 16 participants were asked to include the news retrieval system into their daily news gathering routine and to judge the performance of the system on a daily basis. Differing from standard interactive information retrieval experiments, the evaluation was split into multiple sessions and performed under an uncontrolled environment, two necessary conditions for a realistic evaluation of a long-term user profiling. This novel approach cannot rely on system-centred evaluation measures as common in information retrieval experiments. Thus, standardised evaluation measures need yet to be developed.

The hypotheses were evaluated by analysing users' feedback which was provided during various stages of the experiment. The analysis of their feedback forms seem to support all hypotheses, suggesting that the introduced system can be effectively used to provide a personalised access to video news data. In fact, a majority of all participants claimed in the exit questionnaire that they would use a commercialised system with the presented features for their daily news gathering.

Future work includes a more thorough selection of concepts in the Linked Open Data cloud to be used for augmenting each story. Currently, every concept that is directly linked with the story's concept is used, resulting in many concepts of less importance. A better selection scheme can lead to stronger links between related stories, hence increasing the effectiveness of the introduced approach.

Acknowledgments

This research was supported by the EC under contract FP6-027122-SALERO.

References

1. Bagga, A., Baldwin, B.: Entity-based cross-document coreferencing using the vector space model. In: Proc. Comput. Ling., pp. 79–85 (1998)
2. Bharat, K., Kamba, T., Albers, M.: Personalized, interactive news on the web. Mult. Syst. 6(5), 349–358 (1998)
3. Bürger, T., Gams, E., Güntner, G.: Smart content factory: assisting search for digital objects by generic linking concepts to multimedia content. In: Proc. HT, pp. 286–287. ACM, New York (2005)
4. Campbell, I., van Rijsbergen, C.J.: The ostensive model of developing information needs. In: Proc. Library Science, pp. 251–268 (1996)
5. Chen, L., Sycara, K.: WebMate: A personal agent for browsing and searching. In: Sycara, K.P., Wooldridge, M. (eds.) Proc. Agents 1998, 9–13, pp. 132–139. ACM Press, New York (1998)
6. Dudev, M., Elbassuoni, S., Luxenburger, J., Ramanath, M., Weikum, G.: Personalizing the Search for Knowledge. In: Proc. PersDB 2008 (2008)
7. Fernández, N., Blázquez, J.M., Fisteus, J.A., Sánchez, L., Sintek, M., Bernardi, A., Fuentes, M., Marrara, A., Ben-Asher, Z.: NEWS: Bringing semantic web technologies into news agencies. In: Cruz, I., Decker, S., Allemang, D., Preist, C., Schwabe, D., Mika, P., Uschold, M., Aroyo, L.M. (eds.) ISWC 2006. LNCS, vol. 4273, pp. 778–791. Springer, Heidelberg (2006)
8. Hancock-Beaulieu, M., Walker, S.: An evaluation of automatic query expansion in an online library catalogue. J. Doc. 48(4), 406–421 (1992)
9. Misra, H., Hopfgartner, F., Goyal, A., Punitha, P., Jose, J.M.: TV News Story Segmentation based on Semantic Coherence and Content Similarity. In: Boll, S., et al. (eds.) MMM 2010. LNCS, vol. 5916, pp. 347–357. Springer, Heidelberg (2010)
10. Järvelin, K., Kekäläinen, J., Niemi, T.: ExpansionTool: Concept-Based Query Expansion and Construction. Information Retrieval 4(3), 231–255 (2001)
11. Kelly, D., Teevan, J.: Implicit feedback for inferring user preference: a bibliography. SIGIR Forum 37(2), 18–28 (2003)

TV News Story Segmentation Based on Semantic Coherence and Content Similarity

Hemant Misra, Frank Hopfgartner, Anuj Goyal, P. Punitha, and Joemon M. Jose

Dept. of Computing Science, University of Glasgow, Glasgow, G12 8QQ, UK
{hemant,hopfgarf,anuj,punitha,jj}@dcs.gla.ac.uk

Abstract. In this paper, we introduce and evaluate two novel approaches, one using video stream and the other using close-caption text stream, for segmenting TV news into stories. The segmentation of the video stream into stories is achieved by detecting anchor person shots and the text stream is segmented into stories using a Latent Dirichlet Allocation (LDA) based approach. The benefit of the proposed LDA based approach is that along with the story segmentation it also provides the topic distribution associated with each segment. We evaluated our techniques on the TRECVid 2003 benchmark database and found that though the individual systems give comparable results, a combination of the outputs of the two systems gives a significant improvement over the performance of the individual systems.

1 Introduction

In most part of the 20th century, consuming news was a solely passive activity. People simply followed news coverage by reading newspapers, listening to radio broadcasts or watching the television news. However, the rise of new technologies has rapidly changed this trend; now-a-days publishers and broadcasters also provide content on the WWW, an increasing percentage of this being video clips [1]. Faced with these developments, processing video clips, television news being one among them, has become an important research area that has attracted a lot of attention. The main focus is to tackle the problems that arise when it is required to retrieve some information from this data. In this context, a basic challenge is to segment videoes into meaningful and manageable *segments* in order to ease the access of the video data. The smallest coherent segment in a video is a shot, a unit that has been constantly filmed using the same camera setting. A simple solution to video segmentation is to divide a video into shots using visual features such as colour, texture and shape. State-of-the-art techniques as evaluated within TRECVid [19] reach a very high performance in detecting shot boundaries. Nevertheless, a more challenging, and also more informative approach, is to segment broadcasts into coherent news stories. Segmenting a news broadcast into such stories is essentially finding the boundaries where one story ends and the other begins.

S. Boll et al. (Eds.): MMM 2010, LNCS 5916, pp. 347–357, 2010.

In this paper, we approach the TV news story segmentation task from lexical content and visual similarity perspectives. Segmenting the teletext stream of a television news into stories is a direct application of text segmentation, an active area of research [12,20].

We evaluate the performance of our approaches on the TRECVid 2003 data collection [18], a standard benchmark used for the story segmentation task. The corpus consists of over 130 hours of news video in MPEG-1 format that was broadcast in the year 1998. The collection has been split into a test set and a development set. In the current work, we use the test set, which enables us to compare our results with the runs submitted to TRECVid. The test set was split into more than 32000 shots with representative key frames provided for each shot. Moreover, each broadcast was manually split into coherent story segments and the corresponding transcripts were provided. In Section 2, we provide an overview of state-of-the-art story segmentation approaches. In Section 3, we birefly explain our LDA based approach for the task of text segmentation the details of which can be found in [14]. Segmenting the text transcripts of the news broadcast using LDA based approach not only provides the story boundaries but also the topic distribution associated with each story. In Section 4, we introduce our feature-based approach for video segmentation where we extract colour features from each key frame and identify anchor person shots. Neighbouring shots from these anchor persons are merged based on their similarity with respect to shot length difference and visual dissimilarity. Using the resulting time points of the detected boundary key frames, we segment the video broadcast into stories. The performance of both the approaches and their combination is evaluated in Section 5. In Section 6 we draw the main conclusons of this study and outline the future directions.

2 Background

Segmenting TV news broadcasts into story units was one of the main tasks within TRECVid 2003 and 2004 evaluations. The task description of these evaluations defines stories as "segments of a news broadcast with a coherent news focus which contain at least two independent declarative clauses". Various approaches using text, audio and video streams or a combination of them have been studied to segment TV news broadcast into stories.

In text segmentation, some approaches rely on word repetition [12] while the others use cue phrases [16] to identify story boundaries. The later approaches use the information that transcript of a TV news broadcast is typically laced with cues words such as *welcome, bye, good morning, thank you, next to follow* etc., to indicate the beginning or end of a story.

O'Connor et al. [15] performed story segmentation by clustering key frames based on their low-level colour feature. In their approach, two shots that are very similar based on their visual appearance but have been shown at two distant moments during the broadcast will not be placed in the same cluster.

Due to the feature-rich nature of TV news broadcast, it is not premature to assume that a broadcasts' video and text streams may contain complementary information and that their combination can yield a performance that is better than the performance of a system which only uses the information from a single stream. Indeed, analyses [2,6] have shown that the most successful runs evaluated within TRECVid rely on both text-based and visual-based segmentation approaches to detect story boundaries. Pickering et al. [17], for instance, extracted key entities such as nouns and verbs from the broadcast transcript, computed a term weighting based on their frequency within the text and combined neighbourig shots to accomplish the task.

Hsu et al. [10] perform a story boundary segmentation experiment and compare the average precision of different combinations of audio, video and text fusions. They report that a combination of all modalities worked best to identify correct story boundaries. However, as Chang et al. [5] argue, a better understanding of relations between information extracted from the text stream and relations extracted from different audio and visual streams is still needed. Chaisorn et al. [4] approach this problem using a bifid approach. First, they employ a learning based approach to identify story boundaries, and then classify each story into semantic categories by employing heuristic rules.

Different from all these approaches, our LDA based approach not only estimates the segment boundaries, it also categorizes the segments based on their topic distribution. Moreover, the only assmption in feature based approach is that a story always begins with an anchor person shot.

3 Text-Based Segmentation

In this section, we briefly describe our recently proposed topic model based approach for story segmentation task [14] which exploits the properties of unsupervised Latent Dirichlet Allocation (LDA) [3,7] topic model to estimate the coherence of a segment, and in turn the segment boundaries. The details of our approach and its analysis can be found in [14]

LDA is a generative unsupervised approach to model discrete data such as text. The two main assumptions in LDA are: 1) every document is represented by a topic distribution, and 2) every topic has an underlying word distribution.

In this work, we have used Gibbs sampling method, as decribed in [7], to train the LDA model on the well known Reuters collection volume 1 (RCV1). The training consists of estimating the topic distribution in each training document, represented by θ, and word distribution in each topic, represented by ϕ. After the burn-in period of the Gibbs sampling, these two parameters are estimated by the following equations:

$$\theta_{dt} = \frac{K_{dt} + \alpha}{\sum_{k=1}^{T} K_{dk} + T\alpha} \tag{1}$$

$$\phi_{tv} = \frac{J_{tv} + \beta}{\sum_{k=1}^{V} J_{tk} + V\beta} \tag{2}$$

where K_{dt} is the number of times a word in document d has been assigned to topic t, J_{tw} is the number of times word w has been assigned to topic t in the whole training corpus and V is number of unique words in the training corpus (vocabulary size) after removing stop-words; number of topics, T, and Dirichlet priors, α and β, are hyper-parameters, and in our experiments their values were 50, 1 and 0.01, respectively.

During testing, the topic distribution of an unseen document can be estimated by the following iterative equation [8,13]:

$$\theta_{dt}^{(n+1)} = \frac{1}{l_d} \sum_{v=1}^{V} \frac{C_{dv} \theta_{dt}^{(n)} \phi_{tv}}{\sum_{t'=1}^{T} \theta_{dt'}^{(n)} \phi_{t'v}} \tag{3}$$

where $\theta_{dt}^{(n)}$ is the value of θ_{dt} at nth iteration, C_{dv} is the number of times vocabulary word v has occured in document d, and l_d is the number of words in the document which are present in the training vocabulary. The words in the document which are not in the training vocabulary are dropped, and are not used for estimating the topic distribution.

The likelihood of a document, given its topic distribution, can be estimated as

$$P(C_d|\theta, \phi) = \prod_{v=1}^{V} \left[\sum_{t=1}^{T} \theta_{dt} \phi_{tv} \right]^{C_{dv}} \tag{4}$$

In this paper, the same methodology which is used to compute the likelihood of an unseen document is applied to compute the likelihood of a segment.

For a given text, a coherent segment containing a single story is expected to have only a few active topics (LDA topics as defined in the LDA framework), whereas an incoherent segment, having more than one story in it, may have several active topics. In [13], the authors showed that likelihood of a coherent document is higher as compared to the likelihood of an incoherent document. This observation is the fundamental premise for our LDA based approach: for a given text, the segmentation which provides the highest likelihood is also going to provide the most coherent segments. The task of finding the highest likelihood, and in turn the most coherent segments, is performed in the framework of dynamic programming (DP).

Lets assume a given text $d = \{w_1 \cdots w_{l_d}\}$ of length l_d. For this text, consider a particular segmentation, S, which is made of m segments, $S = \{S_1 \cdots S_m\}$, where S_i has n_i words in it. Further, let w_i^j be the jth word token in S_i, such that $W_i = \{w_i^1 \cdots w_i^{n_i}\}$. Therefore, $\sum_{i=1}^{m} n_i = l_d$, $d = \{W_1 \cdots W_m\}$ and W_i is dependent only on S_i. The likelihood of segment S can be given by

$$P(S|d) = P(d|S)P(S)/P(d) \tag{5}$$

where $P(d|S)$ is the probability of the document d under segmentation S and P(S), considered as a penalty factor, is a prior over segmentations. $P(d)$ is same for all the possible segmentations of a documents and hence can be dropped. Therefore

$$P(S|d) \propto \left[\prod_{i=1}^{m} P(W_i|S) \right] P(S) \propto \left[\prod_{i=1}^{m} P(W_i|S_i) \right] P(S) \propto \left[\prod_{i=1}^{m} \prod_{j=1}^{n_i} P(w_i^j|S_i) \right] P(S)$$

The optimal segmentation is the one that maximises this likelihood, that is, $\hat{S} = \underset{S}{\text{argmax}}\ P(S|d)$, and can be obtained by DP which is typically employed to solve the problem of shortest path in many applications. The likelihood of a segment, $P(W_i|S_i)$, is obtained by (4), where the term C_{dv} is replaced by the word frequency occurence in a segment. That is, for each possible segment, (3) is used to compute its θ and subsequently (4) is employed to estimate its likelihood.

A DP algorithm has two passes, a forward-pass followed by a trace-back. In the forward-pass of our DP, for each segment described by a begin word (B) and an end word (E), the likelihood is computed by (4). This likelihood is accummulated and for each E node, the information about the B node which gives the highest score (in orther words, the B node which is the best starting node for this E node) is stored. On reaching the document end, during trace back, the information about the best starting node is used to get segmentation (segment boundaries) which gives the maximum-likelihood path. In our case, $P(S) = (l_d)^{-m*p}$, where $p = 3$ was empirically found to give the best results on another dataset.

4 Feature-Based Segmentation

In this section, we focus on exploiting various content features to segment news broadcasts into corresponding story segments. In most new broadcasts, e.g. from CNN, Al Jazeera or BBC, stories are often introduced by an anchor person. This also applies to the TRECVid 2003 corpus. The first step in our feature-based story segmentation approach is therefore to identify the first anchor person shot in the video. An analysis revealed that the first anchor person shot usually appears within the initial 25–55 seconds of each broadcast. Since anchor persons are usually filmed in a studio setting with similar visual appearance in each broadcast, identifying these shots is a pattern matching task. Utilising this observation, we first identify the first possible anchor person key frame. In the beginning, we consider each shot in the first 25–55 seconds of video to be the possible anchor person shot candidate. We hence need to identify the key frames which appear more often than any other key frame in the broadcast. We start by computing the visual distance of the MPEG-7 colour structure feature between every candidate within this range and the remaining key frames of the broadcast. Since some shots might be re-appearing shots belonging to the same story, we skip a few shots Δk which may be repeated shots in the neighbourhood of the anchor person shot. The candidate frame with the lowest average visual similarity is considered to be the first anchor person key frame.

The next task is to identify other anchor person shots within the video. In order to classify a shot as an anchor person shot, we also take the neighbouring three shots on both sides, a region of support, into account. This region of support is used to determine whether the anchor person introduces a new story or not. Accordingly, a shot will be treated as story boundary candidate only if the neighbouring shots differ significantly from each other. Unfortunately, no ground truth data exist which can be used to evaluate our anchor person detection approach. Therefore, our evaluation is focused on the actual story boundary detection task, which we treat as a classification task. Twenty sample videos from both CNN and ABC videos of the TRECVid 2003 corpus are used to train an SVM for each collection. Ground truth provided within TRECVid is used to identify true story boundaries in training samples. The following features are used to train an SVM to identify anchor person shots:

- **Distance from Anchor Person Template:** We compute the visual distance between the previously identified template and the current key frame using the MPEG-7 Colour Feature.
- **Semantic Text Similarity:** Following Kolb [11], we compute the semantic similarity between the transcript of the left region of support and the right region of support. We assume that the transcript is similar on both sides if both transcripts form part of the same story.
- **Shot Length Distance:** We compute the absolute difference between the numbers of key frames in the left and the right region of the support. Action-loaded news like sports reports are expected to have more key frames than calmer news, e.g. reports about political party agendas. Therefore, it is an effective feature to distinguish between stories.
- **Average Visual Dissimilarity:** We determine the average difference of the MPEG-7 colour structure feature between the shots from the left and the right region of support. This value can identify the shots which are visually similar to the neighbourhood and so are very less probable to start a new story.
- **Minimum Visual Dissimilarity:** We compute the minimum difference between the shots from the left and right region of support using the colour structure feature. This value is useful to detect when a shot is repeated in a news story, as the minimum distance will be very low in this case.

Despite the assumption that any story starts with the anchor person, it is not always true that a story ends with the appearance of the next anchor person. Hsu et al. [9] argue that within an anchor person shot, there can be a possible presence of a story boundary, as the anchor person continues with the previous story and changes to the new story only towards the mid of the shot. It could also happen that an anchor person introduces stories without any supporting video clips. This gives rise to possible, intra-shot story boundaries. Hence, it is required to split and merge anchor person shots accordingly.

In order to detect such boundaries, we first extract two frames per second of all anchor person shots. As shown in Figure 1, we first split each frame into four regions, with R_1 and R_2 being the first and second quadrant, respectively. We assume that in these two quadrants, anchor person shots will contain the face of the anchor person and a graphic or video indicating the topic of the actual story. Consequently, the visual appearance of the anchor person quadrant will be similar over all frames of the anchor person shot, while the visual appearance of the other quadrant will change whenever a new story begins. Therefore, we determine the eigen difference E_1 and E_2 for both quadrants. If either E_1 or E_2

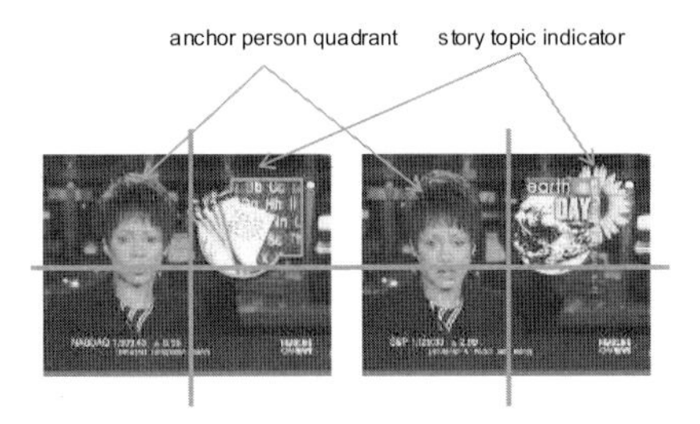

Fig. 1. Example of an intra-shot story boundary

is under a predefined threshold while the other value is above a threshold, we define this frame as a story boundary.

5 Results and Analysis

5.1 Boundary Detection Task

Following the TRECVid guidelines, we evaluate the segmentation performance of both approaches using the precision P_{seg} and recall R_{seg} metrics as defined by (6) and (7). Moreover, we compute the F_1 values using both metrics.

$$P_{seg} = \frac{|\text{determined boundaries}| - |\text{wrong boundaries}|}{|\text{determined boundaries}|} \tag{6}$$

$$R_{seg} = \frac{|\text{detected reference boundaries}|}{|\text{reference boundaries}|} \tag{7}$$

As outlined by Hsu et al. [9], boundaries are correctly detected when a determined boundary lies within five seconds of an actual reference story boundary. Otherwise, the boundary is considered to be wrong. Table 1 shows the independent metrics for both ABC and CNN videos as well as for the combination of both datasets. In the remainder of this section, we will denote these metrics as "baseline" results. As can be seen, the overall performance of both approaches for both datasets is similar.

The main weakness of the feature-based approach seems to be the actual detection of anchor person shots. Whenever an anchor person shot has been missed,

Table 1. Precision, Recall and F_1 measures for both approaches

	CNN			ABC			ABC & CNN		
	R_{seg}	P_{seg}	F_1	R_{seg}	P_{seg}	F_1	R_{seg}	P_{seg}	F_1
Feature-based	0.33	0.69	0.44	0.27	0.69	0.38	0.30	0.70	0.41
LDA	0.30	0.70	0.42	0.32	0.52	0.40	0.31	0.62	0.41
LDA (adapted)	0.31	0.71	0.43	0.38	0.58	0.45	0.34	0.65	0.44

a potential story boundary will be ignored, hence resulting in a drop in precision and recall. Moreover, stories that do not start with an anchor person shot will be missed as well, which is a drawback of our feature based approach. The potential drawback of the LDA approach is that if the test data is from a different domain and as a consequence there is a vocabulary mismatch, those words which did not appear during training will be dropped from the estimations. Therefore, a percentage of content words is lost. To alleviate this problem of vocabulary mismatch between Reuters data used for LDA training and the TRECVid transcripts used for evaluation, we propose to train the LDA model with combined Reuters and TRECVid development data. The results of this LDA adaptation is shown in the last row of Table 1. As can be seen, the performance of the LDA method improves, suggesting that a bigger in-domain adaptation data may have improved the performance even further. A quick analysis of the segmented output reveals that on most occasions the boundaries are estimated correctly or missed by a sentence or two. It is observed that the short segments are typically missed and it is because LDA requires some minimum amount of data for reliable estimation. For two example broadcasts, Figures 2 and 3 show the boundaries, in terms of word number in the transcript, identified by both approaches, as well as the actual boundaries. Figure 2 reveals that most of the time, the boundaries in the ABC broadcast which are identified by both approaches are correct. In the CNN broadcast shown in Figure 3, however, various boundaries have been missed. The reason for this miss is that CNN stories are rather short which is a problem for our text based approach. Both figures illustrate that the two approaches do not identify the same boundaries all the time. This complementarity can be exploited by combining the results of both approaches. It supports the general assumption [6] that a combination of different modalities, text and visual features in our case, can improve the accuracy of story segmentation approaches. Therefore, we fuse detected boundaries from both approaches using the "or" operator. Boundaries from both approaches that are within a one second time window distance from each other are merged to form one single boundary. This buffer will reduce the number of false positives. As Table 2 reveals, this fusion results in a huge improvement in both recall and F_1 measures in comparison to the baseline results shown in Table 1. Precision goes down slightly, indicating that the relative number of wrong boundaries has marginally increased.

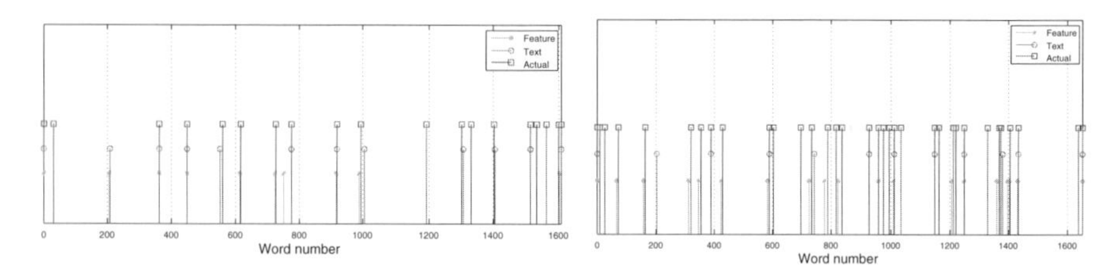

Fig. 2. Performance of both approaches in detecting story boundaries (ABC footage)

Fig. 3. Performance of both approaches in detecting story boundaries (CNN footage)

Table 2. Precision, Recall and F_1 measures

	CNN			ABC			ABC & CNN		
Combination	R_{seg}	P_{seg}	F_1	R_{seg}	P_{seg}	F_1	R_{seg}	P_{seg}	F_1
Feature + LDA	0.51	0.67	0.58	0.52	0.57	0.54	0.52	0.62	0.56
Feature + LDA (adapted)	0.52	0.67	0.58	0.56	0.60	0.58	0.54	0.64	0.58

In comparison with state-of-the-art approaches evaluated within TRECVid 2003, our simple approach ranks in the upper field of all submissions. In addition to the other approaches, however, our LDA based method can also be exploited to categorise detected stories. This categorisation is shown in the following section.

5.2 Story Categorisation Task

All the results previously published in the literature typically concentrated on the segmentation performance (either some error metric or time complexity). Though estimating the segment boundaries is important, if the segments can be identified by a topic (or topic distribution), this information can have profound impact in several other applications such as discourse analysis and information retrieval. LDA being a topic model is in a position to output this information along with the segment boundaries. In this section, we show an example output of the text segmentation phase. To save space, long sentences were terminated by "..." to show continuation beyond the printed words.

"*the holy grail of hiv research is to develop a safe and ... when you have an epidemic like this the way to put an ... a few potential vaccines are in human trials but final results are ... this year s conference represents a major change in emphasis and mood it is somber because hiv continues to be such an elusive foe george strait abc news geneva* **ESTIMATED BOUNDARY is CORRECT: TOPIC 43 has highest probability (0.39)** *now for news back home there is a new face in the ... she s a friend of the lewinsky family and she is telling ... she has testified before kenneth starr s grand jury she has also given an interview to newsweek magazine here is abc s karla davis abc news has confirmed that dale young a forty seven year old ... it is just the most unfortunate sense of timing tomorrow in another washington courtroom team clinton will argue presidential adviser bruce ... the one person not scheduled to be in court is monica lewinsky she and her new legal team still have not reached a deal ... karla davis abc news washington*" **ESTIMATED BOUNDARY is CORRECT: TOPIC 28 has highest probability (0.47)**

The top 10 words of TOPIC 43 and TOPIC 28, obtained after LDA training, are printed below for reference:

TOPIC 43: *'health' 'medical' 'mother' 'hospital' 'people' 'church' 'drug' 'heart' 'doctors' 'disease'*

TOPIC 28: *'pay' 'lead' 'type' 'sep' 'today' 'investigation' 'evidence' 'trial' 'case' 'lewinsky'*

From this example, we notice that the top topic associated with each segment is mostly relevant to the words present in that segment.

6 Conclusions

In this paper, we investigated and compared two approaches for segmenting TV news broadcast into stories: an LDA based method for text segmentation and a low-level feature-based approach for video segmentation. LDA has been previously demonstrated as an approach comprabale to the state-of-the-art approaches for the task of text segmentation [14]; it also outputs the topic distribution of segments. An analysis of the identified story boundaries revealed the complementarity of both the approaches, suggesting that they can be combined to form a more precise segmentation. Indeed, a simple fusion using an "or" operator already leads to significant improvement in performance. With respect to precision and recall, these results are above average in comparison with systems evaluated within TRECVid, outperformed by a few approaches only. While these best performing approaches are tailored to pre-defined rules, e.g., the appearance of cue phrases in the transcript, we base our approach on one assumption only, that is stories always start with an anchor person shot. Our approach is therefore a more general solution to tackle the television news segmentation task. Unlike other approaches, the proposed method computes topic distributions jointly with segmentation, thus allowing to collect information about the thematic content of each segment. This information can be used to keep track of recurring topics. In future work, we aim at including other multimedia domains, such as the audio layer of the news broadcast since TRECVid results support the effectiveness of considering this domain in a segmentation task.

Acknowledgments

This research was supported by the EC under contract FP6-027122-SALERO.

References

1. Ahlers, D.: News Consumption and the New Electronic Media. The Harvard International Journal of Press/Politics 11(1), 29–52 (2006)
2. Arlandis, J., Over, P., Kraaij, W.: Boundary error analysis and categorization in the TRECVID news story segmentation task. In: Leow, W.-K., Lew, M., Chua, T.-S., Ma, W.-Y., Chaisorn, L., Bakker, E.M. (eds.) CIVR 2005. LNCS, vol. 3568, pp. 103–112. Springer, Heidelberg (2005)
3. Blei, D.M., Ng, A.Y., Jordan, M.I.: Latent Dirichlet allocation. In: Dietterich, T.G., Becker, S., Ghahramani, Z. (eds.) Advances in Neural Information Processing Systems (NIPS), vol. 14, pp. 601–608. MIT Press, Cambridge (2002)
4. Chaisorn, L., Chua, T.-S., Lee, C.-H., Tian, Q.: A hierarchical approach to story segmentation of large broadcast news video corpus. In: ICME 2004, pp. 1095–1098 (2004)

5. Chang, S.-F., Manmatha, R., Chua, T.-S.: Combining Text and Audio-Visual Features in video Indexing. In: ICASSP 2005 – Proceedings of Acoustics, Speech, and Signal Processing Conference, March 2005, pp. 1005–1008 (2005)
6. Chua, T.-S., Chang, S.-F., Chaisorn, L., Hsu, W.: Story boundary detection in large broadcast news video archives: techniques, experience and trends. In: MM 2004, pp. 656–659. ACM, New York (2004)
7. Griffiths, T.L., Steyvers, M.: Finding scientific topics. Proceedings of the National Academy of Sciences 101(suppl. 1), 5228–5235 (2004)
8. Heidel, A., an Chang, H., shan Lee, L.: Language model adaptation using latent Dirichlet allocation and an efficient topic inference algorithm. In: Proceedings of EuroSpeech, Antwerp, Belgium (2007)
9. Hsu, S.-F., Chang, W.H., Huang, C.-W., Kennedy, L., Lin, C.-Y., Iyengar, G.: Discovery and Fusion of Salient Multi-modal Features towards News Story Segmentation. In: IS&T/SPIE Electronic Imaging, San Jose, CA (2004)
10. Hsu, W.H., Kennedy, L.S., Chang, S.-F., Franz, M., Smith, J.R.: Columbia-IBM News Video Story Segmentation in TRECVID 2004. In: TREC (2004)
11. Kolb, P.: DISCO: A Multilingual Database of Distributionally Similar Words. In: KONVENS 2008 (2008)
12. Kozima, H.: Text segmentation based on similarity between words. In: Meeting of the Association for Computational Linguistics, Ohio, USA, pp. 286–288 (1993)
13. Misra, H., Cappé, O., Yvon, F.: Using LDA to detect semantically incoherent documents. In: Proceedings of CoNLL, Manchester, UK, pp. 41–48 (2008)
14. Misra, H., Yvon, F., Jose, J.M., Cappé, O.: Text segmentation via topic modeling: An analytical study. In: Proceedings of CIKM, Hong Kong, China (2009)
15. O'Connor, N., Czirjek, C., Deasy, S., Marlow, S., Murphy, N., Smeaton, A.: News Story Segmentation in the Físchlár Video Indexing System. In: ICIP 2001 (2001)
16. Passonneau, R.J., Litman, D.J.: Discourse segmentation by human and automated means. Comput. Linguist. 23(1), 103–139 (1997)
17. Pickering, M.J., Wong, L., Rüger, S.: ANSES: Summarisation of news video. Image and Video Retrieval 2788, 481–486 (2003)
18. Smeaton, A.F., Kraaij, W., Over, P.: TRECVID 2003 – An Overview. In: TRECVid 2003 (2003)
19. Smeaton, A.F., Over, P., Kraaij, W.: Evaluation campaigns and TRECVid. In: MIR 2006, pp. 321–330. ACM Press, New York (2006)
20. Utiyama, M., Isahara, H.: A statistical model for domain-independent text segmentation. In: Meeting of the Association for Computational Linguistics, Bergen, Norway, pp. 491–498 (2001)

Query-Based Video Event Definition Using Rough Set Theory and High-Dimensional Representation

Kimiaki Shirahama[1], Chieri Sugihara[2], and Kuniaki Uehara[2]

[1] Graduate School of Economics, Kobe University,
2-1, Rokkodai, Nada, Kobe, 657-8501, Japan
[2] Graduate School of Engineering, Kobe University,
1-1, Rokkodai, Nada, Kobe, 657-8501, Japan
shirahama@econ.kobe-u.ac.jp, chieri@ai.cs.scitec.kobe-u.ac.jp,
uehara@kobe-u.ac.jp

Abstract. In videos, the same event can be taken by different camera techniques and in different situations. So, shots of the event contain significantly different features. In order to collectively retrieve such shots, we introduce a method which defines an event by using "rough set theory". Specifically, we extract subsets where shots of the event can be correctly discriminated from all other shots. And, we define the event as the union of subsets. But, to perform the above rough set theory, we need both positive and negative examples. Note that for any possible event, it is impossible to label a huge number of shots as positive or negative. Thus, we adopt a "partially supervised learning" approach where an event is defined from a small number of positive examples and a large number of unlabeled examples. In particular, from unlabeled examples, we collect negative examples based on their similarities to positive ones. Here, to appropriately calculate similarities, we use "subspace clustering" which finds clusters in different subspaces of the high-dimensional feature space. Experimental results on TRECVID 2008 video collection validate the effectiveness of our method.

1 Introduction

A video archive contains a large amount and various kinds of videos. When retrieving events in the video archive, the biggest problem is that users are interested in a great variety of events which we cannot assume in advance. Thus, the following two types of existing methods are not so efficient. First, "model-based event definition" methods like [1,2] define an event by a pre-constructed model. But, it is impossible to pre-construct models for all events which may interest users. Second, "concept-based event definition" methods like [5,6] define an event by combining concepts in a pre-defined vocabulary (i.e. ontology). But, all events cannot be necessarily represented by concepts in the vocabulary.

Compared to the above methods, "similarity-based event definition" methods like [3,4] define an event as a set of shots (or shot sequences) which contain

S. Boll et al. (Eds.): MMM 2010, LNCS 5916, pp. 358–369, 2010.

similar features to example shots. So, any event can be defined by providing example shots. But, meaningful events cannot be defined only by similarities. For example, in the event "a car moves", an example shot where a red car moves may be more similar to a shot where a person wears a red cloth than a shot where a white car moves. Like this, similarity-based event definition methods cannot avoid retrieving many irrelevant shots.

The main reason for the above problem is that similarity-based event definition methods only use "positive examples" where an interesting event is shown. So, they cannot discriminate relevant and irrelevant features for the event. To overcome this, we develop a "query-based event definition" method. Here, in addition to positive examples, we use "negative examples" where the event is not shown. And, by contrasting positive examples with negative ones, we can discriminate relevant and irrelevant features to define the event.

So far, only a few researchers have proposed query-based event definition methods [7,8]. Compared to these methods, our method has the following two important advantages. First, [7,8] select negative examples by using random sampling. But, this may cause that shots of an interesting event can be wrongly selected as negative examples. On the other hand, to achieve accurate negative example selection, we select negative examples based on their similarities to positive examples. Second, [7,8] define an event by using SVM classifiers. But, the performance of an SVM classifier depends heavily on the choice of a kernel and parameters. Since we cannot prepare validation data to determine the optimal kernel and parameters for each event, the SVM classifier is not suitable for query-based event definition. In contrast, our method needs no validation data and builds classifiers (classification rules) based only on the discernibility between positive and negative examples.

2 Issues in Query-Based Event Definition

We address the following three issues in query-based event definition:

Large variation of features in the same event: Depending on various factors such as camera techniques, object movements, locations and so on, shots of the same event contain significantly different features. Fig. 1 shows three shots of the event "a car moves in the town". Here, since *shot 1* takes a moving car in a tight shot, a large amount of motion is extracted from all parts of *shot 1*. Also, since *shot 2* takes a car moving in a suburban area, few edges are extracted from the upper part where the sky is shown. On the other hand, since *shot 3* takes a car moving in an urban area, many vertical edges are extracted from the upper and middle parts where buildings are shown. Thus, we assume that shots of the same event are distributed in different subsets (subspaces) in a feature space.

To find the above subsets, we use "rough set theory" which is a set-theoretic classification method based on indiscernibility relations among examples [13]. Specifically, in our case, rough set theory examines whether positive examples can be discernible from negative examples with respect to available features. Then, multiple classification rules called "decision rules" are extracted. Here,

Fig. 1. Example of shots which show the same event but contain different features

each decision rule certainly identifies a subset including positive examples. For example, for the event in Fig. 1, a subset including shots taken in suburban areas like *shot 2* is characterized by the decision rule consisting of many blue-colored pixels in the upper part, many gray-colored pixels in the bottom part and a large amount of motion in the middle part. Therefore, by unifying such subsets, we can cover the whole set of positive examples.

However, a traditional rough set theory can deal only with categorical data [13], while features extracted from each shot are represented in various formats, such as continuous value, histogram, time series and so on. Note that crucial errors inevitably occur by discretizing a feature into a small number of categorical values. That is, the same categorical value is frequently assigned to semantically different shots. Thus, by using the idea of the recently proposed rough set theory for continuous data [14], we propose a rough set theory which can deal with various formats of features. Specifically, we define the indiscernibility relation between positive and negative examples based on their similarity for each feature.

Difficulty of collecting negative examples: For an event, a user can prepare a small number of positive examples by searching previously watched videos or by using on-line video search engines like YouTube. But, the user cannot appropriately prepare negative examples. The reason is that a set of negative examples is just the complement of a set of positive examples. So, it is impossible to collect a variety of negative examples by manually checking a huge number of shots. In addition, collecting negative examples is frequently biased due to user's subjectivity. Hence, query-based event definition should be performed in the condition, where a small number of positive examples and a large number of unlabeled examples (i.e. shots in a video archive) are available.

As a most simple approach, one may think that all unlabeled examples can be regarded as negative, because almost all of them do not show an interesting event. But, in this approach, several shots of the event are wrongly regarded as negative. As a result, the recall of the event retrieval significantly decreases. Thus, it is crucial how to collect negative examples suitable for the event definition.

To this end, we use "partially supervised learning" which builds a classifier from positive and unlabeled examples [9,10,11]. That is, negative examples are selected from unlabeled examples. Regarding this, most of existing partially supervised learning methods select negative examples based on the statistical distribution of positive examples. For example, methods in [10] and [11] use SVM and Naive Bayse to estimate the distribution of positive examples, respectively. But, such methods work well only when a sufficient number of positive examples are available for estimating the true distribution. On the other hand, the method in [9] selects negative examples based on similarities between positive and unlabeled examples. And, it is validated as effective when only a small number of positive examples are available. Thus, we use the method in [9] in our query-based event definition.

High-dimensional feature space: In general, examples are represented by various features such as color, edge, motion and so on. That is, they are represented in a high-dimensional feature space. So, we have to consider the so-called "curse of dimensionality", where similarities among examples are nearly equal to each other, due to noises in many irrelevant dimensions (i.e. features) [12]. Therefore, in partially supervised learning, we have to distinguish relevant and irrelevant features to appropriately calculate similarities between positive and unlabeled examples.

Unlabeled examples show various events and are characterized by different features. Thus, we detect features specific to each unlabeled example, and calculate its similarities to positive examples only by using these features. To this end, we use "subspace clustering" which finds clusters of unlabeled examples in different subspaces of the high-dimensional feature space [12]. That is, each cluster is associated with a different subset of features. For example, a cluster of unlabeled examples where the sky is shown is characterized by color and edge features in the upper part. Also, a cluster of unlabeled examples where an object moves on the road is characterized by the motion in the middle part and the color in the bottom part. Like this, for each unlabeled example, we can detect the specific subset of features by finding the cluster including this example.

3 Query-Based Event Definition Method

First of all, we briefly explain our shot representation. In particular, we consider spatial locations of features to accurately characterize semantic contents. For example, many blue-colored pixels are extracted from the upper part of a shot where the sky is shown, while such pixels are extracted from the bottom part of a shot where the sea is shown. To utilize these spatial locations of features, we partition the keyframe of each shot into 30 regions as shown in Fig. 2 (a). And, from each region, we extract a color histogram, edge histogram and histogram of visual words. Thereby, we can obtain the total 90 features for the shot, as shown in Fig. 2 (b). This is a 90-dimensional shot representation where each dimension is represented by a histogram. In what follows, by using this shot representation,

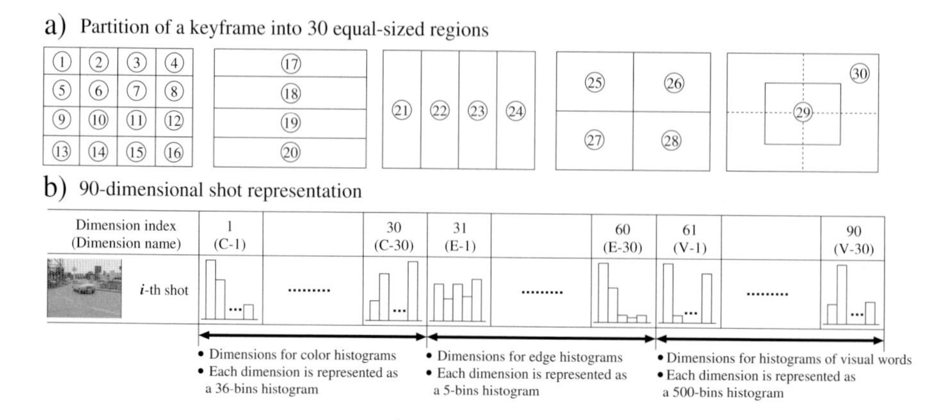

Fig. 2. Illustration of our 90 dimensional shot representation

we present our partially supervised learning method using subspace clustering and event definition method using rough set theory.

3.1 Partially Supervised Learning Method

Given positive examples for an event, we collect negative examples based on two steps shown in Table 1. In the first step called "reliable negative example selection", we select reliable negative examples as unlabeled examples which are unlikely to be positive. That is, reliable negative examples are completely dissimilar to positive examples. For example, for the event "a car moves in the town", reliable negative examples should include shots where the mountain is shown, shots where the beach is shown, and so on. But, only by using positive and reliable negative examples, we cannot estimate a meaningful boundary between positive and negative examples. So, in the second step called "negative example enlargement", we select "additional negative examples" as unlabeled examples which are more similar to positive examples than reliable negative examples. For the above example event, additional negative examples should include shots where a person walks in the mountain, shots where the town is taken from the air, and so on. In this way, we aim to select negative examples from which rough set theory can extract effective decision rules for retrieving the event.

The two-step framework in Table 1 is based on the method proposed in [9]. But, we extend it for the following two points. First, although the method in [9] targets text data where each feature is a word frequency, we extend it to deal with our shot representation where each feature is a histogram. Second, to overcome the curse of dimensionality, we use subspace clustering. Below, we mainly explain points extended from [9]. Note that, for the simplicity, we denote positive, negative, reliable negative, additional negative and unlabeled examples as "p-examples", "n-examples", "rn-examples", "an-examples" and "u-examples", respectively.

Table 1. Overview of our partially supervised learning method

Input: P (set of p-examples), U (set of u-examples),
Output: N (set of n-examples)
/* **Reliable negative example selection** */
1. Detect a set of positive features PF
2. Extract a set of rn-examples RN based on PF
/* **Negative example enlargement** */
3. $N = RN$
4. **while** true **do**
5. Cluster N into k clusters using subspace clustering PROCLUS
6. Extract a set of an-examples AN based on P and k clusters of N
7. **If** $|AN| == 0$, **then** break
8. $N = N \cup AN$
9. **end while**
10. return N

In the 1st line in Table 1, in order to accurately select rn-examples, we detect a set of "positive features" PF which are strongly associated with p-examples. For example, for the event "a car moves in the town", the color feature in the 20th region in Fig. 2 (a) may be selected as a positive feature, because it characterizes the gray-colored road shown in the bottom part. So, if a u-example do not match with such positive features, it should be regarded as an rn-example. To detect positive features, we measure the association of one feature with p-examples based on similarities among p-examples for this feature. Specifically, for each feature f, we group p-examples into clusters with similar histograms by using histogram intersection as a similarity measure. And, we count the number of p-examples $n_P(f)$ in the largest cluster. Also, by applying this largest cluster to u-examples, we count the number of u-examples $n_U(f)$ included in this cluster. Then, we evaluate how much f is associated with p-examples as follows:

$$H(f) = \frac{n_P(f)}{max_P} - \frac{n_U(f)}{max_U}, \tag{1}$$

where max_P and max_U are the largest value of $n_P(f)$ and the one of $n_U(f)$ among all features, respectively. They are used to normalize $n_P(f)$ and $n_U(f)$. Thus, $H(f)$ becomes larger if the largest cluster includes a larger number of p-examples and a smaller number of u-examples. So, if $H(f)$ is larger than the average of $H(j)$ for all features, we regard f as a positive feature. After that, in the 2nd line in Table 1, we use the same ranking-based approach to [9], in order to calculate the similarity between a u-example and the set of p-examples in terms of PF. And, if the similarity is smaller than the average similarity for all u-examples, we regard the u-example as an rn-example.

In the n-example enlargement from 3th to 9th line in Table 1, we select an-examples as u-examples which are significantly similar to rn-examples. Note that since rn-examples show a variety of events and contain different features, an an-example is not similar to all rn-examples. Let us recall the above example. Here, an-examples which are shots where a person walks in the mountain, are similar only to rn-examples which are shots where the mountain is shown. Considering such a variety of rn-examples, we firstly group rn-examples into clusters, and calculate the similarity between a u-example and each cluster. Particularly, since our shot representation is high-dimensional, we use subspace clustering "PROCLUS" proposed in [12]. PROCLUS iteratively improves k clusters where bad clusters such as the ones with few rn-examples are substituted with new clusters by randomly selecting cluster centers. In each cluster, if the average similarity among rn-examples for one feature is larger than the statistical expectation, this feature is associated with the cluster. As a result, the cluster represents a subspace consisting of its associated features.

Then, for i-th cluster of rn-examples, we compute the centroid C_i in the subspace consisting of associated features F_i. Also, we compute the centroid of p-examples C_P in the subspace consisting of positive features PF. Then, we examine whether a u-example u can be regarded as an an-example:

$$Sim_{F_i}(u, C_i) > \mu_i, \quad (2)$$

$$Sim_{F_i}(u, C_i) - Sim_{PF}(u, C_P) > \gamma_i, \quad (3)$$

where $Sim_{F_i}(u, C_i)$ is the similarity between u and C_i in terms of F_i. Specifically, we calculate $Sim_{F_i}(u, C_i)$ as the average of similarities for all features in F_i, where the similarity for each feature is calculated by histogram intersection. Similarly, $Sim_{PF}(u, C_P)$ is calculated as the similarity between u and C_P in terms of PF. Also, μ_i and γ_i are respectively average values of equations (2) and (3) for rn-examples in i-th cluster. Thus, u is selected as an an-example if it is not only sufficiently similar to i-th cluster, but also much more similar to i-th cluster than to the set of p-examples. Finally, as shown in the 7th and 8th lines, we consider already selected an-examples as rn-examples, and iterate the above n-example enlargement step until no an-example is selected.

3.2 Event Definition Based on Rough Set Theory

Given p-examples and n-examples, by using rough set theory, we aim to extract decision rules for discriminating shots of an event from all other shots. Let p_i and n_j be i-th p-example ($1 \leq i \leq M$) and j-th n-example ($1 \leq j \leq N$), respectively. And, p_i^k and n_j^k represent p_i's and n_j's histograms in k-th feature ($1 \leq k \leq 90$), respectively. Below, we extend the traditional rough set theory for categorical features [13] to the one for features represented by histograms.

First, we represent p-examples and n-examples in the form of table, as shown in Fig. 3 (a). This table is called "decision table". In Fig. 3 (a), two p-examples p_1 and p_2 and two n-examples n_1 and n_2 are given for the event "a car moves in the town". Each row represents an example. The rightmost column indicates whether an example is positive ("P") or negative ("N"), while the other columns indicate features represented by histograms. Like this, the decision table provides available information for discriminating between p-examples and n-examples.

Then, for each pair of p_i and n_j, we extract "discriminative features" which are useful for discriminating them. For example, by comparing p_1 to n_1 in Fig. 3 (a),

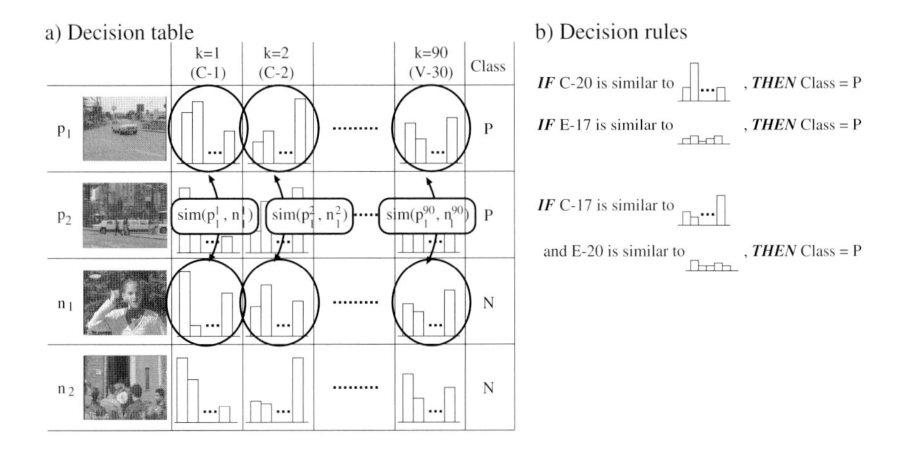

Fig. 3. Example of a decision table and decision rules for "a car moves in the town"

we can extract the color feature in 17th region as discriminative. It is because the sky is shown in 17th region in p_1, which is characterized by many blue-colored pixels. On the other hand, trees are shown in 17th region in n_1, which is characterized by many green-colored pixels. Thus, by using the color feature in 17th region, we can discriminate between p_1 and n_1.

To extract such discriminative features, we calculate the similarity $sim(p_i^k, n_j^k)$ between p_i and n_j for k-th feature. In particular, we use histogram intersection as a similarity measure. Fig. 3 (a) illustrates the process of extracting discriminative features between p_1 and n_1. In this way, we collect the following set of discriminative features $f_{i,j}$ between p_i and n_j:

$$f_{i,j} = \{k \mid sim(p_i^k, n_j^k) < \beta^k\}, \tag{4}$$

where β^k is a pre-defined threshold for k-th feature. $f_{i,j}$ means that when at least one feature in $f_{i,j}$ is used, p_i can be discriminated from n_j.

Next, we extract sets of features which are needed to discriminate p_i from all n-examples. This is achieved by simultaneously using at least one feature in $f_{i,j}$ for all n-examples. That is, we take a conjunction of $\vee f_{i,j}$ as follows:

$$df_i = \wedge\{\vee f_{i,j}|\ 1 \leq j \leq N\} \tag{5}$$

Suppose that the set of discriminative features between p_1 and n_1 is $f_{1,1} = \{C\text{-}17, C\text{-}20, E\text{-}17\}$ and the one between p_1 and n_2 is $f_{2,1} = \{C\text{-}20, E\text{-}17, E\text{-}20\}$. Here, for the simplicity, we use the notation which consists of a capital letter representing the feature name and a hyphenated digit representing the region. For example, *C-17* represents the color feature in 17th region (see Fig. 2 (b)). By using this notation, we compute $df_1 = (C\text{-}17 \vee C\text{-}20 \vee E\text{-}17) \wedge (C\text{-}20 \vee E\text{-}17 \vee E\text{-}20)$. And, df_1 is simplified into $df_1^* = (C\text{-}20) \vee (E\text{-}17) \vee (C\text{-}17 \wedge E\text{-}20)$ [1]. As a result, we

[1] This simplification is achieved by using the distributive law $A \wedge (B \vee C) = (A \wedge B) \vee (A \wedge C)$ and the inclusion relation $A \vee (A \wedge B) = A$.

can know that p_1 can be discriminated from all n-examples n_1 and n_2, by using *C-20*, *E-17* or the set of *C-17* and *E-20*. Each of these represents a "reduct" which is a minimal set of features needed to discriminate p_1 from all n-examples

From each reduct, we construct a decision rule in the form of *IF-THEN* rule. For example, from the above three reducts, we can construct decision rules shown in Fig. 3 (b). Here, the conditional part of each decision rule is obtained by describing a reduct with p_1's histograms and similarities. That is, such a decision rule indicates a subset where p_1 can be correctly identified. Then, we gather decision rules extracted for all p-examples. And, we merge similar decision rules into one decision rule, which indicates a subset where multiple p-examples can be correctly identified. Finally, we retrieve shots which match with a larger number of decision rules than a pre-defined threshold.

4 Experimental Results

We test our query-based event definition method on TRECVID 2008 video archive, containing 71,872 shots in 438 videos [15]. We evaluate the performance of our method for the following three events, *Event 1:* a person opens a door, *Event 2:* a person talks on the street and *Event 3:* a car moves in the town.

Table 2 summarizes the result for the above three events. As shown in the second column, our partially supervised learning method *PSL* is compared to two different n-example selection methods, *Manual* and *Random*. In *Manual*, n-examples are manually selected while they are randomly selected in *Random*. In order to achieve the fair comparison among *Manual*, *Random* and *PSL*, each event is defined by using the same p-examples as in the third column. Additionally, the fourth column shows that in both of *Random* and *PSL*, we select the same number of n-examples (i.e. 50). Also, the fifth column presents precisions calculated from 300 shots retrieved by our method, where numbers of relevant shots are shown in parentheses. And, as seen from the rightmost column, we compare precisions by our method to the ones by SVM. SVM has resulted in top performances in TRECVID for the past few years [15]. Here, in both of our method and SVM, we use the same p-examples and manually selected n-examples.

As can be seen from Table 2, precisions depend significantly on negative examples. In particular, except for *Event 1*, precisions by *Manual* are much larger than the ones by *Random* and *PSL*. Also, for *Event 3*, the precision by *PSL* is much larger than the one by *Random*. On the other hand, for *Event 1*, the precision by *PSL* is smaller than the one by *Random*. From this result, we find that the performance of *PSL* depends on the number of true positive shots. Specifically, the number of true positive shots for *Event 3* is relatively large while the one for *Event 1* is very small. And, for *Event 3*, *PSL* can accurately select n-examples by analyzing features in shots. On the other hand, *Random* wrongly selects some true positive shots as negative, since it does not analyze any feature in shots. But, for *Event 1*, the n-example enlargement in *PSL* does not reach to n-examples which are close to p-examples. Compared to this, due to

Table 2. Results of our query-based event definition method

	n-example selection	# of p-examples	# of n-examples	P@300 (# of rel.)	P@300 by SVM
Event 1	*Manual*	9	16	0.070 (21)	0.060 (18)
	Random	9	50	0.087 (26)	–
	PSL	9	50	0.070 (21)	–
Event 2	*Manual*	11	16	0.087 (26)	0.060 (18)
	Random	11	50	0.050 (15)	–
	PSL	11	50	0.050 (15)	–
Event 3	*Manual*	9	14	0.217 (65)	0.223 (67)
	Random	9	50	0.127 (38)	–
	PSL	9	50	0.170 (51)	–

the randomness and the small number of true positive shots, *Random* can select n-examples close to p-examples without selecting any true positive shots as negative. Thus, in order to handle a case where the number of true positive shots is very small, it may be effective to extend *PSL* by incorporating the mechanism involving the randomness, such as genetic algorithm.

From the fifth and sixth columns in Table 2, we can see that the overall performance of our method is better than that of SVM. Especially, this validates the effectiveness of rough set theory, because both of our method and SVM are tested on the same condition except for classification algorithms. We find a significant difference between the retrieval result by our method and the one by SVM. Fig. 4 shows three retrieved shots for *Event 3* by using either our method or SVM. As seen from Fig. 4 (a), our method can retrieve shots characterized by different shot sizes, such as tight shots like *Shot 1*, medium shots like *Shot 2* and long shots like *shot 3*. Also, the sky is shown in *Shot 1* and *Shot 3*, while buildings are displayed in large regions of *Shot 2* and *Shot 3*. Like this, by using rough set theory, we can cover a large variation of features in an event.

Compared to our method, SVM tends to retrieve shots which are similar to p-examples only for some features. For example, Fig. 4 (b) shows three retrieved shots by SVM, where *Shot 4* is true positive while *Shot 5* and *Shot 6* are false positive. Here, *Shot 5* and *Shot 6* are retrieved only because they have features characterizing roads, that is, many gray-colored pixels in bottom parts. On the other hand, since our method examines whether each shot matches with many decision rules or not, the above kind of shots are not retrieved.

Finally, we closely examine decision rules extracted by rough set theory. Fig. 5 shows the relation between one p-example *Pos 1* for *Event 3* and features involved in decision rules. In Fig. 5, as one feature is involved in a larger number of decision rules, the region corresponding to this feature becomes darker. That is, we can see that many decision rules involve the edge histogram in 19-th region (see Fig. 2 (a)). This characterizes the road where few edges are extracted. Also, many decision rules involve the histogram of visual words in 18-th region, which characterizes cars. In addition, many rules involve histograms of visual words in

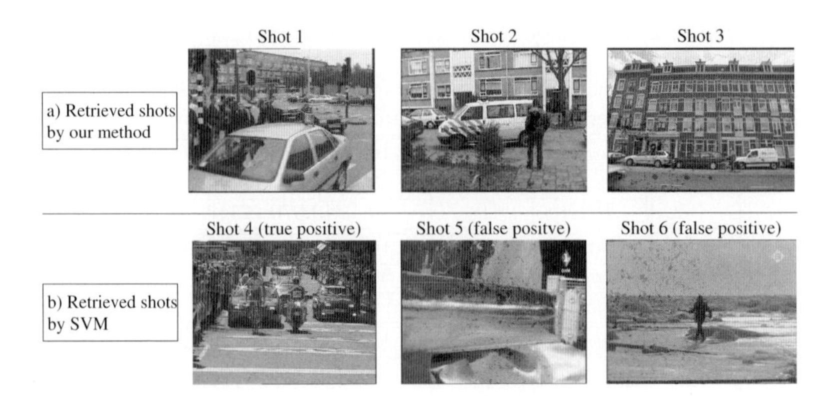

Fig. 4. Three retrieved shots for *Event 3* by our method and SVM

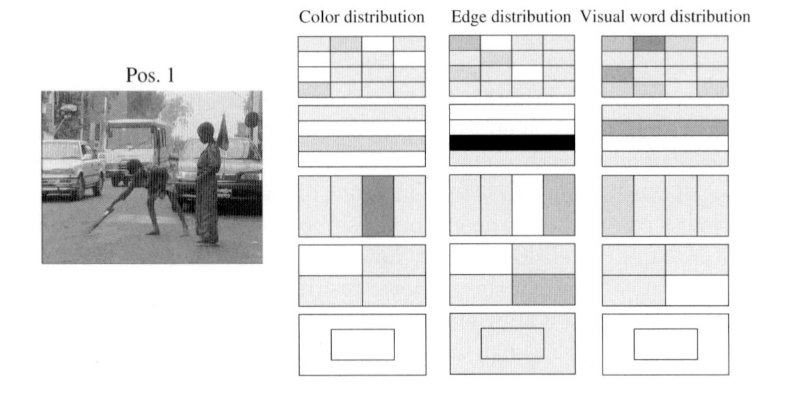

Fig. 5. Illustration of features involved in decision rules for *Event 3*

1-st and 2-nd regions, which characterize street trees and buildings. Like this, rough set theory can extract decision rules which characterize essential objects for the event.

5 Conclusion and Future Works

In this paper, in order to retrieve any interesting event in a video archive, we introduced a query-based event definition method which defines the event from positive and negative examples. To implement this, we address the following three issues. First, considering the difficulty of manually collecting negative examples, we select negative examples by using partially supervised learning. Second, to appropriately calculate similarities in a high-dimensional feature space, we use subspace clustering which finds subspaces characterized by different sets of features. Third, considering that shots of the same event contain significantly different features, we extract multiple definitions (decision rules) of the same event by using rough set theory. Experimental results on TRECVID 2008 video collection validate the effectiveness of our method.

Acknowledgments. This research is supported in part by Strategic Information and Communications R&D Promotion Programme (SCOPE) by the Ministry of Internal Affairs and Communications, Japan.

References

1. Haering, N., Qian, R., Sezan, M.: A Semantic Event-Detection Approach and Its Application to Detecting Hunts in Wildlife Video. IEEE Transactions on Circuits and Systems for Video Technology 10(6), 857–868 (2000)
2. Snoek, C., Worring, M.: Multimedia Event-based Video Indexing Using Time Intervals. IEEE Transactions on Multimedia 7(4), 638–647 (2005)
3. Peng, Y., Ngo, C.: EMD-Based Video Clip Retrieval by Many-to-Many Matching. In: Leow, W.-K., Lew, M., Chua, T.-S., Ma, W.-Y., Chaisorn, L., Bakker, E.M. (eds.) CIVR 2005. LNCS, vol. 3568, pp. 71–81. Springer, Heidelberg (2005)
4. Kashino, K., Kurozumi, T., Murase, H.: A Quick Search Method for Audio and Video Signals based on Histogram Pruning. IEEE Transactions on Multimedia 5(3), 348–357 (2003)
5. Naphade, M., et al.: Large-Scale Concept Ontology for Multimedia. IEEE Multimedia 13(3), 86–91 (2006)
6. Ebadollahi, S., Xie, L., Chang, S., Smith, J.: Visual Event Detection Using Multidimensional Concept Dynamics. In: Proc. of ICME 2006, pp. 881–884 (2006)
7. Natsev, A., Naphade, M., Tešić, J.: Learning the Semantics of Multimedia Queries and Concepts from a Small Number of Examples. In: Proc. of ACM MM 2005, pp. 598–607 (2005)
8. Tešić, J., Natsev, A., Smith, J.: Cluster-based Data Modeling for Semantic Video Search. In: Proc. of ACM MM 2007, pp. 595–602 (2007)
9. Fung, G., Yu, J., Ku, H., Yu, P.: Text Classification without Negative Examples Revisit. IEEE Transactions on Knowledge and Data Engineering 18(1), 6–20 (2006)
10. Yu, H., Han, J., Chang, K.: PEBL: Web Page Classification without Negative Examples. IEEE Transactions on Knowledge and Data Engineering 16(1), 70–81 (2004)
11. Liu, B., Dai, Y., Li, X., Lee, W., Yu, P.: Building Text Classifiers Using Positive And Unlabeled Examples. In: Proc. of ICDM 2003, pp. 179–188 (2003)
12. Aggarwal, C., Procopiuc, C., Wolf, J., Yu, P., Park, J.: Fast Algorithms for Projected Clustering. In: Proc. of SIGMOD 1999, pp. 61–72 (1999)
13. Komorowski, J., Pawlak, Z., Polkowski, L., Skowron, A.: Rough Sets: A Tutorial. In: Pal, S., Skowron, A. (eds.) Rough-Fuzzy Hybridization: A New Trend in Decision Making, pp. 3–98. Springer, Heidelberg (1999)
14. Leung, Y., Fischer, M., Wu, W., Mi, J.: A Rough Set Approach For the Discovery of Classification Rules in Interval-valued Information Systems. International Journal of Approximate Reasoning 47(2), 233–246 (2008)
15. Smeaton, A., Over, P., Kraaij, W.: Evaluation campaigns and TRECVid. In: Proc. of MIR 2006, pp. 321–330 (2006)

Story-Based Retrieval by Learning and Measuring the Concept-Based and Content-Based Similarity

Yuxin Peng and Jianguo Xiao

Institute of Computer Science and Technology
Peking University
pengyuxin@pku.edu.cn

Abstract. This paper proposes a new idea and approach for the story-based news video retrieval, i.e. clip-based retrieval. Generally speaking, clip-based retrieval can be divided into two phases: feature representation and similarity ranking. The existing methods only adopt the content-based features and pairwise similarity measure for clip-based retrieval. The main deficiencies are: (1) In feature representation, the concept-based features is still not used to represent the content of video clip; (2) In similarity ranking, the learning-based method is not considered to rank the similar clips with the query. To address the above issues, in this paper, on one hand, we consider jointly the concept-based and content-based features to represent adequately the news story; on the other hand, we consider jointly the learning classifier and pairwise similarity measure to rank effectively the similar stories with the query. Both are the main novelty of this paper. The model construction of learning classifier for story-based retrieval is our focus, which is constructed as follows: given one query story, we can use its all keyframes as the set of positive examples of its topic, and the retrieval data set in which most of the keyframes are irrelevant to the topic as the candidates of negative examples. The multi-bag SVM is employed to compute the score of all keyframes in the data set, and then the stories in the data set are ranked according to the average score of their keyframes, which reflects their similarity with the query story. We compare and evaluate the performance of our approach on 1334 stories from TRECVID 2005 benchmark, and the results show our approach can achieve superior performance.

Keywords: Story-based Retrieval, Learning, Concept-based Features.

1 Introduction

Clip-based video retrieval plays a major role in video analysis and retrieval. In broad, we can categorize the query-by-example video retrieval techniques into shot-based retrieval and clip-based retrieval. Compared to shot-based retrieval, clip-based retrieval is relatively more meaningful since a clip usually conveys a semantic event which consists of more meaningful and concise information. In news video, clip is normally referred to as "news story", which is generally a segment of news broadcast with a coherent news focus containing some shots. A news story, like clip, usually

S. Boll et al. (Eds.): MMM 2010, LNCS 5916, pp. 370–378, 2010.

conveys one meaningful event. In this paper, we focus on the story-based retrieval in the news video.

Existing approaches on clip-based retrieval (story-based retrieval) can be divided into two categories: some researches focus on the rapid identification of similar clips [1-2, 13], while the others focus on the similarity ranking of video clips [3-8]. In [1], fast algorithms are proposed by deriving signatures to represent the clip contents. The signatures are basically the summaries or global statistics of low-level features in clips. The similarity of clips depends on the distance between signatures. The global signatures are suitable for matching clips with almost identical content but little variation, such as commercial retrieval. In [2], an index structure based on multi-resolution KD-tree is proposed to further speed up clip retrieval. In [3-8], clip-based retrieval is built upon the shot-based retrieval. Besides relying on shot similarity, clip similarity is also dependent on the inter-relationship such as the granularity, temporal order and interference among shots. In [4], shots in two clips are matched by preserving their temporal order, which may not be appropriate since shots in different clips tend to appear in various orders due to the editing effects. Some sophisticated approaches for clip-based retrieval are proposed in [6, 7] where different factors including granularity, temporal order and interference are taken into account. In [6], a cluster-based algorithm is employed to match the similar shots. In [7], the maximum matching is employed to filter irrelevant video clips, while the optimal matching is utilized to rank the similarity of clips. Both algorithms compute the clip similarity by guaranteeing the one-to-one mapping among video shots. In [8], EMD is further proposed to measure the clip similarity by many-to-many matching among video shots. In addition, Wu *et al* [9] use the visual duplicate and speech transcript to measure the similarity of story for the news novelty detection. Hsu *et al* [14] adopt the visual duplicates and semantic concepts to study the topic tracking in news video.

In summary, the existing methods only adopt the content-based features such as color feature [3, 6, 7, 8], texture feature [6], motion feature [7], and text feature [9], and only adopt the pairwise similarity measure [1-8] for clip-based retrieval. The main deficiencies are: (1) In feature representation, the concept-based feature is not used to represent the content of clip; (2) In similarity ranking, the learning-based method is not considered to rank the similar clips with the query. Both (1) and (2) can be the major factors to affect the performance of clip-based retrieval. To address the above issues, in this paper, on one hand, we consider jointly the concept-based and content-based features to represent adequately the news story; on the other hand, we consider jointly the learning classifier and pairwise similarity measure to rank effectively the similar stories with the query. Both are the main novelty of this paper.

The model construction of learning classifier for story-based retrieval is our focus, and our motivation is as follows: News stories from various sources and channels conveying the same topic usually share some common and similar content, which means the intra-story and inter-story have the common and similar shots, such as the focus of the topic like the people, object or scene. Figure 1 illustrates one example, which shows three different reported stories of one topic in TRECID 2005 data (each row represents one story, and each keyframe image represents a subshot). Since the topic is "Hu Jintao visited South America", President Hu Jintao with similar scene will repeatedly occur in many shots of intra-story and inter-story, which is captured as the focus. Under this situation, given one query story, we can use its all keyframes as the set of positive examples of its topic, and the retrieval data set with most of the keyframes irrelevant with the topic as the candidates of negative examples. In this way, we can use a constructed training data set to train the learning classifier, and use the classifier to decide the relevance score of all stories in the data set with the query. That is, we first use the multi-bag SVM [12] to compute the score of each keyframe in the data set, and then the stories in the data set are ranked according to the average score of their all keyframes, which reflects their similarity with the query story. Note that, unlike the query-by-example image and shot retrieval, which generally has only one query image or shot, query-by-example story retrieval uses a story as query, which is composed of some shots and keyframes, and directly provides a set of positive examples on the query topic. So we can directly train the classifier and apply it for the story-based retrieval. We compare and evaluate the performance of our approach on 1334 stories from TRECVID 2005 benchmark, and the results show our approach can achieve superior performance.

Fig. 1. Three stories with one topic "Hu Jintao visited south America"

2 Our Approach

2.1 Feature Representation

In the feature representation, the novelty of our approach mainly lies on: we employ jointly the content-based and concept-based features to represent adequately the news story for story-based retrieval. There are many works on the concept-based and content-based features, however, they are not still adopted jointly for story-based retrieval. In this section, we use the baseline features on Columbia374 [10, 11].

2.1.1 Content-Based Features

We adopt the visual-based features including edge direction histogram (EDH), Gabor (GBR), and grid color moment (GCM) [10, 11] to represent the keyframes of each story, which is described as follows:

- GCM: Divides the keyframe images into 5x5 grids, calculates the mean, standard deviation, and the third root of the skewness of each color channel in LUV color space, resulting in a 225-dimensional color feature.
- EDH: A histogram-based feature with 73 bins: 72 bins for edge direction and 1 bin for non-edge points. EDH detects edge points with a Canny filter and calculates the gradient and direction of each edge point with a Sobel operator, resulting in a 73-dimensional edge feature.
- GBR: uses the mean and standard deviations of the output of a two-dimensional Gabor filter by using the combinations of four scales and six orientations, resulting in a 48-dimensional texture feature.

2.1.2 Concept-Based Features

Besides the low-level visual features in Section 2.1.1, we also employ the concept-based feature for story-based retrieval. Concept-based feature is based on the prediction scores of Columbia374 concept detectors [10, 11]. The three features are described as follows:

- GCM_P374: is a 374-dimensional feature consisting of the scores predicted by the models of the Columbia374 concepts, which are trained on TRECVID 2005 training keyframes with GCM feature.
- EDH_P374: Similar with GCM_P374, EDH_P374 is based on the prediction scores by the models trained on EDH.
- GBR_P374: Similar with GCM_P374, GBR_P374 is based on the prediction scores by the models trained on GBR.

2.2 Similarity Ranking

For the ease of understanding, we use the following notations in this section:

- Let $X = \{x_1, x_2, ..., x_m\}$ be a query story with m keyframes, and x_i represents a keyframe in X.
- Let $Y = \{y_1, y_2, ..., y_n\}$ be a data set with n keyframes, and y_j represents a keyframe in Y.

2.2.1 Learning Classifier

In our approach, the learning model is constructed from X and Y, where $X = \{x_1, x_2, ..., x_m\}$ is used as the set of positive samples, and $Y = \{y_1, y_2, ..., y_n\}$ is used as the candidates of negative samples. Note that the data set Y generally include a few positive samples of the relevant stories with the query X, however, its number is far less than the total number of samples in Y, since most of stories and keyframes in Y are irrelevant to the query. The novelty of our approach mainly lies in the model construction of learning classifier for story-based retrieval, which is constructed as follows:

1. In training phase, for each query, the multi-bag SVM [12] are trained, which can use fully the training data of each SVM model to improve the performance of prediction. For each SVM model, all keyframes in query story $X = \{x_1, x_2, ..., x_m\}$

are used as positive samples. We randomly select (without repetition) a subset of keyframes which is ten times as the size of positive samples from $Y = \{y_1, y_2, ..., y_n\}$ as the negative samples. This is reasonable since only a few stories in the dataset are relevant with the query, and in consequence the ratio of positive(relevant) keyframes in the data set Y is very low, which assures that the positive samples in Y are hardly selected as negative samples. The experimental results show the approach can achieve good performance in Table 1 of Section 3.

2. In the test phase, the 10 SVM models are used to predict the keyframes in the data set Y. Each keyframe y_j in Y get a score from each model, and the similarity score of each y_j is the average value of the output scores on 10 SVM models, where 10 SVM models are adopted for the average fusion on the different negative sample sets. Based on the final score $score(y_j)$ of each keyframe y_j, the similarity score of each story Y_k is defined as follows:

$$score(Y_k) = \frac{1}{|Y_k|} \sum_{y_j \in Y_k} score(y_j) \tag{1}$$

where y_j is a keyframe of story Y_k in the data set Y, and $|Y_k|$ is the number of keyframe of Y_k.

2.2.2 Similarity Measure

In this paper, we focus on the study of ranking capability on story-based retrieval, which is capable of ranking the stories in the data set according to their similarity values with the query, just like the learning-based method for story ranking. So we adopt our optimal matching (OM) algorithm in [7] to measure the similarity between two stories, which can maximize the total weight of matching under the constraint of one-to-one keyframes mapping. Given a query story X and a story Y_k in Y, one complete weighted bipartite graph is constructed based on the similarity of keyframes (vertex) between X and Y_k, where we adopt the cosine similarity to compute the similarity value between two keyframes with GCM, EDH and GBR features defined in Section 2.1.1. The output of OM is a weighted bipartite graph G_{OM} where one keyframe in X can match with at most one keyframe in Y_k and vice versa. The similarity of X and Y_k is assessed based on the total weight in G_{OM} as follows:

$$Similarity(X, Y_k) = \frac{\sum Sim(x_i, y_j)}{|X|} \tag{2}$$

where $Sim(x_i, y_j)$ represents the weight of edge (x_i, y_j) in G_{OM}. The similarity is normalized by the number of keyframes in the query story X.

3 Experiments

We adopt the same experimental data set in [9], which contains 1334 stories selected from the TRECVID 2005 cross-lingual news video corpus. The TREC Video Retrieval Evaluation (TRECVID) is an open and metrics-based evaluation on video analysis and retrieval, and the TRECVID 2005 news videos are used from five different sources (CCTV4, NTDTV, CNN, NBC and MSNBC) with Chinese and English, which sum up to about 127 hours. The 1334 stories are segmented by the story boundary detector from CMU Informedia[9]. The length of stories ranges from 10 seconds to 1700 seconds, with an average of 146.2 seconds. The number of keyframes (subshots) in each story varies from 1 to 404, with an average of 19.4 keyframes.

In the experiment, the stories belonging to the same topic are labeled with the relevance, which is used for the experimental evaluation. We adopt the same ground truth of story topic in [9], which originally uses it for the story novelty detection. Under the guidance of TDT (topic detection and tracking), 33 topics are manually labeled and annotated on the 1334 stories. Among the 33 topics, 6 topics only have Chinese news videos, 10 topics only appear in the English channels, while the remaining 17 topics have stories reported in both Chinese and English(refer to [9] for details). The number of stories in each topic ranges from 5 to 203, with an average of about 43.5 stories per topic. In addition, the stories belonging to one or several topics, and the number of topics annotated on each story varies from 1 to 3, with an average of about 1.1 topics per story. We compare the following methods for the evaluation.

- I Learning classifier using concept-based feature: Employ the learning classifier in Section 2.2.1 and the concept-based feature in Section 2.1.2 for the story-based retrieval;
- II Learning classifier using content-based feature: Employ the learning classifier in Section 2.2.1 and the content-based feature in Section 2.1.1 for the story-based retrieval;
- III Optimal matching using concept-based feature: Adopt optimal matching method in [7], which is briefly described in Section 2.2.2, and the concept-based features in Section 2.1.2 for the story-based retrieval;
- IV Optimal matching using content-based feature: Adopt optimal matching method in [7], and the content-based features in Section 2.1.1 for the story-based retrieval.

In methods I and II, we use the same SVM classifier of RBF kernel with the default parameters, and the *libSVM* tool for the experiment. We adopt average precision(*AP*) to evaluate the returned story list for each query story. For each relevant story in the returned list, we calculate a precision value for the subset of stories before (and including) it. Average precision is the average of precision values calculated at each of the relevant stories. To get a fair and comprehensive evaluation, we use each of all 1334 stories as query, and calculate the mean average precision (*MAP*), which is the mean of the *AP* values over all 1334 stories as an overall experimental result.

Table 1 illustrates the detailed comparison of the four methods. The following conclusions can be obtained:

- In similarity ranking, our proposed learning-based approach, on both concept-based features and content-based features, achieve the better performance than the OM-based similarity measure [7], which shows the learning-based approach is effective for story-based retrieval and ranking. Compared with the OM-based similarity measure, our learning-based approach obtains about 38.9% relative improvement (0.207 versus 0.149) on the average fusion of concept-based features, and 9.6% relative improvement (0.171 versus 0.156) on the average fusion of content-based features respectively.
- In feature representation, compared with the content-based feature, our concept-based feature achieve about 21.1% relative improvement (0.207 versus 0.171) on learning-based method, and has 4.7% relative decrease (0.149 versus 0.156) on the OM-based similarity measure. Totally, our concept-based feature is effective for story-based retrieval and ranking on learning-based method.
- In summary, our method Ⅰ (the learning-based method using concept-based features) achieves the best performance of all four methods. Note that the learning-based method and concept-based feature are ignored by the existing methods for story-based retrieval and ranking, and we consider the two major factors and propose the new approach, which achieves good performance. Our method Ⅰ obtains about 21.1%(0.207 versus 0.171), 38.9%(0.207 versus 0.149) and 32.7%(0.207 versus 0.156) relative improvements compared with the method Ⅱ-Ⅳ respectively.

We further fuse the final results by the average fusion (the average results in Table 1, including 0.207, 0.171, 0.149 and 0.156 in methods Ⅰ-Ⅳ), which are shown in Table 2. We can see that methods Ⅴ-Ⅶ do not achieve better performances compared with their separate results, for example, the result of average fusion is only 0.196 in method Ⅴ, 0.156 in method Ⅵ and 0.177 in method Ⅶ. The reasons may be that the mutual supplement of their separate results is not good. We also vary the weights of their separate results from 0 to 1 with 0.1 as step for the linear fusion scheme. However the fusion results can not still be better than the method Ⅰ on 0.207. How to effectively fuse the different results will be one of our future works.

Table 1. The performance comparison of four methods

Feature	Learning Classifier		Similarity Measure	
	Ⅰ Concept-based	Ⅱ Content-based	Ⅲ Concept-based	Ⅳ Content-based [7]
GCM	0.195	0.166	0.143	0.158
EDH	0.186	0.151	0.138	0.143
GBR	0.188	0.152	0.151	0.143
Average	0.207	0.171	0.149	0.156

Table 2. The results on average fusion of four methods

Methods	MAP
Ⅴ Average fusion of methods Ⅰ and Ⅱ	0.196
Ⅵ Average fusion of methods Ⅲ and Ⅳ	0.156
Ⅶ Average fusion of Ⅴ and Ⅵ	0.177

Table 3. The average time costs of four methods (in seconds)

Feature	Learning Classifier				Similarity Measure	
	Concept-based		Content-based		Concept-based	Content-based [7]
	Train	Test	Train	Test		
GCM(225-d)	2.5	23.7	1.5	18.0	1.0	0.6
EDH(73-d)	3.1	18.9	0.7	7.7	0.9	0.4
GBR(48-d)	2.7	25.5	0.4	4.3	0.9	0.4
Average time	2.8	22.7	0.9	10.0	0.9	0.5

In addition, we give the experimental comparison of time cost on the four methods in Table 3. The experiment is carried out on a machine with Intel quad-core 3GHz CPU and 16GB memory. For the ease of comparison, only one CPU core is used for each method. The time cost of each method is the average of time on all 1334 stories. Note that the learning-based method has the training and the test (retrieval) time, while the similarity measure has only the retrieval time. As shown in Table 3, we can see: (1) The learning-based method has the higher time cost compared with the OM-based similarity measure. (2) The concept-based feature has the higher time cost compared with the content-based features, which is reasonable since the concept-based feature has higher dimension (374 dimensions). Note that the number of feature dimension is a major factor to affect the time cost. For example, the GBR feature (48-dimensions) achieves the lowest time cost on content-based features in the OM-based similarity measure (0.4s) and the learning method (4.7s in total). In addition, the training time is lower than the test time in the learning-based method since the size of training sample set is much smaller than that of the test set.

4 Conclusion

We have proposed a new approach for story-based retrieval. On one hand, we employ the concept-based and content-based feature to represent adequately the news story; on the other hand, we employ the multi-bag SVM classifier and OM-based similarity measure to rank effectively the stories. We compare and evaluate the performance of our approach on 1334 stories from TRECVID 2005 benchmark, and the results have shown our approach can achieve superior performance.

Currently we only adopt the visual-based and concept-based feature. In the future, other effective features including the multimodel features will be used jointly to further improve the performance on story-based retrieval.

Acknowledgements

The work described in this paper was fully supported by the National Natural Science Foundation of China under Grant No. 60873154 and 60503062, the Beijing Natural Science Foundation of China under Grant No. 4082015, and the Program for New Century Excellent Talents in University under Grant No. NCET-06-0009.

References

1. Cheung, S.C., Zakhor, A.: Efficient Video Similarity Measurement with Video Signature. IEEE Trans. on Circuits and Systems for Video Technology (CSVT) 13(1) (January 2003)
2. Yuan, J., Duan, L.-Y., Tian, Q., Xu, C.: Fast and Robust Short Video Clip Search Using an Index Structure. In: ACM International Workshop on Multimedia Information Retrieval (MIR) (October 2004)
3. Chen, L., Chua, T.S.: A Match and Tiling Approach to Content-based Video Retrieval. In: IEEE International Conference on Multimedia and Expo. (ICME), pp. 417–420 (2001)
4. Jain, A.K., Vailaya, A., Xiong, W.: Query by Video Clip. Multimedia System 7, 369–384 (1999)
5. Lienhart, R., Effelsberg, W.: A Systematic Method to Compare and Retrieve Video Sequences. Multimedia Tools and Applications 10(1) (January 2000)
6. Liu, X., Zhuang, Y., Pan, Y.: A New Approach to Retrieve Video by Example Video Clip. In: ACM Multimedia Conference, MM (1999)
7. Peng, Y., Ngo, C.-W.: Clip-Based Similarity Measure for Query-Dependent Clip Retrieval and Video Summarization. IEEE Trans. on Circuits and Systems for Video Technology (CSVT) 16(5) (2006)
8. Peng, Y., Ngo, C.-W.: EMD-Based Video Clip Retrieval by Many-to-Many Matching. In: Leow, W.-K., Lew, M., Chua, T.-S., Ma, W.-Y., Chaisorn, L., Bakker, E.M. (eds.) CIVR 2005. LNCS, vol. 3568, pp. 71–81. Springer, Heidelberg (2005)
9. Wu, X., Hauptmann, A.G., Ngo, C.-W.: Novelty Detection for Cross-Lingual News Stories with Visual Duplicates and Speech Transcripts. In: ACM Multimedia Conference, MM (2007)
10. Yanagawa, A., Chang, S.-F., Kennedy, L., Hsu, W.: Columbia University's Baseline Detectors for 374 LSCOM Semantic Visual Concepts. Columbia University ADVENT Technical Report #222-2006-8, March 20 (2007)
11. Yanagawa, A., Hsu, W., Chang, S.-F.: Brief Descriptions of Visual Features for Baseline TRECVID Concept Detectors. Columbia University ADVENT Technical Report #219-2006-5 (July 2006)
12. Tesic, J., Natssev, A., Smith, J.R.: Cluster-based data modeling for semantic video search. In: ACM International Conference on Image and Video Retrieval, CIVR (2007)
13. Naphade, M.R., Yeung, M.M., Yeo, B.L.: A Novel Scheme for Fast and Efficient Video Sequence Matching Using Compact Signatures. In: SPIE: Storage and Retrieval for Media Databases, pp. 564–572 (2000)
14. Hsu, W.H., Chang, S.-F.: Topic Tracking across Broadcast News Videos with Visual Duplicates and Semantic Concepts. In: International Conference on Image Processing (ICIP), Atlanta, GA (October 2006)

Camera Take Reconstruction

Maia Zaharieva, Matthias Zeppelzauer, Christian Breiteneder, and Dalibor Mitrović

Vienna University of Technology, Interactive Media Systems Group
Favoritenstr. 9-11/188-2, A-1040 Vienna, Austria
{zaharieva,zeppelzauer,breiteneder,mitrovic}@ims.tuwien.ac.at
http://ims.tuwien.ac.at

Abstract. In this paper we focus on a novel issue in the field of video retrieval stemming from film analysis, namely the investigation of film montage patterns. For this purpose it is first necessary to reconstruct the original film sequences, i.e. the camera takes. For the decision whether or not two shots occurring anywhere in a film stem from the same take we use edge histograms and local feature tracking. Evaluation results on experimental film material (where montage patterns are of great importance) show a very good performance of the algorithm proposed.

1 Introduction

Conventional videos such as Hollywood movies and TV-series usually follow specific editing rules (e.g., cross-cutting and shot reverse shot [1]) resulting in well-defined patterns of shot editing within a scene. Documentaries, experimental and art house films challenge the conventional filmmaking by the use of unusual (non-narrative) camera and editing techniques [3]. Currently, the study of such techniques is a tedious manual process performed by film experts.

In this paper we present a new topic in the domain of video retrieval, namely the identification of editing techniques and montage patterns. Furthermore, we introduce a novel approach for the reconstruction of the original film shooting sequences or the camera takes. A *camera take* is defined as a single, continuously-recorded performance with a given camera setup. In the editing process the camera takes are cut into multiple shots and joined together to form a complete movie, i.e. a camera take is a sequence of one or more consecutively recorded video shots. Semantically related and temporally adjacent shots build a video scene. Shots originating from the same camera take can be temporally distributed over the entire movie (see Figure 1).

The reconstruction of camera takes yields relationships of shots that proceed at the same place and time. This high-level structural information is beneficial for tasks such as scene segmentation and analysis of montage patterns, editing style, and motion rhythm. Furthermore, reconstructed camera takes allow for compact video representation and nonlinear browsing. The reconstruction is based on the temporal continuity of shots. It does not require the video content to be similar over the entire camera take. For example, several shots cut out from a camera take that contains a long camera pan can have highly dissimilar content. Methods

S. Boll et al. (Eds.): MMM 2010, LNCS 5916, pp. 379–388, 2010.

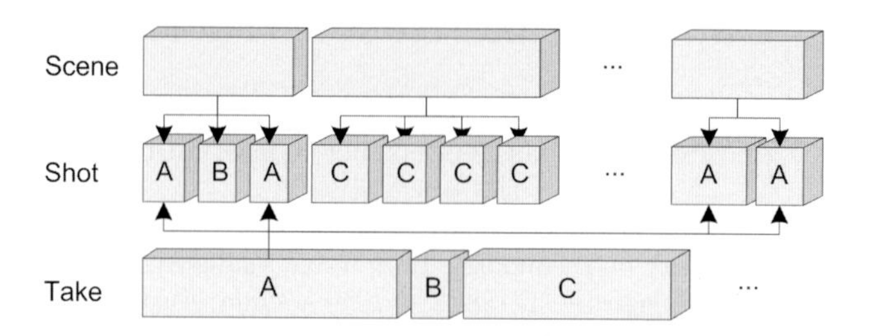

Fig. 1. Camera takes vs. video scenes

based on keyframes and image features may not find similarities among the shots. The presented approach is able to associate the shots with each other.

Various applications for video analysis and retrieval can benefit from the camera takes reconstruction. Examples include:

- *Flashback / -forward detection*: A flashback is defined as a shot that is presented out of chronological order [1]. The detection of camera takes implies the reconstruction of the original chronological order and, thus, allows for a straightforward flashback detection.
- *Montage pattern and rhythm analysis*: The rhythmic relations between two shots indicates highly semantical information. Cinematic rhythm derives from different film techniques such as shot duration, motion, sound rhythm, and montage patterns. For example, the use of alternating close-ups with shorter shots creates a more intense dialog or conflict sequence.
- *Film analysis and reconstruction*: The reconstruction of the montage schema allows for the identification of incomplete copies and altered versions of the original film material.
- *Video summary*: The association among shots of the same camera take can be further used to create a more compact video summary for non linear browsing.

The remainder of this paper is organized as follows. In Secton 2 we give an overview over related research. Section 3 describes the two stage algorithm for camera take reconstruction. Section 4 presents the experimental results. We conclude in Section 5 and give an outlook for further research.

2 Related Work

Current work on video structure analysis focusses mainly on scene detection and classification. Recent approaches on scene detection and classification group shots into a scene if they are content-correlated and temporally close to each other [2,7,8,9,11]. Content correlation is usually determined based on color information. An essential disadvantage of this approach is that false color matches between shots of different scenes result in falsely combined shots. Dynamic scenes often possess different color information which impedes the process of keyframes selection for reliable shot representation. Motion information is often neglected within the process of scene detection. Ngo et al. use motion information for the selection and formation of keyframes as representative for the shot [7]. However, motion is no further used as matching criterion. Rasheed et al. merge shots together that have high motion activity and small shot length to enable high scene dynamics [8]. However, the assumption that shots of the same scene follow the same dynamics holds only for very limited scenarios.

Recently, Truong et al. address the extraction of film takes [10]. The authors apply merge-and-split clustering techniques to group similar shots based on color histograms. A substantial assumption of the approach is that at most one shot is presented from a single camera take. This assumption holds for a great part of Hollywood movies but fails for the most documentary and experimental films. A further limitation of the approach is its inapplicability to shots with extensive camera and/or object motion (e.g. action shots) due to the restrictions of the selected shot representation. Finally, the task of camera take extraction is reduced to a shot similarity detection.

In contrast to existing approaches, we strongly rely on motion information. Motion smoothness between frames of the same camera take allows for the reliable recognition of consecutive shots. Thus, shots are linked together without the problem of appropriate keyframe selection or shot representation. Furthermore, since shots of the same camera take can be temporally apart from each other in the edited film, the reconstruction of camera takes captures information, which is lost by a scene detection algorithm.

3 Camera Take Detection

The core element of the algorithm for camera take detection is the motion smoothness analysis between different shots. However, since motion tracking in a long video can become computationally expensive, we introduce an intermediate step to limit the number of candidates for camera takes. To determine possible camera takes we use a fast and yet reliable similarity measure based on edge histograms. Following, we analyze the motion smoothness based on local feature tracking. Figure 2 gives an overview over the workflow of the algorithm.

3.1 Continuity Analysis

For the detection of candidate camera takes we first construct the set of all continuity regions for a given shot S_x. The continuity region CR between two shots S_x and S_y is defined as the union of the last n frames of S_x and the first n frames from S_y:

$$CR^{S_x,S_y} = \{f_{a-n+1}^{S_x}, f_{a-n+2}^{S_x}, ..., f_a^{S_x}, f_1^{S_y}, f_2^{S_y}, ..., f_n^{S_y}\} \quad (1)$$

where a denotes the number of frames of S_x and f the respective frames in S_x and S_y. S_y represents any other shot from the film. Thus, for a given shot S_x

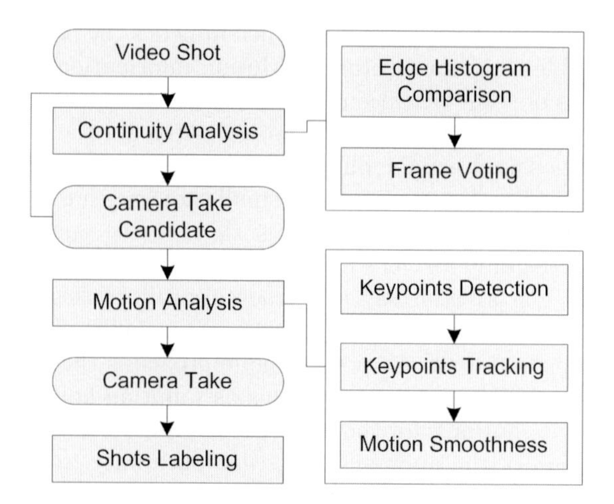

Fig. 2. Algorithm workflow

a set of continuity regions (with common S_x last frames) is constructed. In our evaluations, n is set to three which results in a continuity region of the length 6 between any two shots.

For every frame from the regions an MPEG-7 edge histogram is computed which has been proved to be effective for image similarity retrieval [6]. The edge histogram captures the distribution of orientations of the edges across blocks of a frame. Each video frame is divided into 16 non-overlapping blocks. For each block we create a local edge histogram with 5 bins (vertical, horizontal, 45 degree, 135 degree, and non-directional edges). Thus, the edge histogram for the entire frame contains $16 \times 5 = 80$ bins [4].

Each frame $f_{a-n+1}^{S_x}, f_{a-n+2}^{S_x}, ..., f_a^{S_x}$ is compared to every frame from the set of continuity regions for S_x that represents a shot different than S_x. Following, frames vote for the shot with the highest similarity score in terms of Euclidean distance. A shot S_y is accepted to be a following shot of S_x if 1) the majority frames from S_x vote for S_y, and 2) there is at least one reverse vote, i.e. at least one frame from S_y votes for S_x. In case, S_y is a following shot of S_x, both are assigned to a new candidate camera take: $CT_i = \{S_x, S_y\}$. For every last shot of the current CT_i the process is repeated until there are no more following shots detected.

3.2 Motion Smoothness Analysis

Motion vector fields estimated for consecutive video frames are slowly varying over both space and time. Therefore, we measure the variations of the motion vectors along the temporal direction in the continuity region of each candidate camera take. Figure 3 shows an example for consecutive shots. The difference between the respective motion vectors is very low and, thus, indicates high motion smoothness. On the contrary, Figure 4 depicts frames that are visually similar

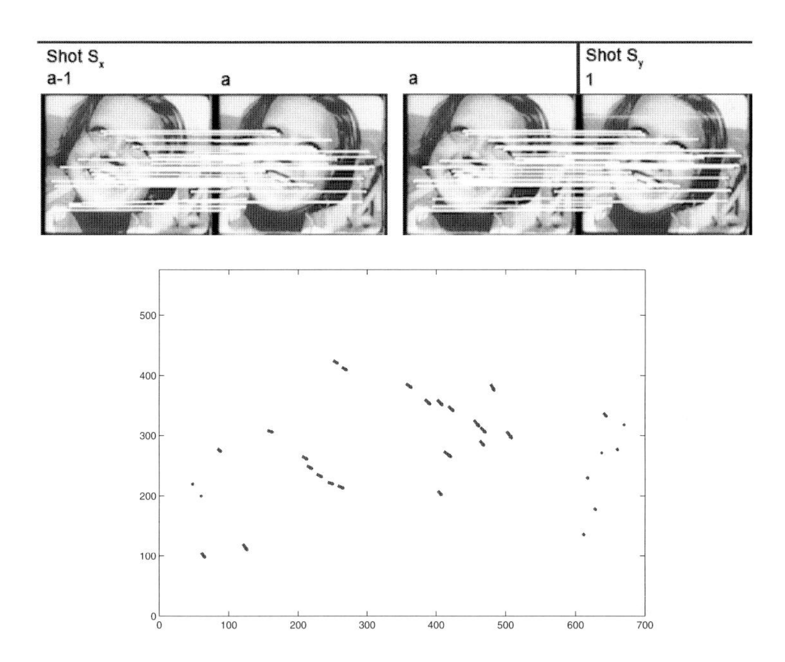

Fig. 3. Motion smoothness for frames of the same camera take (first row: feature tracking; second row: differences between the respective motion vectors)

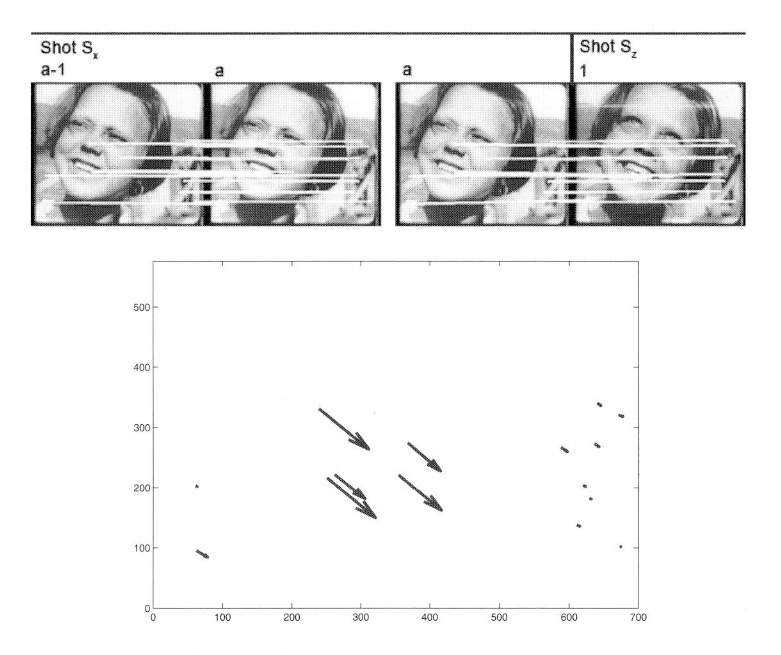

Fig. 4. Motion smoothness for frames of similar but not consecutive shots (first row: feature tracking; second row: differences between the motion vectors)

but belong to different, temporally non consecutive shots. The slight move of the girl's head results in significantly larger differences in the motion vectors.

For motion detection and tracking we apply local feature tracking based on SIFT (Scale Invariant Feature Transform) matching [5]. However, other motion tracking methods can be applied as well. We limit the number of extracted SIFT features per frame to 500. The resulting feature descriptors are matched by identifying the first two nearest neighbors in terms of Euclidean distances. A descriptor is accepted if the nearest neighbor distance is below a predefined threshold. The value of 0.8 was determined experimentally and used through the evaluation tests described in Section 4. Finally, only camera takes with smooth motion vectors are accepted.

4 Experiments

Subject of the evaluation are experimental monochromic documentaries from the late 1920s. Noteworthy is the low quality of the material as a result of multiple storage, copying, and playback over the last decades. The results achieved show that the process of camera take detection performs robustly to various artifacts such as scratches, dirt or frame shrinking.

4.1 Camera Take Detection

The first experiment focusses on the evaluation of camera take detection. The explored movie consists of 1.768 shots (95.678 frames). Our algorithm detected 119 camera takes of two and more shots. The results were evaluated manually by experts. 92.44% of all detected camera shots were correct (see Table 1).

The lack of motion and the same visual appearance of shots may cause false positive detection of camera takes for identical, static shots. Another reason for incorrect detected camera takes is the dissolve editing technique. The gradually replacement and high degree of similarity between the shots falsely assigns them to the same camera take (see Figure 5 for an example).

Table 1. Performance on camera take detection

True positives	110	92.44%
False positives	5	4.20%
Ambiguous camera takes	4	3.36%
Detected camera takes	119	100.00%

Fig. 5. False positive camera take due dissolve (the same scene is shoot from two different perspectives)

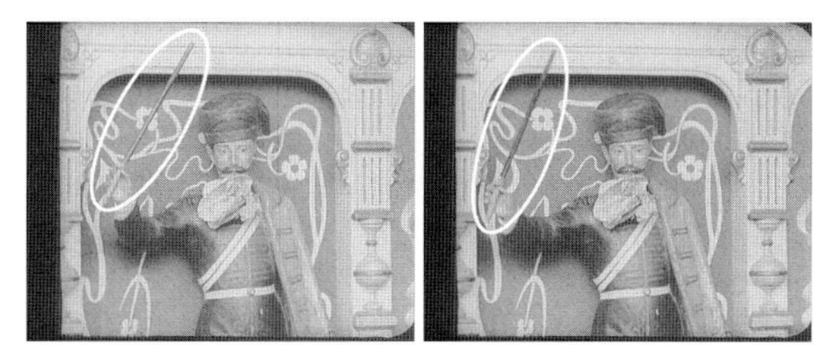

Fig. 6. Ambiguous shot

Additional four detected camera takes could not be verified due to ambiguity. An example for such shots is presented in Figure 6. The shot depicts a figure in a shooting gallery on a fair. Due to the repetitive movement of the timbal in the right hand it is not possible to definitely determine if 1) multiple shots are part of the same camera take or 2) it is always the same shot on different positions.

4.2 Montage Reconstruction

The next experiment aims at the reconstruction and the analysis of the original montage schemas. Montage schemas describe the assemblage of a film through editing. They allow for the analysis of editing techniques and montage patterns. Furthermore, the reconstruction of montage schemas is essential for the analysis of archive film material where the original versions (filmstrips) do often no longer exist. The remaining copies are usually backup copies from film archives that are often incomplete due to bad storage, mold, and film tears.

We investigate three different film sequences. We first detect camera takes. In the next step, we assign labels to the shots of the same camera take.

The first sequence presents workers building a railway. The whole sequence of 19 shots (204 frames) originates from three cross-cut camera takes. Our algorithm successfully detected and assigned the respective shots (see Figure 7).

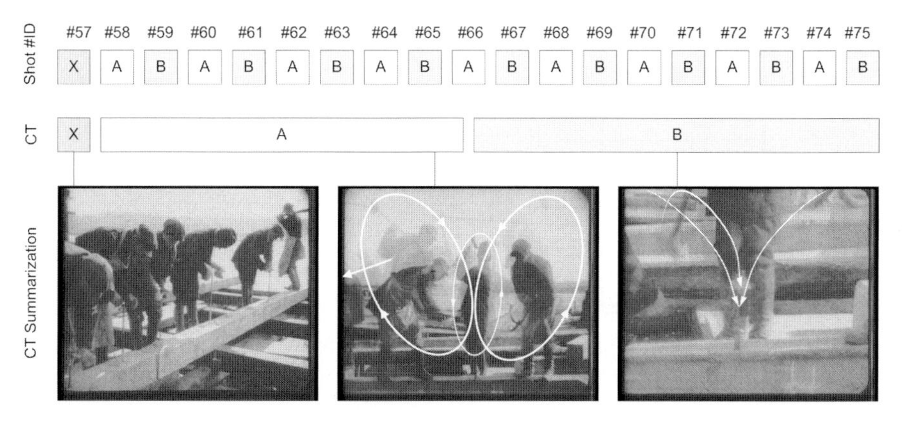

Fig. 7. Detected camera takes in sequence 1 (white arrows show dominant motion)

Since we investigate experimental video material, not all of the resulting montage schemas comply with conventional editing patterns. Figure 8 presents the detected montage pattern in a sequence of 31 shots (1539 frames). It exhibits an unusual editing technique that is not reconstructable with other common scene

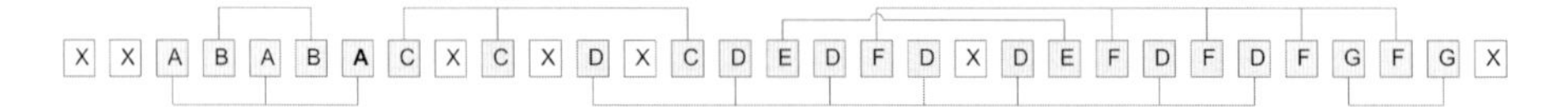

Fig. 8. Detected camera takes in sequence 2 (an X denotes a single shot camera take)

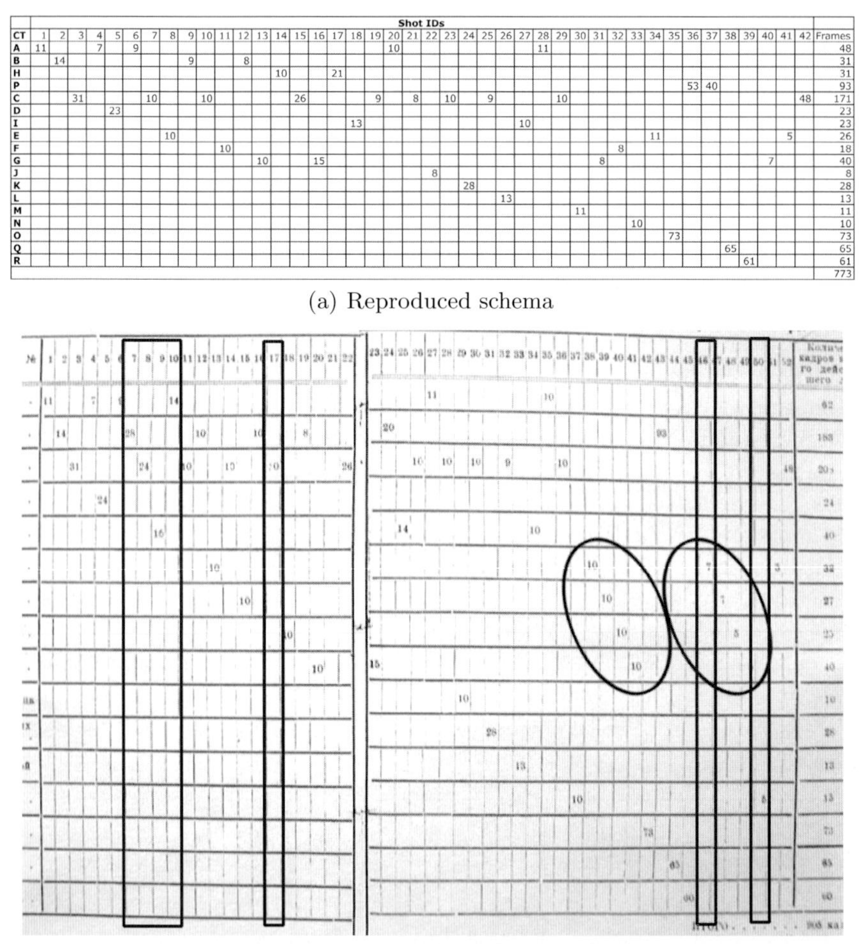

	Shot IDs																				
CT	1	2	3	4	5	6	7	8	9	10	11	12	13	14	15	16	17	18	19	20	21
A	11			7		9														10	
B		14							9			8									
H														10			21				
P																					
C			31				10			10					26				9		8
D					23																
I																		13			
E								10													
F											10										
G													10			15					
J																					
K																					
L																					
M																					
N																					
O																					
Q																					
R																					

CT	22	23	24	25	26	27	28	29	30	31	32	33	34	35	36	37	38	39	40	41	42	Frames
A							11															48
B																						31
H																						31
P															53	40						93
C		10		9				10													48	171
D																						23
I						10																23
E													11							5		26
F											8											18
G										8									7			40
J	8																					8
K			28																			28
L					13																	13
M									11													11
N												10										10
O														73								73
Q																	65					65
R																		61				61
																						773

(a) Reproduced schema

(b) Extract from the original schema (black box frames indicate some of the detected missing shots, oval frames: rearranged shots)

Fig. 9. Montage schema for sequence 3

detection algorithms. The interpretation of such patterns is a research subject for film experts.

The evaluation of the third sequence (42 shots, 773 frames) was motivated by the discovery of the original montage schema from the mid 1920s. It shows the experiment of the filmmaker to graphically chart the montage of shots within a scene (see Figure 9(b)). The reconstructed montage schema indicates missing and rearranged shots (see Figure 9(a)). Currently, it is not clear whether the original film complied with the discovered schema and if the nowadays available copy is a full version of the original film. Notwithstanding, the results demonstrate the reliability of the algorithm and its applicability in a scenario where the original montage schema is not available.

5 Conclusion and Outlook

In this paper we presented a novel application for the reconstruction of camera takes. We applied the proposed algorithm on a test set of experimental documentaries. Presented results demonstrate the reliability of the algorithm and outline its applicability for manifold application scenarios such as montage pattern analysis or comparison of different film cuts.

Reliable camera take detection provides a new perspective to the domain of film analysis. From a technical point of view, it allows for the comparison of different film cuts and the analysis of montage patterns that do not follow conventional editing rules. Moreover, further analysis of the motion smoothness between two shots can provide information about missing frames from the original camera take.

From a semantical point of view, reconstructed camera takes capture information that can be missed by conventional scene detection algorithms. By the analysis of motion smoothness within a given continuity region, the proposed method does not require appropriate shot representation or keyframe and feature selection. Moreover, two shots to be grouped are not required to be visually similar for the whole shot length. Highly dynamical shots (e.g. action) or large camera motion often result in great dissimilarity in the visual perception. However, motion smoothness analysis can still detect consecutive shots due to the smooth transition present in a continuos camera take. This information can be further used to improve the process of video representation and retrieval.

Acknowledgment. This work was partly supported by the Vienna Science and Technology Fund (WWTF) under grant no. CI06 024, "Digital Formalism: The Vienna Vertov Collection".

References

1. Bordwell, D., Thompson, K.: Film Art: An Introduction, 8th edn. McGraw Hill Book Co., New York (2008)
2. Chasanis, V.T., Likas, A.C., Galatsanos, N.P.: Scene detection in videos using shot clustering and sequence alignment. IEEE Transactions on Multimedia 11(1), 89–100 (2009)

3. Dancynger, K.: The technique of film and video editing: history, theory, and practice, 4th edn. Focal Press (2007)
4. ISO-IEC: Information Technology - Multimedia Content Description Interface - part 3: Visual. No. 15938-3, ISO/IEC, Moving Pictures Expert Group (2002)
5. Lowe, D.G.: Distinctive image features from scale-invariant keypoints. Int. Journal of Computer Vision 60(2), 91–110 (2004)
6. Manjunath, B.S., Ohm, J.R., Vasudevan, V.V., Yamada, A.: Color and texture descriptors. IEEE Transactions on Circuits and Systems for Video Technology 11, 703–715 (2001)
7. Ngo, C.W., Pong, T.C., Zhang, H.J.: Motion-based video representation for scene change classification. Int. Journal of Computer Vision 50(2), 127–142 (2002)
8. Rasheed, Z., Shah, M.: Scene detection in holywood movies and tv shows. In: IEEE Conference on Computer Vision and Pattern Recognition (CVPR 2003), vol. 2, pp. II– 343–II–348 (2003)
9. Rasheed, Z., Shah, M.: Detection and representation of scenes in videos. IEEE Transactions on Multimedia 7(6), 1097–1105 (2005)
10. Truong, B.T., Venkatesh, S., Dorai, C.: Extraction of film takes for cinematic analysis. Multimedia Tools and Applications 26(3), 277–298 (2005)
11. Zhao, L., Yang, S.Q., Feng, B.: Video scene detection using slide windows method based on temporal constrain shot similarity. In: IEEE Int. Conference on Multimedia and Expo. (ICME 2001), pp. 1171–1174 (2001)

Semantic Based Adaptive Movie Summarisation

Reede Ren, Hemant Misra, and Joemon M. Jose

Information Retrieval Group
University of Glasgow
18 Lilybank Gardens, Glasgow, UK, G12 8RZ
{reede,hemant,jj}@dcs.gla.ac.uk

Abstract. This paper proposes a framework for automatic video summarization by exploiting internal and external textual descriptions. The web knowledge base Wikipedia[1] is used as a middle media layer, which bridges the gap between general user descriptions and exact film subtitles. Latent Dirichlet Allocation (LDA) detects as well as matches the distribution of content topics in Wikipedia items and movie subtitles. A saliency based summarization system then selects perceptually attractive segments from each content topic for summary composition. The evaluation collection consists of six English movies and a high topic coverage is shown over official trails from the Internet Movie Database[2].

Keywords: Content-based video summarisation, latent Dirichlet allocation.

1 Introduction

Movie summarization aims to create a succinct representation of a long film. Since a movie is a complex media with rich contents, there are various interpretations for a film story. This leads to the research problem of adaptive video summarization or personalized summarization. For example, a user writes down his review of a film and asks for a summary which imitates the statement [12]. Money *et al.* [12] regard such a technique as an essential tool for the management of personal media archives as well as for the web-based business. The major challenge in adaptive video summarization is how-to model video content properly to facilitate the projection from video contents to unknown user requirements. Ronfrad *et al.* [15] manually create textual descriptions to interpret video shots and thus align video segments with semantics. This work shows the probability of content decomposition in a continuous video stream. Kawai *et al.* [8] exploit internal and external textual descriptions, *i.e.* video subtitles and electronic program guide (EPG) data. This approach successfully links video segments with textual descriptions, however, is not robust enough for movie summarization. Kawai *et al.* even report that the performance will worsen, if the EPG document comes from a different content provider [8].

[1] www.wikipedia.com
[2] IMDB, www.imdb.com

S. Boll et al. (Eds.): MMM 2010, LNCS 5916, pp. 389–399, 2010.

In this paper, we propose the usage of Wikipedia as the middle media layer between user statements and internal textual descriptions, *i.e.* closed captions and subtitles. As a common understanding shared by numerous viewers, a Wikipedia movie item can be easily reshaped to match a specific preference. Moreover, the exploitation of Wikipedia alleviates the uncertainty caused by user statements. This results in a common evaluation platform and facilitates the development of an adaptive summarization system. Latent Dirichlet Allocation (LDA) is used to identify semantic topic distributions in both textual descriptions. These topics are matched to identify content coherent video segments. A psychological attention-based system is finally adapted for summary composition, which catches the most attractive video clips in each content coherent video segment. In summary, a general framework is proposed for automatic adaptive video summarization, which projects content topics from a Wikipedia movie item to video subtitles and selects the most attractive part in each content topic.

The remainder of this paper is organized as follows. A review of movie summarisation is provided in Section 2. The system framework is found in Section 3. The module of content modeling is addressed in Section 4, including the description of Wikipedia movie item, movie subtitles and the LDA based content analysis. Section 5 is for the attention based topic selection. Experimental results and conclusions are reported in Section 6 and 7, respectively.

2 Related Work

The literature of movie summarisation can be broadly categorized into three groups, namely scene-based, event-based and saliency-based. Despite the complexity in content modeling, scene-based methods try to index a movie by dividing it into visual scenes [9]. Sundaram *et al.* [17] propose that a scene is a coherent video segment of chromaticity, lighting and ambient sound. They develop a finite memory model to allocate scene boundary. Cao *et al.* [1] align audio and visual scenes to remove false alarm. However, this coherence hypothesis can hardly be fulfilled, especially in action films [16]. Event-based approaches collect shots that satisfy some priori knowledge. Chen *et al.* [2] detect dialogs and violent events by audio features. Smeaton *et al.* [16] classify shots by embedded actions to create action film trailers. Saliency-based summarisation is an exploration of computational psychology to content analysis [14]. Hanjalic *et al.* [6] assert that the interesting video content should be attractive to viewers and that the attractiveness of a video segment reflects its importance in content representation. In [4,14], psychological attention from audio and visual streams are combined to estimate content importance. The main drawback of these approaches is the absence of a clear link to semantics. Little semantic information can be garnered from proposed video structures, neither scene nor event. Kawai *et al.* [8] match EPG sentences with closed captions by a Bayesian belief network and an AdaBoost classifier. A semantic video segmentation is therefore completed by projecting EPG sentences onto a video stream. However, Kawai *et al.* do not analyze the embedded semantics and their approach entirely relies on the matching between textual terms. This results in a low topic coverage, when EPG texts come from a different content provider [8]. In addition, EPG is a low-level content description, which contains too many unnecessary details for an efficient summarization.

3 System Framework

In this section, we address the framework of semantic movie summarisation (Figure 1). We divide this system into two parts, back end and front end. The front end accepts user statements and matches the description with a Wikipedia movie item. The back end involves three components, namely content modeling, topic selection and adaptive summary composition. In content modeling (Section 4), we collect movie subtitles and the related Wikipedia item. The LDA algorithm is used to estimate content topic distributions in both subtitles and the Wikipedia item. We match these content topics to allocate story boundaries in the video stream. The same approach is also used in front end to match user descriptions and Wikipedia items. Topic selection computes the distribution of attention intensity and removes redundant video segments. Adaptive summary composition adjusts summary duration by managing shot candidates to meet the prior requirement on summary length [13].

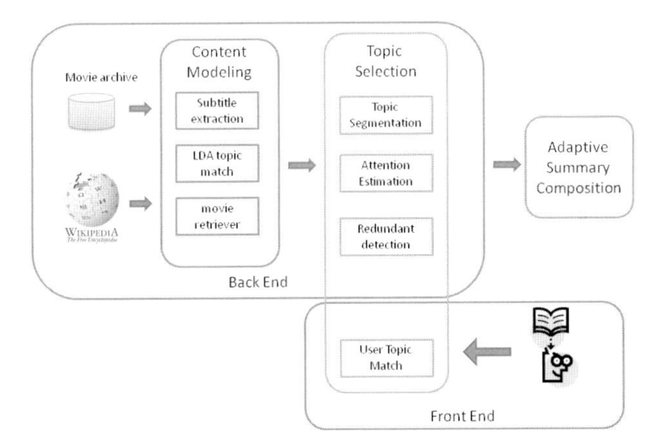

Fig. 1. Semantic Movie Summarisation Framework

4 Content Modeling

In this section, we present the method to exploit implicit semantics among two external knowledge sources, *i.e.* Wikipedia and movie subtitles.

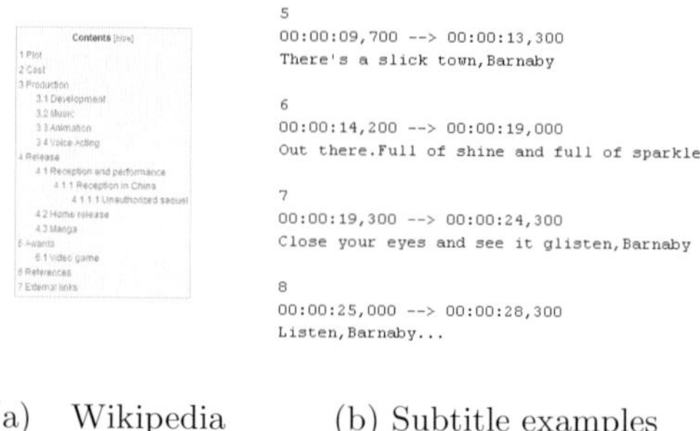

(a) Wikipedia entry (b) Subtitle examples

Fig. 2. External Knowledge for the Film WallE

4.1 Wikipedia Movie Item

The Wikipedia movie item is a structured data, which starts with a short description to clarify the background of movie production. A table of contents (Figure 2(a)) is then shown to list hyper-links to sections. Eight sections are usually involved, namely *plot, cast and characters, production, theme, reception, see also, reference* and *external link*. Section *Plot* is a brief of the film story. It is made up by a few paragraphs, each of which describes a major scene in the movie. This provides a valuable reference for content-based video segmentation. We suggest that projecting the structure of a plot onto video segments would lead to an effective scene segmentation for movies.

4.2 Subtitle

Subtitles are a specific type of closed caption, which can be extracted from DVD textual streams or collected from the web, *e.g.* www.shooter.com. Some examples are shown in Figure 2(b) from the film WallE. A subtitle includes three lines: (1) the sequence number; (2) a time stamp which defines life hold; and (3) audio contents or scene descriptions.

4.3 LDA Based Subtitle Segmentation

Similar to the approach in [8], we divide the subtitles into coherent segments but follow semantic topics from a Wikipedia plot rather than from EPG sentences. The LDA algorithm identifies semantic topic distributions in both the Wikipedia plot and subtitles. As a probabilistic generative model, LDA is able to process general textual description robustly and to take unseen user statements.

Topic Detection and Model Training. In the LDA framework, there are two steps to generate a document: (1) choose a topic distribution, $\theta_{dt}, t = 1...T$, for a document from a Dirichlet distribution of order T; (2) for each word occurrence in the document, a topic, z_i, is chosen from this distribution and a word is selected from the chosen topic. Given the topic distribution, each word is drawn independently from every other word using a *document specific* mixture model. The probability of w_i, the ith word token in document d, is defined as follows.

$$P(w_i|\theta_d, \phi) = \sum_{t=1}^{T} P(z_i = t|\theta_d) P(w_i|z_i = t, \phi) = \sum_{t=1}^{T} \theta_{dt}\phi_{tw_i} \tag{1}$$

where $P(z_i = t|\theta_d)$ is the probability that the tth topic was chosen for the ith word token and $P(w_i|z_i = t, \phi)$ is the probability of word w_i given topic t. The likelihood of document d is a product of such terms and can be defined as:

$$P(C_d|\theta_d, \phi) = \prod_{v=1}^{V} \left[\sum_{t=1}^{T} (\theta_{dt}\phi_{tv}) \right]^{C_{dv}} \tag{2}$$

where C_{dv} is the count of word v in d and C_d is the word-frequency in d. Two parameter sets require learning from a training collection: (1) the topic distribution in each document d ($\theta_{dt}, t = 1...T, d = 1...D$) and (2) the word distribution in each topic ($\phi_{tv}, t = 1...T, v = 1...V$). We use Gibbs sampling [5] and the RCV1 document collection which includes 27,672 documents (news items), to estimate distributions of θ and ϕ. α and β are hyper-parameters which define the non-informative Dirichlet priors on θ and ϕ, respectively. In experiments, α is 1 and β is 0.01. The estimations for θ and ϕ are derived from the counts of hypothesized topic assignments as follows.

$$\phi_{tv} = \frac{J_{tv} + \beta}{\sum_{k=1}^{V} J_{tk} + V\beta} \tag{3}$$

$$\theta_{dt} = \frac{K_{dt} + \alpha}{\sum_{k=1}^{T} K_{dk} + T\alpha} \tag{4}$$

where J_{tv} is the number of times that word v is assigned to topic t and K_{dt} is the number of times that topic t is assigned to some word token in document d. The distribution of words is therefore revealed in each topic as well as the distribution of topics in each train document.

The generative model of LDA can also estimate the topic distribution of an unseen document. We use an iterative procedure [7,10] to estimate topic distributions in Wikipedia plots and subtitles[3]. The update rule is given by:

$$\theta_{dt} \leftarrow \frac{1}{l_d} \sum_{v=1}^{V} \frac{C_{dv}\theta_{dt}\phi_{tv}}{\sum_{t'=1}^{T} \theta_{dt'}\phi_{t'v}} \tag{5}$$

where l_d is the document length in terms of number of content words. In experiments, we find the convergence can be achieved in less than $10 - 15$ iterations.

[3] Subtitles need segmentation as described in the next section.

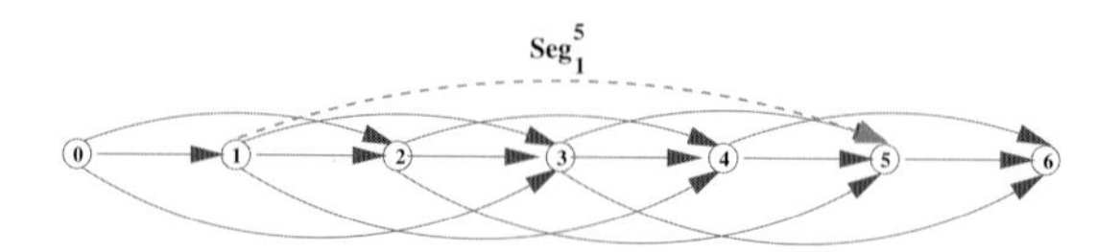

Fig. 3. Example of nodes and segments in dynamic programming

Subtitle Segmentation. Subtitles are a continuous media without a clear topic boundary. We explore all possible segmentations and use Dynamic programming (DP) to find an optimized solution. In addition, we remove stop words and stem all words before DP. In the setup, each possible segment is defined by two nodes, the begin node (B) and the end node (E) as shown in Figure 3. A segment $\mathbf{Seg_1^5}$ (dotted line) is from begin node B_1 (excluding B_1) to end node E_5 (including E_5). Node 0 is treated as null node. In short, each sentence start is a possible B node and each sentence end is a possible E node.

The algorithm for subtitle segmentation is as follows [18]. Assume that a text document $d = w_1 w_2 \cdots w_{l_d}$ has l_d word tokens. A particular segmentation S made up of m consecutive segments can be defined by $S = S_1 S_2 \cdots S_m$. The likelihood of segmentation S is defined as

$$P(S|d) = \frac{P(d|S)P(S)}{P(d)} \tag{6}$$

where $P(d|S)$ refers to the score of segmentation S of the text d; and $P(S)$ is a prior over segmentation which corresponds to a penalty factor. In experiments, the penalty factor is set as $\log P(S) = -p \log(l_d)$, where $p = 3$. If the segment S_i contains n_i word tokens and if w_i^j is the jth word token in segment S_i, we can define W_i as $W_i = w_i^1 \cdots w_i^{n_i}$. Therefore, $d = W_1 \cdots W_m$ and $l_d = \sum_{i=1}^{m} n_i$. Under these assumptions, W_i and S_i have one to one correspondence. Assuming that segments are independent of each other, Equation 6 can be rewritten as[4]:

$$P(S|d) \propto \left[\prod_{i=1}^{m} P(W_i|S)\right] P(S) \propto \left[\prod_{i=1}^{m} P(W_i|S_i)\right] P(S) \propto \left[\prod_{i=1}^{m}\prod_{j=1}^{n_i} P(w_i^j|S_i)\right] P(S)$$

The most likely segmentation $\hat{S}$ is found by maximizing $P(S|d)$, namely $\hat{S} = \underset{S}{\operatorname{argmax}}\ P(S|d)$. This is because the log-likelihood for a coherent segment is typically higher than that for incoherent cases, as several topics are active for a incoherent segment[5]. In the forward-pass of DP, the score of Seg_B^E is computed for each node pair B and E. The path that maximizes the cumulative score from the first to the last node is searched, and for each E node the value of

[4] For a given document d, $P(d)$ is constant for all the segmentations and can be dropped from the equation.

[5] A coherent segment indicates the begin and end nodes from the same story whereas an incoherent segment implies that the begin and end nodes are not from the same story.

the best start node B is stored. Then, the information about the best start node is used in the trace back to find the path that maximizes the score and in turn, the segment boundaries. Algorithm details can be found in [11], where we compare our approach with several other methods, *e.g.* pLSA and TextTiling. In experiments, LDA-based text segmentation demonstrates the best performance among all.

Content Topic Matching. Topic distributions in each paragraph of a Wikipedia plot and in each segment of subtitles are estimated by Equation 5. These topic distributions are matched by minimizing the negative KL divergence. We do not rely on the sequential information for topic matching. The change in topic description sequence therefore will not affect the matching effectiveness.

5 Attention-Based Topic Selection

In this section, we address attention-based topic selection. Attention is a psychological measurement for saliency distribution, which has been widely used for sports highlight detection and video summarisation [4,6,14]. We keep the attention fusion framework of multi-resolution autoregressive [14], but adapt the collection of audio-visual salient features for movie summarisation. Shots in a content topic are scored by audio-visual attention intensities. The top 20% shots in each topic are selected as candidate for summary composition.

Visual Saliency. Region-of-interest (ROI) is used to estimate visual saliency. A ROI denotes a visual area which attracts most attention in a visual frame [6]. A pyramid is created to decompose a visual frame into a number of spatial resolutions. Three features, gray intensity, color contrast and motion intensity, are computed at every pyramid layer. These feature-based stimuli are accumulated by a salient map. The brightest area in the salient map are detected as a ROI [14], whose size is used to score visual saliency.

Audio Saliency. Three features are used for audio saliency, maximum average Teager energy (MTE), mean instant amplitude (MIA), and mean instant frequency (MIF). MTE, MIA and MTF describe audio energy distribution in time and frequency domains. For an audio frame m of length N, the MTE is the dominant signal modulation energy (Equation 7).

$$MTE(m) = \max_{1 \leq k \leq K} \frac{1}{N} \sum_{n=1}^{N} \psi_d(s * h_k) \tag{7}$$

where $n \in [(m-1)*N+1, mN]$ is the sample index, h_k is the impulse response for the kth band-pass Gabor filter, and $\psi()$ is the linear Teager-Kaiser differential energy operator [3]. The filter $j(m) = \underset{m}{\text{argmax}}\ MTE(m)$ is demodulated to derive MIA and MIF on time and frequency domain, respectively.

6 Experiment

The evaluation collection includes six movies. Official trailers from IMDB are gathered as ground truth. This is because film producers publish these trailers on web sites to attract potential customers. These trails include most of interesting aspects in a film and are therefore good enough for personal archive management. As a conclusion, we think these trails satisfy the requirements for semantic video summarization [12]. We cut a trailer into shots and match key frames between a trailer and the source video. If more than 60% of visual frames in a trailer shot are found in a source video shot, we mark the source video shot as a key shot. It must be noted that not all shots in a trailer can be found in a movie. The evaluation collection is listed as follows.

- Wall.E(2008) 98 min. The trailer is 94 sec long and includes 76 shots, among which 58 shots are found in the movie.
- Sex and the City(2007) 138 min. The trailer is 57 sec long and includes 31 shots, among which 26 shots are found in the movie.
- KungFu Panda (2008) 92 min. The trailer is 71 sec long and includes 23 shots, among which 17 shots are found in the movie.
- Transformers (2007) 144 min. The trailer is 118 sec long and includes 47 shots, among which 32 shots are found in the movie.
- The Love Guru (2008) 87 min. The trailer is 150 sec long and include 66 shots, among which 60 shots are found in the movie.
- Quantum Of Solace (2008) 108 min. The trailer is 148 sec long and include 101 shots, among which 97 shots are found in the movie.

The number of key shots in a summary is calculated as topic coverage. Paul *et al.* [13] assert that a high coverage shows the effectiveness of content modeling as well as of the summarization system. To simplify the evaluation strategy, we condense all candidate shots to meet the requirement on summary length until all shots are of 1 sec in length. Then, shots with low attention intensity will be removed. On one hand, candidates with a high attention intensity will be slightly enlarged by inserting more visual frames from the source video, if the summary size is larger than the sum of candidate size. On the other hand, candidates with low attention score will be gradually removed when summary duration decreases, *e.g.* from 5% to 1% of a source video. In addition, we keep the top 20% shots in every topic in order to fulfill the requirement in [13]. The largest summary size is about 20% of the source.

We take the saliency based summarization [14] as the baseline, which collects the 20% most attractive shots without considering video contents. Table 1 lists two experimental results: (1) the number of subtitle segments estimated by using LDA and (2) topic coverage rate reached by the baseline and by the new approach. The number of subtitle segments shows how many topics we consider in the summary composition. The diversity in content topics results in a significant improvement in topic coverage. This shows the effectiveness of the LDA-based content modeling. We think the implicit semantics learnt by LDA

Table 1. Topic coverage for summaries of 20% source videos

Movie	Subtitle Segment Number	Saliency based Topic Coverage	Topic Coverage	Improvement
Wall.E	22	0.650	0.862	+32.6%
KungFu Panda	11	0.400	0.765	+91.3%
Transformers	24	0.785	0.937	+19.2%
Sex and the City	18	0.542	0.823	+51.8%
The Love Guru	14	0.620	0.820	+32.3%
Quantum Of Solace	47	0.750	0.840	+12.0%
mean	-	0.625	0.841	+39.8%

is essential which justifies the selection of content topics. Although being an effective feature for content selection, the saliency is unequally distributed across a video: some meaningful scenes are not so variant and some active shots may not be so meaningful. For example, in the trailer of KungFu Panda, most shots are about the story how the panda masters KungFu, which are of relatively low attention intensity. This will fail the saliency based summarization. However, in the Wikipedia plot, the fight with Tai Lung is only one sentence and the process for the panda to learn KungFu covers three paragraphs. The distribution of content topics will correct the mis-selections based on saliency. In summary, the LDA-based semantic analysis justifies the selection of key content topics and successfully projects the general description of a Wikipedia plot to a continuous video stream. This ensures topic diversity in a summary video and significantly improves the effectiveness of video summarization.

Figure 4 displays the change of topic coverage rate with summary length from 20% to 1% of a source video. The percent of video size is used as x-axis. Topic coverage rate is at about log-like speed deceasing with the number of removed candidate shots during the period from 4% to 1%. This decreasing speed is significantly slower than the linear. This observation is welcomed in the summary composition and shows that attention-based topic selection is effective in keeping key shots.

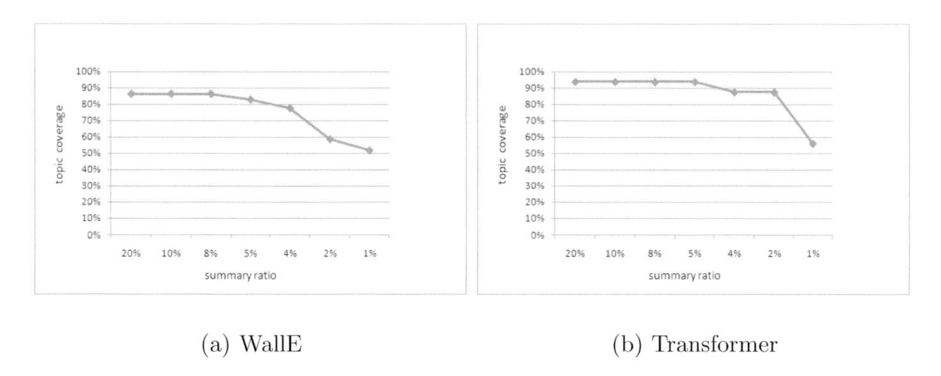

(a) WallE (b) Transformer

Fig. 4. Topic coverage at different summary ratio from 20% to 1% of the original video

7 Conclusion

This paper presents an approach for adaptive movie summarization. The web knowledge base, Wikipedia, is used as a middle layer to bridge the descriptive gap between general user statements and definite movie subtitles. This is because (1) a user statement is usually too general to be matched on movie subtitles and (2) movie subtitles contains too many details and noises. On one hand, a Wikipedia plot stands for a common understanding on a film story. Such a textual description can be easily reshaped to meet personal preferences. On the other hand, a Wikipedia plot contains all necessary semantic aspects. These details can be directly projected onto subtitles. Moreover, the usage of Wikipedia data avoids the uncertainty caused by user statements. This leads to a common evaluation platform for adaptive video summarization. In addition, other justified resources could be also used as the middle media layer. Our approached can be easily improved to exploit textual resources.

In this work, the algorithm of LDA is used to estimate content topic distributions and to segment movie subtitles. The advantage of this approach is the exploitation of implicit semantics among Wikipedia plots and subtitles. This improves the robustness in topic matching and allows the content projection from a general description. The matching between a Wikipedia plot and subtitles results in a semantic video segmentation. This is useful for video indexing as well as for summary composition. However, LDA is associated with a high computational complexity and the RCV1 document collection may not be good enough for the training in the case of movie subtitles. In future, we will try other approaches, e.g. latent semantic analysis (LSA) and use subtitles from a large movie collection for the training. In addition, LDA can also be used to match user requirements and Wikipedia content topics, and some other efficient approaches are available such as textual term based Bayesian brief network.

Acknowledgement

The research leading to this paper was supported by European Commission under contracts FP6-045032 (SEMEDIA) and FP6-027122 (Salero).

References

1. Capus, C., Brown, K.: Fractional fourier transform of the aussian and fractional domain signal support. Vision, Image and Signal Processing 150(2), 99–106 (2003)
2. Chen, L., Rizvi, S.J., Otzu, M.: Incorporating audio cues into dialog and action scene detection. In: Proceedings of SPIE Conference on Storage and Retrieval for Media Databases, pp. 252–264 (2003)
3. Evangelopoulos, G., Maragos, P.: Multiband modulation energy tracking for noisy speech detection. IEEE Transactions on Audio, Speech, and Language Processing 14(6), 24–2038 (2006)

4. Evangelopoulos, G., Rapantzikos, K., Potamianos, A., Maragos, P., Zlatintsi, A., Avrithis, Y.: Movie summarization based on audiovisual saliency detection. In: ICIP 2008, San Diego, CA, October 2008, pp. 2528–2531 (2008)
5. Griffiths, T.L., Steyvers, M.: Finding scientific topics. Proceedings of the National Academy of Sciences 101(supl. 1), 5228–5235 (2004)
6. Hanjalic, A., Xu, L.: Affective video content repression and model. IEEE Trans on Multimedia 7(1), 143–155 (2005)
7. Heidel, A., Chang, H.-a., Lee, L.-s.: Language model adaptation using latent Dirichlet allocation and an efficient topic inference algorithm. In: European Conference on Speech Communication and Technology, Antwerp, Belgium (2007)
8. Kawai, Y., Sumiyoshi, H., Yagi, N.: Automated production of tv program trailer using electronic program guide. In: CIVR, pp. 49–56 (2007)
9. Li, Y., Lee, S.-H., Yeh, C.-H., Kuo, C.-C.: Techniques for movie content analysis and skimming: tutorial and overview on video abstraction techniques. IEEE Signal Processing Magazine 23(2), 79–89 (2006)
10. Misra, H., Cappé, O., Yvon, F.: Using LDA to detect semantically incoherent documents. In: Conference on Computational Natural Language Learning, Manchester, U.K. (2008)
11. Misra, H., Yvon, F., Jose, J., Cappe, O.: Text segmentation via topic modeling: An analytical study. In: CIKM 2009 (2009)
12. Money, A.G., Agius, H.: Video summarisation: A conceptual framework and survey of the state of the art. J. Vis. Comun. Image Represent. 19(2), 121–143 (2008)
13. Over, P., Smeaton, A.F., Awad, G.: The trecvid 2008 rushes summarization evaluation. In: TVS 2008, Vancouver, British Columbia, Canada, pp. 1–20. ACM, New York (2008)
14. Ren, R., Swamy, P.P., Jose, J.M., Urban, J.: Attention-based video summarisation in rushes collection. In: TVS, pp. 89–93 (2007)
15. Ronfard, R., Tran-Thuong, T.: A framework for aligning and indexing movies with their script. In: IEEE International Conference on Multimedia and Expo., Baltimore, USA, July 2003, pp. 21–24 (2003)
16. Smeaton, A.F., Lehane, B., O'Connor, N.E., Brady, C., Craig, G.: Automatically selecting shots for action movie trailers. In: MIR 2006, pp. 231–238. ACM, New York (2006)
17. Sundaram, H., Chang, S.-F.: Determining computable scenes in films and their structures using audio-visual memory models. In: ACM Multimedia, pp. 95–104. ACM, New York (2000)
18. Utiyama, M., Isahara, H.: A statistical model for domain-independent text segmentation. In: Meeting of the Association for Computational Linguistics, pp. 491–498 (2001)

Towards Annotation of Video as Part of Search

Martin Halvey and Joemon M. Jose

Department of Computing Science, University of Glasgow, 18 Lilybank Gardens, Glasgow, G12 8QQ, United Kingdom
{halvey,jj}@dcs.gla.ac.uk

Abstract. Search for multimedia is hampered by both the lack of quality annotations and a quantity of annotations. In recent years there has been a growth in multimedia search services that emphasis interactivity between the user and the interface. Some of these systems present an as yet untapped resource for providing annotations for video. In this paper, we investigate the use of a new innovative grouping interface for video search to provide additional annotations for video collections. The annotations provided are an inherent part of the search interface, thus providing less overhead for the user in providing annotations. In addition we believe that the users are more likely to provide high quality annotations as the annotations are used to aid the users search. Specifically we investigate the annotations provided as part of two evaluations of our system; the results of these evaluations also demonstrate the utility and benefit of a grouping interface for video search [8]. The results of the analysis presented in this paper demonstrate the benefit of this implicit approach for providing additional high quality annotations for video collections.

Keywords: Annotation, video, search, grouping, implicit, tagging, exploitation.

1 Introduction

With the improving capabilities and the falling prices of current hardware systems, there are increasing potential to store and search for videos. It is possible for individuals to create their own digital libraries from videos and images created through digital cameras and camcorders, and use a number of resources to place these collections on the web. However, the current state of the art systems that are used to organise and retrieve these videos are not able to deal with such large and rapidly increasing volumes of video. Current video retrieval systems rely on textual descriptions or methods that use the low-level descriptors, to retrieve relevant videos for users. To date neither of these methods has proved sufficient to overcome the problems associated with video search. The difference between the low-level data representation of videos and the high level concepts that people associate with video, commonly known as the semantic gap, provides difficulties for using these low-level features. While these low-level features are used in some state of the art systems, most video search systems rely only on query by text. Query by text is used by many large scale online

S. Boll et al. (Eds.): MMM 2010, LNCS 5916, pp. 400–410, 2010.

video search engines, such as YouTube[1] or Blinkx[2]. However, query by text relies on the availability of satisfactory textual descriptions of a video and its content. Most of these online video search systems rely on annotations provided by users to provide sufficient descriptions of videos. It is common for users to have diverse perceptions about the same video and as such will annotate that video differently. This can result in synonyms, polysemy and homonymy, which make it particularly difficult for other users to retrieve the same video [6]. It has also been found that users are hesitant to provide a large numbers of annotations [7]. To overcome these problems these videos require additional and superior annotations.

In order to help provide these annotations we investigate the use of a grouping interface for video search to provide additional annotations for video collections. When using this grouping interface, the users organise and conceptualise their search, by adding relevant videos to groups in a workspace. These groups relate to some semantic concept which forms part of the user's solution for the search task that they are carrying out. As these annotations are provided as part of the search process they overcome the tedium for users involved in providing annotations and do not rely on the users expending an additional effort to annotate the collection. In addition as the users are providing annotations to help their own search they are more likely to provide high quality annotations. While the overall goal of this system is not to provide annotations, it is a supplementary benefit for the use of this system. These usage based annotations are extremely useful to other users, as in numerous domains such as the web, medical image retrieval, digital libraries etc. these tags derived from usage data can be used as part of the retrieval index to retrieve high quality multimedia files. In previous work we have evaluated the usability and the benefit of the grouping paradigm for video search [8]. Two user evaluations were carried out in order to determine the usefulness of this grouping paradigm for assisting users. The first evaluation involved users carrying out broad tasks on YouTube, and gave insights into the application of our interface for search on a vast online video collection. The second evaluation involved users carrying out focused tasks on the TRECVID 2007 video collection [12], allowing a comparison over a local collection, on which we could extract a number of content-based features. The results of those evaluations showed that the use of our system results in an increase in user performance and user satisfaction, showing the benefit of a grouping paradigm for video search for various tasks in a variety of diverse video collections. In this paper we examine the log files from these user evaluations to see how users organised their search results and how this organisation could be used to annotate the respective collections. The remainder of this paper is organised as follows: in the following section we will describe the grouping interface that was used for the evaluations outlined above. Subsequently, in Section 3 we will briefly describe the evaluations that were executed using this system and how the user interaction with the system could be used to annotate video collections. Finally, we will provide a discussion of our work and some conclusions.

---

[1] http://www.youtube.com
[2] http://www.blinkx.com/

2 Grouping Interface (ViGOR)

2.1 Search Functionality

The search interface that was used for our evaluations provide facilities that enable the user to both search and organise video search results successfully. The interface comprises of a search panel (A), results display area (B) and workspace (C). The users enter a text based query in the search panel to begin their search. The result panel is where users can view the search results (a). Additional information about each video shot can be easily retrieved by placing the mouse cursor over a video keyframe for longer than 1.5 seconds, which will result in any text associated with that video being displayed to the user (e). If a user clicks on the play button the highlighted video shot will play in a popup panel. Users can play, pause, stop and navigate through the video as they can on a normal media player.

Groups can be created by clicking on the create group button. Users must then select a textual label for the group and can potentially add any number of annotations to the group, but each group must have at least one annotation. Drag-and-drop techniques allow the user to drag videos into a group or reposition the group in the workspace (b). Groups can be deleted, minimised and moved around the workspace using a number of buttons (f). Any video can belong to multiple groups simultaneously, meaning that a video can have multiple labels from multiple groups. The workspace is designed to accommodate a potentially infinite number of groups. Each group can also be used as a starting point for further search queries. The description above describes the basic functionality of the grouping interface; two slightly different versions of the grouping interface were used for evaluation on two different datasets i.e. YouTube and TRECVID. When being used in conjunction with YouTube the interface offers three expansion options for each group (see Figure 1 (c, d)): 1) related videos; 2) videos from the same user 3) and text expansion which is the result of a new search using text extracted from the selected videos. When being used for the TRECVID 2007 collection the interface offered three different expansion options for each group: 1) similar colour; 2) similar shapes, this was retrieved using edge histograms 3) and similar homogenous texture.

In addition to our system outlined briefly above, a number of different innovative search interfaces which involve some sort of grouping or organisation have been developed to assist users while searching for multimedia. PicturePiper [3] provides a means to allow users access to images on the web related to a topic of interest. CueFlik [4] is a web-based image search system that allows users to create their own rules for ranking images based on visual features. Users can then re-rank potentially relevant results according to these rules. EGO [13] is a tool for the organisation of image collections. The main component of EGO is a workspace; the workspace serves as an organisational ground for the user to construct groupings of images. A recommendation system available in EGO observes the user's actions, which enables EGO to make suggestions of potentially relevant images based on a selected group of images. ImageGrouper [10] is another interface for digital image search and organisation,

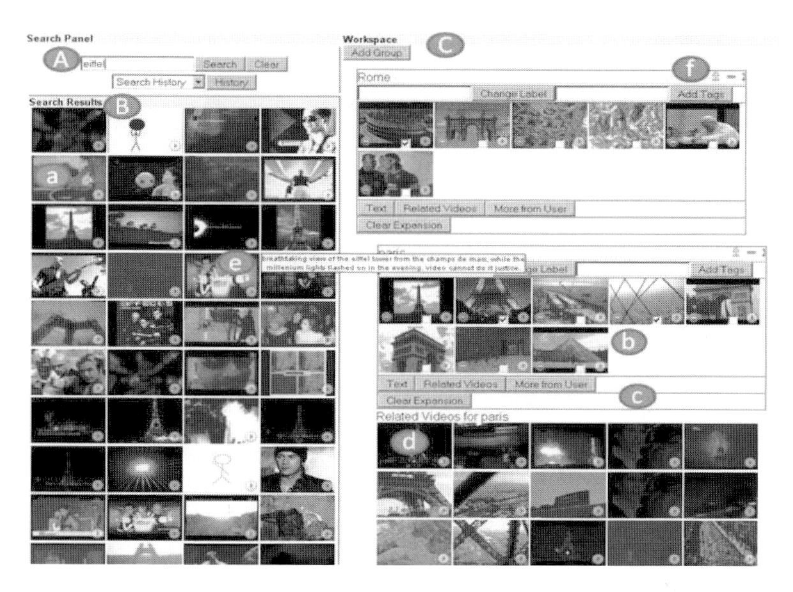

Fig. 1. Screen shot of grouping interface

using ImageGrouper it is possible to search, annotate, and organise images by dragging and grouping images on the workspace. The MediaGLOW system [5] presents an interactive workspace that allows users to organise photographs. Users can group photographs into stacks in the workspace; these stacks are then used to create neighbourhoods of similar photographs automatically. The FacetBrowser [14] is a video search interface that supports the creation of multiple search "facets", to aid users carrying out complex video search tasks involving multiple concepts. Each facet represents a different aspect of the video search task.

As can be seen there are a number of innovative systems that are harnessing the potential of users to organise multimedia search results to assist their search. The organisational features available in many of the systems outlined above can potentially be used to assist annotation of multimedia documents, in the following subsection we demonstrate specifically how the grouping interface that has been described here can be used for annotation of multimedia documents.

2.2 Support for Annotation

The support for annotation in the group interface is provided through the grouping functionality. In the system videos which have some sort of link between them are stored in a group, this group also must have a title which assigns a semantic label to all of the videos of the group and in a way signifies or identifies the link between the videos in the group. In addition as videos can be members of multiple groups it is possible to associate multiple labels with a single video, and also to create links between groups and labels that are assigned to groups.

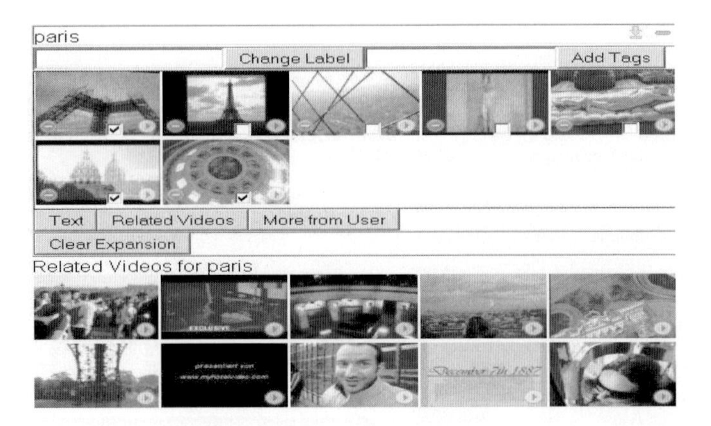

Fig. 2. Example of a user created group for Paris

This approach is similar in some ways to the GroupMe! [1]. GroupMe! allows users to organise and arrange Web resources into groups. This content can be inspected by users immediately as resources are visualised in these groups. These groups are then used as the basis for folksonomy-based ranking strategies that exploit the group structure in their system. However, the annotation that occurs while using our grouping interface for video search is a by product of user interaction, the main goal of our group interface is to aid users conceptualise and organise their search results. In this way our approach is more similar to the work of van Ahn et al. on the ESP game [2]. In their system the annotation is brought about by harnessing the knowledge of people involved in a different task to annotation. In order to measure the effectiveness of the grouping interface for search and retrieval we conducted two user-centred evaluations [8]. As part of these evaluations we recorded user interactions with the system and collections in a number of log files. These log files form the basis for our investigation of the potential of our grouping interface for providing annotations, details of these evaluations and our results are given in the following section.

3 User Evaluations

3.1 YouTube Evaluation

Evaluation Set Up

For the YouTube evaluation four simulated work task situations were created in order to provide broad, ambiguous, open ended tasks for the users. These tasks were related to different topics and multiple aspects of these topics. A between subjects design was adopted for this evaluation. Two interfaces were evaluated; the first was the grouping interface. The second interface mimicked the functionality of YouTube. Users could search via text and when a video was playing users were presented with lists of related videos and videos from the same user, in the same way that YouTube does, this also mimicked the functionality available through the group expansions explained above (see Section 2.1). The order of tasks was varied; this was to avoid any order or learning effect associated with the tasks. Each participant was given five minutes training on their search system and was allowed to carry out training tasks. Users had a maximum of 20 minutes to complete each of these tasks. 16 participants took part in our evaluation, they were randomly divided into two groups of 8 and each group used one of the systems. Thus we have a group of 8 users that used the grouping interface to search for and annotate videos for four tasks each. Full details of the evaluation can be found in [8].

Results

Annotation Rate

Our analysis began by looking at the number of videos that were annotated by the users. In total the users added 1218 videos to groups in approximately 640 minutes (8 users * 4 topics * 20 minutes maximum per topic). On average users added 38.065 videos to groups per search session. Of the videos that were added to the groups 976 (80.131%) were unique; some overlap is to be expected as users were carrying out the same topic, but the relatively low level of overlap is encouraging. In total 2430 terms were added to the videos, this is an average of 2.025 terms in each group label. 152 of these terms were unique, with many being repeated as numerous videos were added to the same group.

Quality of Annotations

One concern that we had was that if users were retrieving videos by mainly using keyword based search that the labels in the groups to which those videos were added would overlap with the annotations that were already attached to that video. To address these concerns we compared the labels assigned via groups with the tags, title and description for each video. It was found that for 426 videos there was overlap with one of these fields with one of the terms that the user assigned in our evaluation, and there was no overlap for 550 videos. Looking at term level, i.e. each individual term in the group name compared with the tags, title and description for each video, it was found that there was overlap with existing annotations for 975 terms and no overlap for 1455 terms.

Following the methodology of Sigburjornsson and van Zwol [11] we have mapped annotations onto the WordNet broad categories. However, unlike their approach we allow tags to be mapped to more than one category as we do not want to infer meaning. Instead we want to compare annotations from different sources to see if similar descriptions of the video content are being created or if we are getting a different description. To that end we compare the distribution of categories from the annotations that already exist on YouTube, the annotations provided by the users using the grouping interface and the queries issued by the users. Other forms of metadata such as titles and descriptions are not investigated here, as previous studies have demonstrated that these types of metadata converge to a different description of multimedia documents in comparison with tags [9]. The queries are investigated as they provide an alternative source of implicit annotations for multimedia documents. In addition we also wish to discover if that annotations provided by using the group interface are closer to queries that are used to retrieve the videos or are closer to user assigned descriptions from YouTube. Figure 3 shows the distribution of the different sources of metadata over the most common WordNet categories. This figure does not show the

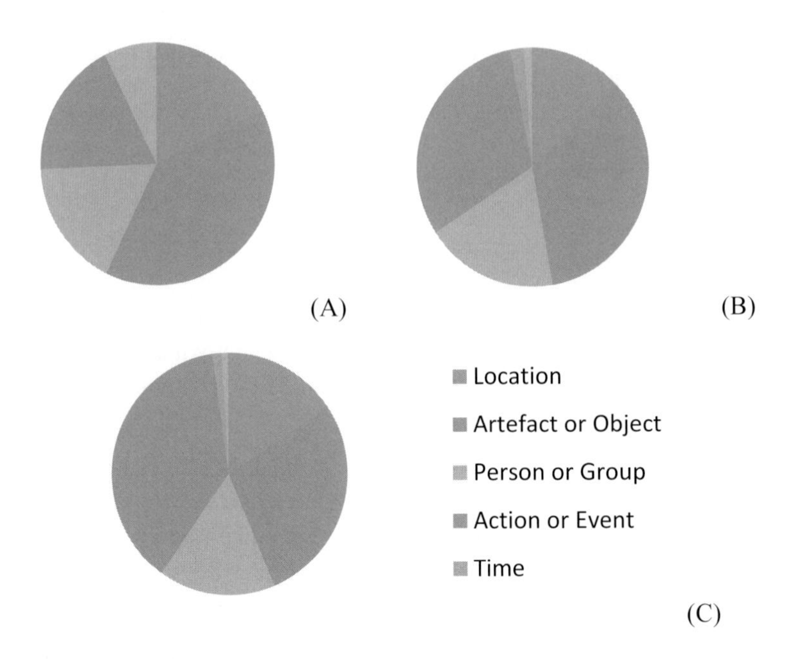

Fig. 3. Most Frequent WordNet categories for queries (A), YouTube tags (B) and user assigned tags (C)

tags that could not be categorised, unlike the work of Sigburjornsson and van Zwol [11] on Flickr[3] tags, the majority of the tags in our study could be categorised using WordNet. The results of the categorisation of the tags assigned during this evaluation are different in all cases to that of Sigburjornsson and van Zwol [11], however, it should be noted that in this case we are investigating a small portion of very specific YouTube videos, where as Sigburjornsson and van Zwol [11] were investigating a large number of Flickr photographs from broad and diverse categories. In our evaluation in all cases we find that objects are the most commonly assigned tags, followed by people or groups. However, it can be seen quite clearly that the descriptions based on the YouTube tags and the group interface tags are more closely aligned, than any description that can be inferred from using query terms. Although the volume of tags that can be assigned to videos from groups that are part of the search process in comparison with YouTube is much smaller, it appears that the description of the videos that is created is very similar. Indeed it seems that the way in which users search for videos and the way in which users organise their results are different. Thus using query terms as annotations may not result in the same description that an object would be assigned by using the description created by tags.

The results of our initial investigation are very positive for a number of reasons. First they show that users can annotate videos at a relatively fast rate as part of the search process. While this rate may not be as fast as other online initiatives such as the ESP game [2], our approach is a seamless part of the natural search process rather than an artificial process external to the media search process, although both have

[3] www.flickr.com

benefits and are valid approaches. In addition it was found that on average around 2 terms were added to each video, again an encouraging result as previous research on online video retrieval has indicated that more annotations results in a video being more likely to be retrieved up to a certain point [7]. Thirdly it was found that while there is some overlap with already existing annotations that quite often users are providing new and different annotations, where overlap exists with existing annotation the possibility of creating a folksonomy exists. Finally it was found that the description created was aligned more closely with other descriptions created using tags, rather than being closer to search terms that were used to retrieve the documents. This indicates that users are giving a more general description of the video content than the search terms used to find the video, but the interaction is still part of the search process meaning that it is in the user's best interest to provide the highest quality of description possible.

3.2 TRECVID Evaluation

The evaluation of our approach for online video search using YouTube was followed up with a second evaluation that evaluated our approach on a collection of broadcast television. In this scenario we did not have user annotations and the needs and problems are slightly different than in an online scenario.

Evaluation Set Up

For the second evaluation a different group of participants carried out focused video search tasks using the TRECVID 2007 collection and tasks. In 2007 the TRECVID collection contained 18,142 shots (over 100 hours) of Dutch magazine television. For the TRECVID 2007 interactive search evaluations there were a total of 24 tasks. For our evaluation we limited the number of tasks that the users carry out to 8. This allowed us to carry out more evaluations, as 24 individual search topics did not have to be carried out for each participant. In order to examine user interactions on different types of tasks we choose the 8 tasks which had the highest number of shots marked as being relevant during TRECVID runs. For our evaluation we adopted a 2-searcher-by-2-topic within subject's Latin Square design. Two interfaces were evaluated; the first was the grouping interface. The second interface allowed the users to query the collection via query by text and query by example. Each participant carried out two tasks using each interface. The order of system usage was varied as was the order of the tasks; this was to avoid any order effect associated with the tasks or with the systems. Each participant was given five minutes training on each system and was allowed to carry out training tasks. Each actual task had a fifteen minute maximum time limit. 16 participants took part in our evaluation. Thus we have a group of 16 users that used the grouping interface to search for videos for 2 tasks each. Full details of the evaluation can be found in [8].

Results

Annotation Rate

Once again our analysis began by looking at the number of videos that were annotated by the users. In total the users added 468 videos to groups in approximately

780 minutes (16 users * 2 topics * 15 minutes maximum per topic). On average users added 14.625 videos to groups per search session. As these tasks were more direct and specific than for the YouTube evaluation it would be expected that users would retrieve less results. In addition for the tasks chosen there were 420 relevant shots in the collections on average (Maximum of 1175 relevant shots, minimum of 210 relevant shots). Of the videos added to groups by the users 330 (78.571%) were unique, some overlap is to be expected as users were carrying out the same topic, but the relatively low level of overlap is encouraging. In total 560 terms were added to the videos, this is an average of 1.196 terms in each group label. 81 of these terms were unique, with many being repeated as numerous videos were added to the same group.

Quality of Annotations

Once again we compared the labels assigned via groups with the already existing annotations for each video for each video. However in this case the already attached annotations for each video had been created using automatic speech recognition and automatic translation from Dutch to English. Some videos in the collection did not have any metadata associated with them. It was found that for 58 videos there was overlap with the metadata with one of the terms that the user assigned in our evaluation, and there was no overlap for the remaining 410 videos. Looking at term level, i.e. each individual term in the group name compared with the metadata for each video, it was found that there was overlap with existing annotations for 83 terms and no overlap for 477 terms. Once again the query terms and user assigned tags were categorised using WordNet. It was found that for these more narrow tasks that the query terms and user assigned tags were more closely aligned than for the YouTube evaluation. However, there were still some differences and we did not have any other user assigned metadata with which to make a comparison. As with the results of for the online video search, the results of our initial investigation for digital libraries are positive. Once again users can annotate videos at a relatively fast rate as part of the search process. While this rate is not as fast as the first evaluation, this is due to the nature of the collections and the tasks. Indeed it is even more encouraging as these tasks were extremely focused and narrow and there were relatively small numbers of relevant videos. Although approximately only 1 term was added to each video via groups, this is not a bad result as in many digital archives and broadcast scenarios only relatively small numbers of people interact with these video collections in comparison with the vast numbers online, any additional annotation or search terms is beneficial. Finally due to the sparseness of the metadata attached to these videos that there was very little overlap with already existing metadata, showing that new and potentially beneficial search terms were being added to each video. In conclusion it has been shown that users can annotate reasonable numbers of videos in a short period of time as part of their search process. This was achieved with no extra over head to the users and with very little overlap between user's assigned annotations and already existing annotations for the videos. In the next section we will discuss our findings and give some directions for future work.

4 Conclusion and Discussion

Our grouping interface is a system that helps users to carry out complex video search tasks by allowing users to organise their results into semantic groups and to share results between multiple groups. Results of user evaluations have shown the benefits of this search paradigm for various video search tasks in a number of scenarios [8]. A supplementary benefit of using this system is that users are adding annotations to the groups that they create as part of the search process and thus to the videos that are part of that group. Having proper labels associated to videos either on the Web or in digital libraries could allow for more accurate image retrieval, could improve the accessibility of sites, and could help users block inappropriate images.

Although the main application of the grouping interface game is not to label images, our main contribution is to show how this system can be used to overcome the well known problem of insufficient annotations and textual descriptions for multimedia in many contexts [7]. What is normally a tedious task for users, i.e. providing annotations or textual descriptions for multimedia, has become a meaningful part of the search process. The results of our analysis of the logs from two evaluations have shown that even small groups of users can annotate relatively large numbers of videos in a short space of time. Indeed despite the fact that these users are carrying out the same tasks the quite often annotate different videos; almost 80% of the videos added to groups in most tasks were unique. One concern might be that if users are retrieving these video through textual searches that the labels that are assigned via groups may already be part of the metadata for that video. However it was found that for the majority of videos that the group labels did not overlap with existing annotations. This is an important finding as previous research has indicated that the more annotations that a video has the more likely it is to be retrieved, up to a certain threshold of annotations [7]. In addition the annotations that were created using the groups match the distribution of other user assigned metadata, rather than any other source of implicit annotation. This indicates that users are providing high quality annotations independent of how they formulate queries as part of the search process, but the annotation is an inherent part of the search process.

There are a number of potential directions for future work. It may be possible to address similar problems in the similar fashion. For example, the grouping interface can be used, with only minor changes, to group and label sound, images or documents. In addition the same process could potentially be used to provide multilingual labels to videos. While it has not been addressed here, in future work we will also look at the effect of this labelling on the retrieval process, i.e. can it bring about an increase in the retrieval performance of users and result in user retrieving more useful videos. In conclusion this interface is a first important step in that we demonstrate how annotation can seamlessly become an explicit part of multimedia the search process, benefiting both the searcher and people who will carry out future searches. Future work in this direction could be a step towards bridging the semantic gap.

Acknowledgements

Thank you to David Vallet, Ivan Cantador and David Hannah for all of their help with this work. This research was supported by the European Commission, under contract IST-FP6-027122 (SALERO).

References

1. Abel, F., Henze, N., Krause, D.: Groupme! In: Proceeding of WWW 2008, pp. 1147–1148 (2008)
2. von Ahn, L., Dabbish, L.: Labelling images with a computer game. In: Proceedings of CHI 2004, pp. 319–326 (2004)
3. Fass, A.M., Bier, E.A., Adar, E.: PicturePiper: using a re-configurable pipeline to find images on the Web. In: Proceedings of UIST 2000, pp. 51–62 (2000)
4. Fogarty, J., Tan, D.S., Kapoor, A., Winder, S.A.J.: CueFlik: interactive concept learning in image search. In: Proceedings of CHI 2008, pp. 29–38 (2008)
5. Girgensohn, A., Shipman, F., Wilcox, L., Turner, T., Cooper, M.: MediaGLOW: organizing photos in a graph-based workspace. In: Proceedings of IUI 2009, pp. 419–424 (2009)
6. Guy, M., Tonkin, E.: Folksonomies Tidying Up Tags. D-Lib Magazine 12(1) (2006)
7. Halvey, M.J., Keane, M.T.: Analysis of online video search and sharing. In: Proceedings of HT 2007, pp. 217–226 (2007)
8. Halvey, M., Vallet, D., Hannah, D., Jose, J.M.: ViGOR: a grouping oriented interface for search and retrieval in video libraries. In: Proceedings of JCDL 2009, pp. 87–96 (2009)
9. Marshall, C.C.: No bull, no spin: a comparison of tags with other forms of user metadata. In: Proceedings of JCDL 2009, pp. 241–250 (2009)
10. Nakazato, M., Manola, L., Huang, T.S.: ImageGrouper: A Group-Oriented User Interface for CBIR and Digital Image Arrangement. J. Vis. Lang. Comput. 14(4), 363–386 (2003)
11. Sigurbjörnsson, B., van Zwol, R.: Flickr tag recommendation based on collective knowledge. In: Proceeding of the WWW 2008, pp. 327–336 (2008)
12. Smeaton, A., Over, P., Kraaij, W.: Evaluation campaigns and TRECVid. In: Proceedings of the 8th ACM international Workshop on MIR, pp. 321–330 (2006)
13. Urban, J., Jose, J.M.: A Personalised Multimedia Management and Retrieval Tool. International Journal of Intelligent Systems 21(7), 725–745 (2006)
14. Villa, R., Gildea, N., Jose, J.M.: A faceted interface for multimedia search. In: Proceedings of the SIGIR 2008, pp. 775–776 (2008)

Human Action Recognition in Videos Using Hybrid Motion Features

Si Liu[1,2], Jing Liu[1], Tianzhu Zhang[1], and Hanqing Lu[1]

[1] National Laboratory of Pattern Recognition Institute of Automation, Chinese Academy of Science, Beijing 100190, China
[2] China-Singapore Institute of Digital Media, 119615, Singapore

Abstract. In this paper, we present hybrid motion features to promote action recognition in videos. The features are composed of two complementary components from different views of motion information. On one hand, the period feature is extracted to capture global motion in time-domain. On the other hand, the enhanced histograms of motion words (EHOM) are proposed to describe local motion information. Each word is represented by optical flow of a frame and the correlations between words are encoded into the transition matrix of a Markov process, and then its stationary distribution is extracted as the final EHOM. Compared to traditional Bags of Words representation, EHOM preserves not only relationships between words but also temporary information in videos to some extent. We show that by integrating local and global features, we get improved recognition rates on a variety of standard datasets.

Keywords: Action recognition, Period, EHOM, Optical flow, Bag of words, Markov process.

1 Introduction

With the wide spread of digital cameras for public visual surveillance purposes, digital multimedia processing has been received increasing attention during the past decade. Human action recognition is becoming one of the most important topics in computer vision. The results can be applied to many areas such as surveillance, video retrieval and human computer interaction etc.

Successful extraction of good features from videos is crucial to action recognition. Yan et al. [1] extend the 2D box feature to 3D spatio-temporal volumetric feature. Recently, Ju Sun et al. [2] propose to model the spatio-temporal context information in a hierarchical way. Among all the proposed features, there is a huge family directly describing motion. For example, Bobick and Davis [3] develop the temporal template which captures both motion and shape. Laptev [4] extracts motion-based space-time features. This representation focuses on human actions viewed as motion patterns. Ziming Zhang et al. [5] propose Motion Context (MC) which captures the distribution of the motion words and thus summarizes the local motion information in a rich 3D MC descriptor. These motion based approaches have shown to be successful for action recognition.

S. Boll et al. (Eds.): MMM 2010, LNCS 5916, pp. 411–421, 2010.

Acknowledging the discriminative power of motion features, we propose to combine period and enhanced histograms of motion words (EHOM) to describe motion in the video. Considering the large variation in realistic videos, our method is more feasible to extract compared to 3D volumes, trajectories, spatio-temporal interest points etc.

Period Features: Periodical motion occurs often in human actions. For example, running and walking can be seen as periodical actions in the leg region. Therefore, a variety of methods use period features to perform action recognition. Cutler and Davis [6] compute an object's self-similarity as it evolves in time. For periodic motion, the self-similarity measure is also periodic, and they apply Time-Frequency analysis to detect and characterize the periodic motion. What's more, Liu et al. [7] also classify periodic motions.

Optical Flow Features: Efros et al. [8] recognize the actions of small scale figures using features derived from optical flow measurements in a spatio-temporal volume for each stabilized human figure. Alireza Fathi et al. [9] develop a method constructing mid-level motion features which are built from low-level optical flow information. Saad Ali et al. [10] propose a set of kinematic features that are derived from the optical flow. All of them achieve good results.

Hybrid Features: We strongly feel that period and optical flow are complementary for action recognition mainly for two reasons. First, optical flow only capture the motion between two adjacent frames thus bringing in local problems, while period can capture global motion in time domain. For example, suppose we want to differentiate walking from jogging. Because they produce quite similar optical flow, it is difficult to distinguish them based on optical flow features alone. Yet, the period feature can easily distinguish them because when somebody jogs, his/her legs move faster. Second, period information is not obvious in several actions such as bending. However, the optical flow of bending with forwarding components and rising up components are quite discriminative. To exploit the synergy, we choose to use hybrid features consisting of both period features (capturing global motion) and optical flow features (capturing local motion) to develop an effective recognition framework.

2 Overview of Our Recognition System

The main components of the system are illustrated in Fig. 1. We first produce a figure-centric spatio-temporal volume (see Fig. 1(a)) for each person. It can be obtained by using any one of detection/tracking algorithms over the input sequence and constructing a fixed size window around it. Afterwards, we divide every frame of the spatio-temporal volume into $m \times n$ blocks to make the proposed algorithm robust to noise and efficient to be computed. By doing this, we also implicitly maintain spatial information in the frame when constructing

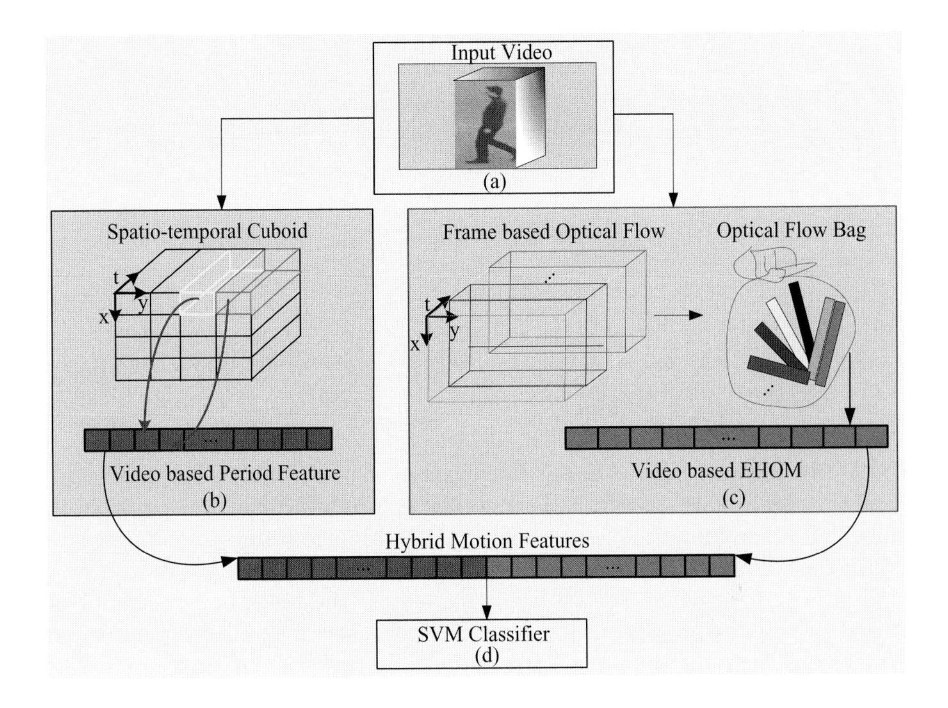

Fig. 1. The Framework of Our Approach

features. As a result, we get $m \times n$ smaller spatio-temporal cuboids consisting of all the blocks at the corresponding location in every frame(Fig. 1(b)). Sec. 3 addresses quasi-period extraction of the cuboid to describe the global motion in time-domain. The feature of all the cuboid are concatenated to form the period feature of the video. Sec. 4 introduces the EHOM feature extraction. Specifically, each frame's optical flow is first assigned a label by k-means clustering algorithm. Based on these labels, Markov process is used to encode the dynamic information (Fig. 1(c)). Then the hybrid features are constructed and fed into the subsequent multi-class SVM classifier (Fig. 1(d)). The experimental results are reported in Sec. 5. Finally, the conclusions are given in Sec. 6.

3 Period Feature Extraction

Based on the spatio-temporal cuboid obtained by dividing the original video, our frequency extraction approach is appearance-based similar to [11]. Fig. 2 shows the block diagram of the module. First, we use probabilistic PCA (pPCA) [12] to detect the maximum spatially coherent changes over time in the objects appearance. The input data that are spatially correlated are grouped together. Different from pixel-wise approaches, pPCA considers these pixels as one physical entity. Hence the method is robust to noise. The final output consists of a

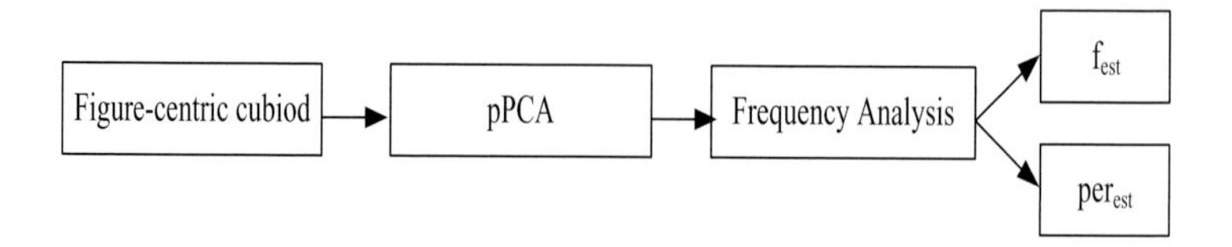

Fig. 2. Block diagram of period extraction module

combination of two indicators: the estimated period f_{est} and the degree of periodicity per_{est}. Next, we will describe the pPCA phase and frequency analysis phase respectively.

pPCA for Robust Periodicity Detection: Let $X^{D\times N} = [x_1x_2....x_N]$ represent the input video, with D the number of pixels in one frame and N the number of image frames. The rows of an aligned image frame are concatenated to form the column x_n. The optimal linear reconstruction $\hat{X}$ of the data is given by: $\hat{X} = WU + \bar{X}$, where $W^{D\times Q} = [w_1w_2...w_Q]$ is the set of orthonormal basis vectors, principal components matrix $U^{D\times Q}$ is a set of Q-dimensional vectors of unobserved variables(see Fig.3(b)) and $\bar{X}$ the set of mean vectors $\bar{x}$. Each eigenvector's corresponding eigenvalue is indicated by $\Lambda = diag(\lambda_1, \lambda_2, ...\lambda_d)$ of the covariance matrix S of the input data X: $S = V\Lambda V^T$, which is calculated by eigenvalue decomposition. The dimension Q is selected by setting the maximum percentage of retained variance we want to preserve in the reconstructed matrix $\hat{X}$.

Frequency Analysis: Periodogram is a typical non-parametric frequency analysis method which estimates the power spectrum based on the Fourier Transform of the autocovariance function. We choose the modified periodogram of the non-parametric class: $P_q(f) = \frac{1}{N}\left|\sum_{n=0}^{N-1} w(n)x(n)\exp(-jn2\pi f)\right|^2$, where N is the frame length, w(n) is the window used and x(n) is principal component vector u_q^T from the pPCA(see Fig. 3(b)). By weighing the spectra $P_q(f)$ with the relative percentages λ_q^* of the retained variance and summing them together, a spectrum is obtained by $\bar{P}(f) = \sum_{q=1}^{Q} \lambda_q^* P_q(f)$, where $\lambda_q^* = \frac{\lambda_q}{\sum_{d=1}^{D}\lambda_d}$.

In order to detect the dominant frequency component in the spectrum $\bar{P}(f)$ (see Fig. 3(c)), we first detect the peaks and local minima which define the peaks' supports. The peaks with a frequency lower than $\frac{f_s}{N}$ are discarded, with f_s being the sampling rate of the video and N the frame length. Afterwards, starting from the lowest found frequency to the highest, each peak is checked against the others for its harmonicity. We require that a fundamental frequency

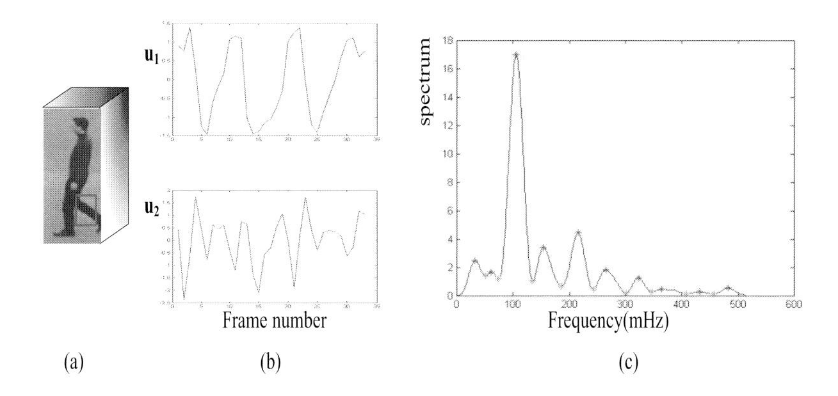

Fig. 3. (a) The spatio-temporal cubic of running is denoted in red. (b) The first 2 principal components of the cubic. (c) The weighted spectrum of running, the peaks is denoted in red and their supports in green.

should have a higher peak than its harmonics and a tolerance of $\frac{f_s}{N}$ is used in the matching process. We select the one group k with the highest total energy to represent the dominant frequency component in the data. The total energy is the sum of the area between the left and right supports $E(.)$ of the fundamental frequency peak f_k^0 and its harmonics f_k^i :

$$f_{est} = \arg\max_{f_k^0} \left\{ E(f_k^0) + \sum_i E(f_k^i) \right\} \tag{1}$$

The estimated frequency f_{est} of Fig. 3(c) is $120mHz$, which means that the motion repeats itself every 8.33 frames. Note that no matter whether the data is periodical or not, as long as there exist some minor peaks in the spectrum $\bar{P}(f)$, the above method may still give a frequency estimate. So we adopt to compare the energy of all peaks found in $\bar{P}(f)$ with the total energy to separate the above cases:

$$per_{est} = \frac{\sum_{k=1}^{K} E_\Delta(f_k)}{\sum_f \bar{P}(f)}, \tag{2}$$

where K is the number of peaks detected and $E_\Delta(f_k)$ as the area of a triangle formed by the peak and its left and right supports. Note that the peak supports should have zero energy for the spectrum of periodic signal. By only using the triangle area for the nominator in eq.(2), we assign a lower per_{est} value for quasi-periodic signal. The obtained per_{est} and f_{est} are then concancated to generate the period component of the hybrid feature.

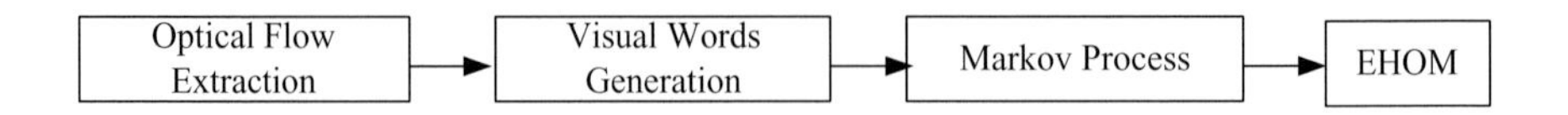

Fig. 4. Block diagram of EHOM extraction module

4 Enhanced Histograms of Motion Words Extraction

As motion frequency is a global and thus coarse description of motion, we adopt a local and finer motion mode descriptor — optical flow as a complement. Fig. 4 shows the block diagram of the module. First, We extract the optical flow of every frame. Then we generate the codebook by clustering all optical flow in training dataset. Afterwards, we would have directly computed the histogram of words occurrences over the entire video sequence based on the obtained visual words, but by doing so the time domain information is lost. For action recognition, however, the dynamic properties of these object components are quite essential, e.g. for the action of standing up or airplane taking off. That is why we go one step further and combine a optical flow based Bags of Words representation with Markov process [13] to get EHOM. It is independent of the length of video and simultaneity maintains both the dynamic information and correlations between words in the video. To our best knowledge, we are the first to consider the relationship between motion words in action recognition.

The Lucas and Kanade [14] algorithm is employed to compute the optical flow for each frame. The optical flow vector field F is then split into horizontal and vertical components of the flow, F_x and F_y. These two non-negative channels are then blurred with a gaussian and normalized. They will be used as our optical flow motion features for each frame. Blurring the optical flows reduces the influence of noise and small spatial shifts in the figure centric volume. For each frame, optical flow features of each block are concatenated to generate a longer vector.

Next, we represent a video sequence as Bags of Words. Our method represents a frame as a single word. In other words, a "word" corresponds to a "frame", and a "document" corresponds to a "video sequence" in our representation. Specifically, given the optical flow vector of every frame in the video, we construct a visual vocabulary with the k-means algorithm and then assign each frame to the closest (we use Euclidean distance) vocabulary word. In fig. 5(a), different colors mean the corresponding frames are assigned to different visual words.

As we mentioned, we go one step further than Bags of Words by considering the relationship between the motion words using Markov process. Before going deep into details, we present some basic definitions in Markov chains. A Markov Chain [15] is a sequence of random observed variables with the Markov property. It is a powerful tool for modeling the dynamic properties of a system. The markov stationary distribution, associated with an ergodic Markov chain, offers a compact and effective representation for a dynamic system.

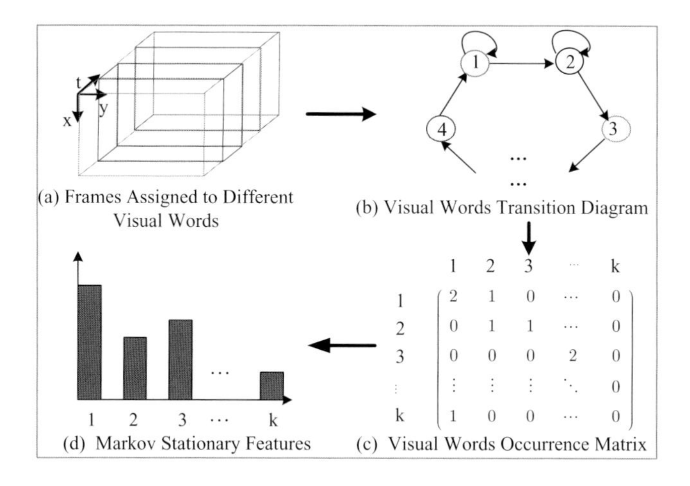

Fig. 5. Construction of E_HOM

Theorem 4.1. Any ergodic finite-state Markov chain is associated with a unique stationary distribution (row) vector , such that $\pi P = \pi$.

Theorem 4.2. 1) The limit $A = \lim_{x \to \infty} A_n$ exists for all ergodic Markov chains, where the matrix

$$A_n = \frac{1}{n+1}(I + P + ... + P^n) \tag{3}$$

2) Each row of A is the unique stationary distribution vector π.

Hence when the ergodicity condition is satisfied, we can approximate A by A_n. To further reduce the approximation error when using a finite n, π is calculated as the column average of A_n.

For consecutive frames in a fixed-length time window with their codebook labels F and F'. we translate the sequential relations between these labels into a directed graph, which is similar to the state diagram of a Markov chain (Fig. 5 (b)). Here we get K vertices corresponding to the K codewords, and weighted edges corresponding to the occurrence of each transition between the words. We further establish an equivalent matrix representation of the graph(Fig. 5 (c)), and perform row-normalization on the matrix to arrive at a valid transition matrix P for a certain Markov chain. Once we obtain the transition matrix P and make sure it is associated with an ergodic Markov chain, we can use eq. 3 to compute π (Fig. 5(d)).

5 Experiments

Here we briefly introduce the parameters used in our experiments. In the period feature extraction phase, pPCA retained variance is 90%, Hanning window is used for periodogram smoothing. If per_{est} is less than 0.4, in other words, the signal is not periodic, we assign the corresponding f_{est} to zero. In EHOM extraction phase, the vocabulary size is set to be 100 and we use $n = 50$ to estimate A by A_n. The length of time window is 20 frames.

For classification, we use support vector machine (SVM) classifier with RBF kernel. We adopt PCA to reduce the dimension of period feature to make it the same as the dimension of EHOM. To prove the effectiveness of our hybrid feature, we test our algorithm on two human action datasets: KTH human motion dataset [16] and Weizmann human action dataset [17]. For each dataset, we perform leave-one-out cross-validation. During each run, we leave the videos of one person as test data each time, and use the rest of the videos for training.

5.1 Evaluating Different Components in Hybrid Feature

We will show that both components in our proposed hybrid feature are quite discriminative. The period features of 6 activities in KTH database are illustrated in Fig. 6. We can see that the bottom three actions have different frequencies in the leg regions (denoted in red ellipses). Specifically, $f_{running} > f_{jogging} > f_{walking}$, where f stands for the frequencies of leg regions. It conforms to the intuitive understanding. Fig. 7 shows the comparison of our proposed EHOM with traditional BOW representation and illustrates that better results are achieved by considering correlations between motion words.

The following experiment is to demonstrate the benefit of combining period and EHOM feature. Fig. 8 shows the classification results for period features, EHOM features and the hybrid of them. The average accuracies are 80.49%, 89.38% and 93.47% respectively. It shows that the EHOM component achieves better result than the period component. We can also draw the conclusion that the hybrid feature is more discriminative than either component alone.

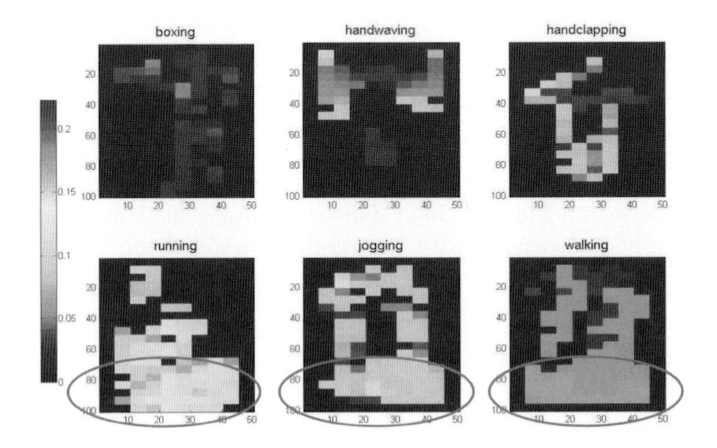

Fig. 6. The frequencies of different actions in KTH database

Methods	Mean Accuracy
BOW	87.25%
EHOM	**89.38%**

Fig. 7. The comparison between BOW and EHOM on KTH dataset

Methods	Mean Accuracy
period feature	80.49%
EHOM feature	89.38%
hybrid feature	**93.47%**

Fig. 8. The comparison of different features about mean accuracy on KTH dataset

5.2 Comparison with the State-of-the-Art

Experiments on Weizmann Dataset: The Weizmann human action dataset contains 93 low-resolution video sequences showing 9 different people, each of which performing 10 different actions. We have tracked and stabilized the figures using background subtraction masks that come with the dataset. In Fig. 9(a) we have shown some sample frames of the dataset. The confusion matrix of our results is shown in Fig. 9(b). Our method has achieved a 100% accuracy.

Experiments on KTH Dataset: The KTH human motion dataset, contains six types of human actions (walking, jogging, running, boxing, hand waving and hand clapping). Each action is performed several times by 25 subjects in four different conditions: outdoors, outdoors with scale variation, outdoors with

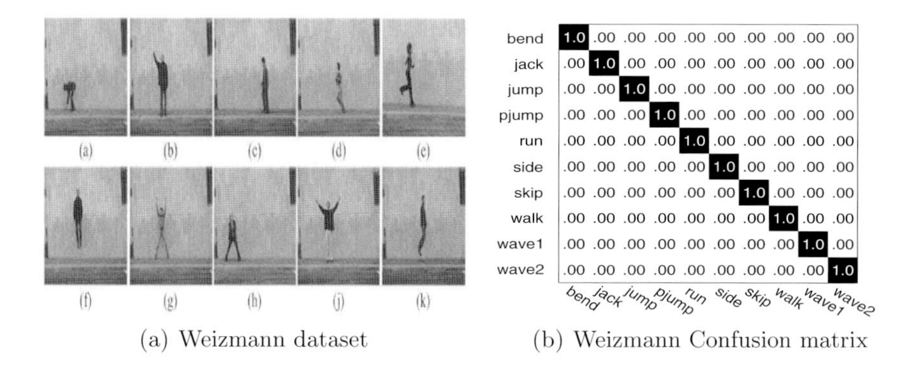

(a) Weizmann dataset (b) Weizmann Confusion matrix

Fig. 9. Results on Weizmann dataset: (a) sample frames. (b) confusion matrix on Weizmann dataset using 100 codewords. (overall accuracy=100%)

Methods	Mean Accuracy
Saad Ali [10]	87.70%
Alireza Fathi [9]	90.50%
Ivan Laptev [18]	91.8%
Our method	**93.47%**

Fig. 10. The comparison of different methods about mean accuracy on KTH dataset

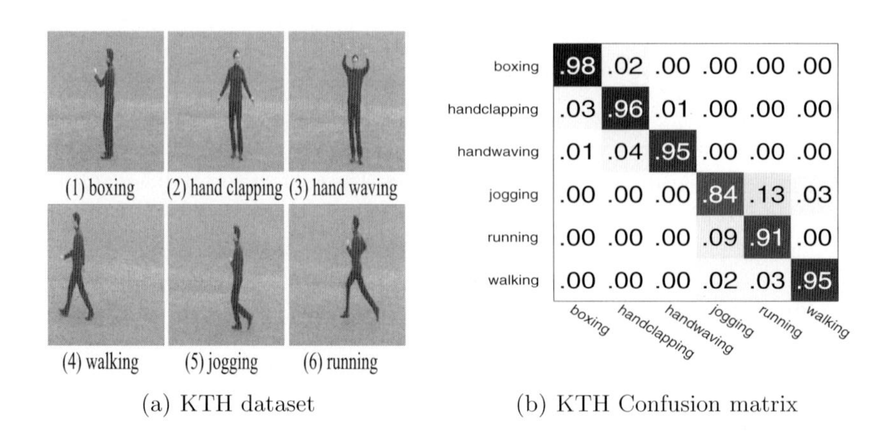

(a) KTH dataset (b) KTH Confusion matrix

Fig. 11. Results on KTH dataset: (a) sample frames. (b) confusion matrix on KTH dataset using 100 codewords. (overall accuracy=93.47%)

different clothes and indoors. Representative frames of this dataset are shown in Fig. 11(a).

Note that the person moves in different directions in the video of KTH database [16], so we divide each video into several segments according to the person's moving direction. Since most of the previously published results assign a single label to each video, we will also report per-video classification on KTH datasets. The per-video classification is performed by assigning a single action label aquired from majority voting. The confusion matrix on the KTH dataset is shown in Fig. 11(b). The most confusion is between the last three actions: running, jogging and walking. We have compared our results with the current state of the art in Fig. 10. Our results outperform other methods. The reason for the improvement is the complementarity of the period and EHOM components in our feature. The other reason is the combination of Bags of Words representation and Markov process keeps the correlation between words and temporary information to some extent.

6 Conclusion

In this paper, we propose an efficient feature for human action recognition. The hybrid feature composed of two complementary ingredients is extracted. As a global motion description in time-domain, period component can capture the global motion in time-domain. As an additional source of evidence, EHOM component could describe local motion information. When generating EHOM, we integrate Bags of Words representation with Markov process to relax the requirement on the duration of videos and maintain the dynamic information. Experiments testify the complementary roles of the two components. The proposed algorithm is simple to implement and experiments have demonstrated its improved performance compared with the state-of-the-art algorithms on the task of action recognition. Since we have already achieved pretty good results in benchmark databases under controlled settings, we plan to test our algorithm in more complicated settings such as movies in future.

References

1. Ke, Y., Sukthankar, R., Hebert, M.: Efficient visual event detection using volumetric features. In: ICCV (2005)
2. Sun, J., Wu, X., Yan, S., Chua, T., Cheong, L., Li, J.: Hierarchical spatio-temporal context modeling for action recognition. In: CVPR (2009)
3. Bobick, A.F., Davis, J.W.: The recognition of human movement using temporal templates. IEEE Transactions on Pattern Analysis and Machine Intelligence 23, 257–267 (2001)
4. Laptev, I., Lindeberg, T.: Pace-time interest points. In: ICCV (2003)
5. Zhang, Z., Hu, Y., Chan, S., Chia, L.-T.: Motion context: A new representation for human action recognition. In: Forsyth, D., Torr, P., Zisserman, A. (eds.) ECCV 2008, Part IV. LNCS, vol. 5305, pp. 817–829. Springer, Heidelberg (2008)
6. Cutler, R., Davis, L.S.: Robust real-time periodic motion detection, analysis, and applications. IEEE Trans. Pattern Anal. Mach. Intell. 22 (2000)
7. Liu, Y., Collins, R., Tsin, Y.: Gait sequence analysis using frieze patterns. In: Heyden, A., Sparr, G., Nielsen, M., Johansen, P. (eds.) ECCV 2002. LNCS, vol. 2351, pp. 657–671. Springer, Heidelberg (2002)
8. Efros, A., Berg, A., Mori, G., Malik, J.: Recognition action at a distance. In: ICCV (2003)
9. Fathi, A., Mori, G.: Action recognition by learning mid-level motion features. In: CVPR (2008)
10. Ali, S., Shah, M.: Human action recognition in videos using kinematic features and multiple instance learning. IEEE Transactions on Pattern Analysis and Machine Intelligence 99 (2008)
11. Pogalin, E., Smeulders, A.W.M., Thean, A.H.C.: Visual quasi-periodicity. In: CVPR (2008)
12. Tipping, M.E., Bishop, C.M.: Probabilistic principal component analysis. J. of Royal Stat. Society, Series B 61 (1999)
13. Li, J., Wu, W., Wang, T., Zhang, Y.: One step beyond histograms: Image representation using markov stationary features. In: CVPR (2008)
14. Lucas, B.D., Kanade, T.: An iterative image registration technique with an application to stereo vision. In: DARPA Image Understanding Workshop (1981)
15. Breiman, L.: Probability. Society for Industrial Mathematics (1992)
16. Schuldt, C., Laptev, I., Caputo, B.: Recognizing human actions: A local svm approach. In: CVPR (2004)
17. Blank, M., Gorelick, L., Shechtman, E., Irani, M., Basri, R.: Actions as space-time shapes. In: ICCV (2005)
18. Laptev, I., Marszalek, M., Schmid, C., Rozenfeld, B.: Learning realistic human actions from movies. In: CVPR (2008)

Bag of Spatio-temporal Synonym Sets for Human Action Recognition

Lin Pang[1,2], Juan Cao[1], Junbo Guo[1], Shouxun Lin[1], and Yan Song[1,2]

[1] Laboratory of Advanced Computing Research, Institute of Computing Technology, Chinese Academy of Sciences, Beijing, China
[2] Graduate University of Chinese Academy of Sciences, Beijing, China
{panglin,caojuan,guojunbo,sxlin,songyan}@ict.ac.cn

Abstract. Recently, bag of spatio-temporal local features based methods have received significant attention in human action recognition. However, it remains a big challenge to overcome intra-class variations in cases of viewpoint, geometric and illumination variance. In this paper we present Bag of Spatio-temporal Synonym Sets (ST-SynSets) to represent human actions, which can partially bridge the semantic gap between visual appearances and category semantics. Firstly, it re-clusters the original visual words into a higher level ST-SynSet based on the distribution consistency among different action categories using Information Bottleneck clustering method. Secondly, it adaptively learns a distance metric with both the visual and semantic constraints for ST-SynSets projection. Experiments and comparison with state-of-art methods show the effectiveness and robustness of the proposed method for human action recognition, especially in multiple viewpoints and illumination conditions.

Keywords: Action Recognition, Spatio-temporal Synonym Sets, Metric Learning.

1 Introduction

Human action recognition is an important technique for multiple real-world applications, such as video surveillance, human-computer interaction, and event-based video retrieval. However, it still remains a challenging task due to the cluttered background, camera motion, occlusion and viewpoint variance, etc.

There are two kinds of traditional methods for action recognition. One is global model based method. For example, Bobick et al. in [1] use motion history images to recognize actions and Blank et al. in [2] represent actions by describing the spatio-temporal shape of silhouettes. However, these methods rely on the restriction of contour tracking and background subtraction. Without prior foreground segmentation, Efros et al. in [3] correlate flow templates with videos and Shechtman et al. in [4] recognize actions by spatial-temporal volume correlation, but these methods are quite sensitive to scale, pose and illumination changes. The other is the local feature based method. Recently, the spatio-temporal local feature based method has been widely

S. Boll et al. (Eds.): MMM 2010, LNCS 5916, pp. 422–432, 2010.

used for human behavior analysis in [5, 6, 7, 8, 9, 10]. By using a Bag-of-Words representation combined with machine learning techniques like Support Vector Machine [5, 7] and graphical models [6], it performs better than the global model based method, especially in the situation with cluttered background and severe occlusion.

In the above spatio-temporal local feature based method, quantizing the local features into visual words is one of the key problem. Among the majority works to date, the visual words are usually obtained by unsupervised clustering methods such as K-means. Niebles et al. in [6] learn the latent topics of the visual words using generative graphical models such as the probabilistic Latent Semantic Analysis (pLSA) model and Latent Dirichlet Allocation (LDA). To get a compact representation, Liu et al. in [10] propose an unsupervised method with Maximization Mutual Information principle to group visual words into video-words-clusters (VWC). Essentially, these clusters of visual words obtained in the unsupervised manner only capture the common visual patterns in the whole training set, thus may not have the most discriminative power in accordance with action category semantics. However, one of the most significant challenges in human action recognition is the different visual appearances of local features within the same action category in cases of viewpoint changes as well as geometric and illumination variance. Meanwhile, actions of different categories may share similar local appearances. An example of intra-class variation and inter-class similarity is shown in Fig.1. It is clear to see that local cuboids a and b with relatively different visual appearances caused by viewpoint change are from the same action category "Running", while cuboids b and c with higher visual similarity are from different action categories. Therefore, spatial-temporal visual vocabularies constructed only by unsupervised visual appearance similarity clustering are limited to handle this gap between visual appearances and category semantics, and the ambiguous vocabularies projection in the visual feature space may hurt the overall classification performance.

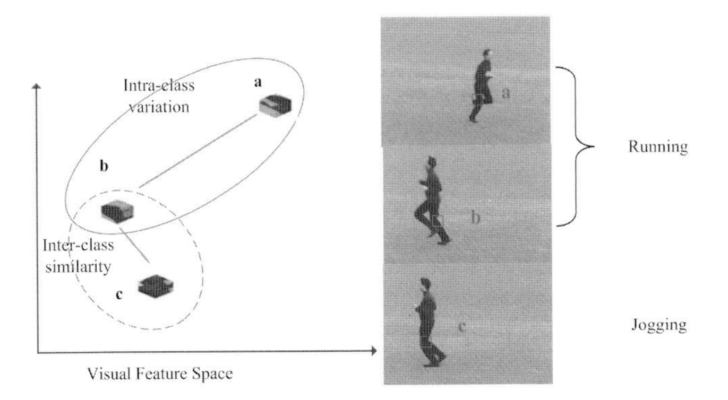

Fig. 1. Intra-class variation and Inter-class similarity of local cuboids in action recognition. Cuboids a and b are from two samples of "Running" with different viewpoints, and cuboid c is from sample of "Jogging".

To address the above issue, motivated by the research in object recognition [11, 12], we propose bag of ***Spatio-temporal Synonym Sets (ST-SynSets)*** for representation of actions to partially bridge the gap between visual appearances and category semantics, where ST-SynSet is a higher level cluster of semantic consistent visual words. The main idea is that though it is hard to measure the semantics of visual words directly, visual words from the same action category with variant visual appearances in cases of scale, viewpoint and illustration change could still keep the similar probability distributions among different categories, which in a way means the semantic similarity. In addition, since the topological proximity of visual words in the visual feature space can't ensure the semantic relevance, vocabulary projection just by visual nearest neighbor mapping lacks the concordance with category semantics. Thus we need to learn a new distance metric for visual vocabulary by integrating the visual and semantic similarity and transfer the visual words to the ST-SynSet space to suppress the errors caused by uncertainty of vocabulary projection. The ST-Synset is different from latent topic in pLSA and LDA in[6] for that it is not a result of a generative model. Without prior assumption of the distribution, the ST-SynSet is the result of a supervised data-mining process of compressing visual words via distributional clustering following the joint distribution of visual words and action categories.

The main contribution lies in two aspects:

First, we propose to cluster visual words which share similar category probability distributions to be ST-Synsets by the Information Bottleneck clustering method, and produce a compact and discriminative representation for actions.

Second, we propose to learn a new distance metric for visual vocabularies based on the synonymy constraints to get a more accurate ST-SynSet projection.

The remainder of the paper is organized as follows. Section 2 presents the proposed method in detail. Experimental results are shown in Section 3, and Section 4 concludes this paper.

2 Spatio-temporal Synonym Sets Based Action Recognition

Fig. 2 shows the flowchart of the proposed action recognition algorithm. First, we adopt the spatio-temporal interest points detector proposed by Dollar et al. in [5] for local feature extraction. This detector produces dense feature points and performs well on the action recognition task [6]. For each cuboid we compute gradient-based descriptor and apply PCA to reduce dimensionality. Then, initial visual vocabularies are constructed by clustering the extracted spatio-temporal local features with K-means algorithm. Second, Sequential Information Bottleneck (SIB) method is implemented to re-cluster the visual words into Spatio-temporal Synonym Sets and informative ST-SynSets are selected by the information score. Third, in order to get a reasonable ST-SynSet projection, we learn a new distance metric for visual vocabulary with the synonym constraints. Finally, we use "Bag of ST-SynSets", the histogram of Spatio-temporal Synonym Sets to represent each action instance, and use Support Vector Machine (SVM) for human action recognition.

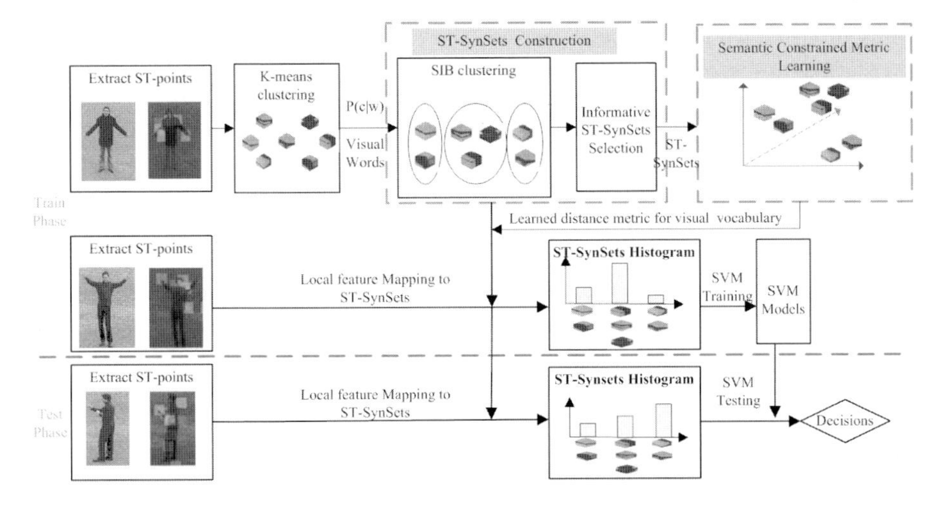

Fig. 2. Flowchart of the proposed action recognition algorithm

2.1 Spatio-temporal Synonym Sets Construction

In the state-of-art method of Bag-of-Words, an action instance is encoded as a histogram of visual words by unsupervised vector quantization of local features. However, action instances usually have significant intra-class variations because of the different attributes of performers (age, gender, clothes, and velocity) and especially different external conditions, such as viewpoints, scales, illuminations and so on .Therefore, the visual words with only the similarity of visual appearances become too primitive to effectively represent the characteristic of each category.

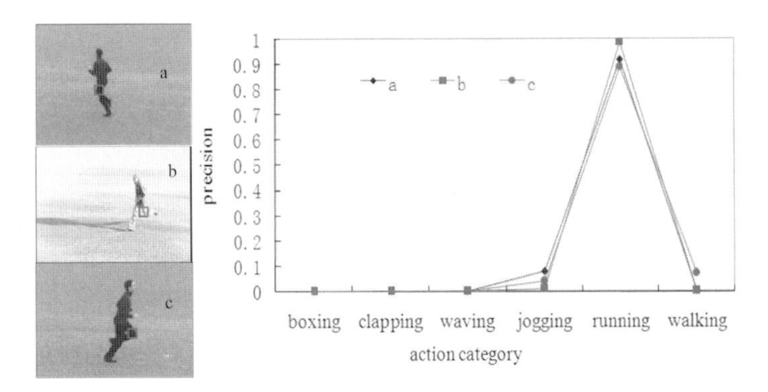

Fig. 3. Category probability distributions of three cuboids with different appearances extracted from "Running "actions in three different scenarios

Motivated by the object recognition method in [12], we find that local features highly correlated with the same human action category may vary in visual appearances under different circumstances but they keep the discriminative ability among different categories. Thus, we define $P\ (c_i \mid w)$ to measure the semantic inference ability of visual word w attributed to particular category c_i. Due to the motion pattern heterogeneity of actions, a number of local features are intrinsic and highly indicative to certain action categories. For example in Fig. 3, three visually different cuboids from "Running " in three different scenarios with large viewpoint variances have similar category probability distributions $P(c \mid w)$ which peak around its belonging classes and denote the semantic similarity of the local cuboids.

Therefore, we define ***Spatio-temporal Synonym Set*** to be the cluster of semantic-consistent visual words which share the similar probabilistic distributions $P(c \mid w)$ among different categories. This higher level representation groups visual words with different visual appearances but similar discriminative power together and thus can partially handle the significant intra-class variations of local features and have more discriminative power to distinguish between action categories.

2.1.1 Distributional Clustering by Information Bottleneck Method

Based on the definition of ST-SynSet, we use the Information Bottleneck (IB) principle [13] which provides a reasonable solution for distributional clustering to cluster the initial visual words to a compact representation for the construction of ST-SynSets. In [10], Liu et al. use the Maximization Mutual Information principle to group visual words into video-words-clusters (VWC). In essence, the visual word clusters obtained in this unsupervised manner mean to capture the common visual patterns with similar distributions among all the video samples in the training set, thus may lack the discriminative power to distinguish different action categories. In our method, we get the most compact representation of visual words and meanwhile maintain as much discriminative information of the categories as possible by clustering the visual words with similar probabilistic distributions $P(c \mid w)$ among different categories using the IB principle.

Given the joint distribution $P(w, c)$ of each visual word w and action category c, the goal of IB principle is to construct the optimal compact representation of visual words set W, namely the ST-SynSets S, such that S preserves as much information of category set C as possible. The IB principle is formulated as the following Lagrangian optimization problem in Equation (1).

$$\underset{S}{Max}\ F(S) = I(S;C) - \beta I(W;S) \tag{1}$$

where $I(S;C)$ and $I(W;S)$ are the mutual information between S and C and between W and S respectively. And the mutual information $I(X; Y)$ is defined as

$$I(X;Y) = \sum_{x \in X, y \in Y} p(x,y) \log \frac{p(x,y)}{p(x)p(y)} \tag{2}$$

Equation (1) means to cluster the visual words into the most compact representation S with the least mutual information with visual words W through a compact bottleneck under the constraint that this compression maintains as much information of the category set C as possible.

In [13], Noam et al. propose a sequential IB clustering algorithm to solve the optimization problem in Equation (1) with the $\beta \to \infty$ limit to generate the hard partitions.

The algorithm starts from an initial random partition S_0 of W, and at each step we draw each w out of its current cluster $S(w)$ and choose for its new cluster by minimizing the score loss caused by merging w to every cluster s_i which is stated in Equation(3). Repeat this process until convergence. To avoid the algorithm being trapped in local optima, we repeat the above procedure for random initializations of S_0 to obtain n different solutions, from which we choose the one that maximize $F(s)$ in Equation(1).

$$d_F(w, s_i) = (P(w) + P(s_i)) \times JS(P(c|w), P(c|s_i)) \tag{3}$$

where c is the variable of category, s_i is the i-th cluster in the current partition S, and $JS(p,q)$ is the Jensen-Shannon divergence [14] which essentially measures the likelihood that the two sample distributions p and q originate from the most likely common source.

2.1.2 Informative ST-SynSet Selection

After distribution clustering, some of the ST-SynSets which have flat and non-salient category probability distributions are not discriminative and have more uncertainty in semantics. Therefore an effective ST-SynSet selection is necessary. We define the maximal mutual information between the ST-SynSet s and each category label c to measure the information score of s, which is formulated in Equation (4).

$$I(s) = \max_c(I(s,c)) = \max_c \sum_{w\in s, c\in C} p(w,c)\log\frac{p(w,c)}{p(w)p(c)} \tag{4}$$

We select the most significant ST-SynSets with the highest $I(s)$ and remove the others with lower $I(s)$. The local appearance samples of cuboids from the three most informative ST-SynSets and their corresponding category probability distributions are shown in Fig 4. From the spatio-temporal patches we can see that they are all signature motion parts of the corresponding action categories.

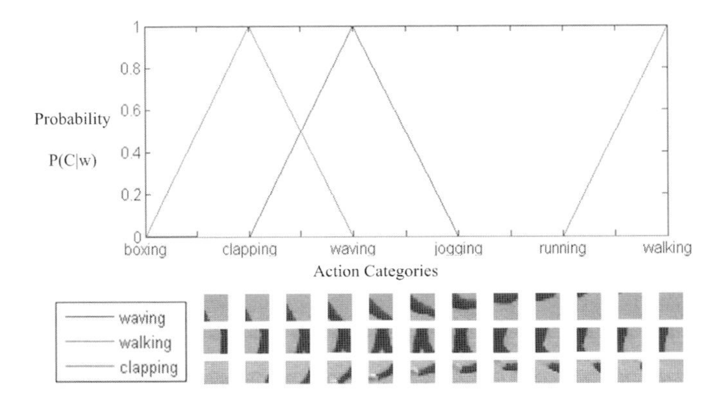

Fig. 4. Local appearance samples of cuboids from the top three informative ST-SynSets and their corresponding category probability distributions

2.2 Distance Metric Learning with Semantic Constraints

The basic mechanism of Bag-of-Words approach involves a step of mapping the local features to visual words according to the distances between the local features and the visual words. In the proposed approach, we need to map the local features to ST-SynSets. It is known that the visual words in the same ST-SynSet may have different visual appearances, while those with similar appearances may represent different category semantics. The inconsistency between semantic space and visual feature space causes the ambiguity and uncertainty of ST-SynSet projection using standard distance metric such as Euclid distance. Therefore, we propose to learn a new distance metric for the visual words by integrating semantic constraints. We use the ST-SynSets structure to get the semantic constraints where we maintain that the visual words of the same ST-SynSet get closer in the new feature space than those in different ST-SynSets.

Given n visual words $\{x_{1...}x_n\}$, with all $x_i \in R^d$, we need to compute a positive-definite $d \times d$ matrix A to parameterize the squared Mahalanobis distance:

$$d_A(x_i, x_j) = (x_i - x_j)^T A(x_i - x_j) \quad i, j = 1.....n \tag{5}$$

Till now, many methods have been proposed for Mahalanobis metric learning [15, 16, 17], and we utilize the information-theoretic metric learning method in [17] because it is fast and effective for the similarity constraints. Given an initial $d \times d$ matrix A_0 specifying the standard metric about inter-point distances, the learning task is posed as an optimization problem that minimizes the LogDet divergence between matrices A and A_0, subject to a set of constraints specifying pairs of examples that are similar or dissimilar.

Here we define the rule of the constraints to preserve small distances for visual words in the same ST-SynSet and large distances for those in different ST-SynSets. The problem is formalized as follows:

$$\begin{aligned} &\min_{A \geq 0} D_{ld}(A, A_0) = trace(AA_0^{-1}) - \log\det(AA_0^{-1}) - d \\ &s.t. \quad d_A(x_i, x_j) \leq u \quad (i, j) \in \mathit{same\ \ ST\text{-}SynSet} \\ &\qquad\ d_A(x_i, x_j) \geq l \quad (i, j) \in \mathit{different\ ST\text{-}SynSets} \end{aligned} \tag{6}$$

A_0 is unit matrix for a standard squared Euclidean distance, and l and u are respectively large and small values, which are given empirically according to the sampled distances of visual words. This problem can be optimized using an iterative optimization procedure by projecting the current solution onto a single constraint per iteration.

With the learned metric matrix A, the distance between local feature and visual word is computed as Equation (5). Then, we use the K-nearest neighbor algorithm to do ST-SynSet projection. The mapping to the ST-SynSet is measured by the majority vote of the feature's K nearest visual words. Here, K is empirically set to be 5.

2.3 Bag of ST-SynSets Action Classification Using SVM

Once the ST-SynSets and the new metric for visual words are obtained, we can describe a given action video using the histogram of ST-SynSets, in the way of "Bag of

ST-SynSets". We use SVM classifier to model each action category. Here, histogram intersection kernel in Equation (7) is used as the kernel.

$$k_{HI}(h_1, h_2) = \sum_{i=1}^{n} \min(h_1(i), h_2(i)) \tag{7}$$

where h_1, h_2 are the histogram of two videos, and $h(i)$ is the frequency of the ith bin.

3 Experiments

We test the proposed method on the KTH human motion dataset [7]. This dataset contains 598 short videos of six types of human actions (walking, jogging, running, boxing, hand waving and hand clapping) performed by 25 persons. The dataset is challenging because there are four different scenarios with variable backgrounds, moving camera and changed scales.

We extract spatio-temporal interest points using the detector proposed by Dollar et al. in [5] and get the corresponding gradient-based descriptors. The detector's scale parameters are empirically set with $\sigma = 2$ and $\tau = 3$ and PCA is applied to get a lower dimension of 100. We build visual vocabulary with the videos of randomly selected 3 persons for each action, and the initial number of visual words is set to 1000. Then we construct ST-SynSets using the SIB algorithm in Section 2.1.1, with a variable number of {20, 50, 100,200, 400} clusters, and after ST-SynSets selection we retain 80 percent of ST-SynSets with higher scores. We adopt the Leave One Out Cross Validation (LOOCV) as [6]. More specifically, at each run of the LOOCV, we use videos of 24 persons to train SVM and the rest for testing, and the average accuracy of 25 runs is reported as the results. The results of the algorithm with different ST-SynSets numbers are shown in Fig 5(a). We can see the optimal accuracy is obtained when the number of ST-SynSets is set to be 100, and after informative ST-SynSets refinement, the total number of ST-SynSets used to represent actions is 80.

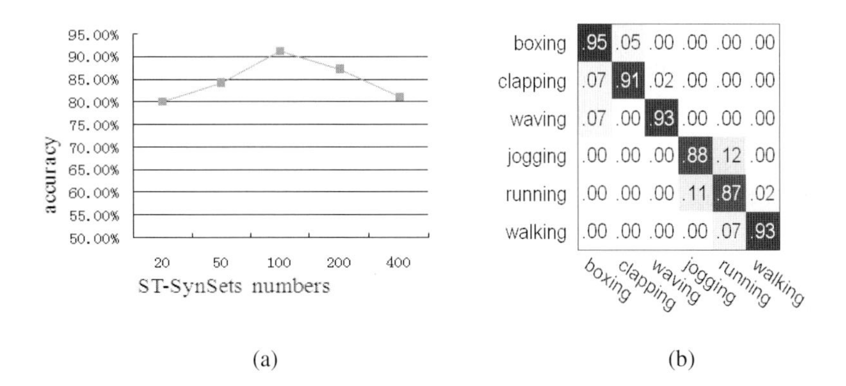

Fig. 5. (a) Average accuracy of the algorithm with different ST-SynSets numbers; (b) Confusion matrix of the method with the best ST-SynSets which are set to be 100 clusters and have an effective ST-Synsets of 80

Confusion matrix of the proposed method with the best ST-SynSets number is shown in Fig 5 (b). We can see most actions are recognized correctly, and the largest confusion is "jogging" vs. "running", which share similar local features and are easily confused.

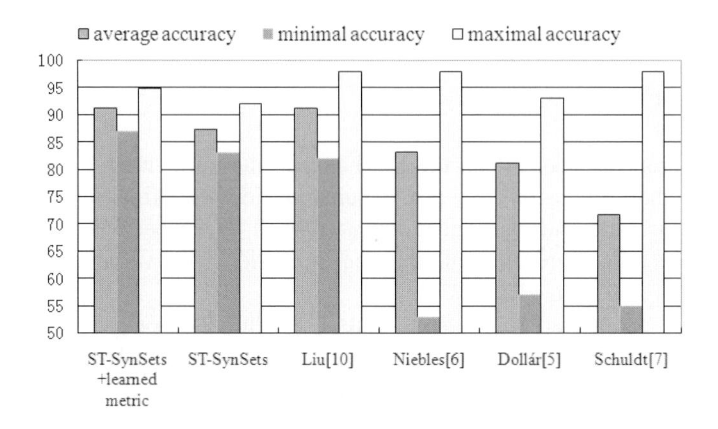

Fig. 6. Performance comparison with other methods

Fig. 6 shows the comparison of accuracies of the popular methods in recent years based on spatio-temporal local features. Although almost all the methods achieve high maximal accuracies for the "walking" action which has slight geometric and viewpoint changes in different scenarios, the proposed method shows superior performance in the minimal accuracy especially for the easily confused actions such as "running". From Fig 6, we can see the proposed method with the ST-SynSet representation and the learned metric achieves the relatively most reliable and stable performance for all types of actions, with the average accuracy of 91.16% and the standard deviation of only 3.13%. The main reason is that this approach fully considers the category semantic information and gets the most discriminative representation visual words clusters to handle the intra-class local variations caused by different viewpoints as well as geometric and anthropometry variances. By clustering the visual words to a semantic meaningful unit with similar discriminative power and by learning the new metric for visual words using synonym constraints, it can learn a category semantic consistent feature space for the visual vocabulary and thus partially bridge the semantic gap in terms of significant intra-class variations and inter-class confusion.

In Fig.7, we show example videos in the four different scenarios from four confusable action categories with their corresponding ST-SynSets histograms when ST-SynSets number is set to be 20. We can see that actions from the same category share the similar ST-SynSets distributions. It is also clear to see from the peaks of these histograms that some ST-SynSets are dominating in particular actions and have the discriminative power among different actions.

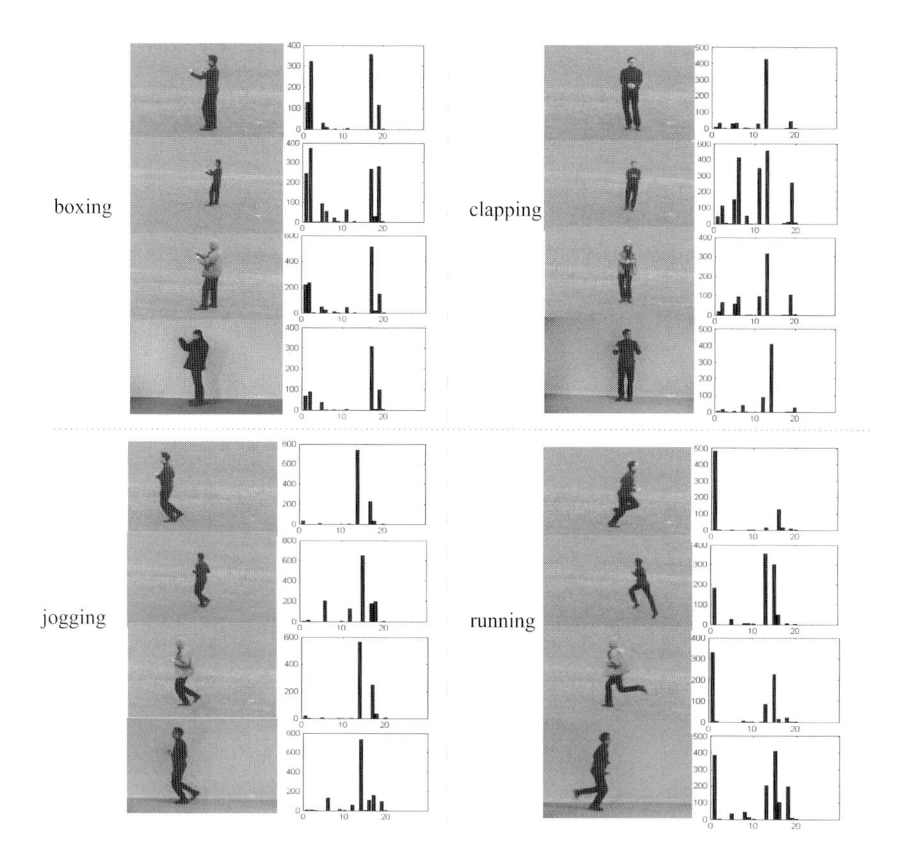

Fig. 7. Examples of ST-SynSets histograms for 4 confusable actions performed in the different scenarios

4 Conclusion and Future Work

In this paper, we propose a higher level representation for action recognition, namely Bag of Spatio-temporal Synonym Sets to bridge the gap between visual appearances and category semantics of human actions. By grouping the semantic-consistent visual words to be ST-SynSet and learning a new distance metric for visual words with both visual and semantic constraints, this approach can partially handle the intra-class variations and inter-class similarities. Experimental results on the KTH dataset show the effectiveness of the proposed method for human action recognition in the scenarios of different viewpoints and illumination conditions.

Because of the good results shown by other methods using the same constrained action database, further comparisons in some more challenging realistic databases with huge variations of anthropometry ,viewpoint, illumination, occlusions and so on are required to highlight the advantage of our method, namely, the robustness and discriminative power to handle the significant intra-class variations. In addition, since the ST-Synset is more compact than visual word, it has relatively stronger extension ability for integrating global geometry and context information. All these will be addressed as our future work.

Acknowledgments. This work was supported by the National Basic Research Program of China (973 Program, 2007CB311100), National High Technology and Research Development Program of China (863 Program, 2007AA01Z416), National Nature Science Foundation of China (60902090, 60802067), Beijing New Star Project on Science & Technology (2007B071), Co-building Program of Beijing Municipal Education Commission.

References

1. Bobick, A.F., Davis, J.W.: The Recognition of Human Movement Using Temporal Templates. IEEE Transactions on Pattern Analysis and Machine Intelligence 23(3), 257–267 (2001)
2. Blank, M., Gorelick, L., Shechtman, E., Irani, M., Basri, R.: Actions as space-time shapes. In: IEEE International Conference on Computer Vision, vol. 2, pp. 1395–1402 (2005)
3. Efros, A.A., Berg, A.C., Mori, G., Malik, J.: Recognizing action at a distance. In: IEEE International Conference on Computer Vision, vol. 2, pp. 726–733 (2003)
4. Shechtman, E., Irani, M.: Space-time behavior based correlation. In: IEEE conference on Computer Vision and Pattern Recognition, pp. 405–412 (2005)
5. Dollár, P., Rabaud, V., Cottrell, G., Belongie, S.: Behavior recognition via sparse spatio-temporal features. In: 2nd joint IEEE international workshop on visual surveillance and performance evaluation of tracking and surveillance, pp. 65–72 (2005)
6. Niebles, J.C., Wang, H.C., Li, F.F.: Unsupervised Learning of Human Action Categories using Spatial-Temporal Words. International Journal of Computer Vision 79, 299–318 (2008)
7. Schuldt, C., Laptev, I., Caputo, B.: Recognizing Human Actions: A Local SVM Approach. In: International Conference on Pattern Recognition, vol. 3, pp. 32–36 (2004)
8. Savarese, S., DelPozo,Niebles, J.C., Li, F.F.: Spatial-temporal correlations for unsupervised action classification. In: IEEE Workshop on Motion and Video Computing (2008)
9. Boiman, O., Irani, M.: Detecting irregularities in images and in video. International Journal of Computer Vision 74(1), 7–31 (2007)
10. Liu, J., Shah, M.: Learning Human Actions via Information Maximization. In: IEEE Conference on Computer Vision and Pattern Recognition, pp. 1–8 (2008)
11. Winn, J., Criminisi, A., Minka, T.: Object Categorization by Learned Universal Visual Dictionary. In: IEEE International Conference on Computer Vision, pp. 1800–1807 (2005)
12. Zheng, Y.T., Zhao, M., Neo, S.Y., Chua, T.S.: Visual synset: towards a higher-level visual representation. In: IEEE Conference on Computer Vision and Pattern Recognition (2008)
13. Slonim, N., Friedman, N., Tishby, N.: Unsupervised document classification using sequential information maximization. In: Proceedings of the 25th ACM SIGIR international conference on research and development in information retrieval, pp. 129–136 (2002)
14. Lin, J.: Divergence Measures Based on the Shannon Entropy. IEEE Transactions on Information Theory 37(1), 145–151 (1991)
15. Xing, E.P., Ng, A.Y., Jordan, M.I., Russell, S.: Distance metric learning with application to clustering with side-information. In: Advances in neural information processing systems, vol. 16, pp. 521–528 (2002)
16. Bar-Hillel, A., Hertz, T., Shental, N., Weinshall, D.: Learning a Mahalanobis Metric from Equivalence Constraints. Journal of Machine Learning Research 6, 937–965 (2005)
17. Davis, J.V., Kulis, B., Jain, P., Sra, S., Dhillon, I.S.: Information-Theoretic Metric Learning. In: International Conference on Machine Learning, pp. 209–216 (2007)

A Novel Trajectory Clustering Approach for Motion Segmentation

Matthias Zeppelzauer, Maia Zaharieva, Dalibor Mitrovic, and Christian Breiteneder

Vienna University of Technology, Interactive Media Systems Group, Favoritenstrasse 9-11, A-1040 Vienna, Austria

Abstract. We propose a novel clustering scheme for spatio-temporal segmentation of sparse motion fields obtained from feature tracking. The approach allows for the segmentation of meaningful motion components in a scene, such as short- and long-term motion of single objects, groups of objects and camera motion. The method has been developed within a project on the analysis of low-quality archive films. We qualitatively and quantitatively evaluate the performance and the robustness of the approach. Results show, that our method successfully segments the motion components even in particularly noisy sequences.

1 Introduction

Motion has a vital importance in human perception. This makes it a qualified attribute for the organization and retrieval of video content. Two types of motion can be distinguished: object motion and camera motion. The detection, description, and segmentation of both is the focus of this paper.

The segmentation of motion in low-quality video with existing methods has shown to be unstable and noisy which led to the development of the proposed approach. We present a robust and efficient clustering scheme for the segmentation of single object motion as well as motion of groups of objects and camera motion. In contrast to other approaches (e.g. [1–3]) we extract motion trajectories by feature tracking directly from the raw video sequence and omit object segmentation. The result of feature tracking is a sparsely populated spatio-temporal volume of feature trajectories. Occlusions and the low quality of the video material lead to numerous tracking failures resulting in noisy and highly fragmented trajectories. The novel clustering scheme directly clusters the sparse volume of trajectories into coherent spatio-temporal motion components belonging to the same objects or groups of objects. The proposed approach takes the following factors into account:

- Robustness to low contrast, flicker, shaking, dirt, etc.
- The input data is a sparse set of fragmented trajectories that are broken off, have different lengths, and varying begin and end times.
- The analyzed time span may be large (shots up to a few minutes length).
- The number of clusters is unknown a priori.

S. Boll et al. (Eds.): MMM 2010, LNCS 5916, pp. 433–443, 2010.

- The resulting clusters have to be temporally coherent (even if the majority of the trajectories breaks off).
- Motion direction and velocity magnitude may change over time inside a cluster (to enable tracking of e.g. objects with curved motion paths and groups of objects that have slightly heterogeneous directions and speeds).
- Efficient computation.
- Flexible selection of trajectory features and similarity measures in order to enable different clusterings and applications.

The organization of this paper is as follows. We present related work in Section 2. The clustering approach is formulated in Section 3. Section 4 introduces the employed video material and presents the experimental setup. Results are discussed in Section 5. We draw conclusions in Section 6.

2 Related Work

Feature trackers are able to provide motion information over large time scales by tracking feature points over multiple successive frames [4]. However, the resulting trajectories are sparse in space and time and have different lengths and varying begin and end times. Consequently, standard methods, such as Mean Shift and K-Means cannot be directly applied for clustering.

Methods for trajectory clustering have been introduced mainly in the field of surveillance [1, 5–9] and video event classification [10]. The methods have different constraints depending on their application domain. Wang and Li present an approach for motion segmentation based on spectral clustering of motion trajectories [8]. A major limitation of their approach is that all feature points must be trackable in all analyzed frames. This is usually not given, especially when working with low-quality video material.

Hervieu et al. perform classification of motion trajectories by an HMM framework for video event classification in sports videos [10]. While the HMM framework is able to handle trajectories of different lengths, the method assumes that the trajectories have similar lifetimes and do not break off (e.g., due to occlusions and tracking failures). A similar assumption is made in [5] where the authors track and cluster motion paths of vehicles. They represent trajectories by global directional histograms which require similar motion trajectories to have similar lifetimes. However, this is not provided for broken trajectories.

Veit et al. introduce an approach for trajectory clustering of individual moving objects [7]. Groups of similarly moving objects, as required in this work, cannot be tracked. Similarly, Rabaud and Belongie cluster motion trajectories on a per-object basis in order to count people in a surveillance video [6].

For clustering of a sparse set of trajectories similarity measures are required that take the different lengths and spatio-temporal locations of the trajectories into account. Buzan et al. employ a metric based on the longest common subsequence (LCSS) to cluster trajectories of different sizes [1]. An asymmetric similarity measure for trajectories of different lengths is proposed by Wang et al. [9]. Their algorithm can handle broken trajectories during clustering due to the asymmetric property of the similarity measure. The measure is not directly applicable in our work because it uses different (spatial) similarity constraints.

There is an important difference between the above mentioned methods and our approach. The presented methods do not consider the temporal location of the trajectories during clustering. They aim at clustering trajectories independently of the time they occur, e.g. in [5]. In this work, we are interested in clustering motion trajectories belonging to the *same* object or group of objects. Therefore, we assume corresponding trajectories to occur within the *same* time and to have similar velocity (direction and -magnitude). Note, that we do not require the trajectories to have similar spatial location, which facilitates tracking of large groups of objects and camera motion (in contrast to e.g. [7]).

3 Trajectory Clustering

The idea behind the proposed scheme is to cluster the entire sparse volume of trajectories directly by iteratively grouping temporally overlapping trajectories. Thus, it is not necessary to split trajectories into sub-trajectories [11] or use global trajectory features [5]. The trajectories are processed in their original representation. The first stage of the algorithm is an iterative clustering scheme that groups temporally overlapping trajectories with similar velocity direction and magnitude. In the second stage (described in Section 3.2) the clusters from the first stage are merged into temporally adjacent clusters covering larger time spans.

3.1 Iterative Clustering

Iterative clustering aims at successively grouping temporally overlapping trajectories. We assume trajectories that perform similar motion at the same time to belong to the same motion component (e.g., object or group of objects). A trajectory t is a sequence of spatio-temporal observations $o_j = < x_j, y_j, f_j >$ with $t = \{< x_j, y_j, f_j >\}$, where x_j and y_j are spatial coordinates and f_j is the frame index of the corresponding observation. The input of the algorithm is a sparse spatio-temporal volume which is represented as a set V containing T trajectories t_i of tracked feature points: $V = \{t_i | i = 1, 2, ..., T\}$.

Clustering starts by the selection of expressive trajectories (representatives of meaningful motion components) for the initialization of clusters. We assume that meaningful motion components span large distances. Therefore, we compute the absolute spatial distance that each trajectory travels during its lifetime. That is the Euclidean distance between the first and last feature point of the trajectory. This measure favors trajectories that belong to an important motion component. Alternatively, the lifetime of the trajectories may be employed as a measure for their expressiveness. However, experiments have shown that the trajectories with the longest lifetimes often represent stationary or slowly moving points, which leads to the selection of inadequate representatives. Consequently, we do not employ the lifetime as an indicator for expressiveness.

We sort the trajectories according to their traveled distances and select the trajectory t_r with the largest distance as representative for the current cluster C_{t_r}. Then all trajectories t_i from the set V are compared to the representative t_r in a pairwise manner. The similarity of trajectories that have no temporal overlap is 0 by definition. Consequently, only temporally overlapping trajectories are compared.

For the pairwise comparison first the temporally overlapping sub-segments of two trajectories t_r and t_i are determined. Following, we extract trajectory features from these sub-segments and a perform similarity comparison. See the description in the following on trajectory features and similarity measures employed in this work. The result of the pairwise comparison of trajectories t_r and t_i is a similarity score $s^{r,i}$.

All trajectories with a score higher than a threshold λ are assigned to the current cluster C_{t_r}:

$$t_i \in C_{t_r} \Leftrightarrow s^{r,i} > \lambda \tag{1}$$

The cluster C_{t_r} is then added to the set S of clusters (which is initially empty).

All trajectories $t_i \in C_{t_r}$ that lie fully inside the cluster are removed from the original set of trajectories V. Trajectories that are temporally not fully covered by the cluster remain in V. That enables trajectories to be assigned to multiple temporally adjacent clusters in further iterations. This is an important prerequisite for the creation of long-term clusters in the second stage of the algorithm.

After updating the set V the next iteration is started by selecting a new representative trajectory t_r from the remaining trajectories in V. The algorithm terminates when no more trajectories are left in V.

The result of iterative clustering is a set of n overlapping clusters $S = \{C_1, C_2, ..., C_n\}$. Each cluster represents a portion of a homogeneous motion component. The temporal extent of the clusters tends to be rather short (it is limited by the temporal extent of the feature trajectories). Consequently, the iterative clustering yields an over-segmentation of the spatio-temporal volume. This is addressed in the second stage (merging), see Section 3.2.

Trajectory features and similarity measures. The proposed iterative clustering scheme allows for the use (and combination) of arbitrary features and similarity measures, for example spatial features compared by Euclidean distance, purely directional features compared by cosine similarity, etc. We compute features adaptively only for the temporally overlapping segments of the compared trajectories. This is different from other approaches, where features are computed a priori for the entire trajectories.

A straight forward way is to directly employ the spatial coordinates of the trajectories as features. For low-quality video the coordinates of the trajectories are often noisy (e.g., due to shaky sequences and tracking failures). For a given segment of a trajectory we compute the dominant direction $\phi = (\Delta x, \Delta y)$ where

$$\Delta x = x_{begin} - x_{end}, \; \Delta y = y_{begin} - y_{end}, \tag{2}$$

and the distance ρ between the first and the last spatial coordinates of the segment:

$$\rho = \sqrt{(\Delta x)^2 + (\Delta y)^2} \ . \tag{3}$$

These features are robust to noise and are location invariant (as required for segmenting motion from the camera and of groups of objects). They represent the velocity direction and magnitude of the trajectories. Dependence on spatial location can easily be integrated by adding absolute coordinates as features.

The presented features require two different metrics for comparison. We employ the cosine metric for the directional features ϕ and a normalized difference for the distance features ρ. The corresponding similarity measures s_ϕ and s_ρ for two trajectories u and v are defined as follows:

$$s_\phi^{u,v} = (\frac{\phi^u \cdot \phi^v}{\|\phi^u\| \cdot \|\phi^v\|} + 1) \cdot \frac{1}{2}, \ s_\rho^{u,v} = 1 - \frac{|\rho^u - \rho^v|}{\max(\rho^u, \rho^v)} \ . \tag{4}$$

The cosine similarity is transformed into the range $[0; 1]$. We linearly combine both similarity measures in order to obtain a single similarity measure $s^{u,v}$ as:

$$s^{u,v} = \alpha \cdot s_\phi^{u,v} + (1 - \alpha) \cdot s_\rho^{u,v} \text{ with } 0 \leq \alpha \leq 1, \tag{5}$$

where α balances the influence of the velocity directions and the velocity magnitudes of the two trajectories.

3.2 Cluster Merging

The goal of cluster merging is to connect clusters that represent the same (long-time) motion component. This is performed by hierarchically merging clusters which share the same trajectories.

The input to this stage is the set of clusters obtained by iterative clustering: $S = \{C_1, C_2, ..., C_n\}$. We start an iteration by sorting the clusters according to their sizes (number of member trajectories) in ascending order. Beginning with the smallest cluster C_i, we search for the cluster C_j which shares the most trajectories with C_i. We merge both clusters when the portion of shared trajectories (connectivity) exceeds a certain threshold μ. The connectivity $c_{i,j}$ between two clusters C_i and C_j is defined as:

$$c_{i,j} = \frac{|C_i \cap C_j|}{\min(|C_i|, |C_j|)} \tag{6}$$

The criterion for merging clusters C_i and C_j into a new cluster $C_i^{'}$ is:

$$C_i^{'} = C_i \cup C_j \Leftrightarrow c_{i,j} > \mu. \tag{7}$$

After merging the clusters C_i and C_j they are removed from the set S and the new cluster is added into an (initially empty) set $S^{'}$. If no cluster C_j fulfills the criterion for merging then $C_i^{'} = C_i$. Following, C_i is removed from S and $C_i^{'}$ is added to $S^{'}$.

Merging is repeated with all remaining clusters in S, until S is empty and $S^{'}$ contains all combined clusters. This makes up one iteration of merging. We perform further iterations by setting $S = S^{'}$ to repeatedly merge newly created clusters until no cluster can be merged any more. Finally, trajectories associated with more than one cluster are assigned to the cluster with the largest temporal overlap. The result of the merging procedure is a smaller set of clusters $S^{'}$ where the clusters represent distinct (long-term) motion components.

The order in which clusters are merged influences the result significantly. We sort the clusters according to their size and begin merging with the smallest clusters. This facilitates that the merging scheme successively generates larger clusters out of small ones (fewer small clusters remain). Furthermore, each cluster is merged with the one having the highest connectivity. This enforces that clusters of the same motion component are merged.

4 Experimental Setup

In this section we present the video material used for the evaluation and the motion analysis framework including pre- and postprocessing steps.

4.1 Video Material

For the evaluation of the proposed method we employ archive film material. The analyzed movies are historical artistic documentaries from the late 1920s. The movies exhibit twofold challenges, that originate from their technical and from their artistic nature. From the technical point of view, the film material is of significantly low quality due to storage, copying, and playback over the last decades. Typical artifacts of archive film material include scratches, dirt, low contrast, flicker, and frame displacements. Such artifacts impede the process of feature tracking and motion recognition, resulting in noisy and broken feature trajectories. Furthermore, frame displacements and significant camera shakes may result in falsely detected motion.

From an artistic point of view, the applied documentary technique is highly experimental. The filmmaker used advanced montage and photographic techniques (e.g., quadruple exposure, reverse filming, etc.) to achieve complex motion compositions. Typical compositions include hammering, camera traveling, and contrapuntal movements (see Figure 1 for examples). Consequently the movies are well-suited for the evaluation of the proposed approach.

4.2 Feature Tracking

We perform motion segmentation for entire shots. Ground truth containing the shot boundaries is provided by film experts. We employ the KLT feature tracker because of its efficiency and its ability to track feature points across large time spans [4]. For most parameters of KLT we use the defaults proposed by the implementation [12]. The search range for tracking is set to 3 to reduce the

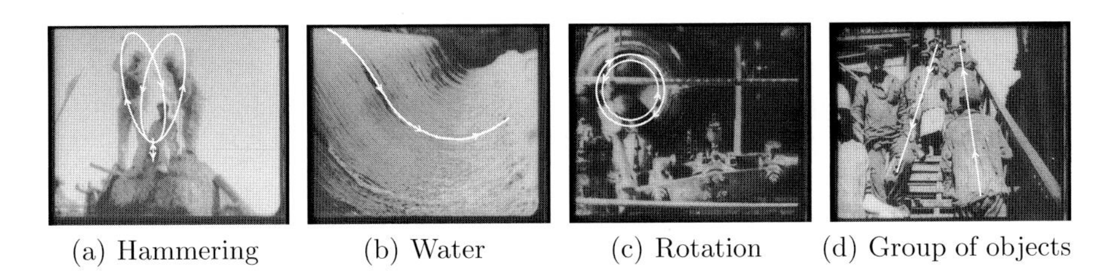

(a) Hammering (b) Water (c) Rotation (d) Group of objects

Fig. 1. Typical motion compositions

Fig. 2. The motion field before (left) and after (right) filtering

number of tracking errors. The minimum distance between selected features is reduced to 5 in order to produce denser motion fields. The number of features is set to 2000. The output of feature tracking is a fragmented set of trajectories in a sparse motion field.

We perform three basic preprocessing steps in order to reduce the noise contained in the motion field. First, we remove trajectories whose lifetime is less than a predefined duration τ ($\tau = 0.5$ seconds in our case). This removes a large number of unstable trajectories.

Second, we detect and remove stationary trajectories. We remove trajectories with a spatial extent in x- and y-direction below a threshold σ which directly corresponds to the amount of shaking in the sequence.

Third, we smooth the trajectories by removing high-frequency components in the discrete cosine spectrum of the spatial coordinates x_j and y_j. This dampens the influence of shaking for the remaining trajectories. Figure 2 shows the effect of preprocessing for a noisy motion field of an entire shot. Most of the stationary and noisy trajectories are removed. The remaining trajectories represent the motion of the airplane in the lower right quarter of the frame.

4.3 Trajectory Clustering

The proposed clustering approach requires three parameters to be set (λ and α for the similarity comparison and μ for cluster merging). The similarity score $s^{u,v}$ as defined in Equation (5) in Section 3.1 ranges from 0 to 1. A value of λ between 0.7 and 0.9 yields satisfactory results in the experiments. The weighting factor α is set to 0.5.

The second parameter μ controls the sensitivity of cluster merging. Due to the high fragmentation of the trajectories the value of μ is chosen rather low to facilitate cluster merging. Values of μ between 10% and 20% of shared trajectories yield the best results in the experiments.

We perform two simple postprocessing steps in order to improve the generated motion segments. First, we detect and remove single outlier trajectories which is necessary since we ignore spatial information during clustering. Second, we remove small clusters (less than 5 trajectories) that usually represent noise.

5 Results

We perform qualitative evaluation by applying our approach to shots with complex motion compositions. In contrast to existing work (e.g. [7, 13]) we additionally evaluate the performance of the approach by a quantitative evaluation.

5.1 Qualitative Evaluation

We have selected approximately 50 shots from different films for evaluating the quality of our approach. Three test sequences are shown in Figure 3.

The first sequence (Figures 3(a) - 3(e)) shows a group of people walking up a hill. The people in the group first move towards the hill (in the lower right quarter), then turn to the left, walk up the hill and finally vanish behind the

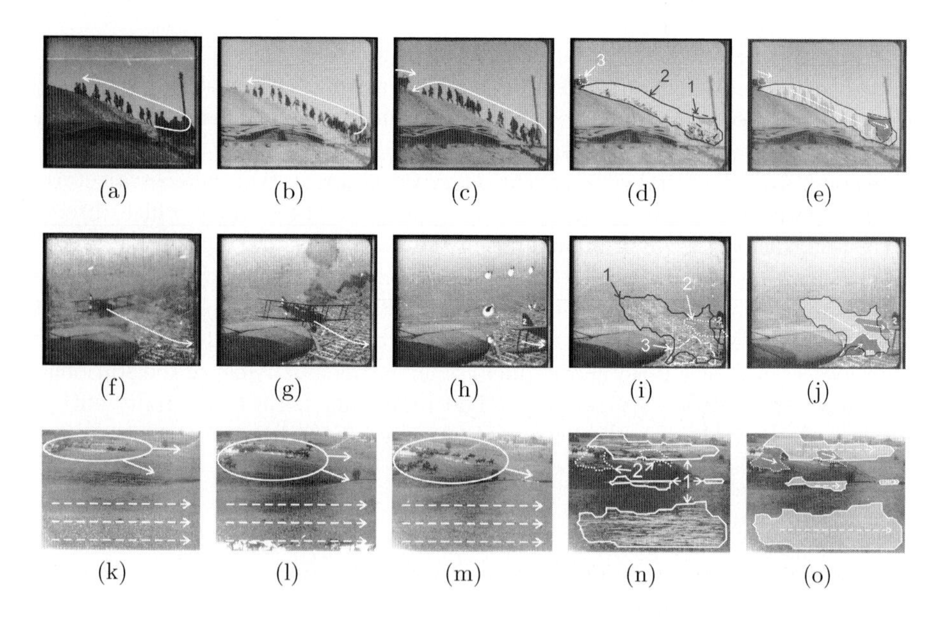

Fig. 3. Segmentation results for 3 test sequences. Columns 1-3 represent keyframes (white annotations mark the dominant motion components). Columns 4 and 5 show the clustered trajectories and the resulting motion segments with their primary direction.

hill. At the end of the sequence a horse enters the scene at the top of the hill in opposite direction (short arrow in Figure 3(c)). From the three keyframes 3(a)-3(c) we observe a large amount of flicker, additionally some frames contain scratches and dirt as in 3(a). Since KLT is sensitive to intensity variations the trajectories frequently break off. However, our approach is able to create temporally coherent motion segments over the entire duration of the shot. The movement of the group of people is represented by segments 1 and 2 (blue and yellow) in Figure 3(d). Segment 1 represents the motion of the people away from the camera and segment 2 captures the people walking up the hill. The third segment (red) represents the horse that appears at the end of the scene from the left. This sequence shows that the approach is able to segment large groups of objects as well as small individual objects.

The second sequence (Figures 3(f) - 3(j)) shows an airplane moving from left to right. The airplane approaches the observing camera and finally passes it. The sequence is shot by a camera that itself is mounted on an airplane, resulting in permanent shaking. Several frames of the shot are heavily blurred (e.g. 3(f)) making feature tracking nearly impossible. The motion of the airplane is represented by segments 1 and 2 (yellow and red). The first segment describes the motion of the airplane from the beginning of the shot to the last quarter of the shot. The second segment continues tracking this motion until the end of the shot. While the two segments are temporally coherent they are not merged by our algorithm because the (noisy) motion field that connects them is too sparse. The third segment (blue) describes an intense camera shake that is not removed during preprocessing. This shot demonstrates the limitations of the approach.

The third sequence shows a herd of horses (surrounded by the white ellipses in Figures 3(k) - 3(m)) moving diagonally into the scene from left to right. At the same time the camera pans to the right (indicated by the dashed arrows). Both motion components are tracked and separated from each other by our approach. The spatially distributed segment (segment 1, yellow) in Figure 3(n) represents the camera motion, while the second segment (red) describes the motion of the herd. Not all individuals of the herd can be tracked robustly by KLT due to the low contrast between the horses and the background. However, the motion trajectories available from tracking are correctly segmented.

Motion segmentation performs well for the presented sequences. Even under noisy conditions the method robustly segments the motion components.

5.2 Quantitative Evaluation

We apply the proposed approach to an entire feature film, in order to perform a quantitative evaluation. The film shows the life of workers of the 1920s and contains a large number of motion studies of physically working people, crowds, industrial machines (e.g., moving pistons), and vehicles (e.g., cars, trains). The film contains 63123 frames (660 shots) and has a duration of approximately one hour (at 18 fps).

The ground-truth is provided a priori by film experts in the form of a protocol that contains the number of motion components per shot and a textual descrip-

Table 1. Percentage of shots containing no false negative (FN), one FN and more than one FN with consideration of all and only the trackable motions, respectively

	0 FN	1 FN	> 1 FN
All motion components	70%	26%	4%
Only trackable motion components	89%	9%	2%

tion of the (groups of) objects, their motion activity and the camera operations. Evaluation is performed manually by comparing the computed motion segments with the ground truth protocol and applying the following rules:

1. A motion component is considered to be correctly detected if one or more clusters exist with similar spatio-temporal locations and similar directions. Otherwise the motion component is considered to be missed.
2. A cluster is considered to be a false positive, if it cannot be assigned to any motion component.

The proposed method is able to segment 60% of all motion components in the film. This low detection rate is a consequence of a poor feature tracking performance. While related literature reports excellent results of KLT for high-quality video [14], the tracker misses 28% of all motion components in the employed video material. The tracker frequently fails for very fast motions, motions in regions with low contrast, and complex scenes of water such as in Figure 1(b). We exclude the motions that KLT misses from the evaluation and yield a significantly higher detection rate of 83% which shows that the proposed method provides high performance when motion tracking is successful.

The false positive rate is relatively high (22%) due to tracking failures and noise. For example, feature points tend to walk along edges resulting in motion components that are wrong but have a significant velocity magnitude. On the other hand, we have configured the system sensitive to small motion components which makes the system prone to noise.

In addition, we test our approach on selected sequences from high-quality video (230 shots from the movie "Lola Runs") and yield a significantly lower false positive rate (3%). The detection rate (for all motion components) is 72% compared to 60% for the low-quality material.

We further evaluate the number of false negatives (motion components that are not correctly segmented) for each shot of the low-quality material. The distribution of false negatives is summarized in Table 1. The approach successfully segments all trackable motion components in 89% of the shots. One trackable motion component is missed in 9% of the shots and only 2% of the shots contain more than one missed component. The greatest potential for improvements lies in the stage of feature detection and tracking. We currently investigate alternative methods, e.g. SIFT which so far shows comparable performance. However this investigation is not the focus of this paper.

The proposed clustering method is computationally efficient. Motion segmentation (excluding feature tracking), requires 10s per shot in average and 110 minutes for the entire film.

6 Conclusions

We have presented a novel clustering scheme that robstly segments sparse and noisy motion trajectories into meaningful motion components. The clustering scheme allows for the selection of trajectory features and similarity measures which makes it well-suited for different types of clusterings and applications. Although the method has been developed for low-quality video, experiments have shown that it is applicable to diverse video material. The low-quality of the employed material mainly influences the feature tracking performance. Where motion is trackable, motion segmentation is successful to a high degree.

Acknowledgments

This work has received financial support from the Vienna Science and Technology Fund (WWTF) under grant no. CI06 024.

References

1. Buzan, D., Sclaroff, S., Kollios, G.: Extraction and clustering of motion trajectories in video. In: Proc. of the Int. Conf. on Pattern Rec., pp. 521–524 (2004)
2. Dagtas, S., Al-Khatib, W., Ghafoor, A., Kashyap, R.: Models for motion-based video indexing and retrieval. IEEE Trans. on Image Processing 9(1), 88–101 (2000)
3. Gevers, T.: Robust segmentation and tracking of colored objects in video. IEEE Trans. on Circuits and Systems for Video Techn. 14(6), 776–781 (2004)
4. Shi, J., Tomasi, C.: Good features to track. In: Proc. of the IEEE Computer Society Conf. on Computer Vision and Pattern Recognition, pp. 593–600 (1994)
5. Li, X., Hu, W., Hu, W.: A coarse-to-fine strategy for vehicle motion trajectory clustering. In: Proc. of the Int. Conf. on Pattern Rec., pp. 591–594 (2006)
6. Rabaud, V., Belongie, S.: Counting crowded moving objects. In: Proc. of the IEEE Comp. Society Conf. on Comp. Vision and Pat. Rec., pp. 705–711 (2006)
7. Veit, T., Cao, F., Bouthemy, P.: Space-time a contrario clustering for detecting coherent motions. In: Proc. of the Int. Conf. on Robot. & Autom., pp. 33–39 (2007)
8. Wang, H., Lin, H.: A spectral clustering approach to motion segmentation based on motion trajectory. In: Proc of the Int Conf on MM & Expo., pp. 793–796 (2003)
9. Wang, X., Tieu, K., Grimson, E.: Learning semantic scene models by trajectory analysis. In: Leonardis, A., Bischof, H., Pinz, A. (eds.) ECCV 2006. LNCS, vol. 3953, pp. 110–123. Springer, Heidelberg (2006)
10. Hervieu, A., Bouthemy, P., Le Cadre, J.P.: Video event classification and detection using 2d trajectories. In: Proc. of the Int. Conf. on Computer Vision Theory and Applications, pp. 110–123 (2008)
11. Bashir, F.I., Khokhar, A.A., Schonfeld, D.: Real-time motion trajectory-based indexing and retrieval of video sequences. IEEE Trans. on MM 9(1), 58–65 (2007)
12. Birchfield, S.: Klt: An implementation of the kanade-lucas-tomasi feature tracker, `http://www.ces.clemson.edu/~stb/klt` (last visited: July 2009)
13. Ewerth, R., Schwalb, M., Tessmann, P., Freisleben, B.: Segmenting moving objects in MPEG videos in the presence of camera motion. In: Proc. of the Int. Conf. on Image Analysis and Processing, pp. 819–824 (2007)
14. Rothganger, F., Lazebnik, S., Schmid, C., Ponce, J.: Segmenting, modeling, and matching video clips containing multiple moving objects. In: Proc. of the IEEE Conf. on Comp. Vis. & Pattern Rec., vol. 2, pp. 914–921 (2004)

New Optical Flow Approach for Motion Segmentation Based on Gamma Distribution

Cheolkon Jung, Licheng Jiao, and Maoguo Gong

Key Lab of Intelligent Perception and Image Understanding of Ministry of Education of China, Institute of Intelligent Information Processing, Xidian University, Xi'an 710071, China
zhengzk@xidian.edu.cn

Abstract. This paper provides a new motion segmentation algorithm in image sequences based on gamma distribution. Conventional methods use a Gaussian mixture model (GMM) for motion segmentation. They also assume that the number of probability density function (PDF) of velocity vector's magnitude or pixel difference values is two. Therefore, they have poor performance in motion segmentation when the number of PDF is more than three. We propose a new and accurate motion segmentation method based on the gamma distribution of the velocity vector's magnitude. The proposed motion segmentation algorithm consists of pixel labeling and motion segmentation steps. In the pixel labeling step, we assign a label to each pixel according to the magnitude of velocity vector by optical flow analysis. In the motion segmentation step, we use energy minimization method based on a Markov random field (MRF) for noise reduction. Experimental results show that our proposed method can provide fine motion segmentation results compared with the conventional methods.

1 Introduction

The development of a powerful moving object segmentation algorithm is an important requirement for many computer vision and ubiquitous systems. In video surveillance applications, motion segmentation can be used to determine the presence of people, car, or other unexpected objects and then start up more complex activity recognition steps. In addition, the segmentation of moving objects in the observed scenes is an important problem to solve for traffic flow measurements or behavior detection during sports activities [1-4].

Up to the present, many significant achievements have been made by researchers in the fields of moving object segmentation. Chang et al. proposed a Bayesian framework that combines motion estimation and segmentation based on a representation of the motion field as the sum of a parametric field and a residual field [5]. Velocity vectors are obtained by affine model, and then motion and segmentation fields are obtained by energy minimization. For energy minimization, iterated conditional mode (ICM) has been used. This method has a shortcoming to input the number of clusters. Luthon et al. proposed motion segmentation algorithm using MRF model [6].

S. Boll et al. (Eds.): MMM 2010, LNCS 5916, pp. 444–453, 2010.

However, this algorithm has shortcomings to set threshold value by user decision and to generate much noise unnecessarily. Aach and Kaup proposed a motion segmentation algorithm of video objects using a statistical approach [7]. They model as a Gaussian distribution the characteristics of pixel difference for background between two consecutive frames. For a given level of significance, the resulting threshold value is theoretically obtained and the frame difference image is threshold so as to yield a change detection mask (CDM). Jung and Kim proposed motion segmentation algorithm using signal detection theory [8]. This algorithm also has poor performance if the number of PDF is more than three.

However, these methods use a Gaussian mixture model (GMM) for motion segmentation. They also assume that the number of PDF is two and has poor performance if the number of PDF is more than three. In this paper, we present a new motion segmentation algorithm which accurately segments motion of moving objects although the number of PDF of velocity vector's magnitude is more than three using gamma distribution. As shown in Fig. 1, the proposed motion segmentation algorithm consists of pixel labeling and motion segmentation steps. First, the pixel labeling step estimates motion by optical flow analysis and assigns a label to each pixel by the magnitude of velocity vectors based on gamma distribution. For the pixel labeling, we have found the number of PDF, and then determined each mean and threshold value. Next, the motion segmentation step removes errors by energy minimization and segments moving regions with motion.

This paper is organized as follows. Pixel labeling by velocity vector is addressed in Section 2. In Section 3, motion segmentation by MRF is explained. Section 4 presents experimental results, and we conclude this paper in Section 5.

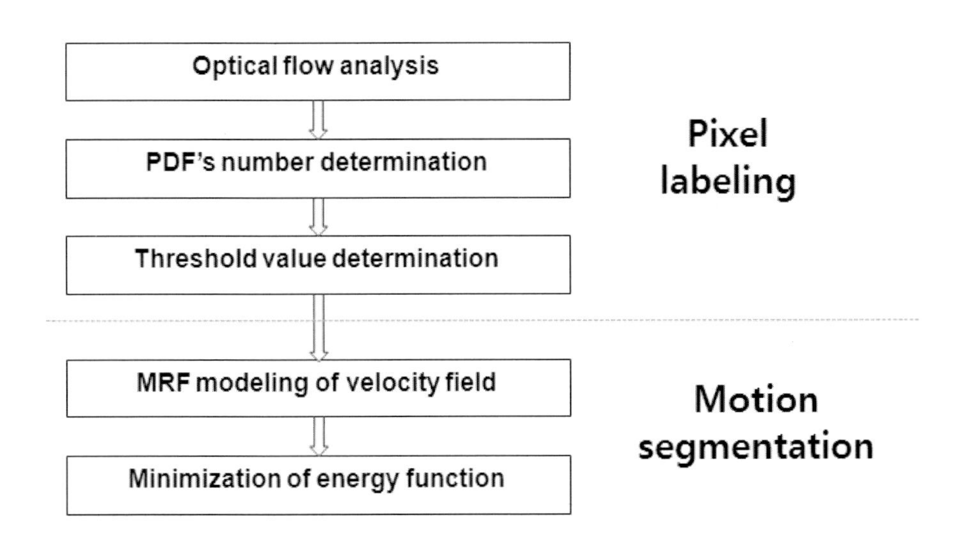

Fig. 1. Block diagram of the proposed method

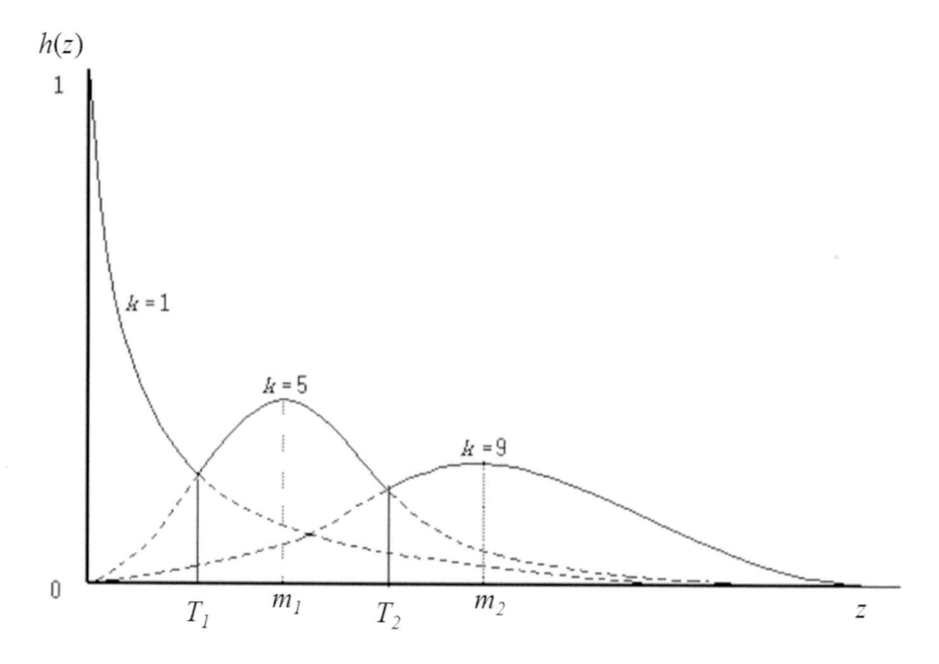

Fig. 2. Distribution of z

2 Pixel Labeling

Optical flow is defined by the velocity field in image plane due to motion of objects in scenes. Let I be the intensity of a pixel (x,y) of an image in time t. In traditional optical flow analysis techniques, the optical flow constraint equation is expressed as [9]:

$$I_x u + I_y v + I_t = 0 \tag{1}$$

where u and v are two components of velocity vector and I_x, I_y, I_t are partial derivatives about x, y, t, respectively. By Horn and Schunck's method [9], the components of the velocity vector is computed by (2) and (3).

$$u^{i+1} = u^i - \frac{I_x(I_x u^i + I_y v^i + I_t)}{\lambda + I_x^2 + I_y^2} \tag{2}$$

$$v^{i+1} = v^i - \frac{I_y(I_x u^i + I_y v^i + I_t)}{\lambda + I_x^2 + I_y^2} \tag{3}$$

where λ is weighted constant and i is iteration number.

If η is threshold value, iteration is stopped by the stop criterion which is expressed as:

$$\sum_{(x,y)} \sqrt{(u^{k+1} - u^k)^2 + (v^{k+1} - v^k)^2} < \eta \tag{4}$$

Let random variable $z(x,y)$ be the magnitude of velocity vector of a pixel (x,y) in an image. Then, random variable $z(x,y)$ is defined as:

$$z(x,y) = \sqrt{u(x,y)^2 + v(x,y)^2} \tag{5}$$

By various experiments, we found that the distribution of z is represented as distribution as shown in Fig. 2. Therefore, we model the distribution of z as mixture gamma distribution. Therefore, the distribution of z, $h(z)$, takes the form as [10]:

$$h(z) = \sum_{k=1}^{M} \delta_k \frac{\mu_k{}^k}{(k-1)!} z^{k-1} e^{-\mu_k z} \tag{6}$$

where k denotes gamma distribution function order and M the maximum value of PDF's number. δ_k is coefficient of each PDF and μ_k is gamma function decaying parameter. Here, δ_k is 1 in k=1,5,9,···, otherwise 0. To make pixel labeling, the correct number of PDF, K, should be determined. Since K is equal to the number of clusters, we use the cluster validity measure proposed by Rose [11]. The principle of this validity is to minimize the within-cluster scatter and maximize the between-cluster separation.

The cluster validity measure, *validity*, is expressed as:

$$validity = w \cdot \frac{\text{intra}}{\text{inter}} \tag{7}$$

where, w is weighted constant. If N is total number of pixel and each cluster is $C_l(1,2, \cdots K)$, $\text{intra} = \frac{1}{\text{N}} \sum_{l=1}^{K} \sum_{z \in C_l} | z - m_l |$ and $\text{inter} = \min(| m_l - m_m |)$ $(l=1,2,\ldots, K-1, m=l+1,\ldots,K)$. Here, m_l means l-th mean. In order to find the label of each pixel, we determine the optimal threshold value T_n. If we assume that $\mu_l=\ldots=\mu_K=\mu$, the optimal threshold value T_n is computed by

$$T_n = \frac{1}{\mu} \sqrt[4]{\frac{4n!}{(4n-4)!}}\,. \tag{8}$$

Here, $T_1 = \frac{1}{\mu}\sqrt[4]{24}$ and $T_2 = \frac{1}{\mu}\sqrt[4]{1680}$. By (8), we assign a label of cluster to each pixel. Label field $L(x,y)$ is expressed as:

$$L(x,y) = l, \quad z(x,y) \in C_l \tag{9}$$

where C_l is l-th cluster. Since z is smaller than 1, z is scaled for computational efficiency by (10) [8].

$$z' = \frac{Z \cdot (z - z_{\min})}{z_{\max} - z_{\min}} \tag{10}$$

where z' is scaled value of z. z_{max} and z_{min} are the maximum and minimum value of z, respectively, and Z is a constant. The scaled value z' instead of z is used for the pixel labeling.

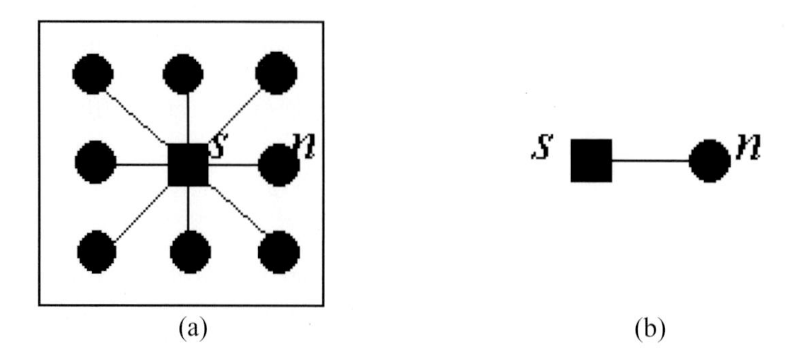

Fig. 3. (a) Neighborhood system. (b) A binary click.

3 Motion Segmentation

If an image is segmented with the threshold T_n in Fig. 2, isolated regions by noise are segmented together. To remove noisy isolated regions, MRF model is used. The MRF model takes advantage of spatial homogeneity of moving objects and removes isolated regions by noise. We have modeled z as MRF and removed the unnecessary regions by noise. Let a random field z be MRF. Neighborhood system N and binary cliques at each point (x,y) are defined as shown in Fig. 3. In the figure, s is a center pixel and n is a neighborhood pixel. If we regard a random field z as MRF, the probability of z is given by a Gibbs distribution that has the following form according to the Hammersley-Clifford theorem [6].

$$P(z) = Q^{-1} \times e^{-U(z)} \tag{11}$$

where Q is a normalized constant called the partition function and $U(z)$ is the energy function. We want to find l that the posteriori probability $p(l|z)$ is maximal. By using the maximum *a posteriori* criterion (MAP), the solution is

$$l^* = \arg \max_l P(l \mid z) . \tag{12}$$

We express relationship between z and l by Bayes rule as follows.

$$P(l \mid z) \propto P(z \mid l) P(l) \tag{13}$$

where $P(l|z)$ is a conditional probability of l in dependence on z and $P(l)$ is *a priori* probability of l. Therefore, we can express Eq. (12) as following form.

$$l^* = \arg \max_l P(l \mid z) = \arg \max_l (P(z \mid l) \cdot P(l)) \tag{14}$$

From Eq. (11), we get

$$\max_{l}(P(l \mid z)) = \min_{l}(U(l \mid z)) \tag{15}$$

Therefore, the maximization of the *a posteriori* probability is equivalent to the minimization of the energy function. The energy function is classically the sum of two terms (corresponding to data-link and prior knowledge, respectively) [6]:

$$U(l \mid z) = U_a(z \mid l) + U_m(l) \tag{16}$$

The link-to-data energy $U_a(z|l)$ (attachment energy) is expressed as

$$U_a(z \mid l) = \frac{1}{2\sigma^2}\sum_{l}(z - m_l)^2 \tag{17}$$

where σ^2 is the observation variance.

The model energy $U_m(l)$ is a regularization term and puts *a priori* constraints on the masks of moving objects, removing isolated points due to noise. Its expression is given by

$$U_m(l) = \sum_{c} V_c(l_s, l_n) \tag{18}$$

where c, s, and n denote a binary clique, a current pixel, and pixel of neighbor, respectively. Here, l_s is a label of s, l_n is a label of n and $V_c(l_s,l_n)$ is a potential function associated with a binary clique, $c=(s,n)$. To put homogeneity constraints into the model, it is defined as,

$$V_c(l_s, l_n) = \begin{cases} -\beta, & \text{if } l_s = l_n \\ +\beta, & \text{if } l_s \neq l_n \end{cases} \tag{19}$$

where the positive parameter β depends on the nature of the clique. To find the minimum of the energy function, iterated conditional modes (ICM) is used [12]. For each pixel s of the current image, the labels from 0 to $K-1$ are tested and the label that induces the minimum local energy in the neighborhood is kept. The process iterates over the image until convergence. Suppose the label of a current pixel in iteration j is denoted as l^j and a prescribed small number is ε. The fixed label of each pixel is achieved if the following condition is satisfied [13]:

$$\sum_{(x,y)} | l^j - l^{j-1} | < \varepsilon \tag{20}$$

A label l is one of $0,1,\ldots,K-1$ and arranged according to the magnitude of velocity vectors. Therefore, motion exists in a pixel with $l>0$ as Eq. (21).

$$motion = \begin{cases} \text{not exist}, & l = 0 \\ \text{exist}, & \text{otherwise} \end{cases} \tag{21}$$

4 Experimental Results

In order to evaluate the performance of the proposed method, four typical image sequences of *Table tennis*, *Claire*, *Street*, and *Smoke* were used in the experiments. All sequences were adjusted to the size of 176x144 pixels. Fig. 4 shows examples of each sequence.

Fig. 5 shows the velocity vectors of each sequence by optical flow analysis. To compute the velocity vectors, we set η =10 in (4). It can be observed that the magnitudes of pixels with large motion are large. Table 1 shows the *validity* of each

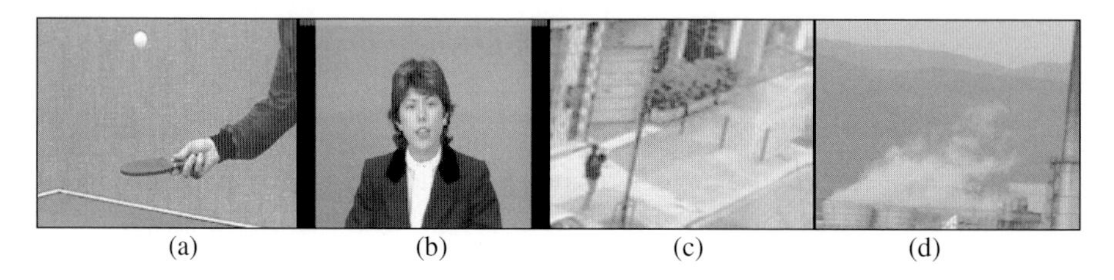

(a) (b) (c) (d)

Fig. 4. Original images. (a) *Table tennis*. (b) *Foreman*. (c) *Street*. (d) *Smoke*.

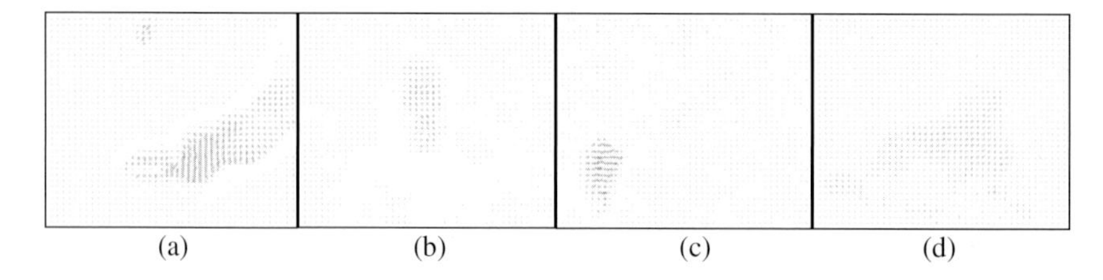

(a) (b) (c) (d)

Fig. 5. Optical flow vectors. (a) *Table tennis*. (b) *Claire*. (c) *Street*. (d) *Smoke*.

Table 1. Validity

K	*Table tennis*	*Claire*	*Street*	*Smoke*
2	0.02895	**0.01135**	**0.00718**	**0.04137**
3	**0.02317**	0.02003	0.01754	0.09762
4	0.03002	0.02037	0.13072	0.17936
5	0.05300	0.08798	0.16737	0.15152
6	0.04324	0.10868	0.27405	0.24072
7	0.06813	0.15721	0.38423	0.20171
8	0.07655	0.11400	0.28709	0.18637
9	0.06501	0.18998	0.51542	0.29961
10	0.07070	0.20081	0.77435	0.23735

sequence according to K. The optimal K is selected by minimizing the *validity* in (7). Therefore, the optimal K is 3 in *Table tennis* sequence, and 2 in the other sequences. By the optimal K, m_l and T_l of (6) are determined. We can assign each pixel to a label by (9). Here, we set $Z = 255$ in (10).

Fig. 6 shows the distribution of z of each sequence. It can be observed that $h(z)$ of *Table tennis* sequence has approximately 3 PDFs and $h(z)$ of the other sequences have 2 PDFs. Also, we can see that $h(z)$ of all test sequences is near mixture gamma distribution as mentioned in Section 2. In the experiments, we set σ=10 and β=1. The iteration number was about 5~10 in (20). Here, we set ε=3. Fig. 7 shows final results of

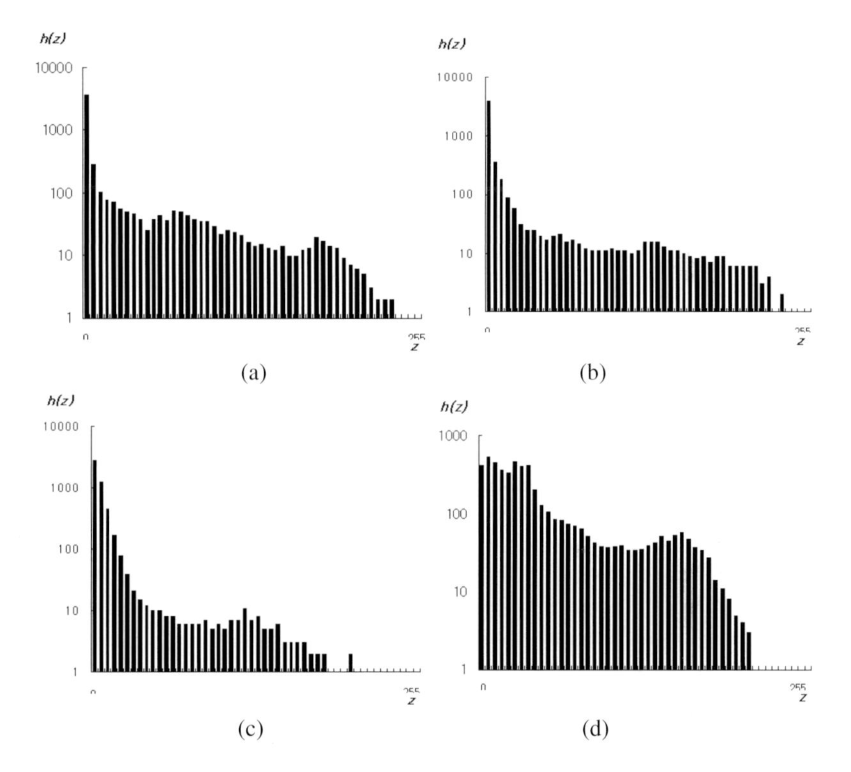

Fig. 6. Histogram of z. (a) *Table tennis*. (b) *Claire*. (c) *Street*. (d) *Smoke*.

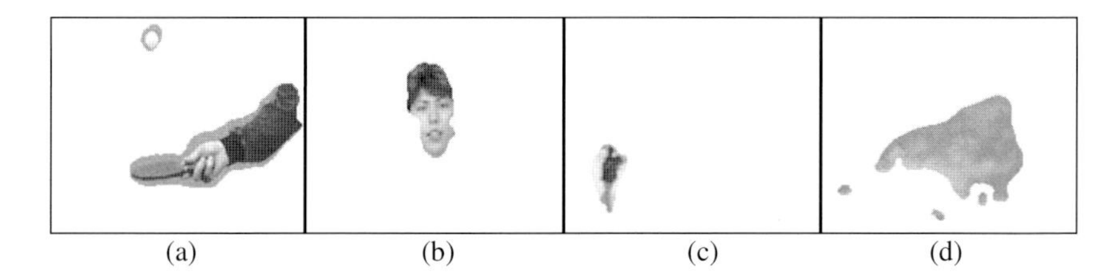

Fig. 7. Motion segmentation results. (a) *Table tennis*. (b) *Claire*. (c) *Street*. (d) *Smoke*.

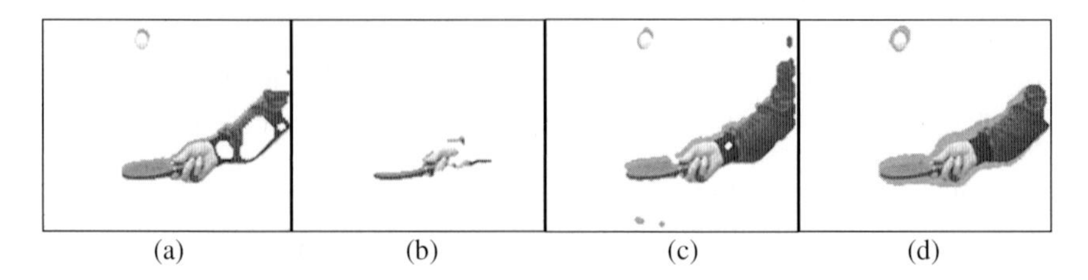

Fig. 8. Motion segmentation results (a) by [6], (b) by [7], (c) by [8], (d) by the proposed method

motion segmentation of each sequence by (21). Fig. 8 shows comparison results between the proposed algorithm and conventional methods. Figs. 8(a)~(d) show the results by motion segmentation algorithm proposed in [6], [7], [8], and this paper, respectively. In Figs. 8(a)~(d), we assumed that σ^2 of each sequence is equal. These results show that the proposed algorithm has better performance than the conventional methods.

5 Conclusions

In this paper, we have proposed a new and accurate motion segmentation algorithm in image sequences based on gamma distribution. The proposed algorithm consists of two steps: pixel labeling using gamma distribution and motion segmentation using MRF. In the pixel labeling step, we have found velocity vectors by optical flow analysis and assign a label to each pixel according to the magnitude of the velocity vectors based on gamma distribution. In the motion segmentation step, we have obtained the final segmentation results through energy minimization based on MRF. We have demonstrated that the proposed algorithm can provide fine segmentation results compared to the conventional methods by experiments. Therefore, it can be used a highly effective object segmentation algorithm for semantic image analysis such as object detection, object recognition, and object-based video coding.

Acknowledgement

This work was supported by the National Basic Research Program (973 Program) of China (No.2006CB705707) and the National High Technology Research and Development Program (863 Program) of China (No.2007AA12Z223, 2008AA01Z125, 2009AA12Z210).

References

1. Spagnolo, P., Orazio, T.D., Leo, M., Distante, A.: Moving object segmentation by background substraction and temporal analysis. Image and Vision Computing 24, 411–423 (2006)
2. Kuo, M., Hsieh, C.H., Huang, Y.R.: Automatic extraction of moving objects for head-shoulder video sequences. Journal of Visual Communication and Image Representation 16, 68–92 (2005)

3. Jung, C., Kim, J.K.: Motion segmentation using Markov random field model for accurate moving object segmentation. In: Proc. of ACM ICUIMC 2008, pp. 414–418 (2008)
4. Kim, M.C., Jeon, J.G., Kwak, J.S., Lee, M.H., Ahn, C.: Moving object segmentation in video sequences by user interaction and automatic object tracking. Image and Vision Computing 19, 245–260 (2001)
5. Chang, M.M., Tekalp, A.M., Sezan, M.I.: Simultaneous motion estimation and segmentation. IEEE Trans. on Image Processing 6, 1326–1333 (1997)
6. Luthon, F., Caplier, A., Lievin, M.: Spatiotemporal MRF approach to video segmentation: Application to motion detection and lip segmentation. Signal Processing 76, 61–80 (1999)
7. Aach, T., Kaup, A.: Bayesian algorithms for adaptive change detection in image sequences using Markov random fields. Signal Processing: Image Communication 7, 147–160 (1995)
8. Jung, C., Kim, J.K.: Moving object segmentation using Markov random field. Korea Information and Communication Society 27, 221–230 (2002)
9. Horn, B.K.P., Schunck, B.G.: Determining optical flow. Artificial Intelligence 17, 185–203 (1981)
10. Barkat, M.: Signal detection & estimation. Artech House (1991)
11. Ray, S., Turi, R.H.: Determination of number clusters in K-means clustering and application in colour image segmentation. In: Proc. of ICAPRDT 1999, pp. 137–143 (1999)
12. Dubes, R.C., Jain, A.K., Nadabar, S.G., Chen, C.C.: MRF model-based algorithms for image segmentation. In: Proc. of ICPR (10th International Conference on Pattern Recognition), vol. 1, pp. 808–814 (1990)
13. Wei, J., Li, Z.: An efficient two-pass MAP-MRF algorithm for motion estimation based on mean field theory. IEEE Trans. on Circuits and Systems for Video Technology 9, 960–972 (1999)

Reducing Frame Rate for Object Tracking

Pavel Korshunov and Wei Tsang Ooi

National University of Singapore, Singapore 119077
pavelkor@comp.nus.edu.sg, ooiwt@comp.nus.edu.sg

Abstract. Object tracking is commonly used in video surveillance, but typically video with full frame rate is sent. We previously have shown that full frame rate is not needed, but it is unclear what the appropriate frame rate to send or whether we can further reduce the frame rate. This paper answers these questions for two commonly used object tracking algorithms (frame-differencing-based blob tracking and CAMSHIFT tracking). The paper provides (i) an analytical framework to determine the critical frame rate to send a video for these algorithms without them losing the tracked object, given additional knowledge about the object and key design elements of the algorithms, and (ii) answers the questions of how we can modify the object tracking to further reduce the critical frame rate. Our results show that we can reduce the 30 fps rate by up to 7 times for blob tracking in the scenario of a single car moving across the camera view, and by up to 13 times for CAMSHIFT tracking in the scenario of a face moving in different directions.

1 Introduction

Object tracking is a common operation in video surveillance systems. However, given an object tracking algorithm, it is unclear what frame rate is necessary to send. Typically, video is sent at the rate of full video camera capacity, which may not be the best option if network bandwidth is limited.

Previously, we have shown in [1] that frame rate can be significantly reduced without object tracking losing the object. We found that the critical frame rate for a given algorithm depends on the speed of tracked object. The simple way to determine the critical frame rate is to run algorithm on a particular video sequence, dropping frames and noticing which rate causes the algorithm to lose the object. Such approach however is not practical, because objects in real surveillance videos move with different speeds, and the critical frame rate therefore should depend on this parameter. We suggest finding critical frame rate using analysis based on the algorithm's key design elements (specific object detection and tracking mechanism) and measured speed and size of the tracked object.

In this paper, we focus on two tracking algorithms, blob tracking algorithm that relies on frame differencing and foreground object detection by Li *et al.* [2] as well as Kalman filter for tracking; and CAMSHIFT algorithm [3], in which objects are

S. Boll et al. (Eds.): MMM 2010, LNCS 5916, pp. 454–464, 2010.

represented as color histograms, and tracking is performed using mean shift algorithm. We present an analytical framework formalizing the dependency between video frame rate and algorithms' accuracy. We estimate critical frame rate using analysis with assumption of known speed and size of the tracked object. Guided by the estimation, we slightly modify these tracking algorithms making them adaptive and more tolerant to the videos with even lower frame rate.

In Section 2, we present analysis of the critical frame rate for object tracking. In Section 3, we demonstrate how the dependency between frame rate and accuracy can be estimated specifically for the blob tracking. We also specify the critical frame rate for this algorithm. In Section 4, we present similar analysis for CAMSHIFT tracking. In Section 5, we show how, using our estimations and measurement of speed and size of the tracked object, we can modify these tracking algorithms adapting them to the reduced frame rate. Section 6 ends the paper with conclusion and future works.

2 General Analysis

We degrade temporal video quality by applying the dropping pattern "drop i frames out of $i+j$ frames", where i is *drop gap*, and j is the number of consecutive remaining frames (see Figure 1). Note that the same frame rate can correspond to two different dropping patterns, for instance, dropping 2 out of 3 frames results in the same frame rate as dropping 4 out of 6 frames. The reason for choosing such dropping pattern is because we found that drop gap is more important factor for the performance of the tracking than simply a frame rate. Therefore, instead of critical frame rate, we focus on finding *critical drop gap*, which would determine the corresponding frame rate.

First, we present an estimation of the critical drop gap for an object tracking algorithm without taking into account the specific method of detection and tracking. For simplicity, consider a video containing a single moving object, which can be accurately tracked by the algorithm. We can notice that dropping frames affects the speed of object. Since video is a sequence of discrete frames, the speed of object can be understood as a distance between the centers of object positions in two consecutive frames, which we call inter-frame speed denoted as Δd. Without loss of generality, we can say that for every object tracking algorithm there exists a $\Delta\tilde{d}$ such that, if object moves for a larger distance than $\Delta\tilde{d}$, the algorithm loses it.

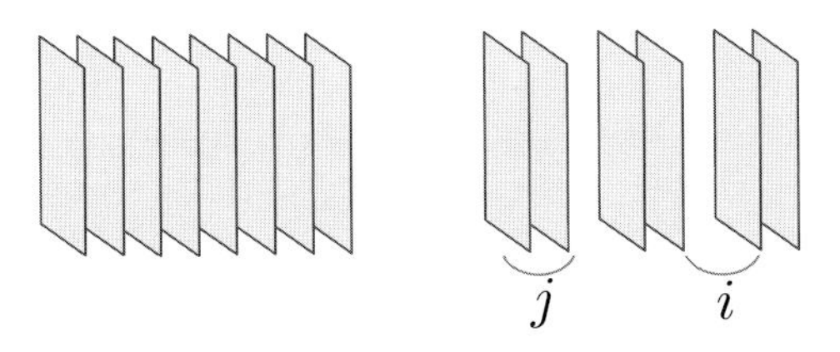

Fig. 1. Dropping i out of $i + j$ frames. i is the drop gap.

Let Δd_0 be the maximal inter-frame speed of the object in the original video, when no frame dropping is applied yet. If we drop frames with drop gap $i = 1$, the new maximum inter-frame speed can be approximated as $\Delta d_1 = 2\Delta d_0$. Then, for general frame dropping pattern, $\Delta d_i = (i+1)\Delta d_0$. Assume we know the original speed of the object and the algorithm's threshold $\Delta\tilde{d}$. Then, we can compute the maximum number of consecutive frames that can be dropped, i.e., critical drop gap $\tilde{i}$, as

$$\tilde{i} = \frac{\Delta\tilde{d}}{\Delta d_0} - 1. \tag{1}$$

3 Blob Tracking Algorithm

For blob tracking algorithm, due to frame differencing detection, the value $\tilde{i}$ depends on the size and the speed of tracked object. If too many consecutive frames are dropped, the object in the current frame appear so far away from its location in the previous frame that the frame differencing operation results in detecting two separate blobs (see Figure 3(b)). Such tracking failure occurs when the distance between blob detected in the previous frame and blob in the current frame is larger than the size of the object itself. Therefore, this distance is the threshold distance $\Delta\tilde{d}$. To determine its value, we need to estimate the coordinates of the blob center in the current frame, which depend on its location and size in the previous frame.

In this analysis, we assume a single object monotonously moving in one direction. Although this assumption considers only a simplified scenario, many practical surveillance videos include objects moving in a single direction towards or away from the camera view. Also, such movements of the object in camera's view as rotating or only changing in size (when object goes away/towards camera view but does not move sideways) do not have a significant effect on frame differencing object detection. We also assume, without loss of generality, that the object moves from left to right with its size increasing linearly. The assumption allows us to consider only changes in coordinate x, and width w. Increase/decrease in size is important because when tracked objects approach or move away from the camera, their size changes. In practice, when object moves in both x and y coordinates, the overall critical drop gap would be the minimum of the two values estimated for corresponding coordinates.

Consider the original video when no frames are dropped. We assume the average distance between fronts of the blob when it shifts from the previous frame to the current frame is Δx^0. We consider the front of the object because it is more accurately detected by frame differencing. When frame differencing is used, the resulted detected blob is the union of the object presented in the previous and current frames (see Figure 3(b)). Therefore, when we drop frames, the width of the blob in the frame following after the drop gap will be larger than that in the original video sequence (see Figure 2 for illustration). However, the front of the blob would be detected in the same way as in the original video.

Since frame dropping affects size of the detected object, we consider average change in size as Δw^0. The superscript indicates the size of the drop gap, which

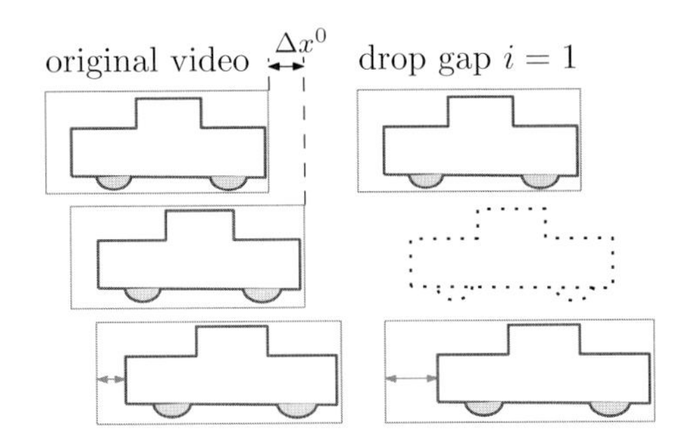

Fig. 2. The schema of the difference between object foreground detection for original video and for video with dropped frames

(a) Detected foreground object with drop gap 14 frames. PETS2001 video.

(b) Binary mask of the frame in 3(a). Effect of drop gap on frame differencing.

Fig. 3. The foreground object detection based on frame differencing

is 0 when frames are not dropped. Assume that x_k^0 is x-coordinate of blob's center in k-*th* frame, then, we can estimate its coordinate in the frame $k+i+1$ as following,

$$x_{k+i+1}^0 = x_k^0 + (i+1)\Delta x^0 - (i+1)\frac{\Delta w^0}{2}. \tag{2}$$

If i frames are dropped after frame k, the detected blob in the $k+i+1$ frame is the union of actual object appearing in frames k and $k+i+1$ (as Figure 2 illustrates). Then, the width difference $(w_{k+i+1}^i/2 - w_k^0/2)$ can be approximated as $(i+1)\Delta x^0/2$. Therefore, the blob's center in the $k+i+1$ frame can be estimated as,

$$x_{k+i+1}^i = x_k^i + (i+1)\Delta x^0 - (i+1)\frac{\Delta x^0}{2} = x_k^0 + (i+1)\frac{\Delta x^0}{2}, \tag{3}$$

since $x_k^i = x_k^0$.

As was mentioned, $\Delta\tilde{d} = |x_{k+\tilde{i}+1}^{\tilde{i}} - x_k^{\tilde{i}}|$, where $\tilde{i}$ indicates the critical drop gap. The failure of the blob tracking implies that $\Delta\tilde{d} = w_k^0$, where value w_k^0 is the width of the blob detected in frame k. Therefore, from equation (3), we obtain $w_k^0 = \Delta\tilde{d} = (\tilde{i}+1)\frac{\Delta x^0}{2}$, from which we can find the critical drop gap to be

$$\tilde{i} = \frac{2w_k^0}{\Delta x^0} - 1. \quad (4)$$

In practice, values w_k^0 and Δx^0 can be determined by either keeping the history of speed and size of tracked object or by estimating their average values for a particular surveillance site.

In addition to the estimation of the critical drop gap for blob tracking, we can estimate the dependency function between accuracy of the algorithm and video frame rate. Such estimation is possible because of the way drop gap affects the accuracy of the frame differencing object detection algorithm used in blob tracking. We can define blob detection error for a particular frame as the distance between blob centers detected in this frame for the degraded video (with dropped frames) and the original video. Then, the average error, denoted as ϵ_{ij}, is the average blob tracking error for all frames in the video. This ϵ_{ij} function can be used as accuracy metric for the blob tracking depicting the tradeoff between tracking accuracy and video frame rate.

Using equations (2) and (3) we can estimate the blob tracking error for $k+i+1$ frame as following,

$$\left|x_{k+i+1}^i - x_{k+i+1}^0\right| = (i+1)\left|\frac{(\Delta x^0 - \Delta w^0)}{2}\right| = (i+1)C, \quad (5)$$

where constant $C \geq 0$ depends on the size and the speed of object in the original video.

Since we apply the dropping pattern "drop i frames out of $i+j$ frames", we need to estimate the blob tracking error for each of the remaining j frames in the video. There is no error in detecting blob for $j-1$ frames that do not have drop gap in front of them, i.e., for these frames, the result of the frame differencing would be the same as in original video with no dropping. Therefore, the average error for all j frames is the error estimated for the frame, which follows the drop gap (equation (5)) divided by j:

$$\epsilon_{ij} = \frac{i+1}{j}\left|\frac{(\Delta x^0 - \Delta w^0)}{2}\right| = \frac{i+1}{j}C. \quad (6)$$

Note the important property of this function that the average error is proportional to i and inversely proportional to j.

We performed experiments to validate the estimation of the average blob tracking error ϵ_{ij}. We use several videos from ViSOR video database, PETS2001 datasets, as well as videos we shot on campus with a hand-held camera (example screenshots in Figure 3(a), Figure 4(c), and Figure 4(d)). Videos include moving cars, person on a bicycle and people walking in a distance. We ran blob tracking algorithm on these videos and applied different dropping patterns. We plot the

(a) Fast moving face shot with a web-cam (CAMSHIFT face tracking).

(b) From database by SEQAM laboratory (CAMSHIFT face tracking).

(c) Shot on campus with hand-held camera (blob tracking).

(d) From VISOR video database (blob tracking).

Fig. 4. Snapshot examples of videos used in our experiments

resulted average error against drop gap i when value j is 1, 3, 6, and 12. The results are shown in Figure 5(a) (original video is 158 frames of 384 $\times$ 288, 30 fps) and Figure 6(a) (original video is 148 frames of 320 $\times$ 256, 30 fps).

Figure 5(a) shows the resulted average tracking error plotted against the drop gap i when value j is 1, 3, 6, and 12. It can be noted from the Figure 5(a) that for each fixed value j the average error is proportional to i. Also, average error is inversely proportional to j, as indicated by the angles of each line in the graph (for instance, angle of the line marked as "j=1" is three times larger than the angle of the line "j=3"). Figure 6(a) demonstrates similar results. These experimental results strongly support our analytical estimation of the average error given in the equation (6). The figures do not reflect the critical drop gap value because even for large drop gaps the blob tracking did not lose the track of the car in this test video sequence.

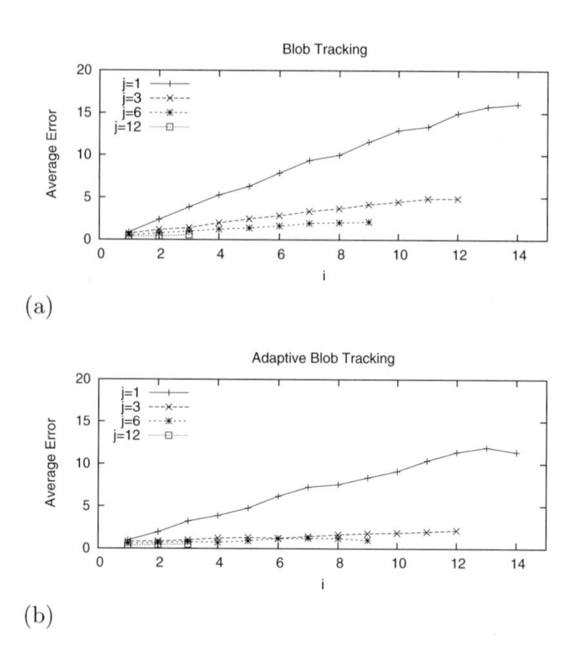

Fig. 5. Accuracy of original and adaptive blob tracking algorithm for PETS2001 video (snapshot in Figure 3(a))

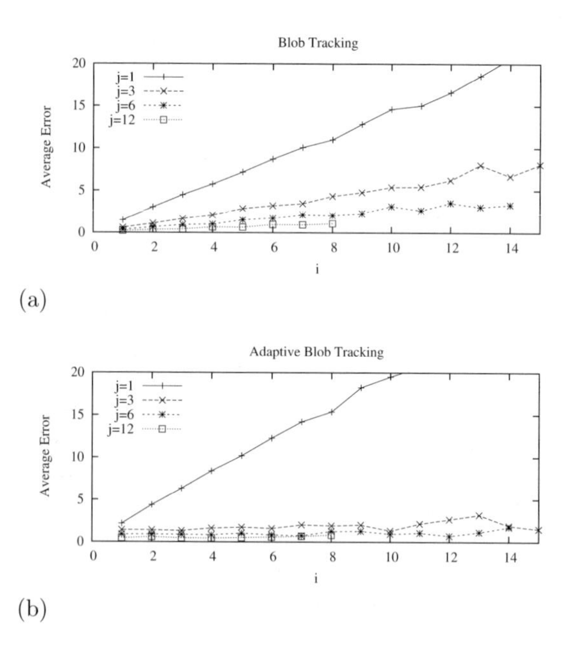

Fig. 6. Accuracy of original and adaptive blob tracking algorithm for VISOR video (snapshot in Figure 4(d))

4 CAMSHIFT Algorithm

CAMSHIFT object tracking [4] relies on color histogram detection and mean shift algorithm for tracking. The algorithm searches for a given object's histogram inside a subwindow of the current frame of the video, which is computed as 150% of the object size detected in the previous frame. Therefore, if the object, moves between two frames from its original location for a distance larger than half of its size, the algorithm will lose the track of the object. Hence, assuming we drop i frames before frame $k+i+1$, the threshold distance $\Delta\tilde{d} = \frac{w_k^0}{2}$, where w_k^0 is the width of the blob detected in frame k. Since CAMSHIFT does not use frame differencing, drop gap does not have an additional effect on object's size. Therefore, we can estimate the center of the blob after drop gap i using the equation (2) instead of equation (3). Hence, the critical drop gap can be derrived as

$$\tilde{i} = \frac{w_k^0}{2\Delta x^0 - \Delta w^0} - 1. \tag{7}$$

Estimating the average tracking error loses its meaning for CAMSHIFT tracking because it uses a simple threshold for detection of the object in the current frame. If the drop gap of the given frame dropping pattern is less than critical drop gap in equation (7), the algorithm continue tracking the object, otherwise it loses it. And the critical drop gap depends on the changes in speed and size of the object.

We performed experiments with CAMSHIFT tracking algorithm to verify our analytical estimation of the critical drop gap (equation (7)). We used several videos of a moving face shot with a simple web-cam, videos of talking heads by SEQAM laboratory and some movie clips (example screenshots in Figure 4(a) and Figure 4(b)). Figure 8(a) (original video is 600 frames of 352 × 288, 30 fps) and Figure 7 (original video is 303 frames of 320 × 230, 30 fps) show average tracking error vs. drop gap for CAMSHIFT tracking and various frame dropping patterns. Figure 7, corresponding to the video of a talking head (see snapshot in Figure 4(b)), demonstrates that tracking algorithm does not lose the face even when drop gap is 14 frames. The reason is because the face in the video does not move around and is always present in the search subwindow of CAMSHIFT

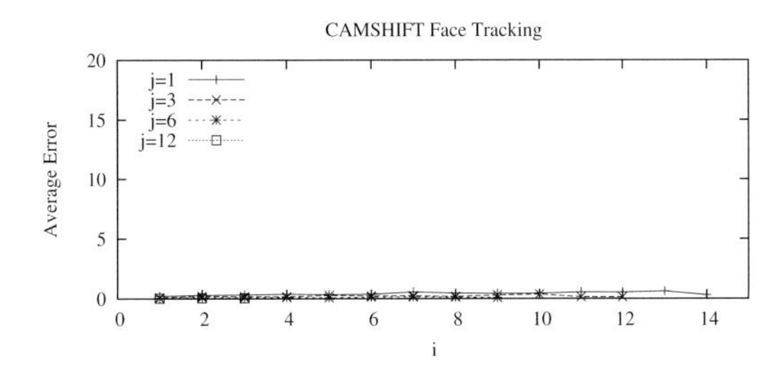

Fig. 7. Accuracy of original and adaptive CAMSHIFT tracking algorithm for video with slow moving face (snapshot in Figure 4(a))

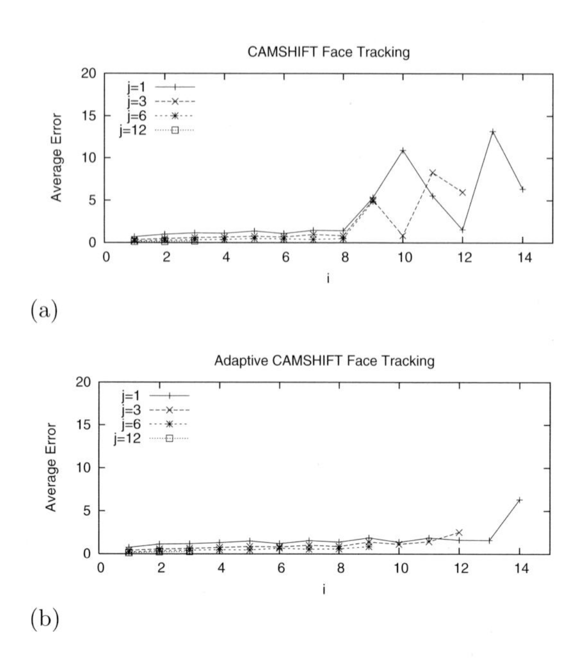

Fig. 8. Accuracy of original and adaptive CAMSHIFT tracking algorithm for video with fast moving face (snapshot in Figure 4(a))

tracker. However, for the experiments shown in Figure 8(a), the video with fast moving head was used (see snapshot in Figure 4(a)). It can be noted that the algorithm does not lose the face until value of drop gap is 8, because for the smaller drop gaps, the face is still within a search subwindow and can be detected by the histogram matching. The fluctuations in the average error for the larger drop gaps appear because the face is either lost by the tracker or, for some large enough gaps, it would move out of the subwindow and move back in, hence the tracker does not lose it. We conducted experiments with more videos and observed that the critical drop gap value is smaller for videos with faster moving faces and larger for videos with slower moving faces. These observations agree with equation (7).

5 Adaptive Tracking

We propose to modify blob tracking and CAMSHIFT algorithms and make them more tolerant to video with low frame rate. We have shown that average error and the critical frame rate of tracking algorithms depend on speed and size of the object in the original video. Therefore, if we record these characteristics for previous frames, the location and the size of object in the frame that follows a drop gap can be approximated. Adjusting to frame dropping in such way allows us to reduce the average error for blob tracking algorithm and increase the critical drop gap for the CAMSHIFT algorithm.

Blob tracking algorithm tracks the detected foreground object using the simplified version of Kalman filter: $x_k = (1 - \alpha)x_{k-1} + \alpha z_k$, where x_k and x_{k-1} represent estimated coordinates of the object in the current and previous frames, z_k is the output of the object detector, and $\alpha \leq 1$ is some constant. When $\alpha = 1$, then the tracker trusts the measurement z_k fully and its average error can be estimated by equation (6). In cases when $\alpha < 1$, the accuracy of the tracking against the frame dropping worsens, due to the larger shifts in blobs' centers for videos with high drop gap. We propose using adaptive Kalman filter [5] to make blob tracking more tolerant to the frame dropping. We apply the filter only to the width of the object, because the front is detected correctly by frame differencing (see Figure 2). The filter can be defined as following,

$$\tilde{w_k} = w_{k-1} + K_k\,(w_{k-1} + u_k) \qquad \tilde{P_k} = P_k + Q_k$$

$$P_k = (1 - K_k)\tilde{P_k} \qquad K_k = \frac{\tilde{P_k}}{(\tilde{P_k} + R_k)},$$

where Q_k and R_k are the process and measurement noise covariances; $\tilde{w_k}$ is the new estimate of the blob's width in the current frame; w_{k-1} is blob's width in the last not dropped frame; u_k is the width measurement provided by the frame-differencing based detector.

Kalman filter depends on correct estimation of the error parameters, Q_k and R_k. By looking at Figure 2, we can set $Q_k = (i\Delta w^0)^2$, which estimates how big the tracked object should be at frame $k + i + 1$ compare to its width before the drop gap at frame k. R_k is essentially the error of the measurement, i.e., the output of the foreground object detector, therefore, $R_k = (w^i_{k+i+1} - w^0_{k+i+1})^2$. Since w^i_{k+i+1} can be estimated as $w^0_k + (i+1)\Delta x^0$ and w^0_{k+i+1} as $w^0_k + (i+1)\Delta w^0$, we can approximate $R_k = (i+1)^2(\Delta x^0 - \Delta w^0)^2$. We obtain the values of Δw^0 and Δx^0 by recording the speed of the object and how fast it grows in size using last two available frames.

To compare how adaptive Kalman filter improves the accuracy of blob tracking, we performed the same experiments varying frame dropping pattern. The average error for blob tracking with adaptive Kalman filter is plotted in Figure 5(b) and Figure 6(b), which can be compared to results with original algorithm in Figure 5(a) and Figure 6(a) respectively. We can note that the accuracy of the adaptive blob tracking algorithm is improved for larger drop gaps (larger frame rate reduction). In both figures, Figure 5(b) and Figure 6(b), the angles of the lines in the graph are not inversely proportional to j anymore, giving fundamentally different bound on the average error. All lines with $j > 1$ are almost parallel to x-axis. It means that Kalman filter adapts very well to the drastic changes in speed and size of the object that occur due to the frame dropping. The constant increase in the average error for $j = 1$, is because, for such dropping pattern, all remaining frames are separated by drop gaps. In this scenario, adaptive Kalman filter accumulates the approximation error of object's size and speed. Therefore, the critical frame rate can be achieved with j that is at least equal to 2. If we take $i = 12$, the original frame rate is reduced by 7 times.

We also modified the CAMSHIFT tracking algorithm, adjusting the size of its search subwindow to the frame dropping. We simply increased the subwindow size in the current frame by $i\Delta x^0$, where i is the drop gap. The average error of this adaptive CAMSHIFT algorithm for the video with fast moving face is shown in Figure 8(b). Comparing with the results of original algorithm in Figure 8(a), we can notice that the adaptive tracker performs significantly better for the larger drop gaps. The experiments show that we can drop 13 frames out of 14 with a tradeoff in small average error. It means that CAMSHIFT algorithm, for this particular video sequence, can accurately track the face with frame rate reduced by 13 times from the original. For the news videos of talking heads, where face does not move significantly around, adaptive algorithm performs with exactly the same accuracy results as the original algorithm. Therefore, Figure 7 illustrates essentially both versions of the algorithm, original and adaptive. These experiments demonstrate that by using analysis to modify CAMSHIFT algorithm, we can improve its performance on videos with fast moving faces, while retaining the original accuracy on videos with slow moving faces.

6 Conclusion

In this paper, we use analysis to estimate the tradeoff between accuracy of two common tracking algorithms and video frame rate. Such estimation depends on the speed and size of the tracked object, and therefore, in practice, such measurements of the object need to be taken (for instance, running average of these values during the last few frames). We also show that slight modifications to existing algorithms can significantly improve their accuracy for the video with larger reductions in frame rate. These findings motivate us to use reasoning for determining critical frame rate (not just running many different experiments) for other video analysis algorithms. The findings also encourage the development of the new object tracking algorithms robust to highly degraded video.

References

1. Korshunov, P., Ooi, W.T.: Critical video quality for distributed automated video surveillance. In: Proceedings of the ACM International Conference on Multimedia, ACMMM 2005, Singapore, November 2005, pp. 151–160 (2005)
2. Li, L., Huang, W., Gu, I.Y., Tan, Q.: Foreground object detection from videos containing complex background. In: Proceedings of the ACM International Conference on Multimedia, ACMMM 2003, Berkeley, CA, USA, November 2003, pp. 2–10 (2003)
3. Bradski, G.R.: Computer vision face tracking as a component of a perceptual user interface. In: Proceedings of the Forth IEEE Workshop on Applications of Computer Vision, WACV 1998, Princeton, NJ, January 1998, pp. 214–219 (1998)
4. Boyle, M.: The effects of capture conditions on the CAMSHIFT face tracker. Technical Report 2001-691-14, Department of Computer Science, University of Calgary, Alberta, Canada (2001)
5. Welsh, G., Bishop, G.: An introduction to the kalman filter. In: Proceedings of SIGGRAPH 2001, Los Angeles, CA, USA, August 2001, vol. Course 8 (2001)

A Study on Sampling Strategies in Space-Time Domain for Recognition Applications

Mert Dikmen[1], Dennis J. Lin[1], Andrey Del Pozo[1], Liang Liang Cao[1], Yun Fu[2], and Thomas S. Huang[1]

[1] Beckman Institute, Coordinated Sciences Laboratory, Department of Electrical and Computer Engineering University of Illinois at Urbana-Champaign
405 N Mathews, Urbana, IL 61801, USA
{mdikmen,djlin,delpozo2,cao4,huang}@ifp.uiuc.edu

[2] Department of Computer Science and Engineering, University at Buffalo (SUNY), 201 Bell Hall, Buffalo, NY 14260-2000, USA
raymondyunfu@gmail.com

Abstract. We investigate the relative strengths of existing space-time interest points in the context of action detection and recognition. The interest point operators evaluated are an extension of the Harris corner detector (Laptev et al. [1]), a space-time Gabor filter (Dollar et al.[2]), and randomized sampling on the motion boundaries. In the first level of experiments we study the low level attributes of interest points such as stability, repeatability and sparsity with respect to the sources of variations such as actors, viewpoint and action category. In the second level we measure the discriminative power of interest points by extracting generic region descriptors around the interest points (1. histogram of optical flow[3], 2. motion history images[4], 3. histograms of oriented gradients[3]). Then we build a simple action recognition scheme by constructing a dictionary of codewords and learning a recognition system using the histograms of these codewords. We demonstrate that although there may be merits due to the structural information contained in the interest point detections, ultimately getting as many data samples as possible, even with random sampling, is the decisive factor in the interpretation of space-time data.

1 Introduction

The amount of data in the digital domain is increasing at an unprecedented rate[1]. However, manual search for arbitrary patterns is dauntingly impractical for respectably sized video corpora due to the sheer amount and high redundancy of the content in videos. It is imperative to develop methods which can index and search the content in the space-time domain. The applications of such

[1] According to Google official statement in May 2009, the upload rate of new content to the popular video sharing site YouTube has recently surpassed 20 hours of content per minute.

S. Boll et al. (Eds.): MMM 2010, LNCS 5916, pp. 465–476, 2010.

methods range from simple semantic search to automatic surveillance of high risk environments.

Human motion analysis can be roughly categorized in three fields: 1) motion analysis of body parts, 2) tracking of human motion and 3) recognition of human activities[5]. Our study concentrates on the last task, where the activity is defined as a short duration repetitive motion pattern. Examples for such visual entities are actions, objects with characteristic motion patterns or events. Fortunately, the constant increase of computational resources has made video processing and pattern analysis more feasible. Yet the ultimate goal of designing a recognition system which can uncover arbitrary query patterns in data residing in space-time domain still remains to be achieved. We build our study on a structure that mimics the experimental nature of Laptev et al.[3] and Nowak et al.[6], except that we work in the video domain instead of the image domain. Here we study the characteristics and usefulness (in comparison to a naive sampling approach) of space-time interest point operators as modules on a standard action recognition system pipeline. The main dataset used for training and testing is the KTH dataset[7], which has been a well established standard for evaluating video action recognition problem. In the action recognition system, initially we find the interest points on the body of the subject performing the action. Next we extract local descriptors on the interest point detections. Then a codeword dictionary is learned for each detector/descriptor combination, and the interest points in each video are histogrammed using the learned dictionary of codewords. In the last step we build a decision system using a multi-class SVM [8].

Extracting features from sampled interest points has several motivations. By sampling the moving parts we retain most of the information about the motion in the scene, but without looking at the entire dataset. The latter has the drawbacks of low signal to noise ratios, the curse of dimensionality, and high computational overhead. Furthermore the bag of features approach[9], where the features are extracted only from the neighborhood of the interest points has demonstrated impressive results in recognition problems in static images. There is reasonable evidence to believe that this approach should be suitable for video domain also, where the interesting parts of the data are even sparser than in the image domain. We concentrate on two established interest point operators; Laptev et al [1] and Dollar et al. [2]. Laptev detector is a straightforward extension of the well known Harris operator used for corner detection in images. Dollar detector utilizes a quadrature pair of time oriented Gabor filters to uncover periodic motion patterns. In addition, we consider a random-sampling based approach.

In the remainder of the paper we first describe the interest point detectors (Section 2) and local region descriptors (Section 3) for capturing features of motion data. We describe the experimental setup of our test bed (Section 4). A brief analysis of interest point detector characteristics is included in Section 5.1, and an overall analysis on several aspects of consideration when working with interest point / descriptor combinations follows in (Section 5).

2 Detectors

2.1 Laptev et al.[1]

This detector is an extension to the well known Harris corner detector. For the space-time case, the scale space $L : \mathbb{R}^2 \times \mathbb{R} \times \mathbb{R}_+^2 \mapsto \mathbb{R}$ is defined by:

$$L(\cdot; \sigma_l^2, \tau_l^2) = g(\cdot; \sigma_l^2, \tau_l^2) * V(\cdot) \tag{1}$$

$$g(x, y, t; \sigma_l^2, \tau_l^2) = \frac{1}{\sqrt{(2\pi)^3 \sigma_l^4 \tau_l^2}} \exp(-(x^2 + y^2)/2\sigma_l^2 - t^2/2\tau_l^2) \tag{2}$$

where $V(x, y, t)$ is the video volume and g is the Gaussian kernel. The time scale parameter τ^2 is handled independently since in general the spatial and temporal dimensions are independent. From the scale space representation the space-time second moment matrix is straightforward to define:

$$\mu(\cdot; \sigma_i^2, \tau_i^2) = g(\cdot; \sigma_i^2, \tau_i^2) * \begin{pmatrix} L_x^2 & L_x L_y & L_x L_t \\ L_x L_y & L_y^2 & L_y L_t \\ L_x L_t & L_y L_t & L_t^2 \end{pmatrix} \tag{3}$$

where the first order derivatives are computed as $L_\epsilon(\cdot, \sigma_l^2, \tau_l^2) = \partial_\epsilon(g * f)$ and $\sigma_i^2 = s\sigma_l^2$ and $\tau_i^2 = s\tau_l^2$. The space-time corners are detected by finding the local maximum of:

$$H = \det(\mu) - k \ \mathrm{trace}^3(\mu) = \lambda_1 \lambda_2 \lambda_3 - k(\lambda_1 + \lambda_2 + \lambda_3)^3 \tag{4}$$

2.2 Dollar et al.[2]

The interest points of Dollar et al. are inspired by Gabor filters in the image domain, which are filters built by multiplying a harmonic function with a 2D Gaussian kernel. In image processing, the harmonic functions are generally a quadrature pair of sin and cos filters tuned to the fundamental frequency of an image texture or a ridge pattern. The sin and cos envelopes decay exponentially in the direction of their spatial frequency, such that the response to the filter is spatially maximized at the center of the particular pattern. The modifications made to this model to form a space-time interest operator are simple yet effective. The frames are filtered with a Gaussian smoothing kernel ($g(x, y; \sigma^2)$) and multiplied with a quadrature pair of even ($h_{ev} = \sin(t; \tau, \omega)$) and odd ($h_{odd} = \cos(t; \tau, \omega)$) harmonic envelopes on the time axis. The response function of the interest point operator is defined as:

$$R(x, y, t) = (V(.) * g(x, y; \sigma^2) * h_{ev}(t))^2 + (V(.) * g(x, y; \sigma^2) * h_{odd}(t))^2 \tag{5}$$

We then find the local maxima of the response function and filter out the faux maxima which have a small magnitude and are most likely produced by noise.

2.3 Random Sampling on the Motion Boundary (RSMB)

We also devise a random sampling method by simple sampling from the motion boundaries obtained by thresholding the pixel-wise difference between two consecutive frames. This sampling method is rather naive, yet it is still more structured than uniformly sampling from all pixels in the video. It is logical in this case since information relevant to action recognition is mostly contained in the regions with motion. Furthermore, since the previous two interest point operators' affinity functions obtain their local maxima on the motion boundary, this random sampling approach is rather close to uniformly sampling from the support of the interest points found with the other approaches. We set the sampling rate to get C samples per frame on the average. This is done by finding all the foreground pixels in a video and then randomly selecting $C \times N$ of them, where N is the number of frames in the video. This ensures that more samples are collected from frames with large amounts of motion and few get selected from frames with little motion. Note that this method is more randomized than selecting a certain fraction of foreground pixels at each frame. The latter ensures that in the case of motion in the frame, at least some samples will be selected in the particular frame, whereas with the random selection over the entire video's foreground pixels, this may not be the case. We opted to use the first approach because there is no hard coded per frame selection rule in the Laptev and Dollar detectors, thus this method of random sampling provides a fairer comparison. For robustness measures, we generate 3 different sets of random point samples in the video and report the average results over these 3 sets.

3 Descriptors

3.1 Histograms of Oriented Gradients and Optical Flow[3]

We construct histograms of the video gradients (HOG) and optic flow (HOF) to describe an interest point. The bins of the histograms are over the orientations and we ignore the magnitude of the image gradients or optic flow. We used eight orientation bins for the HOG descriptor and five bins for HOF.

We define a fixed size volume centered at each interest point. and then subdivide it into n_s and n_t blocks. HOG and HOF descriptors are computed for each block. The histograms of every block of the volume are concatenated to produce one HOG and one HOF descriptor for the interest point. The dimension of the volume used were $7 \times 7 \times 11$ pixels, and we set n_s and n_t to 3 and 2 respectively.

3.2 Motion History Images (MHIST)[4]

Motion History Images have first been proposed for representation and recognition of simple actions, where the subjects are thought to be well localized and scale adjusted. In our implementation we use the MHIST as a local region descriptor in which we extract the motion history image of an $(X \times Y \times T)$ region centered around the sampling point. This process for obtaining foreground pixels is far more robust than background subtraction in noisy datasets, and in datasets obtained by hand held cameras. The drawback is that the non moving pixels of the foreground objects are not registered. However, since the descriptors are designed to capture motion characteristics, this is not a significant drawback. Formally a motion history image H of a $(X \times Y \times T)$ sized region R centered around a point (x_0, y_0) at time t_0 in the video volume $V(x, y, t)$ is described as follows:

$$D(x, y, t) = |V(x, y, z) - V(x, y, t-1)| \tag{6}$$

$$h(x, y, t) = \begin{cases} 0, & D(x, y, t) < k \\ 1, & D(x, y, t) \geq k \end{cases} \tag{7}$$

$$H_{x_0, y_0, t_0}(\hat{x}, \hat{y}) = \max_{0 \leq \hat{t} \leq \tau} (1 - \tfrac{\hat{t}}{\tau}) h(x_0 + \hat{x}, y_0 + \hat{y}, t_o + \hat{t}), \ |\hat{x}| \leq \tfrac{X}{2}, \ |\hat{y}| \leq \tfrac{Y}{2} \tag{8}$$

4 Experiments

We take the KTH action dataset as our baseline dataset for comparing the action recognition performance of different detector-descriptor combinations. The KTH dataset consists of 6 actions (boxing, hand clapping, hand waving, jogging, running, walking), performed by 25 actors in 4 different settings (indoors, outdoors, outdoors with different clothing and outdoors with camera zooming motion). All actors perform each action in each setting about 4 times, yielding obtain 400 shots per action and 2400 samples in total. For robust results, we split the dataset in 3 disjoint partitions by the actor identities ({1,..,8},{9,..,17}, {18,..,25}). We train the system on one partition, set the classifier parameters using a second partition, and utilize the third partition for testing. In total this gives us 6 combinations of training, validation, and test sets, and we run the analysis on all of them. Unless otherwise noted, all performance figures reported are the geometric mean of the performance values over the 6 folds. The recognition system is trained by combining the decision outputs of 15 one vs one SVMs with a global SVM as final classifier, which uses these 15 decision values as feature input. We test two different kernel types for the SVMs (linear and radial basis function) and pick the best performing kernel/parameter combination in each case.

(a)

(b)

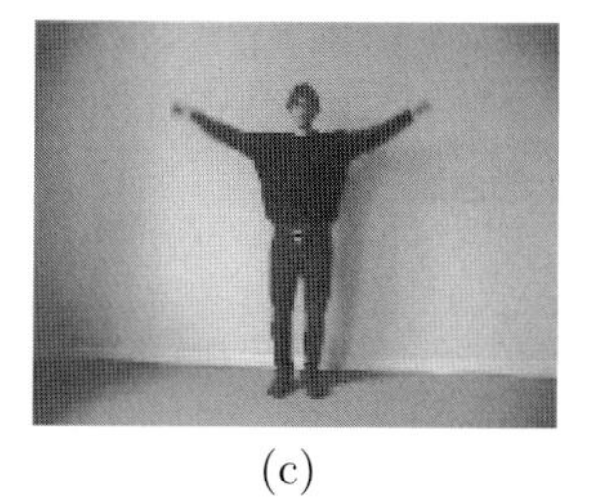

(c)

Fig. 1. Examples from the KTH dataset: a) Boxing (outside), b) Running (outside with different clothing), c) Handwaving (indoors)

(a) Laptev (b) Dollar (c) Random

Fig. 2. Examples of interest points found with different detectors at their native detection thresholds. All the interest points are accumulated up to 10 frames prior to the one shown. The scale of the marker denotes recency.

5 Analysis

5.1 Analysis of the Detectors

The rate at which the detectors produce interest points varies due to different formulations of the response functions (Figure 2). The Laptev detector favors corners in 3 dimensional video data. These corners are produced by strong changes in direction and/or speed of the motion; thus, the detections are quite sparse. The Dollar detector is sensitive to periodic motions. These include but are not limited to space-time corners. Therefore at the native operating point, the Dollar detector tends to fire a bit more than the Laptev detector. The random detector fires the most; however, this is purely due to the rate that we have purposefully set. It can be seen from Table 1 that actions in which the subject traverses the frame produces significantly more interest points than the stationary actions. This is justified because in stationary actions only the upper limbs of the body are moving, where as in translation the entire body moves.

Another aspect of consideration is the repeatability of the interest points under sufficiently similar conditions. To test an upper bound on this, we partition the KTH Dataset into 100 partitions per action (25 actors and 4 settings). Each partition has about 4 shots where the actor performs the same action in an identical setting and manner. We extract a detector signal by counting the number of interest point detections at each frame of the shots[2]. We find the maximum normalized cross correlation value

Table 1. Median number of detections per frame for each shot in an action class

	Laptev	Dollar	Random
Boxing	1.67	11.36	40.73
Handclapping	1.23	8.49	40.00
Handwaving	1.50	11.41	41.12
Jogging	4.49	19.96	66.65
Running	4.49	20.92	78.78
Walking	4.96	14.50	61.88

[2] We exclude the shots from the "outdoors with camera zoom" settings from the experiments in this section because the camera motion introduces artificial inconsistencies to the fundamental periodicity of the motion.

Table 2. Average correlation values over the similar shots per action class

	Laptev Native Rate		Dollar N. R.	Random N. R.
	Laptev	Random	Dollar	Random
Boxing	0.224±0.054	0.182±0.029	0.190±0.048	0.220±0.074
Handclapping	0.246±0.067	0.189±0.033	0.197±0.038	0.305±0.098
Handwaving	0.203±0.034	0.180±0.027	0.213±0.060	0.361±0.089
Jogging	0.510±0.057	0.417±0.066	0.586±0.141	0.750±0.066
Running	0.573±0.063	0.492±0.064	0.637±0.115	0.795±0.055
Walking	0.480±0.079	0.334±0.045	0.395±0.122	0.715±0.078

for each shot pair in a partition and take the average over the whole set for one action. A more robust way of measuring repeatability would be to take locations of the interest points in the consideration. However, the detections are very sparse and the camera is not stabilized. A much larger training data would be needed to infer reliable statistics.

It is desirable that the output of an ideal detector be highly correlated over the shots taken in a similar setting. The number of detected points per time instance we use in the normalized cross correlation metric provides an upper bound for this correlation measurement. As it can be seen in Table 2, the RSMB detector produces the most consistent detections. However this may be due to the fact that by design RSMB is forced to detect certain number of interest points and as this number grows, sampling over time will be more dense. Indeed when the detection rate is lowered to match the detection rate of Laptev detector, the RSMB detector does perform less consistently.

Some actors perform some actions very differently. One person may wave with the entire arm while another person can wave just by bending the forearm by the elbow. However the intra-person variation of an action should be small regardless of the setting. Based on this assumption, we test the repeatability of the interest points over the same actors. We take the average normalized cross correlation value of all shots from the same actor and action and average this value over the actors. The results are tabulated in Table 3. As it was seen before the densely sampling random detector demonstrates the most consistent detections. But when the sampling rate is limited, Laptev interest point detector actually outperforms random. This suggests that the interest point operators are able to pick up some structural characteristics in the space-time data. It is

Table 3. Average of the correlation values over the same actors

	Laptev Native Rate		Dollar N. R.	Random N. R.
	Laptev	Random	Dollar	Random
Boxing	0.209±0.027	0.181±0.016	0.189±0.019	0.200±0.031
Handclapping	0.217±0.024	0.178±0.040	0.195±0.020	0.252±0.042
Handwaving	0.188±0.015	0.178±0.011	0.198±0.025	0.317±0.065
Jogging	0.478±0.030	0.396±0.041	0.487±0.076	0.659±0.051
Running	0.554±0.042	0.470±0.043	0.538±0.056	0.715±0.053
Walking	0.436±0.047	0.315±0.029	0.341±0.085	0.628±0.066

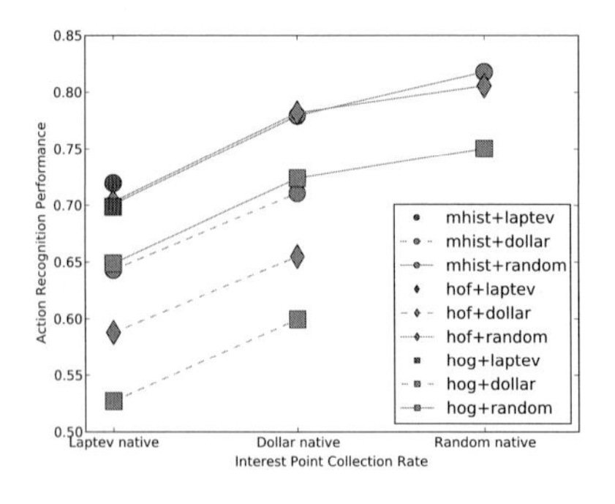

Fig. 3. Recognition rate with respect to the number of interest points collected. The dictionary size is 1000 codewords, while the histograms of codewords are raw unnormalized counts. At the lowest rate of interest point collection Laptev detector is more informative than random point selection but the recognition performance of random sampling consistently improves with the increasing number of points.

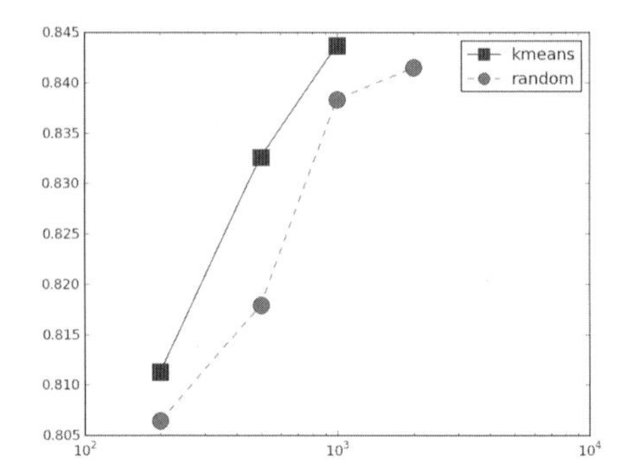

Fig. 4. Effect of dictionary building method to the overall performance. Horizontal axis (log-scale) shows the dictionary size, while vertical axis shows the average recognition accuracy. K-means is more effective in designing dictionaries.

also worth noting that in both experiments the "boxing" action class showed the poorest cross correlation values for all detectors. This is an expected result because it is the action that shows the least periodicity in the sequences of the dataset. Actors often pause of change the frequency of their arm swings.

5.2 Number of Interest Points

Next, we measure the effect of number of interest points on the recognition performance. As mentioned in Section 5.1 the Laptev detector gives the sparsest set of outputs. We configure the Dollar and RSMB detectors to match the rate of the Laptev detector by raising the threshold for the Dollar detector and lowering the collection rate for the RSMB detector. Next we match the native rate of the Dollar detector with the RSMB. Finally we run the system with the entire set of random interest points from RSMB. The end results (Figure 3) are consistent with the findings of Nowak et al.[6] in the sense that, while structurally collected interest points may carry slightly more information than random point collection, it is ultimately the number of samples that defines the recognition performance. It is worth noting that the number of points is the single system parameter which has the biggest impact on the recognition performance (Table 6).

5.3 Descriptor Types

Results on Table 6 demonstrate that explicitly encoding motion information (MHIST, HOF) outperforms the shape based descriptor (HOG). When all other factors are equal, the Motion History descriptor generally is the top performer.

5.4 The Dictionary

We trained dictionaries for our vector quantization codebook using Lloyd's algorithm. We randomly initialize the centers and optimize the objective for 300 iterations. (Many of the dictionaries converged significantly faster.) For each training fold the a separate dictionary is learned. We do not learn a separate dictionary for each action class pair. This probably degrades the one vs one SVM performance but allows for a more straightforward comparison. To further facilitate the comparison, we use the dictionary trained on the maximum number of available points even when we are considering reduced rate detection results.

The effect of dictionary building process is shown in Figure 4, where we compare the results using the dictionaries built by k-means with dictionaries built by simply random sampling of the feature descriptors extracted from the dataset. It is clear that k-means outperforms random descriptor sampling at any dictionary size. The descriptor used for this experiment was motion histograms, while the binning method was just using raw counts.

After the dictionaries are built we vector quantize all of the interest points in a video shot and generate a histogram of the code words. The histogramming process presents several choices for post processing. One method is to leave the counts as is and use the "raw" histogram as the feature presented to the classifier. Another method is the normalize the histogram bins to sum to one such that the histogram resembles a probability mass function. The last method we considered is to convert the histogram to a binary vector by thresholding. We choose the threshold for each bin individually, maximizing the mutual information between the class labels and bin label. This technique of histogram preprocessing has demonstrated strong performance in several cases [10]. As Table 4 shows, the

Table 4. Effect of histogramming method. Table shows performances of each histogramming method according to the dictionary size (leftmost column) and the detection rate of the interest points (top row). The detector whose results are shown is RSMB while the feature used is Motion History Images. The "raw" histogram refer to unaltered counts. The "norm" histogram has the bin sums normalized to 1. The "mt.i" histogram has each bin thresholded in a way to maximize mutual information. The feature used is Motion History Images. The results for the other detector/descriptor combinations are similar.

		Rate of the Random Detector		
		Laptev N.R.	Dollar N.R.	Random N.R.
1000	Raw	0.681	**0.804**	0.844
	Norm	**0.708**	0.799	0.840
	Mt.I.	0.594	**0.804**	**0.848**
500	Raw	0.675	**0.792**	**0.833**
	Norm	**0.701**	**0.791**	0.825
	Mt.I.	0.626	0.783	0.819
200	Raw	0.671	0.772	0.811
	Norm	**0.701**	**0.779**	**0.818**
	Mt.I.	0.614	0.720	0.756

Table 5. Effect of interest point operator and histogramming method on the Weizmann dataset. As in the KTH dataset, the mutual information binning works the best when the dictionary size and the number of interest points are large.

		Laptev Native Rate			Dollar N. R.		Random N. R.
		Laptev	Dollar	Random	Dollar	Random	Random
1000	Raw	0.373	0.336	0.304	0.380	0.590	**0.676**
	Norm	0.611	0.271	0.478	0.477	0.562	**0.652**
	Mt.I.	0.140	0.311	0.318	0.442	0.579	**0.713**

Table 6. Effect of various parameters on action recognition. Leftmost column denotes the dictionary size. The top row signifies the per frame rate of interest point detection. In each case the detectors are modified to match the rate of a sparser interest point operator. The best performing feature-interest point combination is the random sampled motion history features at as many locations as possible. The best performing dictionary size is also the one with the highest number of codewords.

		Laptev Native Rate			Dollar N. R.		Random N. R.
		Laptev	Dollar	Random	Dollar	Random	Random
1000	Motion History	0.741	0.590	0.681	0.756	0.804	**0.844**
	Hist. of Optical Flow	0.722	0.527	0.694	0.705	0.804	**0.828**
	Hist. of Oriented Gradients	0.715	0.504	0.623	0.681	0.748	**0.784**
500	Motion History	0.717	0.595	0.675	0.734	0.792	**0.833**
	Hist. of Optical Flow	0.704	0.523	0.691	0.684	0.797	**0.816**
	Hist. of Oriented Gradients	0.701	0.529	0.668	0.656	0.768	**0.794**
200	Motion History	0.691	0.588	0.671	0.720	0.772	**0.811**
	Hist. of Optical Flow	0.707	0.475	0.622	0.670	0.736	**0.777**
	Hist. of Oriented Gradients	0.673	0.480	0.609	0.640	0.719	**0.759**

mutual information method equals the other techniques at the largest detection and dictionary sizes but is otherwise inferior. Perhaps this is because the thresholding process throws out information, and that the effect is less pronounced as the number of bins grow.

For verification of the results, we repeated the experiments on the Weizmann dataset [11]. In this dataset there are 9 actions performed by 9 actors. All of the actions are performed once by each actor, except for 3 individual cases where the actors performed the actions twice. We exclude these cases from the training and testing for consistency. We performed the experiments similarly over three fold cross validation scheme where 1/3 of the dataset was used for training and the remaining 2/3 for testing. The results in Table 5 are consistent with those for the KTH data.

6 Conclusions

We have presented a study of a general action recognition system using a bag of features approach. The main criticism of the interest point operators has been the fact that they produce a very sparse set of samples, which is undesirable for most statistical learning methods relying on robust statistics. This low sample size problem seems to be a real obstacle for gathering enough evidence about the visual data as demonstrated in the recognition experiments, and also confirmed by the recent study of Wang et al. [12] in which they found a dense grid sampling operator to perform the best. Amongst the interest point detection schemes we considered, randomly sampling the motion boundary yielded the best results. This is not surprising because random sampling approach has the ability to gather arbitrary number of samples from the data, and the more samples the better is the ability of a learning method to distinguish relevant patterns. On the other hand, objective function driven interest point operators are looking for points that satisfy the local maximum condition by comparing the objective value at least to 26 immediate pixel neighbors in 3D, which evidently is even more difficult to satisfy than in the 2D counterpart where only 8 immediate neighbors would be considered. However there seems to be some structural information, though little, embedded in the interest point detections. Figure 3 points out that in the low sample case Laptev detector gives the best performance for any type of feature descriptor probably because the interest point detections themselves contain some information about the action classes being classified.

This paper covers some of the important design variables that need to be considered while building an action recognition system. Certainly many improvements to this basic system have been been proposed and demonstrated improvements to the state of the art [13,14]. The underlying principles of sampling, however, should transfer to the more sophisticated approaches.

References

1. Laptev, I.: On space-time interest points. International Journal of Computer Vision 64(2), 107–123 (2005)
2. Dollár, P., Rabaud, V., Cottrell, G., Belongie, S.: Behavior recognition via sparse spatio-temporal features. In: VS-PETS (October 2005)
3. Laptev, I., Lindeberg, T.: Local descriptors for spatio-temporal recognition. In: MacLean, W.J. (ed.) SCVMA 2004. LNCS, vol. 3667, pp. 91–103. Springer, Heidelberg (2006)
4. Davis, J., Bobick, A.: The representation and recognition of action using temporal templates. In: IEEE Conference on Computer Vision and Pattern Recognition, pp. 928–934 (1997)
5. Aggarwal, J., Cai, Q.: Human motion analysis: A review. CVIU 73, 428–440 (1999)
6. Nowak, E., Jurie, F., Triggs, B.: Sampling strategies for bag-of-features image classification. In: Leonardis, A., Bischof, H., Pinz, A. (eds.) ECCV 2006. LNCS, vol. 3954, pp. 490–503. Springer, Heidelberg (2006)
7. Schuldt, C., Laptev, I., Caputo, B.: Recognizing human actions: a local svm approach. In: Proceedings of the 17th International Conference on Pattern Recognition, ICPR 2004, August 2004, vol. 3, pp. 32–36 (2004)
8. Chang, C.C., Lin, C.J.: Libsvm: a library for support vector machines (2001)
9. Csurka, G., Dance, C.R., Fan, L., Willamowski, J., Bray, C.: Visual categorization with bags of keypoints. In: Sebe, N., Lew, M., Huang, T.S. (eds.) ECCV 2004. LNCS, vol. 3058, pp. 1–22. Springer, Heidelberg (2004)

10. Nowak, E., Jurie, F.: Vehicle categorization: Parts for speed and accuracy. In: ICCN 2005: Proceedings of the 14th International Conference on Computer Communications and Networks, Washington, DC, USA, pp. 277–283. IEEE Computer Society Press, Los Alamitos (2005)
11. Gorelick, L., Blank, M., Shechtman, E., Irani, M., Basri, R.: Actions as space-time shapes. Transactions on Pattern Analysis and Machine Intelligence 29(12), 2247–2253 (2007)
12. Wang, H., Ullah, M.M., Klaser, A., Laptev, I., Schmid, C.: Evaluation of local spatio-temporal features for action recognition. In: British Machine Vision Conference (September 2009)
13. Jhuang, H., Serre, T., Wolf, L., Poggio, T.: A biologically inspired system for action recognition. In: IEEE 11th International Conference on Computer Vision, ICCV 2007, October 2007, pp. 1–8 (2007)
14. Yuan, J., Liu, Z., Wu, Y.: Discriminative 3D subvolume search for efficient action detection. In: IEEE Conference on Computer Vision and Pattern Recognition, Miami, FL (June 2009)

Fire Surveillance Method Based on Quaternionic Wavelet Features

Zhou Yu[1,2], Yi Xu[1,2], and Xiaokang Yang[1,2]

[1] Institute of Image Communication and Information Processing, Shanghai Jiaotong University, Shanghai 200240, China
[2] Shanghai Key Lab of Digital Media Processing and Transmissions, Shanghai 200240, China
{zhouyuyz,xuyi,xkyang}@sjtu.edu.cn

Abstract. Color cues are important for recognizing flames in fire surveillance. Accordingly, the rational selection of color space should be considered as a nontrivial issue in classification of fire elements. In this paper, quantitative measures are established using learning-based classifiers to evaluate fire recognition accuracy within different color spaces. Rather than dealing with color channels separately, the color pixels are encoded as quaternions so as to be clustered as whole units in color spaces. Then, a set of quaternion Gabor wavelets are constructed to establish an comprehensive analysis tool for local spectral, spatial and temporal characteristics of fire regions. Their quaternion Gaussian kernels are used to represent the spectral distribution of fire pixel clusters. In addition, a 2D band-pass filter kernel contained in the quaternion Gabor extracts spatial contours of fire regions. Another 1D temporal filter kernel is enforced to capture random flickering behavior in the fire regions, greatly reducing the false alarms from the regular moving objects. For early alerts and high detection rate of fire events, smoke region is also recognized from its dynamic textures in the proposed fire surveillance system. Experimental results under a variety of conditions show the proposed vision-based surveillance method is capable of detecting flame and smoke reliably.

Keywords: Fire Surveillance, Quaternionic features, Flame recognition, Smoke detection.

1 Introduction

Early fire detection is vital to insure human's safety and prevent hazards before fire gets out of control. Conventional sensor-based fire alarm systems detect chemicals by either ionization or photometry, overlooking the physical presence of combustion material at the location of the sensor. This requirement makes these systems dependent on the distance to fire source as well as sensors' positional distribution. To provide quick responses to fire and gain capability in monitoring open spaces, vision based fire surveillance systems are developed to efficiently identify the danger of flame or smoke with early alerts in areas previously deemed to be impractical for sensor-based systems.

S. Boll et al. (Eds.): MMM 2010, LNCS 5916, pp. 477–488, 2010.

Color cues are important for recognizing flames in fire surveillance. Various color spaces have been used to perform flame pixel classification, including RGB[1][2], YUV[3], HIS[4] and YCbCr[5]. Fire color models are trained in the specified color space for flame detection. However, the color space is selected in an ad hoc manner without foundation of quantitative analysis.

Most of current fire detection methods focus on the feature description of flames [1-6]. The superior ones take into account the disordered spatial features and temporal flickering features to characterize the fire regions. Shape complexity is commonly used to depict the spatial fire features [7][8]. However, it is difficult to figure out the contour finely for fire regions. In addition, rather a long time is needed to detect temporal periodicity of fire flickering behavior [5][7]. Thus, the classification approach with low computational cost is preferred. Smoke is another important element to indicate fire appearance. It is assumed that smoke's color varies within the range from black-grayish to black [2]. However, the smoke color might reflect the color of the background due to semi-transparent property. The shape of the smoke is also considered in some research works [9]. These approaches are dependent on the assumption that smokes have distinct contours, which is not the fact for the thin smokes. Texture features are effective to extract smoke regions. Cui's approach [10] demonstrated high detection accuracy but should extract at least 48 features from each candidate smoke region for final discrimination.

Dynamic characteristics of flame and smoke are important clues to reduce search range of candidate fire regions. Frame differencing is commonly used to filter out the moving regions. But it still needs other decision rules to remove unwanted dynamic features and noises. Flames and smokes are always pointed at the top and spread upward by hot airflows. This is an important visual feature of fire. Yuan performed the orientation analysis using integral image to find candidate fire regions [11]. The block-wise orientation analysis in this method is not quite precise.

In this paper, a unified detection scheme of flames and smoke is presented. It is highlighted in three aspects: (1) Quantitative measures are established to evaluate the flame classification accuracy within the commonly-used color spaces, including RGB, HIS, YCbCr and LAB color spaces. (2) A set of quaternion Gabor wavelets are constructed to establish an comprehensive analysis tool for local spectral, spatial and temporal characteristics of fire regions, where the color cues, contour cues and turbulent motion cues are treated altogether in the filtering process. (3) Motion history image (MHI) analysis combined with statistical texture description and temporal wavelet transform is used to reliably recognize smoke regions.

The content is organized as follows. Section 2 evaluates the impact of color space selection, e.g. RGB, HIS, YCbCr and LAB, on flame element classification. In learning-based clustering process, quaternion representation is adopted to cluster color pixels as units in the color spaces. Section 3 presents a novel quaternionic Gabor filtering method to detect fire region. These quaternion Gabors are capable of conducting spectral analysis together with spatio-temporal wavelet filtering, resulting in reliable fire region detection. A cost-effective framework of smoke detection is given in Section 4. Then section 5 presents experimental comparison with the state-of-the-art methods. Finally, conclusion remarks are drawn in Section 6.

2 Color Space Selection for Flame Detection

Color cues are important for recognizing flames in fire surveillance. In previous works, various color spaces are selected to perform flame classification in an ad hoc way, including RGB[1][2], YUV[3], HIS[4] and YCbCr[5] color space. In this section, we establish a foundation of quantitative analysis of flame recognition accuracy in these commonly-used color spaces. Two of the most popular classifiers based on machine learning model, i.e. Adaboost and SVM, are used to extract the flame-colored regions. It is interesting to show that comparable detection rates and false alarm rate are achieved in these four color spaces if the fire color model is trained with parameter optimized in the specified color space.

Most flame pixels have color components of red, orange and yellow. We collect fire pixels from 19 fire video sequences and compute the spectral distribution of these pixels. Fig.1 illustrates the fire color distribution in RGB, HIS and Lab color spaces.

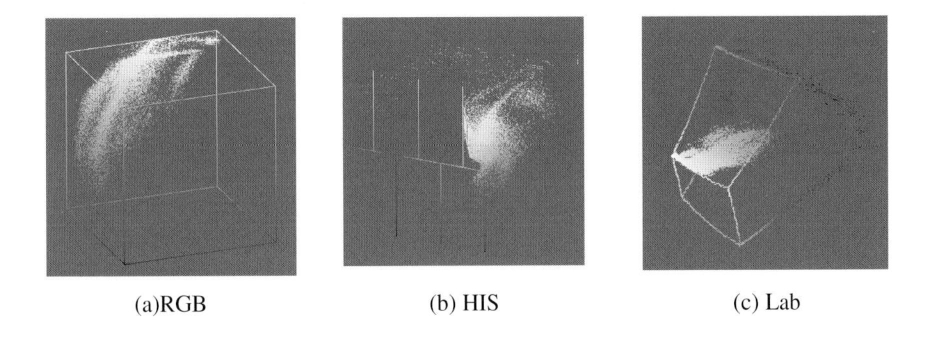

(a)RGB (b) HIS (c) Lab

Fig. 1. Spectral distribution of fire pixels gathered from 19 video sequences

Most former researches select HIS, YCbCr and Lab color spaces rather than RGB color space, assuming more independence among the three channels results in better classification performance. Here we rely on the quantitative analysis to evaluate the flame classification results under different selection of color spaces. We gather non-fire color samples from 15 videos of indoor and outdoor scenes, and classify them against fire colors via Support Vector Machine (SVM) and Adaboost. Each color pixel is represented by a pure quaternion[12], i.e. $x.i + y.j + z.k$, where $i^2 = j^2 = k^2 = -1$, $i.j = -j.i = k$ and imaginary parts x, y, z denote the color channels. Rather than treating each color channel independently, quaternion encodes color pixel as an entity, well preserving interrelationship between three color channels. Quantitative analysis for the classification results is listed in Table 1 and Table 2, respectively obtained by SVM and Adaboost approach. It is observed that comparable detection rates and false alarm rates are achieved in different color spaces if the fire color model is trained with parameter optimized in the specified color space. Thus, the selection of any commonly-used color spaces is reasonable in fire surveillance task. In the following content, we adopt quaternion to represent RGB color vector without additional explanation.

Table 1. Classification results using SVM

Color Spaces	RGB	HIS	YCbCr	Lab
Accuracy	95.4%	96.9%	95.7%	95.9%
False Alarm	10.67%	10.86%	9.46%	10.62%

Table 2. Classification results using Adaboost

Color Spaces	RGB	HIS	YCbCr	Lab
Accuracy	96.17%	96.17%	96.21%	96.04%
False Alarm	9.96%	10.18%	8.86%	10%

We conduct experiments to classify fire color produced by different inflamers using color models trained above. Different kinds of inflamers generate fires with different colors. Still, we can segment fire-color regions (labeled with green pixels) from non-fire color regions successfully no matter what inflamer is, as shown in Fig.2.

(a) Fire of ignited paper

(b) Iron fire

(c) Forest fire

Fig. 2. Classification Results of fire colors generated by different inflamers

3 Construction of Quaternionic Wavelet Filters

In vision-based fire surveillance system, detection of inherent features is fundamental for the whole surveillance performance. Spectral distribution is dominantly exploited as an essential static property to recognize existence of fire. In the state-of-the-art methods, flickering analysis is effective to depict dynamic features of fire elements [4][5]. As for the current fire surveillance systems, they conduct spectral, spatial and temporal analysis sequentially. Multiple threshold constraints should be established by learning-based techniques or empirical experience. In this section, we construct a set of quaternion Gabor wavelets to establish an comprehensive analysis tool for local spectral, spatial and temporal characteristics of fire regions, where the color cues, contour cues and turbulent motion cues are treated altogether in the filtering process. As a result, unitary threshold constraint is available in such an analysis scheme. These quaternion Gabors consist of three components, namely quaternion Gaussian kernel, bandpass filter kernel and temporal filter kernel, respectively responsible for spectral analysis, spatial contour filtering and temporal flickering detection,

3.1 Quaternion Gaussian Kernel for Spectral Analysis

As shown in Fig.3, 20 quaternion Gaussian kernels are utilized to represent the spectral distribution of fire pixels in RGB color space. Given mean value m_q^i and standard deviation σ_q^i $(i=1,\ldots,20)$, we can formulate the quaternion Gaussian kernels as,

$$G_q^i(\mu, m_q^i, \sigma_q^i) = \exp\left(-\frac{(\mu - m_q^i)^2}{2.(\sigma_q^i)^2}\right), \ (i=1,\ldots,20) \tag{1}$$

where m_q^i is a pure quaternion and located at ith cluster center of fire colors. Around each cluster center, the spectral probability density is represented by a quaternion Gaussian kernel with standard deviation σ_q^i. When the quaternion color vector μ falls into any one Gaussian envelope, we can estimate the probability of current pixel to have a fire color. Otherwise, we remove current pixel from candidate fire regions.

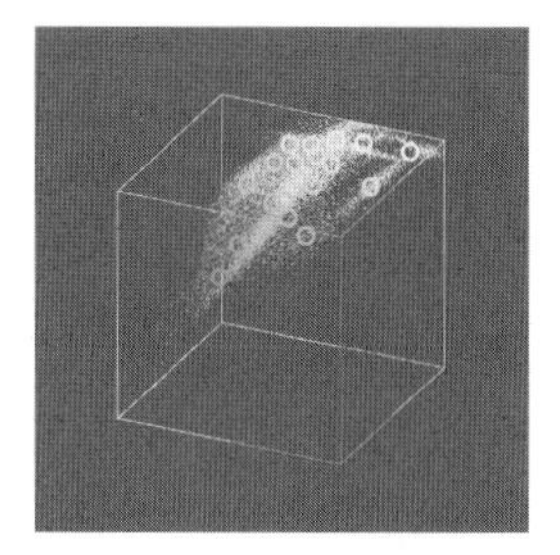

Fig. 3. 20 cluster centers of fire colors in RGB space, circled with different colors

As listed in Table 3, parameters of 20 quaternion Gaussian kernels are provided. In the extensive tests of fire region classification, it is noted that non-fire pixels usually fall into 3 Gaussian envelopes (marked with red cells in Table 3), increasing false alarms in the fire surveillance. These quaternion Gaussians contain the color of sky, road and brilliant yellow objects. In the following classification tests, we assign these Gaussian kernels with lower confidence to greatly reduce the false alarms.

3.2 Band-Pass Filter Kernel for Spatial Contour Filtering

Fire is a complex natural phenomenon involving turbulent flow. The turbulent flames due to an uncontrolled fire expose a flickering characteristic. At the boundary of a flame region, pixels appear as fire elements at a certain flickering frequency. Considering extensive variations of the size and contour orientation of fire regions, we set up a set of band-pass filter kernels to conduct multi-scale and multi-orientation contour analysis. They are formulated as the real part of complex Gabor filter,

$$G_{u,v}(x,y) = \frac{f_m^2}{2^u \sigma^2} \exp\left(-\frac{f_m^2}{2^{u+1}\sigma^2}(x^2+y^2)\right)\left[\cos\left(\frac{f_m}{2^{0.5u}}\left(x\cos\frac{v}{N} + y\sin\frac{v}{N}\right)\right) - \exp\left(-\frac{\sigma^2}{2}\right)\right] \tag{2}$$

Table 3. Parameters of Quaternion Gaussian Kernels

m_q^1	σ_q^1	m_q^2	σ_q^2	m_q^3	σ_q^3
0.24i+0.76j+0.95k	0.06183	0.42i+0.81j+0.98k	0.04769	0.13i+0.65j+0.93k	0.07479
m_q^4	σ_q^4	m_q^5	σ_q^5	m_q^6	σ_q^6
0.39i+0.62j+0.78k	0.10804	0.44i+0.72j+0.9k	0.06713	0.35i+0.7j+0.97k	0.0535
m_q^7	σ_q^7	m_q^8	σ_q^8	m_q^9	σ_q^9
0.86i+0.94j+0.94k	0.08257	0.11i+0.34j+0.71k	0.12228	0.27i+0.63j+0.87k	0.07861
m_q^{10}	σ_q^{10}	m_q^{11}	σ_q^{11}	m_q^{12}	σ_q^{12}
0.22i+0.48j+0.94k	0.06608	0.71i+0.78j+0.87k	0.08201	0.41i+0.5j+0.9k	0.09518
m_q^{13}	σ_q^{13}	m_q^{14}	σ_q^{14}	m_q^{15}	σ_q^{15}
0.52i+0.82j+0.92k	0.05826	0.33i+0.87j+0.93k	0.05341	0.43i+0.89j+0.93k	0.04847
m_q^{16}	σ_q^{16}	m_q^{17}	σ_q^{17}	m_q^{18}	σ_q^{18}
0.28i+0.59j+0.98k	0.04847	0.65i+0.93j+0.95k	0.06505	0.36i+0.8j+0.91k	0.04967
m_q^{19}	σ_q^{19}	m_q^{20}	σ_q^{20}		
0.12i+0.41j+0.87k	0.08967	0.52i+0.92j+0.96k	0.05528		

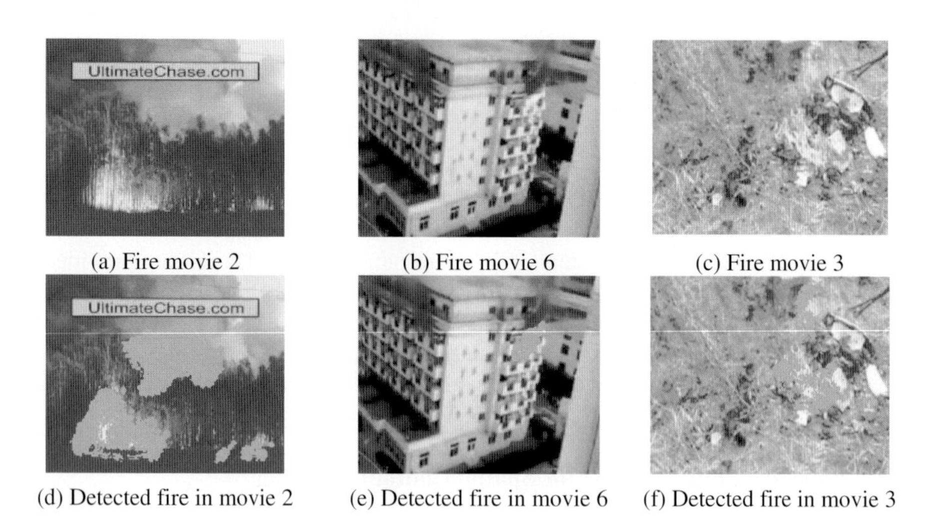

(a) Fire movie 2 (b) Fire movie 6 (c) Fire movie 3

(d) Detected fire in movie 2 (e) Detected fire in movie 6 (f) Detected fire in movie 3

Fig. 4. Fire region detection using spatial filtering (labeled with green color in subfigures (d)-(f))

$$6\sigma \frac{f_m}{2^{0.5u}} = M \tag{3}$$

where (x, y) is the 2D spatial location of current color pixel, f_m denotes the central frequency of the filter with the smallest scale. Parameter u indicates the number of scales and N is the orientation number of the filters. The formulation in (3) is enforced to obtain octave bandwidth filters. Parameter M is a constant, enumerating the number of sinusoidal oscillations within the spatial Gaussian envelope. Denotation σ

is the standard deviation of the Gaussian envelope. Because of the substantial power at low frequencies in natural signals, DC sensitivity is eliminated by the term in the square bracket to avoid a positive bias of the response.

In the experiments, we conduct spatial filtering for fire region detection at **4** scales and **8** orientations. In Fig.4, intermediate filtering results with strong responses are illustrated as candidate fire regions.

3.3 Temporal Filter Kernel for Flickering Detection

In this section, we analyze flickering behavior of fire using temporal filtering method. Similar to the spatial filter kernel, temporal wavelet filters are formulated as the product of a Gaussian window and a sinusoid function. To obtain octave bandwidth property, we construct a set of filters according to (4) and (5),

$$G_{\sigma_T}(t) = \frac{k_T^2}{\sigma_T^2}\exp\left(-\frac{k_T^2 t^2}{2\sigma_T^2}\right)\left[\cos\left(k_T t\right) - \exp\left(-\frac{\sigma_T^2}{2}\right)\right], T = 0,\ldots S-1 \quad (4)$$

$$6\sigma_T k_T = M_T \quad (5)$$

where t indicates the frame time of current color pixel, S is the temporal scale number of the filters. Given an arbitrary filter in the set, which has a Gaussian envelope with standard deviation σ_T and central frequency k_T , constant number M_T of sinusoidal oscillations is observed within the spatial Gaussian envelope. Also, DC sensitivity is eliminated to avoid a positive bias of the response.

(a) (b) (c)

Fig. 5. Fire area detection results (a) not using analysis of temporal features (b) using analysis of temporal features

Finally, we can form a set of quaternion Gabor wavelets to perform comprehensive analysis of spectral, spatial and temporal features of fire regions. The representation of these quaternion Gabors can be formulated as,

$$G_W = G_q^i . G_{u,v}(x, y) \otimes G_{\sigma_T}(t) \quad (i = 1,\ldots,20, u = 0,\ldots,3, v = 0,\ldots,7, T = 0,\ldots,3) \quad (6)$$

where symbol '$\otimes$' denotes tensor product operator.

The related fire detection results are shown in Fig.6. The flowchart of our fire detection scheme is depicted in Fig.7.

In our fire detection framework, one threshold constraint is established for the level crossings in the comprehensive analysis of fire regions. It is noted that the search range of fire regions is greatly reduced by a preprocessing step of dynamic object

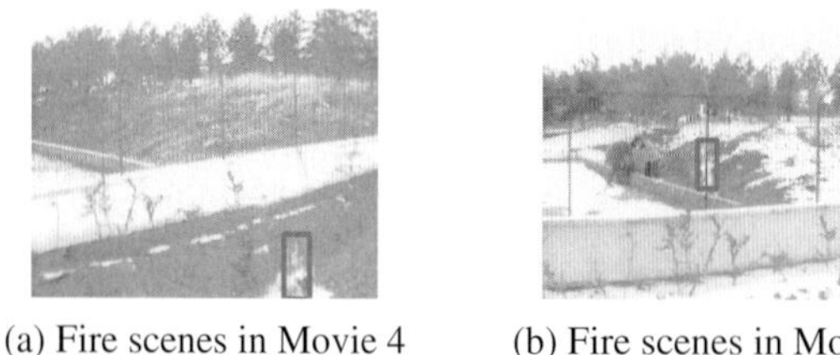

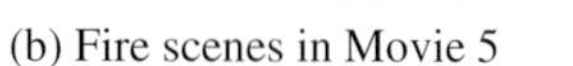

(a) Fire scenes in Movie 4 (b) Fire scenes in Movie 5 (c) Fire scenes in Movie 6

Fig. 6. Fire detection using comprehensive analysis of spectral, spatial and temporal features based on quaternion Gabor filtering

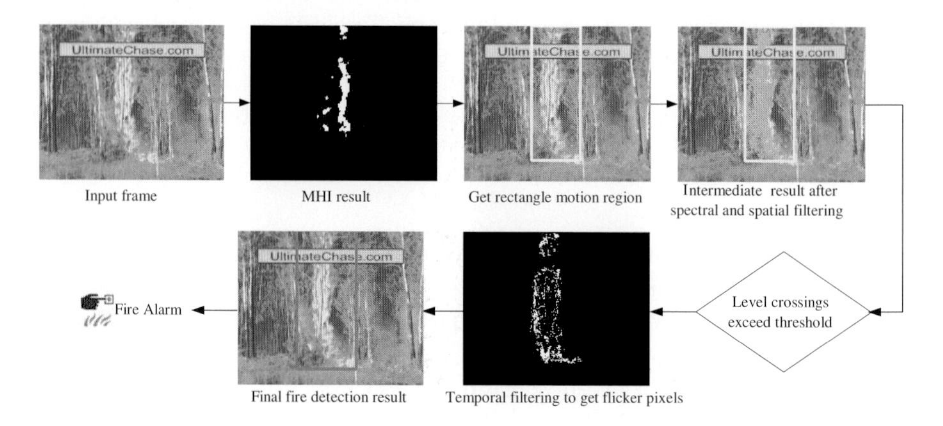

Fig. 7. Our fire Detection framework

extraction. The computation time is reduced to 20% of the time needed in the operation without motion filtering. The motion areas are extracted by Motion History Image (MHI) approach [13][14]. The following discussion of smoke detection is also based on MHI motion filtering method.

4 Smoke Region Detection

As mentioned in section 3.3, we extract moving objects as candidate fire regions using MHI method. Smokes always move upwards because of hot airflow. Hence, we indicate a moving object as smoke candidate if its main orientation points to the positive vertical component. Spectral features are not considered in our smoke detection framework, since semi-transparent smokes always reflect the color of background. Similar to flame regions, flicker process inherent in fire can be used as a clue for smoke detection. An instance of smoke flicker is shown in Fig.8. A similar technique modeling flicker behavior proposed in section 3.3 is developed to detect smoke flicker. Then, we use the GLCM-based texture descriptors in Cui's work [9], i.e. Entropy, Contrast, Angular Second Moment, Inverse Difference Moment and Image pixel correlation to match smoke regions with high confidence.

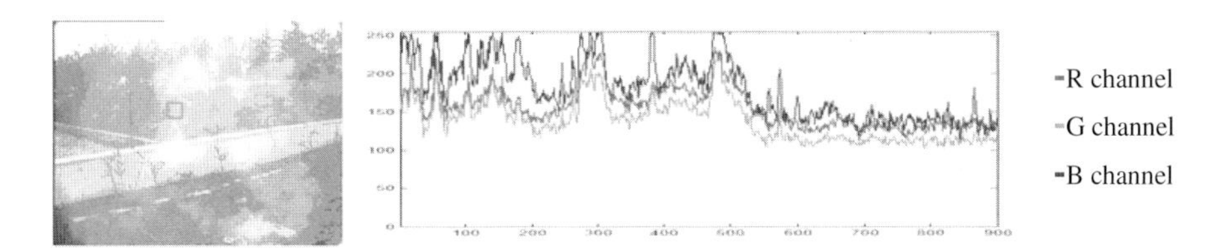

Fig. 8. An instance of smoke flicker (a) Smoke pixel at position (134, 85) (b) Temporal flicker behavior at the smoke pixel

These 5 GLCM features together with the mean value of the candidate smoke region are used to train a robust smoke texture model. As shown in Fig.9, candidate smoke regions are first extracted based on MHI method and flicker analysis. Then they are segmented into small blocks to check if the textures of these blocks are smoke-like ones, where the trained SVM classifier distinguishes smoke textures from other ones.

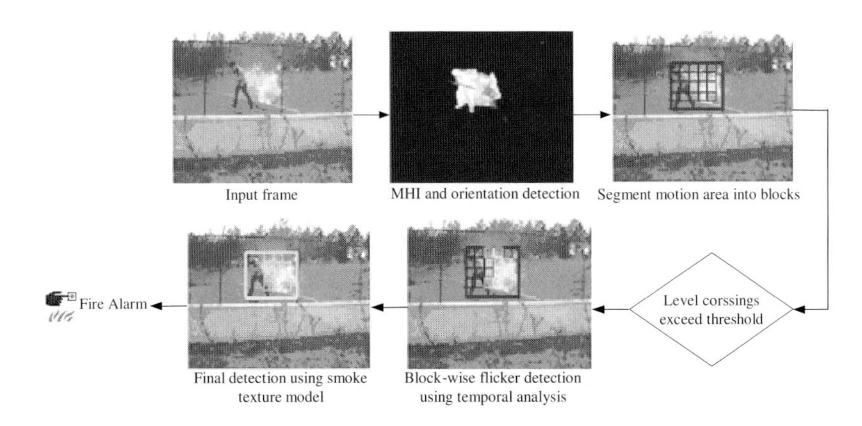

Fig. 9. Smoke Detection framework

5 Experimental Results

The proposed flame detection framework using Quaternionic features is implemented on a PC with an Intel Core Duo CPU 1.86GHz processor. We testify our framework on 12 sample videos. Fire movies 3, 6 are downloaded from Internet. The other fire movies are obtained from the following website: http://www.ee.bilkent.edu.tr/~signal/VisiFire/. The database of ordinary scenes is available at URL: http://homepages.inf.ed.ac.uk/rbf/CAVIAR/.

Some detection results are shown in Fig.10, where fire region is bounded by the red rectangle. It is observed in subfigures (h)~(l) that false alarms are alleviated in the cases of flashlight, illumination variations and clutters in the fire color regions. As compared with the state-of-the-art methods, including Toreyin's Method [4] and Ko's method [5] (Only fire detection results of movie 1,2,5,8 are available in [5] for these

(a) Movie 1 (b) Movie 2 (c) Movie 3 (d) Movie 4 (e) Movie 5 (f) Movie 6
(g) Movie 7 (h) Movie 8 (i) Movie 9 (j) Movie 10 (k) Movie 11 (l) Movie 12

Fig. 10. Test videos and our flame detection results, Movie description: (a) Burning tree; (b) Forest fire; (c) Burning paper; (d) Fire in garden; (e) Burning box; (f) Building fire; (g) Burning bin; (h) Car crash in tunnel; (i) Lighting dynamo; (j) Warm-toned store; (k) Hall of pedestrians; (l) Shining patio

two methods.), our framework outperforms these methods both in true positive and false positive, as listed in Table 4. True positive means the rate of correctly detecting a real fire as a fire and false positive means the rate of recognizing a non-fire as a fire, while missing means the rate of not recognizing a real fire.

Table 4. Comparison of fire detection results

Movie	**Frame number**	**Toreyin's Method**		**Ko's method**		**Our method**	
		TP	FP	TP	FP	TP	FP
1	245	82.6	4.9	77.7	0.0	100.0	0.0
2	199	63.7	10.0	97.9	0.0	95.2	0.0
3	128					83.6	0.0
4	545					93.6	1.3
5	1200	7.1	44.1	56.3	0.0	97.2	0.0
6	635					95.5	3.4
7	696					78.2	0.0
8	375	100.0	0.0	97.5	1.5	100.0	0.0
9	143					100.0	0.0
10	284					100.0	0.0
11	2226					100.0	0.0
12	1147					100.0	6.9

For an image of 200*160 pixels, the processing time per frame in flame detection scheme is about 50ms-200ms. For such an image, our smoke detection method takes about 90ms-200ms per frame. As shown in Fig.11, the motion regions which are

(a) Smoke movie 1 (b) Smoke movie 2 (c) Smoke movie 3

Fig. 11. Test videos and our smoke detection results

detected as smoke is bounded in the yellow frame. Subfigures (b) and (c) show that our method can detect thin smoke, which is available for early alerts for fire accidents.

6 Conclution

Instead of selecting color spaces for flame analysis in an ad-hoc manner, in this paper, quantitative measures are established to evaluate the flame classification accuracy within the commonly-used color spaces, including RGB, HIS, YCbCr and LAB color spaces. It is interesting to find that using different kinds of color space is all reasonable if the fire color model is trained with parameter optimized in the specified color space. More important, a new fire surveillance framework available for flame and smoke detection is presented, where a comprehensive analysis tool is established for local spectral, spatial and temporal characteristics of fire regions based on quaternion wavelet filtering method. Compared with the state-of-the-art method, better detection rate and lower false alarm rate is acquirable in our fire detection scheme.

Acknowledgments. This work was supported in part by Research Fund for the Doctoral Program of Higher Education of China (200802481006), NSFC-60902073, NCET-06-0409, and Cultivation Fund of the Key Scientific and Technical Innovation Project of MOE (706022).

References

1. Ko, B.C., Hwang, H.J., Lee, I.G., Nam, J.Y.: Fire surveillance system using an omnidirectional camera for remote monitoring. In: IEEE 8th International Conference on Computer and Information Technology Workshops, pp. 427–432 (2008)
2. Celik, T., Demirel, H.: Fire detection in video sequences using a generic color model. Fire Safety Journal 44(2), 147–158 (2009)
3. Chen, T., Wu, P., Chiou, Y.: An early fire-detection method based on image processing. In: IEEE International on Image Processing, pp. 1707–1710 (2004)
4. Toreyin, B.U., Dedeoglu, Y., Gudukbay, U., Cetin, A.E.: Computer vision based method for real-time fire and flame detection. Pattern Recognition Letters 27(1), 49–58 (2006)
5. Ko, B.C., Cheong, K.H., Nam, J.Y.: Fire detection based on vision sensor and support vector machines. Fire Safety Journal 44(3), 322–329 (2009)
6. Kopilovic, I., Vagvolgyi, B., Sziranyi, T.: Application of panoramic annular lens for motion analysis tasks surveillance and smoke detection. In: 15th International Conference on Pattern Recognition, vol. 4, pp. 714–717 (2000)
7. Ho, C.C.: Machine vision-based real-time early flame and smoke detection. Measurement Science and Technology 20(4), 045502(13pp) (2009)
8. Liu, C.B., Ahuja, N.: Vision based fire detection. International Conference on Pattern Recognition 4, 134–137 (2004)
9. Cui, Y., Dong, Zhou, H.: An early fire detection method based on smoke texture analysis and discrimination. In: 2008 Congress on Image and Signal Processing, vol. 3, pp. 95–99 (2008)

10. Chen, T.H., Yin, Y.H., Huang, S.F., Ye, Y.T.: The smoke detection for early fire-alarming system base on video processing. In: International Conference on Intelligent Information Hiding and Multimedia Signal Processing, pp. 427–430 (2006)
11. Yuan, F.: A fast accumulative motion orientation model based on integral image for video smoke detection. Pattern Recognition Letters 29(7), 925–932 (2008)
12. Hamilton, W.R.: On quaternions, or on a new system of imaginaries in algebra. Philosophical Magazine 25(3), 489–495 (1844)
13. Davis, J.: Recognizing movement using motion histograms. Technical Report 487, MIT Media Lab (1999)
14. Bradski, G.R., Davis, J.W.: Motion segmentation and pose recognition with motion history gradients. Machine Vision and Applications 3(3), 174–184 (2002)

Object Tracking and Local Appearance Capturing in a Remote Scene Video Surveillance System with Two Cameras

Wenming Yang[1,2], Fei Zhou[1,2], and Qingmin Liao[1,2]

[1] Visual Information Processing Lab., Graduate School at Shenzhen,
Tsinghua University, Shenzhen, China
[2] Tsinghua-PolyU Biometric Joint Lab., Shenzhen, China
yangelwm@163.com, flying.zhou@163.com, liaoqm@sz.tsinghua.edu.cn

Abstract. Local appearance of object is of importance to content analysis, object recognition and forensic authentication. However, existing video surveillance systems are almost incapable of capturing local appearance of object in a remote scene. We present a video surveillance system in dealing with object tracking and local appearance capturing in a remote scene, which consists of one pan&tilt and two cameras with different focuses. One camera has short focus lens for object tracking while the other has long ones for local appearance capturing. Video object can be located via just one manual selection or motion detection, which is switched into a modified kernel-based tracking algorithm absorbing both color value and gradient distribution. Meanwhile, local appearance of object such as face is captured via long focus camera. Both simulated and real-time experiments of the proposed system have achieved promising results.

Keywords: Remote scene, video surveillance, object tracking, local appearance capturing.

1 Introduction

Recently, intelligent video surveillance system has attracted more and more attention from different researchers[1] [2] [3], because it possesses great potential in both civilian and military security applications. However, most of existing video surveillance systems either work on close-up view with short surveillance distance or work on remote scene surveillance with incapability of providing local appearance of object. However, local appearance of object often plays a crucial role in content analysis, object recognition and forensic authentication. For example, it is an urgent need for video surveillance system to provide distinguishable local appearance or features such as human face when intruder appears in surveillance scene. In addition, many of real-time video surveillance systems are only motion detection, not object tracking indeed. So these systems just can deal with surveillance by static camera, and fail once camera becomes moving. Especially, tracking in a remote scene and capturing local appearance simultaneously become a challenging task.

S. Boll et al. (Eds.): MMM 2010, LNCS 5916, pp. 489–499, 2010.

Various tracking algorithms have been proposed to overcome the difficulties that arise from noise, occlusion and changes in the foreground objects or in the surrounding such as Kalman Filter(KF), Particle Filter(PF), Active Shape Model(ASM), meanshift, and their variants[4], [5], [6], [7]. Among these algorithms, meanshift has achieved considerable success in object tracking due to its simplicity and robustness[8], [9]. However, it takes feature spatial distribution into little consideration, so that object and background with partly similar color or texture can not be distinguished.

Yang[10] proposed an object tracking method in joint feature-spatial spaces, feature spatial distribution such as pixel position has been taken into consideration, and experiments on several video sequences achieved satisfactory results. However, in Yang's method, time-consuming iteration always is a problem since it perhaps leads to local optimum. In addition, the use of feature spatial distribution is not adequate in Yang's method, more than one kind of feature need to be introduced simultaneously.

Collins[11] proposed an online tracking features selection method from linear combination of RGB values. This method can adaptively select the top N features, and meanshift algorithm was utilized. The median of N locations produced by meanshift is selected as the final object location. Liang[4] pointed the scale adaptation problem of literature[11] out, giving a corresponding resolution. And an evaluating principle is suggested to perform feature selection. Unfortunately, in their work[4],[11], feature selection was focused on while feature spatial distribution was ignored.

In this paper, we design a remote scene video surveillance system for object tracking and local appearance capturing with two cameras, and present a probabilistic tracking framework integrating feature value combination and feature spatial distribution. In the proposed system, we introduce: 1) a video surveillance system construction consisting of two cameras and a pan&tilt; 2) a 1-dimension feature combining hue, saturation and value; 3) a new formula to compute weight image, avoiding the interference from cluttered background; 4) a modified kernel-based tracking approach absorbing both color value and gradient of sample pixels, which is performed on weight image, not the whole image.

The structure of this paper is as follows: Section 2 presents hardware structure of the proposed system and the computing method of relationship of two views from two cameras. Section 3 shows feature value combining method, new computing method of weight image and a modified kernel-based tracking approach. Both simulated and real-time experimental results are shown in Section 4. Finally, Section 5 concludes the paper with a discussion.

2 The Proposed Surveillance System Structure

2.1 The Hardware Structure

To track object in remote scene and capture local appearance of this object, two kinds of focus lens are necessary in video surveillance system: one is short focus for tracking while the other one is long focus for capturing local appearance. Also two image outputs are necessary for performing different processing algorithms. Meanwhile, to deal with the condition of moving camera, a pan&tilt at least is need to rotate the camera. The ideal case of imaging device is to find a camera with two different focus lens of the same optic axis and two according video outputs, and a pan&tilt used to rotate the camera to tracking object. Unfortunately, no such camera can be found. Since the distance of surveillance is between 50m and 150m, so we select two cameras with different focus lens to simulate ideal camera with different focus lens. For convenience, we name the short focus camera as panorama camera and name the long focus camera as close-up camera in the following context.

For the proposed video surveillance system, some points need to be addressed:

(i) Though the panorama camera has a relatively broad view, no camera can cover overall angles of view in a remote scene without rotation.

(ii) The tracked object, which is usually in a remote scene, needs to capture local appearance in the close-up camera during the surveillance process.

(iii) It is troublesome to align the views of two cameras at the same center exactly. The common case is that the close-up view corresponds to some part of the panoramic view, rather than the central part.

Considering above issues, we fix the two cameras on the same pan&tilt, i.e., the position relationship of two cameras' view is unchangeable in the running. And a controller and computer are employed to adjust the pan&tilt according to the tracking results in panoramic view. Our system is illustrated in Fig. 1.

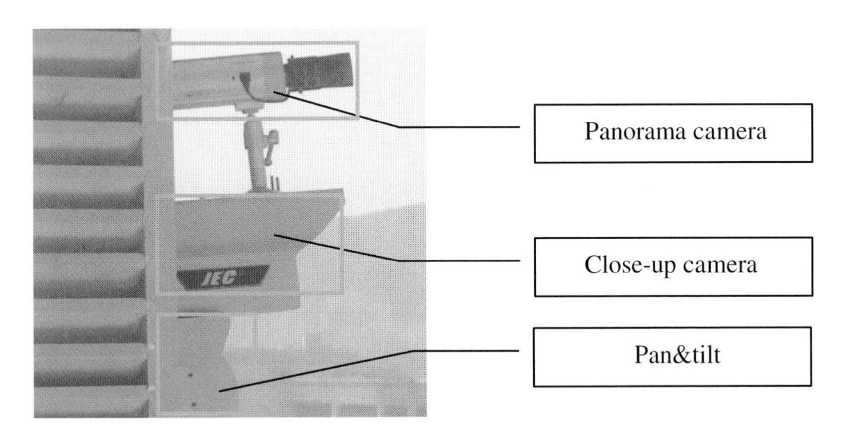

Fig. 1. Hardware structure of the proposed system

2.2 Computing the Relationship of Views from Two Cameras

When an object is found to appear at a certain position in panoramic view, the controller should be informed the direction in which the pan&tilt is rotated for the close-up view, since these two views are not aligned to the same center. This means, though the position relationship of the views of two cameras is unchangeable, the relationship still remains unknown. However, this relationship is of utmost importance to make object remain in close-up view.

A reasonable assumption is made that there is no rotational diversity between two coordinate systems from different views, and it can be guaranteed readily while cameras are fixed. Therefore, translation and scale are the parameters that we take into consideration. We adopt convolution between panoramic view image and close-up view image with varying resolutions to estimate position $\mathbf{x}$ in the panoramic view image to which the close-up view image corresponds:

$$\mathbf{x} = \arg\max_{\mathbf{x}} \left(\frac{\mathbf{P}(\mathbf{x}) * (\mathbf{G}_\sigma - Avg(\mathbf{G}_\sigma))}{Area(\mathbf{G}_\sigma)} \right) \tag{1}$$

where $\mathbf{G}_\sigma$ is a scaled version of close-up view image with variable scale parameter σ, and $\mathbf{P}(\mathbf{x})$ which has the same width and height as $\mathbf{G}_\sigma$ is the image blob located at the position $\mathbf{x}$ of panoramic view image. $Avg(\mathbf{G}_\sigma)$ is the average of pixel values in $\mathbf{G}_\sigma$, and $Area(\mathbf{G}_\sigma)$ is the area of $\mathbf{G}_\sigma$. The symbol * indicates spatial convolution. And convolution is performed on gray images.

$Area(\mathbf{G}_\sigma)$ is used as a normalization term in order to remove the influence of $\mathbf{G}_\sigma$'s size. This convolution can be treated as run a mask (i.e. $\mathbf{G}_\sigma - Avg(\mathbf{G}_\sigma)$) through the panoramic view image. Like some other digital masks, e.g. Sobel and LoG, the average of the above mask is zero. Considering that $\mathbf{x}$ only needs to be

(a) Panoramic view (b) Close-up view

(c) σ=0.1 (d) σ=0.2 (e) σ = 0.4

Fig. 2. Two gray images from the proposed system and the comparison of their convolution results at different parameters

computed once after cameras are fixed, we search exhaustively in the spatial-scale space to guarantee global optimum of $\mathbf{x}$, without considering the computational efficiency.

Fig. 2(a) and Fig. 2(b) show two gray images from panoramic view and close-up view, respectively. The size of the two images is 640×480 pixels. Fig. 2 (c), Fig. 2(d) and Fig. 3(e) are the convolution results with σ=0.1, 0.2 and 0.4. In our experiment, peak value of convolution results gets maximum, when σ=0.2. Therefore, the location of peak value in Fig.2 (d) is the estimated as position $\mathbf{x}$.

3 Tracking Approach

3.1 Color Components and Combination

HSV color model had been used in many image retrieval algorithms for having identical color discrimination with human vision systems. Hue is used as the only feature of object in traditional camshift, in which 1-dimension hue histogram is computed. Lacking of saturation and value, the tracker appears far from robustness. To overcome this disadvantage, many literatures [12], [13] prefer to employ higher-dimension feature spaces.

However, high-dimension space is not suitable to our applications, since our tracked object in the remote scene is relative small, containing few samples pixels in image. Using few samples to cover high-dimension feature space will lead to "curse of dimensionality" without doubt. Therefore, we adopt a 1-dimension feature combining hue, saturation and value with different weights. Hue is non-uniform quantized into 32 portions, while saturation and value are non-uniform quantized into 6 portions respectively:

$$L = w_h Q_h + w_s Q_s + w_v Q_v \tag{2}$$

$$Q_h = 32 \text{, } Q_s = Q_v = 6 \tag{3}$$

$$w_h : w_s : w_v = 16 : 2 : 1 \tag{4}$$

where Q_h, Q_s and Q_v are the quantitative level of hue, saturation and value respectively, i.e.,. w_h, w_s and w_v are the weight of hue, saturation and value, respectively. The range of L is from 0 to 511.

3.2 A New Formula of Computing Weight Image

Weight image is an effective tool for estimating the position and the deformation of object in simple scene. It is generated in traditional camshift by using histogram back projection to replace each pixel with the probability associated with that color value in the target model.

$$I_o(\mathbf{x}) = \eta \cdot b(\mathbf{x}) \tag{5}$$

where $I_o(\mathbf{x})$ is the value of $\mathbf{x}$ in weight image, and $b(\bullet)$ is a function to map pixels into its color feature space has. η is a constant to normalize the value from 0 to 511.

Camshift algorithm works well in simple scenes, but fails in cluttered scenes due to the contamination of weight image. A new formula of computing weight image is proposed to make the weight image more reliable. This formula is established under the observation that the size and the location of tracked object changes smoothly in successive frames.

Before the computation of weight image, we first analyze the connected regions of original weight image $I_o(\mathbf{x})$ using the method in [14], and then a map of size of connected regions is generated by

$$J(\mathbf{x}) = sizeof(R_i \mid \mathbf{x} \in R_i) \tag{6}$$

where R_i is the i-th connected region, and $sizeof(\bullet)$ is a function to map pixels into the size of the connected region which it belongs to. The new formula can be written as:

$$I_n(\mathbf{x}) = C_n \cdot b(\mathbf{x}) \cdot \exp\left(-\alpha \left\| J(\mathbf{x}) - J_t \right\| - \beta \left\| \mathbf{x} - \mathbf{C} \right\|\right) \tag{7}$$

where J_t is the number of pixels in the target model region, and $\mathbf{C}$ is the centroid of tracking result in last frame. α and β are the adjustable parameters. This formula means to give different weights to $b(\mathbf{x})$. The pixels in object tend to get boosted weights while those in background tend to get suppressed weights.

Fig. 3 illustrates the comparison of weight images from different methods and parameters. Fig. 3(a) is the input frame. Fig. 3(b) is the weight image generated by traditional camshift. Fig. 3(c) and Fig. 3(d) are the weight images generated by proposed formula with different parameters.

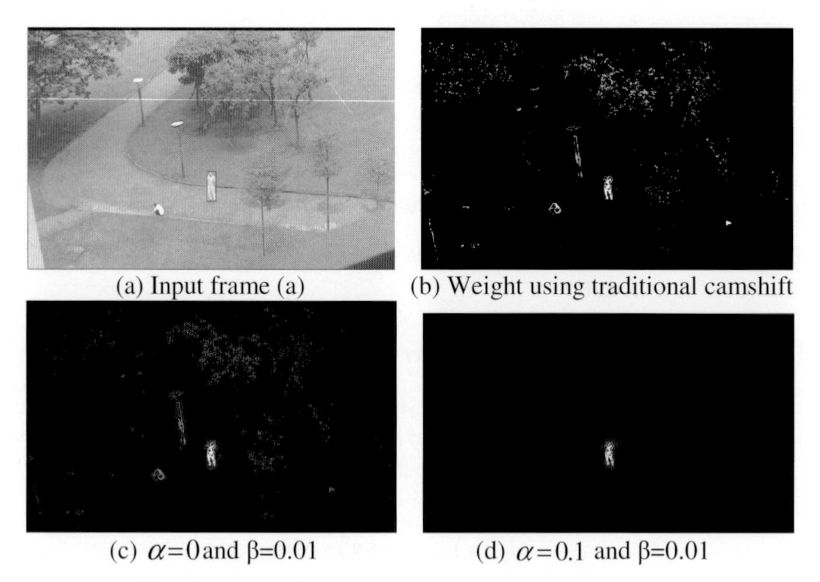

(a) Input frame (a) (b) Weight using traditional camshift

(c) α=0and β=0.01 (d) α=0.1 and β=0.01

Fig. 3. Comparisons of different weight images

3.3 A Modified Kernel-Based Tracking Approach Absorbing Color Value and Gradient Distribution

Although Collins' method[11] and Liang's method[4] selected and updated the features set, the tracking results mainly relied on "shift" of mass center of target image region, which is numerically unstable. Yang[10] proposed an efficient tracking algorithm via meanshift and new similarity measure, one of the advantages of this algorithm is to embed the Euclid distance between feature pixels and region center in meanshift. The new similarity measure is as follows:

$$J(I_x,I_y)=\frac{1}{MK}\sum_{i=1}^{K}\sum_{j=1}^{M}w(\|\frac{(x_*-x_i)}{\tau}\|^2)w(\|\frac{y-y_j}{\tau}\|^2)\kappa(\|\frac{u_i-v_j}{h}\|^2) \tag{8}$$

where I_x, I_y are sample points in model image and target image respectively, K is width of object region and M is height, x and y are the 2D coordinates of model image and target image, u and v are the selected feature vector for model image and target image, x_* and y are the center of sample points in the model image and the current center of the target points, respectively. $\kappa(x)$ is the Radial-Basis Function(RBF) kernel, h is bandwidth, $w(x)$ is another RBF kernel in the spatial domain, and τ is bandwidth. Yang's method[10] actually integrated feature value with the Euclid distance between feature pixels and region center. It provides a fine kernel-based tracking framework. However, there are three points that need to be strengthened in Yang's method.

Firstly, it's not enough that only the Euclid distance between feature pixels and region center is absorbed into kernel-based tracking algorithm. The more precise information of feature spatial distribution needs to be introduced. Since gradient is one of most discriminative spatial features, so we can define gradient similarity and embed it into similarity measure in literature[10]. Let p and q are the corresponding gradients of model image and target image, respectively. Then new similarity measure is defined as follows:

$$J(I_x,I_y)=\frac{1}{MK}\sum_{i=1}^{K}\sum_{j=1}^{M}w(\|\frac{(x_*-x_i)}{\tau}\|^2)w(\|\frac{y-y_j}{\tau}\|^2)\kappa(\|\frac{|u_i-v_j|+|p_i-q_j|}{h}\|^2) \tag{9}$$

In equation (9), u and v are 1-dimension feature combining hue, saturation and value with different weights(see section 3.1) for model image and target image, respectively.

Secondly, computing method of tracking algorithm in literature[10] is a typical time-consuming iteration, which perhaps leads to local optimum. It can be assumed that the tracked object will appear in the position with large weight in weight image because the color change of object remains smoothly and slowly in several successive frames intervals. Thus we compute similarity measure in these pixels with large weight achieving global optimum and reducing time consumption.

Thirdly, since width and height of the object is not same usually, the bandwidth τ for x coordinate and y coordinate should be also different. Actually in our approach, computing of inscribed circle from square in Yang's method[10] is changed into computing of inscribed eclipse from rectangle.

4 Experiments

Both simulated experiments on MPEG testing sequences and real-time outdoor experiments are performed to verify the proposed approach. In all experiments, image gradient, which is obtained by the employment of *Sobel* operator after a Guassian smoothing operation, is the horizontal and vertical gradient of L in equation (2).

The first experiment, the *Ball* sequence for MPEG is tested, the contrast between object(white ball) and background(brown wall) is high. The object is initialized with a manually selected rectangular region in frame 0. The tracking results for frame 3, 16 and 26 are shown in Fig. 4.

Fig. 4. The results of simulated experiment on *Ball* sequence

In the second experiment, a complex video sequence *Library* is used to test. The size of image is 720×480 pixels. The contrast between object and background changes in a wide range and occlusion exists. Especially, contrast is rather low when pedestrian in white passed by white pillar. The object is automatically detected by Gaussian Mixture Modeling (GMM) of background. Fig. 5 shows the tracking results when pedestrian passed by white pillar. Though being similar in color to white pillar, the pedestrian is located accurately.

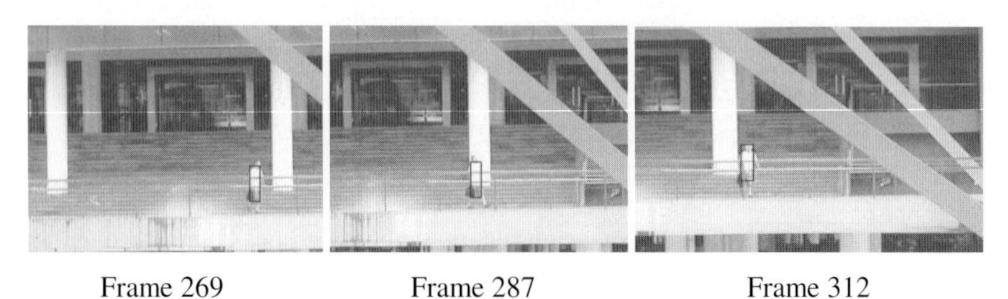

Fig. 5. The results of simulated experiments on the *Library* sequence

In the third experiment, we test the proposed surveillance system for two real-time scenes named by *Bridge* and *Lawn*. The distance between camera and object for *Bridge* scene is about 70m. The size of panoramic and close-up frame is the same, 640×480 pixels. Fig. 6 shows the overall results of our system for Bridge scene with software interface. Image on the left is from panorama camera, while that in the right is from close-up camera. Although it seems very small in image, the object is tracked accurately(see blue rectangle in Fig. 6). Adaboost algorithm[15] is used to detect human face(see red rectangle in Fig. 6). Certainly, many other works, e.g., face recognition and gait recognition, can be tried on the close-up image.

Fig. 6. The real-time outdoor experiments results for *Bridge* scene

Fig. 7. The real-time outdoor experiments results for *Lawn* scene

The distance between camera and object for *Lawn* scene is about 150m. Fig. 7 gives a tracking result for frame 21. Our system provides a framework and platform for object recognition and content analysis, etc in video surveillance for remote scene.

Please note that the pan&tilt is rotated by controller to make object remain in close-up view, therefore the object is almost at the same position in panoramic view all the time.

5 Discussion and Conclusions

In this paper, we established a remote scene surveillance system with a panoramic view for object tracking as well as a close-up view for local appearance capturing, aiming at enhancing the practical functionalities of video surveillance system. In tracking algorithm, we proposed a modified object tracking approach integrating color feature combination with feature pixel spatial distribution. Namely, The HSV color value of pixels and gradient of pixels both are introduced into kernel-based tracking framework. A method to compute the position relationship of these two views is proposed. In addition, a new formula of computing weight image is explored to avoid the interference from cluttered background. Both simulated and real-time experimental results demonstrated that the proposed tracking approach and surveillance system work well.

We will focus on improving the quality of close-up image and the operationality of pan&tilt in future research. Also, real-time online face recognition, gait analysis and recognition will be tried on the basis of our remote scene video surveillance framework. In addition, we set manually several fixed bandwidths and thresholds in experiments, their adaptive selection will be studied in the future work.

Acknowledgments. The research leading to this paper was partially supported by China Postdoctoral Science Foundation(No.20090450353) and Guangdong Nature Science Foundation of China(No.9451805702003923). The authors would like to thank the three anonymous referees whose comments and suggestions have greatly improved this paper.

References

1. Foresti, G.L.: Object Recognition and Tracking for Remote Video Surveillance. IEEE Transactions on Circuits and Systems for Video Technology 9, 1045–1062 (1999)
2. Bue, A.D., Comaniciu, D., Ramesh, V., Regazzoni, C.: Smart cameras with real-time video object generation. In: Proceedings of International Conference on Image Processing, pp. 429–432 (2002)
3. Chen, T.-W., Hsu, S.-C., Chien, S.-Y.: Automatic Feature-based Face Scoring in Surveillance Systems. In: IEEE International Symposium on Multimedia, pp. 139–146 (2007)
4. Liang, D., Huang, Q., Jiang, S., et al.: Mean-shift Blob Tracking with Adaptive Feature Selection and Scale Adaptation. In: IEEE International Conference on Image Processing, San Antonio, United States, pp. 369–372 (2007)
5. Chang, C., Ansari, R., Khokhar, A.: Multiple Object Tracking with Kernel Particle Filter. In: IEEE Computer Society Conference on Computer Vision and Pattern Recognition, San Diego, vol. 1, pp. 566–573 (2005)

6. Shiu, Y., Kuo, C.-C.J.: A Modified Kalman Filtering Approach to On-Line Musical Beat Tracking. In: IEEE International Conference on Acoustics, Speech and Signal Processing, vol. 2, pp. 765–768 (2007)
7. Lee, S.-W., Kang, J., Shin, J., et al.: Hierarchical Active Shape Model with Motion Prediction for Real-time Tracking of Non-rigid Objects. IET Comput. Vis. 1(1), 17–24 (2007)
8. Comaniciu, D., Meer, P.: Mean Shift: A Robust Approach Toward Feature Space Analysis. IEEE Transaction on Pattern Analysis and machine Intelligence 24(5), 603–619 (2002)
9. Comaniciu, D., Ramesh, V., Meer, P.: Kernel-based Object Tracking. IEEE Transaction on Pattern Analysis and machine Intelligence 25(5), 564–577 (2003)
10. Yang, C., Duraiswami, R., Davis, L.: Efficient Mean-Shift Tracking via a New Similarity Measure. In: IEEE Computer Society Conference on Computer Vision and Pattern Recognition, San Diego, vol. 1, pp. 176–183 (2005)
11. Collins, R.T., Liu, Y., Leordeanu, M.: Online selection of discriminative tracking features. IEEE Transaction on Pattern Analysis and Machine Intelligence 27(10), 1631–1643 (2005)
12. Perez, P., Hue, C., Vermaak, J., Gangnet, M.: Color-Based Probabilistic Tracking. In: Heyden, A., Sparr, G., Nielsen, M., Johansen, P. (eds.) ECCV 2002. LNCS, vol. 2350, pp. 661–675. Springer, Heidelberg (2002)
13. Maggio, E., Cavallaro, A.: Multi-Part Target Representation for Color Tracking. In: Proceedings of International Conference on Image Processing, pp. 729–732 (2005)
14. Salembier, P., Oliveras, A., Garrido, L.: Antiextensive Connected Operators for Image and Sequence Processing. IEEE Transactions on Image Processing 7, 555–570 (1998)
15. Jones, M., Viola, P.: Rapid Object Detection using a Boosted Cascade of Simple Features. In: IEEE Conference on Computer Vision and Pattern Recognition, pp. 511–518 (2001)

Dual Phase Learning for Large Scale Video Gait Recognition

Jialie Shen[1,⋆], HweeHwa Pang[1], Dacheng Tao[2], and Xuelong Li[3]

[1] School of Information Systems, Singapore Management University
[2] School of Computer Engineering, Nanyang Technological University, Singapore
[3] Birkbeck College, University of London, UK

Abstract. Accurate gait recognition from video is a complex process involving heterogenous features, and is still being developed actively. This article introduces a novel framework, called GC^2F, for effective and efficient gait recognition and classification. Adopting a "refinement-and-classification" principle, the framework comprises two components: 1) a classifier to generate advanced probabilistic features from low level gait parameters; and 2) a hidden classifier layer (based on multilayer perceptron neural network) to model the statistical properties of different subject classes. To validate our framework, we have conducted comprehensive experiments with a large test collection, and observed significant improvements in identification accuracy relative to other state-of-the-art approaches.

1 Introduction

Increasing demand for automatic human identification in surveillance and access control systems has spurred the development of advanced biometric techniques. Biometrics that are exploited in such identification systems include shoeprint [9], iris [11], fingerprint [12] and palm print [13]. However, many of those techniques have limited effectiveness and feasibility, due to 1) poor robustness on low-resolution videos, and 2) cumbersome user interactions in the course of capturing the biometric parameters. In contrast, human gait holds great promise for biometric-based recognition. Gait analysis offers a number of unique advantages including uniqueness [14] and unobtrusiveness.

Motivated by the potential of automatic gait identification and classification, many techniques have been developed. Some of the earliest studies on this topic have targeted medical and behavioral applications. For example, gait categorization was used in [15] to identify personal friends. It has also been proposed as a basis for clustering patients into treatment groups [16]. Existing gait analysis techniques generally fall into two classes – model-based versus appearance-based.

– Model-based schemes: Such schemes aim to model the human body structure and movement patterns. They can be used to construct generic models for extracting gait parameters, such as kinematic values.

⋆ Jialie Shen is supported by the Lee Foundation Fellowship for Research Excellence (SMU Research Project Fund No: C220/T050024), Singapore.

S. Boll et al. (Eds.): MMM 2010, LNCS 5916, pp. 500–510, 2010.

- Appearance-based schemes: This category of techniques attempt to recognize gait from sequences of binary images of the moving body. The major foci of the techniques include 1) how to track silhouettes, 2) how to study the property of the silhouettes for the purpose of extracting more effective features, and 3) how to design corresponding identification algorithms.

The basic idea of model-based approaches is to recognize human gaits based on predefined structural templates of the human body. One of the earliest work in this domain was done by Niyogi and Adelson [20]. Their gait pattern identification system is based on a 2D skeleton of the human body and joint angles of the lower body, which can be estimated as features for the purpose of gait recognition. In [21], Bregler developed a probabilistic compositional framework to integrate abstractions of different granularity levels, using various statistical schemes including mixture models, EM, and Hidden Markov Model. The framework can be use for tracking and recognition tasks over gait video sequences. Its performance was evaluated with a small test collection containing 33 sequences on 5 different subjects with the gait classes: running, walking and skipping. The achieved classification accuracy varied from 86% to 93%, depending on the initial settings of the framework.

The above approaches generally considered only either dynamic or static information of the body as gait feature. Based on the observation that a combination of the two could improve recognition performance, Wang et al. developed a feature fusion scheme using different combination rules [8]. Experiments on a small test collection (20 subjects) illustrate the potential of their scheme. In [1], Liu et al. designed a simple template using the average of the silhouettes of each gait cycle as gait signature. As each gait cycle could contain a few shuffled frames, their template could not capture information about gait dynamics; this weakness leads to significant performance degradation. More recently, Guo et al. proposed a family of feature subset selection schemes for capturing discriminative gait features [18]. They built their techniques on the notion of mutual information (MI), after comparing it with different statistical methods including feature correlation and One-Way ANOVA. Experiment results showed that they were able to achieve about 95% identification accuracy. This is the current state of the art for the model-based approach.

Obviously, the effectiveness of gait recognition hinges on the ability to extract discriminative features from the videos. Human gait involves various kinds of features, such as head features, upper body features and dynamic gait features. They offer different discriminative capabilities and play different roles in the recognition task, and it is impossible to achieve good performance with just one group of features. Therefore, a fundamental research issue is to *combine* features effectively. While there has been considerable effort in developing feature extraction methods [2, 19], less attention has been given to designing proper schemes for fusing features. In addition, there is a lack of a unified learning framework that combines feature extraction and classification. Without taking class distribution into account, the features generated by current methods may not be sufficiently informative for the recognition task. The main contributions of our study are summarized as follows:

- We have developed a novel dual phase classification framework, called GC^{2}F (**G**ait **C**lassifier based on **C**omposite **F**eatures), for accurate gait class recognition. This

framework allows multiple kinds of features, as well as high-level discriminative information, to be combined effectively within a unified learning framework. In addition, an efficient learning strategy, called backprop.ECOC, is designed for effective feature combination.
- We have carried out a detailed empirical study and deep analysis of the experiment results based on large test collections. The testbed contains 600 short indoor video sequences on 35 subjects. The experiment results show that our framework achieves superior performance over existing solutions.

The rest of the paper is structured as follows: Section 2 introduces the architecture of our proposed system. Section 3 explains the experiment configuration, while Section 4 presents a detailed analysis of the experiment results. Finally, Section 5 draws some conclusions and discusses future research directions.

2 System Architecture

Inspired by neural network architecture, our proposed GC^{2}F gait identification system adopts the "multiple refinement" principle. As depicted in Figure 1, the system contains three functional layers – a feature generation layer (input layer), a hidden classifier layer (middle layer), and a decision layer (output layer). The first layer comprises several basic classifiers. Each classifier contains a set of Gaussian Mixture Models (GMM), with one GMM for each gait class. In the current setting, we have four basic classifiers, corresponding to four different kinds of gait features. The output of each basic classifier is a probability histogram, which captures the probability of the input object belonging to different gait classes according to feature f. The hidden classifier layer includes one node for each gait class. Each of those nodes embeds a neural network, and is connected with

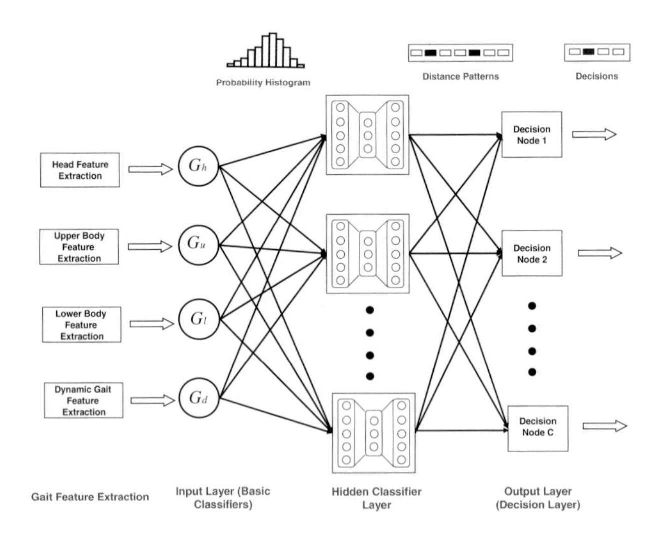

Fig. 1. Structure of GC^{2}F Gait Classification Framework

all the basic classifiers. The hidden classifiers use the statistical features from the input layer to derive the distance patterns between gait classes. Those patterns enable the decision layer to carry out the final categorization process. The following sections provide detailed descriptions of the three layers, their relationships, and the learning algorithms.

2.1 Low Level Gait Feature (Gait Parameter) Extraction

Feature generation is an important prerequisite for gait recognition. There are different parameters for the human gait, corresponding to various positions of the human body. The GC^{2}F framework considers four feature categories that are extracted from the given video sequences. They include Head features (HF), Upper body features (UBF), Lower body features (LBF) and Dynamic gait features (DGF). Detail information is given in Table 1.

Table 1. Summary of different feature types considered in this study

Feature Type	Detail Description	Dimension
Head Features (HF)	Head width, Head height, Head area, Head x offset, Head y offset, x center of neck, y center of neck, Neck width, Neck height	9
Upper Body Features (UBF)	Torso height, Torso width, x center of torso, y center of torso, x pelvis rotation, y pelvis rotation, x center of left elbow, y center of left elbow, x center of right elbow, y center of right elbow, x center of right palm y center of left palm	12
Lower Body Features (LBF)	Thigh length, Shin length, Leg width at hip, Leg width at knee(upper), Leg width at knee (lower), Leg width at ankle, Hip y offset, Foot width, Foot height, x center of left foot, y center of left foot, x center of right foot, y center of right foot	12
Dynamic Gait Features (DGF)	Gait phase, Gait frequency, Knee rotation, Neck rotation, Hip rotation, Ankle rotation, Gait speed, Elbow rotation	8

2.2 Statistical Features Generated by Independent Gaussian Mixture Models

The feature generation layer of GC^{2}F contains a set of basic classifiers, which are intended to model the statistical characteristics of each feature. The classifiers are constructed on Gaussian Mixture Models (GMM). The main reason for choosing GMM is due to its demonstrated effectiveness, simplicity and flexility [27]. As a semi-parametric technique for data modeling and analysis, GMMs can seamlessly combine the advantages of both the KNN and quadratic classifiers into a single structure, and they have been used successfully in many classification and recognition tasks.

A basic classifier employs multiple GMMs, with each GMM independently built with training sets from one gait category. (The training examples are manually selected from the test collection.) In a GMM, the probability of class c is a random variable drawn from the probability distribution for a particular feature f. Given a parameter set Θ_f^s and input feature X_f, the probability of the input object belonging to gait class c, p_{cf}, is calculated using a mixture of multivariate component densities:

$$G_{cf} = p_{cf}(X_f|\Theta_f^s) = \sum_{j=1}^{J} w_{fj}^c \phi_{cf}(X_f|\,\boldsymbol{\mu}_{fj}^c, \boldsymbol{\Sigma}_{fj}^c) \quad (1)$$

where $X_f = \{x_{1f}, x_{2f}, ..., x_{df}\}$ is a input vector containing gait feature f extracted from the video sequence. The Gaussian density is used as the multivariate component in this study, according to GMM $\Theta_f^s = \{w_{fj}^c, \boldsymbol{\mu}_{fj}^c, \boldsymbol{\Sigma}_{fj}^c|, \ 1 < j < J\}$, where w_{fj}^c, $\boldsymbol{\mu}_{fj}^c$ and $\boldsymbol{\Sigma}_{fj}^c$ denote, respectively, the mixture weights, mean vectors and covariance matrices. In addition, $p_{cf}(X_f| \ \boldsymbol{\mu}_{fj}^c, \boldsymbol{\Sigma}_{fj}^c)$ is the probability of a class label c based on feature f extracted from the gait sequence (X_f is the related feature vector input), and is easily calculated using the Gaussian density function ϕ_{cf}:

$$\phi_{cf}(X_f| \ \boldsymbol{\mu}_{fj}^c, \boldsymbol{\Sigma}_{fj}^c) = \frac{1}{(2\pi)^{d/2}(\boldsymbol{\Sigma}_{fj}^c)^2} exp[-\tfrac{1}{2}(X_f - \boldsymbol{\mu}_{fj}^c)^t \boldsymbol{\Sigma}_{fj}^c (X_f - \boldsymbol{\mu}_{fj}^c)] \quad (2)$$

where $\{\boldsymbol{\mu}_{fj}^c, \boldsymbol{\Sigma}_{fj}^c\}$ are associated parameters. The output of each basic classifier is a probability histogram: $PH_f = \{p_{1f}, p_{2f}, ..., p_{cf}, ..., p_{Cf}\}$, where $\sum_1^C p_{cf} = 1$. The histogram models the probability distribution of the input object belonging to different gait classes using feature f.

To train the GMMs, we use an EM algorithm to estimate the model parameters [28]. (The EM is a data-driven optimization method for deriving unknown parameters.) At the same time, the K-means algorithm is used to obtain initial parameters for the mixture model. In essence, the process of estimating the parameter set is an iterative procedure that searches for an optimal parameter configuration Θ_f^s via a maximum likelihood estimation: $(\Theta_f^s)' = \underset{\Theta_f^s}{argmax} \ p_{cf}(\{X_f|\Theta_f^s\})$. The EM-based estimation process terminates when the MAP (maximum posterior) parameter set is found.

2.3 Hidden Classifier Layer of Artificial Neural Networks

The function of the hidden classifier layer is to derive likelihood values between the input object and the gait class labels. The choice of classifiers has important influences on performance. Our proposed system employs multilayer perceptron neural networks because of their ability to model complex mappings between input and output patterns. The second layer of GC^{2}F consists of C multilayer perceptron neural networks. Similar to the first layer, each of them is trained independently through supervised learning, and corresponds to one gait class.

Multilayer Perceptron-based Classifier. Each classifier is a three-layer neural network, consisting of an input layer, a hidden layer and an output layer. The number of neurons in each layer is configured prior to the training process. The interconnected neurons serve as elementary computational elements for information processing. The activation value s_i of a neuron is computed from inputs from its predecessor layer. The first step is to calculate the local field net_i of neuron i,

$$net_i = \sum_{n=1}^{N} x_n w_{ni} + w_{0i} \quad (3)$$

where N denotes the number of neurons in the predecessor layer, $W_i = \{w_{0i}, w_{1i}, ..., w_{Ni}\}$ is the weight vector for neuron i, and w_{0i} is the bias parameter. The combined input is passed through a non-linear function to produce the activation value s_i. We apply the popular sigmoid logistic function here:

$$s_i = \frac{1}{1 + exp[-net_i]} \tag{4}$$

After the neural network is trained, its output can be calculated with Equation 3.

Learning with Backprop.ECOC. The backprop.ECOC algorithm is developed for training our system. The algorithm combines two well known training schemes – back-propagation algorithm [29] and error-correcting output code (ECOC) [30, 31]. Assume that we have a neural network and a training set $R = (x_1, t_1), ..., (x_p, t_p)$ consisting of p pairs of input and target patterns. After training with learning pattern R, the corresponding output pattern o_i and target pattern t_i should be identical ideally. However, there might exist difference in real case and this value can be used to estimate the distance pattern between label and input distance histograms. We apply the backpropagation learning algorithm with gradient descent to search for the minima of the error (objective) function in the space of weights. The relative error function E is defined as:

$$E = \frac{1}{2}\sum_{i=1}^{I}(o_i - t_i)^2 \tag{5}$$

where p denotes the number of learning examples in the training set R, and i iterates over the training samples to aggregate the difference between the calculated output o_i and the predefined target output pattern t_i. The following partial derivatives give the adjustments to the neuron parameters, in order to effect gradient descent:

$$\frac{\partial E}{\partial w_{ij}} = \frac{\partial E}{\partial s_i}\frac{\partial s_i}{\partial w_{ij}} \tag{6}$$

where

$$\frac{\partial s_i}{\partial w_{ij}} = \frac{\partial s_i}{\partial net_i}\frac{\partial net_i}{\partial w_{ij}} \tag{7}$$

We apply error correcting output codes (ECOC) to encode different classes and to further reduce training errors [23]. It consists of two steps: training and classification. In the first step, we define the coding matrix M where $M \in \{-1, 1\}^{C \times n}$, and n is the length of the code. Each row of M is the base codeword created for one class. The creation process of matrix M can be treated as a group of learning problems - separating classes, one per each column. The classification is carried out by comparing the base codeword for each class with the classifier outputs. Before training, each gait class is allocated one codeword t_i using M. The backprop.ECOC scheme is summarized in Algorithm 1.

Given the probability histograms for different gait features and the trained neural network $NN_c^{'}$, the likelihood value of gait class c is computed as $ld_c = \|o_i^{'} - t_c\|$, where $o_i^{'}$ is the output vector generated by $NN_c^{'}$. The set of likelihood values $ld = \{ld_1, ld_2, ..., ld_C\}$, produced by the hidden classifiers, are fed into the decision layer.

Algorithm 1. backprop.ECOC learning algorithm to train multilayer feedforward network

Input: t_c: ECOC codeword for gait class c.
NN_c: multilayer feedforward network.
Output: Trained multilayer feedforward network $NN_c^{'}$.
Description:
1: Initialize the weights of NN to small random values;
2: Select an input training example (x_c, t_c) and feed it into the network;
3: Output result o_i;
4: Compute the gradient of output units using
5: $\frac{\partial E}{\partial s_i} = \frac{1}{2}\frac{\partial((o_i - t_c)^2)}{\partial s_i} = o_i - t_i$;
6: and compute the gradient of other units with
7: $\frac{\partial E}{\partial s_i} = \sum_{l \in succ(i)} \frac{\partial E}{\partial s_l}\frac{\partial s_l}{\partial s_i}$;
8: Update the weights according to the following equation;
9: $\Delta w_{ij}(t) = \varepsilon \frac{\partial E}{\partial w_{ij}}(t)$;
10:
11: **if** $E > \Lambda$, where Λ is predefined **then**
12: Go back to step 2;
13: **else**
14: Output trained neural network $NN_c^{'}$;

2.4 Decision Layer and Gait Recognition with GC^{2}F

Taking the set of likelihood values $ld = \{ld_1, ld_2, ..., ld_C\}$ from the hidden classifiers, each node in the decision layer implements a classification function,

$$de_c = \phi_c(ld) \tag{8}$$

where ϕ_c is a discriminative function and de_c denotes the distance between the input object and gait class c. Similar to the hidden classifier layer, we select a sigmoid function to model the discriminative function ϕ_c. Thus, the output of decision node c is express as:

$$de_c = \frac{1}{1 + exp[\sum_{c=1}^{C} w_{ic} ld_c]} \tag{9}$$

where w_{ic} is the weight of the connection from node i in the hidden classifier layer to decision node c. w_{ic} can be estimated via the backpropagation learning algorithm. Once the whole system is trained, it can perform gait identification. Given an image sequence, the first step is to extract different gait features. As summarized in Table 1, we consider four different types of low-level human gait parameters corresponding to different parts of the human body, including head and neck, upper body, lower body, and dynamic gait. Following that, the basic classifiers in the first layer generate a probability histogram for each feature type, which is input to the hidden classifier layer. Using the probability histograms, the neural networks in the second layer compute likelihood values for the various gait classes. Based on their outputs, the decision nodes in the last layer derives class distances that quantify the closeness between the class labels and the input object. Finally, the object is assigned to the class with the minimum distance, using $c^* = \underset{1 \geq c \geq C}{argmin}\, de_c$.

3 Experiment Configuration

This section explains the experiment configuration for our performance evaluation, which covers test collection, evaluation metrics and parameter setting for different components of the GC^2F framework. The test machine is a Pentium(R) D, 3.20GHz PC running the Microsoft Windows XP operating system. We compare the performance of a range of gait signature generation methods for model-based approach, including our proposed GC^2F and state-of-the-art techniques like GUO and its variants [18], WANG [8], and BJ [22].

Our test collection contains 600 short indoor video sequences on 35 subjects, which we selected manually from the Southampton HiD Gait database[1]. All subjects walk either from the left to the right, or from the right to the left. Each video was filmed at a rate of 25 frames per second, at a resolution of 720 $\times$ 576 pixels. The competing systems are evaluated on their recognition rates at top n results (R@n). For example, R@5 means the percentage of subjects identified correctly in the top 5 results.

4 Experiment Results and Analysis

In this section, we present evaluation results on the identification effectiveness of our proposed system. We also study the effects of different kinds of gait feature combinations on the proposed system.

4.1 Effectiveness Study

The first set of experiments is a comparative study on the identification accuracy of the various gait recognition schemes. Table 2 shows experiment results obtained with the test collection described in Section 3. Among the competing schemes, BJ performs the worst; for example, its recognition rates at top 5 and 7 results are 86.17% and 86.52%. WANG gives better accuracy, outperforming BJ by a margin of 4% in some cases. Meanwhile, GUO and its variants achieve further improvements, with recognition rates that are very close to those reported in [18]. GC^2F is the overall winner, giving an average improvement of 1.75% over its closest competitor. While the improvements achieved are lower for some of the test cases, the performance gain is consistent across recognition rates measured at different number of top results.

4.2 Importance of Combining Gait Features

In this section, we investigate how different gait feature combinations influence the recognition rate. The methodology is as follows. First, we prepare 14 combinations of gait features, including HF+UBF+LBF+DGF, HF+UBF+LBF, UBF+LBF+DGF,

[1] http://www.gait.ecs.soton.ac.uk/database

Table 2. Identification Accuracy Comparison of Different Gait Recognition Methods. For GUO-MI, GUO-ANOVA and GUO-Cor methods, dimension of feature set is 45.

Rank	Gait Recognition Methods (%)					
	GC²F	GUO-MI	GUO-ANOVA	GUO-Cor	WANG	BJ
1	**97.45**	96.21	95.89	95.41	88.25	85.56
2	**97.87**	96.28	96.01	95.58	88.39	85.64
3	**97.81**	96.64	96.14	95.67	88.45	85.79
4	**98.15**	96.85	96.23	95.87	89.54	85.91
5	**98.61**	97.01	96.51	96.06	89.75	86.17
6	**98.72**	97.11	96.78	96.32	89.87	86.29
7	**98.83**	97.34	96.99	96.54	89.91	86.52
8	**98.97**	97.57	97.27	97.01	90.09	86.79
9	**99.14**	98.01	97.54	97.19	90.21	86.87

Table 3. Identification Accuracy Comparison of Different Gait Feature Combinations with GC²F

Feature Combination	Recognition Rate with Different Number of Top Results							
	R@1	R@2	R@3	R@4	R@5	R@6	R@7	R@8
HF+UBF+LBF+DGF	97.45	97.87	97.81	98.15	98.61	98.72	98.93	99.07
UBF+LBF+DGF	88.65	88.73	88.95	89.21	89.36	89.59	90.27	91.01
HF+UBF+LBF	86.15	86.72	86.81	87.15	87.53	87.96	88.95	89.12
HF+LBF+DGF	82.67	82.93	83.27	83.56	83.91	84.39	84.73	85.16
UBF+LBF	78.67	79.13	79.97	80.17	80.46	81.29	81.65	82.01
LBF+DGF	71.21	71.43	71.97	72.06	72.27	72.54	73.23	73.89
HF+LBF	71.51	71.55	71.83	72.01	72.29	72.72	73.08	73.21
HF+LBF	71.29	71.56	71.87	71.97	72.17	72.61	73.16	73.75
HF+DGF	69.17	69.49	69.45	69.95	70.91	71.43	72.85	73.18
UBF	68.12	68.62	69.09	70.02	70.86	71.17	71.49	73.07
LBF	67.45	68.09	69.31	69.57	70.29	70.56	71.53	72.07
DGF	44.05	44.76	46.35	47.80	48.91	49.13	50.54	49.79
HF	42.45	43.59	44.72	45.32	45.91	46.13	47.03	48.17

HF+LBF+DGF, UBF+LBF, HF+UBF, HF+DGF, LBF+DGF, HF+LBF, UBF+DGF, HF, UBF, LBF and DGF. The combinations are then used to build different versions of GC²F.

Table 3 summarizes recognition rates for the various feature combinations. The last four rows demonstrate that GC²F performs very poorly with just a single type of gait feature. In particular, HF and DGF achieve less than 50% accuracy. This proves that good identification rate cannot be achieved with just one feature type. Interestingly, UBF and LBF appear to be more effective than HF and DGF. We also observe a steady improvement in accuracy as more features are integrated. For example, UBF+LBF provides an additional 9.8% in accuracy over UBF or LBF alone, while adding HF gives another 5.3% gain. Among all the feature combinations, HF+UBF+LBF+DGF offers the best performance, delivering about 10% better accuracy over the closest competitor (UBF+LBF+DGF).

5 Conclusion

Due to its special characteristics, gait recognition is becoming more attractive recently as effective biometric techniques. In this article, we propose and evaluate a novel gait recognition system, called GC^2F. In contrast to previous solutions, GC^2F allows multiple kinds of features and high level discriminative information to be effectively combined within a unified learning framework. The system is developed on the "refine-and-classify" principle, and has a multilayer architecture. The first layer is made up of basic classifiers, constructed from the Gaussian Mixture Model (GMM). The outputs of the basic classifiers take the form of probability histograms that model the probability relationship between the input object and various gait classes. The second layer employs multilayer perceptron neural networks to model the statistical properties of the gait classes. To reduce error in the trained system, we develop an efficient learning algorithm called backprop.ECOC. We have carried out a detailed empirical study and analysis of GC^2F with a large test collection of human gait videos. The experiment results show that the system achieves superior identification accuracies over existing solutions.

References

1. Liu, Z., Sarkar, S.: Simplest representation yet for gait recognition: Averaged silhouette. In: Proceedings of International Conference on Pattern Recognition, pp. 211–214 (2004)
2. Liu, Z., Sarkar, S.: Improved gait recognition by gait dynamics normalization. IEEE Trans. Pattern Anal. Mach. Intell. 28(6), 863–876 (2006)
3. Zhou, X., Bhanu, B.: Integrating Face and Gait for Human Recognition at a Distance in Video. IEEE Transactions on Systems, Man, and Cybernetics, Part B 37(5), 1119–1137 (2007)
4. Huang, P.: Automatic gait recognition via statistical approaches for extended template features. IEEE Transactions on Systems, Man, and Cybernetics, Part B 31(5), 818–824 (2001)
5. Liu, Z., Sarkar, S.: Effect of silhouette quality on hard problems in gait recognition. IEEE Transactions on Systems, Man, and Cybernetics, Part B 35(2), 170–183 (2005)
6. Sarkar, S., Phillips, P.J., Liu, Z., Vega, I.R., Grother, P., Bowyer, K.W.: The humanid gait challenge problem: Data sets, performance, and analysis. IEEE Trans. Pattern Anal. Mach. Intell. 27(2), 162–177 (2005)
7. Wang, L., Tan, T., Ning, H., Hu, W.: Silhouette Analysis-Based Gait Recognition for Human Identification. IEEE Trans. Pattern Anal. Mach. Intell. 25(12), 1505–1518 (2003)
8. Wang, L., Ning, H., Tan, T., Hu, W.: Fusion of static and dynamic body biometrics for gait recognition. IEEE Trans. Circuits Syst. Video Techn. 14(2), 149–158 (2004)
9. de Chazal, P., Flynn, J., Reilly, R.B.: Automated processing of shoeprint images based on the fourier transform for use in forensic science. IEEE Trans. Pattern Anal. Mach. Intell. 27(3), 341–350 (2005)
10. Lee, L., Grimson, W.E.L.: Gait analysis for recognition and classification. In: Proceedings of IEEE Int. Conf. Automatic Face and Gesture Recognition, pp. 148–155 (2002)
11. Daugman, J.: How iris recognition works. IEEE Transactions on Circuits and Systems for Video Technology 14(1), 21–30 (2004)
12. Ross, A., Dass, S.C., Jain, A.K.: Fingerprint warping using ridge curve correspondences. IEEE Trans. Pattern Anal. Mach. Intell. 28(1), 19–30 (2006)

13. Jain, A.K., Feng, J.: Latent palmprint matching. IEEE Transactions on Pattern Analysis and Machine Intelligence 99(1) (2008)
14. BenAbdelkader, C., Cutler, R., Nanda, H., Davis, L.: Eigengait: Motion-based recognition of people using image self-similarity. In: Bigun, J., Smeraldi, F. (eds.) AVBPA 2001. LNCS, vol. 2091, pp. 284–294. Springer, Heidelberg (2001)
15. Cutting, J., Kozlowski, L.: Recognizing friends by their walk: Gait perception without familiarity cues. Bulletin Psychonomic Soc. 9(5), 353–356 (1977)
16. Murray, M., Drought, A., Kory, R.: Walking pattern of normal men. J. Bone and Joint Surgery 46-A(2), 335–360 (1964)
17. Wagg, D.K., Nixon, M.S.: On automated model-based extraction and analysis of gait. In: Proc. of the Sixth IEEE International Conference on Automatic Face and Gesture Recognition (FGR 2004), pp. 11–16 (2004)
18. Guo, B., Nixon, M.S.: Gait feature subset selection by mutual information. IEEE Transactions on Systems, Man, and Cybernetics, Part A 39(1), 36–46 (2009)
19. Li, X., Maybank, S.J., Yan, S., Tao, D., Xu, D.: Gait components and their application to gender recognition. IEEE Transactions on Systems, Man, and Cybernetics, Part C 38(2), 145–155 (2008)
20. Niyogi, S., Adelson, E.: Analyzing and recognizing walking figures in xyt. In: Proc. of IEEE International Conference on Computer Vision and Pattern Recognition (CVPR), pp. 469–474 (1994)
21. Bregler, C.: Learning and recognizing human dynamics in video sequences. In: Proc. of IEEE International Conference on Computer Vision and Pattern Recognition (CVPR), pp. 568–574 (1997)
22. Bobick, A., Jonhson, A.: Gait recognition using static activity-specific analysis. In: Proc. of IEEE International Conference on Computer Vision and Pattern Recognition, CVPR (2001)
23. Dietterich, T., Bakiri, G.: Solving Multiclass Learning Problems via Error-Correcting Output Codes. Journal of Artificial Intelligence Research 2 (1995)
24. Cunado, D., Nixon, M., Carter, J.: On the learnability and design of output codes for multiclass problems. Comput. Vis. Image Understand. 90(1) (2003)
25. Müller, B., Reinhardt, J., Strickland, M.T.: Neural Networks: An Introduction (Physics of Neural Networks). Springer, Heidelberg (1995)
26. MacKay, D.: Information Theory, Inference, and Learning Algorithms. Cambridge University Press, Cambridge (2003)
27. McLachLan, G., Peel, D. (eds.): Finite Mixture Models. Wiley Interscience, Hoboken (2000)
28. Dempster, A., Laird, N., Rubin, D.: Likelihood from incomplete data via the em algorithm. Journal of the Royal Statistical Society, Series B 39(1), 1–38 (1977)
29. Haykin, S.: Neural Networks: A Comprehensive Foundation. Macmillan Publishing, New York (1994)
30. Kong, E., Dietterich, T.: Error-correcting output coding corrects bias and variance. In: Proc. of the 12th IEEE International Conference on Machine Learning, pp. 313–321 (1995)
31. Crammer, K., Singer, Y.: On the learnability and design of output codes for multiclass problems. Machine Learning 47(2-3) (2002)

Semantic Concept Detection for User-Generated Video Content Using a Refined Image Folksonomy

Hyun-seok Min, Sihyoung Lee, Wesley De Neve, and Yong Man Ro

Image and Video Systems Lab, Korea Advanced Institute of Science and Technology (KAIST), Yuseong-gu, Daejeon, 305-732, Republic of Korea
{hsmin,ijiat,wesley.deneve}@kaist.ac.kr, ymro@ee.kaist.ac.kr

Abstract. The automatic detection of semantic concepts is a key technology for enabling efficient and effective video content management. Conventional techniques for semantic concept detection in video content still suffer from several interrelated issues: the semantic gap, the imbalanced data set problem, and a limited concept vocabulary size. In this paper, we propose to perform semantic concept detection for user-created video content using an image folksonomy in order to overcome the aforementioned problems. First, an image folksonomy contains a vast amount of user-contributed images. Second, a significant portion of these images has been manually annotated by users using a wide variety of tags. However, user-supplied annotations in an image folksonomy are often characterized by a high level of noise. Therefore, we also discuss a method that allows reducing the number of noisy tags in an image folksonomy. This tag refinement method makes use of tag co-occurrence statistics. To verify the effectiveness of the proposed video content annotation system, experiments were performed with user-created image and video content available on a number of social media applications. For the datasets used, video annotation with tag refinement has an average recall rate of 84% and an average precision of 75%, while video annotation without tag refinement shows an average recall rate of 78% and an average precision of 62%.

Keywords: Folksonomy, semantic concept detection, tag refinement, UCC.

1 Introduction

Nowadays, end-users can be considered both consumers and producers of multimedia content. This is for instance reflected by the significant growth in the amount of user-generated image and video content available on social media applications. An example of a highly popular website for the consumption of user-created video content is 'YouTube' [1], an online video sharing service that has nine million visitors every day, with 50,000 videos being uploaded on a daily basis [2]. Online user-created video content is currently estimated to have a size of approximately 500,000 terabytes, while it is predicted that the amount of user-created video content will continue to increase to 48 million terabytes by the end of 2011 [3]. Similar observations can be made for Flickr, a highly popular website for image sharing.

S. Boll et al. (Eds.): MMM 2010, LNCS 5916, pp. 511–521, 2010.

The availability of vast amounts of user-created image and video content requires the use of efficient and effective techniques for indexing and retrieval. However, users typically do not describe image and video content in detail [4]. This makes it difficult to bring structure in the image and video collections of users. Therefore, the automatic detection and annotation of semantic concepts can be seen as a key technology for enabling efficient and effective video content management [5-7]. Although tremendous research efforts have already been dedicated to advancing the field of automatic semantic concept detection, robust methods are not available yet.

Semantic concept detection still suffers from several major problems: the semantic gap, the imbalanced data set problem, and the use of a constrained concept vocabulary [8] [9]. Traditional approaches typically use a training database and classifiers in order to detect a limited number of semantic concepts [8-12]. These approaches make use of low-level features that are extracted from the video content and that are subsequently mapped to semantic concepts. However, people typically perceive a gap between the meaning of low-level features and the meaning of semantic concepts. As such, it is hard to model a high number of semantic concepts using a training database and classifiers [13]. The imbalanced data set problem refers to concepts that may occur infrequently. Consequently, the detection performance for rarely occurring concepts may be low due to the unavailability of a high number of training samples.

Current social media applications such as Flickr and YouTube provide users with tools to manually tag image and video content using their own vocabulary. The result of personal free tagging of image and video content for the purpose of retrieval and organization is called a folksonomy [14] [15]. The term ‘folksonomy’ is a blend of the words ‘taxonomy’ and ‘folk’, essentially referring to sets of user-created metadata.

A folksonomy typically contains a high number of images that have been manually annotated by users with a wide variety of tags. Therefore, a folksonomy may provide enough training data to learn any concept. However, tags in an image folksonomy are frequently characterized by a significant amount of noise. The presence of tag noise can be attributed to the fact that users may describe images from different perspectives (imagination, knowledge, experience) and to the fact that manual tagging is a time-consuming and cumbersome task. For example, batch tagging may cause users to annotate images with concepts not present in the image content.

This paper discusses semantic concept detection for user-created video content using an image folksonomy. Relying on the collective knowledge available in an image folksonomy is a promising approach to overcome the interrelated issues that still taunt conventional semantic concept detection. In addition, we discuss a method that allows reducing the number of noisy tags in an image folksonomy. This tag refinement method essentially uses tag co-occurrence statistics. Preliminary results show that our method is able to reduce the level of noise in an image folksonomy. Further, we demonstrate that the use of an image folksonomy allows for the effective detection of an unlimited number of concepts in user-generated video content.

The remainder of this paper is organized as follows. Section 2 presents an overall overview of the proposed system for semantic concept detection, while Section 3 describes how to refine tags in an image folksonomy. In addition, a method for computing the similarity between a video shot and images in a refined folksonomy is outlined in Section 4. Experimental results are subsequently provided in Section 5. Finally, conclusions are drawn in Section 6.

2 System Overview

Fig. 1 shows the proposed system for annotating user-created video content. Our method mainly consists of two modules: a module responsible for image folksonomy refinement and a module responsible for semantic concept detection. Given a target concept *w*, the folksonomy refinement module collects images that are representative for concept *w*. The functioning of this module will be explained in more detail in Section 3. To determine whether a concept *w* is present in a video segment, we measure the similarity between the video segment and the folksonomy images that are representative for concept *w*.

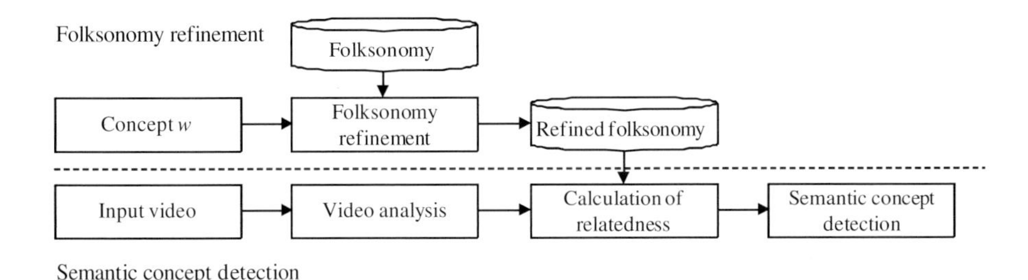

Fig. 1. Proposed system for annotating user-created video content by means of a folksonomy

3 Image Folksonomy Refinement

3.1 Definition of Correct and Incorrect Tags

It is well-known that manual tagging of image content is a time consuming and cumbersome task. The use of batch tagging may reduce the effort required by manual image tagging. However, this often results in image content annotated with incorrect tags [4]. Moreover, user-driven image tagging also introduces a personal perspective.

Image		
User-defined tags	village, valley, grass, horses, flowers, hills, water, lake, pond, waterfalls	church, valley, temple, tower, light, street, night, palm, tree, star

Fig. 2. Example images with user-defined tags (retrieved from Flickr). Underlined tags represent tags that are regarded as correct by human visual perception.

In our research, we assume that users assign two types of tags to image content: correct and incorrect tags. Correct tags are tags that are representative for the visual semantics of the image content, while so-called incorrect or noisy tags are not visually related to the image content. Fig. 2 shows a number of example images retrieved from Flickr, annotated with user-provided tags. Correct tags are underlined, differentiating these tags from the incorrect tags.

3.2 Measuring Confidence between a Concept and Folksonomy Images

Techniques for reducing the number of noisy tags in an image folksonomy are important in order to allow for precise search and tag-based data mining techniques [17]. To mitigate the amount of noise in an image folksonomy, we explore the relationship between different tags. In particular, the co-occurrence of several tags is taken into account. For example, the presence of the tag 'valley' typically implies that the presence of the tag 'hills' is also relevant (as shown in Fig. 2).

In general, user-generated images are created in regular situations. In other words, user-generated images rarely contain artificially organized scenes. For example, an image depicting the concepts 'sphinx' and 'water' is rare compared to an image containing the concepts 'beach' and 'water'. Consequently, we assume that the semantic concepts depicted in an image are correlated. As such, if noisy tags are present in the set of tags for a particular image, it ought to be possible to differentiate the noisy tags from the correct tags by analyzing the tag co-occurrence statistics.

We define the notion of concept confidence to represent the degree of cohesiveness between an image I and an associated semantic concept w. In this work, the confidence value is obtained by making use of the tags assigned to images. The concept confidence value is normalized, thus varying between 0 and 1. If an image I is not related to the concept w, then its confidence value will be close to 0. Likewise, if a confidence value is close to 1, then we assume that image I is highly related to concept w. The confidence value for an image I and a concept w is measured as follows:

$$Confidence(w, \mathbf{T}) = \frac{\sum_{t \in \mathbf{T}} relation(w,t)}{|\mathbf{T}| - 1}, \tag{1}$$

where $\mathbf{T}$ represents the set of user-supplied tags for image I, where $|\mathbf{T}|$ represents the number of tags in $\mathbf{T}$, and where *relation()* denotes a function that maps tags on relation information. In this paper, we measure relation information between tags using tag co-occurrence statistics. In particular, the proposed method measures how often a particular tag co-occurs with another tag. This can be expressed as follows:

$$relation(w,t) := \frac{\left|\mathbf{I}^{\{w \cap t\}}\right|}{\left|\mathbf{I}^{\{t\}}\right|}, \tag{2}$$

where $\mathbf{I}^{\{w \cap t\}}$ denotes the set of images annotated with both concept w and tag t, where $\mathbf{I}^{\{t\}}$ represents the set of images annotated with tag t, and where $|\cdot|$ counts the number of elements in a set.

3.3 Refinement of Folksonomy Images

Let I be an image in the folksonomy $\mathbf{F}$ and let $\mathbf{T}$ be the set of user-defined tags associated with image I. Given the target concept w, images need to be found that are related to concept w. To find related images, the proposed refinement method uses the concept confidence computed using Eq. (1). This process can be described using the following equation:

$$\mathbf{F}_{w,refined} = \{ I \mid Confidence\,(w, \mathbf{T}) \geq threshold \}, \tag{3}$$

where w represents a concept to be detected, where $\mathbf{T}$ denotes the set of tags for image I, where *confidence()* is a function that measures concept confidence (as defined in Eq. (1)), and where *threshold* is a value that determines whether an image is related to the given target concept w or not.

4 Semantic Concept Detection

In our research, concept detection is based on measuring the similarity between a video shot and refined folksonomy images. In this section, we first describe how to extract visual features from a video shot. Next, we explain how to measure visual similarity between a video shot and a single folksonomy image. We then proceed with a discussion of how to measure the relatedness between a video shot and a set of folksonomy images all related to the same concept, enabling semantic concept detection.

4.1 Shot-Based Extraction of Visual Features

A shot in a video sequence is composed of visually similar frames. In addition, a shot is typically used as the basic unit of video content retrieval. Therefore, semantic concept detection is carried out at the level of a shot. In order to extract visual features, a video sequence $\mathbf{S}$ is first segmented into N shots such that $\mathbf{S} = \{s_1, s_2, \ldots, s_N\}$, where s_i stands for the ith shot of video sequence $\mathbf{S}$ [18]. Next, low-level visual features such as color and texture information are extracted from several representative key frames. The extracted low-level visual features are described using the following MPEG-7 color and texture descriptors [20][21]: the Color Structure Descriptor (CSD), the Color Layout Descriptor (CLD), the Scalable Color Descriptor (SCD), the Homogeneous Texture Descriptor (HTD), and the Edge Histogram Descriptor (EHD). Besides color and texture information, we also extract spatial information from the key frames. This is done using the rectangle division technique described in [19]. Finally, we denote the set of visual features for shot s_i as $\mathbf{X}_i = \{x_{i,1}, x_{i,2}, \ldots, x_{i,L}\}$, where $x_{i,l}$ represents the lth low-level feature extracted from shot s_i and where L is the total number of low-level features.

4.2 Similarity Measurement between a Video Shot and a Folksonomy Image

The proposed method uses a refined image folksonomy for the purpose of detecting semantic concepts in user-created video content. Specifically, the presence of a concept w in a particular video shot is estimated by making use of the visual similarity between the video shot and a set of folksonomy images that contain concept w.

Let I be an image in the refined folksonomy $\mathbf{F}_{w,refined}$. Similar to a shot in a video sequence, a set of low-level visual features is extracted for image I. This set of low-level features is represented by $\mathbf{X}_f$. The visual similarity between a video shot and the fth image in the image folksonomy can be measured as follows:

$$Sim(\mathbf{X}_i, \mathbf{X}_f) = 1 - \frac{1}{L} \cdot \sum_{l=1}^{L} \sqrt{(x_{i,l} - x_{f,l})^2}, \tag{4}$$

where $x_{i,l}$ denotes the lth low-level visual feature of shot s_i and where $x_{f,l}$ denotes the lth low-level visual feature of the fth image in the refined image folksonomy.

4.3 Semantic Concept Classification

Given a target concept w, a refined image folksonomy can be denoted as a finite set of images $\mathbf{F}_{w,refined} = \{I_1, I_2, \ldots, I_F\}$. The refined image folksonomy only contains images related to the concept w. To estimate the presence of concept w in a particular video shot, we measure the visual similarity between the video shot and all of the refined folksonomy images. The visual similarity or relatedness between a video shot s_i and the complete set of folksonomy images related to concept w is measured as follows:

$$Relatedness(\mathbf{X}_i, \mathbf{F}_{w,refined}) = \frac{1}{F} \cdot \sum_{f=1}^{F} Sim(\mathbf{X}_i, \mathbf{X}_f), \tag{5}$$

where $\mathbf{X}_i$ denotes the visual feature set of video shot s_i, where $\mathbf{X}_f$ denotes the low-level visual feature set of folksonomy image I_f, and where *Sim()* is defined in Eq. (4). If the relatedness value is higher than a pre-determined threshold, then the shot contains concept w.

5 Experiments

In this section, we first describe our experimental setup, including the construction of the image and video datasets used. We then present our experimental results.

5.1 Experimental Setup

A number of experiments were performed in order to verify the effectiveness of the proposed method for semantic concept detection in user-created video content. Our image folksonomy is constructed using the MIRFLICKR-25000 image collection [22]. This dataset consists of 25,000 images downloaded from 'Flickr' using its public API. Each image in the data set is annotated with tags provided by anonymous users. The average number of tags per image is 8.94. Also, the data set contains 1386 tags that have been assigned to at least 20 images.

The video annotation performance was tested for 6 target concepts: 'street', 'tree', 'architecture', 'water', 'terrain', and 'sky'. Further, 70 different user-generated video sequences were retrieved from 'YouTube', resulting in a set of 1,015 video shots that need to be annotated. In order to evaluate the performance of the proposed semantic concept detection technique, the ground truth for all video shots was created in a manual way. In particular, three participants independently selected ground truth concepts by relying on their visual perception. Fig. 3 shows a number of extracted key frames and the corresponding ground truth concepts.

Key frames		
Ground truth	architecture, tree, sky	terrain, sky, architecture
Key frames		
Ground truth	street, water, sky	water, sky

Fig. 3. Extracted key frames and corresponding ground truth concepts

5.2 Experimental Results

We compare the accuracy of our semantic concept detection method with a technique that does not make use of a refined image folksonomy. The performance of the different concept detection methods is measured using the traditional 'recall' and 'precision' metrics. The corresponding definitions can be found below:

$$recall = \frac{N_{TP}}{N_{True}} \tag{6}$$

$$precision = \frac{N_{TP}}{N_{TP} + N_{FP}} \tag{7}$$

In the equations above, N_{TP} denotes the number of true positives, N_{FP} represents the number of false positives, and N_{True} is the number of positive samples (i.e., the total number of samples in the ground truth annotated with a particular target concept).

Given a target concept *w*, the annotation technique that does not make use of refinement uses all images in the image folksonomy that have been tagged with concept *w*. Consequently, the annotation performance is affected by folksonomy images with visual semantics that do not contain concept *w*. Fig. 4 summarizes our experimental results. Specifically, video annotation with refinement has an average recall rate of 84% and an average precision of 75%, while video annotation without refinement has an average recall rate of 78% and an average precision of 62%.

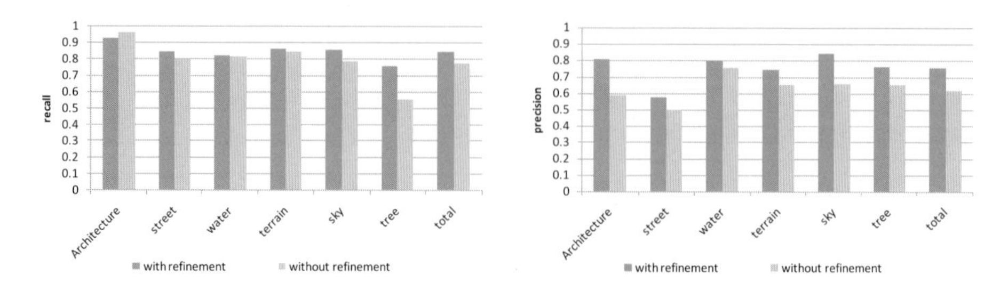

Fig. 4. Performance of semantic concept detection

Fig. 5 shows two example images that have been filtered by the proposed method. Although the user-defined tags contain the target concepts (street and tree), the content of the images does not contain these target concepts. Due to the presence of images with incorrect tags, the performance of the folksonomy-based annotation method degrades when tag refinement is not used. However, the proposed method is able to detect and remove folksonomy images with incorrect tags by making use of concept confidence values.

Concept	street	tree
Folksonomy	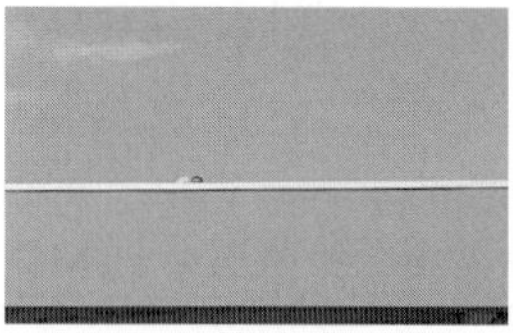	
User-defined tags	buh, brescia, colori, fdsancorastorta, ben, muibien, street	amanda, tattoo, cherries, harrypotter, tree, shamrock, girls, explored
Concept confidence value	0.1713	0.1301

Fig. 5. Example images (street and tree) filtered by the proposed method

Refined folksonomy		
User-defined tags	hollywood, florida, beach, resort, roadtrip, sea, clouds	grand canyon, sunset, sun, clouds, hill, landscape, blue, peaceful
Concept confidence value	0.4339	0.4756

Fig. 6. Example images that are recommended for the sky by the proposed method

Further, even if the user-defined tags do not contain the target concept, the actual content of the images may still include a particular target concept. The proposed folksonomy refinement process is able to find such images by exploring the tag relations. Fig. 6 shows two example images that are recommended for the concept 'sky'. While the corresponding tag sets of the two images do not include the concept 'sky', tags that are highly related to the concept 'sky' were used to annotate the images. Highly related tags include 'clouds' and 'beach' for the image to the left, and 'blue' and 'sunset' for the image to the right.

However, while the video annotation method with refinement mitigates the impact of noise in an image folksonomy, we have observed that the performance of our method is low for concepts such as 'tree' and 'street'. Indeed, although the images in the refined image folksonomy are all related to the target concepts 'tree' and 'street', the diversity of the respective image sets is high in terms of visual similarity and tags used. This observation is for instance illustrated in Fig. 7 for the concept 'tree'.

Refined folksonomy		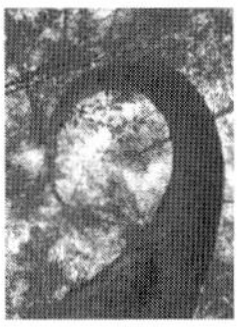	
User-defined tags	tree, cliff, windswept, wales	tree, pine, spiral	Snow, albero, foglie, tree
Concept confidence value	0.5108	0.5000	0.5410

Fig. 7. Example images and corresponding confidence values for the concept 'tree'

6 Conclusions and Future Work

This paper discussed a new method for semantic concept detection in user-created video content, making use of a refined image folksonomy. Tag refinement mitigates the impact of tag noise on the annotation performance thanks to the use of tag co-occurrence statistics. That way, the proposed annotation method can make use of folksonomy images that are better related to the concept to be detected. Our experimental results show that the proposed method is able to successfully detect various semantic concepts in user-created video content using folksonomy images.

Despite the use of tag refinement for the purpose of tag noise reduction, the refined image folksonomy may still contain a diverse set of images that are all related to the same target concept. Future research will conduct more extensive experiments in order to study the aforementioned observation in more detail. In addition, future research will focus on improving the concept detection accuracy by also making use of a video folksonomy.

References

1. YouTube, http://www.youtube.com/
2. 7 things you should know about YouTube (2006), http://www.educause.edu/ELI/7ThingsYouShouldKnowAboutYouTu/156821
3. Ireland, G., Ward, L.: Transcoding Internet and Mobile Video: Solutions for the Long Tail. In: IDC (2007)
4. Ames, M., Naaman, M.: Why We Tag: Motivations for Annotation in Mobile and Online Media. In: ACM CHI 2007, pp. 971–980 (2007)
5. Wang, M., Hua, X.-S., Hong, R., Tang, J., Qi, G.-J., Song, Y.: Unified Video Annotation via Multi-Graph Learning. IEEE Trans. on Circuits and Systems for Video Technology 19(5) (2009)
6. Wang, M., Xian-Sheng, H., Tang, J., Richang, H.: Beyond Distance Measurement: Constructing Neighborhood Similarity for Video Annotation. IEEE Trans. on Multimedia 11(3) (2009)
7. Yang, J., Hauptmann, A., Yan, R.: Cross-Domain Video Concept Detection Using Adaptive SVMs. In: Proceedings of ACM Multimedia, pp. 188–197 (2007)
8. Chen, M., Chen, S., Shyu, M., Wickramaratna, K.: Semantic event detection via multimodal data mining. IEEE Signal Processing Magazine, Special Issue on Semantic Retrieval of Multimedia 23(2), 38–46 (2006)
9. Xie, Z., Shyu, M., Chen, S.: Video Event Detection with Combined Distance-based and Rule-based Data Mining Techniques. In: IEEE International Conference on Multimedia & Expo. 2007, pp. 2026–2029 (2007)
10. Jin, S.H., Ro, Y.M.: Video Event Filtering in Consumer Domain. IEEE Trans. on Broadcasting 53(4), 755–762 (2007)
11. Bae, T.M., Kim, C.S., Jin, S.H., Kim, K.H., Ro, Y.M.: Semantic event detection in structured video using hybrid HMM/SVM. In: Leow, W.-K., Lew, M., Chua, T.-S., Ma, W.-Y., Chaisorn, L., Bakker, E.M. (eds.) CIVR 2005. LNCS, vol. 3568, pp. 113–122. Springer, Heidelberg (2005)
12. Wang, F., Jiang, Y., Ngo, C.: Video Event Detection Using Motion Relativity and Visual Relatedness. In: Proceedings of the 16th ACM International Conference on Multimedia, pp. 239–248 (2008)
13. Jain, M., Vempati, S., Pulla, C., Jawahar, C.V.: Example Based Video Filters. In: ACM International Conference on Image and Video Retrieval (2009)
14. Ramakrishnan, R., Tomkins, A.: Toward a People Web. IEEE Computer 40(8), 63–72 (2007)
15. Al-Khalifa, H.S., Davis, H.C.: Measuring the Semantic Value of Folksonomies. Innovations in Information Technology, 1–5 (2006)
16. Lu, Y., Tian, Q., Zhang, L., Ma, W.: What Are the High-Level Concepts with Small Semantic Gaps? In: Proceedings of IEEE Conference on Computer Vision and Pattern Recognition, CVPR (2008)
17. Xirong, L., Snoek, C.G.M., Worring, M.: Learning Tag Relevance by Neighbor Voting for Social Image Retrieval. In: Proceeding of the 1st ACM International Conference on Multimedia Information Retrieval, pp. 180–187 (2007)

18. Min, H., Jin, S.H., Lee, Y.B., Ro, Y.M.: Contents Authoring System for Efficient Consumption on Portable Multimedia Device. In: Proceedings of SPIE Electron. Imag. Internet Imag. (2008)
19. Yang, S., Kim, S.K., Ro, Y.M.: Semantic Home Photo Categorization. IEEE Trans. on Circuits and Systems for Video Technology 17(3), 324–335 (2007)
20. Ro, Y.M., Kang, H.K.: Hierarchical rotational invariant similarity measurement for MPEG-7 homogeneous texture descriptor. Electron. Lett. 36(15), 1268–1270 (2000)
21. Manjunath, B.S., et al.: Introduction to MPEG-7. Wiley, New York (2002)
22. Huiskes, M.J., Lew, M.S.: The MIR Flickr Retrieval Evaluation. In: ACM International Conference on Multimedia Information Retrieval (MIR 2008), Vancouver, Canada (2008)

Semantic Entity-Relationship Model for Large-Scale Multimedia News Exploration and Recommendation*

Hangzai Luo[1], Peng Cai[1], Wei Gong[1], and Jianping Fan[2]

[1] Shanghai Key Lab of Trustworthy Computing, East China Normal University
[2] Department of Computer Science, University of North Carolina at Charlotte

Abstract. Even though current news websites use large amount of multimedia materials including image, video and audio, the multimedia materials are used as supplementary to the traditional text-based framework. As users always prefer multimedia, the traditional text-based news exploration interface receives more and more criticisms from both journalists and general audiences. To resolve this problem, we propose a novel framework for multimedia news exploration and analysis. The proposed framework adopts our semantic entity-relationship model to model the multimedia semantics. The proposed semantic entity-relationship model has three nice properties. First, it is able to model multimedia semantics with visual, audio and text properties in a uniform framework. Second, it can be extracted via existing semantic analysis and machine learning algorithms. Third, it is easy to implement sophisticated information mining and visualization algorithms based on the model. Based on this model, we implemented a novel multimedia news exploration and analysis system by integrating visual analytics and information mining techniques. Our system not only provides higher efficiency on news exploration and retrieval but also reveals extra interesting information that is not available on traditional news exploration systems.

1 Introduction

News reports are the major information source for most people. As a result, news reports have significant impact on routine life of the masses. From watching and analyzing news, the users not only have fun but also learn new technologies, make investment decisions, understand the trend of the people's thought and even international affairs. Consequently, there is an urgent demand for deep analysis of large-scale multimedia news collections. However, news websites still adopt a framework with hierarchical browsing, headline recommendation and keyword-based retrieval. This framework may bring more hits to the websites, but can hardly help users to deeply analyze and explore the large collection of multimedia news reports. Criticisms have been given by journalists and general audiences [1] for a long time. Therefore, it is essential to provide effective and efficient exploration and retrieval on large-scale news collections, especially for multimedia news reports carrying text, image and video.

Several systems have been put online to address the problem [2,3,4]. Several approaches are proposed to analyze special types of information in news reports [5,6].

* This work is supported by Shanghai Pujiang Program under 08PJ1404600 and NSF-China under 60803077.

S. Boll et al. (Eds.): MMM 2010, LNCS 5916, pp. 522–532, 2010.

However, these systems and approaches still cannot satisfy the news analysis and exploration requirements due to several problems. First, they can only extract and present very limited information of large-scale news collections to the users and generally lack exploration tools. These approaches summarize the large-scale news collection into a very compact summarization and present only the summarization to the users. Even though the summarization is a good general overview of the whole news collections, it can carry few details. Second, they can only summarize and present one type of medium. Most of them work on textual information. Only 10×10 [2] displays the summarization via news pictures. As more and more news reports adopt multimedia, summarization on single medium can hardly present enough attractive information effectively and efficiently.

Based on the above observations, a novel multimedia news summarization and exploration approach is needed to satisfy the user's demand on news exploration and analysis applications. The new approach must summarize the large-scale news reports in multimedia format, and provide powerful exploration and retrieval tools on multimedia data to help users explore and analyze the large volume of information. However, to develop such a system, several challenging problems must be resolved.

The first problem is how to **extract and model the semantics of the large-scale multimedia data mathematically**. Any exploration, retrieval or summarization algorithm needs to process, analyze and transform the semantics of the raw news collection. But the raw news reports may be in different forms and media. Therefore, there is an urgent need for a computable semantic model that is able to uniformly model different types and media of news reports. Currently, multimedia documents are generally modeled as raw visual features [7] and semantic concepts [8], and text documents are modeled as document vectors. These models are suitable for document level semantic modeling, but may not be suitable for other granularity. However, event and entity level modeling is needed for news exploration and analysis applications as the same event or entity may be reported by many different news reports.

The second problem is how to **evaluate and extract interesting information from the semantic model**. Even though the semantic model carries most information of the large-scale news collection, most of it is uninteresting and unattractive. Only very limited information can catch the eye of the users. To implement efficient and effective news exploration and analysis applications, the system must present only the interesting and attractive information to the users and filter out others. However, it is a very difficult task because the "interestingness" property of a news report is always subjective. There are algorithms that can evaluate document level interestingness, such as tf-idf[9], Pagerank[10] and HITS [11]. But the news exploration and analysis applications prefer event and entity level computation as discussed above. How to evaluate the event and entity interestingness is still a challenging problem.

The third problem is how to **organize and display the information to enable most efficient human-computer information exchange**. The total amount of the news information reported everyday is very large. As a result, users may need to spend a long time to find an interesting news report. However, the user's patience is limited. A user may submit at most 3-5 queries for a task. Therefore, if the system cannot present interesting reports to the users in these 3-5 queries, the user may no longer use the system. Consequently, it's very important to enable the users to find the news report of interest as fast as possible. To resolve this problem, the system must provide highly efficient human-computer information exchange approaches, so that the user can receive as much information as possible for each query. As the vision is the most efficient perception for people to accept and process information, different visualization approaches are proposed to improve human-computer information exchange efficiency [12,6]. But these techniques are designed for special summarization information. To visualize the large-scale multimedia semantics efficiently, new algorithms must be developed.

The above problems are inter-related and cannot be resolved independently. The central problem is the semantic model. It is the basis of interestingness weighting and visualization algorithms. Thus its design must consider the requirements of both the semantic extraction algorithms and the interestingness weighting and visualization algorithms. To resolve this problem, we propose the semantic entity-relationship model to model the multimedia semantics of large-scale news collections. The proposed model is able to model the multimedia semantics in a uniform framework and is flexible enough to implement sophisticated information mining and visualization algorithms. By integrating a novel interestingness weighting algorithm and hyperbolic visualization technique with this model, we have implemented a visual recommendation system for multimedia news exploration and analysis. The system not only provides more efficient news exploration and analysis services but also disclose interesting information that is not available on traditional news exploration systems. In this paper we will introduce the proposed models and algorithms.

2 Semantic Entity-Relationship Model

As discussed above, a proper semantic model must be proposed to model the multimedia semantics of the large-scale news collection. Because a news event or entity may be reported in different news with different formats, the model must be able to model the semantics at event or entity level. In addition, the relationship among events and entities are important news semantics. Each news report can be summarized as several semantic entities and their relationships. Based on this observation, we propose the semantic entity-relationship model to model the news semantics.

The semantic entity is defined as any entity that has stable and real semantics during a period of time. The semantic entity is the extension of traditional named entity. It can be a person, organization, number, event, product or mathematical equation. For example, "United States", "General Motor", "2009/4/17", "Purchase Manager's Index", "Boeing 747" and "subprime mortgage crisis" are all semantic entities. Further more, each semantic entity can have different names and visual, textual or audio properties, as shown in Figure 1. One can find that the primary semantics of a news report can be carried by its semantic entities. In addition, the semantic entities may carry not only the text semantics but also the multimedia semantics from relevant images and video shots. Therefore, the semantic entity is suitable for news semantic modeling.

Aside from the semantic entities, the primary information in news reports are the relationships among semantic entities. Consequently, news semantic model must carry such information. Therefore, the news semantics can be modeled as a graph with semantic entities as nodes and relationships as edges. The relationship among semantic

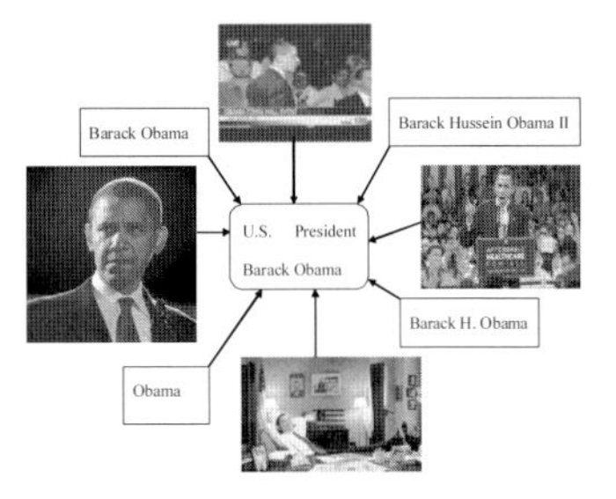

Fig. 1. The semantic entity "U.S. President Barack Obama" is represented as different text strings, images and video shots in different news reports

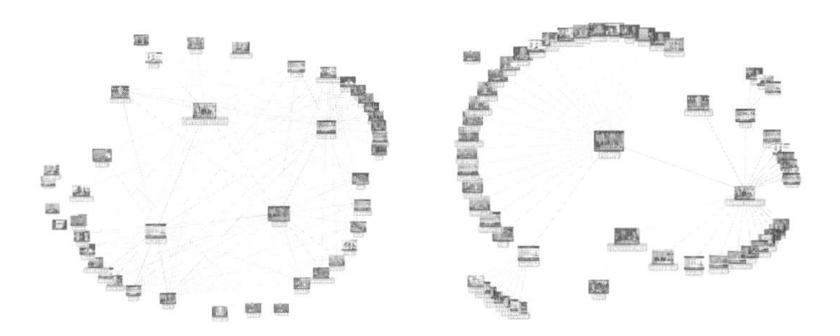

Fig. 2. Examples of semantic entity-relationship model

entities are defined by the verbs in sentences. Ideally, the relationship can be modeled by typed, directed edges between two semantic entities. However, the syntax parsing algorithms cannot achieve high accuracy on large-scale data. Therefore, we simplify the problem by defining the relationship as the co-occurrence of two semantic entities: if two semantic entities appear in a sentence, image or video shot at the same time, then a relationship is recognized between the two semantic entities. Based on this simplification, our semantic entity-relationship model can be represented as a unidirectional graph with semantic entities as nodes and relationship as edges. To preserve the semantics of verbs, we treat verbs as abstract semantic entities. An example of semantic entity-relationship model is given in Figure 2.

One may argue that a semantic entity or relationship may not be attractive if it appears only a few times in a large-scale news collection. To resolve this problem, we enable each node/edge to carry a weight in the semantic entity-relationship graph. The weight represents the interestingness of the associated semantic entity or relationship. Therefore, the semantic entity-relationship model can be represented as a weighted unidirectional graph:

$$D = \{(k_i, w_D(k_i)), (r(k_i, k_j), w_D(r(k_i, k_j)))\} \quad (1)$$

where k_i represents a semantic entity, $w_D(k_i)$is its weight, $r(k_i, k_j)$ represents a relationship, and $w_D(r(k_i, k_j))$ is its weight. Obviously, the interestingness weight may be different for the same semantic entity or relationship for different purpose or users.

Consequently, there may be different semantic entity-relationship model for the same collection of news reports. The simplest weighting algorithm is to use the frequency as the weight. We call such semantic entity-relationship model the raw model. It is the basis to compute more sophisticated models.

One can find that the semantic entity-relationship model preserves most semantics of the news reports. It can model the semantics of not only a single news report but also a collection of news reports. In addition, the multimedia semantics can be modeled uniformly in the model. However, no matter how many nice properties it has, the model is useless without efficient extraction algorithm and interestingness weighting algorithm. To resolve this problem, we propose an algorithm to extract semantic entity-relationship model from a large-collection of multimedia news reports. We will introduce the raw model extraction algorithm in this section. The interestingness weighting algorithm will be discussed in the next Section. As the semantic entity-relationship model is complex, it cannot be extracted in one simple step. To extract the semantic entities from news reports, the first step is to segment the semantic entities from different media. Apparently, the semantic entity segmentation algorithms in different media are different. Below we will discuss the problems separately.

2.1 Text Semantic Entity Detection

As the semantic entity is similar to the named entity, a potential approach for detecting it is to adopt algorithms for named entity detection. The most advanced named entity detection algorithm is the CRF (Conditional Random Field) -based algorithm [13]. However, because the CRF-based algorithm can only utilize short context information and part of linguistic features, it performs very poor on new semantic entities. On the contrary, the new semantic entities have significant different statistical properties than common strings. Therefore, a novel algorithm that can take the advantages of both approaches are needed to detect the semantic entity. However, the CRF-based algorithm is a sequence annotation approach, but the statistical approach is a typical classification approach. They have totally different methodology. As a result, they cannot be integrated easily. To resolve this problem, we propose an approach that integrates the CRF intermediate probabilities, statistical features and linguistic features as the multi-modal features to perform classification. Our algorithm trains a binary classifier for consecutive strings $s = (w_1, ..., w_n)$, where w_i is a word. For a s in a sentence $\xi = (..., c_{-2}, c_{-1}, s, c_1, c_2, ...)$, where c_j is a context word, several features are extracted for classification. The features used in our system are listed in Table 1. In Table 1, $B_l(s)$ is the probability that a semantic entity starts between c_{-1} and w_1 in text stream based on the observation ξ. It can be computed via a CRF model [13]. So does $B_r(s)$. $C_l(s)$ is the probability that c_{-1} leads a semantic entity. Some word may inherently have higher probability to lead a semantic entity, such as "the". Therefore, $C_l(s)$ may carry useful information for semantic entity detection. So does $C_r(s)$. The four features discussed above can be computed in advance via manually annotated training data. $R(s)$ is the internal relevance of words of s: $R(s) = \frac{n \times f(s)}{\sum_{i=1}^{n} f(w_i)}$. Where $f(x)$ is the frequency that string x occurs in recent news report sentences. $R(s)$ measures the internal relevance of words in s. If $R(s)$ is close to 0, it means the occurrence of$s$ is just by chance. Therefore, only s with high $R(s)$

Table 1. Features for semantic entity detection

Group	Feature(s)	Explanation
Sequence	$B_l(s)$ and $B_r(s)$	Left and Right Boundary Probability
Statistical	$E_l(s)$ and $E_r(s)$	Left and Right Context Entropy
Statistical	$R(s)$	Internal Relevance
Linguistic	$C_l(s)$ and $C_r(s)$	Left and Right Context Leading Probability

Table 2. Context examples of "swine flu"

see 2009 **swine flu** outbreak	lying to get **swine flu** drug
known as " **swine flu** ", is due	**Swine flu** is a type of
new strain of **swine flu** will develop	suspected **swine flu** victim in quarantine

can be a semantic entity. $E_l(s)$ is the word context entropy measured at the position -1: $E_l(s) = -\sum_{x\in\Omega(s,c_{-1})} p(x)\log(p(x))$. Where $\Omega(s,c_{-1})$ is the set of words that may appear in front of s (i.e. at position -1), $p(x)$ is the probability that word x appears in front of s. $E_l(s)$ can be computed via a large collection of recent news reports. So does $E_r(s)$. A semantic entity may appear in different context, as shown in Table 2. The more complex the context are, the higher probability will s be a semantic entity. Therefore, the context entropy $E_l(s)$ and $E_r(s)$ carry useful information for semantic entity detection and can be used as the features for classification. Because the statistical features measure the properties of recent news reports, there is no need to train $R(s)$, $E_l(s)$ and $E_r(s)$ in advance. They are extracted from the data to be processed.

We have evaluated our algorithm on datasets with different languages. The results are shown in Figure 3. One can find that our algorithm outperforms CRF algorithm on text semantic entity detection on both Chinese and Spanish data.

2.2 Multimedia Semantic Entity Extraction and Coreference

With the above algorithm, semantic entities in text sentences can be detected. These semantic entities carry abundant semantic information and are thus suitable for semantic modeling. However, a semantic entity may be represented as not only text strings but also images and video shots. Therefore, visual representations of the semantic entities must be extracted for complete semantic entity representation. There are two major sources of visual properties of semantic entities. The first one is the video shots of TV news broadcasts. The second one is the informative images of web news reports. As the two sources have different properties, we use different algorithms to extract the visual representation of semantic entities from different sources.

For TV news video clips, video shot is the basic units for video content representation. Thus it can be used as the basic unit for semantic modeling. However, the video shot must be associated properly with the text semantic entities before we can model

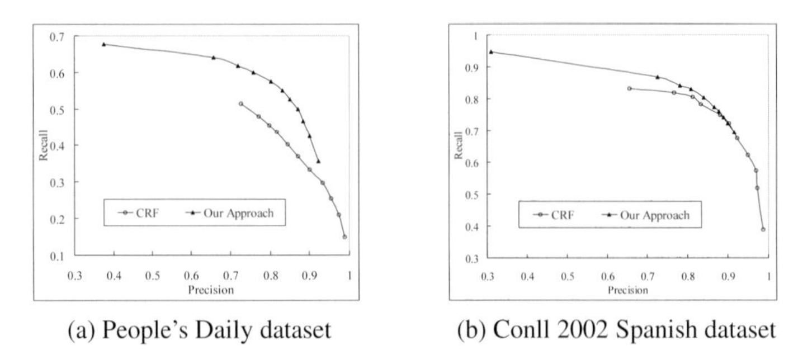

(a) People's Daily dataset (b) Conll 2002 Spanish dataset

Fig. 3. Text semantic entity detection performance

the semantic entity via video shots. One potential approach is to perform semantic classification to extract the representative semantic keywords from the video shots. To do so, we have proposed a salient object-based algorithm in previous works [14]. However, the semantic keywords extracted via semantic classification are not enough to fully represent the semantics of video shots. To resolve this problem, we propose to use the synchronized ASR (automatic speech recognition) script or closed caption text to associate video shots with proper text semantic entities. First, each video shot is associated with an ASR script or closed caption text sentence that is synchronized in the audio channel with this video shot. Second, the text semantic entities in the ASR script or closed caption sentence can be extracted via the algorithm addressed in above subsection. Finally, the video shot is used as the video representation of all text semantic entities in the ASR script or closed caption sentence. It is obvious that a text semantic entity may appear many times in the same or different news reports. Therefore, many video shots may be associated with the same text semantic entity via above approach. As a result, it is very important to select the most representative video shots for a text semantic entity and remove irrelevant ones. Consequently, the video shots must be weighted via their visual and semantic properties. To resolve this problem, we have proposed a set of algorithms to evaluate the representativeness of video shots via its visual semantic objects, repeat frequency/pattern and other visual properties [14]. Our algorithm will assign low weights to unrepresentative shots (like anchor shots) and high weights to representative ones. With this algorithm, only the few video shots with the highest weights are selected as the video representation of the text semantic entity.

Many web news reports may have associated picture or video. These informative multimedia materials carry abundant visual information of the news event thus are suitable to represent the visual properties of the relevant semantic entities. However, a news web page may embed many pictures and video clips that are ads., page structural elements and images for other news reports. Apparently, the informative images and video clips must be detected from the large number of embedded multimedia materials before they can be adopted to interpret the visual properties of the semantic entities. The most relevant techniques are web data extraction techniques [15]. However, as web data extraction algorithms are designed for the semi-structured data in web pages (such as a HTML formatted table), they may not adapt to informative multimedia elements that are mostly short HTML tags. As a result, it is more attractive to develop approaches

Table 3. The number of samples in informative image extraction dataset

	English Pages		Chinese Pages	
	Training	Test	Training	Test
Irrelevant Images	15036	15002	11481	11364
Informative Images	531	560	786	757
Ratio	28.3	26.8	14.6	15.0

extracting informative multimedia elements via only the inherent features of the HTML. There are 4 types of features that may imply whether a HTML multimedia tag is an informative element or not. Type I is the element's URL features, such as the file type (gif, jpg, flv, etc.), length of URL, special chars (+, -, ?, etc.), and special words (button, arrow, logo, etc.). Type II is the other attributes of the HTML tag, such as width, height and aspect ratio, alignment, alt, etc. Type III is the relationship between the element URL and the page URL, such as: if they are in the same Internet domain; if they carry the same word. Type IV is the surrounding tag properties of the HTML tag, such as: if it has an link (surrounded by A tag) or is an list item (surrounded by LI tag). More details of the features can be found at `http://memcache.drivehq.com/infimg/`

After the representative features are extracted, the next problem is that the number of irrelevant multimedia elements is far more than that of informative multimedia elements in news web pages. As shown in Table 3, the ratio between the irrelevant images and the informative images is up to 28.3 in our dataset. Consequently, existing machine learning algorithms will have reduced performance on these imbalanced training data [16]. To resolve this problem, we propose an automatic filtering algorithm to eliminate most irrelevant elements that are obviously not informative before classification.

For the i-th dimension of feature f_i, the most simple form of rule is $R_j \equiv (f_i = d)$. It matches any sample that its i-th dimension is equal to d. Similarly, we can define rules in the form of $R_j \equiv (f_i \geq c)$ and $R_j \equiv (f_i \leq c)$. In addition, multiple rules can be combined to form a complex rule: $R_k \equiv (R_{j_1} \wedge R_{j_2})$. It matches any sample that matches R_{j_1} and R_{j_2} simultaneously. Obviously, many rules can be extracted from a training set Ω but most of them are useless. Because our purpose is to filter out most negative samples and reduce the ratio between negative and positive samples, only those rules that match large amount of negative samples and little positive samples are useful for us. Therefore, we first select rules that satisfy:

$$\alpha(R_j) = \frac{|\Omega^+(R_j)|}{|\Omega^-(R_j)|} < \varepsilon \frac{|\Omega^+|}{|\Omega^-|} \tag{2}$$

where $\Omega^+(R_j)$ is the set of positive samples in Ω that matches with R_j, and $\Omega^-(R_j)$ is the set of negative samples in Ω that matches with R_j. Ω^+ is the set of positive samples in Ω, and Ω^- is the set of negative samples in Ω. $0 < \varepsilon < 1$ is a constant. In our experiments we select $\varepsilon = 0.1$. Any rules satisfying Eq. (2) will match higher proportion of negative samples than positive samples in the training set. Therefore, if we eliminate all matching samples from the training data, the proportion of negative samples in the training data will be reduced.

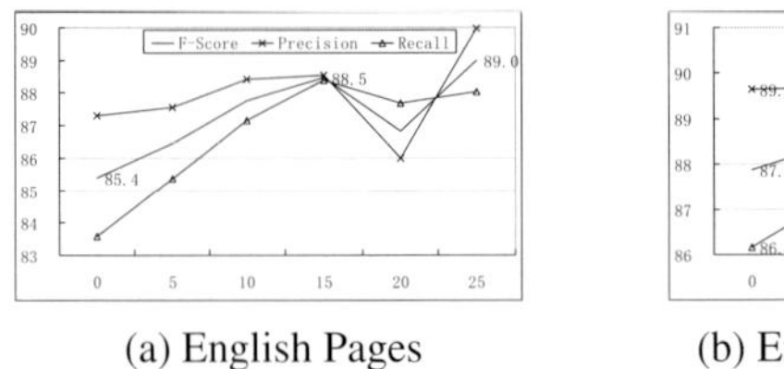

(a) English Pages

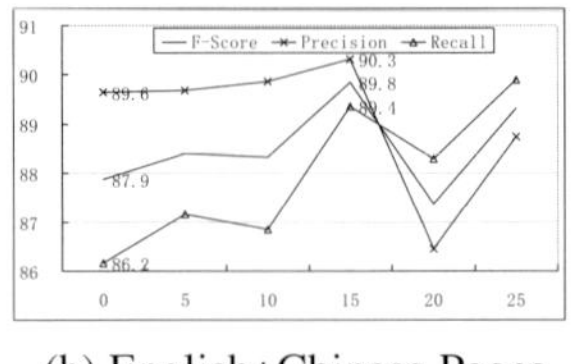

(b) English+Chinese Pages

Fig. 4. Overall classification performance

Even though rules selected via Eq. (2) is guaranteed to reduce the negative/positive samples ratio in training data, not all of them are suitable for filtering. Some rules may be a sub rules of other rules (i.e. $\Omega(R_{j_1}) \subseteq \Omega(R_{j_2})$). Some rules may match too few samples thus may be just an overfit. To resolve this problem, we iteratively select the best rule and apply to the training data via $\widehat{R} = \arg\max_{R_j} \{|\Omega^-(R_j)|\}$, i.e. we select the rule that matches the most number of negative samples. After one rule is selected, it is applied to the training data: $\Omega \leftarrow \Omega - \Omega\left(\widehat{R}\right)$. Then another rule can be selected with the modified training data. The iteration continues until $\left|\Omega\left(\widehat{R}\right)\right| < \lambda$, where λ is a constant. We use $\lambda = 10$ in our experiments.

After the training data is filtered with above algorithm, a SVM classifier can be trained from the remaining data. By integrating our filtering algorithm and the SVM classification algorithm, the classification performance can be improved. We have evaluated our algorithm on the dataset shown in Table 3. The overall classification performance is given in Figure 4. The horzontal axis is the number of rules used in the filtering step. One can find that the overall classification performance can be improved via our filtering algorithm. After the informative pictures and video clips are detected, they can be associated with the alt text and title text in HTML files. Then the same approach introduced at the beginning of this subsection can be applied to associate the pictures and video clips with the text semantic entities.

3 Interestingness Weighting

With semantic entities extracted, their relationships can be extracted by counting the co-occurrence frequencies of semantic entities in images, video shots and text sentences. As a result, the raw semantic entity relationship model can be extracted. However, the raw model is not very attractive, as frequent entities and relationships may not always interesting. For example, the semantic entities "USA" and "Obama"may have very high weights in the raw model. But most users have already known them thus may not be interested. Only those unknown entities or relationships may be interesting. Consequently, a good interestingness weighting algorithm needs to evaluate how well a user may know the semantic entities and relationships. To do this, the user's personal knowledge can be used to predict the semantic entities and relationships. If one entity or relationship can be predicted accurately, then it is already known by the user and is uninteresting at all. Only unpredictable entities and relationships are interesting for the user. However, to use this approach, 3 problems must be resolved: (1) how to model the user's knowledge mathematically; (2) how to perform the prediction; (3) how to extract the user's knowledge model. We will discuss these problems separately below.

As the user's knowledge about news is composed of a set of news events, it is indeed a news report collection. Therefore, the user's knowledge can also be modeled by our semantic entity relationship model. If we use K to represent the user's knowledge model and D_r to represent the raw semantic entity relationship model for the news collection of interest, the next step is to predict D_r via K. By observing the semantic entity relationship model carefully, one can find that it can be normalized to a probability distribution model, i.e. the weights are normalized to item probabilities. If D_r and K are two distributions, then their difference can be computed via KL-divergence: $d(D_r \parallel K) = \Sigma_{x \in \{k_i, r(k_i,k_j)\}} w_{D_r}(x) \log \frac{w_{D_r}(x)}{w_K(x)}$. Apparently, the difference is the most interesting part for the user. Therefore, by decomposing $d(D_r \parallel K)$ and normalizing the result, the interestingness weights of each semantic entity or relationship can be computed. More details can be found in our previous publication [17].

The remaining problem is how to estimate K. It sounds impossible to estimate K accurately at first glance, because users may learn a piece of information from may uncontrollable information sources, like a friend. However, by checking the route of information spread carefully, one can find that almost all important pieces of information are rooted from a publicly available report, e.g. a web report, a blog or a video report, no matter how the user knows it directly. Therefore, the most important knowledge of K can be estimated from the historical public reports. Certainly, the user's personal preference, memory ability and other properties have significant impact on K. We have discussed these problems and proposed detailed solutions to estimate K via the historical public reports and the user's properties in our previous work [18].

4 Knowledge Visualization and Visualization Based Retrieval

With the above algorithms, a semantic entity relationship model carrying appropriate interestingness weights can be extracted from the multimedia news collection of interest. However, it is obvious that no user has interest to examine this model directly, because it is just abstract data. To enable the users explore and retrieve the model more efficiently and effectively, hyperbolic visualization technique is used in our system to display the semantic entity relationship model [17,18,14]. As our semantic entity relationship model has nice mathematical structure, sophisticated computation can be implemented and thus advanced applications can be developed. We have developed personalized news recommendation, visualization based visual/text retrieval and exploration and other applications in our previous works [14,19].

5 Conclusion

In this paper, the semantic entity relationship model is proposed to represent the semantic information of large-scale multimedia news collections. Algorithms to extract the model from the raw news report data is also proposed. The proposed model can represent both text and multimedia semantic information in a uniform framework. In addition, the proposed model inherently enables cross-media semantic analysis on multimedia data like video clips and web pages. Therefore, the difficult visual semantic analysis task on video and image can be simplified. As the proposed model has nice mathematical structure, advanced applications can be developed based on it.

References

1. Wagstaff, J.: On news visualization (2005), http://www.loosewireblog.com/2005/05/on_news_visuali.html
2. Harris, J.: Tenbyten, http://tenbyten.org/10x10.html
3. Tagcloud, http://www.tagcloud.com/
4. Weskamp, M.: Newsmap, http://www.marumushi.com/apps/newsmap/index.cfm
5. Gastner, M., Shalizi, C., Newman, M.: Maps and cartograms of the 2004 us presidential election results (2004), http://www.cscs.umich.edu/~crshalizi/election/
6. Yang, J., Hauptmann, A.G.: Annotating news video with locations. In: Sundaram, H., Naphade, M., Smith, J.R., Rui, Y. (eds.) CIVR 2006. LNCS, vol. 4071, pp. 153–162. Springer, Heidelberg (2006)
7. Borgne, H.L., Guerin-Dugue, A., O'Connor, N.: Learning midlevel image features for natural scene and texture classification. IEEE Trans. on CSVT 17(3), 286–297 (2007)
8. Chang, S., Manmatha, R., Chua, T.: Combining text and audio-visual features in video indexing. In: IEEE International Conference on Acoustics, Speech, and Signal Processing (ICASSP), Philadelphia, pp. 1005–1008 (2005)
9. Jones, K.S.: A statistical interpretation of term specificity and its application in retrieval. Journal of Documentation 28, 11–21 (1972)
10. Page, L., Brin, S., Motwani, R., Winograd, T.: The pagerank citation ranking: Bringing order to the web, http://dbpubs.stanford.edu:8090/pub/1999-66
11. Kleinberg, J.M.: Authoritative sources in a hyperlinked environment. Journal of the ACM 46, 668–677 (1999)
12. Hetzler, E.G., Whitney, P., Martucci, L., Thomas, J.: Multi-faceted insight through interoperable visual information analysis paradigms. In: IEEE Symposium on Information Visualization, p. 137 (1998)
13. McCallum, A., Li, W.: Early results for named entity recognition with conditional random fields, feature induction and web-enhanced lexicons. In: Natural language learning at HLT-NAACL, pp. 188–191 (2003)
14. Luo, H., Fan, J., Satoh, S., Yang, J., Ribarsky, W.: Integrating multi-modal content analysis and hyperbolic visualization for large-scale news video retrieval and exploration. Signal Processing: Image Communication 23, 538–553 (2008)
15. Zhai, Y., Liu, B.: Structured data extraction from the web based on partial tree alignment. IEEE Transactions on Knowledge and Data Engineering 18, 1614–1628 (2006)
16. Orriols, A., Bernad'oMansilla, E.: The class imbalance problem in learning classifier systems: A preliminary study. In: GECCO Workshops, pp. 74–78 (2005)
17. Luo, H., Fan, J., Yang, J., Ribarsky, W., Satoh, S.: Exploring large-scale video news via interactive visualization. In: IEEE Symposium on Visual Analytics Science and Technology, pp. 75–82 (2006)
18. Luo, H., Fan, J., Yang, J., Ribarsky, W., Satoh, S.: Analyzing large-scale news video databases to support knowledge visualization and intuitive retrieval. In: IEEE Symposium on Visual Analytics Science and Technology (2007)
19. Luo, H., Fan, J., Keim, D.A., Satoh, S.: Personalized news video recommendation. In: Multimedia Modeling, pp. 459–471 (2009)

PageRank with Text Similarity and Video Near-Duplicate Constraints for News Story Re-ranking

Xiaomeng Wu[1], Ichiro Ide[2], and Shin'ichi Satoh[1]

[1] National Institute of Informatics
2-1-2 Hitotsubashi, Chiyoda-ku Tokyo 101-8430, Japan
{wxmeng,satoh}@nii.ac.jp
[2] Graduate School of I.S., Nagoya University
Furo-cho, Chikusa-ku Nagoya 464-8630, Japan
ide@is.nagoya-u.ac.jp

Abstract. Pseudo-relevance feedback is a popular and widely accepted query reformulation strategy for document retrieval and re-ranking. However, problems arise in this task when assumed-to-be relevant documents are actually irrelevant which causes a drift in the focus of the reformulated query. This paper focuses on news story retrieval and re-ranking, and offers a new perspective through the exploration of the pair-wise constraints derived from video near-duplicates for constraint-driven re-ranking. We propose a novel application of PageRank, which is a pseudo-relevance feedback algorithm, and use the constraints built on top of text to improve the relevance quality. Real-time experiments were conducted using a large-scale broadcast video database that contains more than 34,000 news stories.

Keywords: PageRank, Video Near Duplicate, News Story Re-ranking, Video Data Mining.

1 Introduction

News videos are broadcast everyday across different sources and times. To make full use of the overwhelming volume of news videos available today, it is necessary to track the development of news stories from different sources, mine their dependencies, and organize them in a semantic way. News story retrieval is a fundamental step for news topic tracking, threading, video summarization, and browsing from among these research efforts. News story retrieval aims at searching for evolving and historical news stories according to the topics, such as *the Trial of Saddam Hussein* and *2006 North Korean nuclear test*.

1.1 Background

News story retrieval is normally studied under the Query by Example theme using the textual features as the underlying cues [3,4,11]. Relevance feedback is a

S. Boll et al. (Eds.): MMM 2010, LNCS 5916, pp. 533–544, 2010.

popular and widely accepted query reformulation strategy for this task. Some researchers have attempted to automate the manual part of relevance feedback, which is also known as pseudo-relevance feedback [1,5,10]. Pseudo-relevance feedback is obtained by assuming that the top k documents in the resulting set containing n results (usually where $k \ll n$) are relevant, and has the advantage in that assessors are not required. However, in pseudo-relevance feedback, problems arise when assumed-to-be relevant documents are actually irrelevant, which causes a drift in the focus of the reformulated query [10]. How to reduce the inappropriate feedback taken from irrelevant documents or how to guarantee the relevance quality is the main focused issue for pseudo-relevance feedback studies.

To tackle this issue, we offer a new perspective that explores the pairwise constraints derived from video near duplicates for news story retrieval and constraint-driven re-ranking. The main points of discussion include: (1) a novel scheme for pseudo-relevance feedback on the basis of near duplicates built on top of text, (2) a novel application of PageRank as a pseudo-relevance feedback algorithm used for constraint-driven re-ranking, and (3) real-time experiments conducted on a large-scale broadcast video database containing more than 34,000 news stories.

1.2 Framework Overview

Our system works on a large-scale broadcast video database. The system uses a news story as the search query and outputs stories depicting the query topic from within the database. A news story is formally defined as a semantic segment within a news video, which contains a report depicting a specific topic or incident. A story is described as a group of shots. Each shot is described by using a set of representative keyframes and closed-captions. Figure 1 depicts our proposed news story retrieval and re-ranking system.

Initially, candidate news stories that are similar to the query are searched using a topic-tracking method based only on the textual information. The reason

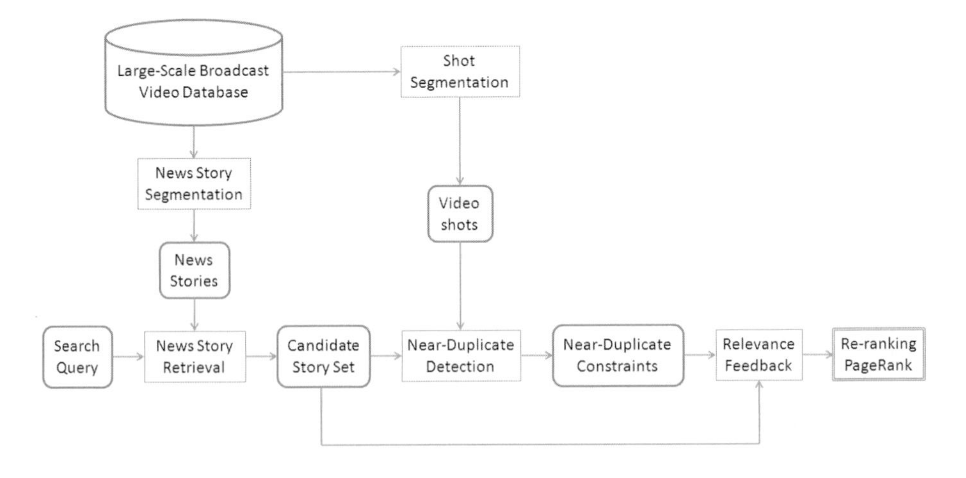

Fig. 1. Proposed news story retrieval and re-ranking system

that we use a text-based method first before using the near-duplicate detection is because the latter requires visual information processing and is computationally far more expensive than textual information processing. On the other hand, the coverage of near duplicates is normally insufficient for conducting a complete and thorough topic tracking compared that of the textual information. After text-based topic tracking, near duplicates are detected from the set of candidate news stories and used to group together the stories that share these near duplicates. We believe that video near-duplicate constraints are highly-useful for guaranteeing a higher relevance quality than the textual constraints. PageRank is then applied as a pseudo-relevance feedback algorithm on the basis of these pairwise constraints, and used to re-rank news stories depicting the same topic.

2 News Story Retrieval Based on Textual Information

News story retrieval is normally studied under the Query by Example theme with the textual features used as the underlying cues [3,4,11]. In news videos, the focal point or content of evolving stories depicting the same topic normally varies slowly with time. The news story retrieval method used in this work should be robust enough for this focal-point variation. In this paper, [3] is used for this purpose, which uses the semantic and chronological relation to track the chain of related news stories in the same topic over time.

A story boundary is first detected from a closed-caption text broadcasted simultaneously with the video. The resemblances between all story combinations are evaluated by adopting a cosine similarity between two keyword frequency vectors generated from two news stories. When the resemblance exceeds a threshold, the stories are considered related and linked. Tracking is achieved by considering the children stories related to the search query as new queries to search for new children stories. This procedure forms a simple story link tree starting from the story of interest, i.e. the search query. Children stories are defined as news stories related to a parent, under the condition that the time stamps of the children stories always chronologically succeed their parent. The link tree can also be considered a set of candidate news stories that is similar to the search query, which is further used for near-duplicate detection.

3 Video Near-Duplicate Detection

In addition to the textual features, in broadcast videos, there are a number of video near duplicates, which appear at different times and dates and across various broadcast sources. Near duplicates, by definition, are sets of shots composed of the same video material used several times from different sources or material involving the same event or the same scene, as shown in Fig. 2. These near duplicates basically form pairwise equivalent constraints that are useful for bridging evolving news stories across time and sources.

After text-based news story retrieval, near duplicates are detected only from the set of candidate news stories. This can not only dramatically reduce the

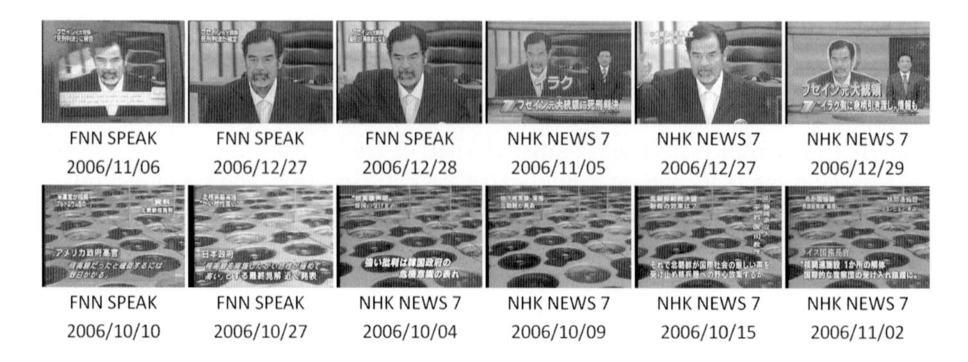

Fig. 2. Near duplicates across different stories of two topics. The label under each image is the program name and the airdate. Above: *Trial of Saddam Hussein*. Below: *2006 North Korean nuclear test*.

computation burden due to visual information processing, but also reduce the probability of potential errors caused by near-duplicate detection. We used an interest-point-based algorithm with a local description for the near-duplicate detection. This algorithm was proposed by Ngo et al. [9] and proved to be robust to variations of translation and scaling introduced due to video editing and different camerawork.

They tested their algorithm using a keyframe database instead of a video archive [9]. We extract multiple keyframes from each video shot to extend it to near-duplicate shot detection. The shot length is equally divided, and the frames at the points of division are selected as the keyframe. This is to tolerate the variation introduced by the camera and object motion. In equation terms, given the shot length L, the $(i \times L/(N+1))^{\text{th}}$ frames are extracted as the keyframe, where $i = 1 \cdots N$. N indicates the number of keyframes and is empirically set to three in this paper. To tolerate the significant impact of video captions, we propose cropping the keyframe beforehand so that only the central part is used for the near-duplicate detection. On the other hand, we also manually excluded anchorperson shots that are not related to the topic while highly possible to be detected as near duplicates. Since anchorperson-shot detection has been extensively studied and many good algorithms were already proposed, this process can be automated if needed.

4 PageRank with Near-Duplicate Constraints

4.1 PageRank

Eigenvector centrality is a measure of the importance of a node in a network. It assigns relative scores to all nodes in the network based on the principle that connections to high-scoring nodes contribute more to the score of the node in question than equal connections to low-scoring nodes. PageRank [2] is a variant of the Eigenvector centrality measure. PageRank (**PR**) is iteratively defined as

$$\mathbf{PR} = d \times \mathbf{S}^* \times \mathbf{PR} + (1 - d) \times \mathbf{p} \tag{1}$$

$\mathbf{S}^*$ is the column normalized adjacency matrix of $\mathbf{S}$, where $\mathbf{S}_{i,j}$ measures the weight between node i and j. d is a damping factor to be empirically determined, and $\mathbf{p} = [\frac{1}{n}]_{n\times 1}$ is a uniform damping vector, where n is the total number of nodes in the network. Repeatedly multiplying $\mathbf{PR}$ by $\mathbf{S}^*$ yields the dominant eigenvector of the matrix $\mathbf{S}^*$. Although $\mathbf{PR}$ has a fixed-point solution, in practice, it can often be estimated more efficiently using iterative approaches. PageRank converges only when matrix $\mathbf{S}^*$ is aperiodic and irreducible. The former is generally true for most applications and the latter usually requires a strongly connected network, a property guaranteed in practice by introducing a damping factor d. It is generally assumed that the damping factor is set around 0.85.

The simplicity and effectiveness of PageRank for text mining were demonstrated through document summarization studies [12,14], which suggested combining PageRank with text similarity. However, except the study [7], little attention has been paid to applying PageRank to the document retrieval task. This is because PageRank usually requires a strongly connected network where the group of good nodes should comprise a majority of all the nodes in this network. This fundamental assumption is feasible for document summarization because people normally seek to summarize a set of documents, all of which are strongly related to a certain topic. However, for document retrieval, the set of documents searched for by the query normally contains errors that are irrelevant to the topic of interest. The potential existence of these assumed-to-be good nodes that are actually bad nodes will have a large impact on the majority distribution of the network, and thus, they cause a drift in the focus of PageRank. This issue is similar to the problem existing in pseudo-relevance feedback studies that were introduced in Sec. 1.1.

4.2 Near-Duplicate Constraints

A variety of algorithms have been proposed for applying near-duplicate constraints to news video retrieval tasks. Zhai et al. [6] linked news stories by combining image matching and textual correlation. Hsu et al. [8] tracked four topics with near duplicates and semantic concepts, and found that near duplicates significantly improve the tracking performance. These two works use video near duplicate and textual information as two independent modalities. Since each modality is individually processed and fusion is based only on the score functions of their processing results, the potential inter-modal relationships between the two modalities have not been thoroughly explored and thus are wasted. Apart from these multimodality fusion studies, Wu et al. [13] presented a system built on near-duplicate constraints, which are applied on top of text to improve the story clustering. This work depends on manual near-duplicate labeling, which is impossible to handle within large-scale databases.

In this paper, we apply PageRank as a pseudo-relevance feedback algorithm and integrate video near-duplicate constraints to guarantee the relevance quality. Given a story used as a search query, candidate stories similar to the query are

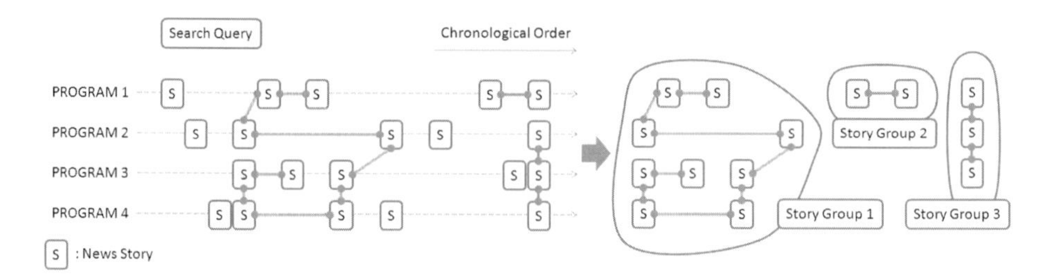

Fig. 3. Group stories sharing near duplicates

searched across various news programs (Fig. 3). Two stories are linked together if they share at least one pair of near duplicates. This kind of two stories can be regarded as having a must-link constraint, which indicates that they discuss the same topic. Stories are then clustered into groups based on these links. We make the following two assumptions.

- **Assumption 1**
  Most stories in the same story group depict the same topic.
- **Assumption 2**
  The largest story group depicts the same topic as the query.

Assumption 1 is feasible in most cases because near duplicates are detected only from the set of candidate stories that are similar to the query, so that the probability of potential errors caused by near-duplicate detection is small. **Assumption 2** is also feasible because most near duplicates are shared between stories depicting the same topic as the query. From another point of view, the noise or outlier topics are normally different from each other so that fewer near duplicates are shared between them.

The experimental results from news story grouping based on ten search queries are listed in Table 1 (Sec. 5.1 for more information on these queries). From #Story (#TP), we can see that most news stories clustered in the largest group depict the same topic as the search query (except for T6). In other words, both **Assumption 1** and **Assumption 2** described above were feasible in our experiment. For T6, the news topic is on a child abduction-murder that occurred in Hiroshima, and the query story was broadcasted on 2005/12/01. The next day, another child abduction-murder was reported near Tokyo. In most news programs, these two news topics were continuously broadcasted. Due to the high similarity between them, the story segmentation method [3] used in this paper failed to segment them from each other, so that the stories of the query topic also contained shots of the noise topic. As a result, stories depicting these two topics were clustered in the same story group based on the near-duplicate constraints.

Given both **Assumption 1** and **Assumption 2** are feasible, it is reasonable to assume that stories in the largest story group would satisfy the PageRank requirement in a strongly connected network where the group of good nodes comprises a majority among all the nodes in this network (Sec. 4.1). Based on

Table 1. Experimental results of news story grouping. #Candidate: number of stories in the candidate story set. #TP: number of relevant stories depicting the same topic as the search query. #Group: number of clustered story groups based on near-duplicate constraints. #Story: number of stories in the largest story group.

Topic Number	#Candidate (#TP)	#Group	#Story (#TP)
T1	106 (43)	11	9 (9)
T2	370 (174)	3	72 (72)
T3	164 (115)	3	75 (71)
T4	35 (13)	3	10 (10)
T5	93 (64)	1	47 (47)
T6	148 (25)	7	28 (4)
T7	119 (60)	7	26 (22)
T8	35 (34)	3	13 (13)
T9	48 (44)	2	30 (30)
T10	65 (54)	3	39 (38)

Fig. 4. Examples of near duplicates depicting *Trial of Saddam Hussein*. #story: number of stories sharing the corresponding near duplicate.

these assumptions, the largest story group is chosen as the relevance feedback. For example, story group 1 shown in Fig. 3 is the largest story group, so it will be used as the relevance feedback and further integrated with PageRank. Examples of near duplicates within the largest story group depicting *the Trial of Saddam Hussein* are shown in Fig. 4.

4.3 Applying PageRank

To define the text similarity, we use term frequency-inverse document frequency (tf-idf) weighting, which is one of the best-known schemes for text mining, to represent each story. To do so, a semantic analysis is first applied to the compound nouns extracted from each story to generate a keyword vector for four semantic classes, *general*, *personal*, *locational/organizational*, and *temporal*. From our experimental results, we found that the keywords from the *temporal* class are normally not helpful for identifying news stories depicting the same topic. Therefore, only the compound nouns of the other three classes are used as the keywords.

For each story, a keyword vector $\boldsymbol{V}$ can be created as follows. The keyword similarity between two stories is thus defined by the cosine similarity shown in Eq. 3.

$$\boldsymbol{V} = (tfidf(t_1), tfidf(t_2), \cdots, tfidf(t_N)) \tag{2}$$

$$\cos(\boldsymbol{V}_i, \boldsymbol{V}_j) = \frac{\boldsymbol{V}_i \cdot \boldsymbol{V}_j}{\|\boldsymbol{V}_i\|\|\boldsymbol{V}_j\|} \tag{3}$$

The set of stories may be represented by a cosine similarity matrix $\mathbf{S}$, where each entry $\mathbf{S}_{i,j} = \cos(\boldsymbol{V}_i, \boldsymbol{V}_j)$ in the matrix is the similarity between the corresponding story pair (s_i, s_j). Different from document summarization studies where sentences are regarded as nodes, we regard one document or one story as one node. Thus, we have the overall relevancy (**PR**) of each story given its similarity to other stories, iteratively defined by using PageRank (Eq. 1).

This is equivalent to using all the stories searched for by the query as relevance feedback, which is highly sensitive to the potential existence of irrelevant stories. To guarantee the relevance quality, we put restrictions on the adjacency matrix $\mathbf{S}$ based on the video near-duplicate constraints. In Sec. 4.2, stories are clustered into groups based on the near-duplicate constraints. Given the largest story group being denoted by $\mathbb{S}$, each entry in the adjacency matrix $\mathbf{S}$ is defined by using Eq. 4. In other words, we only regard stories in $\mathbb{S}$ as high-quality nodes, and for each node s_j in the whole network, we only use the connection between s_j and these high-quality nodes $\{s_i : s_i \in \mathbb{S}\}$ as a vote of support to iteratively define the overall relevancy of s_j.

$$\mathbf{S}_{i,j} = \begin{cases} \cos(\boldsymbol{V}_i, \boldsymbol{V}_j) & (s_i \in \mathbb{S}) \\ 0 & (s_i \notin \mathbb{S}) \end{cases} \tag{4}$$

5 Experiments

5.1 Database

We tested our system using a large-scale broadcast video database comprised of actually broadcasted videos from 2005/10/19 to 2007/01/19. These videos were broadcasted from six different news programs produced by three different Japanese TV stations. Closed-captions were segmented into stories using the algorithm developed by Ide et al. [3], and the videos were segmented into shots by comparing the RGB histograms between adjacent frames. The stories and shots were used as the basic units of analysis. The keywords were derived from a list of compound nouns extracted from the closed-captions [3], while the keyframes were derived by equally dividing the shot and selecting the points of division. The set of near-duplicate pairs was detected using the algorithm developed by Ngo et al. [9]. The database was comprised of 34,279 news stories (compared to around 800 news stories used by Wu et al. [13]).

Ten search queries were selected for experimentation, as listed in Table 2, including five Japanese and five foreign news stories. The design of these queries is based on the biggest topics of the important domestic and international news stories from 2005/10/19 to 2007/01/19. The candidate stories that were similar to these queries were searched for across the six news programs, and the near duplicates were detected from the set of candidate news stories. The duration

Table 2. Ten search queries selected for experimentation

Topic Number	Topic	Duration	Domestic / Foreign
T1	*Trial of Saddam Hussein (1)*	15 months	Foreign
T2	*Architectural forgery in Japan*	2 months	Domestic
T3	*Fraud allegations of Livedoor*	2 months	Domestic
T4	*Trial of Saddam Hussein (2)*	2 months	Foreign
T5	*7 July 2005 London bombings*	1 month	Foreign
T6	*Murder of Airi Kinoshita*	1 month	Domestic
T7	*Murder of Yuki Yoshida*	1 month	Domestic
T8	*Murder of Goken Yaneyama*	1 month	Domestic
T9	*2006 North Korean missile test*	1 month	Foreign
T10	*2006 North Korean nuclear test*	1 month	Foreign

within which the search was conducted varied from 1 to 15 months. Our experiments on news story re-ranking were conducted based on our proposed algorithm discussed in Sec. 4.

5.2 News Story Re-ranking

The stories searched for using the ten queries were ranked based on PageRank with video near-duplicate constraints. To evaluate the performance, we compared our algorithm with the four baseline algorithms listed as follows. Note that **BL2**, **BL3**, and **BL4** can also be considered traditional algorithms under the pseudo-relevance feedback theme.

- **Baseline 1 (BL1)**
  The cosine similarity between each news story and the original query is evaluated and used for the story ranking.

- **Baseline 2 (BL2)**
  After ranking the stories using **BL1**, the top-k ($k = 10, 20, 30$) stories were chosen as the relevance feedback. The Rocchio algorithm [1], which is a classic algorithm for extracting information from relevance feedback, is used to re-rank the stories. The Rocchio algorithm formula is defined by using Eq. 5, where $\boldsymbol{V}_m$ is the expanded query, $\boldsymbol{V}_0$ is the original query, $\mathbb{D}_{rel}$ and $\mathbb{D}_{irr}$ are the sets of relevant and irrelevant stories, respectively, and $\alpha = 1$, $\beta = 1$, and γ are weights. In this baseline, we allow only positive feedback, which is equivalent to setting $\gamma = 0$. Note that $|\mathbb{D}_{rel}| = k$. The cosine similarity between each news story and the expanded query is then evaluated and used for story re-ranking.

$$\boldsymbol{V}_m = \alpha \times \boldsymbol{V}_0 + \beta \times \frac{\sum_{\boldsymbol{V}_i \in \mathbb{D}_{rel}} \boldsymbol{V}_i}{|\mathbb{D}_{rel}|} - \gamma \times \frac{\sum_{\boldsymbol{V}_i \in \mathbb{D}_{irr}} \boldsymbol{V}_i}{|\mathbb{D}_{irr}|} \tag{5}$$

- **Baseline 3 (BL3)**
  Apply PageRank to all stories searched for by the query.

– **Baseline 4 (BL4)**
After ranking the stories using **BL1**, the top-k ($k = 10, 20, 30$) stories are chosen as the relevance feedback. PageRank is used for the story re-ranking. Each entry in the adjacency matrix $\mathbf{S}$ is defined by using Eq. 6, where $\mathbb{D}_{rel}$ is the set of top-k stories.

$$\mathbf{S}_{i,j} = \begin{cases} \cos(\boldsymbol{V}_i, \boldsymbol{V}_j) & (s_i \in \mathbb{D}_{rel}) \\ 0 & (s_i \notin \mathbb{D}_{rel}) \end{cases} \quad (6)$$

5.3 Experimental Results and Discussions

An evaluation using the average precision ($AveP$) was performed using Eq. 7. In Eq. 7, r denotes the rank, N the number of stories searched, $rel()$ a binary function on the relevance of a given rank, and $P()$ the precision at a given cut-off rank. N_{rel} denotes the number of relevant stories with $N_{rel} \leq N$. Table 3 lists the results, where PRND denotes our proposed algorithm using PageRank with Near-Duplicate constraints and MAP is the Mean Average Precision.

$$AveP = \frac{\sum_{r=1}^{N}(P(r) \times rel(r))}{N_{rel}} \quad (7)$$

Table 3. Experimental results of story re-ranking ($AveP$: %)

	BL1	BL2_10	BL2_20	BL2_30	BL3	BL4_10	BL4_20	BL4_30	**PRND**
T1	82.6	69.92	67.58	80.37	65.99	69.73	69.22	82.02	**96.38**
T2	87.92	87.47	91.35	93	67.61	92.06	92.41	94.35	**97.45**
T3	88.4	96.92	97.05	97	97.33	97.89	97.54	97.78	**98.36**
T4	81.68	94.85	90.79	90.99	85.93	95.3	80.35	62.91	**100**
T5	93.46	97.37	97.67	97.63	96.08	97.11	98.12	97.95	**99.42**
T7	85.36	96.86	98.12	98.15	93.91	98.21	99.09	99.3	**99.42**
T8	99.46	99.83	**100**	**100**	**100**	**100**	**100**	**100**	100
T9	98.28	98.62	98.74	98.64	98.62	**98.97**	98.85	98.74	98.87
T10	94.92	97.09	97.22	97.13	95.13	97.27	**97.58**	96.95	97.34
MAP	89.87	92.55	93.63	95.01	85.81	94.22	94.11	95.2	**98.25**

From Table 3, we can see that our proposed re-ranking algorithm outperformed the baselines for most topics, and the MAP of PRND is higher than all baselines. The main reason for this is that our re-ranking algorithm based on PageRank improves the informativeness and representativeness of the original query (**BL1**), and near-duplicate constraints guarantee a higher relevance quality than the textual constraints used in traditional pseudo-relevance feedback algorithms (**BL2**, **BL3**, and **BL4**).

In particular, for T1, T2, and T4, we can see they had a higher level of improvement than the other topics. By observing the results, we found that ranking based on the original query tends to associate higher rank to stories closer to the query in terms of their airdates. Therefore, stories temporally closer to the query (normally within one week in our experiments) tend to comprise the majority from among the top-k stories chosen as the relevance feedback in **BL2** and **BL4**. This is because the focal point of the evolving stories depicting the same topic normally varies over time, so the stories temporally distant from the query tend to have less similarity to the query. For long topics like T1, T2, and T4, stories that are relevant but temporally distant from the query tend to acquire less contribution or a lower vote of support from the top-k relevant stories. On the other hand, evolving stories always repeatedly use the same representative shots, even if their airdates are distant from each other. Figure 2 shows two examples, from which we can see that the detected near-duplicates guaranteed a larger coverage of evolving stories in terms of airdate. This leads to increased coverage of textual information for the relevance feedback when applying PageRank based on these near-duplicate constraints. This can explain the greater improvement of T1, T2, and T4 compared to the other topics. This is also considered one of the contributions of our proposed re-ranking algorithm. For another long topic T3, because the focal-point variation is small in this case, the problem described above was not reflected in this experiment.

6 Conclusion

This paper focuses on news story retrieval and re-ranking, and offers a new perspective by exploring the pairwise constraints derived from video near duplicates for constraint-driven re-ranking. Compared to some other similar works, we use the constraints built on top of text to improve the relevance quality. As a future work, an experiment is under development for evaluating the actual time necessary to process the dataset during each phase of our proposed system. Also, we are developing an experiment for integrating the near-duplicate constraints into pseudo-relevance feedback, and comparing it to PageRank. Another future work is quantitatively evaluating the performance of near-duplicate detection algorithm, and checking up on whether the false alarms will have a large impact on the proposed system.

References

1. Rocchio, J.: Relevance Feedback in Information Retrieval. The SMART Retrieval System (1971)
2. Brin, S., Page, L.: The Anatomy of a Large-Scale Hypertextual Web Search Engine. Computer Networks 30, 107–117 (1998)
3. Ide, I., Mo, H., Katayama, N., Satoh, S.: Topic Threading for Structuring a Large-Scale News Video Archive. In: Enser, P.G.B., Kompatsiaris, Y., O'Connor, N.E., Smeaton, A., Smeulders, A.W.M. (eds.) CIVR 2004. LNCS, vol. 3115, pp. 123–131. Springer, Heidelberg (2004)
4. Mo, H., Yamagishi, F., Ide, I., Katayama, N., Satoh, S., Sakauchi, M.: Key Image Extraction from a News Video Archive for Visualizing Its Semantic Structure. In: Aizawa, K., Nakamura, Y., Satoh, S. (eds.) PCM 2004. LNCS, vol. 3331, pp. 650–657. Springer, Heidelberg (2004)

5. Qin, Z., Liu, L., Zhang, S.: Mining Term Association Rules for Heuristic Query Construction. In: Dai, H., Srikant, R., Zhang, C. (eds.) PAKDD 2004. LNCS (LNAI), vol. 3056, pp. 145–154. Springer, Heidelberg (2004)
6. Zhai, Y., Shah, M.: Tracking news stories across different sources. In: ACM Multimedia, pp. 2–10 (2005)
7. Zhang, B., Li, H., Liu, Y., Ji, L., Xi, W., Fan, W., Chen, Z., Ma, W.-Y.: Improving web search results using affinity graph. In: SIGIR, pp. 504–511 (2005)
8. Hsu, W.H., Chang, S.-F.: Topic Tracking Across Broadcast News Videos with Visual Duplicates and Semantic Concepts. In: ICIP, pp. 141–144 (2006)
9. Ngo, C.-W., Zhao, W., Jiang, Y.-G.: Fast tracking of near-duplicate keyframes in broadcast domain with transitivity propagation. In: ACM Multimedia, pp. 845–854 (2006)
10. Song, M., Song, I.-Y., Hu, X., Allen, R.B.: Integration of association rules and ontologies for semantic query expansion. Data Knowl. Eng. 63, 63–75 (2007)
11. Lin, F., Liang, C.-H.: Storyline-based summarization for news topic retrospection. Decis. Support Syst. 45, 473–490 (2008)
12. Wan, X., Yang, J.: Multi-document summarization using cluster-based link analysis. In: SIGIR, pp. 299–306 (2008)
13. Wu, X., Ngo, C.-W., Hauptmann, A.G.: Multimodal News Story Clustering With Pairwise Visual Near-Duplicate Constraint. IEEE Transactions on Multimedia 10, 188–199 (2008)
14. Otterbacher, J., Erkan, G., Radev, D.R.: Biased LexRank: Passage retrieval using random walks with question-based priors. Inf. Process. Manage. 45, 42–54 (2009)

Learning Vocabulary-Based Hashing with AdaBoost

Yingyu Liang, Jianmin Li, and Bo Zhang

State Key Laboratory of Intelligent Technology and Systems
Tsinghua National Laboratory for Information Science and Technology
Department of Computer Science and Technology, Tsinghua University
Beijing 100084, China
liangyy08@mails.tsinghua.edu.cn,
{lijianmin,dcszb}@mail.tsinghua.edu.cn

Abstract. Approximate near neighbor search plays a critical role in various kinds of multimedia applications. The vocabulary-based hashing scheme uses vocabularies, i.e. selected sets of feature points, to define a hash function family. The function family can be employed to build an approximate near neighbor search index. The critical problem in vocabulary-based hashing is the criteria of choosing vocabularies. This paper proposes a approach to greedily choosing vocabularies via Adaboost. An index quality criterion is designed for the AdaBoost approach to adjust the weight of the training data. We also describe the parallelized version of the index for large scale applications. The promising results of the near-duplicate image detection experiments show the efficiency of the new vocabulary construction algorithm and desired qualities of the parallelized vocabulary-based hashing for large scale applications.

1 Introduction

Approximate nearest neighbor search plays a critical role in various kinds of multimedia applications, including object recognition [13], near-duplicate image detection [10], content-based copy detection [11,16]. In such applications, typically, the multimedia objects are represented as sets of elements (e.g., local feature points), between which the similarity can be evaluated via searching the nearest neighbors of each element.

The recent explosion of multimedia data has led the research interest into large scale multimedia scene. The various typical approximate near neighbor search algorithms, such as ANN [4] and LSH[5,8], show high performance in relatively small datasets, but do not fit in the large scale scene. For example, the popular Euclidean locality sensitive hashing based on p-stable distributions (E2LSH)[5] typically requires hundreds of bytes for each point. Also, instead of all points in the buckets, it performs a refinement step to return only those within a distance threshold, which requires loaded data in the memory. These shortcomings prevent it from usage in large datasets.

The bag-of-features (BOF) image representation [17] is introduced in this context. Each feature point in the dataset is quantized by mapping to the ID

S. Boll et al. (Eds.): MMM 2010, LNCS 5916, pp. 545–555, 2010.

of the nearest one in a selected set of feature points called a visual vocabulary. The vector quantization approach can be interpreted as an approximate near neighbor search: the space is partitioned into Voronoi cells by the vocabulary, and points are treated as neighbors of each other if they lie in the same cell. The BOF approach can deal with large scale datasets for its efficient and space-saving property. However, it is an approximation to the direct matching of individual feature points and somewhat decreases the performance [10].

The vocabulary-based hashing scheme [12] combines the merits of BOF and LSH. Vocabularies are employed to define a hash function family in which each function maps an input point to the ID of the nearest one in the corresponding vocabulary. The function family is incorporated into the locality sensitive hashing scheme in [9] to build an index for approximate near neighbor search. This approach shows better performance than BOF and LSH, and it is efficient for large databases. In this vocabulary-based hashing scheme, the vocabularies define the hashing functions and thus determine the index, so they play a key role and should be carefully designed. The vocabulary construction algorithm in [12] first generates random vocabularies and then selects from them according to two criteria. However, it is time-consuming since it must generate a large amount of random vocabularies to select effective vocabularies.

In this paper, a new approach utilizing AdaBoost [7] is proposed for the vocabulary construction in vocabulary-based hashing. An index quality criterion is designed for AdaBoost to adjust the weight of the training data. We also describe the parallelized version of the index for large scale applications. Near-duplicate image detection experiments are carried out to demonstrate the effectiveness and efficiency of the approach. The results show that the new vocabulary construction algorithm is significantly more efficient than that in [12], and the parallelized vocabulary-based hashing shows desired qualities for the large scale scene.

This paper is organized as follows. The vocabulary-based hashing index is briefly reminded in Section 2. Section 3 presents the proposed vocabulary construction algorithm and Section 4 describes the parallelization. Experimental results are provided in Section 5. Section 6 concludes the paper.

2 Vocabulary-Based Hashing

In this section we briefly describe the vocabulary-based hashing scheme proposed in [12].

Denote a hash function family mapping a domain S into U as $\mathcal{H} = \{h : S \rightarrow U\}$. [12] propose to use feature point vocabularies to define hash functions by partitioning the space into Voronoi cells. Formally, A hash function $h \in \mathcal{H}$ is defined as

$$h(q) = \arg\min_{0 \leq i < t} D(q, w_h^i), w_h^i \in V_h$$

where

$$V_h = \{w_h^i, 0 \leq i < t\}$$

is a vocabulary associated with h, t is the size of the vocabulary and $D(q, w)$ is the Euclidean distance between points q and w.

Here we remind the hashing index scheme using a given function family [9]. First for a given parameter k, define a function family $\mathcal{G} = \{g : S \rightarrow U^k\}$ such that $g(p) = (h_1(p), \dots, h_k(p))$, where $h_i \in \mathcal{H}$. Then for a given parameter L, choose L functions $g_1, \dots, g_L$ from $\mathcal{G}$. During the construction of the index, each data point p is stored in the buckets $g_j(p)$, for $j = 1, \dots, L$. To find neighbors for a query point q, search all buckets $g_1(q), \dots, g_L(q)$ and return all the points encountered. Thus, the functions $g_1, \dots, g_L$ define a hashing index and different hashing function family $\mathcal{H}$ leads to different index. The vocabulary-based hashing index is constructed by employing the vocabulary-based hash functions. For simplicity, we call $V_g = (V_{h_1}, \dots, V_{h_k})$ a vocabulary associated with $g = (h_1, \dots, h_k)$ and let $V = (V_{g_1}, \dots, V_{g_L})$.

3 Vocabulary Construction

As mentioned above, the vocabularies play a key role in the scheme. The basic idea for the vocabulary construction in [12] is to select vocabularies of best quality from randomly generated ones. It is time-consuming because sufficient amount of random vocabularies are required for selection. We propose an algorithm that utilizes AdaBoost [7] to speed up the construction. The AdaBoost approach needs a criterion for representing the quality of a point being indexed to adjust the weight of the training data. So the first subsection focus on designing the criterion. Then we describe the vocabulary construction algorithm and provide an analysis of the AdaBoost approach in this context.

3.1 Index Quality Criterion

As noted in [12,10], a high-quality search index should return ground truth points and filter noise points with high probability at the same time. For example, in the typical application of similar image search, the retrieved neighbors are used for voting. Here the true positive neighbors can be regarded as useful information while the false positive neighbors bring noise into the voting. Thus we define the index quality of a point to be the signal/noise ratio of the returned neighbors if the point is used as a query. Formally, denote the dataset as $\mathcal{P}$. Suppose p and q are near neighbors if $D(p, q) < R$. Let

$$\begin{aligned} TP_g(q) &= \{p : g(p) = g(q), D(p, q) < R\} \\ T(q) &= \{p : D(p, q) < R\} \\ P_g(q) &= \{p : g(p) = g(q)\}. \end{aligned}$$

Assume there is only one true positive neighbor. Its possibility of being returned is $|TP_g(q)|/|T(q)|$, which can be used as the measure for information brought in. Further assume the weight of the noise brought in by one returned neighbor is w. We define the index quality of q in g to be

$$\hat{v}_g(q) = \frac{|TP_g(q)|/|T(q)| + 1}{w|P_g(q)| + 1} \tag{1}$$

where $|\cdot|$ is the number of points in the set. Note that better designs are possible but left for future work.

Here we discuss the setting of w in detail. We simplify the analysis by making the following assumptions. First, the noise brought in by n returned neighbors will counteract the useful information brought in by true neighbors and thus $w=1/n$. Second, the data points come from N_o multimedia objects (e.g., images or videos) and the returned neighbors scatter among these objects uniformly and independently. Then if two or more of the n returned points belong to the same object, the information of the true neighbors will be counteracted. Thus, we expect that the n noisy points belong to different multimedia objects with high probability:

$$\frac{\prod_{i=1}^{n-1}(N_o - i)}{N_o^{n-1}} > 1 - \epsilon.$$

Since $\ln(1+x) \approx x$ with small x, approximately we have,

$$\sum_{i=1}^{n-1}(\frac{i}{N_o}) < \epsilon.$$

Setting $\epsilon=0.05$, we have

$$w = \frac{1}{n} \approx \sqrt{\frac{10}{N_o}}. \tag{2}$$

The formula (1) and (2) define the criterion $\hat{v}_g(p)$. We tune it to fit the AdaBoost scheme as follows: if $\hat{v}_g(p)$ is among the smallest $|\mathcal{P}|/10$ ones, then $v_g(p)=-1$, indicating that p is not well-indexed in g and needs more emphasis; otherwise, $v_g(p)=1$. Formally,

$$I_g(p,q) = \begin{cases} 1 & \text{if } \hat{v}_g(p) > \hat{v}_g(q) \\ 0 & \text{otherwise} \end{cases} \tag{3}$$

$$v_g(p) = \begin{cases} -1 & \text{if } \sum_{q\in\mathcal{P}} I_g(p,q) < |\mathcal{P}|/10 \\ 1 & \text{otherwise} \end{cases} \tag{4}$$

3.2 AdaBoost for Vocabulary Construction

Now we have the criterion $v_g(p)$ describing the quality of a point p indexed in the hash table g and thus can utilize AdaBoost in vocabulary construction. During the construction of $V = (V_{g_1}, \ldots, V_{g_L})$, we compute $\{v_{g_i}(p), p \in \mathcal{P}\}$ after V_{g_i} is constructed. Then the weights of the points are adjusted accordingly, emphasizing points with low index quality in the construction of $V_{g_{i+1}}$.

The algorithm is described in **ConstructVocabulary**$(\mathcal{P}, t, k, L, C)$: $\mathcal{P}$ is the training dataset; $\mathcal{W}$ is the weight for points in $\mathcal{P}$; t, k and L are index parameters; C is a parameter indicating the number of repetitions, typically $C=10$.

We first briefly analyze the AdaBoost approach in the context of vocabulary construction, which is analogous to that in the classification context. During the

Procedure. ConstructVocabulary$(\mathcal{P}, t, k, L, C)$

1 For each $p \in \mathcal{P}$, assign the weight $\mathcal{W}_1(p) = 1/|\mathcal{P}|$

2 For $i=1$ to L

1) V_{g_i} = **ConstructVg**$(\mathcal{P}, \mathcal{W}_i, t, k, C)$

2) Compute $\{v_{g_i}(p), p \in \mathcal{P}\}$

3) α_i= **ComputeAlpha**$(\mathcal{W}_i, v_{g_i})$

4) For each $p \in \mathcal{P}$, $\mathcal{W}_{i+1}(p) = \mathcal{W}_i(p)\exp\{-\alpha_i v_{g_i}(p)\}$

5) $Z_i = \sum_{p\in\mathcal{P}} \mathcal{W}_{i+1}(p)$

6) For each $p \in \mathcal{P}$, $\mathcal{W}_{i+1}(p) = \mathcal{W}_{i+1}(p)/Z_i$

3 Return $V = (V_{g_1}, \ldots, V_{g_L})$

---

analysis, we specify the subprocedures **ConstructVg** and **ComputeAlpha**. Let $N=|\mathcal{P}|$, $v_i(p)=v_{g_i}(p)$. We have

$$\begin{aligned}\mathcal{W}_{L+1}(p) &= \mathcal{W}_L(p)\frac{\exp\{-\alpha_L v_L(p)\}}{Z_L}\\ &= \mathcal{W}_1(p)\frac{\exp\{-\sum_{j=1}^{L}\alpha_j v_j(p)\}}{\prod_{j=1}^{L} Z_j}\\ &= \frac{\exp\{-\sum_{j=1}^{L}\alpha_j v_j(p)\}}{N\prod_{j=1}^{L} Z_j}.\end{aligned}$$

As $\sum_{p\in\mathcal{P}} \mathcal{W}_{L+1}(p) = 1$,

$$\begin{aligned}\prod_{j=1}^{L} Z_j &= \frac{1}{N}\sum_{p\in\mathcal{P}}\exp\left\{-\sum_{j=1}^{L}\alpha_j v_j(p)\right\}\\ &\geq \frac{1}{N}\sum_{p\in\mathcal{P}}\left(1-\sum_{j=1}^{L}\alpha_j v_j(p)\right)\\ &= 1-\frac{1}{N}\sum_{j=1}^{L}\alpha_j\sum_{p\in\mathcal{P}} v_j(p).\end{aligned}$$

Assume that the query shares the same distribution with the dataset, which leads to $E\big[v_j(q)\big] \approx \frac{1}{N}\sum_{p\in\mathcal{P}} v_j(p)$, then

$$\begin{aligned}\prod_{j=1}^{L} Z_j &\geq 1-\sum_{j=1}^{L}\alpha_j\frac{1}{N}\sum_{p\in\mathcal{P}} v_j(p)\\ &\approx 1-\sum_{j=1}^{L}\alpha_j E\big[v_j(q)\big].\end{aligned}$$

Thus, $1-\prod_{j=1}^{L} Z_j$ serves as a lower bound for $\sum_{j=1}^{L} \alpha_j E\big[v_j(q)\big]$, which indicates the quality of the index defined by $g_1, ..., g_L$. So we turn to minimize $\prod_{j=1}^{L} Z_j$. A greedy approach is adopted, i.e. incrementally minimize Z_j from $j=1$ to L. So the problem becomes

$$(g_j, \alpha_j)^* = \arg\min_{g,\alpha} \sum_{p\in\mathcal{P}} \mathcal{W}_j(p) \exp\{-\alpha v_g(p)\}$$

We greedily select g_j and then optimize α_j. An approximate solution of g_j will be

$$g_j = \arg\max_{g} \sum_{p\in\mathcal{P}} \mathcal{W}_j(p) v_g(p)$$

which leads to the following **ConstructVg**.

Subprocedure. ConstructVg$(\mathcal{P}, \mathcal{W}, t, k, C)$

1 Compute the mean m of $\mathcal{P}$ and its bounding box $\mathcal{B}$, i.e. the minimum and maximum value in each dimension
2 For $i=1$ to C, $j=1$ to k
Draw t random points $p_{j,s}^i (0 \le s < t)$ within $\mathcal{B}$ uniformly and independently
3 Centralize to

$$w_{j,s}^i = p_{j,s}^i - \frac{1}{t}\sum_{s=0}^{t-1} p_{j,s}^i + m$$

4 Let

$$V_{h_j}^i = \{w_{j,s}^i\}, V_g^i = (V_{h_1}^i, \ldots, V_{h_k}^i)$$

5 Return V_g^i that maximizes $\sum_{p\in\mathcal{P}} \mathcal{W}(p) v_g(p)$

The objective Z_j becomes a function of α. It has nice analytical properties, and many algorithms exist for the optimization. We use Newton's algorithm in **ComputeAlpha**. During the experiments, we observe that setting the parameter $T=20$ will be sufficient for the procedure to converge.

Subprocedure. ComputeAlpha$(\mathcal{W}, v_g)$

1 $\alpha^{(0)} = 1.0$
2 For $i = 1$ to T

$$\alpha^{(i)} = \alpha^{(i-1)} + \frac{\sum_{p\in\mathcal{P}} \mathcal{W}(p) v_g(p) \exp\{-\alpha^{(i-1)} v_g(p)\}}{\sum_{p\in\mathcal{P}} \mathcal{W}(p) v_g^2(p) \exp\{-\alpha^{(i-1)} v_g(p)\}}$$

3 Return $\alpha^{(T)}$

3.3 Implementation for Large Datasets

As in [12], we adopt a hierarchical approach for large datasets: the dataset is partitioned into t_1 subsets and vocabulary construction is performed for each subset. More specifically, during the vocabulary construction step, the dataset $\mathcal{P}$ is first clustered into t_1 points, which form the first level vocabulary $\widehat{V} = \{\widehat{w}^i, 0 \le i < t_1\}$ for all $h \in \mathcal{H}$. Then we hash points on $\widehat{V}$ and each bucket forms a subset $\mathcal{P}_i$. The algorithm **ConstructVocabulary** uses each $\mathcal{P}_i$ as input dataset to construct vocabularies $V_i = (V_{i,g_1}, \ldots, V_{i,g_L}), V_{i,g} = (V_{i,h_1}, \ldots, V_{i,h_k}), V_{i,h} = \{w^s_{i,h}, 0 \le s < t_2\}$, which form the second level. During the search step, we hash the query point on $\widehat{V}$, find which $\mathcal{P}_i$ it falls in, and use V_i to find its approximate near neighbors.

4 Parallelization

The vocabulary-based hashing can be parallelized naturally for the tables work in parallel. The parallelized version of the index is illustrated in Figure 1($L=2$)[12]. The query is sent to each table and further forwarded to the corresponding bucket. Points in those buckets are then returned. The search time in vocabulary-based hashing consists of two parts: hashing on the first level $\widehat{V}$ needs $O(t_1)$ if brute-force search is adopted; hashing on V_i needs $O(t_2kL)$. After parallelization the second part of the search time is reduced to $O(t_2k)$. Also, each table can hold more points and thus the index can deal with larger datasets. Additionally, a single table can be further split into two or more tables which still work in parallel and thus can be deployed on machines without large memory. Our experiments in the next section show its benefits.

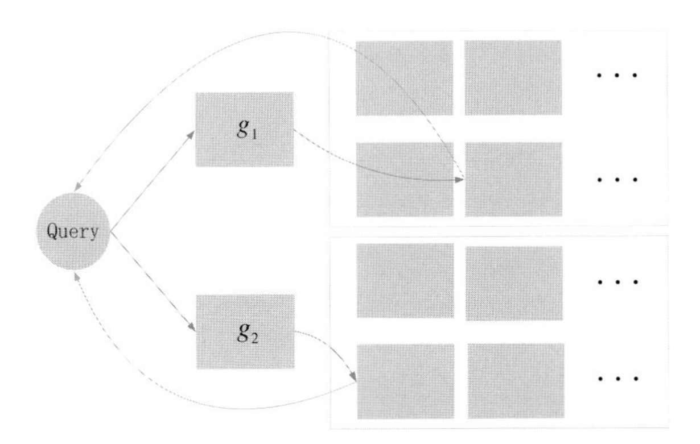

Fig. 1. Parallelized vocabulary-based hashing[12]

5 Experiments

5.1 Settings

We evaluate our index scheme in the typical application of near-duplicate image detection. The experiments are carried out on a Intel(R) Xeon(R) machine with 16GB of memory. The parallelized index is also distributed on other machines with Intel(R) Pentium(R) 4 CPU and 4G memory. The algorithms are implemented in C++.

Fig. 2. Examples of near duplicates. From left to right, first row: original, change gamma to 0.5, crop 70%; second row: scale down by 5 times, frame, rotate 90°.

Datasets. We construct the vocabularies on *Flickr60k* and evaluate on the *Holidays* and *Flickr1M* datasets [10,2]. The *Holidays* (1491 images) is divided into 500 groups, each of which represents a distinct scene or object. The *Flickr60k* (67714 images) and *Flickr1M* (1 million images) are two distinct datasets downloaded arbitrarily from Flickr[1].

For vocabulary construction, we use a dataset *Flickr60k* distinct from the testbed, in order to show more accurately the behavior in large scale scenes, where the dataset itself is too large, or it is updated incrementally so that we do not have the entire dataset at hand for the construction.

For evaluation, we construct the testbed similar to the web scale context from *Holidays* and *Flickr1M*. The first image of each group in *Holidays* is selected to form the query set. Transforms are applied to each query image and the generated duplicates are added into the *Flickr1M* to form the test dataset. The transforms are similar to those in [11,14], and are implemented using ImageMagick[3]. They are listed below and the number in brackets next to each operation denotes the number of near-duplicate images generated.

SIFT [13] descriptors extracted from the images by the software in [2] are used in the experiments.

1. Exact duplicate [1].
2. Changing contrast [2]: (a) change contrast with default parameters in ImageMagick; (b) increase constrast by 3x.
3. Changing intensity [2]: intensity (a) decreased by 50%; (b) increased by 50%.
4. Changing gamma [2]: change gamma to (a) 0.5 or (b) 2.0.
5. Cropping [3]: crop the image by (a) 10% or (b) 50% or (c) 70%, preserving the center region.
6. Framing [1]: Add an outer frame to the image, where the size of the frame is 10% of the framed image.
7. Scaling [2]: scale the image down by (a) 2 or (b) 5 times.
8. Rotating [2]: Rotate image by (a) 90°, (b) 180°.
9. Inserting text [1]: insert the text at the center of the image.
10. Changing format [1]: change the image format JPEG to GIF.

Evaluation Measure. To evaluate the performance of the index, the near neighbor retrieved from the index are used to perform a vote on the images as in [10]. Note that in practice there is usually a post-verification step of the top n positions, especially in large scale scenes where the voting results need further refinements. So rate of true positives returned in the top n positions after voting (perf@n) serves as a suitable performance measure [15,10,6,12]. As there are 17 duplicate images, we choose perf@20 for our evaluation.

5.2 Results

Effectiveness. Figure 3 shows the effectiveness of the index while evaluating on the test data subsets of different sizes. VBH_Ada is the proposed approach with the index parameters $t_1 = 20000, t_2 = 2, k = 8, L = 8$. BOF is the bag-of-features approach with a codebook size of 20000. VBH_Ada outperforms BOF and shows better scalability since it is a refinement of BOF. VBH_Rs and VBH_Rn are vocabulary-based hashing with the same index parameters as VBH_Ada. VBH_Rs employs the construction approach in [12], generating 100 random vocabularies. VBH_Rn uses random vocabularies with words drawn from the bounding box of the dataset uniformly and independently at random. The better performance and scalability of VBH_Ada and VBH_Rs indicates that the construction algorithms do contribute to the effectiveness of the index.

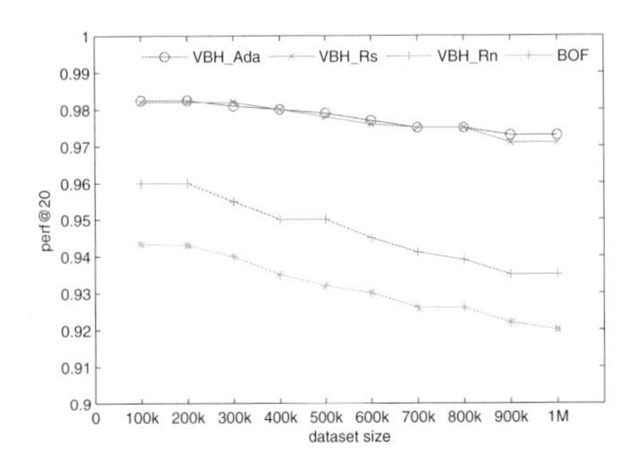

Fig. 3. Effectiveness

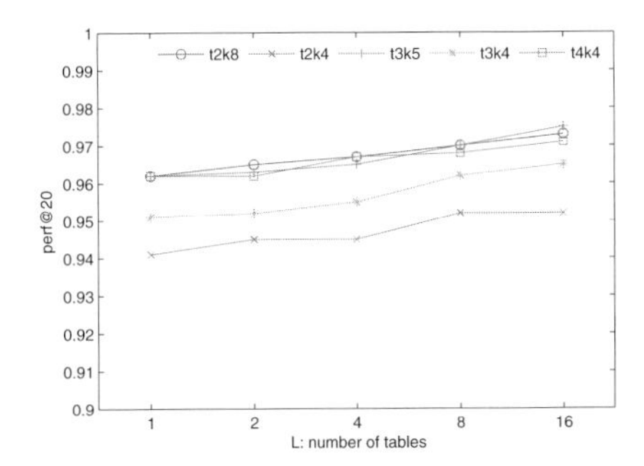

Fig. 4. Parameters

Parameters. Figure 4 shows the performance of vocabulary-based hashing on the 1M dataset with different index parameters t_1, t_2, k, L. All the settings use $t_1 = 20000$, and $txky$ means $t_2 = x, k = y$. Indexes with similar number t_2^k shows similar performance and larger number leads to better results for larger t_2^k indicates finer partition of the feature space. Also the performance increase with table number L.

Time. The vocabulary construction time for the approach in [12] and our proposed approach is presented in table 1. Both approaches are applied on *Flickr60k* to construct vocabularies with index parameters $t_1 = 20000, t_2 = 2, k = 8, L = 16$. The approach in [12] generates 100 random vocabularies for selection. The proposed approach consumes significantly less time since the AdaBoost method avoids the need for generating a large amount of random vocabularies. This makes it more practical in the large scale scene.

Table 1. Vocabulary construction time

Method	Time
[12]	154hr
proposed	21hr

Table 2. Feature extraction and search time per query (Flickr1M dataset)

Method	Feature extraction	Search
serial VBH	0.51s	8.97s
parallelized VBH	0.51s	1.27s
BOF	0.51s	9.16s

The search time for a query in the near-duplicate detection task is presented in table 2. Although the vocabulary-based hashing approach does more computation while searching the near neighbor, it filters out much more noisy points than BOF, thus the total query time does not increase. Further, after parallelization, the approach consumes much less time than the serial version.

Space. The dataset consists of 2072M SIFT descriptors, which occupy a space of 323G. They are impractical to load into memory and thus can not be indexed by some typical structures like E2LSH. Our index keeps only one integer for one point in each table, so each table occupies 8G space. And the parallelized version can be deployed across typical PCs without large amount of memory. If global descriptors are used, such as GISTIS[6], two orders of magnitude more images can be handled, approaching to web scale applications.

6 Conclusion

This paper proposed a new vocabulary construction algorithm for vocabulary-based hashing index. Experiment results show its efficiency which makes it more practical for large scale applications. We also described the parallelized version of the index scheme. Near-duplicate image detection experiments on a dataset with 1M images show its effectiveness and efficiency in the large scale scene.

Acknowledgments

This work was supported by the National Natural Science Foundation of China under the grant No. 60621062 and 60605003, the National Key Foundation R&D Projects under the grant No. 2007CB311003. The authors would like to thank Herve Jegou and Matthijs Douze for providing the datasets, and thank three anonymous reviewers and Jinhui Yuan for their helpful comments.

References

1. Flickr, `http://www.flickr.com`
2. Holidays dataset, `http://lear.inrialpes.fr/people/jegou/data.php`
3. Imagemagick, `http://www.imagemagick.org`
4. Arya, S., Mount, D., Netanyahu, N., Silverman, R., Wu, A.: An optimal algorithm for approximate nearest neighbor searching fixed dimensions. J. ACM (1998)
5. Datar, M., Immorlica, N., Indyk, P., Mirrokni, V.S.: Locality-sensitive hashing scheme based on p-stable distributions. In: SCG (2004)
6. Douze, M., Jégou, H., Singh, H., Amsaleg, L., Schmid, C.: Evaluation of gist descriptors for web-scale image search. In: CIVR. ACM, New York (2009)
7. Freund, Y., Schapire, R.E.: A decision-theoretic generalization of on-line learning and an application to boosting. J. of Computer and System Sciences (1997)
8. Gionis, A., Indyk, P., Motwani, R.: Similarity search in high dimensions via hashing. In: VLDB (1999)
9. Indyk, P., Motwani, R.: Approximate nearest neighbors: towards removing the curse of dimensionality. In: STOC (1998)
10. Jégou, H., Douze, M., Schmid, C.: Hamming embedding and weak geometric consistency for large scale image search. In: Forsyth, D., Torr, P., Zisserman, A. (eds.) ECCV 2008, Part I. LNCS, vol. 5302, pp. 304–317. Springer, Heidelberg (2008)
11. Ke, Y., Sukthankar, R., Huston, L.: Efficient near-duplicate detection and subimage retrieval. In: MM (2004)
12. Liang, Y., Li, J., Zhang, B.: Vocabulary-based hashing for image search. In: MM (to appear, 2009)
13. Lowe, D.G.: Distinctive image features from scale-invariant keypoints. In: IJCV (2004)
14. Meng, Y., Chang, E., Li, B.: Enhancing dpf for near-replica image recognition. In: Proceedings of IEEE Computer Vision and Pattern Recognition (2003)
15. Nister, D., Stewenius, H.: Scalable recognition with a vocabulary tree. In: CVPR (2006)
16. Poullot, S., Buisson, O., Crucianu, M.: Z-grid-based probabilistic retrieval for scaling up content-based copy detection. In: CIVR (2007)
17. Sivic, J., Zisserman, A.: Video Google: A text retrieval approach to object matching in videos. In: ICCV (2003)

Mediapedia: Mining Web Knowledge to Construct Multimedia Encyclopedia

Richang Hong, Jinhui Tang, Zheng-Jun Zha,
Zhiping Luo, and Tat-Seng Chua

Computing 1, 13 Computing Drive, 117417, Singapore
{hongrc,tangjh,zhazj,luozhipi,chuats}@comp.nus.edu.sg

Abstract. In recent years, we have witnessed the blooming of Web 2.0 content such as Wikipedia, Flickr and YouTube, etc. How might we benefit from such rich media resources available on the internet? This paper presents a novel concept called *Mediapedia*, a dynamic multimedia encyclopedia that takes advantage of, and in fact is built from the text and image resources on the Web. The *Mediapedia* distinguishes itself from the traditional encyclopedia in four main ways. (1) It tries to present users with multimedia contents (e.g., text, image, video) which we believed are more intuitive and informative to users. (2) It is fully automated because it downloads the media contents as well as the corresponding textual descriptions from the Web and assembles them for presentation. (3) It is dynamic as it will use the latest multimedia content to compose the answer. This is not true for the traditional encyclopedia. (4) The design of *Mediapedia* is flexible and extensible such that we can easily incorporate new kinds of mediums such as video and languages into the framework. The effectiveness of *Mediapedia* is demonstrated and two potential applications are described in this paper.

Keywords: Web Knowledge, Multimedia Encyclopedia.

1 Introduction

The word "encyclopedia" comes from the classical Greek and was first used in the title of a book in 1954 by *Joachimus Fortius Ringelbergius* [7]. The encyclopedia as we recognize it was developed from the dictionary in the 18th century. However, it differs from dictionary in that each article in encyclopedia covers not a word, but a subject. Moreover, it treats the published article in more depth and conveys the most relevant accumulated knowledge on the subject. Thus encyclopedia is a wealth of human knowledge and has been widely acknowledged.

Most early encyclopedias are laid out using plain text with some drawings or sketches [7]. Their presentation is also somewhat plain and not so vivid. This aroused because of the initiation of print and photograph technologies at that time. In the 20th century, the blooming of multimedia technology has fostered the progress of encyclopedia. One landmark development in encyclopedia is the production of Microsoft's Encarta, which was published on CD-ROMs and supplemented with videos and audio files as well as high quality images. Recently

S. Boll et al. (Eds.): MMM 2010, LNCS 5916, pp. 556–566, 2010.

Fig. 1. The manually grouped Flickr's top 60 images accompanied with the disambiguation entries for "apple" on Wikipedia (Yellow dots denote some other images in top 60). We can see the diversity and somewhat noisy nature of Flickr images and the inherent ambiguity of the concept.

web based encyclopedias such as Wikipedia emerge by leveraging on the hypertext structure and the attributes of user contributed contents. Although there are some criticisms on Wikipedia's bias and inconsistencies, it is still the most popular encyclopedia due to the timeliness of its contents, its online accessibility and it is free of charge. To date, Wikipedia contains more than 13 million articles, of which about 2.9 million are in English[1]. Considering the success of Wikipedia, is it the ultimate form of encyclopedia, or are there other ways to construct a more interesting, useful and attractive encyclopedia?

As we know, Web 2.0 content such as Flickr, Zoomr, YouTube, etc. allow users to distribute, evaluate and interact with each other in the social network. Take Flickr as an example, it contains more than 3.6 million images as of June 2009 and many of the images are in high resolution. Thus the characteristics of Web 2.0 enrich the resources available online. Is it then possible to utilize these rich multimedia repositories to offer the dynamic meaning of concepts as well as new concepts by ways of automatically assembling them for multimedia encyclopedia? Actually projects such as Everything, Encarta and Wikipedia have included some images, audios, and even videos. However, they only appear in a limited number of but not all entries, and they may not be the latest and most representative. Moreover, the presentation is somewhat tedious and unattractive as it focuses mostly on textual description with multimedia contents used mostly as illustrations.

In this paper, we propose a multimedia encyclopedia called *Mediapedia* which is automatically produced and updated by leveraging on the online Web 2.0 resources. The novel form of encyclopedia interprets the subject in a more intuitive and vivid way. The key characteristics of *Mediapedia* that distinguishes it from other encyclopedias are that: (1) the presentation is in form of video;

[1] http://en.wikipedia.org/wiki/Wikipedia

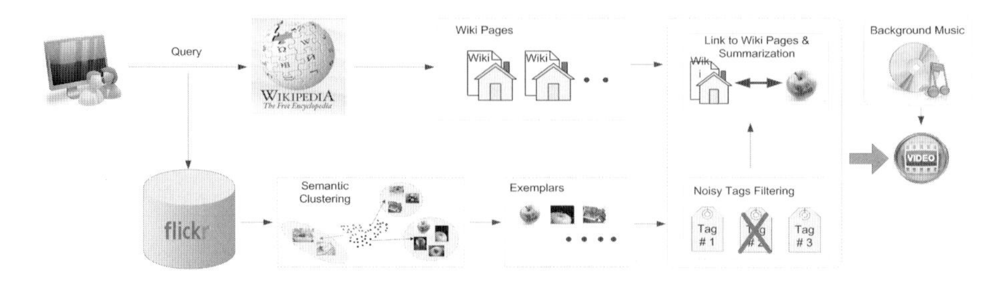

Fig. 2. The system framework of the proposed *Mediapedia*. It mainly consists of: (1) image clustering for producing exemplars; (2) association of exemplars to Wikipedia and (3) multimedia encyclopedia presentation.

(2) it is fully automatically produced, (3) it is updated dynamically; and (4) the whole framework is flexible and extensible that facilitates more potential applications. Through *Mediapedia*, the users can choose to view the concept based on its most "common" meaning or in "diverse" form, which affects the duration of presentation for the concept. When users input a query, the system first crawls the diverse images from Flickr and generates the exemplar images; it then associates the exemplars with Wikipedia summary after noise tag filtering; and finally it automatically produces the multimedia encyclopedia for the concept with synchronized multimedia presentation.

Although *Mediapedia* is promising and desirable, we have to face many challenges. As an example, Fig. 1 illustrates the manually grouped Flickr's top 60 images accompanying with disambiguation entries for the concept "apple" on Wikipedia. We can see that the retrieved images are diverse and somewhat noisy. The disambiguation page on Wikipedia identifies different senses or sub-topics of the concept. Thus we have to find the exemplars from the piles of images and associate them with the corresponding concise Wiki description. We conclude the challenges as follows. (1) How to make the tradeoff between "typicality" and "diversity". In other words, face with the list of retrieved images, which ones are more typical for characterizing the concept and to what extent they are sufficient for showing the diversity. (2) Where and how to discover the corresponding textual contents and prune them to describe the image exemplars. (3) How to present the multimedia content (e.g., text, image and audio) to ensure coherence and elegance of the multimedia encyclopedia. (4) The final and most important challenge is why do we do this work? Are there any potential applications based on this work? In next section, we will answer the "how to construct" question to tackle the challenges (1), (2) and (3). Section 3 evaluates the performance of *Mediapedia*. We describe several potential applications in Section 4, and conclude the paper in Section 5.

2 How to Construct

This section describes the system framework and the algorithms involved. Figure 2 illustrate the proposed framework for *Mediapedia*. We first elaborate on the image clustering for producing the exemplars. We then discuss the association of exemplars with Wikipedia that aims to associate the exemplars with user contributed contents on Wiki [9][17][18]. We finally assemble the exemplars and the concise descriptions to produce the multimedia encyclopedia, where images, transcripts and the background music are presented in an attractive and vivid way. We describe the detailed algorithms in the following subsections.

2.1 Image Clustering for Producing Exemplars for Concepts

Considering the attributes of images from Flickr, a question naturally arise is how to efficiently present the representative images for the concept to users? Some works have been proposed to organize the retrieved images into groups for improving user experiences[1][3]. However, these works are based on traditional clustering algorithm, and although they can produce a more organized result, how to present the clusters to users is still challenging. The studies on finding exemplars from piles of images can be seen as a further step to solving the problem, where the most popular way may be the k-centers algorithm such as [3]. Frey *et al.* proposed affinity propagation to discover exemplars from a set of data points and it has been found to be more effective than the classical methods [4]. It can also be considered as an effective attempt to tackle the problem of finding image exemplars [12][5], which is consistent with the first of the four challenges. Here, we take advantage of the Affinity Propagation (AP) algorithm [4] to acquire exemplars for presentation.

We denote a set of n data points as $\mathcal{X} = \{x_1, x_2, \cdots, x_n\}$ and the similarity measure between two data points as $s(x_i, x_j)$. Clustering aims at assembling the data points into $m (m < n)$ clusters, where each cluster is represented by an "exemplar" from $\mathcal{X}$. Two kinds of messages are propagated in the AP algorithm. The first is the "responsibility" $r(i,k)$ sent from data point i to data point k, which indicates how well k serves as the exemplar for point i taking into account other potential exemplars for i. The second is the "availability" $a(i,j)$ sent from data point k to data point i, which indicates how appropriate for point i to choose point k as exemplar taking into account the potential points that may choose k as their exemplar. The messages are iterated as:

$$r(i,k) \leftarrow s(i,k) - \max_{k' \neq k}\{a(i,k') + s(x_i, x_{k'})\}, \quad (1)$$

$$a(i,k) \leftarrow \min\{0, r(k,k) + \sum_{i' \notin \{i,k\}} \max\{0, r(i',k)\}\}. \quad (2)$$

where the self-availability is updated in a slightly different way as:

$$a(k,k) \leftarrow \sum_{i' \neq k} \max\{0, r(i',k)\}. \quad (3)$$

Upon convergence, the exemplar for each data point x_i is chosen as $e(x_i) = x_k$ where k maximizes the following criterion:

$$\arg\max_k \; a(i,k) + r(i,k) \quad (4)$$

2.2 Association of Exemplars to Wikipedia Pages

The AP algorithm facilitates the finding of exemplars for each cluster. Given the exemplars, we have to face the second challenge of where and how to obtain the corresponding textual description and prune it for those exemplars. As we know, most uploaded images contain a large number of user contributed tags in social media network. However, the tags tend to contain a lot of noise, and are inadequate to help users understand the inherent meaning of the images. An intuitive way is to associate the exemplars with Wikipedia by leveraging the exemplar's tags. We thus need to remove the noisy tags first and then analyze the correlation of remaining tags of each exemplar with its corresponding Wiki pages. We also need to prune the Wiki pages by summarization techniques to produce a brief description in *Mediapedia*.

Noisy Tag Filtering. Since we aim to describe the exemplars with their corresponding text descriptions in Wikipedia by leveraging on the tags of the exemplars, the quality of the tags should therefore be of high quality. In other words, we need to remove insignificant tags such as typo, number, model ID and stop-words, etc., from the tag list. WordNet[2] is a popular lexical database and has been widely used in eliminating noisy tags [8]. Here we list the tags in their respective word group and removing those tags that do not appear in WordNet. We denote the tags after noise filtering as $\mathcal{T} = \{t_{ij}, 1 \leq i \leq m, 1 \leq j \leq N(t_i)\}$, where j indicates the tag in the group t_i and $N(t_i)$ indicates the total number of tags in t_i.

We then utilize the Normalized Google Distance (NGD) [10] between the concept and its associated tags in each cluster as a metric for the semantic relationship between them. Since NGD is a measure of semantic inter-relatedness derived from the number of hits returned by Google search engine, it can be used to explore the semantic distance between different concept-tag pairs. Given the concept q and tag t_{ij}, the NGDbetween them is defined as:

$$ngd(q, t_{ij}) = \frac{\max\{\log f(q), \log f(t_{ij})\} - \log f(q, t_{ij})}{\log M - \min\{\log f(q), \log f(t_{ij})\}} \tag{5}$$

Here, M is the total number of retrieved web pages; $f(q)$ and $f(t_{ij})$ are the number of hits for the concept q and tag t_{ij} respectively; and $f(q, t_{ij})$ is the number of web pages containing both q and t_{ij}. We then use the NGD metric to rank the tags with respect to the concept q. Note that some tags may appear in more than one cluster since a concept usually possesses a variety of presentations and the AP algorithm only groups the images with feature similarity.

Linking to Wikipedia. As aforementioned, for each concept, we obtain several exemplar images via clustering the retrieved Flickr images. We then collect the corresponding documents from Wikipedia by Wiki dump[3]. By considering these

[2] http://wordnet.princeton.edu/
[3] http://en.wikipedia.org/wiki/Dump

exemplar images as characterizing the various aspects of the query in a visual manner while the Wiki documents provide textual descriptions of different senses of a concept, we thus argue that the seamless combination of the visual and textual description can help users better comprehend the concept[15]. However, it is nontrivial to automatically associate each image with the corresponding document.

Here we resort to Latent Semantic Analysis (LSA) method to calculate the similarity between each image and the documents, while the similarity is in turn used to associate the images with the documents [14]. LSA is an approach for automatic indexing and information retrieval that maps document as well as terms to a representation in the so-called latent semantic space. The rationale is that documents which share frequently co-occurring terms will have a similar representation in the latent space. We give the necessary mathematical equations and methodology for LSA here. More detailed theoretical analysis can be found in [14].

LSA represents the association of terms to documents as a term-document matrix:

$$\mathcal{R} = \begin{pmatrix} r_{11} & \cdots & r_{1n} \\ \vdots & \ddots & \vdots \\ r_{m1} & \cdots & r_{mn} \end{pmatrix} = [d_1, \cdots, d_n] = [t_1, \cdots, t_m]^T \tag{6}$$

where element r_{ij} denotes the frequency of term i occurs in document j and properly weighted by other factors [16]. d_i is the i-th column of $\mathcal{R}$ corresponding to the i-th document, and t_j is the j-th row of $\mathcal{R}$ corresponding to the j-th term. The documents and terms are then re-represented in a k-dimensional latent vector space. This is achieved using the truncated singular value decomposition as follows:

$$\mathcal{R} = \mathbf{U}_k \mathbf{\Sigma}_k {\mathbf{V}_k}^T \tag{7}$$

where $\mathbf{U}_k$ and $\mathbf{V}_k$ are orthonormal matrices consisting of the left and right singular vectors respectively, and $\mathbf{\Sigma}_k$ is a diagonal matrix containing the singular values.

Given a concept q consisting of the tags of a image, its corresponding representation in k-dimensional latent semantic space is $q^T\mathbf{U}_k$, and the documents are represented as columns of $\mathbf{\Sigma}_k {\mathbf{V}_k}^T$. The similarities between the concept and documents are calculated as:

$$sim = (q^T\mathbf{U}_k)(\mathbf{\Sigma}_k {\mathbf{V}_k}^T) \tag{8}$$

where the i-th column of sim is the similarity between the concept and the i-th document.

2.3 Encyclopedia Presentation and User Interface

Given the exemplars and the associated Wiki pages, the next problem is how to present the multimedia content to ensure coherence and elegance of the multimedia encyclopedia. Since the articles from Wikipedia are highly structured

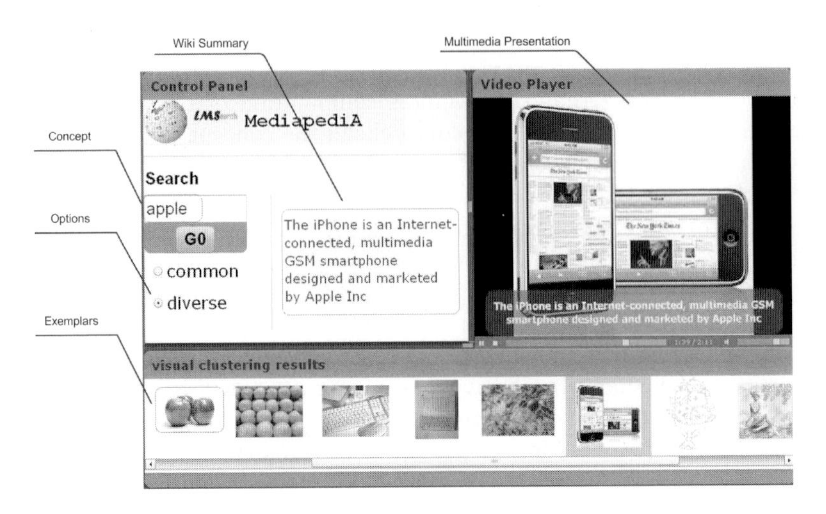

Fig. 3. The user interface of *Mediapedia*. The five annotations indicate the functionalities of the corresponding parts.

and cover as many sub-topics about the concept as possible, it is infeasible to directly incorporate the Wiki content into *Mediapedia*. A concise summary of the concept would be more desirable.

There are many methods focusing on the summarization of individual document or multiple documents. In the scenario of web documents, the performance of those methods can be improved by leveraging the hypertext structure including anchor text, web framework etc. Ye *et al.* proposed to summarize the Wikipedia pages by utilizing their defined Wiki concepts and inforbox structure through the extended document concept lattice model [6]. Moreover, it produces summary with various lengths to suit different users' needs. We therefore employ the summarization method in [6] to provide a concise description for the concept where the length of summary can be controlled by the users.

In the presentation of encyclopedia, we claim that video or well structured audio visual presentation is more attractive and vivid as compared to plain text even with hypertext links to sound, images and motion. In this study, we employ the APIs from *imageloop*[4] for presenting the multimedia encyclopedia by way of slideshow which consecutively displays the images ordered by the size of clusters. The display effect is constrained to the options of *imageloop*. In order to demonstrate the performance of exemplars generation and Wiki page association, we design the user interface to show the composed slideshow accompanying with the exemplars and text summary from Wiki on the same panel. Figure 3 illustrates the user interface. We embed background music to *Mediapedia* for enhancing its presentation. In our rudimentary system, the background music is randomly selected from a pool of *Bandari Rhythms*[5].

4 http://www.imageloop.com/en/api/index.htm
5 http://en.wikipedia.org/wiki/Bandari_music

3 Evaluation of Mediapedia

In this section, we briefly evaluate the performance of the generation of exemplars and their links to Wikipedia pages respectively. More detailed analysis of the performance on individual components can be referred to the literatures [4][11][6][14]. We choose 20 concepts, most of which are from the "NUS-WIDE" concept list [8] and crawl the top ranked 1000 returned images for each concept. For simplicity, we adopt the 225-D block-wise color moment as the visual features. The similarity between two images is considered as the negative distance between their feature vectors [4]. The damping factor in AP algorithm is set to 0.5 and all the preferences, i.e., the diagonal elements of the similarity matrix are set to the median of $s(i,k), i \neq k$. After clustering, we rank the clusters according to their size. Figure 4(a) illustrates the top 12 exemplars for the concept "apple" generated by the AP algorithm.

Fig. 4. Evaluation on the modules. (a) the top 12 exemplars for the concept "apple" generated by affinity propagation. (b) two exemplar images for the concept "apple" and their associated Wiki summaries with the option of two sentences.

For the association of exemplars to Wiki pages, we download the Wikipedia entries for each concept as well as the disambiguation entries for those concepts that contain ambiguity. Here, the concepts are directly used as queries for downloading. We utilize NGD to filter some irrelevant tags with respect to the concept. We then apply LSA to the relationship matrix constructed by the tags and the documents. For each image, the value of its relevance to Wiki pages is accumulated by the normalized similarity between each tag-document pair, which is similar to the metric of Normalized Discounted Cumulative Gain (NDCG). Note that if the concept is without ambiguity (by referring to Wikipedia), the exemplar images and Wiki pages are directly associated. For concise presentation, Wiki pages are summarized by the method in [6]. Figure 4(b) illustrates the two exemplar images associated with the summary from Wiki pages. Note that the number of the sentences is optional and here we show the summaries with two sentences only. The tags in yellow font are removed by the step of noisy tag filtering.

We briefly introduce the process of testing and present some experimental results for the concept "apple". Since the individual component has been evaluated in other works[4][11][6][14], we only briefly test the overall performance of this system. In this paper, we define three metrics for user based evaluation: 1) experience,

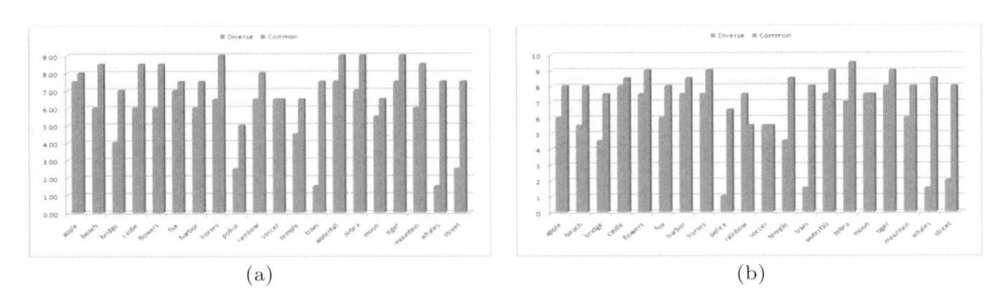

Fig. 5. User based evaluation on the overall performance: (a) experience and (b) informative

do you think the system is interesting and helpful? 2) informative, do you feel the system can properly describe the concept (taking account of both "typicality" and "diversity")? Four students are involved in the user study. For the metrics, each student is asked to score the system performance on 10 levels ranging from 1 (unacceptable) to 10 (enjoyable). Figure 5 demonstrates the user based evaluation. We can see that the overall performance is not as good as expected and should be improved further. On the one hand, the images from Flickr are too diverse and noisy to be ideally grouped. On the other hand, the rational of our study is based on the assumption that the distribution of images from Flickr can automatically make a tradeoff between "typicality" and "diversity". However, in some cases, this assumption may not agree with the truth. Thus we can see for the option of "common", the performance is acceptable while should be improved for the option of "diverse".

4 The Potential Applications

This section answers the final challenge about why we do this work. Similar to encyclopedia which has been widely used all around our lives, *Mediapedia* can also be applied to many fields: as the coinage *Mediapedia* indicates, we can lookup *Mediapedia* for interpretation with multimedia form. Imaging the scenario of child education, such a vivid and colorful encyclopedia would be very attractive. Here, we simply present another two potential applications to show how we can benefit from *Mediapedia*.

4.1 Definitional Question-Answering

Definitional QA in multimedia is emerging following the blooming of internet technology and the community contributed social media. Chua *et al.* presented a survey on Multimedia QA and the bridge between definitional QA in text and multimedia [2]. Analogous to textual definitional QA, it is a way of Multimedia QA by incorporating some defined key-shots, which equate the sentences in textual QA and indicate the key sub-topics of the query, into the answers. The idea has been approved successfully in event driven web video summarization, i.e., multimedia QA for events [11]. *Mediapedia* can be viewed as another way for multimedia QA.

The attributes of intuitiveness and vividness of *Mediapedia* facilitate the application in multimedia QA. Compared to the process of determining key-shots first and then combining them into the answer, the way of *Mediapedia* would be more direct and flexible. Moreover, we can apply some image processing techniques such as the film digital effects to produce more coherent and elegant video answers.

4.2 Tourist Site Snapshot

There are a huge number of images related to some popular tourist sites on Flickr. Moreover, more photos are taken on some landmark buildings or scenes. Take "Paris" as an example, the "Eiffel Tower" can be deemed as one of the famous landmarks. Thus in our system, the images from those sites will be grouped into comparatively larger clusters by the AP algorithm. After association to the entry on Wikipedia, the produced slideshow will comprise of views of the popular landmark and their corresponding descriptions. This will provide informative snapshots of the tourist sites to the users.

It is more exciting that many images from the social sharing websites offer geo-information which specifies the specific world location of the photo taken [13]. Considering the fact that Flickr alone carries over one hundred million geo-tagged photos (out of a total of 3 billion photos), the use of these images would improve the performance of selecting the exemplar images and provide better tourist site snapshots, even tourism recommendation.

5 Discussions and Conclusions

This paper presented a novel concept named *Mediapedia* which aims to construct multimedia encyclopedia by mining web knowledge. The *Mediapedia* distinguishes itself from traditional encyclopedia in its multimedia presentation, full automated production, dynamic update and the flexible framework where each module is extensible to potential applications. In the proposed system, we employed the AP algorithm in producing the exemplars from image pool, while using LSA to associate exemplars to Wiki pages and utilizing document lattice model to perform Wiki pages summarization. We finally assembled them for multimedia encyclopedia. Two potential applications are described in detail.

This study can be deemed as an attempt at constructing *Mediapedia* by leveraging on web knowledge. The experimental results, however, were not as good as expected and should be improved further. This may be aroused by the assumption that the distribution of images from Flickr can automatically make a tradeoff between "typicality" and "diversity". However, this is not the truth for all the concepts. Improvement can be made by taking into account the tags in producing the exemplar, and leveraging the images embedded in Wikipedia to facilitate better association and so on. An alternative approach to tackle the problem is to start our *Mediapedia* from Wikipedia, by identifying different senses of the concept by Wikipedia first and then associating them with images/audios. Thus, the framework proposed in this study is still evolving. The main contribution of this work can be seen as an attempt to construct multimedia encyclopedia by mining the rich Web 2.0 content. Moreover, we provide an interesting system which performance is acceptable. In future works, we will improve the performance of *Mediapedia* by incorporating interactivity and exploring another design alternative.

References

1. Deng, C., Xiaofei, H., Zhiwei, L., Wwei-Ying, M., Ji-Rong, W.: Hierarchical clustering of WWW image search results using visual, textual and link information. In: Proc. of ACM MM 2004, pp. 952–959 (2004)
2. Tat-Seng, C., Richang, H., Guangda, L., Jinhui, T.: From Text Question-Answering to Multimedia QA. To appear in ACM Multimedia Workshop on Large-Scale Multimedia Retrieval and Mining, LS-MMRM (2009)
3. Gonzalez, T.: Clustering to minimize the maximum intercluster distance. Theoretical Computer Science 38(2), 293–306 (1985)
4. Frey, B., Dueck, D.: Clustering by Passing Messages Between Data Points. Science 315(5814), 972 (2007)
5. Yangqing, J., Jingdong, W., Changshui, Z., Xian-Sheng, H.: Finding Image Exemplars Using Fast Sparse Affinity Propagation. In: ACM International Conference on Multimedia, ACM'MM 2008, Vancouver, BC, Canada (2008)
6. Shiren, Y., Tat-Seng, C., Jie, L.: Summarization Definition from Wikipedia. In: ACL-IJCNLP, Singapore, August 2-7 (2009)
7. Carey, S.: Two Strategies of Encyclopaedism. Pliny's Catalogue of Culture: Art and Empire in the Natural History. Oxford University Press, Oxford (2003)
8. Tat-Seng, C., Jinhui, T., Richang, H., Haojie, L., Zhiping, L., Yantao, Z.: NUS-WIDE: A Real-World Web Image Databased From National University of Singapore. In: ACM International Conference on Image and Video Retrieval, Greece, July 8-10 (2009)
9. Meng, W., Xian-Sheng, H., Yan, S., Xun, Y., Shipeng, L., Hong-Jiang, Z.: Video Annotation by Semi-Supervised Learning with Kernel Density Estimation. In: ACM'MM (2006)
10. Rudi, C., Paul, V.: The Google Similarity Distance, ArXiv.org or The Google Similarity Distance. IEEE Trans. Knowledge and Data Engineering 19(3), 370–383 (2007)
11. Richang, H., Jinhui, T., Hung-Khoon, T., Shuicheng, Y., Chong-Wah, N., Tat-Seng, C.: Event Driven Summarization for Web Videos. To appear in ACM Multimedia Workshop on Social Media, WSM (2009)
12. Zheng-Jun, Z., Linjun, Y., Tao, M., Meng, W., Zengfu, W.: Visual Query Suggestion. To appear in ACM'MM, Beijing, China (2009)
13. Yan-Tao, Z., Ming, Z., Yang, S., Hartwig, A., Ulrich, B., Alessandro, B., Fernando, B., Tat-Seng, C.: Neven Hartmut: Tour the World: building a web-scale landmark recogntion engine. In: Proceedings of CVPR 2009, Miami, Florida, US, June 20-25 (2009)
14. Ding, C.H.Q.: A Similarity-Based Probability Model for Latent Semantic Indexing. In: Proc. of SIGIR (1999)
15. Haojie, L., Jinhui, T., Guangda, L., Tat-Seng, C.: Word2Image: Towards Visual Interpretation of Words. In: Proceedings of ACM'MM 2008 (2008)
16. Dumais, S.: Improving the retrieval of information from external sources. Behavior Research Methods, Instruments and Computers 232, 229–236 (1991)
17. Meng, W., Kuiyuan, Y., Xian-Sheng, H., Hong-Jiang, Z.: Visual Tag Dictionary: Interpreting Tags with Visual Words. To appear in ACM Multimedia Workshop on Web-Scale Multimedia Corpus (2009)
18. Jinhui, T., Shuicheng, Y., Richang, H., Guojun, Q., Tat-Seng, C.: Inferring Semantic Concepts from Community-Contributed Images and Noisy Tags. To appear in ACM'MM 2009 (2009)

Sensing Geographical Impact Factor of Multimedia News Events for Localized Retrieval and News Filtering

Xu Zhang[1,2], Jin-Tao Li[1], Yong-Dong Zhang[1], and Shi-Yong Neo[3]

[1] Institute of Computing Technology, Chinese Academy of Sciences, Beijing 100190, China
[2] Graduate School of the Chinese Academy of Sciences, Beijing 100039, China
[3] National University of Singapore, 3 Science Dr, Singapore 117543
{zhangxu,jtli,zhyd}@ict.ac.cn,
neoshiyo@comp.nus.edu.sg

Abstract. News materials are reports on events occurring in a given time and location. Looking at the influence of individual event, an event that has news reported worldwide is strategically more important than one that is only covered by local news agencies. In fact, news coverage of an event can accurately determine the event's importance and potential impact on the society. In this paper, we present a framework which extracts the latent impact factor of events from multimedia news resources by a geographical approach to support: (a) localized retrieval for end users; and (b) pre-screening of potential news elements that should be filtered for use by web monitoring agencies.

Keywords: Multimedia News Impact Factor, Retrieval, Filtering.

1 Introduction

Information retrieval especially in the domain of multimedia news is increasingly important with the ever increasing amount of multimedia data. With the rise of multimedia news search engines such as Yahoo!, Google and MSN, it has become extremely easy for end-users to gain access to the wide range of published multimedia news available on the Web. This has however created a scenario where the end-user is subjected to an extensive amount of unrelated and undesired information. In particular, regulating agencies are also facing difficulties in filtering undesirable news materials, especially those in non-text formats such as videos. Consequently, to handle this problem and enhance news video retrieval, some form of advanced video processing is needed.

Most prior researches in video processing rely strongly on features within video source only for retrieval [1], [2]. A recent approach [3] demonstrated that features from external information such as blogs and online news articles may also be useful in improving retrieval. In this paper, we propose another useful facet of news, the geographical news coverage, or what we term as the event impact factor. News reports consist of essential events occurring in a given time and location, with different news agencies having its own degree of involvement. A piece of news that has a worldwide coverage is termed more strategically important than one that is only covered by local news

S. Boll et al. (Eds.): MMM 2010, LNCS 5916, pp. 567–576, 2010.

agencies. In fact, the news coverage of an event has direct implications to its impact on the society. Based on this observation, we propose a framework to extract and utilize the derived event impact factor as a major enhancement for: (a) supporting localized multimedia news retrieval; (b) acting as a gauge for filtering of sensitive news for data governing agencies.

Sensing the Impact factor. The impact factor of an event can be inferred from the news coverage of the event. In reality, a unique event can have different video presentations and be narrated in different languages for different locations or lingual groups. In order to determine the news coverage for an event, it is necessary to find all the related news materials reporting about the event. We therefore perform clustering of the collected news materials to obtain coherent groups of news for different events. The process is similar to topic detection and tracking (TDT) [4]. Using the groups obtained, the news coverage of an event can then be inferred through the strategic influence of each news agency. For example: "[Straits Times, Channel 5] Singapore", "[人民日报(People's Daily), CCTV] China", "[CNN, NYT, MSNBC] US". Subsequently, the impact factor matrix is derived for each cluster of news materials to support retrieval and filtering.

Utilizing the Impact factor for localized retrieval and news filtering. To leverage the impact factor matrix, we first need to know an event's actual occurrence location. We determine this by space and frequency analysis of location entities mentioned in news articles and news videos. There might be more than one location entity mentioned, but through the analysis, confidence score can be assigned to each entity. One useful heuristic rule to aid in confidence scoring is that news reports will commonly begin by highlighting the event location. At present, news search engines localize searches through the detection of IP address. Our proposed impact factor can similarly be exploited like IP address to further rank news accordingly. The rationale is that impact factor scores can effectively depict an event's importance. In retrieval we can allow users to indicate several locations which they are interested so as to view the highly important news in these locations. As for news filtering, we can forbid users from a targeted location "T" in gaining access to sensitive news materials, for example politics-related news videos. As an illustration, a particular event E which occurs in "T" may have a large amount of news coverage in many countries (high impact factor), but significantly low impact factor in "T" itself. This scenario can signal that news regarding the event could be sensitive in nature for "T". The intuition for this is built upon a reverse psychologically effect where local media in "T" have likely done preventive measures to block sensitive news.

2 Sensing Event Impact Factor

The influence of an event is location sensitive, as people living in different parts of the world are likely to see only news which is made known to them through their local media broadcasters. Thus, it makes perfect sense to rank news according to the location sensitive impact factor. Figure 1 shows the overall framework for event impact factor sensing. The first step to understand the news coverage is to obtain semantic groups of news reports related to an event. News reports of the same event can be in the form of video, text or audio, and presented differently across multiple agencies even in different languages. It is necessary to pre-cluster related news materials on a per-event basis. Catering to this, we construct an event space to model events and utilize an unsupervised temporal multi-stage clustering algorithm to obtain coherent groups of news. With this grouping, it is possible to infer the coverage of an event and its impact by analyzing the strategic influence of the news agencies reporting the event.

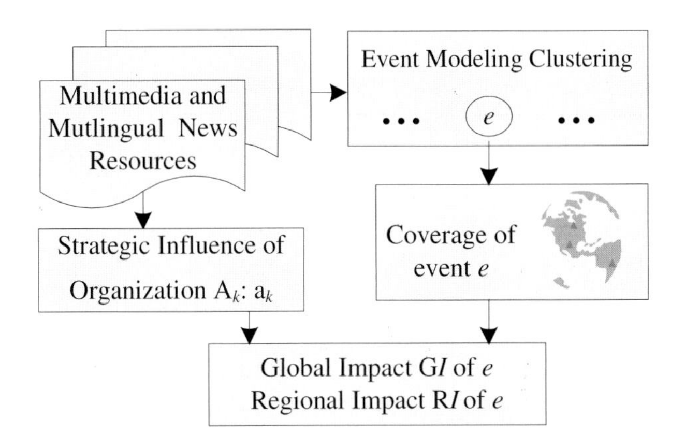

Fig. 1. Framework for Sensing Impact Factor

2.1 Multimedia News Event Modeling and Clustering on News Video and Online Articles

There are subtle differences between news and events. While an event is the occurrence of an incident, news is the report on a series of events. Regardless of the media format in publication for a news report, it will contain various event aspects such as: *Location*, *Time*, *Subject*, *Object*, *Quantity*, *Action* and *Description*, etc. Thus by translating a piece of news onto a distinct point in a multi-dimensional event space, where each dimension is an event aspect, we can effectively use news materials to model events. For this task, we have chosen the event feature space and methods according to the observed effectiveness in [3]. The features used in our event space consist of: (a) text entities from speech; (b) high level features (HLF); (c) near duplicate (ND) information. They are extracted from videos and online news articles. In our model, we define essential text event entities as follows: **Location** {country, city, county, places of interest, etc}, **Time** {video timestamp or specific date mentioned, etc}, **Object** {tangible like car, people, intangible like war, oil prices}, **Subject** {person's name, organization, etc}, **Quantity** {numerical}, **Action** {death, birth, murder, etc} and **Descriptions** {other deeds}.

Apart from text entities, HLF is also utilized, which are extracted from the visual content of videos through support vector machine analysis on low level features. As shown in [5], the use of HLF allows news to have more semantic information. Coupled with this, our model is further enhanced by ND information [6], which can easily provide the linkage between different news video events. This is because reports of the same event can sometimes have the same footage or slightly different footages.

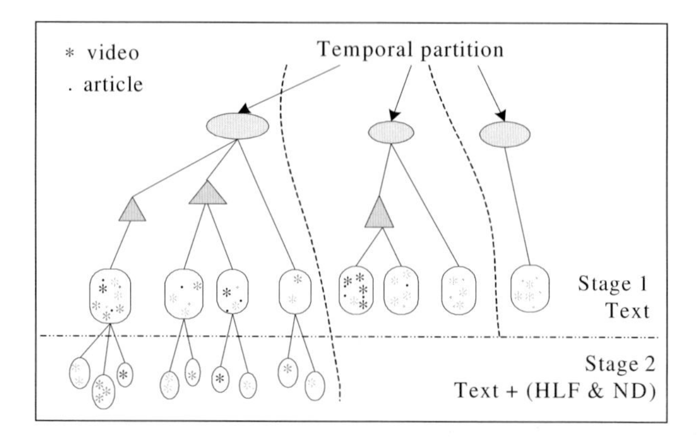

Fig. 2. Temporal Multi-stage Event Clustering

In order to obtain coherent groups of news from news video and online articles, an unsupervised temporal multi-stage clustering algorithm as shown in Figure 2 is used. The multi-stage clustering framework uses text during the first stage of clustering and a combination of text, HLF and ND for the second stage. The aim of the first stage is to identify possible events using text, which is one of the best feature for event detection and tracking. However, text from speech transcripts may be erroneous and insufficient. Therefore we supplement it with text from parallel news articles. This combination enables key entities in events (from text) to be correctly extracted for effective event clustering. In the second stage, HLF and visual features are utilized to refine the initial clusters. News resources are then divided into temporal partitions before clustering. The aim of temporal partitions is to reduce the computation time, and provide a smaller clustering space from which better clustering results can be obtained. More details can be obtained from [3].

2.2 Strategic Influence of News Agencies and Event Impact Factor

As news materials are copyrighted, details of publishing agencies are usually appended with the news elements. Considering such a characteristic, we introduce a two-stage approach to map events into their news coverage areas with respect to agencies by: a) identify the agencies and their strategic area of influence; and b) map the news reports or videos in clusters from Section 2.1 to corresponding agencies and tabulate their impact scores. We use a list containing over 1000 news agencies for computing the impact scores. This list is a subset of which Google online news [7] performs their crawling. The approximate breakdown for the new agencies across countries is given in Figure 3.

In many countries, events that are reported by authoritative agencies are relatively more influential than those reported by smaller agencies. The strategic penetration depth (or circulation) of the agency also plays a significant role in determining the amount of potential readers and viewers. This strategic reach of an agency can be

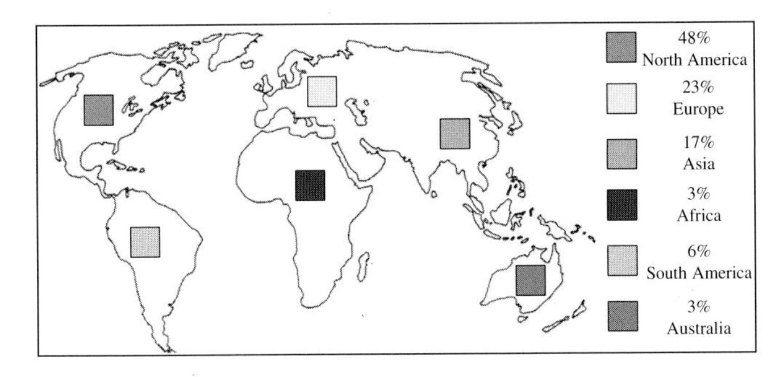

Fig. 3. Strategic breakdown of news agencies at marco-continental level

estimated through its local viewer/readership population (at county, province or country level, depending on the area of circulation). Another observation about authoritative agencies is that they tend to produce larger amount of news reports. It is therefore possible to evaluate the agency strategic importance (ASI) by counting the daily reports. We approximate this value by the number of reports of each news agency during a certain period. The ASI a_k of each news agency A_k is defined in equation 1 to provide the underlying strategic importance.

$$a_k = n_{A_k} \cdot p_{A_k} \tag{1}$$

where n_{A_k} is the total number of reports issued by A_k during the given period and normalized to [0,1] by standard normalization; p_{A_k} is the population normalized to [0,1] in the location of A_k which is extracted together with agencies.

With the individual ASI a_k, it is then possible to measure the impact of an event appropriately with respect to various locations. In particular, we consider the impact factor of an event *e* at an individual country-level since this is most appropriate for many applications. The regional impact factor $RI_{e|L}$ of *e* at location *L* is given in equation 2.

$$RI_{e|L} = \lambda \sum_i (a_i \cdot n_e) + (1-\lambda) \sum_j (a_j \cdot len_e) \tag{2}$$

where n_e is the total of text articles about event *e* (for text-based news) and len_e is the length of videos about event *e* (for video news), both of which are normalized to [0,1] by standard normalization. Since video news usually emphasizes events that are most interesting or important due to expensive and limited airtime, we give higher weight to video news. Empirically the weight λ for text news is set as 0.3, weight for video news as 0.7. The impact of an event to various countries is tabulated in the form of an impact factor table. In addition, a global impact factor GI_e of *e* can be calculated by aggregating the scores of all reporting locations. Figure 4 shows the impact of event "Sichuan Earthquake" collated from news report during the period 12th to 22nd of May2008 normalize to 4 shading intensity with the region of highest impact in black.

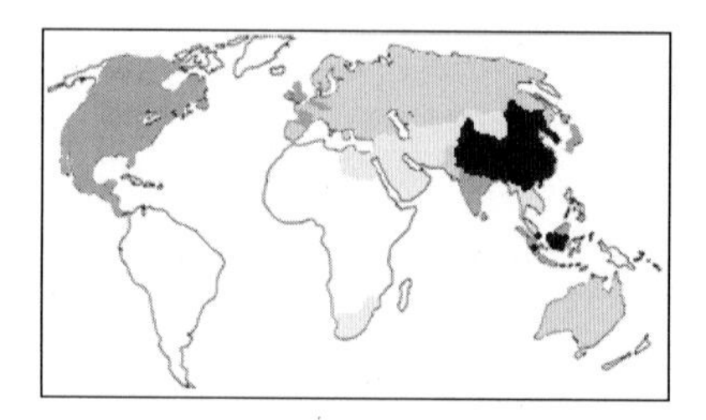

Fig. 4. Impact of event "Sichuan Earthquake"

3 Utilizing Event Impact Factor

After getting the impact factor table, a prior step to extract the event occurrence location is carried out by spatial and frequency analysis of location text entities from each news cluster. With the available event feature space, the features are then employed for: (a) localized news retrieval; and (b) pre-screening of potential news elements that should be filtered for data monitoring agencies.

3.1 Getting the Event's Occurrence Location

Content of news usually contains more than one location entity, which makes it challenging to determine the event's occurrence location. We make decisions with the help of text pragmatics and video cinematic in a rule-based fashion considering the following two characteristics. First, it is observed that news is often presented in a top-down fashion where highlights of news are usually presented at the beginning. This characteristic increases the likelihood that the actual event location appears in the beginning rather than at the end. Second, the actual location entity is likely to be mentioned more frequently than other location entities for emphasis.

Combining the above two characteristics, we propose a *spatial-freq* approach to determine event's location of occurrence. Let $LE=\{L_1,..., L_i,...L_n\}$be all location entities in a news article or video, the event's occurrence location L_e is defined as follows:

$$L_e = \arg\max_{L_i} \left(\beta \cdot r_{L_i} + (1-\beta) \cdot f_{L_i}\right) \tag{3}$$

where r_{L_i} is the reciprocal of number of words before the first appearance of location entity L_i, and f_{L_i} is the frequency of L_i in news content; β is set as 0.6 to emphasize the relative position of L_i. Recognizing the first appearance of L_i requires high complexity in checking for each word. Thus, we use Aho-Corasick algorithm[8] which is designed for fast matching of strings and finding the first possible substrings for our problem.

3.2 Localizing Geo-based News Retrieval

Current online news search engines facilitate localized news searching through users' IP addresses and display news from their localized news agencies. (For example:

CNN and NYT for US, "Straits Times" for Singapore, etc) This rationale is straightforward as news agencies situated near the location of the user are likely to deliver news which is of interest to that user. However, if there is no news coverage of an event by that local broadcaster, it is possible for users to miss that event. With the knowledge of an event location together with its global and regional impact factors, our system aims to further enhance localized retrieval with respect to user's interest, location and IP. Given an event e with its occurrence location L_e, global impact GI_e and regional impact $RI_{e|L}$, the modified ranking score $\overline{s}_{R_{e|L}}$ for news report R_e (article or video) about e at location L is shown in equation 4.

$$\overline{s}_{R_{e|L}} = \theta \cdot \left(\frac{RI_{e|L}}{GI_e} \right) + (1-\theta) \cdot s_{R_e} \tag{4}$$

where s_{R_e} is the original ranking scores generated by initial retrieval algorithm on a scale from 0 to 1 (1 is most relevant) for each news report, the weight θ for impact-based ranking score is empirically set as 0.4 for aggregation.

3.3 News Filtering for Network Monitoring

The idea of online news monitoring is to restrict users within a location to news materials which could be sensitive in nature, as monitoring organizations do not have access control over overseas news agencies. Traditional news filtering systems performs filtering using a predefined list of key words and various matching strategies. That approach can only detect sensitive text in news according to a preference list, and have much limitation on news video. With the discovered event impact factor, it is now possible to estimate the sensitivity $S_{e|L}$ of an event e with respect to location L as in Equation 5.

$$S_{e|L} = \frac{GI_e - RI_{e|L}}{GI_e} \tag{5}$$

where the regional impact $RI_{e|L}$ of e in L and its global impact GI_e can be accessed through the impact factor table generated in Section 2. The larger the value of $S_{e|L}$ is, the event e is more sensitive in the location of L. This criterion is based on the assumption that events should be reported more in its happening places.

4 Experiments

In this section, we will present the experimental results to the proposed tasks of retrieval and filtering. We leverage news video corpus from TRECVID 2006[5] testing dataset, consisting of about 160 hours of multilingual news videos in English, Arabic and Chinese. In addition, we crawled 463,000 news articles through various online news archives, including the Highbeam Research database [9] which archived over 1000 sources. These archived news articles are reported during the same period with TRECVID news video.

4.1 Clustering Performance

Clustering performance is crucial as it will have a predecessor effect on the later applications. To evaluate the clustering performance, manual assessment is done to measure the clustering quality using human annotators. As it is too labor-intensive to screen through all clusters, a subset is selectively chosen for evaluation. This subset stretches over a 7-day period equivalent to approximately 15% of the testing corpus. During evaluation, the human annotators are asked to access inter-cluster and intra-cluster correctness. Three runs are designed to test the effectiveness of: (1) text features, (2) high level features and (3) near duplicate information. The results are tabulated in Table 1.

Table 1. Performance of clustering

TRECVID2006	Baseline	A	B	C
Prec.	0.376	0.418	0.437	0.478

where Baseline: (text event entities with temporal partitions)
A) baseline+ HLF
B) baseline + near duplicate
C) baseline+ HLF + near duplicate

From Table 1, we can draw the following conclusions. First, improvements can be seen with the addition of HLF and near duplicated information. In particular, near duplicate information seem more effective. This can be attributed to the fact that an event can have similar footages across multiple news report and this makes near duplicate information important. The run which utilizes both HLF and near duplicate information yields the best performance, demonstrating the complementary boosting effect.

4.2 Effects of Localizing Geo-based Retrieval

This series of experiment is designed to investigate the effects brought by the enhancement of localization. Ten users with prior knowledge of news video retrieval are selected for the experiments. The users are asked to use the retrieval system and evaluate the retrieved results on a scale of 1 to 5 (5 as with best performance). The following are two sample questions from the evaluation questions list: *How relevant is the news to the input query? Does the news reflect the location you are interested?*

For comparison, we make use of the previous news video system as in [3]. The results are tabulated in Table 2, which shows that users prefer our system with the localization effect. Interactive user interface is suggested by users as part of our future system enhancement.

Table 2. User rating (average)

	[3] without impact factor	[3] with impact factor
Score	3.97	4.64

4.3 Sensitive News Video Filtering

The last series of test is designed to access the effectiveness of the system in ranking sensitive news video. For this test, we select 5 locations of interest: China, India, Saudi Arabia, Russia, and Japan. For each location, the system ranks news videos according to the sensitivity score *S* generated by equation (5). Top n (Max: n= 100) News videos with $S_{e|L}$ above threshold *T* is selected as potentially to be filtered, where *T* is determined by Gaussian estimation as $\overline{S}_{e|L} + 3\delta_s$, where $\overline{S}_{e|L}$ is mean of $S_{e|L}$, $\delta_{S_{e|L}}$ is variance of $S_{e|L}$. As there is no ground truth available to access the performance, we employ 5 human assessors to manually judge the sensitivity. Each human assessor will view the news video clip of the top *n* list and evaluate the sensitivity by three categories: *sensitive*, *not sensitive*, and *unable to judge*; depending on the impact on politics, military or economic at that location. The purpose of this manual assessment is to compare between the judgments of sensitive news videos from our system and from human perception. Subsequently, the agreement values are measured in term of *Fleiss' kappa* for measuring agreements[10].

Table 3. Kappa User Agreement (k)

TRECVID2006	Highest	Lowest	Average
Inter User	0.89	0.51	0.74
User vs. System	0.78	0.46	0.63

From Table 3, we can conjecture that the agreement between users is quite high (an average of 0.74). This is also true for the agreement score between user and our system on sensitive news (an average of 0.63). Although little news which is not sensitive to users is potentially filtered by our system, this is much because of the threshold is too strict. Better estimation of threshold can improve the results. Overall, this signifies that our system is able to filter and detect sensitive news quite reasonably.

5 Conclusion

In this paper, we presented the idea of mining geographical impact factor of news events to support news video retrieval. Two applications are proposed based on our framework: (a) localized news retrieval for end users; and (b) pre-screening of potential news elements for data monitoring agencies. Preliminary results using the TRECVID 2006 dataset and archived news articles demonstrate the effectiveness and usability of our framework. For future work, we are looking into analyzing different opinions across multiple news agencies to summarize and provide more various aspects of news events.

Acknowledgements

This work was supported by National Basic Research Program of China (973 Program, 2007CB311100), National High Technology and Research Development

Program of China (863 Program, 2007AA01Z416), National Nature Science Foundation of China (60873165、60802028), Beijing New Star Project on Science & Technology (2007B071) and Co-building Program of Beijing Municipal Education Commission.

References

1. Chang, S.-F., Winston, J.W., Kennedy, L., Xu, D., Yanagawa, A., Zavesky, E.: Columbia University TRECVID-2006 Video Search and High-Level Feature Extraction. Gaithersburg (2006)
2. Campbell, M., Hauboldy, A., Ebadollahi, S., Joshi, D., Naphade, M.R., Natsev, A.P., Seidl, J., Smith, J.R., Scheinberg, K., Tešić, J., Xie, L.: IBM Research TRECVID-2006 Video Retrieval System
3. Neo, S.-Y., Ran, Y., Goh, H.-K., Zheng, Y., Chua, T.-S., Li, J.: The use of topic evolution to help users browse and find answers in news video corpus. In: Proceedings of the 15th international conference on Multimedia (2007)
4. James, A., Ron, P., Victor, L.: On-line new event detection and tracking. In: Proceedings of the 21st annual international ACM SIGIR conference on Research and development in information retrieval (1998)
5. `http://www-nlpir.nist.gov/projects/trecvid/`
6. Zheng, Y.-T., Neo, S.-Y., Chua, T.-S., Tian, Q.: The use of temporal, semantic and visual partitioning model for efficient near-duplicate keyframe detection in large scale news corpus. In: Proceedings of the 6th ACM international conference on Image and video retrieval (2007)
7. `http://news.google.com/`
8. Aho, A.V., Corasick, M.J.: Efficient string matching: an aid to bibliographic search. Commun. ACM 18(6), 333–340 (1975)
9. `http://www.highbeam.com/web/`
10. Landis, J., Koch, G.: The measurement of observer agreement for categorical data. Biometrics (1977)

Travel Photo and Video Summarization with Cross-Media Correlation and Mutual Influence

Wei-Ta Chu[1], Che-Cheng Lin[1], and Jen-Yu Yu[2]

[1] Department of Computer Science and Information Engineering, National Chung Cheng University, Chiayi, Taiwan
wtchu@cs.ccu.edu.tw, john72831@yahoo.com.tw
[2] Information and Communication Research Lab, Industrial Technology Research Institute, Hsinchu, Taiwan
KevinYu@itri.org.tw

Abstract. This paper presents how cross-media correlation facilitates summarization of photos and videos captured in journeys. Correlation between photos and videos comes from similar content captured in the same temporal order. We transform photos and videos into sequences of visual word histograms, and adopt approximate sequence matching to find correlation. To summarize photos and videos, we propose that the characteristics of correlated photos can be utilized in selecting important video segments into video summaries, and on the other hand, the characteristics of correlated video segments can be utilized in selecting important photos. Experimental results demonstrate that the proposed summarization methods well take advantage of the correlation.

Keywords: Cross-media correlation, photo summarization, video summarization.

1 Introduction

Recording daily life or travel experience by digital videos or photos has been widely accepted in recent years, due to popularity and low cost of digital camcorders and cameras. Large amounts of videos and photos are especially captured in journeys, in which people are happy to capture travel experience at will. However, massive digital content burdens users in media management and browsing. Developing techniques to analyze travel media thus has drawn more and more attention.

There are at least two unique challenges in travel media analysis. First, there is no clear structure in travel media. Unlike scripted videos such as news and movies, videos captured in journeys just follow the travel schedule, and the content in video may consist of anything people willing or unwilling to capture. Second, because amateur photographers don't have professional skills, the captured photos and videos often suffer from bad quality. The same objects in different photos or video segments may have significant appearance. Due to these characteristics, conventional image/video analysis techniques cannot be directly applied to travel media.

People often take both digital cameras and digital camcorders in journeys. They usually capture static objects such as landmark or human faces by cameras and capture

S. Boll et al. (Eds.): MMM 2010, LNCS 5916, pp. 577–587, 2010.

evolution of events such as performance on streets or human's activities by camcorders. Even with only one of these devices, digital cameras have been equipped with video capturing functions, and on the other hand, digital camcorders have the "photo mode" to facilitate taking high-resolution photos. Therefore, photos and videos in the same journey often have similar content, and the correlation between two modalities can be utilized to develop techniques especially for travel media.

In our previous work [1], we investigate content-based correlation between photos and videos, and develop an effective scene detection module for travel videos. The essential idea of this work is to solve a harder problem (video scene detection) by first solving an easier problem (photo scene detection) accompanied with cross-media correlation. In this paper, we try to further take advantage of cross-media correlation to facilitate photo summarization and video summarization. We advocate that summarizing a media can be assisted by other media's characteristics and the correlation between them.

Contributions of this paper are summarized as follows.

- We explore cross-media correlation based on features resisting to significant visual variations and bad quality. Two-level cross media correlations are investigated to facilitate the targeted tasks.
- We advocate that the correlated video segments influence selection of photos in photo summaries, and in the opposite way, the correlated photos influence selection of video segments in video summaries.

The rest of this paper is organized as follows. Section 2 gives literature survey. Section 3 describes the main idea of this work and the components developed for determining cross-media correlation. Photo summarization and video summarization are addressed in Section 4. Section 5 gives experimental results, and Section 6 concludes this paper.

2 Related Works

We briefly review works on home video structuring and editing. Then, studies especially about highlight generation and summarization are reviewed as well. Gatica-Perez et al. [2] cluster video shots based on visual similarity, duration, and temporal adjacency, and therefore find hierarchical structure of videos. On the basis of motion information, Pan and Ngo [3] decompose videos into snippets, which are then used to index home videos. For the purpose of automatic editing, temporal structure and music information are extracted, and subsets of video shots are selected to generate highlights [4] or MTV-style summaries [5]. Peng et al. [6] further take media aesthetics and editing theory into account to perform home video skimming.

For summarizing videos, most studies exploit features such as motion and color variations to estimate the importance of video segments. However, different from scripted videos, drastic motion changes in travel videos don't imply higher importance, because motion may be caused by hand shaking. Similarly, drastic color changes may be from bad lighting conditions or motion blur. In this paper, we exploit correlation between photos and videos to define importance of photos and video segments.

3 Cross-Media Correlation

In travel media, a photo scene or a video scene means a set of photos or video shots that were captured in the same scenic spot. To faithfully represent a journey by a photo summary or a video summary, we have to consider important parts of media and fairly select data from each scene to generate summaries. Scene boundaries of photos and videos are therefore important clues to the proposed summarization modules. We will determine cross-media correlation first, and briefly review video scene detection utilizing correlation [1].

Figure 1 shows the flowchart of finding cross-media correlation between a photo set and a video. Note that all video segments captured in the same journey are concatenated as a single video stream according to the temporal order.

- Photo Scene Detection

There are large time gaps between photos in different scenic spots because of transportation. This characteristic can be utilized to cluster photos into several scenes. We check time gaps between temporally adjacent photos, and claim a scene boundary exists between two photos if their time gap exceeds a dynamic threshold [7]. The method proposed in [7] has been widely applied in photo clustering, and has been proven very effective. After this time-based clustering, photos taken at the same scenic spot (scene) are clustered together.

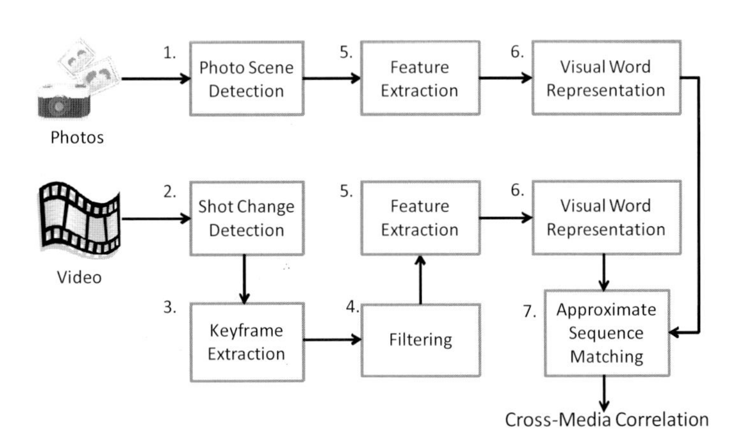

Fig. 1. Flowchart for finding cross-media correlation

- Keyframe Extraction

For the video, we first segment it into shots based on difference of HSV color histograms in consecutive video frames. To efficiently represent each video shot, one or more keyframes are extracted. We adopt the method proposed in [8], which automatically determines the most appropriate number of keyframes based on an unsupervised global k-means algorithm [9]. The global k-means algorithm is an incremental deterministic clustering algorithm that iteratively performs k-means clustering while increasing k by one at each step. The clustering process ends until the clustering results converge.

- Keyframe Filtering

Video shots with blurred content often convey less information, and would largely degrade the performance of correlation determination. To detect blurred keyframe, we check edge information in different resolutions [10]. The video shots with blurred keyframes are then put aside from the following processes.

Video shot filtering brings two advantages to the proposed work. First, fewer video shots (keyframes) are needed to be examined in the matching process described later. Moreover, this kind of filtering reduces the influence of blurred content, which may cause false matching between keyframes and photos.

- Visual Word Representation

After the processes above, correlation between photos and videos is determined by matching photos and keyframes. Image matching is an age-old problem, and is widely conducted based on color and texture features. However, especially in travel media, the same place may have significantly different appearance, which may be caused by viewing angles, large camera motion, and overexposure/underexposure. In addition, landmarks or buildings with apparent structure are often important clues for image matching. Therefore, we need features that resist to luminance and viewpoint changes, and are able to effectively represent local structure.

We extract SIFT (Scale-Invariant Feature Transform) features [11] from photos and keyframes. The DoG (difference of Gaussian) detector is used to locate feature points first, and then orientation information around each point is extracted to form 128-dimenional feature descriptors.

SIFT features from a set of training photos and keyframes are clustered by the k-means algorithm. Feature points belong to the same cluster are claimed to belong to the same *visual word*. Before matching photos with keyframes, SIFT features are first extracted, and each feature point is quantized into one of visual words. The obtained visual words in photos and keyframes are finally collected as visual word histograms. Based on this representation, the problem of matching two image sequences has been transformed into matching two sequences of visual word histograms. According to the experiments in [1], we present photos and keyframes by 20-bin visual word histograms.

Conceptually, each SIFT feature point represents texture information around a small image patch. After clustering, a visual word presents a concept, which may correspond to corner of building, tip of leaves, and so on. The visual word histogram presents what concepts compose the image. To discover cross-media correlation, we would like to find photos and keyframes that have similar concepts.

- Approximate Sequence Matching

To find the optimal matching between two sequences, we exploit the dynamic programming strategy to find the longest common subsequence (LCS) between them. Given two visual word histogram sequences, $X = \langle \boldsymbol{x}_1, \boldsymbol{x}_2, ..., \boldsymbol{x}_m \rangle$ and $Y = \langle \boldsymbol{y}_1, \boldsymbol{y}_2, ..., \boldsymbol{y}_n \rangle$, which correspond to photos and keyframes, respectively. Each item in these sequences is a visual word histogram, i.e., $\boldsymbol{x}_i = h[j]$, $0 \leq j \leq N-1$, where N is the number of visual words. The longest common subsequence between two subsequences X_m and Y_n is described as follows.

$$LCS(X_m, Y_n) = \begin{cases} LCS(X_{m-1}, Y_{n-1}) + 1, & \text{if } x_m = y_n, \\ \max(LCS(X_{m-1}, Y_n), LCS(X_m, Y_{n-1})), & \text{otherwise,} \end{cases} \quad (1)$$

where X_i denotes the ith prefix of X, i.e., $X_i = \langle \boldsymbol{x}_1, \boldsymbol{x}_2, ..., \boldsymbol{x}_i \rangle$, and $LCS(X_i, Y_j)$ denotes the length of the longest common subsequence between X_i and Y_j. This recursive structure facilitates usage of the dynamic programming approach.

Based on visual word histograms, the equality in eqn. (1) occurs when the following criterion is met:

$$x_i = y_j \text{ if } \sum_{k=0}^{N-1} |(h_i(k) - h_j(k))| < \delta, \quad (2)$$

where h_i and h_j are the visual word histograms corresponding to the images x_i and y_j. According to this measurement, if visual word distributions are similar between a keyframe and a photo, we claim that they are conceptually "common" and contain similar content.

- Video Scene Detection

Figure 2 shows an illustrated example to conduct video scene detection based on cross-media correlation. The double arrows between photos and keyframes indicate matching determined by the previous process, and are representation of the so-called cross-media correlation. If a video shot's keyframe matches the photo in the ith photo scene, this video shot is assigned as in ith video scene as well. For those video shots without any keyframe matched with photos, we apply interpolation and nearest neighbor processing to assign them. Details of visual word histogram matching and scene detection processes please refer to [1].

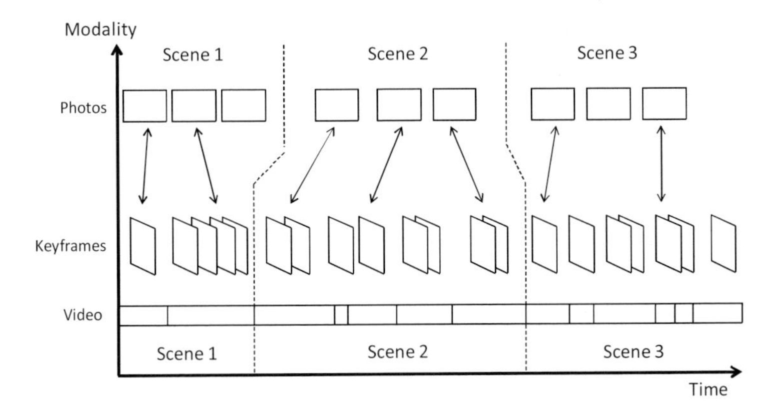

Fig. 2. Illustration of video scene detection

4 Photo Summarization and Video Summarization

On the basis of photo scenes and video scenes, each of which corresponds to a scenic spot, we develop summarization modules that consider characteristics of the correlated media. In a word, how video content evolves affects the selection of photos in photo summary. On the other hand, how photo being taken affects the selection of video segments in video summary. This idea is totally different from conventional approaches, such as attention modeling in photo summarization and motion analysis in video summarization.

4.1 Local Cross-Media Correlation

Matching based on visual word histogram and the LCS algorithm comes from two factors: First, the matched photos and keyframes contain objects with similar concepts, e.g., both images contain large portion of grass and tree, or both images contain artificial objects. Second, the matched images were taken in the same temporal order, i.e., a photo at the beginning of a journey unlikely matches with a keyframe at the end of a journey.

Correlation determined by this process suffices for scene boundary detection. However, to select important data as summaries, finer cross-media correlation is needed to define importance value of each photo and keyframe. In this work, we call the correlation described in Section 3 *global cross-media correlation*, which describes matching in terms of visual concepts. In this section, we need further analyze *local cross-media correlation* to find matching in terms of objects.

For the photos and keyframes in the same scene, we perform finer matching between them by the SIFT matching algorithm [11]. Let s_i denote a feature point in the photo p_m, we calculate the Euclidean distance between s_i and each of the feature points in the keyframe k_n, and find the feature point t_{j^*} that are nearest to s_i. That is,

$$j^* = \arg\min_{j=1,2,\ldots,J} d(s_i, t_j). \tag{3}$$

Similarly, we can find the second nearest feature point $t_{j^\dagger}$ to s_i. The feature point s_i is claimed to match with the point t_{j^*} if

$$\frac{d(s_i, t_{j^*})}{d(s_i, t_{j^\dagger})} < \gamma, \tag{4}$$

where the threshold γ is set as 0.8 according to the suggestion in [11].

For the photo p_m and the keyframe k_n, we claim they contain the same object, such as a building or a statue, if the number of matched feature points exceeds a predefined threshold τ. This threshold can be adjusted dynamically according to the requirements of users. The experiment section will show the influence of different thresholds on summarization performance.

We have to emphasize that local cross-media correlation is determined based on SIFT feature matching rather than visual word histograms. Visual word histograms describe global distribution of concepts (visual words), while feature points describe local characteristics that more appropriate whether two images have the same building or other objects.

4.2 Photo Summarization

The idea of defining each photo's importance comes from two perspectives. The first factor directly comes from the determined local cross-media correlation. When a view or an object is both captured in photos and videos, the captured content must attract people more and is likely to be selected into summaries. We propose the second factor by considering the characteristics of videos to define photo's importance. When people take a closeup shot on an object, this object must attract people more and is likely to be selected into summaries. Therefore, a photo's importance is set higher if it matches with a keyframe that is between a zoom in action and a zoom out action, or is between a zoom in action and camera turning off.

To detect zoom in and zoom out actions, we first find motion vectors and motion magnitudes based on the optical flow algorithm. A keyframe is equally divided into four regions, i.e., left-top, right-top, left-bottom, and right-bottom regions. If motion vectors in all four regions point to the center of the keyframe, a zoom out action is detected. If all motion vectors diverge from the center of the keyframe, a zoom in action is detected.

Two factors defining importance values can be mathematically expressed as follows.

- Factor 1:

The first importance value of the photo p_m is defined as

$$PT_{1,m} = \frac{I_{1,m}}{\max_{i=1,2,\ldots,M} I_{1,i}}, \tag{5}$$

where M is the number of photos in this dataset. The value $I_{1,m}$ is calculated as

$$I_{1,m} = \begin{cases} L_1(h_{p_m}, h_{k_n}), & \text{if the photo } p_m \text{ matches with the keyframe } k_n, \\ 0, & \text{otherwise,} \end{cases}$$

where h_{p_m} and h_{k_n} are visual word histograms of the photo p_m and the keyframe k_n, respectively. The value $L_1(\cdot,\cdot)$ denotes L_1-distance between two histograms.

- Factor 2:

The second importance value of the photo p_m is defined as

$$PT_{2,m} = \frac{I_{1,m} \times ZoomIn(p_m)}{\max_{i=1,2,\ldots,M} I_{1,i} \times ZoomIn(p_i)}, \tag{6}$$

where $ZoomIn(p_m) = 1$ if the keyframe k_n that matches with p_m locates between a zoom in action and a zoom out action, or between a zoom in action and camera turning off. The value $ZoomIn(p_m) = 0$, otherwise.

Note that two importance values are normalized, and then integrated by linear weighting to form the final importance value of p_m:

$$PT_m = \alpha \times PT_{1,m} + \beta \times PT_{2,m}. \tag{7}$$

Currently, the values α and β are set as 1.

In photo summarization, users can set the desired number of photos in summaries. To ensure the generated summary contain photos of all scenes (scenic spots), we first pick the most important photo in each scene to the summary. After the first round, we sort photos according to their corresponding importance values in descending order, and pick photos sequentially until the desired number is achieved.

According to the definitions above, only photos that are matched with keyframes have importance values larger than zero. If all photos with importance values larger than zero are picked but the desired number hasn't achieved, we define the importance value of a photo p_i not picked yet by calculating the similarity between p_i and its temporally closest photo p_j that has nonzero importance value, i.e.,

$$T_i = L_1(h_{p_i}, h_{p_j}). \tag{8}$$

We sort the remaining photos according to these alternative importance values in descending order, and pick photos sequentially until the desired number is achieved.

4.3 Video Summarization

Similar to photo summarization, we advocate that photo taking characteristics in a scene affect selection of important video segments in video summaries. Two factors are also involved with video summary generation. The first factor is the same as that in photo summarization, i.e., video shots whose content also appears in photos are more important. Moreover, a video shot in which many keyframes match with photos is relatively more important. Two factors defining importance values can be mathematically expressed as follows.

- Factor 1:

The first importance value of a keyframe k_m is defined as

$$KT_{1,m} = \frac{I_{1,m}}{\max_{i=1,2,\ldots,M} I_{1,i}}, \tag{9}$$

where *M* is the number of keyframes in this dataset. The value $I_{1,m}$ is calculated as

$$I_{1,m} = \begin{cases} L_1(h_{k_m}, h_{p_n}), & \text{if the keyframe } k_m \text{ matches with the photo } p_n, \\ 0, & \text{otherwise,} \end{cases}$$

where h_{k_m} and h_{p_n} are visual word histograms of keyframe k_m and the photo p_n, respectively.

- Factor 2:

The second importance value of the keyframe k_m is defined as

$$KT_{2,m} = \frac{I_{2,m}}{\max_{i=1,2,\ldots,M} I_{2,i}}, \tag{10}$$

where the value $I_{2,m}$ is the sum of visual word histogram similarities between keyframes at the same shot as k_m and their matched photos. That is,

$$I_{2,m} = \sum_{j=1}^{J} L_1(h_{k_j}, h_{p_{j^*}}). \tag{11}$$

This expression means there are J keyframes in the shot containing k_m, and the notation p_{j^*} denotes the photo matched with the keyframe k_j.

These two importance values are integrated by linear weighting to form the final importance value of k_m:

$$KT_m = \alpha \times KT_{1,m} + \beta \times KT_{2,m}. \tag{12}$$

In video summarization, users can set the desired length of video summaries. To ensure the generated summary contain video segments of all scenes (scenic spots), we first pick the most important keyframe of each scene. Assume that the keyframe k_i is selected, we determine length and location of the video segment S_i corresponding to k_i as

$$S_i = \left(\frac{t(k_{i-1}) + t(k_i)}{2}, \frac{t(k_i) + t(k_{i+1})}{2}\right), \tag{13}$$

where $t(k_i)$ denotes the timestamp of the keyframe k_i, and k_{i-1} and k_{i+1} are two nearest keyframes that are before and after k_i, and with nonzero importance values. Two values in the parentheses respectively denote the start time and end time of the segment S_i.

We pick keyframes and their corresponding video segments according to keyframe's importance values until the desired length of video summary is achieved. If all keyframes with nonzero importance values are picked but the desired length hasn't achieved, we utilize a method similar to that in eqn. (8) to define remaining keyframe's importance, and pick appropriate number of keyframe accordingly.

5 Experimental Results

We collect seven sets of travel media captured in seven journeys for performance evaluation. Each dataset includes a video clip and many photos. The video is encoded in MPEG-1 format, and the resolution is 480×272. Each photo is normalized into 400×300 in experiments. The first five columns of Table 1 show information about the evaluation data.

To objectively demonstrate summarization results, we ask content owners of these data to manually select subset of keyframes and photos as the ground truth of summaries. In generating video summaries or photo summaries, we set the number of keyframes or photos in manual summaries as the targeted number to be achieved. For example, in generating the video summary for the first dataset, 98 keyframes and their corresponding video segments should be selected in the video summary. We measure summarization results by precision values, i.e.,

$$\text{Precision} = \frac{\#\text{ correctly selected keyframes}}{\#\text{ selected keyframes}}. \tag{14}$$

Note that precision and recall rates are the same due to the selection policy.

Figure 3(a) shows precision/recall rates of video summarization for seven datasets under different matching thresholds τ, while Figure 3(b) shows precision/recall rates of photo summarization. In Section 4.1, a photo is claimed to has local cross-media correlation with a keyframe if matched SIFT points is larger than the threshold τ. Generally, we see that using five or ten matched points as the threshold we can obtain better summarization results, i.e., looser thresholds draw slightly better performance. The second dataset has the worst performance, because photos and videos in this dataset don't have high content correlation as that in others, and the content in them is involved with large amounts of natural scenes such that local cross-media correlation based on SIFT matching cannot be effectively obtained. This result conforms that cross-media correlation really impacts on the proposed photo and video summarization methods.

We also conduct subjective evaluation by asking content owners to judge summarization results. They give a score from five to one, in which a larger score means higher satisfaction. Table 2 shows results of subjective evaluation. Overall, both video and photo summarization achieves more than 3.7. The worse performance on the second dataset also reflects in this table.

Table 1. Information of evaluation data

Dataset	# scenes	length	#kf	#photos	#kf in manual sum.	#photos in manual sum.
S1	6	12:57	227	101	98	48
S2	4	15:07	153	30	32	12
S3	5	8:29	98	44	71	11
S4	5	11:03	176	62	97	21
S5	3	16:29	136	50	103	15
S6	2	5:34	67	23	43	12
S7	6	15:18	227	113	112	32

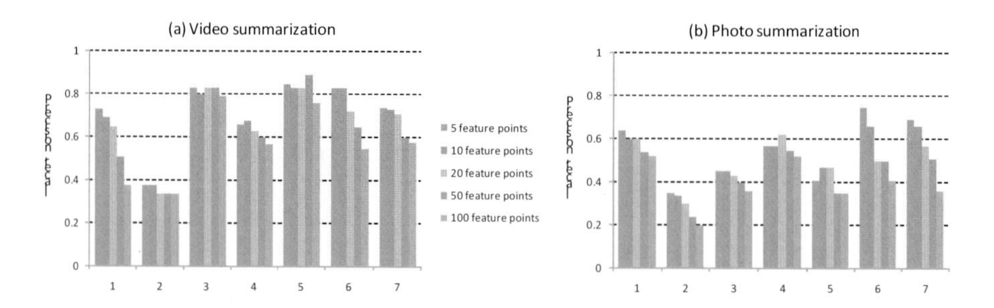

Fig. 3. Performance of (a) video summarization and (b) photo summarization

Table 2. Subjective evaluation on summarization results

	S1	S2	S3	S4	S5	S6	S7	Overall
Video sum.	4	2	4	3	5	4	4	3.7
Photo sum.	5	2	4	3	4	5	4	3.8

6 Conclusion

Two novel ideas have been presented in this paper. Because photos and videos captured in journeys often contain similar content, we can find correlation between them based on an approximate sequence matching algorithm. After that, we first solve an easier problem (photo scene detection), and then solve a harder problem (video scene detection) by consulting with the correlation. To summarize photos and videos, we further exploit cross-media correlation and propose that photo summarization is influenced by the correlated video segments, and contrarily video summarization is influenced by the correlated photos. We respectively consider two factors based on correlation to conduct summarization, and demonstrate the effectiveness of the proposed methods.

In the future, extensive experiments will be conducted to demonstrate the effectiveness and limitation of the proposed methods. For those datasets with less content correlation, more elaborate techniques should be integrated to accomplish the targeted tasks. Moreover, we will try to extend this work to other domains, in which different modalities have high content correlation.

Acknowledgments

This work was partially supported by the National Science Council of the Republic of China under grants NSC 98-2221-E-194-056 and NSC 97-2221-E-194-050.

References

1. Chu, W.-T., Lin, C.-C., Yu, J.-Y.: Using Cross-Media Correlation for Scene Detection in Travel Videos. In: ACM International Conference on Image and Video Retrieval (2009)
2. Gatica-Perez, D., Loui, A., Sun, M.-T.: Finding Structure in Home Videos by Probabilistic Hierarchical Clustering. IEEE Transactions on Circuits and Systems for Video Technology 13(6), 539–548 (2003)
3. Pan, Z., Ngo, C.-W.: Structuring Home Video by Snippet Detection and Pattern Parsing. In: ACM International Workshop on Multimedia Information Retrieval, pp. 69–76 (2004)
4. Hua, X.-S., Lu, L., Zhang, H.-J.: Optimization-based Automated Home Video Editing System. IEEE Transactions on Circuits and Systems for Video Technology 14(5), 572–583 (2004)
5. Lee, S.-H., Wang, S.-Z., Kuo, C.C.J.: Tempo-based MTV-style Home Video Authoring. In: IEEE International Workshop on Multimedia Signal Processing (2005)
6. Peng, W.-T., Chiang, Y.-H., Chu, W.-T., Huang, W.-J., Chang, W.-L., Huang, P.-C., Hung, Y.-P.: Aesthetics-based Automatic Home Video Skimming System. In: Satoh, S., Nack, F., Etoh, M. (eds.) MMM 2008. LNCS, vol. 4903, pp. 186–197. Springer, Heidelberg (2008)
7. Platt, J.C., Czerwinski, M., Field, B.A.: PhotoTOC: Automating Clustering for Browsing Personal Photographs. In: IEEE Pacific Rim Conference on Multimedia, pp. 6–10 (2003)
8. Chasanis, V., Likas, A., Galatsanos, N.: Scene Detection in Videos Using Shot Clustering and Symbolic Sequence Segmentation. In: IEEE International Conference on Multimedia Signal Processing, pp. 187–190 (2007)
9. Likas, A., Vlassis, N., Verbeek, J.J.: The Global K-means Clustering Algorithm. Pattern Recognition 36, 451–461 (2003)
10. Tong, H., Li, M., Zhang, H.-J., Zhang, C.: Blur Detection for Digital Images Using Wavelet Transform. In: IEEE International Conference on Multimedia & Expo., pp. 17–20 (2004)
11. Lowe, D.: Distinctive Image Features from Scale-Invariant Keypoints. International Journal of Computer Vision 60(2), 91–110 (2004)

An Augmented Reality Tourist Guide on Your Mobile Devices

Maha El Choubassi, Oscar Nestares, Yi Wu, Igor Kozintsev, and Horst Haussecker

Intel Corporation
2200 Mission College Blvd
Santa Clara, CA 95052, USA
{maha.el.choubassi,oscar.nestares,yi.y.wu,igor.v.kozintsev, horst.haussecker}@intel.com

Abstract. We present an augmented reality tourist guide on mobile devices. Many of latest mobile devices contain cameras, location, orientation and motion sensors. We demonstrate how these devices can be used to bring tourism information to users in a much more immersive manner than traditional text or maps. Our system uses a combination of camera, location and orientation sensors to augment live camera view on a device with the available information about the objects in the view. The augmenting information is obtained by matching a camera image to images in a database on a server that have geotags in the vicinity of the user location. We use a subset of geotagged English Wikipedia pages as the main source of images and augmenting text information. At the time of publication our database contained 50 K pages with more than 150 K images linked to them. A combination of motion estimation algorithms and orientation sensors is used to track objects of interest in the live camera view and place augmented information on top of them.

Keywords: Mobile augmented reality, image matching, SIFT, SURF, location and orientation sensors, optical flow, geotagging.

1 Introduction

In the past few years, various methods have been suggested to present augmented content to users through mobile devices [1,2,3,4,5]. Many of the latest mobile internet devices (MIDs) feature consumer-grade cameras, WAN and WLAN network connectivity, location sensors (such as Global Position System - GPS) and various orientation and motion sensors. Recently, several applications like Wikitude (www.wikitude.org) for G1 phone and similar applications for iPhone have been announced. Though similar in nature to our proposed system, these solutions rely solely on the location and orientation sensors, and, therefore, require a detailed location information about points of interest to be able to correctly identify visible objects. Our system extends this approach by using the image matching techniques both for recognition of objects and for precise placement of augmenting information.

S. Boll et al. (Eds.): MMM 2010, LNCS 5916, pp. 588–602, 2010.

In this paper, we demonstrate *a complete end-to-end* mobile augmented reality (MAR) system that consists of Intel® Atom^{TM} processor-powered MIDs and Web-based MAR service hosted on a server. On the server, we store a large database of images crawled from geotagged English Wikipedia pages that we update on a regular basis. The MAR client application is running on a MID. Figure 1 demonstrates a snapshot of the actual client interface. In this example the user has taken a picture of a Golden Gate bridge in San Francisco. The MAR system uses the location of the user along with the camera image to return top 5 candidate matching images from the database on the server. The user has an option of selecting the image of interest, which retrieves a corresponding Wikipedia page with an article about the Golden Gate bridge. A transparent logo is then added to the live camera view "pinned" on top of the object (Figure 1). The user will see the tag whenever the object is in the camera view and can click on it later on to return to the retrieved information.

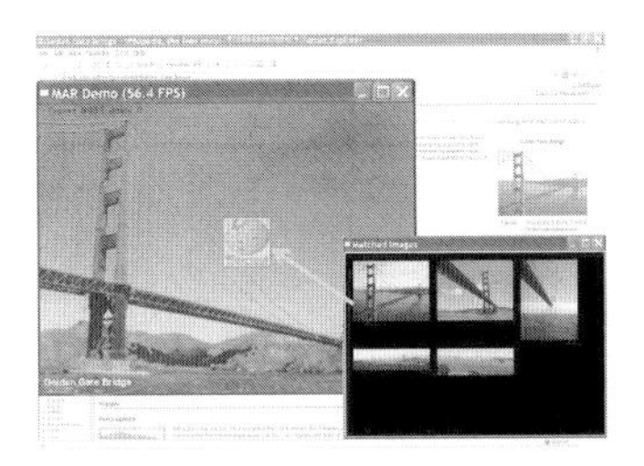

Fig. 1. A snapshot of our MAR system

The following are the main contributions of this paper:

- In [6], we relied on location and orientation sensors to track objects. In this paper, we additionally use the image content for tracking objects using motion estimation, thus improving the precision of placement of augmenting information in the live camera view.
- We created a representative database sampled from the large Wikipedia database to perform systematic testing of the MAR performance in a realistic setting.
- We propose a new method to improve the matching algorithm based on histograms of minimum distances between feature descriptors and present the results for our sample database.
- We propose a simple and promising way to extend our database of images by combining text and GPS information and various sources of images like Google image search and Wikiepdia. As a result we enhance the accuracy of image matching of our system.
- We implemented a fully-functional client application and optimized its performance for Intel® Atom^{TM} processor.

The organization of our paper is as follows. In Section 2, we give an overview of our MAR system. In Section 3, we describe the small sample Wikipedia database.

In Section 4, we present an enhanced version of the matching algorithm based on our testing results on this database. In Section 5, we extend our matching algorithm by investigating histograms of minimum distances between descriptors. In Section 6, we supplement our database by images from Google image search and show better image matching results on our sample database. In Section 7, we illustrate our algorithms with experimental results. In Section 8, we explain some of the implementation and code optimization aspects. Finally, we conclude in Section 9.

2 System Overview

In this section, we present the overview of MAR system as illustrated in Figure 2. The system is partitioned into two components: client MID and a server. The MID continuously communicates with the server through WAN or WLAN network.

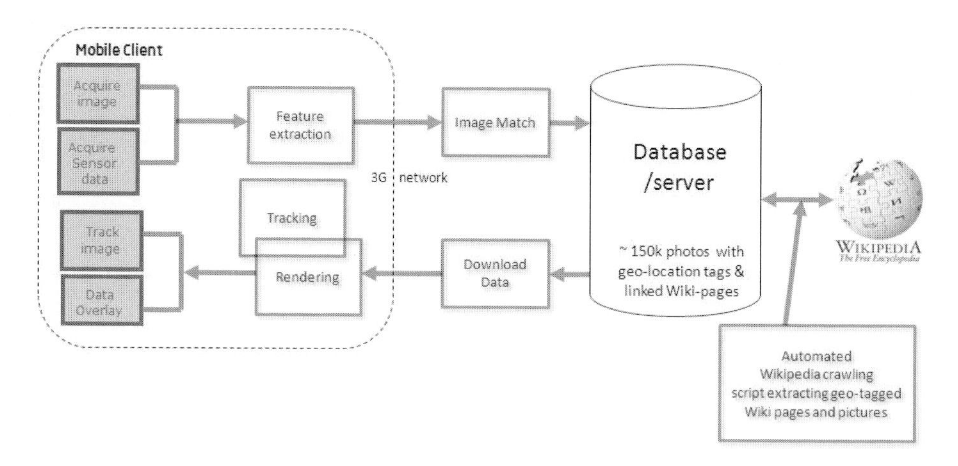

Fig. 2. MAR system Diagram with the image acquisition, feature extraction, rendering, and tracking on the client side, and the database and image matching on the server side

2.1 Client

- **Query acquisition.** Client MID devices contain cameras, orientation and location (such as GPS) sensors. Client continuously acquires live video from the camera and waits for the user to select the picture of interest. As soon as a picture is taken, sensor data from location and orientation sensors is written to the images EXIF fields.
- **Feature extraction.** Visual features are then extracted from the picture. We use 64-dimensional SURF [7] which are fast to compute compared to SIFT [8]. Client sends extracted feature vectors and recorded sensor data to

the server for searching matched images and related information. The reason of sending the compact visual features instead of full resolution images is to reduce network bandwidth, hence reduce latency.
- **Rendering and overlay.** Once the matched information is found, related data including matched image thumbnail, wikipage link, etc. will be downloaded to the client via WAN or WLAN network. Client will render the augmented information such as wiki tag related to the query object and overlay it on the live video. Our device has orientation sensors: a compass, an accelerometer, and a gyroscope. As the user moves the device around, the augmented information representing the query will be pinned to the position of the object. This way, multiple queries can be interacted with separately by simply pointing the camera at different locations.
- **Tracking.** When a query is made, the direction of the MID is recorded using the orientation sensors on the device. Client will continuously track the movement of orientation sensor [6]. However, tracking using orientation sensors is not very precise. We extend our tracking method to also include the visual information. The image based stabilization is based on aligning neighbor frames in the input image sequence using a low parametric motion model. The motion estimation algorithm is based on a multi-resolution, iterative gradient based strategy [9], optionally robust in a statistical sense [10]. Two different motion models have been considered, pure translation (2 parameters) and pure camera rotation (3 parameters).

2.2 Server

- **Database collection.** We crawled Wikipedia pages, in particular those with GPS information associated with them [6]. We downloaded the images from these pages to our server, extracted visual features from images and built our image database. At the time of publishing [6], the database had 50K images. Currently, it has over 150K images and it is constantly growing. This fact further emphasizes the need and the importance of applications exploring large resources of images.
- **Image match.** Once the server receives query from client MID, the server will perform image match. For image matching, we combine both the GPS information and the image content. More explicitly, we restrict the database of candidate matching images to those within certain search radius from the user GPS location. Next, we use computer vision techniques to match the query image to the GPS-constrained set of images. Downscaling the database from 150K images to a much smaller candidate image set after GPS filtering enhances the performance of image-based matching algorithms. If the number of nearby candidates is reasonable (< 300) we perform brute-force image matching based on the ratio of the distance between the nearest and second nearest neighbor descriptors [8]. Otherwise, for scenarios with highly dense nearby images or with no GPS location available, the database is large and we use indexing as proposed by Nister & Stewenius in [11]. Details of the matching algorithm are presented in the following sections.

3 Building a More Realistic Database

In our previous paper [6], we showed quantitative results on the standard ZuBuD test set [12] with 1005 database images and 115 query images. Also, we only had qualitative results for a small benchmark of images of iconic landmarks in an effort to represent MAR. Although the results on the ZuBuD set are informative, they do not precisely reflect the performance of the actual MAR system. To better represent MAR, we select 10 landmarks around the world: *the triumph arc* (France), *the Pisa tower* (Italy), the *Petronas towers* (Malaysia), *the colosseum* (Italy), *the national palace museum* (Taiwan), *the golden gate bridge* (California, USA), *the Chiang Kai-Shek memorial hall* (Taiwan), *the capitol* (Washington D.C., USA), *the palace of justice* (France), and *the Brandenburg gate* (Germany). For each landmark, we follow the steps of the MAR system. First, we filter the 150K images database on the server to a smaller set that includes only the images within a radius of 10 miles[1]. Next, we randomly pick one of the landmark images in the set as a query image, and keep the others as the database images. We now have a set of 10 query images and 10 databases. The total number of images in all the databases is 777. Please refer to Figure 3 for samples from these databases. Each row corresponds to one landmark and the leftmost image is the chosen query image. In the leftmost column, we show the number of relevant images out of the total number of images in the database. For example for the triumph arc landmark, the database has 5 images of the arc out of all the 27 images inside it. Moreover, the characteristics of these databases vary. For instance, the size of the database is only 5 for *the national palace museum*, while it is 200 for the justice palace. Also, the database corresponding to the golden gate bridge has 12 relevant images out of 19 images, while that of *the justice palace* has only 1 relevant image among 200 images. Finally, we manually annotate the databases. Simulating MAR on these databases, we obtain representative quantitative results for its performance since such databases are sampled from the Wikipedia data on our server, i.e, the actual database queried by MAR. Note that images labeled by the same landmark might look different depending on the lighting conditions, the viewpoint, and the amount of clutter.

4 Matching Enhancement by Duplicates Removal

4.1 Original Matching Algorithm

In order to recognize the top matching database images, our algorithm inspects each database image at a time and compares it to the query image. More specifically, for each SURF keypoint in the query image, the algorithm finds the nearest and the second nearest neighbor keypoints in the database image based on the L_1 distance between descriptors. Next, it computes the distances ratio and decides whether the query keypoint matches the nearest keypoint in the image database as in [8]. After the algorithm detects the matching pairs between the

[1] Our MAR system uses this radius also.

Fig. 3. Sample images from the GPS-constrained databases for 10 landmarks

query image and all the database images, it ranks these images in a descending order according to the number of such pairs.

4.2 Duplicates Removal

Running MAR on the data collected in Section 3, we identified one problem in the matching algorithm that degrades MAR's performance: having multiple matching keypoints in the query image corresponding to one keypoint in the database image magnifies false matches and degrades the matching accuracy. We adjusted the algorithm to remove duplicate matches and improve matching accuracy at no computational cost as shown below.

Example. In Figure 4, we see the top 5 images returned by MAR for the golden gate bridge query. The features used are SURF with a threshold[2] of 500. Clearly, the top 1 image is a mismatch.

In Figure 5, we see that many of the matches between the query image and the top 1 image are actually duplicates: for example, the 14 keypoints in the query image correspond to only one keypoint in the database image. Moreover, these matches are actually false. It turns out that many of the 60 matches are due

[2] The threshold is a parameter in the SURF algorithm. The larger this parameter is, the less sensitive is the keypoints detector and the smaller is the number of keypoints.

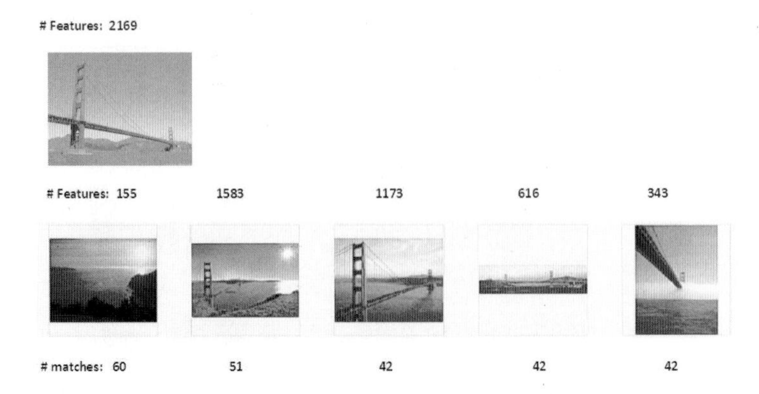

Fig. 4. Top 5 matches retrieved by MAR for the golden gate bridge query

Fig. 5. Right: query image. Left: database image. Illustration of a typical problem of the matching algorithm: one keypoint in the database image has duplicate matches in the query image (14 matches in this example).

to duplicates. Even when these matches are false, they are highly amplified by their multiplicity, which eventually affects the overall retrieval performance. We believe that this problem particularly arises in cases of strong imbalance between the number of keypoints in the database and query images (155 versus 2169). The imbalance forces many keypoints in the query image to match one single point in the database image. The standard ZuBuD data set does not have this issue since its images are more or less uniform and have comparable numbers of keypoints. Contrarily, Wikipedia and web images in general are highly variant, which emphasizes the necessity of having representative sample databases as in Section 3. To solve this problem, our adjusted algorithm prohibits duplicate matches. Whenever a keypoint in the database image has multiple matches in the query image, we only pick the keypoint with the closest descriptor. Applying the new algorithm to the golden gate query, the number of matches between the query image and the previously top 1 image decreases from 60 to 21, and the new top 5 images are shown in Figure 6. We recognize that duplicate matches may still be correct in particular for repetitive structures, however as it is shown for this particular example and in the more extensive results in Section 7, substituting

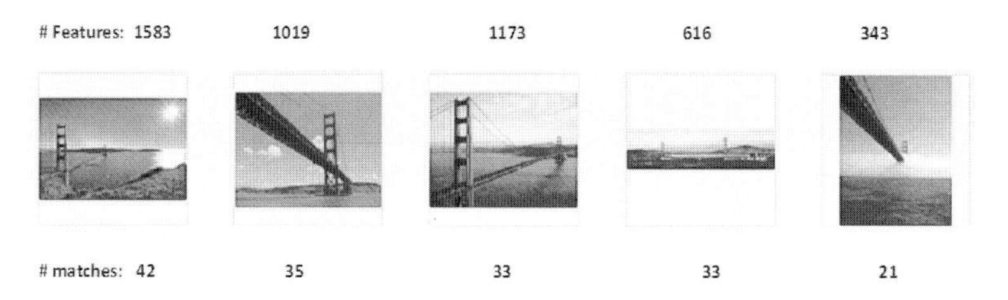

Fig. 6. Top 5 matches retrieved by the adjusted algorithm for golden gate bridge query

the duplicates with the "best" matching descriptor, i.e., the nearest, improves the overall performance of our system.

5 Matching Enhancement by Histograms of Distances

For each keypoint in the query image, the current matching algorithm computes the minimum distance and the second minimum distance between its descriptor and the descriptors of the database image keypoints. Next, the ratio between the distances is used to decide about the match as explained in Section 4.1. Instead of merely relying on the distances ratio, we propose further exploring other statistics based on descriptors' distances. More explicitly, we compute the histogram of minimum distances between the query image and the top ten retrieved database images. Our purpose is to extract more information from these distances about the similarity/dissimilarity between the query and the database images. Next, we examine these histograms in order to remove obviously mismatching images. The cost of this approach is not high, since the distances are already computed and hence we are only leveraging their availability. We applied this approach to 10 query images and their corresponding GPS-constrained databases. We obtained promising results.

5.1 Algorithm

Let Q be the the query image. We first apply our existing matching algorithm and we retrieve the top 10 database images D_1, D_2, ..., D_{10}. Next, we consider each pair (Q, D_i) at a time and build the histogram H_i of minimum distances from keypoints in the query image Q to the database image D_i. There is no additional overhead since these distances were already calculated by the existing matching algorithm. For each histogram H_i, we obtain its empirical mean M_i and and skewness S_i:

$$M_i = \frac{1}{n}\sum_{j=1}^{n} H_{i,j},$$

$$S_i = \frac{\frac{1}{n}\sum_{j=1}^{n}(H_{i,j} - M_i)^3}{\left(\frac{1}{n}\sum_{j=1}^{n}(H_{i,j} - M_i)^2\right)^{3/2}}.$$

The smaller the skewness is, the closer to symmetric is H_i. Our main assumptions are:

1. If M_i is large then many of the descriptors pairs between Q and D_i are quite distant and hence are highly likely to be mismatches. Therefore, image D_i must not be considered a match for image Q.
2. If S_i is small (close to zero), then the histogram H_i is almost symmetric. Having many descriptors in Q and D_i that are "randomly" related, i.e., not necessarily matching, would result in this symmetry. We expect this scenario when the the two images Q and D_i don't match and hence there is no reason for the histogram to be biased.

Based on these assumptions, we remove database images that have very low skew S_i. We also cluster the images based on the means M_1, M_2, ..., M_{10} into two clusters (we used $k-$means). We remove the images that belong to the cluster with the higher mean. For the experimental results, please refer to Section 7.2. Figure 7 displays the block diagram of our adjusted matching algorithm.

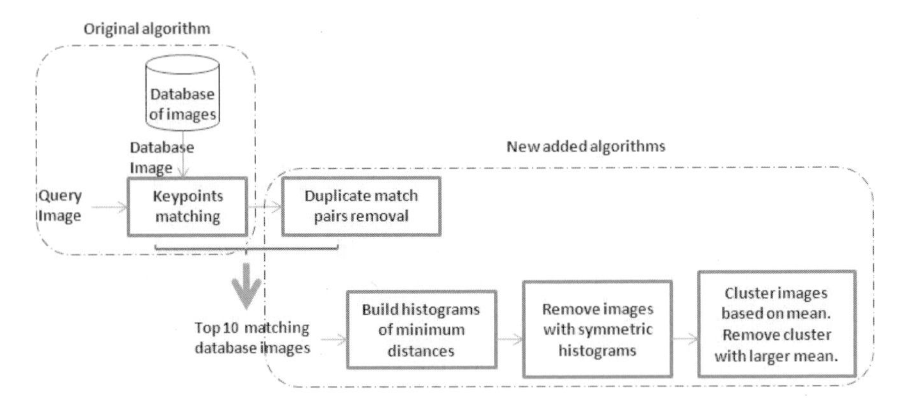

Fig. 7. Block diagram of our algorithm for image matching with the duplicates removal and enhancement based on histograms added

6 Matching Enhancement by Database Extension

As mentioned before, the images on our server are downloaded from Wikipedia. Although Wikipedia is a good source for information in general, it is not the best source for images. As we see in Figure 3 there is only one image that matches *the justice of palace* query out of 200 images in the database. However, not only the matching algorithm but also the number of matching images in the database impacts the performance accuracy. Moreover, the query image might be taken from different views and under different light conditions. Having a small number of matching images, e.g., one image in the day light for *the justice of palace* query, would limit the performance of MAR. For this reason, we prefer if the database has more matching images to the query and we aim at extending it by combining GPS-location, geotagged Wikipedia pages, text in these pages, and the large resources of web images. For each Wikipedia page, we don't only download its images, but we also use its title as a text query on Google image search engine. Next, we pick the first few retrieved images and associate them with the same Wikipedia page. As we will see in Section 7.3, the performance of our MAR system improves significantly.

7 Experimental Results

7.1 Duplicates Removal

To test the impact of duplicate matches removal, we ran both the original and the new matching algorithms (see Section 4) on the sample databases of Section 3 for various SURF thresholds. In Figure 8, we display the average of the top 1, top 5, and average precision over the 10 databases. Clearly, removing the duplicate matches improves the performance at almost the same running time.

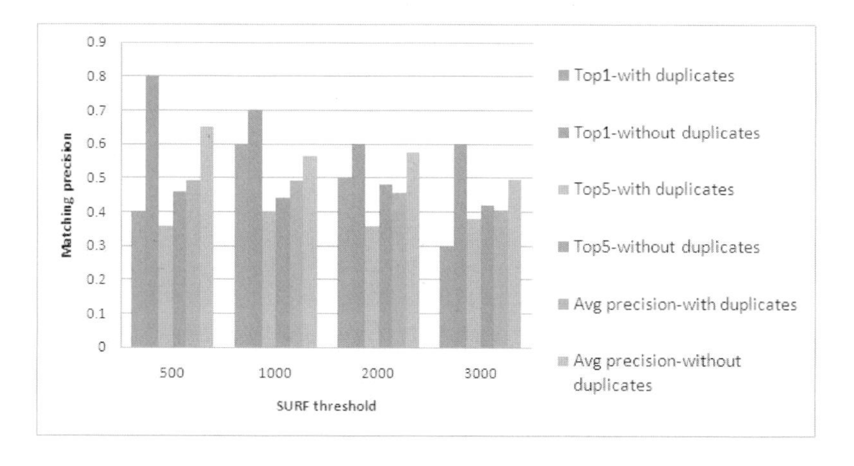

Fig. 8. Top 1, top 5, and average precision results for the 10 landmarks of Section 3 vs. the SURF threshold

7.2 Matching Enhancement through Histograms of Distances

In Figure 9, we display the top 10 retrieved images for each of the 10 databases in Section 3 by the updated matching algorithm of Section 4.2 with duplicates removal. We also used the histograms of distances to further refine the match as explained in Section 5.1. Based on the histograms analysis, images labeled with a red square in the figure are rejected and those tagged with a blue square are not rejected. From these experiments, we see that portraits and statues of people can be clearly distinguished and rejected since most of the queries are those of buildings. Such images in general have a very low skew, i.e., a very symmetric histogram. They also may have a large mean.

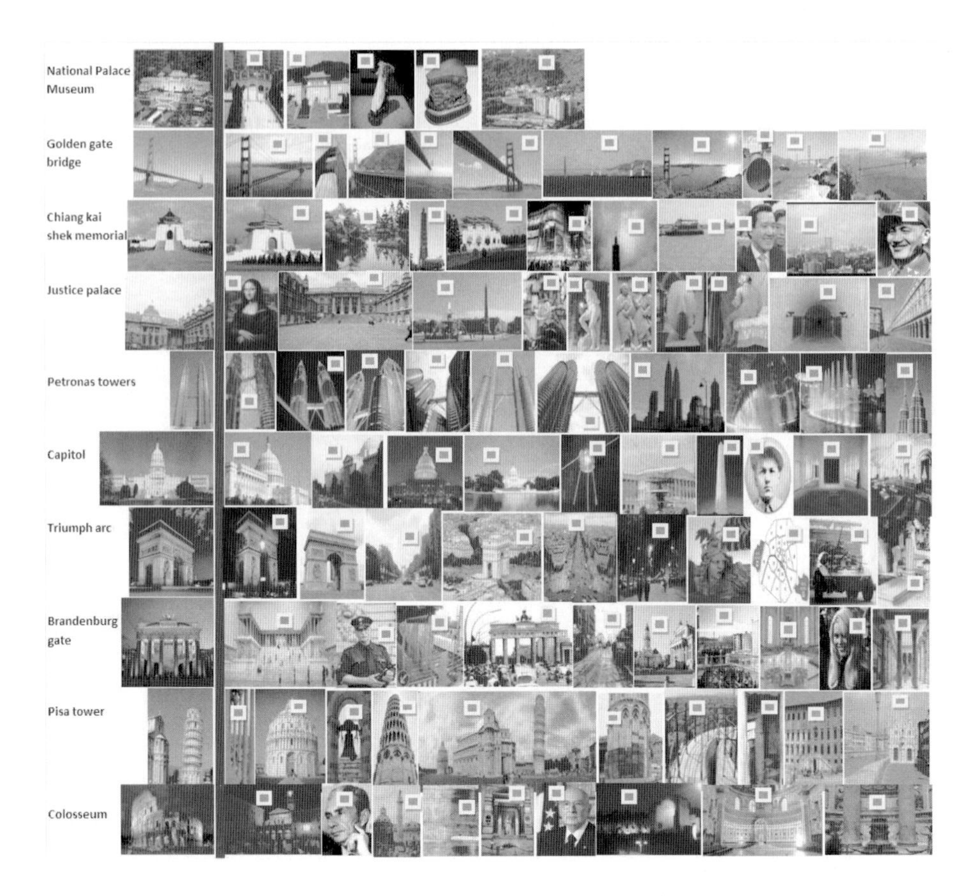

Fig. 9. Results for matching enhancement based on histograms of distances. Images before the red line are the queries.

7.3 Matching Enhancement through Database Extension

For each of our 10 landmarks, we download the top 21 images returned by Google image search when queried with the title of the landmark Wikipedia page and add them to the corresponding landmark database. Next, we apply our improved matching algorithm with no duplicates and using histograms of distances and display the improved results in part (a) of Figure 10. Part (b) illustrates the top 10 images returned by MAR for query images under new light conditions (golden gate bridge at night and triumph arc during the day). The top returned images have similar lighting conditions due to the sensitivity of SURF features to light conditions, which applies to other popular features such as SIFT. Hence we need a diverse set of images in the database in order to compensate for the limitation of visual features. Finally, in the actual scenario Google images relevant to all geotagged Wikipedia pages must be added to our databases, while in part (a) we only added the images corresponding to the query

Fig. 10. Results for matching enhancement based on database extension, images before the red line are the queries. (a) All the 10 landmarks. (b) New light conditions for query: night for golden gate bridge, day for triumph arc. (c) Add also a large number of mismatching images to the database (427 mismatching images and 21 google images to 174 database.

landmark. For this reason, in order to make sure that the improvement in the matching accuracy is not due to the bias in the number of matching vs. mismatching images, we added 427 various mismatching images to *the Brandenburg gate* database in addition to the 21 google images (original size 174 images). In part (c), we view the top 10 retrieved images. Clearly, we still have a significant improvement in performance due to the database extension. Note that for parts (b) and (c) we only use the improved matching algorithm excluding duplicates but without refinement based on histograms of distances. Indeed, the main point of the results in part (b) is to stress the sensitivity to light conditions and the need for diverse databases, while the purpose of the results of part (c) is to prove the success of database extension. Both points are clear from the results independently of the particular refinements of the matching algorithm.

8 Implementation and Optimization

8.1 Code Optimization

The original SURF feature extraction code is based on OPENCV implementation (`http://sourceforge.net//projects/opencv/library`) and match algorithm as described in Section 2.2. We further identified the hot spots in the MAR source codes and employed the following optimization:

- Multi-threaded all the hotspots, including interesting point detection, key-point description generation and image match.
- Data and computation type conversion. In the original implementation, double and float data types are used widely, and floating point computations as well. We quantized the keypoint descriptor from 32-bit floating point format to 8-bit char format. We also converted many floating point computation to fixed point computations in key algorithms. By doing that, not only the data storage is reduced by $4X$, but also the performance is improved by taking advantage of the integer operations. The image recognition accuracy isn't affected from our benchmark results.
- Vectorization. We vectorized the image match codes using SSE intrinsic to take advantage of 4-way SIMD units. Significant speedups are observed compared to original implementation.

8.2 Tracking

The tracking algorithm explained in Section 2.1 has been optimized by: 1) using a simplified multi-resolution pyramid construction with simple 3-tap filters; 2)

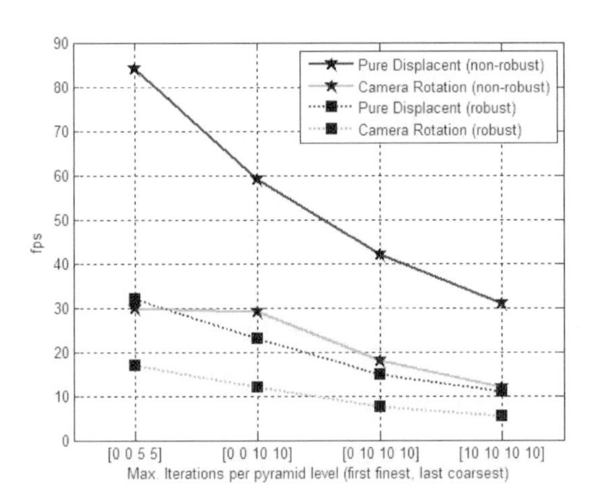

Fig. 11. Performance measured in fps of the image-based stabilization method on an Intel® Atom™ processor based platform (1.6GHz, 512KB L2 cache) for several choices of model, estimation method, resolution levels and iterations per level

using a reduced linear system with gradients from only 200 pixels in the image instead of from all the pixels in the images; 3) using SSE instructions for the pyramid construction and the linear system solving; 4) using only the coarsest levels of the pyramid to estimate the alignment. Performance was measured on an Intel® Atom™ system (1.6GHz, 512KB L2 cache) using VGA (640×480) input video and different options. The results are shown in Figure 11, which displays the measured frames per second (fps) for different models (displacement/camera rotation), estimation method (robust/non-robust) and resolution levels and iterations per level used. For pure displacement model, using non-robust estimation and running 5 iterations in the levels 3 and 4 of the multi-resolution pyramid (being level 1 the original resolution) the performance is over 80 fps.

9 Conclusion

In [6], we presented MAR with a fully functional prototype on a MID device with Intel® Atom™ processor inside. In this paper, we described new improvements to the matching and tracking algorithms in addition to the design of the system and its database. We also built a new realistic testing database. With all these improvements, MAR demonstrates the powerful capabilities of future mobile devices by integrating location sensors, network connectivity, and computational power.

References

1. Pradhan, S., Brignone, C., Cui, J.H., McReynolds, A., Smith, M.T.: Websigns: hyperlinking physical locations to the web. Computer 34, 42–48 (2009)
2. Lim, J., Chevallet, J., Merah, S.N.: SnapToTell: Ubiquitous Information Access from Cameras. In: Mobile and Ubiquitous Information Access (MUIA 2004) Workshop (2004)
3. Zhou, Y., Fan, X., Xie, X., Gong, Y., Ma, W.Y.: Inquiring of the Sights from the Web via Camera Mobiles. In: 2006 IEEE International Conference on Multimedia and Expo., pp. 661–664 (2006)
4. Takacs, G., Chandrasekhar, V., Gelfand, N., Xiong, Y., Chen, W.C., Bismpigiannis, T., Grzeszczuk, R., Pulli, K., Girod, B.: Outdoors augmented reality on mobile phone using loxel-based visual feature organization (2008)
5. Quack, T., Leibe, B., Van Gool, L.: World-scale mining of objects and events from community photo collections. In: Proceedings of the 2008 international conference on Content-based image and video retrieval, pp. 47–56. ACM, New York (2008)
6. Gray, D., Kozintsev, I., Wu, Y., Haussecker, H.: WikiReality: augmenting reality with community driven websites. In: International Conference on Multimedia Expo., ICME (2009)
7. Bay, H., Tuytelaars, T., Van Gool, L.: SURF: Speeded Up Robust Features. In: Leonardis, A., Bischof, H., Pinz, A. (eds.) ECCV 2006. LNCS, vol. 3951, pp. 404–417. Springer, Heidelberg (2006)
8. Lowe, D.G.: Distinctive Image Features from Scale-Invariant Keypoints. International Journal of Computer Vision 60(2), 91–110 (2004)

9. Lucas, B.D., Kanade, T.: An iterative image registration technique with an application to stereo vision, pp. 674–679
10. Nestares, O., Heeger, D.J.: Robust multiresolution alignment of MRI brain volumes, pp. 705–715
11. Nister, D., Stewenius, H.: Scalable Recognition with a Vocabulary Tree. In: IEEE Computer Society Conference on Computer Vision and Pattern Recognition (2006)
12. Shao, H., Svoboda, T., Van Gool, L.: ZuBuD: Zurich Buildings Database for Image Based Recognition. Technique report No. 260, Swiss Federal Institute of Technology (2003)

Transfer Regression Model for Indoor 3D Location Estimation

Junfa Liu, Yiqiang Chen, and Yadong Zhang

Institute of Computing Technology, Chinese Academy of Sciences
100190, Beijing, China
{liujunfa,yqchen,zhangyadong}@ict.ac.cn

Abstract. Wi-Fi based indoor 3D localization is becoming increasingly prevalent in today's pervasive computing applications. However, traditional methods can not provide accurate predicting result with sparse training data. This paper presented an approach of indoor mobile 3D location estimation based on TRM (Transfer Regression Model). TRM can reuse well the collected data from the other floor of the building, and transfer knowledge from the large amount of dataset to the sparse dataset. TRM also import large amount of unlabeled training data which contributes to reflect the manifold feature of wireless signals and is helpful to improve the predicting accuracy. The experimental results show that by TRM, we can achieve higher accuracy with sparse training dataset compared to the regression model without knowledge transfer.

Keywords: 3D Location, Transfer Regression Model, Semi-Supervised Learning, Manifold Regularization.

1 Introduction

Nowadays, location information is an important source of context for multimedia service systems, especially for mobile service systems. For instance, in a LBS (Location Based Services) system of video on demand, users in different location will receive different video program or same video with different qualities, and more accurate location will promise better video quality.

Wi-Fi based location systems are becoming more and more popular for some advantages of WLAN (Wireless Local Access Network). Today, GPS provides localization for users outdoors, but it can not provide precise indoor location because of the technical constraints. Historically, we have seen indoor location systems based on infra-red, ultrasound, narrowband radio, UWB, vision, and many others in 2D location [1]. But deployments of high precision indoor location systems have not involved large coverage areas due to the cost of equipping the environment, meanwhile, the accuracy or deployment cost is unsatisfactory. On the other hand, many systems which use the 802.11 WLAN as the fundamental infrastructure is more and more popular. Considering the expanding coverage of Wi-Fi signals and Wi-Fi modules are equipped in the mobile devices widely, existed WLAN usually is selected as the manner for indoor localization system to minimize the cost of deployment and infrastructure.

S. Boll et al. (Eds.): MMM 2010, LNCS 5916, pp. 603–613, 2010.

Most systems use RSS (Received Signal Strength) as Wi-Fi feature. Those methods employ machine learning or statistical analysis tools. The algorithms of those learning methods usually contain two phases: offline training phase and online localization phase. Ferris et al. in [2] proposed a framework using Gaussian process models for location estimation. Nguyen et al. [3] apply kernel methods and get more accurate result for location prediction. Pan et al. [4] realized a system based on manifold regularization as a semi-supervised algorithm to mobile-node location and tracking. This method can reduce greatly the calibration effort.

However, those works addressed 2D location problems. Today, indoor 3D location is becoming more and more important. As we all know, many public or proprietary services are always based on the geospatial locations and therefore named as Location Based Services (LBS). As the rapid developments of ubiquitous computing systems as well as wireless networking, location sensing techniques, LBS is increasingly considered as a very important service. However, most existing LBS can only deal with 2D geospatial data and offer related 2D geospatial services. They lack in supporting 3D datasets as well as 3D services. In the mean time, 3D information is rapidly increasing, partly due to the fast data acquisition techniques, which can bring people more vivid presentations of the real world .To use the existing and newly emerging 3D datasets, implement 3D LBS systems turn to be more and more necessary.

This paper presented TRM (Transfer Regression Model) and its application in indoor 3D location estimation. TRM is essential a regressive model which also learns the mapping function between signal space and location space. It has the advantage that can import the other dataset in training progress when the training dataset is rather sparse, which will refine the regressive model and improve the prediction accuracy.

The rest of this paper is organized as follows. We describe related research works in Section 2, and present in details the theory of TRM and TRM based 3D location in Section 3. Section 4 shows the experiments. Finally, we give our conclusions and future works in Section 5.

2 Related Works

Some systems on 3D location [5, 6, 7, 8] have been developed to realize 3D LBS, the most significant achievements in the 3D research area concerning key issues of 3D GIS, which is intensive and covers all aspects of the collecting, storing and analyzing real world phenomena. Chittaro et al. in [6] presented an approach to give evacuation instructions on mobile devices based on interactive location-aware 3D models of the building. This system uses RFID technology to determine user's position in the real building. It needs place RFID tags in needed part of the building. RFID tags have a range of about 4 meters and send their signal to the RFID reader every 500 milliseconds. But few systems can provide accurate and satisfactory 3D location and easily be implemented especially in indoor environment.

There are also some pure 3D location systems such as SpotON [9], an indoor 3D location sensing technology based on RF signal strength, can be used to locate indoor but it is too expensive compared with WLAN based systems and it alone is almost certainly not the ultimate solution in the problem space. Pedestrian Localization [10], a foot-mounted inertial unit, a detailed building model, and a particle filter be combined to provide absolute positioning in a 3D building, but the foot-mounted inertial unit is expensive and each building needs a detailed building model which is time-consuming.

From those documents described above, 3D location estimation is still an on going problem. Furthermore, the current systems always require changing the infrastructure which sequentially leads to high cost. Intuitively, we can train dedicated 2D location model in each floor, but it would bring huge human labors to collect the data and label the data.

In this case, we propose the method of 3D location based on TRM. TRM is designed to train effective regressive model for the target floor with sparse training data by reusing the large amount of data collected from the other floor.

3 TRM for 3D Location Estimation

In this section, we first demonstrate the 3D location problem. Then, TRM a regression model with the capability to import knowledge from the third party of dataset is introduced. Then, TRM based 3D location estimation is described.

3.1 Indoor 3D Location Problem

Let us define some parameters formally. Suppose there are totally r APs deployed in the wireless environment, and they can be detected both on the third and fourth floors. The RSS values can be represented as a row vector $x = (x_1, x_2, \ldots, x_r) \in R_r$, where x_i stands for the RSS value received from AP $_i$ and sometime we fill the value with -100db if no actual signal strength is detected. The value of -100db is the lowest signal strength that can be detected in our experiment.

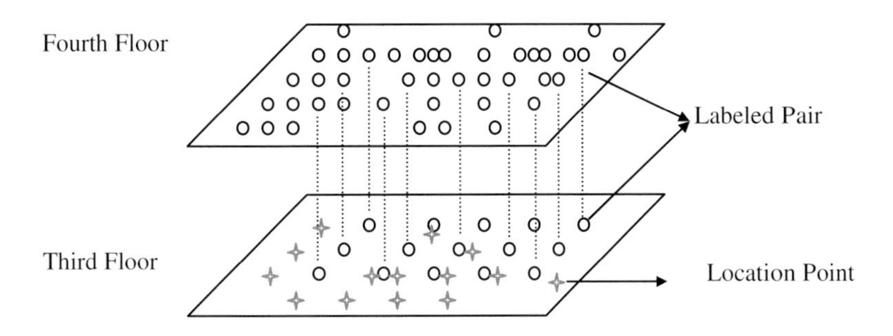

Fig. 1. The 3D location problem

We carry out the 3D location experiment on the third and fourth floor in a building. As shown in Fig. 1, the black circles represent the position with labeled coordinates while the red stars represent the position we want to know the coordinates. There are some labeled pairs, and each pair means two points with same physical coordinates but from different floors. In the case of 2D location estimation, we may train two prediction model $y = f_3(x)$ and $y = f_4(x)$ respectively for third floor and fourth floor. x is the Wi-Fi signal vector and y a two dimensional vector represents the coordinates of the location point. The problem is: If we just use the labeled data on the third floor to train the location estimation model f_3, the prediction accuracy is very limited for the training data is so sparse. So can we reuse the plentiful data of the fourth floor to improve the prediction accuracy for model f_3? If the old data can be reused well, the labeled data and human calibration effort would be reduced greatly. This is always a meaningful direction to build 3D location model with less manual cost.

Now, the 3D location problem is to train a regressive model f_3 with three datasets X,Y,Z, that is: $f_3 : X \rightarrow Y \mid Z$. X is the Wi-Fi signal set collected in third floor, Y is the coordinates of the location points, and Z is the Wi-Fi signal set collected in fourth floor. As we know, some regressive model such as SVR (Support Vector Regression) and BP (Back Propagation) network can not perform such learning task which is based on three datasets, so we propose the novel learning framework of Transfer Regression Model.

3.2 Transfer Regressive Model

This section proposes TRM (Transfer Regression Model), a kind of regression method for learning from sparse training data. Limited training data will lead to reduction of regression accuracy for traditional regression method. TRM can adopt large number of unlabeled data based on Manifold Regularization and transfer useful knowledge from third part of dataset when training the regression model. For TRM is original from MR (Manifold Regularization), we give a brief introduction of the theory of regularization and manifold regularization.

3.2.1 Regularization for Regression Problems

Regularization is a well built mathematical approach for solving ill-posed inverse problems. It is widely used in machine learning problem [11], and many popular algorithms such as SVM, splines and radial basis function can be broadly interpreted as the instance of regularization with different empirical cost functions and complexity measures.

Here we give a brief introduction of regularization on regression learning, whose detail can be referred in [11]. Given a set of data with labels $\{x_i, y_i\}_{i=1,2,\dots,l}$, the standard regularization in $RKHS$ (Reproducing Kernel Hilbert Space) is to estimate the function f^* by a minimizing:

$$f^* = \arg\min_{f \in \mathrm{H}_K} \frac{1}{l} \sum_{i=1}^{l} V(x_i, y_i, f) + \gamma \|f\|_K^2 \tag{1}$$

Where V is some loss function, and $\|f\|_K^2$ is the penalizing on the $RKHS$ norm reflecting smoothness conditions on possible solutions. The classical Representer

Theorem states that the solution to this minimization problem exists in $RKHS$ and can be written as:

$$f^*(\boldsymbol{x}) = \sum_{i=1}^{l} \alpha_i \boldsymbol{K}(\boldsymbol{x}_i, \boldsymbol{x}) \tag{2}$$

Substituting this form in the problem above, it is transformed to optimize α^*. In the case of the squared loss function of $V(x_i, y_i, f) = (y_i - f(x_i))^2$. We can get final RSL (Regularized Least Squares) solution:

$$\alpha^* = (\boldsymbol{K} + \gamma l \boldsymbol{I})^{-1} \boldsymbol{Y} \tag{3}$$

3.2.2 Manifold Regularization

M. Belkin, et al [12] extended the standard framework to manifold regularization, which is described as the following form:

$$f^* = \arg\min_{f \in \mathrm{H_K}} \frac{1}{l} \sum_{i=1}^{l} V(x_i, y_i, f) + \gamma_A \|f\|_K^2 + \gamma_I \|f\|_I^2 \tag{4}$$

The differences between manifold regularization and standard framework lie in two aspects. The first is the former incorporate geometric structure of the marginal distribution P_X in the minimizing, which is reflected by the third item $\|f\|_I^2$. While minimizing, the coefficient γ_I controls the complexity of the function in the intrinsic geometry of P_X and γ_A controls the complexity of the function in ambient space. M. Belkin, et al [12] employ Laplacian manifold to represent the geometric structure embedded in high dimensional data. The second difference is the importing of unlabeled data $\{X_i\}_{i=l+1}^{i=u}$ by manifold regularization. While the lost function is calculated only by the labeled samples as before, when calculating the third penalizing item $\|f\|_I^2$, manifold regularization imports large mount of unlabeled samples that reflect the manifold distribution structure. So the solution changed with a new form:

$$f^*(x) = \sum_{i=1}^{l+u} \alpha_i^* K(x, x_i) \tag{5}$$

And the corresponsive solution of α^* is in Equation (6). $J = diag(1,1,...,0,0)$ with the first l diagonal entries as 1 and the rest 0. $\mathbf{L}$ is the Laplacian graph. We can notice that if γ_I=0 which means there is no unlabeled data in training, the Equation (6) will be identical with Equation (3).

$$\alpha^* = \left(JK + \gamma_A lI + \frac{\gamma_I l}{(u+l)^2} LK \right)^{-1} Y \tag{6}$$

3.2.3 Transfer Regression Model

We extend framework of manifold regularization to TRM, which can transfer the knowledge of the third party of dataset to the training model while importing large amount of unlabeled data.

We first have a survey on our transfer problem in Fig.1. The target is to build the regressive relationship $y = f(x)$ between variables of X and Y. X and Y correspond to two sparse training data. There is another variable Z, and Y is the function of Z, that is $y = h(z)$. In the task of 3D location estimation, X means the signal vectors collected on the third floor and Z means the signal vectors collected on the fourth floor and Y means the physical coordinators of the corresponding location points. There are some data pairs in X and Z, and these data have same function value in Y, which represent the same physical coordinates in different floor. Our purpose is to find schedules to reuse the data Z to improve the regressive model of X and Y.What we can depend on is the data pairs in X and Z, and the relationship of Z and Y .

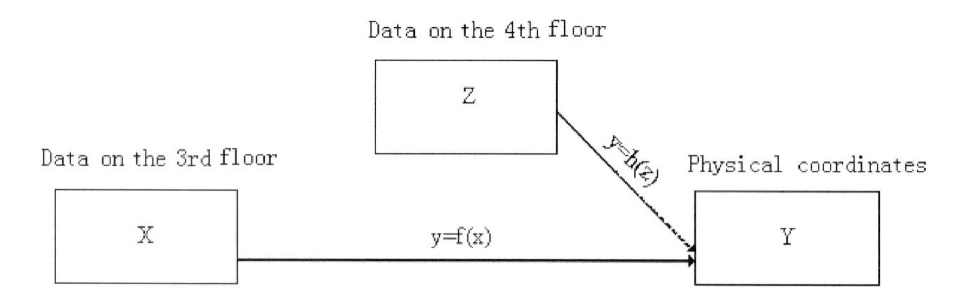

Fig. 2. Transfer problem in location estimation

Obviously, the traditional regression model such as BP (Back Propagation) network and SVM (Support Vector Machine) can not perform regressive training based on three datasets. In this paper, based on the framework of MR, we designed the schedule of TRM to improve $y = f(x)$ by importing data Z and optimize the regression model $y = f(x)$ and $y = h(z)$ at the same time. The strategy is to make the output of h and f have identical output while keep normal constraints of model h and f .

Now we construct the optimization problem:

$$(f^*,g^*,h^*)=\arg\min_{f\in \mathrm{H}_K}\frac{1}{l}\sum_{i=1}^{l}V(x_i,y_i,f)+\gamma_{A1}\|f\|_K^2+\gamma_{I1}\|f\|_I^2$$
$$+\frac{1}{l}\sum_{i=1}^{l}V(z_i,y_i,h)+\gamma_{A2}\|h\|_K^2+\gamma_{I2}\|h\|_I^2 \tag{7}$$
$$+\frac{1}{l}\sum_{i=1}^{l}(h(z_i)-f(x_i))^2$$

Sequentially, according to the classical Representer Theorem, the solution to this minimization problem exists in $RKHS$ (Reproducing Kernel Hilbert Space). We construct the regressive function as follows:

$$f^*(x)=\sum_{i=1}^{l}\alpha_i K_1(x_i,x) \tag{8}$$

$$h^*(z)=\sum_{i=1}^{l}\omega_i K_2(z_i,z) \tag{9}$$

Substitute them to (7), we can obtain the optimization problem of α and ω:

$$(\alpha^*,\omega^*)=\arg\min\frac{1}{l}(Y-J_1K_1\alpha)^T(Y-J_1K_1\alpha)+\gamma_{A1}\alpha^TK_1\alpha+\frac{\gamma_{I1}}{(u_1+l)^2}\alpha^TK_1L_1K_1\alpha$$
$$+\frac{1}{l}(Y-J_2K_2\omega)^T(Y-J_2K_2\omega)+\gamma_{A2}\omega^TK_2\omega+\frac{\gamma_{I2}}{(u_2+l)^2}\omega^TK_2L_2K_2\omega$$
$$+\frac{1}{l}(J_1K_1\alpha-J_2K_2\omega)^T(J_1K_1\alpha-J_2K_2\omega) \tag{10}$$

For h and f are both based on the labeled pairs, which means the number of the labeled data is the same, so $J_1=J_2=J$, $u_1=u_2=u$.

After derivation, we can represent α and ω as:

$$\alpha=\left(2JK_1+\gamma_{A1}lI+\frac{\gamma_{I1}l}{(u+l)^2}L_1K_1\right)^{-1}(Y+JK_2\omega) \tag{11}$$

$$\omega=\left(2JK_2+\gamma_{A2}lI+\frac{\gamma_{I2}l}{(u+l)^2}L_2K_2\right)^{-1}(Y+JK_1\alpha) \tag{12}$$

Combine two formulations above, and let

$$A=\left(2JK_1+\gamma_{A1}lI+\frac{\gamma_{I1}l}{(u+l)^2}L_1K_1\right)^{-1} \tag{13}$$

$$B = \left(2JK_2 + \gamma_{A2} lI + \frac{\gamma_{I2} l}{(u+l)^2} L_2 K_2 \right)^{-1} \tag{14}$$

Then, the solutions are:

$$\alpha^* = (I - AJK_2BJK_1)^{-1}(AY + AJK_2BY) \tag{15}$$

$$\omega^* = (I - BJK_1AJK_2)^{-1}(BY + BJK_1AY) \tag{16}$$

We can observed from the Equation (15) that the solution of α^* is based on the parameters A, K_1, Y, B, K_2, the later two parameters is computed upon the dataset of Z, which means the third party data have contribution to the final regression model.

In addition, if the data Z have no contribution for the training, such as when $Z = X$, then $K_1 = K_2 = K$, $L_1 = L_2 = L$, $\alpha = \omega$. We can get:

$$\alpha^* = \omega^* = \left(JK + \gamma_{A1} lI + \frac{\gamma_{I1} l}{(u+l)^2} LK \right)^{-1} Y \tag{17}$$

The solution is identical as the solution of MR in Equation (6). It just means that MR (Manifold Regularization) is a special case of TRM, which can train regression model based on three dataset.

Once α^* is computed, the model f in Equation (8) is then obtained, and it can be applied as a regressive model to predict the physical coordinates for the new input point on the third floor. For the model contains location and Wi-Fi information of the fourth floor, it performs better than that without knowledge transfer. It can be observed in the followed experiment section.

4 Experiments

4.1 Experiment Setup

Our experiments are carried out based on real Wi-Fi data set. The data were collected within the building of ICT (Institute of Computing Technology). To test the ability of the transfer model, we collected Wi-Fi signal data in third and fourth floor in a 34.8*83.4m area respectively. The physical structures are same for the two floors as shown in Fig. 3.

We prepared a large amount of data in the third floor which is enough to provide high location accuracy, and just acquired a small dataset which can only provide low location accuracy, as shown in Table 1.

In the third floor, there are totally 23 APs are detected, and we choose 10 of them to pick up the RSS. There are 500 labeled data are selected as training data.

In the fourth floor, the same APs as in third floor are selected. There are also 500 labeled data are acquired corresponding to those of third floor. So total 500 labeled pairs are prepared and 12700 unlabeled data are prepared in the training data set.

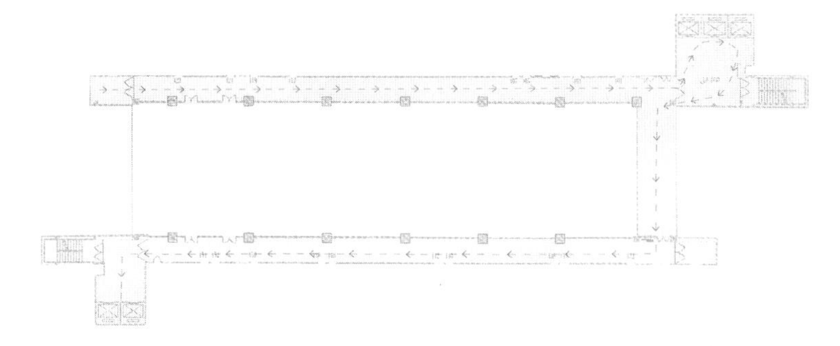

Fig. 3. The shared physical layout of third and fourth floor

4.2 TRM Based Location Estimation

Two experiments are carried out that train the prediction model with and without transferring knowledge from the data of the fourth floor.

- With transfer: In this case, three datasets are involved in the training procedure of TRM.
- Without transfer: In this case, just two datasets X, Y from third floor are involved, and the dataset Z of fourth floor is ignored. As mentioned before, if there is no contribution of dataset Z, TRM will be transformed into MR which is a regressive model without knowledge transfer.

For a given location, we use ED (Error Distance) parameter to evaluate the predicting result. In the experiment. If $y_0 = f(x_0)$ is the predicted position for the given input x_0, and y_t is the true position at that location, the prediction error will be the Euclidean distance between y_0 and y_t, that is $\text{ED} = \| y_0 - y_t \|$. If ED is less than a threshold ED_0, the result is regarded as a correct one, or a wrong one. Three thresholds are selected to evaluate our prediction accuracy.

In the experiments, some parameters of TRM need to be specified first. We test them as in our previous work [13]. Finally, they are set as: the number of neighbors $k = 6$, and γ_A=0.014362, γ_I=0.2852, K is calculated using Gaussian Kernel function, and the kernel parameter is set as 0.935. Under such case of the parameters, we get better results than other configure case.

Table 1. Experiments of transfer knowledge

	Prediction Accuracy		
	ED_0 =1m	ED_0 =2m	ED_0 =3m
With Transfer	65.2%	78.3%	89.6%
Without Transfer	51.5%	66.7%	85.3%
Improved Percent	26.6%	17.4%	5%

The regression results are shown in Table 1. From Table 1, the prediction accuracy is improved in different degree corresponding to different error distance threshold.

We also compute the prediction accuracy under different AP selection, and the strategy to select the APs in our experiment is according to the average signal strength that collected at all points. Generally speaking, strong signal strength means the AP is detected well, which is beneficial for location estimation. Finally, we compare the result of TRM and other two regressive model SVR and BP neural network. For SVR and BP model, the data from fourth floor is ignored.

Fig. 4 gives the comparison under different AP selection. It can be concluded that, on the whole, TRM outperforms both SVR and BP.

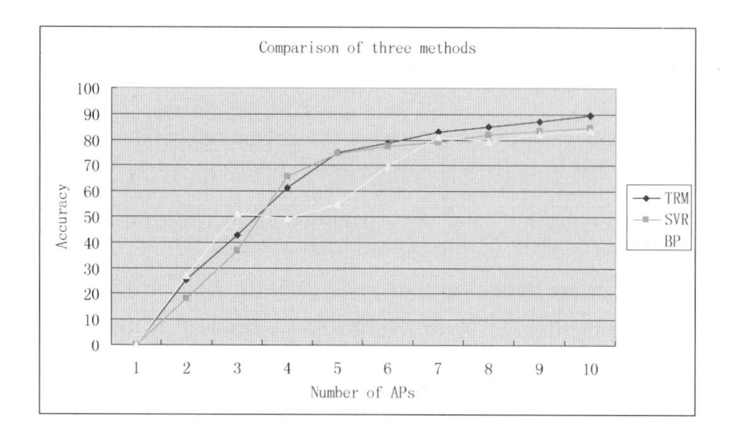

Fig. 4. Comparison of three methods under different AP selection

5 Conclusion and Future Works

In this paper, we present a novel transfer regressive model for indoor 3D location problem. TRM can effective import knowledge from the third party of dataset. The experiments of applying TRM to 3D location problem show that it is a good model to transfer the knowledge of location and Wi-Fi signal between various floors of the building. What's more, only a small labeled training data is needed for the target floor, which reduces calibration effort greatly.

In the future, we will consider the transfer learning on different devices, the situation that the well trained model can be transferred to a new type of wireless device. For instance, we use the notepad computer first to get a location model with high accuracy. Once we want to estimate location with a mobile cell phone, no large amount of data needed any more to train the new model, for our transfer model will utilize the previous data collected by the notebook computer.

Acknowledgements

This work is supported by National High-Technology Development '863' Program of China (2007AA01Z305) and Co-building Program of Beijing Municipal Education Commission. Thank Qiong Ning, Xiaoqing Tang for Wi-Fi data collection.

References

1. Hightower, J., Borriello, G.: Location systems for ubiquitous computing. Computer 34(8), 57–66 (2001)
2. Ferris, B., Haehnel, D., Fox, D.: Gaussian Processes for Signal Strength-Based Location Estimation. In: Proceedings of Robotics: Science and Systems (2006)
3. Nguyen, X., Jordan, M.I., Sinopoli, B.: A kernel-based learning approach to ad hoc sensor network localization. ACM Transaction on Sensor Networks 1(1), 134–152 (2005)
4. Pan, J.J., Yang, Q., Chang, H., Yeung, D.Y.: A Manifold Regularization Approach to Calibration Reduction for Sensor-Network Based Tracking. In: Proceedings of AAAI (2006)
5. Zlatanova, S., Rahman, A.A., Pilouk, M.: 3D GIS: current status and perspectives. In: Proceedings of the Joint Conference on Geo-spatial theory, Processing and Applications, Ottawa, July 8-12, p. 6. CDROM (2002)
6. Chittaro, L., Nadalutti, D.: Presenting evacuation instructions on mobile devices by means of location-aware 3D virtual environments. In: Proceedings of MobileHCI 2008 Amsterdam, Netherlands, September 2-5 (2008)
7. Chittaro, L., Burigat, S.: 3D location-pointing as a navigation aid in Virtual Environments. In: Proceedings of the working conference on Advanced visual interfaces, May 25-28 (2004)
8. Zlatanova, S., Verbree, E.: Technological Developments within 3D Location-based Services. In: International Symposium and Exhibition on Geoinformation 2003 (invited paper), Shah Alam, Malaysia, October 13-14, pp. 153–160 (2003)
9. Hightower, J., Boriello, G., Want, R.: SpotON: An indoor 3D Location Sensing Technology Based on RF Signal Strength, University of Washington CSE Report #2000-02-02 (February 2000)
10. Woodman, O., Harle, R.: Pedestrian localisation for indoor environments. In: Proceedings of the 10th international conference on Ubiquitous computing, pp. 114–123. ACM, New York (2008)
11. Evgeniou, T., Pontil, M., Poggio, T.: Regularization Networks and Support Vector Machines. Advances in Computational Mathematics 13, 1–50 (2000)
12. Belkin, M., Niyogi, P., Sindhwani, V.: Manifold Regularization: A Geometric Framework for Learning from Labeled and Unlabeled Examples. Journal of Machine Learning Research 7, 2399–2434 (2006)
13. Sun, Z., Chen, Y.Q., Qi, J., Liu, J.F.: Adaptive Localization through Transfer Learning in Indoor Wi-Fi Environment. In: Proceedings of Seventh International Conference on Machine Learning and Applications, pp. 331–336 (2008)

Personalized Sports Video Customization for Mobile Devices

Chao Liang[1,2], Yu Jiang[1], Jian Cheng[1,2], Changsheng Xu[1,2], Xiaowei Luo[3], Jinqiao Wang[1,2], Yu Fu[1], Hanqing Lu[1,2], and Jian Ma[4]

[1] National Labortory of Pattern Recognition, Institute of Automation, Chinese Academy of Sciences, 100190 Beijing, China
[2] China-Singapore Institute of Digital Media, 119615, Singapore
[3] School of Automation, Wuhan University of Technology, 430070 Wuhan, China
[4] Nokia Research Center, 100176, Beijing, China
{cliang,jcheng,csxu,jqwang,yfu,luhq}@nlpr.ia.ac.cn, zhejiangyu@hotmail.com, luo.xiaowei@hotmail.com, jian.j.ma@nokia.com

Abstract. In this paper, we have designed and implement a mobile personalized sports video customization system, which aims to provide mobile users with interesting video clips according to their personalized preferences. With the B/S architecture, the whole system includes an intelligent multimedia content server and a client interface on smart phones. For the content server, the web casting text is utilized to detect live events from sports video, which can generate both accurate event location and rich content description. The annotation results are stored in the MPEG-7 format and then the server can provide personalized video retrieval and summarization services based on both game content and user preference. For the client interface, a friend UI is designed for mobile users to customize their favorite video clips. With a new '4C' evaluation criterion, our proposed system is proved effective by both quantitative and qualitative experiments conducted on five sports matches.

Keywords: Sports video, personalized customization, mobile phone.

1 Introduction

With the advance of 3G technology, mobile multimedia, such as mobile TV and video on demand (VOD) services, is growing in popularity and promises to be an important driver for the wireless consumer industry. Among various video genres, sports video is extremely popular with mobile users. Equipped with such multimedia-enable mobile devices, sports fans nowadays can freely enjoy their favorite matches at anytime in anyplace.

However, due to the characteristics of mobile devices and sports video, mobile-based sports video service still faces some challenges: 1) massive video data and limited accessible bandwidth increase the burden of wireless network, which may cause the network jam and service delay; 2) interesting video clips only account for a small portion of the total sport match, so that transmitting the whole

S. Boll et al. (Eds.): MMM 2010, LNCS 5916, pp. 614–625, 2010.

match is neither economical nor necessary to mobile consumers; 3) due to the diverse preferences, traditional broadcast mode, in which game video or highlight collections are made by studio professionals, cannot meet audiences' personalized appetites. For example, a Barcelona's fan may prefer to enjoy Messi's shots in the Euro-Cup football matches. In such condition, the ability to detect semantic events and provide personalized video customization is of great importance.

For accurate event location and rich semantic description, text-facilitated sports video analysis has been validated as an effective method under current state of the art. Babaguchi *et al.* [1] proposed a multimodal strategy using closed caption (CC) for event detection and achieved promising result in American football games. Because CC is a direct transcript from speech to text, its colloquial words and poorly-defined structure will cause the difficulty for automatic text parsing. Xu *et al.* [2] utilized match report and game log to facilitate event detection in soccer video. Since text records and video events were matched based on the global structures, their method is not able to perform live event detection.

Sports video customization includes personalized retrieval and summarization, where the former focus on the particular person or event, while the latter concerns the whole situation. Zhang *et al.* [3] designed a relevance feedback strategy to retrieve suitable video clips in accordance with user request on both high-level semantics and low-level visual features. Babaguchi *et al.* [4] proposed personalized video retrieval and summarization with user preference learning from their operation for browsing. Although experiments conducted on PCs validate the effectiveness of the above approaches, their usability on mobile devices is still worth considering. Take the relevance feedback as an example, its complex operation and time-consuming interaction are likely to annoy mobile users.

In view of the extensive application prospect, mobile video service is attracting wide attention from both academy and industry. In [5], highlight detection technique is studied to extract and deliver interesting clips from the whole sports match to mobile clients. In [6] and [7], IBM and NTT DoCoMo have also developed their primary video summarization system for mobile clients. This paper presents a novel personalized sports video customization system for the mobile device. The main contributions of our work are summarized as follows:

- An user-participant multi-constraint 0/1 knapsack problem is raised to model personalized mobile video customization, where the video content and user preference can be synthetically considered and well balanced;
- An client interface is elegantly designed to facilitate mobile users customizing their favorite sports video with various personalized preferences;
- A '4C' criterion is proposed to evaluate the effectiveness of the personalized mobile video service from consistency, conciseness, coverage and convenience.

The rest of the paper is organized as follows. First, the system framework is presented in Section 2. Then, the technical details of the sports video analysis and data preparation modules are described in section 3 and 4, respectively. After that, personalized video customization is introduced in section 5 and user client design is discussed in section 6. Finally, experimental results are reported in Section 7 and conclusion and future work are given in Section 8.

2 System Framework

Fig. 1 illustrates the B/S framework of the proposed personalized video customization system. The multimedia server takes the responsibility of video content analysis and data processing. It aligns the sports video with web-casting text to detect various semantic events. Then a multimedia database is created to store the sports video data as well as their corresponding MPEG-7 description. Once the server receives a customization request, it will searches through the database and tailor the optimal set of video clips to the particular user by solving a constraint optimization problem. The mobile client allows the user to customize their favorite sports video clips with personalized preference on video content under limited device resources.

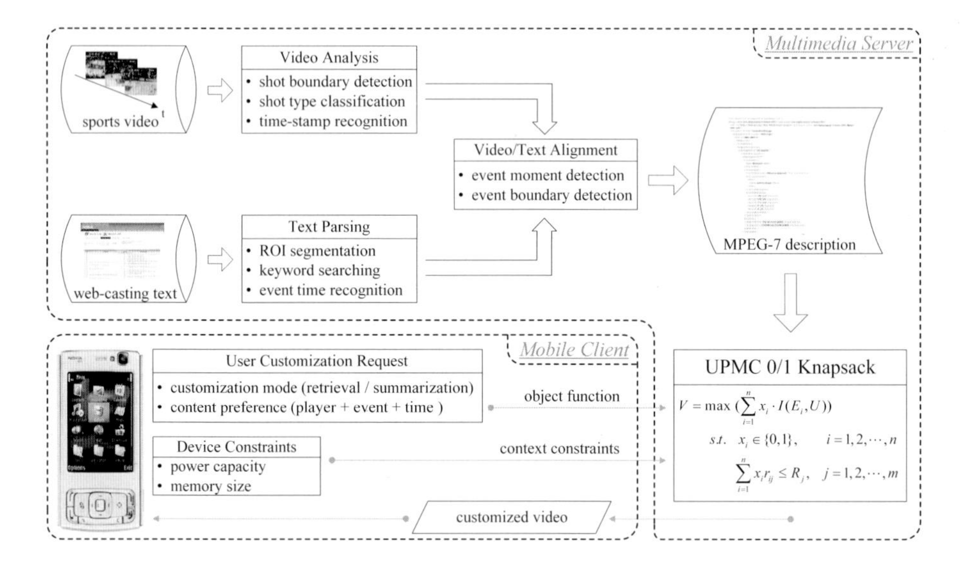

Fig. 1. System framework of mobile-based personalized video customization

3 Sports Video Analysis

High-level semantics extraction serves as the basis of our personalized sports video customization. Only with detailed content description, including involved player (who), event development (what) and happening moment (when), can we tailor the most interesting video clips to the particular audience. In the server end, we adopt our previous work, the web-casting text based method [8] to annotate sports video. Compared with other approaches, the advantage of our method lies in two sides: 1) the web-casting text provides detailed semantic description of video content and can be automatically extracted; 2) the timestamp-based method enables live event detection and thus can provide timely video service.

The basic idea of web text based video analysis is to detect semantic events through matching timestamps extracted from video content and its related text description. Therefore, live web text parsing, live video analysis and the timestamp based alignment constitute the major components of the approach.

Web text parsing is critical for the high-level semantics extraction. As shown in Fig. 2, the web text contains well-structured semantic description of sports matches, including the event time, current score, involved players and events. To live resolve the event information, the background program keeps sending request to the website server to get the latest webpage regularly, and extracts the match content descriptions using rule-based keyword matching method automatically.

Meanwhile, live video analysis algorithm is also conducted on the corresponding sports video. First, a static region detection method is applied to segment the overlaid digital clock region. Then the edge feature's changing is employed to describe the digits' temporal pattern and locate the digits area. The template of the 0 9 digits are captured after the SECOND digit area is located. At last, the game time is recognized by template matching. A realistic example is illustrated in Fig. 2, where text event happening at 11:37 is aligned with related video frames with clock digits indicating the same moment.

After detecting the event moment in the video, we need further identify a temporal range containing the whole event evolvement process. Due to the common production rule in sports video broadcasting, event occurrence is always accompanied by a group of specific shots transition (see examples in Fig. 3). Hence, the event start/end boundaries can be effectively detected by utilizing such temporal transition patterns. For reader who want to know the technical details, we recommend our previous work [8] for further reference.

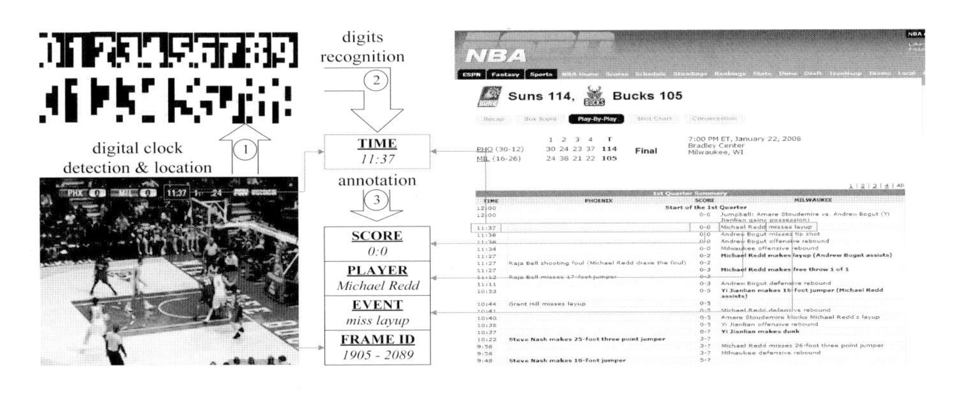

Fig. 2. Web-casting text based sports video analysis

Fig. 3. Typical temporal shot transition patterns for (a) shot event and (b) foul event

4 Data Formulation

The incorporation of web-casting text into sports video analysis, which links low-level features with high-level semantics in a cross-media way, generates the accurate event location and detailed description. For efficient data usage and management, we store the derived video annotation result in standard MPEG-7 XML format. As shown in Fig. 4, the XML file is organized in a hierarchical structure, where the whole match includes several quarters and each quarter contains a group of events. For each event in the match, the XML file records the detailed information including its involved player(s), event type and subtype, current score and related video segment.

Fig. 4. An example of MPEG-7 sports video description XML file

5 Personalized Mobile Video Customization

Since users are usually interested in watching the video content that matches his/her preference, the multimedia server is expected to provide as much relevant video content as possible to meet users' requirements. However, in the particular context of mobile multimedia, various client-side resource constraints, such as viewing time, memory size and power capacity, limit the allowed data transmission quantity. Therefore, the multimedia server is expected be able to select the optimal set of video segments within various client-side resource constraints.

5.1 Event Importance Computation

To evaluate the selected video segment group, event importance computation is indispensable. Both event influence on the match and its relevance to user request are taken into account. For influence computation, event rank and occurrence time are two main factors [4]. For relevance measurement, the semantic consistency on involved players and event types are considered.

1) ***Event Rank*:** The rank of event is directly determined by its influence to the game state. In a common sports match between two teams, say team A and B, the game state is always in one of three state, which are 'two teams tie', 'A team leads' and 'B team leads'. For the last two states, the change of leading degree can be further divided into two classes, increasing and decreasing the scoring gap. Based on the game state and its degree variation, game events in our system are assigned to various influence ranks (in Table 1).

Table 1. Event importance rank list

Rank Level	Event Description
1	score events and their inductive foul events that can change the game state from one situation to another.
2	score events and their inductive foul events that retain the game state but change the scoring gap in the match.
3	offense events that failed to score a goal or common foul events.
4	all other events that are not in the Rank 1 to 3

Then, the rank-based event importance $I_r(E_i)$ $(0 \leq I_r \leq 1)$ is defined as:

$$I_r(E_i) = 1 - \frac{R_i - 1}{3} \cdot \alpha \tag{1}$$

where R_i $(0 \leq R_i \leq 4)$ represent the rank level of event E_i and α $(0 \leq \alpha \leq 1)$ denotes the adjustable parameter to control the effects of rank difference on event importance computation.

2) ***Event Occurrence Time*:** In the sports match, offense events occurring at the final stage of the game are usually fatal to the component because they have little time to save the match. Thus, such events are of great importance to the final result. The event occurrence time based importance $I_t(E_i)$ $(0 \leq I_t \leq 1)$ is define as follows:

$$I_t(E_i) = 1 - \frac{N - i}{N} \cdot \beta \tag{2}$$

where N is the total event number, i is the index number of event E_i, and β $(0 \leq \beta \leq 1)$ is the adjustable parameter to control the effects of event occurrence time on event importance computation.

3) ***Event Relevance*:** Different to the type rank and occurrence time that are decided by event itself, the event relevance reflects the semantic consistency between event content and user's preference. This index plays a main role in our personalized customization because it effectively introduces the user's request in event importance evaluation. Through increasing the importance score of semantically related events, our system can finally tailor the appropriate video clip to the particular user. Based on the video annotation result, the event relevance based importance $I_u(E_i)$ $(0 \leq I_u \leq 1)$ is defined as:

$$I_u(E_i, U) = \gamma^{Dist_{person}(E_i, U)} \cdot (1 - \gamma)^{Dist_{event}(E_i, U)} \tag{3}$$

where functions $Dist_{person}$ and $Dist_{event}$ measure the semantic consistence between event E_i and user request U on the subjects of involved players and event types, and adjustable parameter γ $(0 \leq \gamma \leq 1)$ denotes the user preference between the above two subjects.

According to the above analysis, the event importance can be formulated as:

$$\begin{aligned} I(E_i, U) &= \lambda \cdot I_m(E_i) + (1 - \lambda) \cdot I_u(E_i, U) \\ &= \lambda \cdot I_r(E_i) \cdot I_t(E_i) + (1 - \lambda) \cdot I_u(E_i, U) \end{aligned} \tag{4}$$

where λ $(0 \leq \lambda \leq 1)$ is the fusion parameter controlling the weights on event influence on the match $I_m(E_i)$ and its semantic consistency to the user $I_u(E_i)$.

5.2 Personalized Mobile Video Customization

In personalized mobile video service, the multimedia server is required to present as much desired video content as possible within various resources constraints imposed by the mobile client. This problem can be modeled as a constraint optimization problem, and have been studied in [9,10]. Merialdo *et al* [9] raised a 0/1 Knapsack Problem to model the viewing time limited TV program personalization. Since the model can satisfy only a single client-side resource constraint at a time, their approach is not appropriate to the multi-constraint mobile video service. To fill the gap, Wei *et al* [10] proposed a Multi-choice Multi-dimension Knapsack strategy to summarize video segments on various abstraction levels and tailor these length-variant video segments to the particular resource-constraint client. However, we argue that it is unnecessary to present all video segments to the user and some excessively short video segments may actually annoy the audience. In light of this, we formulated our personalized mobile video customization as a user-participant multi-constraint 0/1 Knapsack problem as follows:

$$\begin{aligned} V = max&\Big(\sum_{i=1}^{n} x_i \cdot I(E_i, U)\Big) \\ s.t. \quad & x_i \in \{0, 1\}, \qquad i = 1, 2, \cdots, n \\ & \sum_{i=1}^{n} x_i r_{ij} \leq R_j, \quad j = 1, 2, \cdots, m \end{aligned} \tag{5}$$

where $I(E_i, U)$ represents the integrative importance value of event E_i under specified user request U, r_{ij} be the j_{th} resource consumption of the i_{th} event, R_j be the client-side resource bound of the j_{th} resource, xi denotes the existence of the i_{th} event in the selected optimal set, and n and m represent the number of events and resource types.

To refine users' personalized requests, we functionally divide their queries into retrieval and summarization. Video retrieval can be regarded as a 'point' query where users explicitly focus on the particular player and/or event type. In contrast, video summarization can be regarded as a 'plane' query where users are interested in the general situation of the match. These two customization modes can be treated in a unified manner with different fusion parameters in event importance computation. When λ is approximating to 0, event importance is mainly decided by user preference, thus only semantically consistent events can be assigned higher importance score and presented to the user. While in the case of λ approximating to 1, events' values are largely up to the match itself, hence the selected events can reflect the global situation of the match.

6 User Client Design

To avoid the difference among various mobile devices, a web-based video customization UI is designed on the server end. Through a specified browser, mobile users can easily visit the website and submit their personalized video customization requirements, including favorite matches, players, event types and allowed viewing time. As for the device context information, such as power capacity

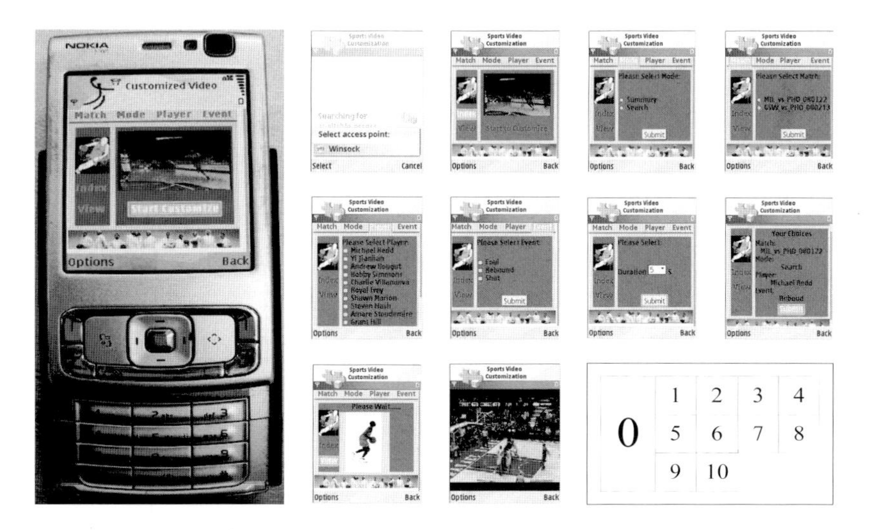

Fig. 5. The client interface on the Nokia N95 cell phone and a customization scenario example. (1) connect multimedia server; (2) enter customization user interface; (3) choose customization type; (4) choose match; (5) choose favorite players; (6) favorite event(s); (7) set viewing time; (8) request confirmation; (9) wait; (10) video playing.

and memory size, are all automatically submitted to the server. Once the user confirms his/her request, the multimedia server will search through the whole database and find the optimal set of video segments to the particular client. To better understand the above operation flow, a realistic example conducted on a Nokia N95 cell phone is illustrated in Fig. 5.

7 Experiment

To verify the effectiveness of our proposed personalized sports video customization system, we conduct both quantitative and qualitative experiments on three NBA 2008 basketball matches and two Euro-Cup 2009 football matches. The corresponding text records are from ESPN for basketball and BBC for football. In average, there are about 400 text events happened in one 100-minute basketball match and 50 text records in one 90-minue football match.

7.1 Semantic Event Location

Similar to [8], we use the event boundary accuracy (BDA) to measure the detected event boundary compared with the manually labeled boundary:

$$BDA = \frac{\tau_{db} \cap \tau_{mb}}{\tau_{db} \cup \tau_{mb}} \tag{6}$$

where τ_{db} and τ_{mb} represent the automatically detected event boundary and manually labeled event boundary, respectively. Since audience is insensitive to the frame-level difference, we adopt a shot rather than a frame (used in [8]) as the basic boundary unit in our evaluation. Table 2 gives the event boundary detection result of four events in basketball and three events in football matches. The average event location precision is reaching 90%, which lays a solid foundation for the personalized video customization. As for the lower BDA value of shot event in basketball matches, it is mainly due to the irregular shot switching in the case of free throw events drawing near the game end.

7.2 Personalized Video Customization

Before we give the experimental results of personalized video customization, we first mention the setting of adjustable parameters for event importance computation. As can be seen from 1 and 2, α and β are control coefficient to consider

Table 2. Event boundary detection result of basketball(B) and football(F) matches

No.	Shot	Foul	Rebound	Block	No.	Goal	Foul	Corner
B1	88.2%	90.5%	90.6%	92.3%	F1	93.4%	90.2%	90.4%
B2	88.5%	92.8%	91.3%	88.7%	F2	90.8%	94.4%	88.6%
B3	85.5%	91.5%	89.1%	90.1%				
Ave.	87.4%	91.6%	90.3%	90.4%	Ave.	92.1%	92.3%	89.5%

how large the difference of event rank and occurrence time affect the event importance. In our following experiment, both of these two coefficients are set as 1, which denotes the above rank and time differences fully affect the event significance. As for the parameter λ in equation 3, it is set as 0.5 denoting the equal preference between event types and involved player. To simplify the calculation, the Dist function is realized as a piecewise function, which outputs 0 when the text event and user request is consistent on the specified item of involved player or event type, and outputs 1 when they are inconsistent and 1/2 for the special case of no user request.

Due to the intrinsic subjectivity of personalized video customization, we carry out a user study to evaluate the performance of the customization system. Our study involves 4 student volunteers, who are all sports fans and familiar to the operations on smart phones. After watching the integral sports matches, students are asked to use the mobile phone to customize specific video clips with their personalized preference. To validate the effectiveness of the video customization algorithm and the usability of the mobile client, a particularly-designed '4C' questionnaire is handed out to the participants to get their feedbacks. Motivated by the pioneering work by He *et al* [11], we define the '4C' criterions as follows:

- **Consistency.** Whether the generated video clip is consistent with user request on content semantics, such as involved players and event types;
- **Conciseness.** Whether the generated video clip capture the main body of the match without including irrelevant events;
- **Coverage.** Whether the generated video clip covers all important events happened in the match under current viewing time limit;
- **Convenience.** Whether the mobile client can facilitate the user conveniently customize his/her favorite sports video clip.

All the above four items are answered on a five-scale Likert scale where 1 represents strongly disagree and 5 denotes strongly agree. The detailed results under various λ s are given in Table 3, where A~D represents four students.

Table 3. Users' 4C evaluation under different fusion parameters

	$\lambda = 0.1$					$\lambda = 0.5$					$\lambda = 0.9$				
	A	B	C	D	Ave.	A	B	C	D	Ave.	A	B	C	D	Ave.
Consistency	4	4	3	5	4.00	3	4	2	4	3.25	4	3	2	3	3.00
Conciseness	4	4	3	3	3.50	4	5	4	3	4.00	5	3	3	4	3.75
Coverage	3	2	3	2	2.50	3	4	3	5	3.75	5	3	3	5	4.00
Convenience	4	3	4	5	4.00	4	4	5	5	4.50	4	3	5	5	4.25

As can be seen from Table 3, when the fusion parameter λ increases from 0.1 to 0.9, users' evaluation on result consistency gradually declines from 4 to 3, while the result coverage, behaves oppositely, rising from 2.5 to 4. The contrast reveals the important role that λ plays in balancing game content and user preference in video customization. When λ is small, the server will lay more emphasis on

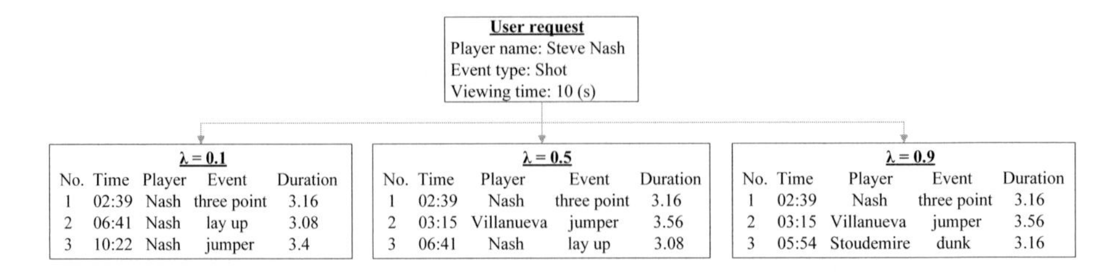

Fig. 6. Change of customized video segments under different fusion parameters

user's request, hence more semantically related events are selected. In contrast, when λ is big, the server is partial to the whole situation of the match, thus the globally interesting events are presented. To better understand the function of λ, a real customization example is illustrated in Fig. 6.

8 Conclusions and Future Work

In this paper, we proposed a personalized sports video customization system for mobile users. Compared with previous work, our system can provides timely multimedia service with the help of automatic live video analysis and meet user personalized preference with a newly proposed user-participant multi-constraint 0/1 knapsack model. Both quantitative and qualitative experiments conducted on basketball and football matches validated the effectiveness of our approach.

In the future, we will investigate an objective criterion to evaluate the personalized video customization result so that different systems/methods can be objectively compared on a common baseline. In addition, besides sports video, more challenging video types, such as movie and news programs, will also be studied so that we can provide more abundant visual enjoyment to our clients.

Acknowledgement

This work is supported by National Natural Science Foundation of China (Grant No. 60833006), Beijing Natural Science Foundation (Grant No. 4072025), and partly supported by Nokia Research Center.

References

1. Babaguchi, N., Kawai, Y., Kitahashi, T.: Event based indexing of broadcasted sports video by intermodal collaboration. IEEE Transaction on Multimedia 4(1), 68–75 (2002)
2. Xu, H., Chua, T.: The fusion of audio-visual features and external knowledge for event detection in team sports video. In: Proc. of Workshop on Multimedia Information Retrieval, pp. 127–134 (2004)

3. Zhang, Y., Zhang, X., Xu, C., Lu, H.: Personalized retrieval of sports video. In: Proc. of Workshop on Multimedia Information Retrieval, pp. 312–322 (2007)
4. Babaguchi, N., Ohara, K., Ogura, T.: Learning personal preference from viewers operations for browsing and its application to baseball video retrieval and summarization. IEEE Transaction on Multimedia 9(5), 1016–1025 (2007)
5. Liu, Q., Hua, Z., Zang, C., Tong, X., Lu, H.: Providing on-demand to mobile deices. In: Proc. ACM Multimedia, pp. 6–11 (2005)
6. Tseng, B., Lin, C., Smith, J.: Video summarization and personalization for pervasive mobile. In: SPIE, pp. 359–370 (2002)
7. Echigo, T., Masumitsu, K., Teraguchi, M., Etoh, M., Sekiguchi, S.: Personalized delivery of digest video managed on mpeg-7. In: ITCC, pp. 216–220 (2001)
8. Xu, C., Wang, J., Lu, H., Zhang, Y.: A novel framework for semantic annotation and personalized retrieval of sports video. IEEE Transaction on Multimedia 3(10), 421–436 (2008)
9. Merialdo, B., Lee, K., Luparello, D., Roudaire, J.: Automatic construction of personalized tv news programs. ACM Multimedia, 323–331 (1999)
10. Wei, Y., Bhandarkar, S., Li, K.: Video personalization in resource-constrained multimedia enviroments. ACM Multimedia, 902–911 (2007)
11. He, L., Sanocki, E., Gupta, A., Grudin, J.: Auto-summarization of audio-video presentation. ACM Multimedia, 489–498 (1999)

3D Thumbnails for Mobile Media Browser Interface with Autostereoscopic Displays

R. Bertan Gundogdu, Yeliz Yigit, and Tolga Capin

Bilkent University, Computer Engineering Department
06800 Ankara, Turkey
{gundogdu,yyigit,tcapin}@cs.bilkent.edu.tr

Abstract. In this paper, we focus on the problem of how to visualize and browse 3D videos and 3D images in a media browser application, running on a 3D-enabled mobile device with an autostereoscopic display. We propose a 3D thumbnail representation format and an algorithm for automatic 3D thumbnail generation from a 3D video + depth content. Then, we present different 3D user interface layout schemes for 3D thumbnails, and discuss these layouts with the focus on their usability and ergonomics.

Keywords: Mobile Multimedia, 3D Thumbnails, 3D User Interfaces.

1 Introduction

Today mobile devices are becoming one of the main means to use multimedia in our daily life. It is now possible to send/receive multimedia messages, watch TV broadcasts, and perform basic videoconferencing on current-generation mobile devices. To further benefit from the advances in mobile multimedia processing hardware and software solutions, new levels of experiences will be required in mobile multimedia.

The European FP7 3DPHONE project [1] aims to develop applications enabling such a new level of user experience, by developing an *end-to-end all-3D imaging mobile phone*. The goal of this work is to build a mobile device, where all fundamental functions are realized in 3D, including media display, user interface (UI), and personal information management (PIM) applications. Various solutions are needed for achieving this all-3D phone experience: including building of mobile autostereoscopic displays, 3D images/video, 3D UIs, 3D capture/content creation solutions, compression, and efficient rendering.

Developing such an all-3D mobile phone requires building a solution that takes into account the entire user experience of a mobile phone, instead of merely putting separate solutions together. For example, 3D enabled phones already exist in the market, and various vendors have started to promote autostereoscopic displays on mobile devices. However, despite users' clear interest in 3D technologies, simply providing 3D display capabilities is not sufficient, as it was demonstrated that the usability of these devices and applications have to be improved. To make the best use of the technologies to support user experience, a new user interaction paradigm (an all-3D phone experience) will be needed, taking advantage of the latest advances on 3D graphics

S. Boll et al. (Eds.): MMM 2010, LNCS 5916, pp. 626–636, 2010.

rendering on mobile handheld platforms. User experience will be driven towards immersive 3D media, through the use of autostereoscopic displays and 3D interaction using sensors. This leads to a need for new concepts in 3D UI development.

We aim to address the question: what are the best design decisions for building an all-3D user experience on a mobile phone? To answer this question, we are currently constructing the hardware and software platforms of a prototype 3D phone device, which together provide an integrated environment allowing the implementation of the 3D based UIs and applications. We use the Texas Instruments' OMAP 34x MDK [2] as the base platform of the 3D phone, and a two-view lenticular-based solution and a quasi-holographic display for output. The prototype also features two cameras for stereo capture, based on SGS Thomson's VS6724 camera.

In this paper, we focus on the particular problem of how to visualize and browse 3D video and images in a media browser application, running on a mobile device with an autostereoscopic display. Particularly, we propose a 3D thumbnail-based approach for representing 3D media. We first present an algorithm for generating 3D thumbnails that preserve the significant parts of the input 3D image. Then, we discuss different 3D user interface layout schemes for a media browser application that makes use of these 3D thumbnails. We then discuss our solutions with the focus on their usefulness, and their ergonomics and comfort.

The paper is organized as follows: Section 2 reviews previous work on thumbnail generation, 3D video representation, and 3D user interface design, which are the essential components of our work. Section 3 describes our proposed approach on 3D thumbnail generation and 3D layouts. In Section 4, we present our initial results and discuss future directions of our research.

2 Previous Work

Our approach combines the three distinct problems: thumbnail generation, 3D video representation, and 3D user interaction steps. Therefore, we discuss each topic under a different subsection.

2.1 Thumbnail Representation

A major part of our approach is creation of thumbnails from 3D video content without losing perceivable elements in the selected original video frame. It is essential to preserve the perceivable visual elements in an image for increasing recognizability of the thumbnail. Our thumbnail representation involves the computation of important elements, and performing non-uniform scaling to the image. This problem is similar to the recently investigated image retargeting problem for mobile displays.

Various automatic image retargeting approaches have been proposed [3]. Firstly, retargeting can be done by standard image editing algorithms such as uniform scaling and manual cropping. However, these techniques are not an efficient way of retargeting: with uniform scaling, the important regions of the image cannot be preserved; and with cropping, input images that contain multiple important objects leads to contextual information lost and image quality degrades. On the other hand, automatic cropping techniques have been proposed, taking into account the visually important parts of the input image, which can only work for single object [4]. Another alternative approach is based on the epitome, in which the image representation is in miniature and condensed version, containing the most important elements of the original image [5]. This technique is suitable even when the original image contains repetitive unit patterns. For creating meaningful thumbnails from 3D images/video, we have adopted a saliency-based system that preserves the image's recognizable features and qualities [3].

2.2 3D Video

A number of 3D imaging and video formats have recently been investigated. These formats can be roughly classified into two classes: N-view video formats and geometry-enhanced formats. The first class of formats describes the multi-view video data with N views. For stereoscopic (two-view) applications, conventional stereo video (CSV) is the most simple format.

In the second class of 3D formats, geometry-enhanced information is added to the image. In the multi-view video + depth format (MVD) [6], a small number of sparse views is selected and enhanced with per pixel depth data. This depth data is used to synthesize a number of arbitrarily dense intermediate views for multi-view displays. One variant of MVD is Layered Depth Video (LDV), which further reduces the color and depth data by representing the common information in all input views by one central view and the difference information in residual views [7]. Alternatively, one central view with depth data and associated background information for color and depth is stored in an LDV representation to be used to generate neighboring views for the 3D display. Geometry-enhanced formats such as MVD or LDV allow more compact methods, since fewer views need to be stored. The disadvantage however is the intermediate view synthesis required. Also, high-quality depth maps need to be generated beforehand and errors in depth data may cause considerable degradation in the quality of intermediate views. For stereo data, the Video-Depth format (V+D) is the special case. Here, one video and associated depth data is coded and the second view is generated after decoding.

Regarding the capabilities of current mobile devices, stereo formats CSV and V+D are the most likely candidates to be used for real-time applications. As the V+D format is more flexible for generating new views, the thumbnail representation is based on this format in this study.

2.3 3D User Interfaces

There are numerous studies on 3D user interfaces, but very few of them target 3D UIs on mobile devices or offer solutions for 3D thumbnail layout problem in a media browser application. One rather early study on how to present information on a small scale display area is a work by Spence [8], which provides a bifocal display solution in which the information is presented in two different resolutions. A high resolution is used for a specific data element to provide details about that element and a low resolution for the rest of the data elements that is enough for high-level view of each element. A 3D version of the bifocal display of Spence may be suitable for a media browser application that uses 3D thumbnails so that the high resolution element will be a 3D thumbnail and the low resolution ones are 2D thumbnails or simplified versions.

A more recent design suitable for mobile devices is Cover Flow® by Apple [9]. It is originally designed to display album covers in the media player application, however, today it is used in many different applications to present photos, videos, web pages, etc. The structure can be considered as two stacks placed on both sides of the item in focus.

Lastly, a web browser extension named CoolIris® [10] incorporates 3D properties in its presentation for photo and video thumbnails. It uses a wall metaphor and places thumbnails on the wall as a grid. The grid has strictly three rows and the number of columns increases with the number of items. The viewport is tilted around the y-axis when scrolling horizontally through the thumbnails and perspective projection gives the depth perception while scrolling. Although the design is basically 2D, the tilting of viewport and perspective projection makes navigation faster by showing more items than a regular 2D grid. Loading items on the fly, instead of using pagination, also helps to increase the efficiency of navigation. A fundamental difference between this approach and Cover Flow® is that this approach does not focus on a particular item.

3 Proposed Solution

3.1 Overview of the System

Our system consists of a 3D thumbnail generation subsystem and a 3D user interface subsystem. Figure 1 shows the overview of our proposed solution.

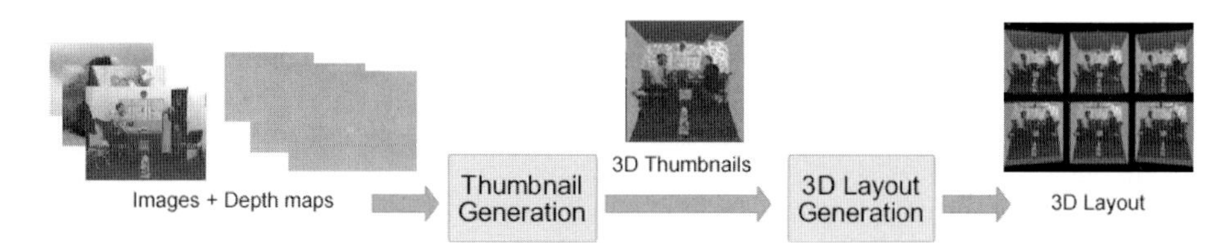

Fig. 1. Overview of the system

3.2 Thumbnail Generation

The input of the thumbnail generation subsystem is a frame from the input 3D video, in the form of a depth image. Thus, two images are input to the system – an RGB color map, and a corresponding depth map approximating the depth of each pixel.

Our thumbnail generation approach is shown in Figure 2. The input color map is first segmented into regions in order to calculate each region's importance, as explained in Section 3.2.1 and 3.2.2. Our method then removes the important parts of the image, later to be exaggerated, and fills the gaps of the background using the technique described in Section 3.2.3. Afterwards, the filled background is resized to standard thumbnail size of 192 x 192. Then, important regions are pasted onto the resized background, as explained in Section 3.2.4. The final 3D thumbnail is generated by constructing a 3D mesh, as described in Section 3.2.5.

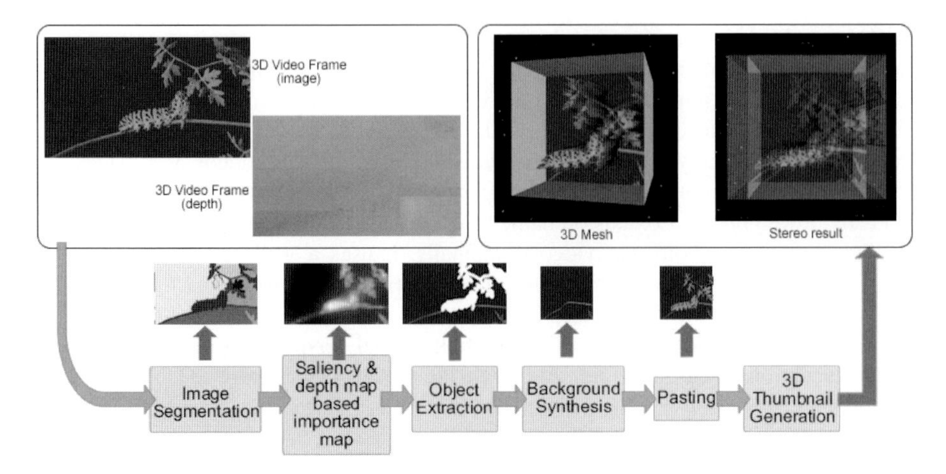

Fig. 2. Overview of the 3D thumbnail generation system

3.2.1 Image Segmentation

We use the mean-shift image segmentation algorithm for separating the input image into regions. In addition to the mean-shift method [11], there are alternative segmentation methods such as graph-based [12] and hybrid segmentation approaches [13]. Pantofaru et al. compares the three methods by considering correctness and stability of the algorithms [13]. The results of this work suggest that both the mean-shift and hybrid segmentation methods create more realistic segmentations than the graph-based approach with a variety of parameters. The results show that these two methods are also similar in stability. As the hybrid segmentation method is a combination of graph-based and mean-shift segmentation methods, and thus is more computationally expensive, we have chosen the mean-shift algorithm for its power and flexibility of modeling.

The mean-shift based segmentation method is widely used in the field of computer vision. This method takes three parameters together with the input image: spatial radius h_s, color radius h_r and the minimum number of pixels M that forms a region. The CIE-Luv color space is used in mean-shift algorithm, therefore the first step is to convert the RGB color map into $L\alpha\beta$ color space [14]. The color space has luminance, red-green and blue-yellow planes. These color planes are smoothed by Gaussian kernels.

Afterwards, for each pixel of the image with a particular spatial location and color, the set of neighboring pixels within a spatial radius h_s, and color radius h_r is determined and labeled. Figure 3 shows the image segmentation results for different values of h_s, h_r and M. In this work, we have chosen the parameters as shown in configuration (b).

3.2.2 Color and Depth Based Saliency Map

The next step computes an *importance map* from the input color map and depth map. The aim of the retargeting algorithm is to resize images without losing the important regions on the input image. For this purpose, we calculate the importance of each pixel, as a function of its *saliency* in the color map and the depth map. We compute the saliency based on color and depth differently, as described below.

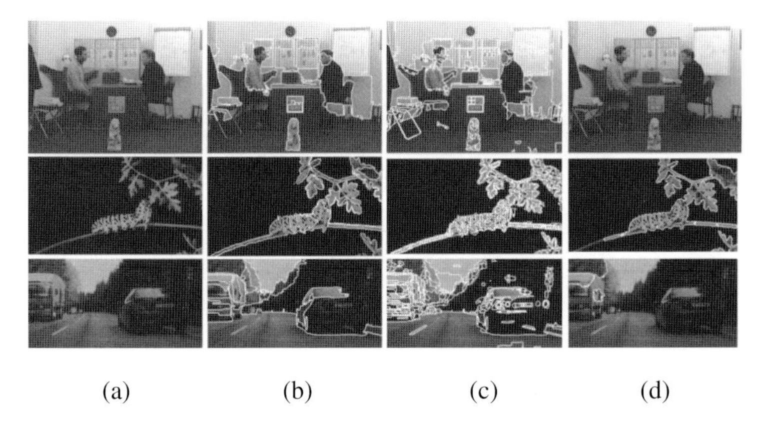

(a) (b) (c) (d)

Fig. 3. Mean Shift Segmentation with different parameters. (a) Original Image; (b) hs = 7, hr = 6 and M = 800, number of regions: 31, 15, 19; (c) hs = 6, hr = 5 and M = 50, number of regions: 214, 268, 246; (d) hs = 32, hr = 30 and M = 150, number of regions: 1, 7, 6.

Computation of Saliency Based on Color Map

Most of the physiological experiments verify that human vision system is only aware of some parts of the incoming information in full detail. The concept of saliency has been proposed to locate the points of interest. In this work, we apply the *graph-based visual saliency* image attention model [15]. There are two steps for constructing the bottom-up visual saliency model: constructing activation maps on certain feature channels and normalization. This method is based on the bottom-up computation framework because in complex scenes that hold intensity, contrast and motion, visual attention is in general unconsciously driven by low-level stimulus.

Feature extraction, *activation* and *normalization of the activation map* are three main steps of the graph-based visual saliency method:

- In the *feature extraction step*, the features such as color, orientation, texture, intensity are extracted from the input image through linear filtering and computation of center-surround differences for each feature type.
- In the *activation step*, an activation map is formed from the feature maps produced in step 1. A pixel with a high activation value is considered significantly different from its neighborhood pixels.
- In the last step, *normalization of the activation map* is performed, by normalizing the effect of feature maps, and summing them into the final saliency value of the pixel based on the color map.

Figure 4 shows sample results that are based on graph-based visual saliency image attention method. The detailed explanation of the algorithm can be found in [15].

Computation of Saliency Based on Depth Map

We observe that depth is another factor to decide whether an object is of interest or should be ignored. In other words, closer objects should be more salient due to proximity to the eye position. Therefore, we also use the depth map to calculate depth

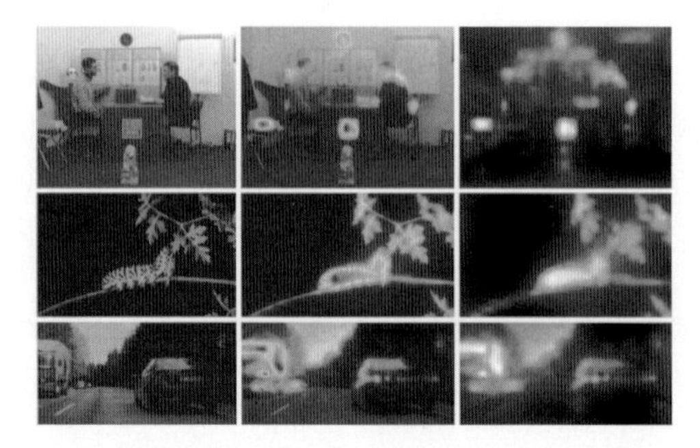

Fig. 4. Graph-based saliency map. a) Original image; b) salient parts of the image (red - most salient); c) resulting saliency map.

saliency for each pixel in the input image. The function below for computing the depth importance map, adapted from the work of Longurst et al. [16], uses a model of exponential decay to get a typical linear model of very close objects. Therefore, in the equation, d and A_D are constants to approximate the linear model by the overall rate of exponential decay [16]:

$$S_D = \frac{1}{d\sqrt{2\pi}}\left(exp - \frac{D^2}{d^2}\right)A_D$$

$$d = 0.6$$

$$A_D = 1.5$$

Computation of Overall Saliency Map

For each region that was computed as the result of the earlier segmentation step, we compute the overall saliency of the region, by averaging the sum of the color-based and depth-based saliency of pixels belonging to the region.

3.2.3 Background Resynthesis

The next steps after extracting important regions from the original color map are to resize the color map that has gaps to the standard thumbnail size, and to fill gaps with the information extracted from the surrounding area. This step adopts Harrison et al.'s inpainting method [17] that reconstructs the gaps with the same texture as the given input image by successively adding pixels from the image. The procedure is capable of reproducing large features from this input image, even though it only examines interactions between pixels that are close neighbors. Decomposing the input image into a feature set is avoided and the method could reproduce a variety of textures, making it suitable for this purpose.

3.2.4 Pasting of Visually Significant Objects

After extracting the important objects from the original color map and background resynthesis, the next step is to paste them onto the new background. We use a constraint-based method to paste each important region due to their overall saliency value from the most salient to least [3]. The goal of the algorithm is to preserve the relative positions of the important regions in order to maintain certain perceptual cues and keep the resized image's layout similar to the original image. To achieve this, there are four constraints: positions of the important objects must stay the same, aspect ratios of the important objects must be maintained, the important objects must not overlap in the retargeted background if they are not overlapping in the original image and the background color of the important region must not change.

By using a decreasing order of the overall saliency, this step reduces the change in position and size of the important objects. Thus, from the most salient object to the least, the algorithm searches whether the four constraints are satisfied or not that are described above, and changes the size and position of the important object according to the original and target color map by calculating aspect ratio.

3.2.5 3D Thumbnail Creation

As a result of the previous step, two channels for the thumbnail are generated – the RGB thumbnail color map, and the corresponding depth map approximating the depth of each pixel. In the 3D thumbnail creation step, these color and depth maps are converted to a 3D mesh representation. This conversion is necessary, as the depth image representation has several drawbacks for the thumbnail case. Thumbnails are expected to be small in size, to allow several of them to appear simultaneously on display. Depth images also introduce image fitting problems such as matching the perspective of the 3D thumbnail and the rendered 3D scene, and the eye-separation value of the cameras. Another drawback is that using a 3D mesh is much more flexible in terms of 3D display usage. One can apply different object reconstruction methods to obtain the meshes and then render the scene for multiple view rendering.

For construction of the 3D mesh, we gather the depth values of every pixel on the thumbnail using the depth image and produce vertices for each pixel. Then we connect these vertices to form the mesh. Depth values of pixels are mapped to the depth values of vertices. To increase the contrast in depth values we use only the minimum and maximum depth values of pixels and normalize them to [0-1] interval. We then simplify the 3D mesh to obtain a model that can be rendered in real time on the mobile device.

3.3 3D User Interface Layout for Viewing Thumbnails

One of the main reasons, and perhaps the most important one, to use thumbnails is to display as many of the items as possible at the same time. Therefore, the thumbnails are expected to be small, but still reasonably large to let the user have an idea on what is "inside" each item. We consider this principle while designing our 3D media browser user interface. We aim to make use of the 3D content and the autostereoscopic 3D display properties, for viewing the generated 3D thumbnails.

We extend the card metaphor to support 3D cards and design two essentially different layouts, and experiment with various variations of them. The first design – i.e. linear design – focuses on one of the items by centering it on the screen and displaying the other items at the back (Figure 5-a). The second design (Figure 5-b) – i.e. grid-based design – treats every item equally forming a grid structure. The following discusses the variations we apply to these main designs:

- *Rotation:* in the linear design, we form a circular list (Figure 5-a) with the items at the back; in the grid-based design, all the items are forming two circular (Figure 5-d) or linear (Figure 5-c) lists that are placed on top of each other.

- *Framed Thumbnails:* we consider the use of frames around thumbnails to enhance the depth perception of 3D content and to establish boundaries between thumbnails (Figure 5-d and Figure 5-e).
- *Depth:* the initial depth values of vertices of the mesh are mapped to [0-1] interval, but in order to provide a better 3D perception on the autostereoscopic display, we scale each item's depth values by different values in a search for the best value (Figure 5-a and Figure 5-b).

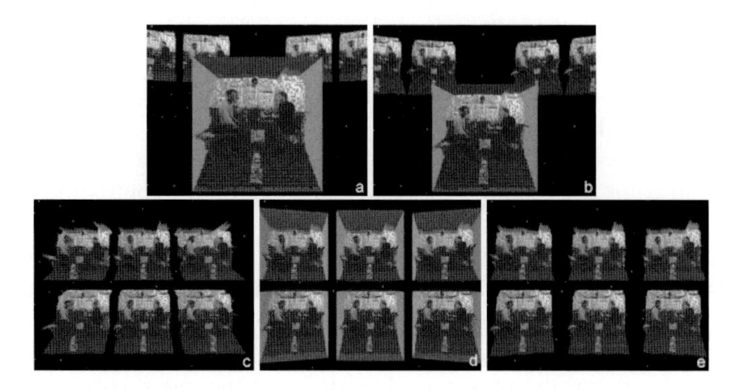

Fig. 5. Design alternatives for 3D user interface layout

4 Discussion

3D Thumbnail Generation
We use the automatic retargeting method for creating 3D thumbnails. We have selected this approach because we aim to preserve the multiple important objects' original position, importance and background color on the given color map. As discussed, there are alternative methods for creating thumbnails such as cropping and scaling. However, the results of earlier evaluation tests for the 2D case, performed by Setlur et al., show that neither cropping nor scaling preserves recognizability as much as the retargeting method [3].

However, the retargeting method has disadvantages and limitations: the semantic relationships between objects aren't preserved; important regions that are resized independently by their overall saliency lead to wrong relative proportions and incorrect handling of repeated textures if the background is complex.

Although we use a simple method for creating 3D thumbnails for our system, the results are reasonable. It is also possible to use computer vision techniques that derive 3D meshes from images and GPU-based computer graphics methods that give 3D view to 2D images. We'll investigate these methods in our future work, and perform subjective and ergonomic evaluation of these alternatives.

3D Layout
We experiment with different 3D layout options, and identify a number of design suggestions for producing a 3D media interface with high usability. Subjects in our initial tests have reported that the use of rotation in grid-based layout provides a cleaner view with less overlap. In the absence of rotation, the depth effect is more

noticeable in thumbnails that are out of focus. Subjects have also reported that in the grid-based design, use of frames increases the 3D perception in some cases, but it also increases the eye strain, thus frameless design was more comfortable than the framed design in grid-based layout. On the other hand, in the linear design, the use of frames in the centered thumbnail yields significantly better perception in 3D than non-frame representation.

For the linear design option, we conclude that the average depth value scaling values in the interval [1.5-2] provide a better depth perception on our auto-stereoscopic displays. The scaling values below this interval are still usable, but do not provide a good depth perception. Scaling values above are also usable until a certain limit, but they increase the screen space needed to show the thumbnail because of the increasing perspective effect. As for the background items in the linear-based design, the thumbnails need to be small in size, however, smaller size than a threshold decreases the depth perception.

As a future work, we plan to conduct several ergonomics tests on the UI part of this study as well as experimenting with different UI designs that use 3D thumbnails.

Acknowledgments

This work is supported by FP7 ICT All 3D Imaging Phone (3DPHONE) project. The depth images shown in this work are courtesy of Fraunhofer Heinrich Hertz Institute.

References

1. European 3DPHONE Project Homepage (2009), `http://www.the3dphone.eu`
2. Texas Instruments Zoom™ OMAP34x MDK (2009), `http://www.omapzoom.org`
3. Setlur, V., Takagi, S., Raskar, R., Gleicher, M., Gooch, B.: Automatic Image Retargeting. In: International Conference on Mobile and Ubiquitous Multimedia, Christchurch, New Zealand, pp. 59–68 (2005)
4. Suh, B., Ling, H., Bederson, B., Jacobs, D.: Automatic Thumbnail Cropping and its Effectiveness. In: ACM Symposium on User interface Software and Technology, Vancouver, Canada, pp. 95–104 (2003)
5. Jojic, N., Frey, B., Kannan, A.: Epitomic Analysis of Appearance and Shape. In: IEEE International Conference on Computer Vision, Nice, pp. 34–41 (2003)
6. Smolic, A., Müller, K., Dix, K., Merkle, P., Kauff, P., Wiegand, T.: Intermediate View Interpolation based on Multi-View Video plus Depth for Advanced 3D Video Systems. In: IEEE International Conference on Image Processing, San Diego, CA, pp. 2448–2451 (2008)
7. Müller, K., Smolic, A., Dix, K., Kauff, P., Wiegand, T.: Reliability-Based Generation and View Synthesis in Layered Depth Video. In: Proc. IEEE International Workshop on Multimedia Signal Processing (MMSP 2008), Cairns, Australia, pp. 34–39 (2008)
8. Spence, R., Apperley, M.: Data Base Navigation: an Office Environment for the Professional. Journal of Behaviour and Information Technology 1(1), 43–54 (1982)
9. Apple, Inc. iPod Touch Description (2009), `http://www.apple.com/ipodtouch`
10. Cooliris, Inc. (2009), `http://www.cooliris.com`

11. Comaniciu, D., Meer, P.: Mean shift: A robust Approach Toward Feature Space Analysis. IEEE Trans. on Pattern Analysis and Machine Intelligence 24(5), 603–619 (2002)
12. Felzenszwalb, P., Huttenlocher, D.: Efficient Graph-Based Image Segmentation. International Journal of Computer Vision 59(2), 167–181 (2004)
13. Pantofaru, C.: A Comparison of Image Segmentation Algorithms. Technical Report, Pittsburgh, PA (2005)
14. Mirmehdi, M., Petrou, M.: Segmentation of Color Textures. IEEE Transactions on Pattern Analysis and Machine Intelligence 22(2), 142–159 (2000)
15. Harel, J., Koch, C., Perona, P.: Graph-Based Visual Saliency. In: Advances in Neural Information Processing Systems, Cambridge, MA, vol. 19, pp. 545–552 (2007)
16. Longhurst, P., Deba Hista, K., Chalmers, A.: A GPU Based Saliency Map For High-Fidelity Selective Rendering. In: International Conference on Computer Graphics, Virtual Reality, Visualisation And Interaction, Cape Town, South Africa, pp. 21–29 (2006)
17. Harrison, P.: A Non-Hierarchical Procedure for Re-Synthesis of Complex Textures. In: Proc. WSCG, pp. 190–197 (2001)

Video Scene Segmentation Using Time Constraint Dominant-Set Clustering

Xianglin Zeng[1], Xiaoqin Zhang[1], Weiming Hu[1], and Wanqing Li[2]

[1] National Laboratory of Pattern Recognition, Institute of Automation, Beijing, China
{xlzeng,xqzhang,wmhu}@nlpr.ia.ac.cn
[2] SCSSE, University of Wollongong, Australia
wanqing@uow.edu.au

Abstract. Video scene segmentation plays an important role in video structure analysis. In this paper, we propose a time constraint dominant-set clustering algorithm for shot grouping and scene segmentation, in which the similarity between shots is based on autocorrelogram feature, motion feature and time constraint. Therefore, the visual evidence and time constraint contained in the video content are effectively incorporated into a unified clustering framework. Moreover, the number of clusters in our algorithm does not need to be predefined and thus it provides an automatic framework for scene segmentation. Compared with normalized cut clustering based scene segmentation, our algorithm can achieve more accurate results and requires less computing resources.

1 Introduction

Recently, video scene segmentation has attracted much attention due to its critical role in video structure analysis. Scenes are the semantic units of a video, and scene segmentation is often formulated as a process of clustering video shots into groups, such that the shots within each group are related to each other from a narrative or thematic point of view. It is a challenging task since a good scene segmentation requires the understanding of the semantic meanings conveyed by the video shots.

Much work has been done on scene segmentation in the last decade. They can be roughly classified into three categories.

• **Shot clustering based approach:** It is well known that video shots belong to the same scene are semantically similar. The similarities between the shots provide a basic clue for the clustering based approach. In [1], the normalized cut algorithm is employed to cluster video shots, and a temporal graph analysis is further used to detect the scenes. Yeung et al. [2] propose a time constraint clustering algorithm for partitioning a video into several story units. Rui et al. [3] propose an intelligent unsupervised clustering algorithm based on the time locality and scene structure.
• **Boundary detection based approach:** In this approach, shot boundaries are considered as the candidates of scene boundaries and the false boundaries are removed by checking the coherence of the similarity between different shots. Rasheed and Shah [4] construct a backward shot coherence measure for scene detection. Lin et al. [5] propose a shot correlation measure for scene segmentation based on dominant color grouping and tracking.

S. Boll et al. (Eds.): MMM 2010, LNCS 5916, pp. 637–643, 2010.

• **Model based approach:** This approach views that to group N shots into K scenes is equivalent to estimating the model parameters $\{\Phi_i\}_{i=1}^{K}$, which represent the boundaries of K scenes. In [6], scene segmentation is formulated as a Bayesian inference problem and solved by the Markov chain Monte Carlo (MCMC) technique. Zhao et al. [7] propose an effective method for video scene segmentation based on best-first model merging.

This paper proposes a time constraint dominant-set clustering algorithm for shot grouping and scene segmentation, in which the number of clusters does not need to be predefined and thus it provides an automatic framework for scene segmentation. The rest of the paper is organized as follows. The proposed method for scene segmentation is described in Section 2. Experimental results are presented and discussed in Section 3, followed by conclusions and remarks in Section 4.

2 Scene Segmentation

Figure 1 shows an overview of the proposed algorithm for scene segmentation. There are three main steps: computing similarity matrix, dominant-set based clustering and scene construction.

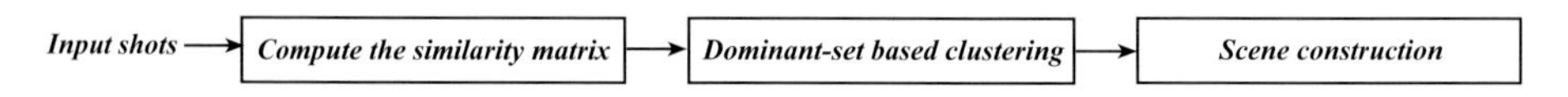

Fig. 1. The flowchart of the proposed algorithm

2.1 Shot Similarity Matrix

Given the input shots, the similarity between every pair of shots is calculated based on both visual evidence and time constraint, and is stored in the similarity matrix W.

Visual Evidence. The visual evidence includes both static feature and motion feature. Specifically, we adopt the autocorrelogram in HSV color space as the static feature for a frame in a shot. Let the autocorrelogram feature of frame k be f^k, and their similarity with respect to the static feature is defined as:

$$ColSim(f^x, f^y) = exp(-\frac{||f^x - f^y||^2}{\delta}) \tag{1}$$

where δ is the scaling parameter. Based on Eq.(1), the similarity between shot i and shot j with respect to static feature is defined as follows.

$$StaSim(shot_i, shot_j) = \min_{f^x \in shot_i} \max_{f^y \in shot_j} (ColSim(f^x, f^y)) \tag{2}$$

where f^x and f^y can be key frames of $shot_i$ and $shot_j$ respectively.

For motion feature, we measure the motion activity of a shot using inter-frame histogram difference. The motion activity of shot i is defined as

$$Mot(shot_i) = \frac{1}{b-a} \sum_{m=a}^{b-1} Dist(m, m+1) \tag{3}$$

$$Dist(m, m+1) = \sum_{k} |hist^{(m)}(k) - hist^{(m+1)}(k)| \quad (4)$$

where $hist^{(m)}$ is the histogram of frame m. From Eqs.(3) and (4), the motion information inside a shot is represented by the average distance between consecutive frames. The similarity between shots based on the motion feature is defined as follows.

$$MotSim(shot_i, shot_j) = \frac{2 \times \min(Mot(shot_i), Mot(shot_j))}{Mot(shot_i) + Mot(shot_j)} \quad (5)$$

As a result, shot similarity with respect to the visual evidence is obtained through a linear combination of the static and motion feature similarities.

$$VisSim(shot_i, shot_j) = w \times StaSim(shot_i, shot_j) + (1-w) \times MotSim(shot_i, shot_j) \quad (6)$$

where $w(w \geq 0.8)$ is a weighting factor for the static feature, because it is more reliable than the motion feature.

Time Constraint. Time information is very important in video structure analysis. As for scene segmentation, temporally adjacent shots likely belong to the same scene and vice versa. Therefore, the time constraint needs to be incorporated into the shot similarity calculation. Here, we adopt the exponential attenuation time constraint, meaning that the similarity between different shots exponentially declines according to their temporal distance.

$$W(shot_i, shot_j) = \begin{cases} exp\,(-\frac{dist(i,j)}{\sigma}) & \text{if } dist(i,j) < D \\ 0 & \text{otherwise} \end{cases} \quad (7)$$

where $dist(i, j)$ is defined as $|\frac{b_i+e_i}{2} - \frac{b_j+e_j}{2}|$, in which b_k, and e_k are the indexes of the beginning frame and ending frame of shot k, and D is a threshold.

Finally, the shot similarity based on both visual evidence and time constraint is:

$$ShotSim(shot_i, shot_j) = VisSim(shot_i, shot_j) \times W(shot_i, shot_j) \quad (8)$$

2.2 Clustering

As described in [8], the dominant-set clustering algorithm begins with the similarity matrix and iteratively bipartitions the shots into dominant set and non-dominant set, therefore, produces the clusters progressively and hierarchically. The clustering process usually stops when all shots are grouped into one of the clusters or when certain criteria are satisfied. We choose to terminate the clustering process when more than 95% shots are clustered so as to avoid forming tiny and meaningless clusters.

2.3 Scene Construction

Although the time constraint is incorporated into the similarity matrix, the resulting shot group still contains shots which are discontinuous in time. In this paper, we adopt an interrelated connection strategy to overcome the discontinuous problem.

Table 1. Scene construction algorithm

1. $Set \quad l \leftarrow m, e \leftarrow last(label(l), m)$
2. $While \quad l \leq e \quad do$
   $if \quad last(label(l), m) > e \quad e \leftarrow last(label(l), m)$
   $l \leftarrow l + 1$
3. $shot_m, shot_{m+1}, \cdots, shot_{m+e}$ construct a scene

Let $last(A, a) = max_{i \geq a, Label(i)=A}\ i$ represent the shot index of the last occurrence in shot group A starting at shot a. Thus the algorithm to construction a scene proceeds in Table 1. The scene boundary is constructed at the index $last(B, b)$ if starting from a. In this way, a shot label occurs in one scene will not occur in other scenes. Due to the time constraint part, this rule will not generate unreasonable scenes, and thus the discontinuous problem is effectively migrated.

3 Experimental Results

3.1 Database

We constructed a challenging database including news video, TV-series, instructional video, movies, and home video. News video data contained three clips from TRECV2006; TV-series data were collected from the teleplay "Friends"; badminton didactical clips constitute the instructional video data; movie data were from movie "If You Are The One"; home videos were download from "youtube". The detail information of the database and the manually labeled scenes are shown in Table 2.

3.2 Evaluation Criterion

In our work, so a tolerance of 3 shots is defined for correct detection. If a scene boundary is detected, but the distance between its boundary and the corresponding labeled boundary is $4 \sim 6$ shots, this detection is considered as wrong matching. All the detections can be divided into four categories: the number of wrong matching N_{WM}; the number of annotated scenes missed N_M; the number of correct detections N_C; the

Table 2. Database

	Clips	Time	Frame	Shot	Scene
News Video	clip1	0:28:20	50,950	249	21
	clip2	0:28:20	50,950	35	4
	clip3	0:28:20	50,950	216	18
TV-series	clip1	0:28:52	43,153	322	16
	clip2	0:22:01	32,903	230	13
Instructional Video	clip1	0:20:05	29,983	88	5
Movies	clip1	0:27:13	40,837	144	7
Home video	clip1	0:20:12	30,775	33	6

number of false detections N_F. Thus the number of scenes annotated by human N_{true} and the number of detected scenes $N_{detected}$ can be represented as:

$$N_{true} = N_C + N_{WM} + N_M, \quad N_{detected} = N_C + N_{WM} + N_F$$

Recall and *precision* are commonly used as the measure for performance evaluation,

$$Precision = N_C / N_{detected}, \quad Recall = N_C / N_{true}$$

however, *recall* and *precision* are two contradictory criterions, so we introduce F_1 as an all-around criterion.

$$\boldsymbol{F_1} = \frac{2 \times Precision \times Recall}{Precision + Recall}$$

3.3 Results

We conduct a comparison experiment between the proposed algorithm and normalize cut based scene detection algorithm [1] on our database. The parameters employed in our algorithm are set as follows $\{\delta = 1, w = 0.85, \sigma = 750, D = 1500\}$, and all the shots are segmented by twin comparison algorithm [9].

The results of our algorithm are detailed in Table 3, from which we can see that our algorithm achieved good performance. The number of the wrong matching and undetected scenes are small. Among all categories of video clips, the segmentation results on the news videos is the best, this is because news video often has a more structural content. Notice that the our algorithm tends to over-segment the videos where no scene

Table 3. Results of our algorithm

	Scenes(Detected)	Correct	Wrong match	Missed	False	P	R	F_1
News Video	30	19	2	0	9	0.63	0.90	0.75
	5	4	0	0	1	0.80	1.00	0.89
	24	16	0	2	8	0.67	0.89	0.76
TV-series	32	13	1	2	18	0.41	0.81	0.54
	22	13	0	0	9	0.59	1.00	0.74
Instructional Video	11	4	1	0	3	0.50	0.80	0.62
Movies	16	5	2	0	9	0.31	0.71	0.43
Home video	4	4	0	2	0	1.00	0.67	0.80

Table 4. Results of normalized cut based algorithm

	Scenes(Detected)	Correct	Wrong match	Missed	False	P	R	F_1
News Video	32	16	1	4	15	0.50	0.76	0.60
	5	4	0	0	1	0.80	1.00	0.89
	31	18	0	0	13	0.58	1.00	0.73
TV-series	32	13	2	1	17	0.41	0.81	0.54
	24	10	3	0	11	0.42	0.81	0.54
Instructional Video	8	4	1	0	3	0.50	0.80	0.62
Movies	10	2	3	2	5	0.20	0.29	0.24
Home video	3	3	0	3	0	1.00	0.50	0.67

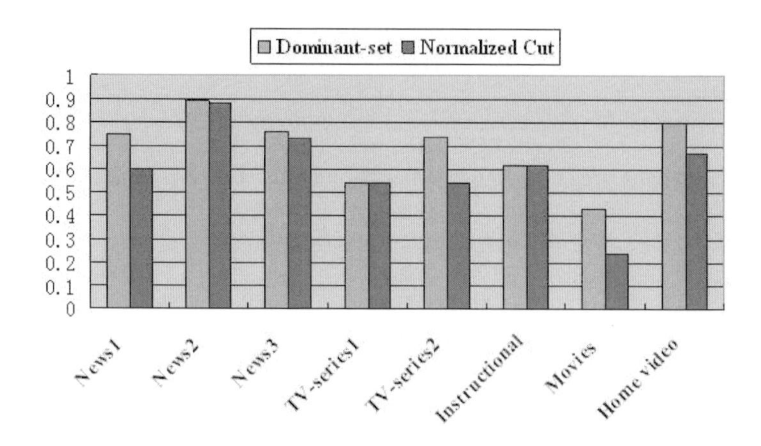

Fig. 2. Comparison of results

exists. From user's perspective, over segmentation is probably better than under segmentation, because it is much easier to merge the over-segmented scenes than to split the under-segmented scenes.

In the normalized cut based algorithm, the number of shot clusters needs to be predefined. However, the number of shot clusters usually does not equal to the number of scenes. We found empirically that a proper value for the cluster number is twice the scene number for the normalized cut clustering algorithm. The results of normalized cut based scene segmentation are shown in Table 4. To clearly compare the results of these two algorithms, the bar charts of the F_1 values are shown in Figure 2. It can be seen that our algorithm outperforms normalized cut based algorithm and, in addition, it does not need any prior knowledge to specify the cluster number. While the number of clusters is very important in the normalized cut based algorithm and greatly impacts its performance. Moreover, since our algorithm does not need to solve the eigen-structure of the similarity matrix, it is more efficient than normalized cut based algorithm, and the computation time needed by normalized cut based clustering is about 4 times of that of the dominant-set clustering.

4 Conclusion

This paper has presented a time constraint dominant-set clustering algorithm for shot grouping and scene segmentation, in which the similarity between shots is based on autocorrelogram feature, motion feature and time constraint. The algorithm offers an automatic framework for scene segmentation with less demand on computing resources and better performance than normalized cut based scene segmentation algorithm.

Acknowledgment

This work is supported by NSFC (Grant No. 60825204, 60672040,) and the National 863 High-Tech R&D Program of China (Grant No. 2006AA01Z453, 2009AA01Z318).

References

1. Ngo, C., Ma, Y., Zhang, H.: Video summarization and scene detection by graph modeling. IEEE Transactions on Circuits and Systems for Video Technology 15(2), 296–305 (2005)
2. Yeung, M., Yeo, B.: Time-constrained clustering for segmentation of video into story units. In: Proceedings of IEEE International Conference on Pattern Recognition, vol. 96, pp. 375–380 (1996)
3. Rui, Y., Huang, T., Mehrotra, S.: Exploring video structure beyond the shots. In: IEEE International Conference on Multimedia Computing and Systems, pp. 237–240 (1998)
4. Rasheed, Z., Shah, M.: Scene detection in hollywood movies and tv shows. In: IEEE International Conference on Computer Vision and Pattern Recognition, pp. 343–348 (2003)
5. Lin, T., Zhang, H.: Automatic video scene extraction by shot groupings. In: Proceedings of IEEE International Conference on Pattern Recognition, vol. 4, pp. 39–42 (2000)
6. Zhai, Y., Shah, M.: A general framework for temporal video scene segmentation. In: IEEE International Conference on Computer Vision, vol. 2, pp. 1111–1116 (2005)
7. Zhao, L., Qi, W., Wang, Y.J., Yang, S., Zhang, H.: Video shot grouping using best-first model merging. In: SPIE symposium on Electronic Imaging Storage and Retrieval for Image and Video Databases (2001)
8. Pavan, M., Pelillo, M.: A new graph-theoretic approach to clustering and segmentation. In: IEEE International Conference on Computer Vision and Pattern Recognition, pp. 3895–3900 (2003)
9. Zhang, H., Kankanhalli, A., Soliar, S.: Automatic partitioning of full-motion video. Multimedia Systems 1(1), 10–28 (1993)

Automatic Nipple Detection Using Shape and Statistical Skin Color Information

Yue Wang, Jun Li, HeeLin Wang, and ZuJun Hou

Institute for Infocomm Research
A*Star (Agency for Science, Technology and Research), Singapore
{ywang,stulj,hlwang,zhou}@i2r.a-star.edu.sg

Abstract. This paper presents a new approach on nipple detection for adult content recognition, it combines the advantage of Adaboost algorithm that is rapid speed in object detection and the robustness of nipple features for adaptive nipple detection. This method first locates the potential nipple-like region by using Adaboost algorithm for fast processing speed. It is followed by a nipple detection using the information of shape and skin color relation between nipple and non-nipple region. As this method uses the nipple features to conduct the adult image detection, it can achieve more precise detection and avoids other methods that only detect the percentage of exposure skin area to decide whether it is an adult image. The proposed method can be also used for other organ level detection. The experiments show that our method performs well for nipple detection in adult images.

Keywords: Pornographic image, adult image, obscene image, nipple detection, naked image detection.

1 Introduction

There are a huge number of adult images that can be freely accessed in multimedia documents and databases through Internet. To protect children, detection and blocking the obscene images and videos received more and more concern. Automatic recognition of pornographic images has been studied by some researchers. Current methods can be briefly classified into two kinds [1]: (1) Skin-based detection and (2) Feature-based detection.

Skin-based methods focus on skin detection. Many skin models have been developed based on color histogram [1], chromatic distribution [2], color and texture information [3][5][6][8][9]. After skin region has been detected, perform one of below detections: (a) Model-based detection [3] which is using a geometrical model to describe the structure or shape of human body; (b) Region-based detection which extracts features for recognition based on the detected skin regions. These features include contour and contour-based features [1][8], shape features [2][6], a series of features [9] from each connected skin region: color, texture, and shape, etc. Feature-based methods focus on using the features directly extracted in the images. These features include normalized central moments and color histogram [4], shape feature

S. Boll et al. (Eds.): MMM 2010, LNCS 5916, pp. 644–649, 2010.

(Compactness descriptor) [7], etc. These methods tend to use a global matching rather than a local matching. All existing methods mentioned above suffer from a fundamental problem that they did not conduct the detection at the organ (object) level. A certain percentage of skin detected over the whole image or a human body does not mean it is a naked adult image. To make a correct judgment, the basic rule is checking whether the female nipples, male and female private parts are exposure into the image. The only paper can be found in literature that detects the sex organ is in [10] for nipple detection. This method conducted the skin detection first, and then performed the nipple detection using self-organizing map neural network. They claimed that the correct nipple detection rate is 65.4%.

This paper focuses on nipple detection in images. It is a fundament step in pornography image detection. Our method is an organ model driven, that means we emphasize the features of organ to be detected. In nipple detection, shape and skin are the most important features for nipple appearance. Therefore, in the real application, both of them should be combined for detection, at least play the same important role. Our method consists of two stages:

(1) Rapid locating for potential nipple region. Adaboost algorithm with Haar-like features is used to rapidly locate the possible nipple regions.
(2) Nipple detection which combines shape and skin statistical information is applied to determine whether the located regions from stage 1 are the real nipples.

The remaining structure of this paper is arranged as follows. Section 2 briefly introduces the Adaboost algorithm with Haar-like features and its application in searching the possible nipple region. Section 3 describes the details of the nipple model for nipple detection. Experimental results and discussion are presented in Section 4. Finally, the conclusion of this paper is presented in Section 5.

2 Rapid Locating for Possible Nipple Region

Adaboost algorithm is first proposed in [11] for fast face detection with the Haar-like features which are based on computing the gray level values within rectangle boxes. The Adaboost algorithm takes as input of a training set of positive and negative samples and then constructs a cascade detector as linear combination of simple and weak classifiers. It can achieve a fast processing speed by using integral image.

In our Adaboost training proceeding, we are using 638 single nipple images as positive samples, 19370 images as negative samples. Some nipple samples from positive training set are shown in Fig. 1(a). The trained Adaboost detector contains 16 stages with 257 weak classifiers.

Adaboost algorithm may not always get a correct detection. We found that there are some false alarms in Adaboost nipple detection. Fig. 1(b) shows two examples. The red boxes indicated in Fig. 1(b) are the results from Adaboost algorithm. There are total 4 regions that have been located as possible nipple regions. However, two of them are false results, the regions of eye and belly button are wrongly detected as nipple. This is because Haar-features calculate the summary grey values within the rectangle box, the detail clues in these boxes are lost. To remove such false nipples regions, a further nipple detection is applied following. The details are showed in next section.

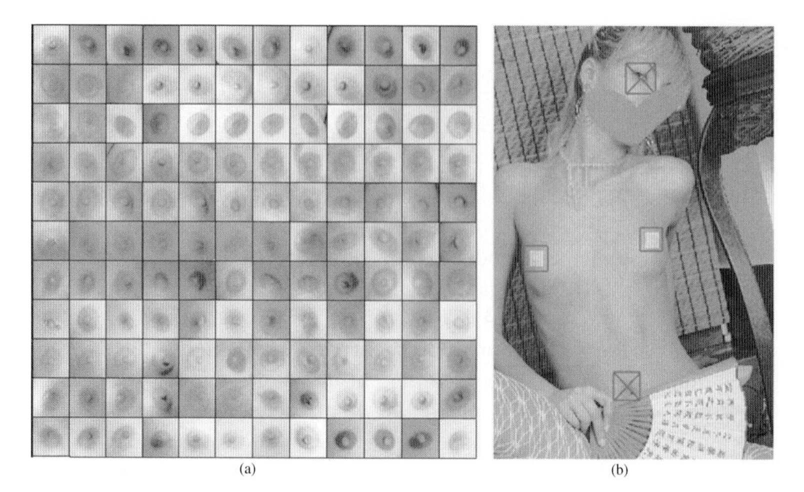

Fig. 1. (a) Some nipple samples from training set. (b) Nipple detection results in Adaboost algorithm. Red boxes without cross sign are the correct results from Adaboost algorithm, while the red boxes with cross sign are the false alarms detected that are removed in section 3.

3 Nipple Detection

A filter which is using the nipple skin statistical information is developed to remove the false nipple detected in Adaboost. First the statistical information of skin color relation between nipple and region surrounding nipple has to be extracted, and then this information is used to filter off the non-nipple region detected in last stage.

3.1 Color Statistical Information Extraction for Nipple Skin

Here we focus on the red and green components in nipple regions, as the statistical information shows that the red and green have a related significant difference between the nipple and non-nipple skin. We are using the same nipple images that have been used as the positive samples in Adaboost training. We first extract the standard deviation of gray values of pixels in nipple and non-nipple skin, and then extract means and distributions of $(R_{sur} - R_{nip})$ and $(G_{sur} - G_{nip}) / (R_{sur} - R_{nip})$, here *nip* and *sur* denote the nipple and non-nipple region. It can be observed that the nipple skin contains more R component compared with non-nipple skin, but it is reversed for G component. The extracted information is used to determine the thresholds used in the nipple detection algorithm presented in next section.

3.2 Nipple Detection Algorithm

The results from Adaboost algorithm are fed into this detection. The color and gray information are used in this stage. It involves few steps list below:

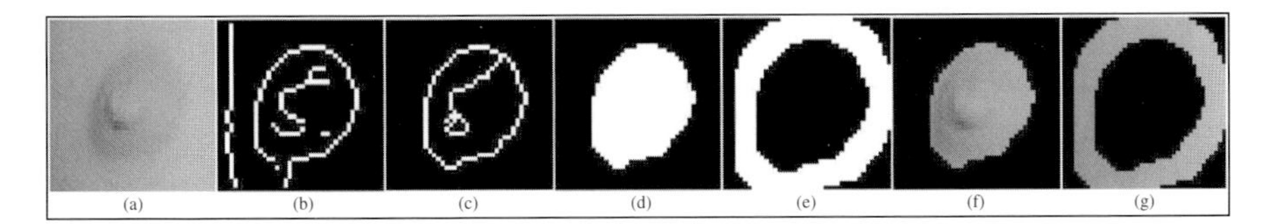

Fig. 2. Create the templates for nipple and ring region surrounding nipple. (a) Input image for stage 1. (b) Canny edge detection. (c) Connect broken boundaries and remove spur. (d)(e) Templates of sampling region for nipple. (f)(g) Sampling region for nipple region.

(1) Conduct Canny edge detection for input from stage 1, see Fig. 2(a)(b).
(2) Clean the small size objects and then fill holes for remained object, see Fig. 2(c).
(3) Get the biggest object from the remained objects and check the ratio of its width and height. If the ratio is not in a certain range, will judge it as a fake nipple region and Stop. Otherwise, assume it is a nipple pattern and go to next step.
(4) Form the templates for sampling the nipple region and the region surrounding nipple, see Fig. 2(d)(e).
(5) The template for the ring region surrounding nipple is created from the templates of nipple by dilating and subtraction. Calculate the standard deviations (std_{nip}, std_{sur}) of those pixels in nipple region and the ring region surrounding nipple in their grey images, respectively. If anyone of (std_{nip}, std_{sur}) exceeds a certain range which is extracted in section 3.1, judge it as a fake nipple and Stop.
(6) Calculate the mean values (M_{nip}, M_{sur}) for R, G, B components of those pixels in nipple and the ring region surrounding nipple in their color images, respectively.
Here denote M_{nip}=(R_{nip}, G_{nip}, B_{nip}) and M_{sur}=(R_{sur}, G_{sur}, B_{sur}). In order to be a true nipple, all of conditions listed below must be satisfied:

(i) $(R_{sur} - R_{nip})/ R_{sur} < Threshold_0$;
(ii) $Threshold_1 > (G_{sur} - G_{nip})/(R_{sur} - R_{nip}) > Threshold_2$;

Based on the extracted statistical information in section 3.1, the $Threshold_0$, $Threshold_1$ and $Threshold_2$ are set to 0, 0.57 and -1.34, respectively.

The red boxes with cross sign in Fig. 1(b) show the removed nipple regions based on above algorithm. From there it can be observed that although some false nipple regions have been located in stage 1 of our method, they can be filtered out in stage 2.

4 Experimental Results and Discussions

The above algorithm has been simulated by using Matlab codes and tested to pornographic images downloaded from Internet. A database of 980 images, which consists of 265 images with 348 labeled nipples and 715 non-nipple images, is used for testing. Fig. 3 shows some results of nipple detection. Table 1 presents the experimental results for this testing database. There are 75.6% of nipples have been corrected detected but 24.4% missing. Those results presented in [10] which are 65.4% and 34.6%, respectively. However, it is hard to compare as we are not using the same dataset for testing.

Table 1. Experimental result for testing

	Number	Percentage (%)
Total images	980	
Nipple images	265	
Non-nipple images	715	
Total Nipples	348	
Detected nipples	263	263 / 348 = 75.6 %
Missing nipples	85	85 / 348 = 24.4 %
False Detection	170	170 / 980 = 0.174 /per frame

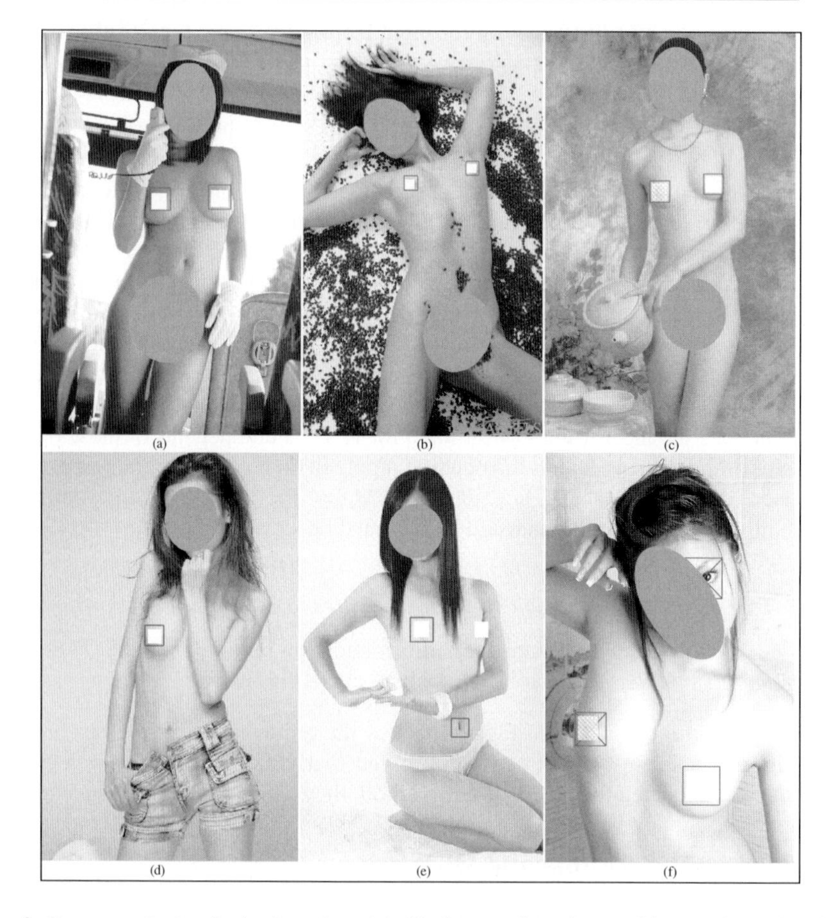

Fig. 3. Some results in nipple detection. (a)-(d) Correct detection, red boxes show the nipple detected; (e) Belly button is wrongly detected as nipple. Right nipple is not located in Adaboost algorithm. (f) Left nipple is judged as a fake nipple in stage 2 due to half of its region is in the shadow. The red boxes with cross sign are determined as the false nipples in stage 2.

The false detection in our method is still need to be reduced, as there are total 170 false alarms have been detected, false rate is 0.174/per frame. Two samples of false detections are showed in Fig. 3(e)(f). It can be observed that the belly button (Fig. 3(e)) sometime is still confusing our algorithm due to the similar shape and appearance with nipple. The shadow (Fig. 3(f)) may also cause a wrong decision in stage 2 of our method due to half of its region is in the shadow.

5 Conclusion

This paper presents a two-stage nipple detection algorithm for adult images inspection. It is using nipple features for organ level detection. In the first stage, Adaboost algorithm is applied to fast locate the potential nipple regions. In the second stage, a nipple model is implemented to further confirm the real nipple from the results of first stage. This nipple model includes the shape and skin color information of nipple and the skin surrounding nipple to effectively detect real nipple in the images. The proposed method was tested for finding the nipple in the real images, the experimental results show the efficient and accuracy of the proposed algorithm.

Our future work will focus on the following aspects to improve our method: (1) A dual-threshold or multi-threshold can be applied in Adaboost to reduce the false detection in stage 1 of proposed method. (2) Color information can be involved in Adaboost. (3) To fully judge whether an image is pornography, the private part of human body must be detected as well. A different model has to be constructed for this purpose.

References

1. Hu, W., Wu, O., Chen, Z., Fu, Z., Maybank, S.: Recognition of Pornographic Web Pages by Classifying Texts and Images. IEEE Transactions on Pattern Analysis and Machine Intelligence 29(6), 1019–1034 (2007)
2. Lee, J.-S., Kuob, Y.-M., Chung, P.-C., Chen, E.-L.: Naked Image Detection Based on Adaptive and Extensible Skin Color Model. Pattern Recognition 40, 2261–2270 (2007)
3. Forsyth, D.A., Fleck, M.M.: Automatic Detection of Human Nudes. International Journal of Computer Vision 32(1), 63–77 (1999)
4. Wang, J.Z., Li, J., Wiederhold, G., Firschein, O.: System for Screening Objectionable Images. Computer Comm. 21(15), 1355–1360 (1998)
5. Zhu, H., Zhou, S., Wang, J., Yin, Z.: An Algorithm of Pornographic Image Detection. In: IEEE Fourth International Conference on Image and Graphics 2007, pp. 801–804 (2007)
6. Zheng, Q.-F., Zeng, W., Wen, G., Wang, W.-Q.: Shape-based Adult Images Detection. In: Proceedings of the Third International Conference on Image and Graphics, pp. 150–153 (2004)
7. Shih, J.-L., Lee, C.-H., Yang, C.-S.: An Adult Image Identification System Employing Image Retrieval Technique. Pattern Recognition Letter 28, 2367–2374 (2007)
8. Yang, J., Fu, Z., Tan, T., Hu, W.: A Novel Approach to Detecting Adult Images. In: 17th International Conference on Pattern Recognition 2004, pp. 479–482 (2004)
9. Arentz, W.A., Olstad, B.: Classifying Offensive Sites Based on Image Content. Computer Vision and Image Understanding 94(1-3), 295–310 (2004)
10. Fuangkhon, P., Tanprasert, T.: Nipple Detection for Obscene Pictures. In: Proc. of the 5th Int. Conf. on Signal, Speech and Image Processing, Greece, August 2005, pp. 315–320 (2005)
11. Viola, P., Jones, M.: Rapid Object Detection Using A Boosted Cascade of Simple Features. In: Proc. of IEEE Conference on Compter Vision and Pattern Recognition, USA, pp. 511–518 (2001)

Asymmetric Bayesian Learning for Image Retrieval with Relevance Feedback

Jun Wu* and Mingyu Lu

School of Information Science & Technology, Dalian Maritime University,
Dalian 116026, China
wujunas8@gmail.com

Abstract. Bayesian learning (BL) based relevance feedback (RF) schemes plays a key role for boosting image retrieval performance. However, traditional BL based RF schemes are often challenged by the small example problem and asymmetrical training example problem. This paper presents a novel scheme that embeds the query point movement (QPM) technique into the Bayesian framework for improving RF performance. In particular, we use an asymmetric learning methodology to determine the parameters of Bayesian learner, thus termed as asymmetric Bayesian learning. For one thing, QPM is applied to estimate the distribution of the relevant class by exploiting labeled positive and negative examples. For another, a semi-supervised learning mechanism is used to tackle the scarcity of negative examples. Concretely, a random subset of the unlabeled images is selected as the candidate negative examples, of which the problematic data are then eliminated by using QPM. Then, the cleaned unlabeled images are regarded as additional negative examples which are helpful to estimate the distribution of the irrelevant class. Experimental results show that the proposed scheme is more effective than some existing approaches.

Keywords: Asymmetric learning, Bayesian, image retrieval, relevance feedback.

1 Introduction

Relevance feedback (RF), as a powerful tool for bridging the gap between the high-level semantic concepts and the low-level visual features, has been extensively studied in content-based image retrieval (CBIR) [1]. RF focuses on the interactions between the user and the search engine by letting the user provide feedback regarding the retrieval results, i.e. the user has the option of labeling a few images returned as either positive or negative in terms of whether they are relevant to the query concept or not. From this feedback loop, the engine is refined and improved results are returned to the user. Early RF schemes are heuristic, which aims to improve the query vector or similarity measure function. However, these methods make strong assumption that the target class has an elliptical shape in the feature space, but it is not hold true in the semantically relevant image retrieval. Later on, researchers began to consider RF as a statistical learning problem, which attempts to train a learner to classify

* Corresponding author.

S. Boll et al. (Eds.): MMM 2010, LNCS 5916, pp. 650–655, 2010.

the images in the database as two classes, i.e. relevant (positive) class and irrelevant (negative) class, in terms of whether they are semantically relevant to the query or not. Support vector machine (SVM) has good performance for pattern classification by maximizing the margin of classification hyperplane and thus has been widely used to design RF schemes [2]. However, training a SVM learner is a very time-consuming process, which is inconsistent with the real time requirement of RF. In contrast, Bayesian learner is very easy to construct, not needing any complicated iterative parameter estimation methods [3].

Duan et al. [4] proposed an adaptive Bayesian RF algorithm, termed as Rich get Richer (RGR), which aims at emphasizing the more promising images and deemphasizing the less promising one by assigning high probabilities to the images similar to the query. However, this method often suffers from the small example problem. To address this problem, Yin et al [5] proposed a hybrid RF method by combining BL and query point movement (QPM) technique. But they ignore another characteristic indwelled in RF, i.e. the asymmetrical distribution between the positive and negative examples. Zhang et al [6] proposed a stretching Bayesian method which assumed that each negative example represents a unique irrelevant semantic class and the unlabeled examples near to the concerned negative example are regarded as additional negative examples. The examples close to the concerned negative example should have a strong chance to belong to the same semantic class, and thus a few irrelevant classes that contain the observed negative examples are emphasized. However, there are a lot of irrelevant classes existing in database and most of them are ignored.

In view of above discussion, an asymmetric Bayesian learning (ABL) scheme is developed in this paper, which investigates three special problems in RF, i.e. real time requirement, small example problem, and asymmetric training example problem [1]. First, we construct a very simple learner based on Bayesian inference so as to avoid a complicated learning process. Moreover, an asymmetric learning strategy is applied to estimate the distribution of the positive and negative class. Finally, a novel semi-supervised learning mechanism is presented for tackling the scarcity of negative examples.

2 The Proposed Scheme

Given the query, we apply Bayesian theory to determine the degree that an image in database is classified as a positive or a negative one according to the prior history of feedbacks provided by the user. Since the probability over the whole database is updated after each feedback, the CBIR system, therefore, able to retrieve as many as positive images and reject as many as negative images from being retrieved. Let P denote the positive example set while N denotes the negative example set, and x denotes a random image in database. We use $\mathrm{p}(\ )$ to denote a probability. Based on Bayesian inference, the following equations hold:

$$\mathrm{p}(P|x) = \mathrm{p}(x|P)\mathrm{p}(P)/\mathrm{p}(x) \tag{1}$$

$$\mathrm{p}(N|x) = \mathrm{p}(x|N)\mathrm{p}(N)/\mathrm{p}(x) \tag{2}$$

Then, the CBIR system can judge whether x is relevant to the query using the leaner:

$$L(x,P,N)=\frac{p(P|x)}{p(N|x)}=\frac{p(x|P)p(P)}{p(x|N)p(N)}\approx\xi\frac{p(x|P)}{p(x|N)}\propto\frac{p(x|P)}{p(x|N)} \tag{3}$$

Here, the response of the learner describes the relevancy confidence of image x to the query, and thus the learner could produces a rank of images according to how confident it believes the images are relevant to the query. Since the number of the positive images is much less than that of the negative images in the database, $p(P)/p(N)$ is treated as a small constant ξ and thus the learner is further simplified as $p(x|P)/p(x|N)$. The class-conditional probability density function $p(x|P)$ and $p(x|N)$ can be approximated by using Gaussian kernels. To simplify the following description, we use c_i (i=1, 2) denotes the class label (c_1=P, c_2=N). We assume that each feature dimension of all examples belonged to c_i class satisfies Gaussian distribution.

$$p(x_k|c_i)=\frac{1}{\sqrt{2\pi}\sigma_k^{(c_i)}}exp\left[-\frac{1}{2}\left(\frac{x_k-\mu_k^{(c_i)}}{\sigma_k^{(c_i)}}\right)^2\right] \tag{4}$$

where x_k is the kth dimension of the feature vector of an image, $\mu_k^{(c_i)}$ and $\sigma_k^{(c_i)}$ are the mean value and the standard deviation of the kth dimension of all examples belonged to c_i class, respectively. Finally, the $p(x|c_i)$ can be determined by using equation (5):

$$p(x|c_i)=\prod_{k=1}^{T}p(x_k|c_i) \tag{5}$$

where T is the number of dimensions of the feature space. Generally, $\mu_k^{(c_i)}$ and $\sigma_k^{(c_i)}$ can be estimated depending upon user labeled images. From Eq. (3)-(5), it can be seen that four parameters of the constructed learner are needed to determine: $\mathbf{\mu}^{(P)}=\{\mu_k^{(P)}\}$, $\mathbf{\sigma}^{(P)}=\{\sigma_k^{(P)}\}$, $\mathbf{\mu}^{(N)}=\{\mu_k^{(N)}\}$, $\mathbf{\sigma}^{(N)}=\{\sigma_k^{(N)}\}$, k=1…T.

Most BL based RF schemes directly estimate $\mathbf{\mu}^{(P)}$ and $\mathbf{\sigma}^{(P)}$ for relevant class using the observed positive examples. However, the positive examples labeled in RF may not be the most representative examples in the potential target class. Hence, the user has to repeat many rounds of feedback to achieve desirable results. Inspired by [5], ABL, by using QPM, attempts to mine a potentially better pattern for representing the target class. But the strategy used in ABL is different from that used in [5] which apply QPM to estimate the parameters for the relevant and irrelevant classes in the same manner. But ABL uses QPM for relevant and irrelevant classes with different purposes.

QPM aims to reformulate the query vector through user's feedback so as to move the query point to a region involving more positive examples in the feature space. Let Q denote the original query, the reformulated query, denoted as Q^*, can be computed by:

$$Q^*=\alpha Q+\beta\sum_{y_k\in P}\frac{y_k}{|P|}-\gamma\sum_{y_k\in N}\frac{y_k}{|N|} \tag{6}$$

where $|\bullet|$ denotes the size of a set, and α, β, and γ are constants used for controlling the relative contribution of each component. Our experiments show that the ABL is not sensitive to the setting of these parameters. Empirically, we set the values of $\alpha = 0.3$, $\beta = 0.6$, and $\gamma = 0.3$. In some sense, Q^* represents the mass centroid of the all possible positive examples, and thus it is reliable to assume that Q^* is also a rational estimate for the mean vector of the assumed Gaussian density of the positive examples. Hence, we set $\boldsymbol{\mu}^{(P)} = Q^*$. Then, based on $\boldsymbol{\mu}^{(P)}$ and *P*, $\boldsymbol{\sigma}^{(P)}$ can be estimated.

Unlike positive examples, each negative example is 'negative in its own way' [1] and the small number of labeled negative examples can hardly be representative the entire irrelevant class. ABL applies a semi-supervised learning mechanism to overcome the scarcity of negative examples. Our approach is based on a fact that, for any given query, negative examples make up an extremely large proportion of the existing database. So a random subset of the unlabeled image set can be selected as the additional negative examples. Furthermore, to improve the data quality, a QPM-based data cleaning method is applied to eliminate the possible positive examples in the selected unlabeled images.

(1) Collecting unlabeled images.

A random subset of the unlabeled images, denoted as N_U, is generated by using random sampling.

$$N_U = Sampling(U) \text{ with } |N_U| = fix(\sigma_S \cdot |U|) \tag{7}$$

where U denotes the unlabeled image set, $Sampling(\bullet)$ denotes random sampling from a certain set, $fix(\bullet)$ denotes the mantissa rounding operator, and $\sigma_S \in [0,1]$ is the sampling scale which is used for controlling the number of examples sampled from U.

(2) Cleaning the selected unlabeled images.

Since Q^* represents the mass centroid of all possible positive examples, the examples close to Q^* should have a strong chance to be positive. Depending upon this assumption, ABL tries to remove the *k* most "similar" examples to Q^* from N_U. A simplified version of Radial Basis Function is applied to measured the similarity between Q^* and the examples in N_U. The cleaned image set, denoted as N^*, can be generated by

$$N^* = N_U - \left\{ \mathrm{x}_i \left| \underset{\forall \mathrm{x}_i \in N_U}{argmax} \; exp\left(-\left\|\mathrm{x}_i - Q^*\right\|_2^2\right) \right.\right\}_{i=1}^{k} \text{ with } k = fix(\sigma_C \cdot |N_U|) \tag{8}$$

where $\sigma_C = \eta \cdot \sigma_S \in [0,1]$ is the cleaning scale used for controlling the number of examples removed from N_U, and η is a constant used for adjusting the relationship between σ_C and σ_S. To eliminate the 'bad' examples as much as possible σ_C may be slightly larger than σ_S because the candidate negative examples are readily available in the database. Empirically, η is set to 1.3. Finally, $\boldsymbol{\mu}^{(N)}$ and $\boldsymbol{\sigma}^{(N)}$ are estimated depending upon $N \cup N^*$.

Essentially, our approach could be regarded as a type of active semi-supervised learning algorithm. In the absence of the teacher, our approach just discards the problematic data after identification instead of asking the teacher for labels as in the standard active learning scenario.

3 Experiments

To demonstrate the effectiveness of the proposed ABL, we compare it with SVM active learning (SVM-AL) [2] and Rich get Richer (RGR) method [4]. 3000 images selected from COREL dataset are used to form the testing image database.

At the beginning of retrieval, the images in the database are ranked according to their Euclidean distances to the query. After user feedback, three learning methods are then used to rerank the images in the database. In each round of RF, the user labels 20 images for the system.

Sampling is the most important step for ABL. Selecting a small number of unlabeled examples might make the improvement trivial, while selecting a large number of unlabeled examples might include non-informative or even poor examples into the training set. To select optimal values of σ_S, various feasible values of each parameter are tested. After 60 experiments, the parameter with the best performance among those experiments is $\sigma_S = 0.1$.

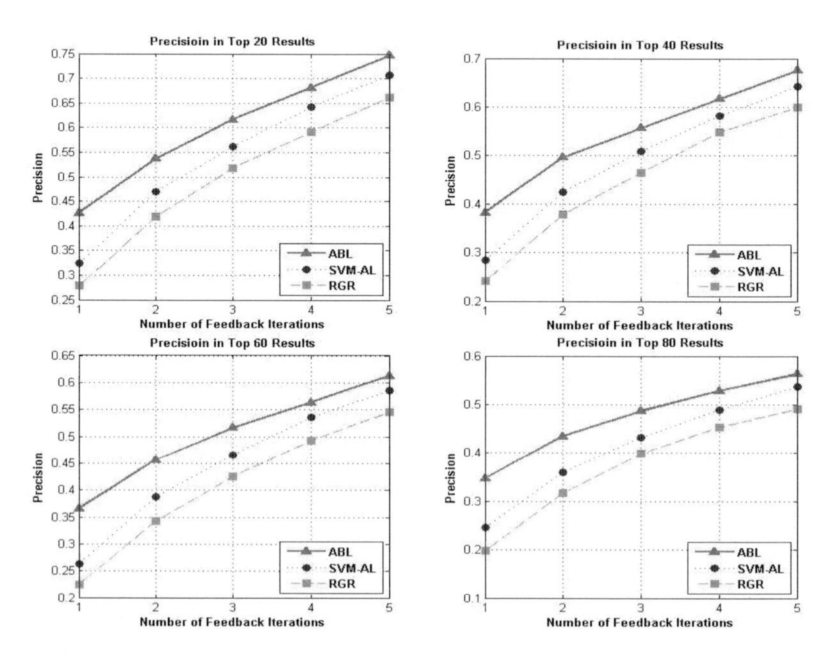

Fig. 1. Performance of the proposed algorithm compared with some existing algorithms

Fig. 1 shows the average precision at the top20, top 40, top 60, and top 80 retrieval results of the three methods. As can be seen, our ABL outperforms the other two methods, especially at the first round of relevance feedback. By iteratively adding the user's feedbacks, the performance difference between ABL and the other two methods gets smaller. Since the number of the labeled images is very limited after the first round of feedback, SVM-AL and RGR can hardly show meaningful results, yet ABL can outperform them obviously since the unlabeled examples are used by it. As the number of feedbacks increases, the performance of SVM-AL and RGR becomes much better. But our ABL continues to perform the best. The above observations show that the proposed asymmetric learning mechanism is effective and ABL method can improve the retrieval performance significantly by using only a few rounds of feedbacks.

4 Conclusions

In this paper, a novel asymmetric Bayesian leaning (ABL) based RF algorithm is presented. There are several key elements in our scheme:

(1) To satisfy the real-time requirement in RF, a very simple learner is constructed based on Bayesian inference. (2) Asymmetric learning strategy is applied to tackle the distribution imbalance between the positive and negative examples. (3) Semi-supervised learning mechanism is introduced in our scheme, and the unlabeled data is helpful to improve the generalization capability of learner. Experimental results illustrate the effectiveness of the proposed algorithm.

Acknowledgments. This research was supported by Natural Science Foundation of China (No.60773084, No. 60603023, No. 60973067) and Doctoral Fund of Ministry of Education of China (No. 20070151009).

References

1. Zhou, X., Huang, T.S.: Relevance Feedback in Image Retrieval: A Comprehensive Review. ACM Multimedia Syst. J. 8, 536–544 (2003)
2. Tong, S., Chang, E.: Support Vector Machine Active Learning for Image Retrieval. In: Proc. ACM Int. Conf. on Multimedia, pp. 107–118. ACM Press, Ottawa (2001)
3. Wu, X.D., Kumar, V., Quinlan, J.R., et al.: Top 10 Algorithms in Data Mining. Knowledge Information Systems 14, 1–37 (2008)
4. Duan, L., Gao, W., Zeng, W., et al.: Adaptive Relevance Feedback Based on Bayesian Inference for Image Retrieval. Signal Processing 85(2), 395–399 (2005)
5. Yin, P., Bhanu, B., Chang, K., et al.: Integrating Relevance Feedback Techniques for Image Retrieval using Reinforcement Learning. IEEE Trans. on Pattern Analysis and Machine Intelligence 27(10), 1536–1551 (2005)
6. Zhang, R., Zhang, Z.: BALAS: Empirical Bayesian Learning in Relevance Feedback for Image Retrieval. Image and Vision Computing 24, 211–223 (2006)

Automatic Visualization of Story Clusters in TV Series Summary

Johannes Sasongko and Dian Tjondronegoro

Faculty of Science and Technology, Queensland University of Technology, Brisbane, Australia
{w.sasongko,dian}@qut.edu.au

Abstract. This paper describes a visualization method for showing clusters of video stories for the purpose of summarizing an episode of a TV series. Key frames from the video story segments are automatically extracted and clustered based on their visual similarity. Important keywords are then extracted from video subtitles to describe the semantic content of each story cluster in the form of tag clouds. The evaluation of the automatic processing has shown promising results, as the generated summaries are accurate and descriptive.

Keywords: Visualization, video summarization, story clusters.

1 Introduction

The popularity of television series provides a business opportunity for "episode guide" websites such as TV.com [1], as well as official websites provided by producers or distributors of various TV series. Conventionally, these websites present a textual summary of each episode in the series, possibly with some screen captures, but both the summary and the images have to be created or selected manually.

A number of different video summarization techniques have been developed in the past, with different points of view and objectives. Based on the form of the summary output, existing techniques are classified into two categories [2]. The first category produces a set of *keyframes*, which are static images representing the contents of a video, while the second category produces a *video skim*, or a shorter video from the original one.

There has been some research in the past on use of speech transcripts in the field of multimedia retrieval. A simple text-based video retrieval method was presented in [3], where speech recognition is used to obtain textual information similar to that of subtitles. The resulting text is scanned for certain keywords that signify specific emotions. However, the use of this technique for video summarization is limited, due to relying only on finding predefined words. A different video summarization method based on keyword extraction was proposed in [4], whereby tf-idf vectors are created for each segment in the subtitles. Important keywords in the video are detected by clustering the tf-idf vectors. The segments where these keywords occur are used for the video summary. This method, however, does not take into account the visual aspect of the video.

S. Boll et al. (Eds.): MMM 2010, LNCS 5916, pp. 656–661, 2010.

In this paper, we propose a hybrid visualization method for summarizing an episode of a TV series. This method combines the visual-based information in the form of keyframes extracted from the video, as well as textual-based information in the form of keywords taken from the episode subtitles. The visualization shows shots from story clusters within the video, combined with a tag cloud of keywords for each cluster and for the whole episode.

2 Framework Description

2.1 Automatic Clustering of Stories

In order to visually separate a video into stories, the system firstly detects the transition between video shots. The shot boundary detection method is based on comparing the histogram of nearby frames using the chi-square test [5] with emphasis on color hue, which has been shown to be effective [6]. In order to speed up the process, frames are sampled at one tenth of the original frame rate. Shot changes at less than two seconds from the preceding ones are ignored because these are generally due to very fast-moving camera shots. The effectiveness of this shot boundary detection method has been extensively tested for rushes videos in the TRECVID database [7], and our preliminary tests showed its suitability for TV series as well.

After the shot units are extracted, keyframes are selected automatically to visually represent each shot. To save processing time, this is done using a simple method whereby for every shot, the system selects the frame at the two-seconds into the shot as the keyframe. Thus any shot that is less than two seconds is deemed too short and not significant enough to be used in the summary.

The actual clustering of stories uses a time-constrained hierarchical clustering similar to the one described in [8]. Our clustering method uses the histogram difference of shot keyframes, calculated using the chi-square test, as the distance metric. Two shots are linked into one cluster if they satisfy these two criteria: (1) the histogram difference between the shots fall below a set threshold determined from experiments; and (2) the shots occur within a set time difference of each other, ensuring that shots far apart in the video are not accidentally clustered together.

Each of the resulting clusters shows a particular story, for example, a conversation. To filter out very short story clusters (noise), clusters that are less than 15 seconds in length are removed. This leaves the clusters that cover significant parts of the episode.

2.2 Automatic Keyword Detection

Keywords in a video are detected from its subtitles. For many recent TV series, this is available in the respective DVD releases, and can be extracted using programs such as SubRip [9] or Avidemux [10]. Subtitle texts are associated with video shots based on their timestamps, and a database of words appearing in the subtitle texts is then built. Stop word removal is used to filter out common words that are not suitable as keywords. The words are also stemmed using the Porter stemmer [11, 12].

To rank the keywords in a particular video the following formula is used:

$$score_t = \frac{n_t}{n} \times \log \frac{1}{F_t}, \qquad (1)$$

where n_t is the occurrence of term t in the episode; n is the occurrence of all terms in the episode; and F_t is the frequency of t in spoken English. A published word frequency database for spoken English is available in [13].

Keyword scores for each story cluster are calculated similarly, except we use a measure like tf-idf in order to compare the word frequency within the cluster with the word frequency in the whole episode. This increases the value of unique keywords within the particular cluster. This tf-idf value is then combined with the inverse word frequency in spoken English. We define the score of a particular term in a cluster as:

$$score_{clust,t} = \frac{n_{clust,t}}{n_{clust}} \times \log \frac{C}{C_t} \times \log \frac{1}{F_t}, \qquad (2)$$

where $n_{clust,t}$ is the occurrence of term t in the cluster *clust*; n_{clust} is the occurrence of all terms in *clust*; C is the number of clusters in the episode; C_t is the number of clusters containing t; and F_t is the frequency of t in spoken English.

2.3 Visualizing the Summaries

In order to show the keywords within the whole video or a particular cluster, we chose to visualize them as a tag cloud of twenty of the highest-scored keywords, sorted alphabetically. The size of the keyword text in the output is scaled based on the score. Therefore, higher-valued keywords are shown in larger font sizes.

Cluster keyframes are shown in thumbnail size below the keywords tag cloud. Each thumbnail is accompanied by a timestamp indicating where the shot appears in the video. When the user clicks on a thumbnail, the full-size picture is displayed.

Combined together, the keywords tag cloud and image thumbnails give users a visual and textual overview of stories and themes within the TV series episode.

3 Results

The method presented in this paper was tested on four popular TV series. The particular series and episodes used in this experiment were selected arbitrarily in order to demonstrate the generality of our method.

The ground truth used in the experiment is partly based on the "episode recaps" found on TV.com [1]. Using this, we determined stories contained in the videos. These stories are then matched with the stories we obtained in the summary.

The first video that we used for experiment is from the series "Doctor Who". This video is characterized by a straightforward plot, with no side stories or flashbacks. The recording in this video uses a lot of close-up shots.

The second video comes from "Battlestar Galactica", which has several parallel plots with characteristically distinct environment backgrounds, taking place in two different planets and a space ship. There are also several flashbacks.

The third video is from the series "Desperate Housewives". This video also has several parallel plots happening around the same time at various locations. There is a recurring flashback that is shown a few times. Compared to the other videos, this video is a lot more visually diverse and is shot with more kinds of backgrounds.

The last video in the experiment dataset comes from "Terminator". The plot in this video involves three timelines: the "past", the "present", and the "future". These three timelines are shown interspersed with each other.

Results and Sample Output. Table 1 shows the result of the story clustering.

Table 1. Accuracy of the story clustering method on the test videos

Vide	Actual	Found	Accurate	Precision	Recall
Doctor Who	18	23	14	77.78%	60.87%
Battlestar Galactica	28	17	15	53.57%	88.24%
Desperate Housewives	26	22	16	61.54%	72.73%
Terminator	23	17	15	65.22%	88.24%

Fig. 1(a) shows keywords from the Doctor Who episode. This episode depicts characters watching the *death* of *planet Earth* due to *heat* from the sun. *Humans* and a*liens* are watching from space, and the plot involves someone tampering with the *sunfilter* of the spaceship windows (causing them to *descend*), endangering the ship *guests*. The story cluster shows the exchange of *gifts* between *guests*. One of the characters gave "the *air* of [his] *lungs*", while another guest gave "the gift of bodily *salivas*". The *Jolco* keyword shown prominently here is a name.

Fig. 1(b) shows tags obtained from the Battlestar Galactica video. In this series, the *fleet* refers to a number of space ships that the Galactica ship protects, and *Cylons* are a type of humanoid robots featured in the series. *Chief*, *Cally*, and *Cottle* are names of some of the ship's crew. The episode plot is about the Galactica "*jumping*" to the wrong location. The story cluster in depicts characters *Helo* and *Kara* (code-named *Starbuck*) talking about a *Cylon* robot named *Sharon* who previously *lied* to them.

Fig. 1(c) shows tags from the Desperate Housewives video. These include names of important characters including *Bree*, *Ian*, *Jane*, *Mike*, *Monique*, *Orson*, and *Zach*. Other relevant tags include *remember*, *memory* (a character has amnesia and lost his memory), and *date* (several couples in the episode are dating). The story cluster in Fig 8 shows a flashback of Mike coming from back a *hardware* store to fix a *leaking sink*. Other tags are closely related to this story and mentioned in conversations, e.g. *damage*, *pipe*, *seeping*, *wash*, and *water*.

Fig. 1(d) comes from the TV series Terminator. Some personal names such as *Roger*, *David*, *Lauren*, and *Sarah* are picked up as keywords. The plot involves a *cyborg* from the *future* coming to *kill* an unborn *baby* who has *immunity* against a certain disease. The story cluster shows a conversation within the episode, with the topic of *cyborgs* and how one of the characters has a "not exactly *legal*" dealing with a *cybernetics company*. The topic of *birdhouses* comes up during small talk.

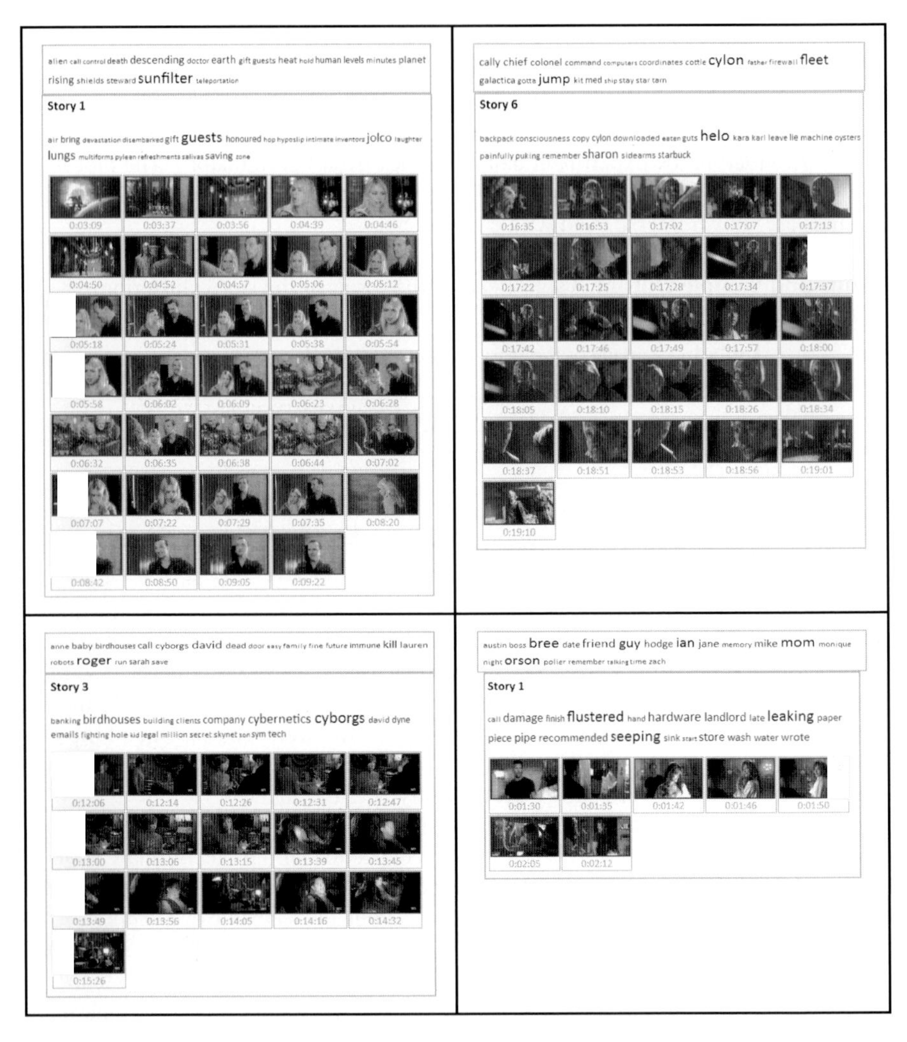

Fig. 1. Keywords and sample story clusters from the four videos (clockwise from top left): (a) Doctor Who; (b) Battlestar Galactica; (c) Desperate Housewives; (d) Terminator

4 Conclusion and Future Work

In this paper we have detailed a method for visualizing story clusters within an episode of TV series. The main parts of this visualization are sets of keyframe clusters and keyword tag clouds for each cluster and for the whole episode. Besides the potential application in an episode guide website, this visualization can also be useful for browsing personal video libraries or for commercial video archiving.

While the main focus of this paper is on TV series, the method can be easily adapted for other types of videos. Because the story clustering is independent of the subtitles, our method is still useful for visualizing videos where the complete subtitles are not available or only available in low quality, for example due to live captioning.

In the future we would like to extend the work to create a summary for of each episode in one season, as well as a summary for the whole season of a particular TV series. This will provide better input to the tagging and will give an opportunity for more interesting visualization features, because the system can pick up common topics and entities (e.g. person, location, organization) throughout the season.

The system would also benefit from using a better scene segmentation method instead of simple clustering. While the story clustering algorithm described here works quite well, sometimes shots in one story exhibit distinct histogram patterns, which the clustering method fails to take into account.

Acknowledgments. This research is funded by the Smart Services CRC.

References

1. TV.com, http://www.tv.com/
2. Truong, B.T., Venkatesh, S.: Video Abstraction: A Systematic Review and Classification. ACM Trans. Multimedia Comput. Commun. Appl. 3(1), 1–37 (2007)
3. Xu, M., Chia, L.T., Yi, H., Rajan, D.: Affective Content Detection in Sitcom Using Subtitle and Audio. In: 12th International Multimedia Modelling Conference, pp. 129–134 (2006)
4. Yi, H., Rajan, D., Chia, L.T.: Semantic Video Indexing and Summarization Using Subtitles. In: Aizawa, K., Nakamura, Y., Satoh, S. (eds.) PCM 2004. LNCS, vol. 3331, pp. 634–641. Springer, Heidelberg (2004)
5. Patel, N.V., Sethi, I.K.: Compressed Video Processing for Cut Detection. IEE Proc.-Vis. Image Signal Process. 153(6), 315–323 (1996)
6. Lupatini, G., Saraceno, C., Leonardi, R.: Scene Break Detection: A Comparison. In: Eighth International Workshop on Research Issues in Data Engineering: Continuous-Media Databases and Applications, pp. 34–41 (1998)
7. Sasongko, J., Rohr, C., Tjondronegoro, D.: Efficient Generation of Pleasant Video Summaries. In: 2nd ACM Trecvid Video Summarization Workshop, pp. 119–123. ACM, New York (2008)
8. Yeung, M.M., Yeo, B.L.: Time-Constrained Clustering for Segmentation of Video into Storyunits. In: 13th International Conference on Pattern Recognition, vol. 3, pp. 375–380 (1996)
9. SubRip DVD subtitles ripper, http://zuggy.wz.cz/dvd.php
10. Avidemux, http://fixounet.free.fr/avidemux/
11. Porter, M.F.: An Algorithm for Suffix Stripping. In: Sparck Jones, K., Willett, P. (eds.) Readings in Information Retrieval. Morgan Kaufmann Multimedia Information and Systems Series, pp. 313–316. Morgan Kaufmann Publishers, San Francisco (1997)
12. Porter, M.: The Porter Stemming Algorithm, http://tartarus.org/~martin/PorterStemmer/
13. Leech, G., Rayson, P., Wilson, A.: Companion Website for: Word Frequencies in Written and Spoken English, http://ucrel.lancs.ac.uk/bncfreq/

From Image Hashing to Video Hashing

Li Weng and Bart Preneel*

Katholieke Universiteit Leuven, IBBT
li.weng@esat.kuleuven.be, bart.preneel@esat.kuleuven.be

Abstract. Perceptual hashing is a technique for content identification and authentication. In this work, a frame hash based video hash construction framework is proposed. This approach reduces a video hash design to an image hash design, so that the performance of the video hash can be estimated without heavy simulation. Target performance can be achieved by tuning the construction parameters. A frame hash algorithm and two performance metrics are proposed.

1 Introduction

Perceptual hashing is a technique for the identification of multimedia content. It works by computing hash values from robust features of multimedia data. Differing from a conventional cryptographic hash [1], a perceptual hash does not vary, even if the content has undergone some incidental distortion. Therefore, a perceptual hash can be used as a persistent fingerprint of the corresponding content. One can tell whether two multimedia files correspond to the same or similar content(s) by comparing their hash values. Since a hash value is much shorter than the original file in size, this approach is more efficient than direct comparison among multimedia files. A perceptual hash can also be used for authentication, such as message authentication codes (MAC) and digital signatures (DS) [1]. In the former case, the algorithm is designed to support a secret key. The hash value is significantly different when a different key is used, so that only entities knowing the key can generate the correct hash. In the latter scenario, a perceptual hash is electronically signed. Compared to conventional MAC and DS, *perceptual* MAC and DS have the advantage that they do not need to be regenerated when multimedia data undergoes incidental distortion.

* This work was supported in part by the Concerted Research Action (GOA) AM-BioRICS 2005/11 of the Flemish Government and by the IAP Programme P6/26 BCRYPT of the Belgian State (Belgian Science Policy). The first author was supported by IBBT (Interdisciplinary Institute for BroadBand Technology), a research institute founded by the Flemish Government in 2004, and the involved companies and institutions (Philips, IPGlobalnet, Vitalsys, Landsbond onafhankelijke ziekenfondsen, UZ-Gent). Additional support was provided by the FWO (Fonds Wetenschappelijk Onderzoek) within the project Perceptual Hashing and Semi-fragile Watermarking.

S. Boll et al. (Eds.): MMM 2010, LNCS 5916, pp. 662–668, 2010.

2 Video Hash Construction

Designing a perceptual hash algorithm is challenging. The hash value must be sensitive to significant content modification, but insensitive to content-preserving processing. These traits are defined as *robustness* and *discriminability*. The former is the ability to resist incidental distortion; the latter is the ability to avoid *collisions*, i.e., different contents result in similar hash values. In addition, the hash value must be sensitive to the secret key.

Since there is no strict boundary between similar and dissimilar contents, threshold-based hash comparison is usually used in practice. When comparing a pair of hash values, a decision is made from two hypotheses: 1) $\mathbb{H}_0$ – *they correspond to different contents*; 2) $\mathbb{H}_1$ – *they correspond to similar contents.*

A distance metric is used to measure the similarity between hash values. Only if the distance d is below a threshold t, the contents are judged as similar. The performance can be quantified by the *detection rate* "p_d" – probability$\{d < t \mid \mathbb{H}_1\}$, and the *false positive rate* "p_f" – probability$\{d < t \mid \mathbb{H}_0\}$. By choosing different values for the threshold, p_d and p_f can be plotted as the receiver operating characteristic (ROC) curve, which characterizes the performance.

Performance evaluation is a difficult task. Deriving the ROC curve generally requires extensive simulation. Typically, a database is used as the ground truth. For each element in the ground truth, several legitimate distortions are applied to produce near-duplicates. The ground truth and the near-duplicates compose an expanded database. Elements in the expanded database are pairwise compared to derive the ROC curve. This procedure is repeated for different keys and different algorithm parameters. Due to excessive computation and the huge storage for test data, the conventional way of performance evaluation may be impractical for video hashing. In order to simplify the task, a framework for video hash construction is proposed in the following.

Since a video is essentially a sequence of frames, a straight-forward way to construct a video hash is to concatenate hash values of video frames. This approach has several advantages: 1) existing image hash algorithms can be used; 2) no need to store an entire video; 3) the video hash performance can be estimated from the image hash performance. In the proposed framework, it is assumed that N frames are extracted from the input video. For each frame, a *frame hash* value $h_{i,i=1,\cdots,N}$ is computed by an image hash algorithm $\mathbb{A}$. The video hash $H = \{h_1|\cdots|h_N\}$ is the concatenation of frame hash values. When two video hash values H_1 and H_2 are compared, *each frame hash h_{i1} of H_1 is compared with the corresponding h_{i2} of H_2; the video files are judged as similar if at least T frame pairs are similar*. The ratio $T/N \in (0.5, 1)$ is a constant, denoted by α. The rules are based on the assumption that if two videos are perceptually different, then each pair of extracted frames are also likely to be perceptually different. Assuming $\mathbb{A}$ has performance $< p_d, p_f >$ and the N frames are independent, the detection rate and the false positive rate of the overall scheme, denoted by P_d and P_f, can be formulated as:

$$P_d = \sum_{k=T}^{N} \binom{N}{k} \cdot p_d^k \cdot (1 - p_d)^{N-k} \tag{1}$$

$$P_f = \sum_{k=T}^{N} \binom{N}{k} \cdot p_f^k \cdot (1 - p_f)^{N-k}. \tag{2}$$

The above equations show that P_d and P_f increase with p_d and p_f, but decrease with α. For reasonable values of p_d and p_f, i.e., $0 < p_f \ll p_d < 1$, P_d increases with N towards 1, and P_f decreases with N towards 0. The speed of convergence is faster if p_d is closer to 1 or p_f is closer to 0. Therefore, target performance can be achieved by choosing a $< p_d, p_f >$ pair and a suitable N.

Equations (1)-(2) can be adapted to cope with dependent frames. The idea is to divide extracted frames into groups and assume elements in each group depend on each other. Since the dependency varies, only the worst case is considered here as a lower bound. Elements within the same group are assumed to have 100% dependency, i.e., they show the same result for a hypothesis test. Note that dependency does not change the threshold α and the decision rules stay the same. The performance can be estimated by replacing N and T in (1)-(2) with $N' = \lfloor N/a \rfloor$ and $T' = \lceil T/b \rceil$, where a and b are respectively the largest and the smallest cardinalities of the groups. By assuming $a = b = a'$, the dependency actually introduces some expansion in the number of extracted frames N. About a' times more frames are needed for the same performance.

In practice, frames can be missing, and new frames might appear. These factors are treated separately in the following. When two video hash values A and B are compared, it was assumed that each pair of frame hash values A_i and B_i correspond to each other. The complexity is $O(N)$. This is not true if the order of frames has changed. For each A_i, the problem is to find the corresponding $B_{i'}$. A naive approach is to compare each A_i with all frame hash values of B and choose the most similar one. This approach assumes that the frames are completely disordered, so it has the highest complexity $O((N+1)N/2)$. If there is only a time shift, then once a single frame is synchronized, all the rest frames are synchronized too. This only increases the complexity to $O(2N-1)$. A more general case, with complexity $O(lN)$, is that each frame can be synchronized by a maximum of l comparisons. After all, desynchronization increases complexity for hash comparison, but does not affect the formulas.

On the other hand, non-repeatability can be modeled by adding a parameter $\beta \in (0.5, 1]$ to (1)-(2). It is the *overlapping ratio* between the extracted frames from two similar videos. When two hash values of similar videos are compared, a frame hash is either compared to a correct counterpart with probability β, or to a "random" one with probability $1-\beta$. Therefore, new formulas can be derived by replacing p_d and p_f in (1)-(2) with $p_d' = \beta p_d + (1-\beta)p_c(t)$ and $p_f' = \beta p_f + (1-\beta)p_c(t)$, where $p_c(t)$ is defined as the *general collision rate*. For hash length n and threshold t, $p_c(t)$ is the probability that arbitrary two hash values' Hamming distance is no less than t:

$$p_c(t) = \sum_{k=t}^{n} \binom{n}{k} \cdot p^k \cdot (1-p)^{n-k}, \tag{3}$$

where p is the probability that two bits coincide, assuming each bit is independent. Since $p_c(t)$ is normally much smaller than p_d and p_f, the effect of non-repeatability is that both P_d and P_f are decreased. Assuming $\alpha = 0.7$, $a' = 10$, $\beta = 0.85$, $n = 144$, $t = 101$, $p = 0.5$, Fig. 1 illustrates the convergence of P_d and P_f by a few examples.

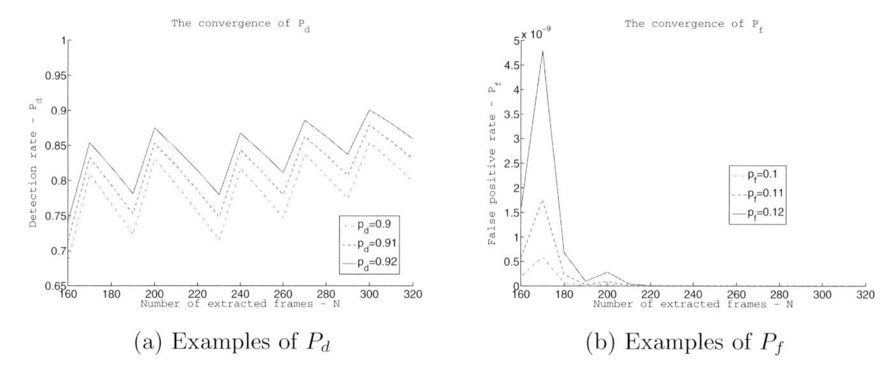

(a) Examples of P_d (b) Examples of P_f

Fig. 1. The convergence of P_d and P_f for practical scenarios

3 A Frame Hash Algorithm

In this section, a frame hash algorithm is proposed. The algorithm is based on statistics. Since an image is essentially a certain allocation of pixels, the content can be characterized by the statistics of pixel values. In order to reduce collision, statistics of different orders are extracted. The maximum order is limited to four as a compromise between performance and complexity. Specifically, the standard deviation, the third- and the fourth-order auto-cumulants are used in the proposed scheme. The latter two are defined as [2]:

$$C_{3X}(k,l) = E[x(n)x(n+k)x(n+l)], \tag{4}$$

$$\begin{aligned} C_{4X}(k,l,m) = {} & E[x(n)x(n+k)x(n+l)x(n+m)] \\ & -C_{2X}(k)C_{2X}(l-m) - C_{2X}(l)C_{2X}(k-m) \\ & -C_{2X}(m)C_{2X}(k-l), \end{aligned} \tag{5}$$

where X is a zero-mean vector, E means expectation, and $C_{2X}(k) = E[x(n)x(n+k)]$. A reason to use higher-order cumulants is that they are immune to Gaussian noise [2]. Additionally, it is known that natural images are non-Gaussian [3]. The assumption that content-preserving processing preserves non-Gaussianity has been validated by applying the fourth-order cumulant in image hashing [4].

The algorithm starts with pre-processing. An extracted frame is first converted to gray and resized to a canonical size 512×384. The output is filtered by an averaging filter and a median filter. Histogram equalization is then performed. These steps limit the computation complexity, remove slight noise, and stabilize the content. The pre-processed image is divided into 256×192 pixel blocks with 50% overlapping. Pixels in each block are scanned into a vector and normalized to be zero-mean. From each vector, the standard deviation, the third- and the fourth-order cumulants are computed. Each statistic is computed twice to increase robustness. The first time is for the original block, the second time is for a transposed version of the original block. The mean value is the output. In order to make the hash compact, only one value is kept for each statistic. Each block is allocated with 16 bits for quantization. Specifically, 3 bits for the standard deviation, 6 bits for the third-order cumulant, and 7 bits for the fourth-order cumulant. Higher-order statistics are are given finer quantization, because they are more representative. There are 9 blocks, so a hash value has 144 bits. There are two steps of randomization. The first is key-based feature extraction. The parameters k, l, and m in (4)-(5) are decided by a secure pseudo-random number generator (PRNG), which accepts a secret key. The second step is dithering. After quantization, the binary output is XORed with a dither sequence generated by the secure PRNG. The distance metric for hash value comparison is the bit error rate (BER), aka the normalized Hamming distance.

The performance of the algorithm has been evaluated by simulation. In the simulation, 432 different natural scene images in the JPEG format are used. They consist of several categories: architecture, landscape, sculpture, objects, humanoid, and vehicle. Each category includes 72 images of three canonical sizes: 1600×1200, 1024×768, and 640×480. There are 10 operations defined to generate distorted but authentic images, each with 5 levels, listed in Table 1.

Table 1. The robustness test

Distortion name	Parameter range, step	Average bit error rates				
1. Rotation	Angle: $2° - 10°, 2°$	0.086	0.142	0.174	0.194	0.210
2. Gaussian noise	Standard deviation: $10 - 50, 10$	0.020	0.026	0.032	0.041	0.051
3. Central cropping	Percentage: $2\% - 10\%, 2\%$	0.078	0.127	0.159	0.185	0.209
4. JPEG compression	Quality factor: $50 - 10, 10$	0.001	0.014	0.010	0.016	0.033
5. Scaling	Ratio: $0.5 - 0.1, 0.1$	0.003	0.008	0.012	0.022	0.055
6. Median filter	Window size: $3 - 11, 2$	0.005	0.012	0.021	0.030	0.036
7. Gaussian filter	Window size: $3 - 11, 2$	0.003	0.003	0.003	0.003	0.003
8. Sharpening	Strength: $0.5 - 0.1, 0.1$[1]	0.022	0.022	0.023	0.023	0.023
9. Salt & pepper	Noise density: $0.01 - 0.05, 0.01$[1]	0.009	0.015	0.021	0.026	0.030
10. Row/col. removal	No. of removals: $5 - 25, 5$	0.009	0.012	0.015	0.017	0.020

In the robustness test, a distorted image is compared with the original by the hash values. For each comparison, a key is randomly chosen from 500 ones. The average BERs are listed in Table 1. The maximum BER is about 0.2, given by rotation and cropping. In the discrimination test, there are 22032 images

[1] These are parameters for Matlab functions fspecial() and imnoise().

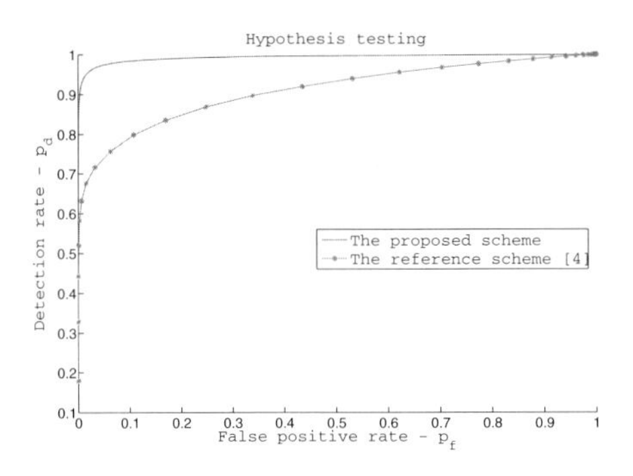

Fig. 2. ROC curves

in total. Any two of them are compared by their hash values. A default key is used. The average BER between different images is 0.33 with standard deviation 0.045. Therefore, the BER margin against incidental distortion is about 0.1. The ROC curve is plotted in Fig. 2. A state-of-the-art image hash algorithm [5] is tested by the same procedure for performance comparison. Its ROC curve for a default key is also plotted in Fig. 2. The figure shows that the proposed scheme has superior performance.

4 New Performance Metrics

Although the ROC curve tells the performance, sometimes it is useful to measure robustness and discriminability separately. In the following, two metrics are proposed for this purpose, defined as the *robustness index* and the *discrimination index*. They are suitable for hash comparison schemes based on the Hamming distance. Before the details, note that the robustness can be represented by the $p_d(t)$ curve, which is the detection rate p_d versus the threshold t. Similarly, the discriminability is represented by the $p_f(t)$ curve.

The discrimination index is basically the difference between the $p_f(t)$ and the $p_c(t)$ curves. The idea is to use the $p_c(t)$ curve as a reference, because it is close to the theoretical (best) $p_f(t)$ curve. Although the general collision rate is larger than the theoretical false positive rate according to the definition, practical schemes hardly achieve it, i.e., p_f is almost always larger than p_c. Therefore, the general collision rate can be used as a reference or even a lower bound of the false positive rate in practice. The $p_c(t)$ curve is derived by assuming $p = 0.5$ in (3) for different values of t. The $p_f(t)$ curve comes directly from the ROC curve. The KL-divergence is used to measure the distance between the two. The discrimination index is defined as

$$D(p_c||p_f) = \sum_t p'_c(t) \frac{p'_c(t)}{p'_f(t)}, \tag{6}$$

where $p'_c(t) = p_c(t)/\sum p_c(t)$ and $p'_f(t) = p_f(t)/\sum p_c(t)$ are normalized versions of $p_c(t)$ and $p_f(t)$. A larger index means better discrimination. The indices of the proposed scheme and the reference scheme are −2.09 and −2.11 respectively. Therefore the proposed scheme has better discrimination. This is consistent with the ROC curve comparison.

Since there is no non-trivial upper bound for p_d, the robustness index is defined differently. The idea is to model $p_d(t)$ by (3). Note that when hash values corresponding to similar contents are compared, p in (3) should be close to 1. Therefore, p can be interpreted as the average detection rate per bit. The robustness index is defined as the p value which results in a $p_c(t)$ curve that is closest to the $p_d(t)$ curve by (6). A larger index means stronger robustness. The indices of the proposed scheme and the reference scheme are 0.96 and 0.99 respectively. Interestingly, the reference scheme shows stronger robustness. It is confirmed by a robustness test. The ROC curves and the indices imply that the superior overall performance of the proposed scheme is due to a good balance between robustness and discriminability.

5 Conclusion

In this work, it is proposed that a perceptual video hash can be constructed from the perceptual hash values of video frames. This framework has the advantage that the performance can be estimated from the performance of the frame hash. A frame hash algorithm based on statistics is proposed. Simulation shows that it outperforms an existing algorithm. As a complement to the ROC curve, two metrics are proposed to measure robustness and discriminability.

References

1. Schneier, B.: Applied Cryptography: Protocols, Algorithms, and Source Code in C, 2nd edn. John Wiley & Sons, Chichester (1996)
2. Mendel, J.: Tutorial on higher-order statistics (spectra) in signal processing and system theory: theoretical results and some applications. Proc. of the IEEE (1991)
3. Krieger, G., Zetzsche, C., Barth, E.: Higher-order statistics of natural images and their exploitation byoperators selective to intrinsic dimensionality. In: Proc. of the IEEE Signal Processing Workshop on Higher-Order Statistics, pp. 147–151 (1997)
4. Weng, L., Preneel, B.: On secure image hashing by higher-order statistics. In: Proc. of IEEE International Conference on Signal Processing and Communications, pp. 1063–1066 (2007)
5. Swaminathan, A., Mao, Y., Wu, M.: Robust and secure image hashing. IEEE Transactions on Information Forensics and Security 1(2), 215–230 (2006)

Which Tags Are Related to Visual Content?

Yinghai Zhao[1], Zheng-Jun Zha[2], Shanshan Li[1], and Xiuqing Wu[1]

[1] University of Science and Technology of China, Hefei, Anhui, 230027, China
[2] National University of Singapore, Singapore, 639798
{yinghai,ssnl}@mail.ustc.edu.cn,
junzzustc@gmail.com, xqwu@ustc.edu.cn

Abstract. Photo sharing services allow user to share one's photos on the Web, as well as to annotate the photos with tags. Such web sites currently cumulate large volume of images and abundant tags. These resources have brought forth a lot of new research topics. In this paper, we propose to automatically identify which tags are related to the content of images, i.e. which tags are content-related. A data-driven method is developed to investigate the relatedness between a tag and the image visual content. We conduct extensive experiments over a dataset of 149,915 Flickr images. The experimental results demonstrate the effectiveness of our method.

Keywords: Flickr, tag, content-relatedness, visual content.

1 Introduction

Online photo-sharing services, such as Flickr [1] and Photobucket [2], encourage internet users to share their personal photos on the web, as well as to annotate the photos with tags (*i.e.*, keywords). Take Flickr for example, it currently cumulates around four billion images as well as billions of tags [1].

These gigantic volume of social tagged images have brought forth many novel research topics. For example, Sigurbjörnsson et al. [7] proposed a tag recommendation strategy based on tag concurrence analysis, while Wu et al. [9] proposed to recommend tags by taking both tag concurrence and image visual content into account. The social tagged images are also used to aid image search. Liu [10] and Li [4] proposed to compute the relevance between tags and images, which in turn facilitated the image search. Although these works investigated the usage of the social images and tags. The facets of tags have been seldom studied.

As reported in [3], the tags associated with images are mainly to describe the image contents and provide other information, such as time stamp, location, and subjective emotion. We argue that the automatic identification of tags which are content-related can aid more intelligent use of the social images and tags and thus facilitate the researches and applications over these resources. To do this, we propose a data-driven method to analyze the relatedness between the tags and the content of the images.

Specifically, different tags have different prior probabilities to be content-related because of the semantic nature of tags. For example, tag *"flower"* is more

S. Boll et al. (Eds.): MMM 2010, LNCS 5916, pp. 669–675, 2010.

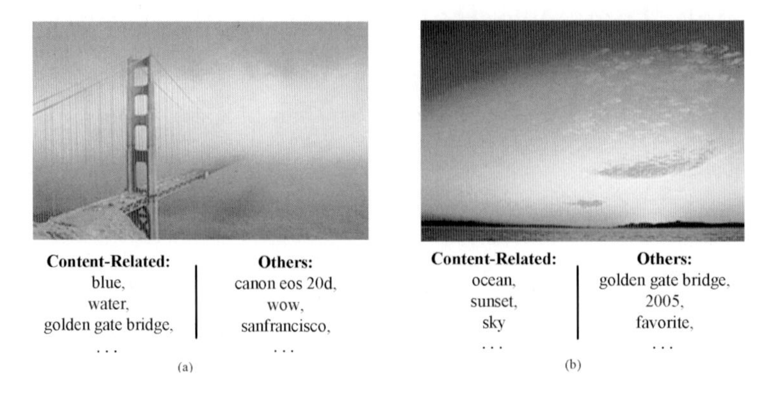

Fig. 1. Two Flickr images and the tags labeled to them

probable to be content-related than *"music"*. Moreover, the content-relatedness between the same tag and the associated images may vary with different images. Fig. 1 shows two images which are annotated with tag *"golden gate bridge"*. However, Fig. 1(a) is a photo of the bridge while (b) is a sunset picture maybe taken on the bridge. Thus, the tag "*golden gate bridge*" is content-related to Fig. 1(a), but is not content-related to Fig. 1(b). Therefore, we investigate the probability that a tag is related to the content of a specific image from the above two aspects.

The rest of this paper is organized as follows. Our method is elaborated in Section 2. Then, the evaluation results over Flickr images are reported in Section 3. Finally, we give the conclusion remarks in Section 4.

2 Tag Content-Relatedness Analysis

Intuitively, (1) if one tag is often used by different users to annotate similar images, this tag is widely accepted to be a proper description for some objective aspects of images content, i.e., this tag is content-related in nature. Moreover, (2) if one tag is labeled to an image by one user, and this image is similar to many other images labeled with this tag by different users, this tag is very likely to be content-related to the specific image. Motivated by these two observations, we propose the following method.

Given a tag t and a set of images $\mathcal{X} = \{x_i\}_{i=1}^N$ that are annotated with t, our target is to derive the scores $\mathcal{P} = \{p_i\}_{i=1}^N$ which measure the content-relatedness between tag t and each image x_i. The probability p_i can be represented as $p_i(r = 1|t, x_i, \mathcal{X}\backslash x_i)$, where $r \in \{1, 0\}$ is an indicator of being content-related or not and $\mathcal{X}\backslash x$ denotes all the other images in $\mathcal{X}$ except x. To reduce the tagging bias of single user in $\mathcal{X}$, each user is limited to contribute only one image to $\mathcal{X}$. Thus, x and $\mathcal{X}\backslash x$ are conditionally independent given t. According to Bayes' formula, the probability $p(r = 1|t, x, \mathcal{X}\backslash x)$ can be derived as follows:

$$p(r = 1|t, x, \mathcal{X}\backslash x) = \frac{p(r = 1|t, \mathcal{X}\backslash x)p(r = 1|t, x)}{p(r = 1|t)}, \tag{1}$$

where $p(r = 1|t)$ indicates how likely t is content-related without knowing any prior knowledge, and is set to be uniform over all tags.

Then, Eq.1 can be formulated as:

$$p(r = 1|t, x, \mathcal{X}\backslash x) \propto p(r = 1|t, \mathcal{X}\backslash x)p(r = 1|t, x), \tag{2}$$

where the item $p(r = 1|t, \mathcal{X}\backslash x)$ indicates the prior probability that t is content-related given the social tagged image resources $\mathcal{X}\backslash x$, and $p(r = 1|t, x)$ expresses the likelihood that t is content-related to x. Overall, $p(r = 1|t, x, \mathcal{X}\backslash x)$ gives out the posterior probability of t being content-related to x with the assistance of social tagged images $\mathcal{X}\backslash x$. For simplicity, we denote $p(r = 1|t, \mathcal{X}\backslash x)$, $p(r = 1|t, x)$ and $p(r = 1|t, x, \mathcal{X}\backslash x)$ as $PrCR$, $LiCR$ and $PoCR$, respectively.

2.1 Probability Estimation

Because one object could be presented from different points of view and one image may only show some local parts of the object, we estimate $p(r = 1|t, \mathcal{X}\backslash x)$ through the local visual consistency over all $x_i \in \mathcal{X}\backslash x$. Here, local visual consistency is a measurement of visual similarities between an image and its K-nearest neighbors. The $p(r = 1|t, \mathcal{X}\backslash x)$ can be estimated as:

$$p(r = 1|t, \mathcal{X}\backslash x) = \frac{1}{KN}\sum_{i=1}^{N}\sum_{j=1}^{K} s(x_i, x_j), \tag{3}$$

where $x_j \in \mathcal{X}\backslash x$ is one of the K-nearest neighbors of x_i, and N=$|\mathcal{X}\backslash x|$ is the number of images in $\mathcal{X}\backslash x$. Moreover, $s(x_i, x_j)$ denotes the visual similarity between x_i and x_j .

Similarly, the likelihood $p(r = 1|t, x)$ can be evaluated through the local visual consistency of image x with respect to its K'-nearest neighbors in $\mathcal{X}\backslash x$:

$$p(r = 1|t, x) = \frac{1}{K'}\sum_{i=1}^{K'} s(x, x_i), \tag{4}$$

where $x_i \in \mathcal{X}\backslash x$, $i = 1, \cdots, K'$, are the K' nearest neighbors in $\mathcal{X}\backslash x$. For simplicity, we let $K = K'$ in our method.

2.2 Visual Similarity

In this subsection, we show the different definitions of visual similarity measurement $s(.)$. It's widely accepted that global features, such as color moments and GIST [6], are good at characterizing scene-oriented (*e.g.*, "*sunset*"), color-oriented (*e.g.*, "*red*") images, while local features, such as SIFT [5], perform better for object-oriented (*e.g.*, "*car*") images. The fusion of multiple features can achieve better representation for image content. Here, we use three similarity definitions that are based on global, local features, and both global and local features, respectively.

f is a feature vector that may be the concatenation of several kinds of global features extracted from image x, the global visual similarity between two images x_i and x_j can be calculated through Gaussian kernel function as:

$$s_g(x_i, x_j) = \exp(-\frac{||f_i - f_j||^2}{\sigma^2}), \tag{5}$$

where σ is the radius parameter of Gaussian kernel.

To computer local visual similarity, bag-of-visual-words method is adopted here [8]. Each image is represented as a normalized visual word frequency vector of dimension D. Then, the local visual similarity between two images could be calculated through the cosine similarity:

$$s_l(x_i, x_j) = \frac{v_i^T v_j}{||v_i||||v_j||}, \tag{6}$$

where v_i and v_j are visual word representations of x_i and x_j, respectively.

Furthermore, a fused visual similarity is obtained through the line combination of global and local visual similarities:

$$s_c(x_i, x_j) = \alpha s_g(x_i, x_j) + (1 - \alpha) s_l(x_i, x_j), \tag{7}$$

where $0 < \alpha < 1$ is the combination coefficient.

3 Experiments

3.1 Data and Methodologies

For experimental data collection, the 60 most popular Flickr tags in April 2009 are selected as seed queries. For each query, the first 2000 images are collected through Flickr image searching. During this process, only the first image is kept for one user. Tag combination and tag filtering operations are conducted to get the tags with more than one word and remove noises. Afterwards, another 207 most frequent tags are selected as queries for further image collection. Finally, 149,915 images are obtained in all.

The following visual features are used to characterize the content of images:

- global features: 6-dimensional color moment in LAB color space and 100-dimentional GIST feature [6] processed by PCA.
- local features: 128-dimensional SIFT descriptors [5].

For parameter setting, the nearest neighbor number, K and K', are both set to 100, and the size of $\mathcal{X}$ is set to 600. Moreover, the size of the visual word codebook D is set to 5000, and the coefficient α in Eq.7 is set to 0.5.

3.2 How Probable the Tag Is Content-Related

In this experiment, we investigate the prior content-relatedness ($PrCR$) probabilities $p(r = 1|t, \mathcal{X}\backslash x)$ for different tags. The $PrCR$ results calculated with

Table 1. The first 10 and the last 10 tags in the *PrCR* based tag sorting results. The tags in bold are examples being ranked at inappropriate places.

Position	*PrCR_g*	*PrCR_l*	*PrCR_f*	Position	*PrCR_g*	*PrCR_l*	*PrCR_f*
1	**winter**	flowers	flowers	51	italy	australia	trip
2	snow	cat	**winter**	52	wedding	trip	canon
3	blue	dog	snow	53	music	nature	japan
4	flower	food	flower	54	europe	vacation	europe
5	beach	**christmas**	green	55	france	wedding	france
6	green	people	cat	56	nyc	birthday	music
7	flowers	snow	beach	57	paris	canon	spain
8	water	green	food	58	spain	new	wedding
9	sky	flower	blue	59	taiwan	california	california
10	cat	city	water	60	california	taiwan	taiwan

global visual similarity, local visual similarity and fused visual similarity are denoted with *PrCR_g*, *PrCR_l*, and *PrCR_f*, respectively. We sort the 60 Flickr tags according to their *PrCR* probabilities in descending order. Due to the limited space, only the first 10 and the last 10 tags are listed in Tab.1.

From the results in Tab.1, we observe that:

- Tags whose semantics are related to concrete objects, colors, or scenes are generally ranked at the top of the results, while tags whose semantics are related to locations, abstract concepts or time, are ranked at the bottom.
- The *PrCR_g* metric succeeds to promote scenes and colors oriented tags, such as "*beach*", and '*blue*", to the top of the list while the *PrCR_l* metric prefers concrete objects related tags, such as "*flowers*", "*dog*".
- *PrCR_f* is benefited from the fusion of global and local similarity measurements, and gets the most reasonable result.

3.3 How Probable the Tag Is Content-Related to Specific Image?

In this section, we evaluate the performance of the proposed method in sorting the image tags according to their content-relatedness probability. For evaluation, 400 images are randomly selected with their tags being manually labeled into two levels: "content-related" or not. We measures the tag sorting performance with mean average precision (MAP). MAP is obtained by averaging the AP scores of the sorted tag lists over all the test images.

We compare the performances of five methods: (1) *Baseline* which are the tagging results in the order of user inputting; (2) *LiCR_g* which sorts tags based on $p(r = 1|t, x)$ with global visual similarity; (3) *LiCR_l* which sorts tags according to $p(r = 1|t, x)$ with local visual similarity; (4) *LiCR_f* which sorts tags according to $p(r = 1|t, x)$ with fused similarity; and (5) *PoCP* which sorts tags according to posterior probability $p(r = 1|t, x, \mathcal{X}\backslash x)$ with fused visual similarity.

The experimental results are shown in Fig. 2 (a). From the results, we can observe that all the last four methods could consistently boost the tag sorting performance comparing to the original input order. Our proposed method *PoCR*

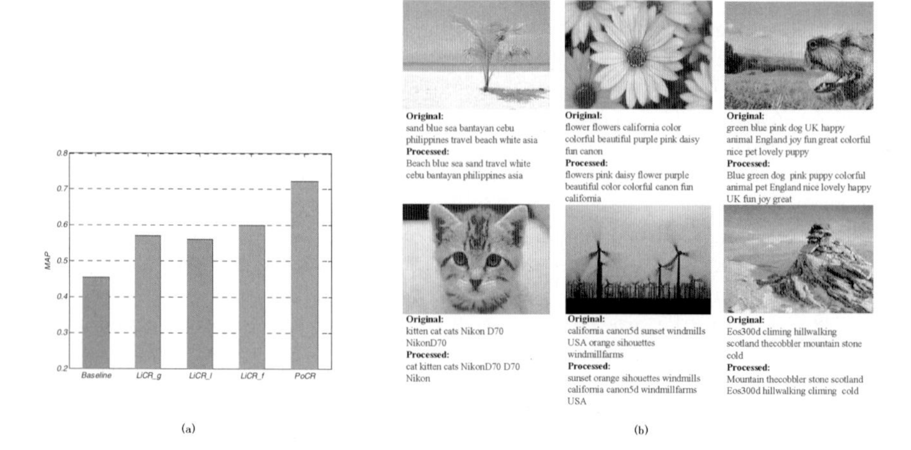

Fig. 2. (a) MAP results of tag content-relatedness analysis for 400 images. (b) Some example results of tag content-relatedness analysis. According to our method, the tags which are content-related to the image are ranked at the top of the processed tag list.

achieves the best performance, and obtains 59.8%, 26.6%, 29.3%, and 20.4% relative improvements compared to *Baseline*, *LiCR_g*, *LiCR_l*, and *LiCR_f* respectively. We illustrate some example images and their original tag lists, sorted tag lists, in Fig. 2 (b).

4 Conclusion

In this paper, we have firstly shown the fact that tags of Flickr images are not always content-related to images. Then, we propose an data-driven approach to evaluate the probability of a tag to be content-related with respect to an image. It's worth noting that our method requires no model training process and could be scalable to large-scale datasets easily. Experiments on 149,915 Flickr images demonstrate the effectiveness of the proposed method.

References

1. Flickr, `http://www.flickr.com/`
2. Photobucket, `http://photobucket.com/`
3. Ames, M., Naaman, M.: Why we tag: motivations for annotation in mobile and online media. In: Proceedings of the SIGCHI conference on Human factors in computing systems, San Jose, California, USA, pp. 971–980 (2007)
4. Li, X., Snoek, C.G., Worring, M.: Learning tag relevance by neighbor voting for social image retrieval. In: Proceeding of the 1st ACM international conference on Multimedia information retrieval, Vancouver, Canada, pp. 180–187 (2008)
5. Lowe, D.G.: Distinctive image features from scale-invariant keypoints. Int. J. Comput. Vision 60(2), 91–110 (2004)

6. Oliva, A., Torralba, A.: Modeling the shape of the scene: A holistic representation of the spatial envelope. Int. J. Comput. Vision 42(3), 145–175 (2001)
7. Sigurbjörnsson, B., van Zwol, R.: Flickr tag recommendation based on collective knowledge. In: Proceeding of the 17th international conference on World Wide Web, Beijing, China, pp. 327–336 (2008)
8. Sivic, J., Zisserman, A.: Video google: A text retrieval approach to object matching in videos. In: Proceedings of the Ninth IEEE International Conference on Computer Vision, pp. 1470–1477 (2003)
9. Wu, L., Yang, L., Yu, N., Hua, X.-S.: Learning to tag. In: Proceedings of the 18th international conference on World wide web, Madrid, Spain, pp. 361–370 (2009)
10. Liu, D., Hua, X.-S., Yang, L., Wang, M., Zhang, H.-J.: Tag ranking. In: Proceedings of the 18th international conference on World wide web, Madrid, Spain, pp. 351–360 (2009)

Anchor Shot Detection with Diverse Style Backgrounds Based on Spatial-Temporal Slice Analysis

Fuguang Zheng, Shijin Li*, Hao Wu, and Jun Feng

School of Computer & Information Engineering, Hohai University, Nanjing, China
lishijin@hhu.edu.cn

Abstract. Anchor shot detection is a challenging and important task for news video analysis. This paper has put forward a novel anchor shot detection algorithm for the situations with dynamic studio background and multiple anchorpersons based on spatio-temporal slice analysis. Firstly, two different diagonal spatio-temporal slices are extracted and divided into three portions, after which sequential clustering is adopted to classify all slices from two sliding windows obtained from each shot to get the candidate anchor shots. And finally, structure tensor is employed, combining with the distribution properties to precisely detect the real anchor shots. Experimental results on seven different styles of news programs demonstrate that our algorithm is effective toward the situations described above. And the usage of spatio-temporal slice can also reduce the computational complexity.

Keywords: Anchor shot detection, dynamic background, multiple anchorpersons, spatio-temporal slice, sequential clustering.

1 Introduction

As the rapid development of the research on Content-Based Video Retrieval (CBVR), people hope to retrieve the programs of their own interest from huge amount of news videos. Typically, a complete news video consists of anchor shots, news story footage and some possible commercial blocks. For the unique structure of the news program, an anchor shot indicates the beginning of a news story. The anchor shot with the following news story represents an integrated news event.

Recent years, many algorithms have been proposed to deal with anchor shot detection. They are mainly based on the assumption that the similarity between anchor shots is quite high, and these methods can be categorized into two classes, i.e., one is based on template matching [1-4] and another is based on the scattered distribution properties of anchor shots [5-8].

The prevalent methods based on template matching always assume that the studio background is static. Zhang et al.[1] constructed three models for an anchor shot: shot, frame, and region. An anchor shot was modeled as a sequence of frame models and a frame was modeled as a spatial arrangement of regions. These models varied for

* Corresponding author.

S. Boll et al. (Eds.): MMM 2010, LNCS 5916, pp. 676–682, 2010.

different TV stations and it was difficult to construct all the possible models for different news videos. Ma et al.[2] proposed an edge detection based method to locate anchor shots, which used the difference of Gaussian (DoG) operator and generalized Hough transform(GHT) to match the contours of anchor persons. In order to improve the detection results of previous algorithms, toward the anchor shots with a small area of dynamic background, Xu et al.[3] proposed to build a simple anchorperson model in the first frame of the anchor shot to deal with picture in picture situation in news program. In our previous work [4], each frame was equally divided into 64 regions, after which a weighted template was built for the anchor shots. The algorithm was effective to detect anchor shots with dynamic background, but it could not cope with the news program with multiple anchorpersons.

Due to the poor generalization ability of the template matching-based algorithms, researchers proposed some new approaches based on the distribution properties of anchor shots. Gao et al. [5] employed graph-theoretical cluster analysis method to classify the shots, the experimental results proved that its performance was quite good, but the complexity of their algorithm was rather high. In [6], the authors created a set of audio and video templates of anchorperson shots in an unsupervised way, then shots were classified by comparing them to all the templates when there was one anchor and to a single best template when there were two anchors. Further more, face detection was also proposed to refine the final detection results [7-8]. However, it is well known that they are quite time consuming. Due to the inherent difficulties of anchor shot detection, Santo et al.[11] proposed to combine multiple algorithms[5,12-13] to provide satisfactory performance for different styles of news programs.

Liu et al. [9] proposed a new method to locate anchor shots based on spatial-temporal slice analysis, with which the authors extracted horizontal and vertical direction slices from news video, and k-means clustering was employed to detect the anchor shots quickly. However, if the anchorperson wasn't seated at the center of the screen, the vertical direction slice wouldn't include the anchorperson's body, and k-means algorithm is not suitable, as we don't know how many clusters those samples should be clustered into. In their method, the cluster with most elements was considered as the candidate anchor shot, so they could just cope with news program hosted by one anchorperson.

In summary, the above mentioned existing methods have some limitations, such as high computational complexity [5,7,8,11], unable to deal with dynamic studio background [1-3,6,9] or multiple anchorpersons[1,4,6,9]. To improve the poor performance caused by these limitations, a novel approach to anchor shot detection based on spatio-temporal slice is proposed in the paper. Our algorithm is based on the following two weak assumptions, which can be satisfied by many different styles of news programs:

(1) Each anchorperson's clothes will not vary in one news program, while his/her position is not restricted to be the same all the time.
(2) The same anchorperson's shots tend to be scattered in a news program at least two times.

The presented algorithm is made up of three steps. Firstly, two different diagonal spatio-temporal slices are extracted and divided into three portions, then sequential clustering is adopted to cluster all the three portioned slices in each sliding windows obtained from each shot to get the candidate anchor shots. And finally, structure tensor is employed to compute the mean motion angle of the slice, combining with the distribution properties to detect the real anchor shots precisely.

2 The Proposed Algorithm

2.1 Extracting Spatio-temporal Slice

Spatio-temporal slice is a collection of scans in the same position of every frame of a video as a function of time. There are many kinds of selection directions, and the typical ones are horizontal, vertical and diagonal. Since the anchorperson may be seated at left, center or right of the screen, we choose two diagonal spatio-temporal slices, then the spatio-temporal slice of each anchor shot could include the body of the anchorperson. Fig.1 shows the two methods we get pixels from a frame, red line between the two green lines shows the position we get pixels. Ignoring the areas outside of the green lines can reduce computational cost and avoid the disturbances of scrolling captions.

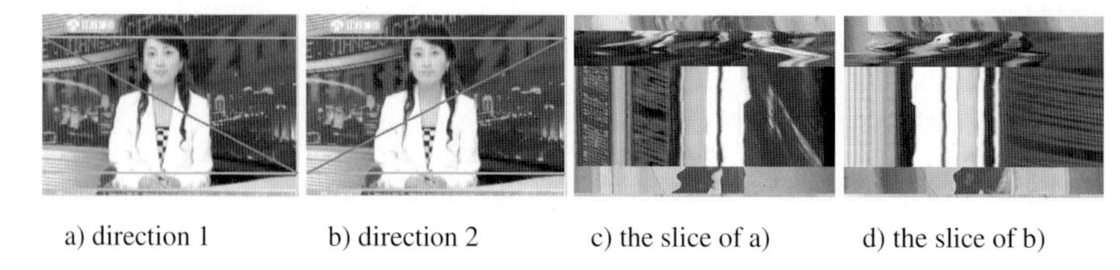

a) direction 1 b) direction 2 c) the slice of a) d) the slice of b)

Fig. 1. Two scanning directions of the spatio-temporal slice and the corresponding slice images

Taking the anchorperson's position and studio background into consideration, we can find two interesting observations of the slice image. On the one hand, anchorperson will appear on the left, center or right part of the slice that is corresponding to the anchorperson's position. On the other hand, the anchorperson's body of an anchor shot in the slice is almost unchanged, but the region with dynamic background is disorder, and static background corresponds to a completely unchanged area. Based on the above mentioned observations, we propose to cluster the nearly unchanged anchorperson portions.

2.2 Sequential Clustering Analysis of Slices

According to the distribution properties of anchor shots mentioned in Section 1, combined with the characteristics of the spatio-temporal slices in the last section, the similarity between sub-images of a slice by a same anchorperson at the same position is quite high. And these sub-images are widely scattered through the news program with a long time span between them, while the other non-anchor shots with high similarity only occur on their neighboring shots.

Because anchorperson only appears on the left, center or right region of the slice, and in order to eliminate the interference of dynamic background, we equally divide the slice into three portions. Then the portion with an anchorperson will have lots of sub-images with high similarity. So we extract two sub-images from every shot with the same size in the same portion, one starts from the beginning of the shot, and another terminates in the end of the shot, because sometimes zooming effect is applied to anchor shots. Another reason is that each anchor shot will contribute equally to the last clusters, no matter how long the shot is and thus avoiding the weather forecast shots. By scrupulously observing those sub-images, the one located at the anchorperson's position is almost occupied by anchorperson's body. So these sub-images are quite similar, no matter whether its background is static or dynamic.

In order to group these anchorperson's sub-images into a cluster, we extract 3 color moments of each sub-image in the HSI color space. The clustering algorithm is aimed at grouping the similar shots into a cluster, and doesn't concern about how many clusters it will generate. Moreover, the number of anchorpersons in different news programs is unknown. For these reasons, sequential clustering algorithm [14] is adopted. The algorithm scans all samples once, if the minimum distance between the current sample and the clusters already exist exceeds a pre-selected threshold (800 in this paper), the sample is constructed as a new cluster, or combine it into the cluster with the minimum distance to it and recalculate the cluster's center. And the clusters whose member amount exceeds a threshold (8 in the paper) are selected as the candidate anchor shots.

2.3 Precisely Labeling Anchor Shots

The sequential clustering algorithm only uses color features, but the texture feature is neglected. As shown in Fig.1, the region with anchorperson's body of the slice is almost unchanged in vertical direction, and the other dynamic regions are in disorder. We adopt the structure tensor feature proposed in [10] to distinguish anchor shot from other non-anchor shots, since there are minor or no motion in the anchor portion, while there are obvious motion in the other non-anchor shot or the dynamic background. Firstly, we calculate each pixel's direction of gray level change Φ in every sub-image. And then each sub-image's total direction φ is computed by Equation (1).

$$\varphi = (\sum_{i=1}^{N} |\Phi_i|) / N \tag{1}$$

Where N is total pixel number of one sub-image. Finally, Equation (2) is employed to get the mean motion angle σ of every cluster.

$$\sigma = \frac{1}{M}\sum_{i=1}^{M} \varphi_i \tag{2}$$

Where M is the number of cluster members. The last decision rules are listed as follows:

(1) Remove the cluster whose mean motion angle σ is less than 80°, and also remove the member whose overall direction φ is less than 80° in the remaining clusters.

(2) If there are some continuous shots in the clusters filtered by step (1), remove these shots, and also remove the clusters whose shot time span between the

first member and last member is too short, as these shots may be interview shots, lecture shots or weather forecast shots.

(3) All the above steps including clustering and post-processing heuristics are conducted on one of the three portions of each slice, so each portion has its own decision, and a shot has six chances to be detected as an anchor shot. We choose the shots judged as anchor shots twice as the final anchor shots, as there are two slices.

3 Experimental Results and Analysis

Our experiments are carried out on the data of 7 representative news video programs collected from our local TV stations and China Central TV station (CCTV). Fig. 2 shows some examples of the key frames of anchorperson shots in these videos. Clip 1 to 4 are from our local TV stations in Jiangsu Province, and Clip 5 and Clip 6 are from CCTV News channel, while Clip 7 is from CCTV International in Chinese channel. Each news program of Clip 1 to 4 and Clip 7 last about 60 minutes, while Clip 5 and Clip 6 last 30 minutes. All the videos are with a frame rate of 25fps.

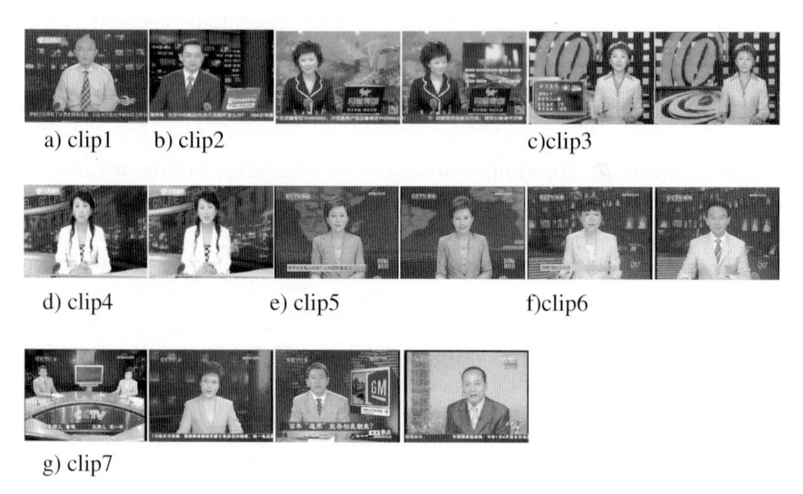

Fig. 2. Some key frames of anchorperson shots in our experimental videos

From Fig.2, it can be found that the background of anchor shots in Clip1 and Clip2 are static. And there are two anchorpersons in Clip2, in addition, at the end of the two news programs followed by weather forecasts. Anchor shots in Clip3 have a partially dynamic background and globally dynamic background in Clip4 and Clip5. Program in Clip6 have two anchorpersons, but several lecture shots in it will disturb the decision. Clip7 is presented by three anchorpersons at different positions, among which there is a partially dynamic background for the first male anchorperson.

The performance of our algorithm is evaluated in terms of Precision and Recall, which are commonly used in CBVR.Table.1 gives the detection results of our algorithm, and those of the method proposed in [9].

Table 1. Experimental results comparison of the proposed algorithm and the method in [9]

Program	Anchor shots	Precision of [9]	Recall of [9]	Precision of our method	Recall of our method
Clip1	18	84%	100%	100%	100%
Clip2	24	100%	84%	100%	96%
Clip3	30	100%	100%	100%	100%
Clip4	24	80%	95%	100%	100%
Clip5	16	94%	100%	100%	100%
Clip6	13	100%	54%	100%	100%
Clip7	46	92%	50%	96%	100%

From Table.1 we can find the precision and recall of our method are quite high, whether it is with static background, partially dynamic background or fully dynamic background. And the performance is also excellent toward programs with single anchorperson or multiple anchorpersons. Compared with the algorithm in [9], some false alarms occur when the studio background is fully dynamic, because at this situation only the vertical slice is useful; moreover, the anchorperson has to be seated at the centre of the screen. And the recall is too low when there are several anchorpersons to report news, it's impossible to detect any other anchorperson by searching anchor shots from the cluster with the most elements in [9]. The hostess in Clip2, the male anchorperson in Clip6 and the two hosts in Clip7 are all missed.

The time cost of our algorithm is lower than the method in [9], because we only extract the spatial-temporal slices in two sliding windows from each shot, but Liu et al. [9] used a sliding window throughout the whole video, and their method was more time consuming. For a video clip of 30 min, our approach needs 150s, while the method in [9] needs 200s, both running on a same PC of Pentium 2.4G, 512M RAM.

4 Conclusions

In this paper a novel algorithm for detecting anchor shot based on the distribution properties which can deal with the situation with dynamic studio background and multiple anchorpersons. By using spatio-temporal slice analysis, video processing is converted into image processing which will significantly reduce the algorithm complexity. The experimental results on six different styles of news video demonstrate that the proposed algorithm is accurate, robust and efficient. But the algorithm is not suitable for the situation that the anchorperson occupies only a very small percent of the screen, such as the anchorperson standing far away from the camera. In the future, we will study more elaborate schemes to detect this kind of situation in a full-length video program. And the detailed comparison with the state of the art algorithms such as those methods in [5, 8] are currently being undertaken by our group.

Acknowledgments. The authors would like to thank Mr. Zhang Haiyong from Jiangsu Broadcast Corporation, China, for his help providing us with the videos used in this paper. This work is partially funded by NSFC (granted number 60673141).

References

1. Zhang, H., Gong, Y., Smoliar, S., et al.: Automatic Parsing of News Video. In: Proceedings of the International Conference on Multimedia Computing and Systems, Boston, MA, pp. 45–54 (1994)
2. Ma, Y., Bai, X., Xu, G., et al.: Research on Anchorperson Detection Method in News Video. Chinese Journal of Software 12, 377–382 (2001)
3. Xu, D., Li, X., Liu, Z., Yuan, Y.: Anchorperson Extraction for Picture in Picture News Video. Pattern Recogn. Lett. 25, 1587–1594 (2004)
4. Zheng, F., Li, S., Li, H., et al.: Weighted Block Matching-based Anchor Shot Detection with Dynamic Background. In: Proceeding of International Conference on Image Analysis and Recognition, pp. 220–228 (2009)
5. Xinbo, G., Xiaoou, T.: Unsupervised Video-shot Segmentation and Model-free Anchorperson Detection for News Video Story Parsing. IEEE Transactions on Circuits and Systems for Video Technology 12, 765–776 (2002)
6. D'Anna, L., Percannella, G., Sansone, C., Vento, M.: A Multi-Stage Approach for News Video Segmentation Based on Automatic Anchorperson Number Detection. In: Proceedings of International Conference on Mobile Ubiquitous Computing, Systems, Services and Technologies, pp. 229–234 (2007)
7. Lan, D., Ma, Y., Zhang, H.: Multi-level Anchorperson Detection Using Multimodal Association. In: Proceedings of the 17th International Conference on Pattern Recognition, pp. 890–893 (2004)
8. Santo, M., De, F.P., Percannella, G., et al.: An Unsupervised Algorithm for Anchor Shot Detection. In: Proceedings of the 18th International Conference on Pattern Recognition, pp. 1238–1241 (2006)
9. Anan, L., Sheng, T., Yongdong, Z., et al.: A Novel Anchorperson Detection Algorithm Based on Spatio-temporal Slice. In: Proceedings of International Conference on Image Analysis and Processing, pp. 371–375 (2007)
10. Chong-Wah, N., Ting-Chuen, P., Hong-Jiang, Z.: Motion Analysis and Segmentation through Spatio-temporal Slices Processing. IEEE Transactions on Image Processing 12, 341–355 (2003)
11. De Santo, M., Percannella, G., Sansone, C., et al.: Combining Experts for Anchorperson Shot Detection in News Videos. Pattern Analysis & Application 7, 447–460 (2004)
12. Bertini, M., Del Bimbo, A., Pala, P.: Content-based Indexing and Retrieval of TV News. Pattern Recognition Letter 22, 503–516 (2001)
13. Hanjalic, A., Lagendijk, R.L., Biemond, J.: Semi-automatic News Analysis, Indexing and Classification System Based on Topics Preselection. In: Proceedings of SPIE: Electronic Imaging: Storage and Retrieval of Image and Video Databases, San Jose (1999)
14. Theodoridis, S., Koutroumbas, K.: Pattern Recognition, 2nd edn. Academic Press, London (2003)

The SLDSRC Rate Control Scheme for H.264

Jianguo Jiang[1,2], Wenju Zhang[1], and Man Xuan[1]

[1] School of Computer & Information, Hefei University of Technology, Hefei 230009, China
[2] Engineering Research Center of Safety Critical Industrial Measurement and Control Technology, Ministry of Education, Hefei 230009, China
jianguoj@163.com, wena_tt@yahoo.cn, eel.xuan@gmail.com

Abstract. A novel slice-layer double-step rate control (SLDSRC) scheme for H.264 is proposed. It not only resolves the problem of inter-dependency between rate-distortion (R-D) optimization (RDO) and rate control (RC), but also improves control accuracy by introducing double-step mechanism, new source rate prediction model, header-bit prediction method. The new rate-quantization (R-Q) model distributes bit rate more reasonable; the novel header-bit prediction method satisfies the requirement of high accuracy at low bit rate. Experimental results show the proposed algorithm heightens the control precision, improves the PSNR and reduces fluctuation of output bit rate, compared to RC algorithm in JVT-H017.

Keywords: Double-step RC, R-Q model, SAQD, control accuracy, PSNR.

1 Introduction

RC, one of the important video coding technologies, has been playing an important role in the video transmission, storage and the hardware design. It has been extensively studied in many standards (MPEG-2, MPEG-4, H.263, H.264) [1-6]. With the existence of chicken and egg dilemma [5], it is very complex to achieve RC for H.264. JVT reference software JM adopts the RC algorithm proposed in JVT-H017 [6]. Based on it, many improved algorithms have been proposed. But most of them do not change the holistic structure of the RC algorithm for H.264 although associate the complexity of the image itself to optimize bit allocation [7-9] for overcoming the weakness that the complexity of video image isn't taken into account in JVT-H017, and there are still some limitations.

2 SLDSRC Algorithm

2.1 Double-Step Encoder Mechanism

The SLDSRC scheme is divided into the preparation step and the encoding step, denoted as p_step and e_step respectively in Fig.1. Each frame is performed through the above two steps.

S. Boll et al. (Eds.): MMM 2010, LNCS 5916, pp. 683–688, 2010.

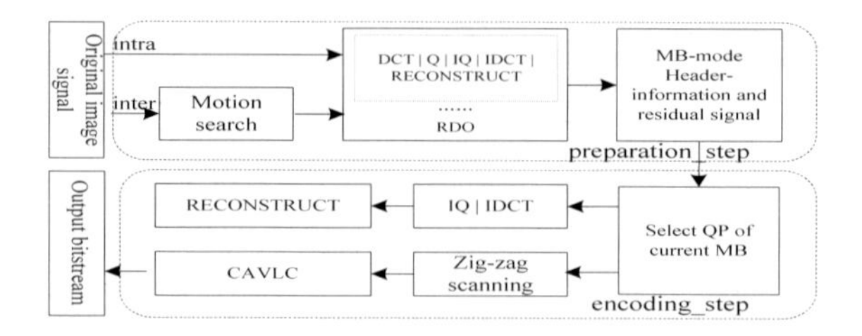

Fig. 1. The framework of SLDSRC

p_step: To encode a new frame, update QP_0 according to QP distribution for slices in a frame and the change trend of image complexity between adjacent frames, and after motion search, perform RDO for all MBs in the current frame for obtaining the header information and residual signal. This step makes preparation for e_step. QP and processing unit in this step are QP_0 and a frame, respectively.

e_step: Search the best QP for each slice in the QP discrete set $[QP_0-2, QP_0+2]$, recorded as QP_c, and then encode the residual signal of each slice obtained in p_step. QP and processing unit in this step are QP_c and a slice, respectively.

Generally, the decrease of the coding gain is not much even though QP_0 and QP_c are different as long as the difference is restricted to a small range, as $|QP_c - QP_0| \leq \delta, \delta \leq 2$. We observe almost identical results for numerous test sequences and show only "foreman" as a representative one in table 1. Experiment results demonstrate that PSNR just decrease about 0.12dB if $|QP_c - QP_0| = 2$, and the decrease is negligible when $|QP_c - QP_0| = 1$.

It should be noted that if a slice is inter-coded, the residual signal for QP_0 is simply re-quantized using QP_c; if it is intra-coded, the residual signal should be updated since QP_c may be different from QP_0 and its adjacent reference units are no longer the reconstructed units for QP_0. For such a case, update residual signal following the same intra-mode determined by the RDO for QP_0.

Table 1. A compare of PSNR for $QP_c - QP_0 \in [-2, +2]$

QPc-QP0 / Rate(kbps)	-2	-1	0	1	2
48	31.61	31.70	31.74	31.68	31.68
64	32.01	32.98	33.00	32.98	32.94
96	34.66	34.75	34.77	34.72	34.71
150	37.65	37.74	37.78	37.75	37.75
250	40.26	40.41	40.44	40.38	40.35
500	44.11	44.26	44.30	44.23	44.22

2.2 The R-Q Model Based On SAQD

The quadratic Laplace-distribution-based R-Q model [5] has been widely used, but this kind of ρ-domain prediction model is less accurate than the q-domain model, and the inaccuracy is mainly due to the roughness of residual-signal complexity denotation, such as MAD-denotation. It only reflects the time-domain residual difference, but can not reflect the state of actual coding bit stream. In frequency-domain [10], in terms of the statistical characteristic of DCT coefficients [11], we develop the sum of absolute quantization distortion SAQD and exploit a better R-Q model.

SAQD is defined as formula (1):

$$\mathrm{SAQD}=\sum_{i=0}^{N-1}\sum_{j=0}^{N-1} d_{i,j}, d_{i,j}=\begin{cases}0, & |X_{i,j}|>\varphi\cdot Qstep \\ |X_{i,j}|, & |X_{i,j}|\le\varphi\cdot Qstep\end{cases} . \tag{1}$$

where $X_{i,j}$ is the DCT coefficient for the position (i, j) before quantization, Qstep, N and φ are the quantization stepsize, MB-height, and the threshold constant, respectively.

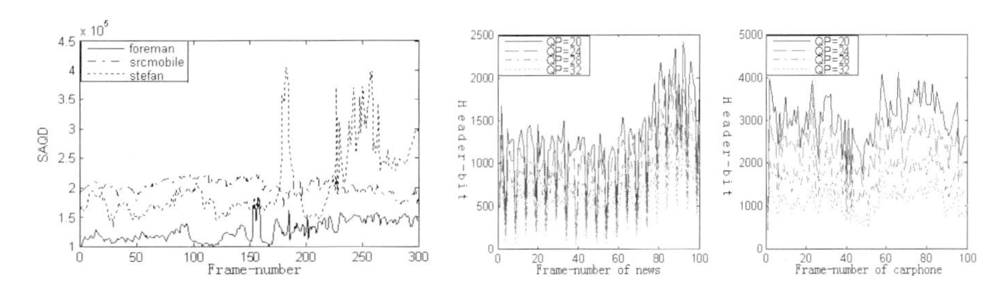

Fig. 2. SAQD variation for three sequences

Fig. 3. Variation of header bits with QP

It's well known that the smaller the residual DCT coefficient is, the greater the QP is, the greater the error between before and after quantization will be, the more detailed information will be lost, resulting in worse image distortion as well as bigger SAQD. Contrarily, more detailed information will be reserved, the smaller the distortion will be and the smaller the SAQD will be. The SAQD variation for three sequences with the QP 28 and the format CIF is shown as Fig.2. We can see the SAQD for srcmobile is bigger than foreman. That's because the complexity of the former image is greater than the latter. Meanwhile, the SAQD stability of the above two sequences are higher than stefan, which precisely reflects the characteristics of intense image movement and large complexity-change of stefan. Therefore, SAQD can be used to denote image complexity accurately.

Due to higher accuracy of SAQD denotation as image complexity, we update the quadratic R-Q model [5] as formula (2):

$$B_s - B_{head} = (\frac{X_1}{\mathrm{Qstep}^2}+\frac{X_2}{Qstep})\cdot SAQD . \tag{2}$$

where B_s, B_{head} are all the bits allocated to coding blocks and the corresponding header bits, respectively. Qstep is quantization stepsize, X_1, X_2 are model parameters, updated by linear regression technique after encoding each slice, which can be referred to JVT-H017 [6].

2.3 Header-Bit Prediction

In order to predict header bits, JVT-H017 treats the average header bits for the encoded units as that of the current encoding unit. It is simple, but not effective. The paper presents an accurate and effective method to predict header bits.

In the RDO process, the encoder determines the encoding mode for every MB by minimizing Lagrange cost function, shown as formula (3):

$$J_{MODE}(S_k, I_k \mid QP, \lambda_{MODE}) = D_{REC}(S_k, I_k \mid QP) + \lambda_{MODE} \cdot R_{REC}(S_k, I_k \mid QP). \quad (3)$$

$$\lambda_{MODE} = 0.85 \times 2^{(QP-12)/3}. \quad (4)$$

where I_k is the encoding mode of MB S_k, and R_{REC} and D_{REC} are the bit rate and distortion of encoded block, respectively.

The larger QP becomes, the larger λ_{MODE} will become, and the larger the importance of R_{REC} in the Lagrange cost function will get, thus more attention will be paid to R_{REC} while the smaller the importance of D_{REC} will fall. For such a case, I_k becomes simple, such as rough division mode, small MV, and therefore the header bits will decrease. Experimental results demonstrate that header bits decrease as QP increases as Fig.3.

Since QP_0 in p_step and QP_c in e_step may be different, the header bits generated in p_step may not be equal to the actual header bits generated in e_step. However, header bits decrease as QP increases, and $QP_c (QP_c \in [QP_0 - 2, QP_0 + 2])$ is equal to QP_0 or almost evenly fluctuates around QP_0 (as shown in experimental results), so the actual header bits in e_step equals to or almost evenly fluctuates around the actually generated header bits, and the number of the fluctuating bits can be counteracted one another. Based on the above, a concise and accurate method is developed for predicting header bits as formulate (5):

$$B_{head\_e,i} = B_{head\_p,i}. \quad (5)$$

where $B_{head\_e,i}$ is the header bits of the current frame, i-th frame, and $B_{head\_p,i}$ is actually generated header bits in p_step.

3 Experiment Results and Analysis

The presented SLDSRC algorithm was implemented on JM10.0 platform for H.264 baseline-profile encoder under constant bit rate constraint. In order to brighten the advantages of the new models and methods in SLDSRC algorithm, we initialize GOP-layer QP and distribute frame-layer bit rate as JVT-H017. Several QCIF sequences are encoded with IPPP format, regarding a frame in intra-frame coding as a slice and a MB in inter-frame coding as a slice. φ in equation (1) is endowed with an experiential value 1.0, X_1, X_2 in equation (2) are initialized with 1.0, 0.0, respectively. $slidwd_{max}$ is set to experiential value 20.

In table 2, it is very clear that the bit rate using our algorithm is more approximate to the signal channel bandwidth, and the PSNR is raised. We know from Fig.4 that the bit rate fluctuation between two successive frames by our algorithm is smaller. Those prove that our algorithm not only makes a reasonable use of signal channel bandwidth, but also obtain higher control accuracy for a single frame, therefore avoids the video buffer underflow and overflow. A compare of the predicted header bits and the

Table 2. A compare of two algorithms in accuracy and PSNR

Test Sequences (QCIF)	Target Bits (kbps)	Inital QP	Control Accuracy (kbps)		Rate Offset (kbps)		PSNR (dB)	
			JVT-H017	Proposed	JVT-H017	Proposed	JVT-H017	Proposed
foreman	48	32	48.10	48.00	0.10	0.00	31.74	32.00(+0.28)
	64	30	64.06	64.00	0.06	0.00	33.00	33.28(+0.28)
	96	26	96.06	96.00	0.06	0.00	34.77	34.98(+0.21)
carphone	48	32	48.03	48.01	0.03	0.01	32.59	32.98(+0.39)
	64	30	64.03	64.00	0.03	0.00	33.81	34.12(+0.31)
	96	28	96.01	96.00	0.01	0.00	35.63	35.90(+0.27)
Hall monitor	48	26	48.06	48.01	0.06	0.01	38.51	38.83(+0.32)
	64	24	60.06	60.01	0.06	0.01	39.60	39.90(+0.30)
	96	23	96.02	90.00	0.02	0.00	40.84	41.09(+0.25)
Mother daughter	48	32	48.04	48.00	0.04	0.00	38.47	38.80(+0.33)
	64	26	64.08	64.01	0.08	0.01	39.98	40.28(+0.30)
	96	23	96.11	96.02	0.11	0.02	41.84	42.11(+0.27)
news	48	32	48.09	48.01	0.09	0.01	35.28	35.65(+0.37)
	64	26	64.03	63.99	0.03	0.01	37.28	37.68(+0.40)
	96	23	96.09	95.98	0.09	0.02	39.77	40.02(+0.25)
salesman	48	32	48.11	48.02	0.11	0.02	36.27	36.48(+0.21)
	64	26	64.09	64.00	0.09	0.00	38.21	38.45(+0.24)
	96	23	96.17	96.02	0.17	0.02	40.46	40.74(+0.28)

actual header bits between two algorithms is shown as Fig.5. It is obvious that the number of predicted bits by our algorithm is closer to the actual header bits than JVT-H017, which reveals our algorithm predicts header bits more accurately.

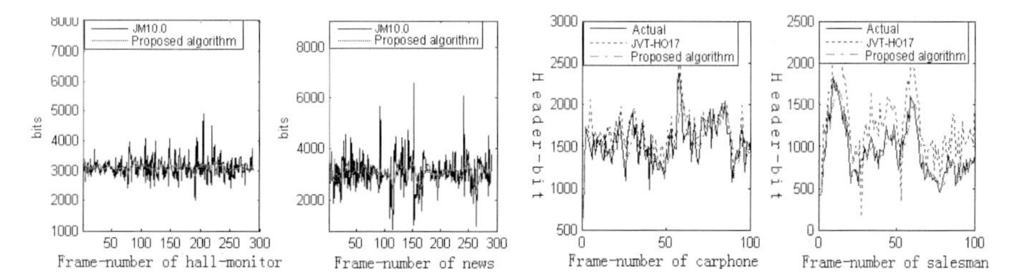

Fig. 4. Compare of bit rate fluctuation

Fig. 5. Compare of header bits

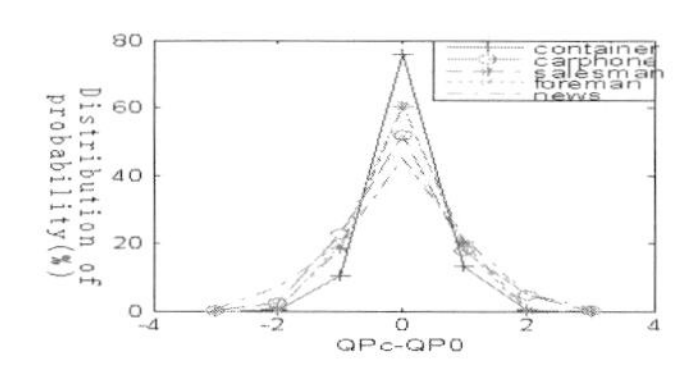

Fig. 6. Probability distribution of $QP_c - QP_0$

Fig.6 is the $QP_c - QP_0$ statistical probability distribution with QP_c ranging from $QP_0 - 3$ to $QP_0 + 3$. We can conclude that the probability of $|QP_c - QP_0| = 3$ is very small, so we set QP_c among the discrete set $[QP_0-2, QP_0+2]$ in our algorithm. It validates the reasonability and feasibility of double-step algorithm.

4 Conclusion

The proposed SLDSRC algorithm adopts double-step coding mechanism to resolves the chicken and egg dilemma radically and introduces new source-bit prediction model, header-bit prediction method and R-Q model to improve control precision and image quality. In addition, the double-step coding style increases coding complexity, however it can not affect coding rate. The algorithm just applied to H.264 baseline-profile encoder, future work will be extended to the main profile and extended profile.

References

1. Lee, H.J., Chiang, T.H., Zhang, Y.Q.: Scalable Rate Control for MPEG-4 Video. J. IEEE Transactions on Circuits and Systems for Video Technology 10(6), 878–894 (2000)
2. Vetro, A., Sun, H., Wang, Y.: MPEG-4 rate control for multiple video objects. J. IEEE Transactions on Circuits and Systems for Video Technology 9(1), 186–199 (1999)
3. MPEG-2 Test Model 5, Doc. ISO/IEC JTC1/SC29 WG11/93-400 (April 1993), http://www.mpeg.org/MPEG/MSSG/tm5
4. Li, Z.G., Xiao, L., Zhu, C., Feng, P.: A Novel Rate Control Scheme for Video Over the Internet. In: Proceedings of ICASSP, Florida, USA, pp. 2065–2068 (2002)
5. Li, Z., Pan, F., Pang, K.: Adaptive basic unit step rate control for JVT, JVT–G012. In: Joint Video Team of ISO/IEC and ITU 7th Meeting, Pattaya, Thailand (2003)
6. Ma, S., Li, Z., Wu, F.: Proposed draft of adaptive rate control, JVT–H017. In: Joint Video Team of ISO/IEC and ITU 8th Meeting, Geneva (2003)
7. Jiang, M., Yi, X., Ling, N.: Improved frame step rate control for H.264 using MAD ratio. In: IEEE International Symposium on Circuits and Systems, British Columbia, Canada, pp. 813–816 (2004)
8. Miyaji, S., Takishima, Y., Hatori, Y.: A novel rate control method for H.264 video coding. In: IEEE International Conference on Image Processing, Genoa, Italy, pp. 11-309–11-312 (2005)
9. Jiang, M., Yi, X., Ling, N.: One enhancing H.264 rate control by PSNR-based frame complexity stimation. In: International Conference on Consumer Electronics, pp. 231–232 (2005)
10. Kim, Y.K., He, Z., Sanjit, K.M.: Low-delay rate control for DCT video coding via p-set source modeling. J. IEEE Trans. Circuits and Systems for Video Technology 11(8), 92–940 (2001)
11. Malvar, H.S., Hallapuro, A., Karczewicz, M., et al.: Low-Complexity Transform and Quantization in H.264/AVC. J. IEEE Trans. on Circuits and Systems for Video Technology 13(7), 598–603 (2003)

Adaptively Adjusted Gaussian Mixture Models for Surveillance Applications

Tianci Huang, Xiangzhong Fang, Jingbang Qiu, and Takeshi Ikenaga

Graduate School of Information, Production, and System, Waseda University
N355, 2-7, Hibikino, Wakamatsu, Kitakyushu, Fukuoka, Japan, 808-0135
bond0060@ruri.waseda.jp

Abstract. Segmentation of moving objects is the basic step for surveillance system. The Gaussian Mixture Model is one of the best models to cope with repetitive motions in a dynamic and complex environment. In this paper, an Adaptively Adjustment Mechanism was proposed by fully utilizing Gaussian distributions with least number so as to save the amount of computation. In addition to that, by applying proposed Gaussian Mixture Model scheme to edge segmented image and combining with data fusion method, the proposed algorithm was able to resist illumination change in scene and remove shadows of motion. Experiments proved the excellent performance.

1 Introduction

Background subtraction is a conventional and effective solution to segment the moving objects from the stationary background But in an actual scene, the complex background such as snowy or windy conditions, make the conventional algorithm unfit for the real surveillance systems. Stauffer and Grimson [1][2] proposed to model each pixel by a mixture of Gaussians. Saeid et al. [3] proposed an improved method based on GMM, but it was not able to cope with illumination change and shadow problem. Huwer et al. [4] proposed a method of combining a temporal difference method with an adaptive background model subtraction scheme to deal with lighting changes. J. Zhan et al. [5] analyzed the foreground by GMM and operated the classification based on SVM method, high accuracy were achieved, but with a high computation cost. To save huge computation load for surveillance system, the Adaptive Adjustment Mechanism was proposed. Moreover, laplacian edge segmented image was utilized in our method as the input of the modified GMM. To improve the quality of segmentation, data fusion mechanism is put forward to make up the lost information. The remaining parts for this paper are arranged as follows. Section 2 introduces the conventional GMM procedure and describes Adaptively Adjustment Mechanism. Section 3 specifies analysis of applying edge-based image segmentation and data fusion scheme. Section 4 and Section 5 present the experimental results and conclusion respectively.

S. Boll et al. (Eds.): MMM 2010, LNCS 5916, pp. 689–694, 2010.

2 Adaptive Adjustment Mechanism for Gaussian Models

2.1 Gaussian Mixture Model

According to the original GMM, the pixel process is considered a time series of vectors for color images. The algorithm models the recent history of each pixel as a mixture of K Gaussian distributions. A match is found if the pixel value is within 2.5 standard deviation of a distribution. If current pixel value matches none of the distributions, the least probable distribution is updated with the current pixel values, a high variance and low prior weight. After the prior weights of the K distributions are updated the weights are renormalised. The changing rate in the model is defined by $1/\alpha$. α stands for learning rate. And parameters for matching distribution are updated. The Gaussians are ordered based on the ratio of ω/σ. This increases as the Gaussian's weight increases and its variance decreases. The first B distributions accounting for a proportion of the observed data are defined as background.

2.2 Adaptive Adjustment Mechanism

Even though K (3 to 5) Gaussian distributions are capable of modeling a multimodal background, the huge number of total Gaussian distributions induced a great computational load for surveillance system.

In fact not all the pixels of the background objects moved repetitively or changes diversely all the time. For the areas where less repetitive motion occurs, such as the ground, houses and parking lot in the scene of Fig. 1(a), it is easy to find that the first and second highest weighted Gaussians (Fig. 1(b) and (c)) are

Fig. 1. (a) The 363th frame from PetsD2TeC2 [6]; (b) M_1st_WG(The Mean of 1st Weighted Gaussians); (c) M_2nd_WG; (d) M_3rd_WG; (e) M_4th_WG; (f) Foreground mask by GMM

adequate to model the mult-possibilities of background variability. So adaptive Adjustment Mechanism was proposed to drop unnecessary Gaussians component which contributed less to the multi-possibilities for modeling background, then adaptive number of distribution could be adopted for different pixels according to their corresponding value changing history. The update of weight, mean and variance for our proposal is based on online EM algorithm [7].

E-step: As the online EM algorithm does, it begins from estimating of the Gaussian Mixture Model by expected sufficient statistics, which is called E-step. Due to the unpredictable possibilities for the complexity of background pixel, and the first L frames is very important for Gaussian models to dominant background component and achieve stable adaptations. And then keep the number of Gaussians models, K, fixed during E-step. Experiments also show these could provide a good estimation which helps to improve the accuracy for M-step process. For initialization part, we define a parameter $N_{i,j}$ to record number of Gaussian models for the pixel at the position *(i,j)* in each frame, also a parameter called *sum_match* to record the sum of matches for a particular Gaussian distribution.

M-step: The L-recent window update equations give priority over recent data therefore the tracker can adapt to changes in environment. When a new pixel value comes, check it against first $N_{i,j}$ Gaussian distributions in turn. If the i_{th} distribution G_i matches, update parameters as M-step in EM does. After that, we compare the value of ω_i/σ_i with value of $\omega_{i-1}/\sigma_{i-1}$. If $\omega_i/\sigma_i > \omega_{i-1}/\sigma_{i-1}$, exchange the order of G_i and G_{i-1} and operate $i = i - 1$, repeat it until $i = 1$ or $\omega_i/\sigma_i \leq \omega_{i-1}/\sigma_{i-1}$. Or else no match found, operate as follows:

$$N_{i,j}^k = N_{i,j}^{k-1} + 1, \ \ if \ \ N_{i,j} < K; \ \ N_{i,j}^k = K, \ \ if \ \ N_{i,j} = K \tag{1}$$

then replace the mean value of the N_i, j_{th} distribution with current pixel. After that the Gaussians are eliminated from least updated ones according to two parameters: value of weight, which represent the time proportions that those colors stay in the scene and *sum_match*, which takes for the percentage of importance in K guassians to dominant background component from history.

$$\omega_k = \frac{\omega_k}{\sum_{i=1}^{N_{i,j}} \omega_i}, \quad k = 1, 2, ...N_{i,j} \tag{2}$$

where $N_{i,j}$ is the number of left Gaussians.

As this adaptive Adjustment Mechanism processes with GMM, the stable value pixels did not need K Gaussians modeling for adaptation. The total number of Gaussians of PetsD2TeC2 with a resolution of 384x288, 2821 frames was experimented shown as Fig. 2. When comes to M-step, obvious decrease occurred. Especially when larger the K is, more unnecessary Gaussians were eliminated.

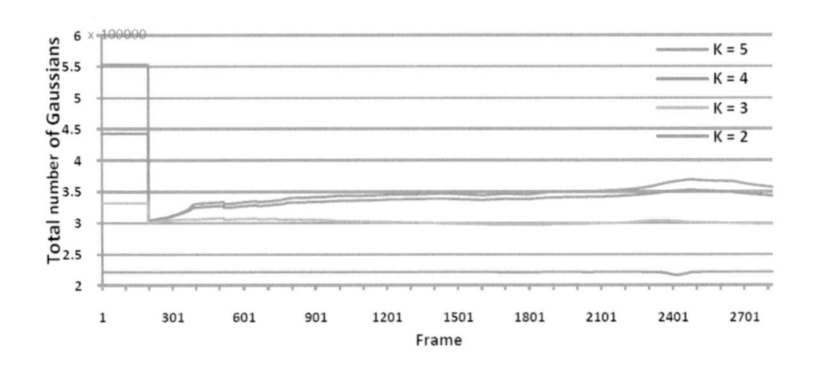

Fig. 2. The total number of Gaussians for each frame based on different value of K

3 Laplacian Edge Detection and Data Fusion

Even though GMM is capable of dealing with complex environment especially unpredictable repetitive motion, two major problems influence the detection accuracy when applying GMM to surveillance system: illumination changes and shadow of moving objects.

In our proposal, the well-known Laplacian edge detection method was utilized since it runs very fast and achieves good results. And then the improved GMM method mentioned above was applied to the mask generated by laplacian edge detection. Because laplacian operator could enhance the effect of edges of object, so the influence from illumination and shadow area were weakened intensively (refer to Fig. 3(c)). Considering this point of advantage, the edge segmented grey level information from video stream was proposed to act as input of improve GMM to avoid illumination influence and shadows. Meantimely, even though edge of motion was clear and shadow of people was removed, inside hole of motion appeared in the detection mask. To solve this problem, we proposed data fusion scheme.

We named the mask by applying GMM on RGB color space as *Mask_RGB*, and the mask by applying GMM on laplacian edge segmented image as *Mask_Edge*. *Mask_RGB* contains all the information of moving objects except repetitive motion, and also misclassified foreground. In the other hand, *Mask_Edge* excludes the misclassified foreground pixels, but it lost information inside of motion. In the proposal, *Mask_Edge* takes an important role as a criterion to confirm the foreground pixels in *Mask_Edge* whether are correctly classified. For a foreground pixel in $Mask\_RGB(i,j)$, neighboring foreground pixels in a 6x6 region centered as $Mask\_Edge(i,j)$ is checked, we define a threshold, which equals to 6 for indoor and 3 for outdoor. And compared with this number we can determine whether $P(i,j)$ should belong to foreground or background.

4 Experimental Results

Our experimental results is based on the 4 outdoor sequences and 3 indoor sequences from [6]. Proposed background modeling method is evaluated by the metric proposed by Black et al. in [8]. Through this method for data fusion, some noise can also be removed. As the Fig. 3 (d) below shows, it is clear that shadow was removed from motion, and illumination changes did not influence

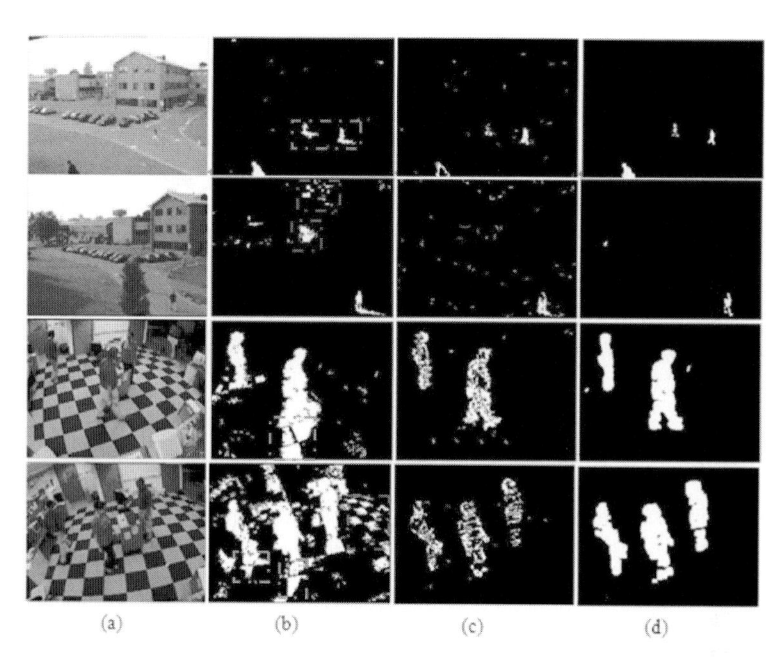

Fig. 3. Comparison of foreground: (a) original frames; (b) Foreground mask of GMM_RGB (shadow noted by green contour, and influence caused by illumination changes by red). (c) Foreground mask of GMM_RGB. (d) Foreground mask by proposed algorithm.

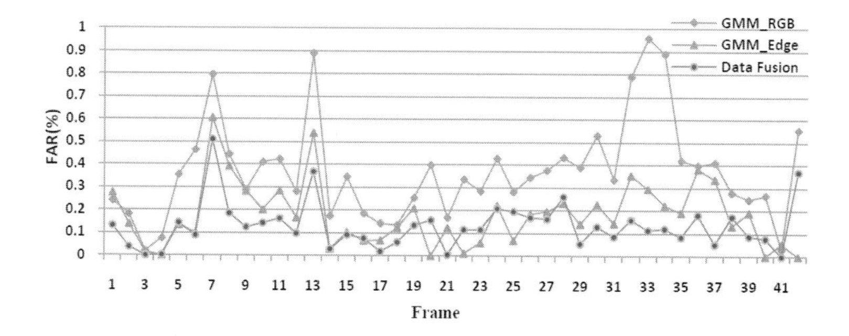

Fig. 4. False alarm rate for every 60th frame of PetsD2TeC2 sequence

the segmentation results. Compared with Fig. 3(c), inside information of moving objects was filled up. For indoor sequence, foreground is a little shattered, if adding the postprocessing filter in final step, results would be better. The FAR (False Alarm Rate) is shown as Fig. 4.

5 Conclusion

This paper presented an effective and efficient algorithm based on Gaussian Mixture model for surveillance system. An adaptive Adjustment Mechanism is proposed to reduce the number of Gaussian distributions. Additionally, aiming at excluding the influence by illumination and shadow problem, we proposed to apply our improved GMM on the laplacian edge segmented image, and a data fusion mechanism is put forward to solve the problem of losing inside motion information. The results of segmentation by proposal consequently proved its effectiveness and efficiency. Experiments show the detection rate and false alarm rate between different methods, which validated the improvement on detection accuracy and segmentation quality.

Acknowledgments. This work was supported by fund from MEXT via Kitakyushu innovative cluster projects and CREST, JST.

References

1. Stauffer, C., Grimson, W.E.L.: Adaptive background mixture models for real-time tracking. In: IEEE Intl. Conference on Computer Vision and Pattern Recognition, pp. 246–252 (2007)
2. Stauffer, C., Grimson, W.E.L.: Learning patterns of activity using real-time tracking, vol. 22, pp. 747–757 (2000)
3. Saeid, F., Hamed Moradi, P., Hamed, B.: Multiple object tracking using improved gmm-based motion segmentation. In: Electrical Engineering/Electronics, Computer, Tele-communications and Information Technology International Conference, vol. 2, pp. 1130–1133 (2009)
4. Huwer, S., Niemann, H.: Adaptive change detection for real-time surveillance applications. In: 3rd IEEE Workshop on Visual Surveillance, pp. 37–45 (2000)
5. Zhan, J., Chen, C.: Moving object detection and segmentation in dynamic video backgrounds. In: IEEE conference on computer vision, pp. 64–69 (2007)
6. Brown, L.M., Senior, A.W., Tian, Y., Connell, J., Hampapur, A., Shu, C., Merkl, H., Lu, M.: Performance evaluation of surveillance systems under varying conditions. In: IEEE International Workshop on Performance Evaluation of Tracking and Surveillance (2005)
7. KaewTraKulPong, P., Bowden, R.: An improved adaptive background mixture model for real-time tracking with shadow detection. In: 2nd European Workshop on Advanced Video Based Surveillance Systems, AVBS 2001 (2001)
8. James Black, T.E., Rosin, P.: A novel method for video tracking performance evaluation. In: International Workshop on Visual Surveillance and Performance Evaluation of Tracking and Surveillance, pp. 125–132 (2003)

Estimating Poses of World's Photos with Geographic Metadata

Zhiping Luo, Haojie Li, Jinhui Tang, Richang Hong, and Tat-Seng Chua

School of Computing, National University of Singapore
{luozhipi,lihj,tangjh,hongrc,chuats}@comp.nus.edu.sg

Abstract. Users can explore the world by viewing place related photos on Google Maps. One possible way is to take the nearby photos for viewing. However, for a given geo-location, many photos with view directions not pointing to the desired regions are returned by that world map. To address this problem, prior know the poses in terms of position and view direction of photos is a feasible solution. We can let the system return only nearby photos with view direction pointing to the target place, to facilitate the exploration of the place for users. Photo's view direction can be easily obtained if the extrinsic parameters of its corresponding camera are well estimated. Unfortunately, directly employing conventional methods for that is unfeasible since photos fallen into a range of certain radius centered at a place are observed be largely diverse in both content and view. Int this paper, we present a novel method to estimate the view directions of world's photos well. Then further obtain the pose referenced on Google Maps using the geographic Metadata of photos. The key point of our method is first generating a set of subsets when facing a large number of photos nearby a place, then reconstructing the scenes expressed by those subsets using normalized 8-point algorithm. We embed a search based strategy with scene alignment to product those subsets. We evaluate our method by user study on an online application developed by us, and the results show the effectiveness of our method.

1 Introduction

Google Maps is a widely used online service to explore world's places. However, current service mainly relies on the geographical metadata of photos result in simply considering the photos nearby a place are exactly related to the geographic content. We can observe the limitations of such application. On one hand, photos taken by non location-aware devices may be wrongly placed on the map by uploaders manually. On the other hand, even the photos are correctly placed (manually by users or automatically from EXIF tags), their viewing directions may not be pointing to the desired region. Therefore, many photos without truly poses in terms of position and view direction are returned.

Prior know the poses in terms of position and view direction of photos, then let Google Maps returns only nearby photos with view direction pointing to the target places is one of the most feasible way to address the above problems.

S. Boll et al. (Eds.): MMM 2010, LNCS 5916, pp. 695–700, 2010.

Assume the geographic metadata is correct, It can be easy to get the photo's position expressed by latitude and longitude coordinate referenced by a world map from the metadata. In this paper we consider Google Maps be the case as its popularity of use. We then can further obtain the photo's view on the map by geo-registering its view direction estimated in the camera coordinate system. To estimate the view direction, we can first estimate the camera rotation and translation via scene reconstruction. When there are significant overlaps among photos [1], rotation and translation can be robustly estimated.

The most used technique for scene reconstruction is described as follows. First compute correspondences among images based on feature matching. Second, employ RANSAC [2] to decide the inliers (the actual correspondences) when tuning the estimation of fundamental matrix [3]. At last, compute the camera extrinsic parameters, such as rotation and translation, based on the assumption of fixed intrinsic parameters.

However, those photos in the Internet that we can easily obtain are largely diverse both in image content and view. We randomly selected 10 photos fallen into the range of 20-meter radius centered at a region within "Acropolis, Athens, Greece", and calculate the inliers among them. unfortunately, Only average 40% inlier rate are obtained by point-to-point matching SIFT [4] features with well configure and under 10^6 iterations in RANSAC for tuning the estimations of fundamental matrices. Since current techniques for scene reconstruction heavily depend on inlier rate so that such a low value cannot satisfy the estimation of camera's extrinsic parameters well. Since scene reconstruction can be well done if there are significant overlaps among photos, We propose to divide the whole photos a number of subsets. In each set, the photos are more convinced to be visually relevant, meanwhile they represent an underlying scene consisting of those photos. Therefore, those underlying scenes might be well reconstructed since their own photos are overlapped so better that will product higher inlier rate. Besides, compared to the total set which is with too complex scene, those underlying scenes out of the whole scene are more robust to be reconstructed and aligned in their own camera coordinate systems. To generate such subsets, we cannot resort to clustering algorithm like k-mean [5] to automatically cluster a number of photo sets because wrong matches based on visual similarity may occur due to the diversity. Meanwhile, although photos may be associated with meaningful textual tags, there is still no warrantee of that photos with the same salient phrase are visually relevant due to the tag noise [6].

In this paper, We build a search infrastructure to generate the subsets of photos with respect to underlying scenes. We embed a scene alignment algorithm using flow into the infrastructure, in particular using the SIFT flow [7] algorithm as we choose SIFT features to perform feature matching. we consider the photos nearby a place as an image database, given a photo that needs to be estimated its view direction, we find the best matched photos with respect to the given photo by searching in the database, thus generate a set of photos with significant overlaps. Note that a photo may occurs in multiple scenes, we choose the scene where that photo obtains the highest inlier rate when reconstructing the scene.

For each scene's reconstruction, we use RANSAC based normalized 8-point algorithm [9] to estimate the camera rotation, translation for simplicity. Although there are more robust algorithms can product higher inlier rate, such as [8], cost too expensive computation for large set of photos. Unfortunately, there is always placing a large amount of photos nearby a place on the world map. We do not perform ground evaluation because it is so difficult to get the ground truth for that whether a photo is exactly shot to the target place. Instead of we developed an online system for world exploration to perform the user study on our method.

2 The Methodology

Our goal is to find a set of matched photos for each given query example, facilitating the estimation of the camera rotation, translation. We describe our method in three steps as follows.

2.1 Search Infrastructure

Because objects present in photos are in different spatial location and captured at different scale in the case of Google Maps, we believe that the conventional methods for building a CBIR (query by example) system are not sufficient in effectiveness to return the most relevant results at most. To design a more robust CBIR, we search the photos by computing alignment energy using the flow estimation method to align photos in the same scene, where a photo example is aligned to its k-nearest neighbors in the photo collection. Since the use of SIFT features gives birth to robust matching across diversity, we employ the SIFT flow [7] algorithm to align them.

We introduce the search infrastructure as follows. We first extract SIFT features of all photos in the database, and index them using k-d tree algorithm. Given a query example of an underlying scene, we aim to find a set of matched photos from the database of a place to reconstruct the scene. The search results is returned by SIFT-based matching, and enhanced by scene alignment. In our method, we adopt k-nearest neighbors to search, and k=50 is used in our case, although other values also can be used. Figure 1 shows the top 10 nearest

Fig. 1. An example of the search results returned by our search paradigm

neighbors of a query (marked with most left box). The number below each photo is the minimum energy of alignment obtained by the SIFT flow algorithm, which is used to rank the search results. As can be seen, this set of photos is promising to robustly estimate the extrinsic parameters of the corresponding cameras.

2.2 Estimation of Rotation, Translation

For each set of photos, we estimate the camera rotation, translation by using normalized 8-point algorithm to estimat the fundamental matrices, and apply RANSAC to detect the inliers for robust estimation of each fundamental matrix. The steps are detailed below.

1. For each pair of photos consists of p and $p^{'}$.
2. Determine the correspondences set C by SIFT features matching between p and $p^{'}$.
3. Randomly choose 8 correspondences and compute an initial fundamental matrix $\mathbf{F}$ using normalized 8-point algorithm, apply RANSAC to detect outliers, and determine inlier rate induced from current $\mathbf{F}$, if the appropriate inlier rate is achieved, the current $\mathbf{F}$ is the robust one, else repeat step 3. The $\mathbf{F}$ can be written as:

$$\mathbf{F} := \mathbf{K}^{'-1}\mathbf{T}\mathbf{R}\mathbf{K}^{-1} \tag{1}$$

where $\mathbf{T}$ is the translation vector and $\mathbf{R}$ is a 3×3 rotation matrix.

Using a pinhole camera model, the 3D view direction V_p of the photo p can be obtained as $V_p = \mathbf{R}^{'} * [0\ 0\ -1]^{'}$, where $'$ indicates the transpose of a matrix or vector. We retain the x and y components of V_p as the 2D view direction of p. The 3D position is $-R' * T$, so that the 2D position is the remains of x and y components too.

2.3 Geo-registration of View Direction

Assume the geographic metadata is correct, we will use it expressed as latitude and longitude coordinate as the photo's position on the map. In this section, we begin geo-register the views of photos on the map by using the metadata and the whole procedure is illustrated in Figure. 3. In a reconstructed scene, for each pair of photos denoted as p and i, their coordinates and directions are denoted as L_{wcs_p} and L_{wcs_i}, V_{wcs_p} and V_{wcs_i} respectively. We clean the geographic metadata and get the GPS coordinates in terms of latitude and longitude as the photo's position on the map, which are marked as L_{geo_p} and L_{geo_i} in the figure. Since the angel demonstrated as ∂_p between the vector representing view direction and the vector linking two positions in the specific coordinate system is fixed. So we can register the directions in the camera coordinate system wcs onto the geographic coordinate system geo referenced by the fixed angels.

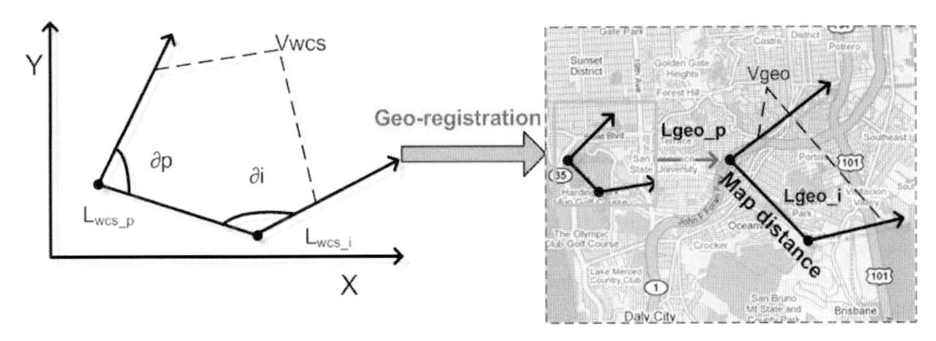

Fig. 2. Procedure of view direction's geo-registration

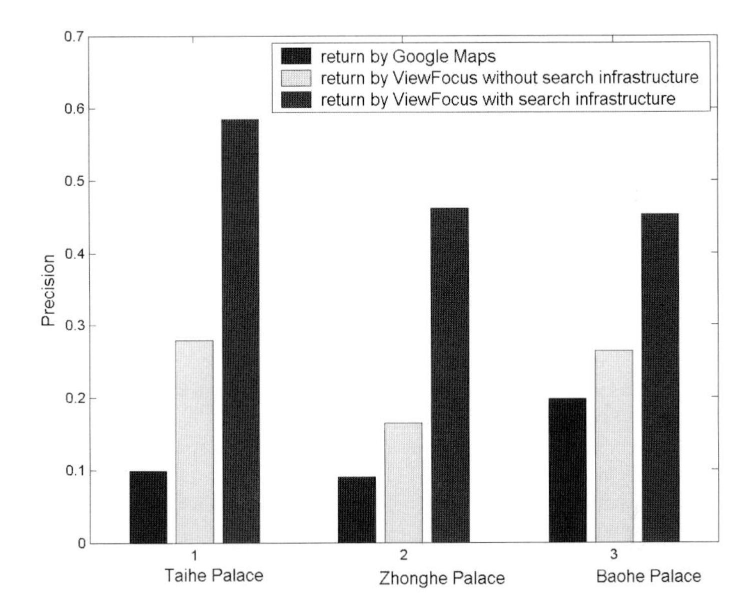

Fig. 3. Comparisons of precision evaluated on *ViewFocus*

3 User Study

Since it is difficult to obtain the ground truth about which photo nearby a place is exactly shot to the target region. We use a developed online system [10] named *ViewFocus* to evaluate our method by user study. This system allows users draw a rectangle on the map interface, then it returns the photos with view direction pointing to the enclosed region. Therefore, we can manually examine how many precise photos are returned so that to perform the user study. We use *precision* as the evaluation metrics. The settings of the evaluation are designed as follows: (a) examine how many related photos nearby the target regions on Google Maps; (b) examine how many related photos returned by the system using view direction filtering without the search infrastructure; and (c) examine how many related photos returned by the system using view direction filtering with the search infrastructure. In order to be more be judgeable about the photos' view directions, we select three famous buildings including "Taihe Palace", "Baohe Palace" and "Zhonghe Palace" in "Forbidden City" in Beijing, China. Therefore, assessors are more easy to justify whether the photos are shot to the target regions. We set the radius of the range centered at the target region to 100-meter in our evaluation. Therefore, there are 312, 383, 365 photos nearby these three buildings respectively. The comparisons of precision are presented in Figure 3. As can be seen, our method outperforms the others.

4 Conclusion

We presented a method to estimate the pose of the world's photos. We proposed a novel search infrastructure based on SIFT flow to generate number of sets of photos for scene reconstruction. Finally, we demonstrated the effectiveness by user study on the method proposed about the view direction and position of a photo.

References

1. Schaffalitzky, F., Zisserman, A.: Multi-view Matching for Unordered Image Sets, or How Do I Organize My Holiday Snaps? In: Heyden, A., Sparr, G., Nielsen, M., Johansen, P. (eds.) ECCV 2002. LNCS, vol. 2350, pp. 414–431. Springer, Heidelberg (2002)
2. Fischler, M.A., Bolles, R.C.: Random sample consensus: a paradigm for model fitting with applications to image analysis and automated cartography. Morgan Kaufmann, San Francisco (1987)
3. Luong, Q., Faugeras, O.: The fundamental matrix: Theory, algorithms and stability analysis. Int. J. Comput. Vision 17(1), 43–75 (1996)
4. Lowe, D.G.: Distinctive Image Features from Scale-Invariant Keypoints. Int. J. Comput. Vision 60(6), 91–100 (2004)
5. Kanunqo, T., Mount, D.M., Netanyahu, N.S., Piatko, C.D., Silverman, R., Wu, A.Y.: An Efficient k-Means Clustering Algorithm: Analysis and Implementation. IEEE Trans. Pattern Anal. Math. Intell. 24(7), 881–892 (2002)
6. Chua, T.-S., Tang, J., Hong, R., Li, H., Luo, Z., Zheng, Y.-T.: NUS-WIDE: A Real-World Web Image Database from National University of Singapore. In: Proc. of ACM Conf. on Image and Video Retrieval, CIVR 2009 (2009)
7. Liu, C., Yuen, J., Torralba, A., Sivic, J.: SIFT Flow: Dense Correspondence across Different Scenes. In: Forsyth, D., Torr, P., Zisserman, A. (eds.) ECCV 2008, Part III. LNCS, vol. 5304, pp. 28–42. Springer, Heidelberg (2008)
8. Urfalioḡlu, O.: Robust Estimation of Camera Rotation, Translation and Focal Length at High Outlier Rates. In: 1st Canadian Conference on Computer and Robot Vision (CRV 2004), pp. 464–471. IEEE Computer Society, Los Alamitos (2004)
9. Hartley, R.: In defence of the eight-point algorithm. IEEE Trans. Pattern Anal. Math. Intell. 19(6), 580–593 (1997)
10. Luo, Z., Li, H., Tang, J., Hong, R., Chua, T.-S.: ViewFocus: Explore Places of Interests on Google Maps Using Photos with View Direction Filtering. To appear in ACM Multimedia 2009 (2009)

Discriminative Image Hashing Based on Region of Interest*

Yang Ou[1], Chul Sur[2], and Kyung Hyune Rhee[3,**]

[1] Department of Information Security, Pukyong National University,
599-1, Daeyeon3-Dong, Nam-Gu, Busan 608-737, Republic of Korea
ouyang@pknu.ac.kr

[2] Department of Computer Science, Pukyong National University
kahlil@pknu.ac.kr

[3] Division of Electronic, Computer and Telecommunication Engineering,
Pukyong National University
khrhee@pknu.ac.kr

Abstract. In this paper, we propose a discriminative image hashing scheme based on Region of Interest (ROI) in order to increase the discriminative capability under image content modifications, while the robustness to content preserving operations is also provided. In our scheme, the image hash is generated by column-wisely combining the fine local features from ROI and the coarse global features from a coarse represented image. Particularly, a small malicious manipulation in an image can be detected and can cause a totally different hash. The experimental results confirm the capabilities of both robustness and discrimination.

Keywords: Discriminative Image Hashing, Region of Interest, Interest Point Detector.

1 Introduction

Traditional data integrity issues are addressed by cryptographic hash functions (e.g.MD5, SHA-1) or message authentication code, which are very sensitive to every bit of the input message. However, the multimedia data always undergoes various acceptable manipulations such as compression. The sensitivity of traditional hash functions could not satisfy these perceptual insignificant changes. The image hashing takes into account changes in the visual domain and emerged rapidly. Particularly, the image hashing should be robust against image Content Preserving Operations (CPOs) while highly discriminative to Content Changing Operations (CCOs).

Recently, an image hashing scheme based on Fourier Mellin Transform (FMT) was proposed in [1] to make the hash resilient to geometric and filtering operations. A novel histogram shape-based image hashing scheme [2] is proposed

* This work was supported by Pukyong National University Research Fund in 2009(PK-2009-50).

** Corresponding author.

S. Boll et al. (Eds.): MMM 2010, LNCS 5916, pp. 701–706, 2010.

more recently by employing the insensitivity of histogram shape to geometric distortions. The experimental results in [2] depicted that it is much more robust than previous approaches [1,3] against CPOs. However, even though most image hashing schemes provide a satisfactory robustness property, the sensitivity to CCOs, i.e, distinguishing certain malicious distortions, is still somewhat indefinite.

In this paper, aiming to achieve a good discriminative capability to CCOs, we propose a discriminative image hashing scheme based on Region of Interest (ROI). We show that the proposed scheme is also robust to various content preserving operations, such as image filtering, strong noises and compression with very high ratio. A solid comparison results of discriminative capability and Receiver Operating Characteristic (ROC) curves are presented among several existing schemes and our schemes. The results indicate that our proposed scheme performs excellent capabilities on both robustness and discrimination.

2 Discriminative Image Hashing

Our scheme is based on Region of Interest in which the local features of an image can be captured. The global and local features of the image are smartly combined and a discriminative hash is generated which is robust against CPOs, as well as sensitive to CCOs. The scheme includes three steps: (1) ROI construction based on interest points; (2) Global and local feature extraction; (3) Image hash generation. Figure 1 shows the coding pipeline of our scheme.

The ROI Construction Based on Interest Points. Since the local image structure around an interest point is rich, a circular region centered by the point can be captured as a ROI which reflects local features. In order to robustly detect interest points, we firstly resize the input image then apply an order statistic filter for denoising. Wavelet transform is performed sequently to downsample the image. The famous Harris corner detector is empoyed to detect interest points in LL sub-band which is a coarse representation of the original image I.

In particular, given n which is the number of ROIs, the first n maximum corner responses are selected as n interest points. Following, restore n points in the same positions in I. Finally, the n circular ROIs are constructed by drawing

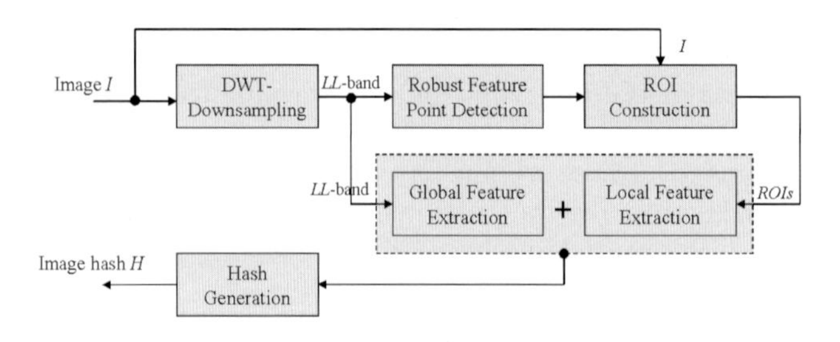

Fig. 1. The coding pipeline of the proposed image hashing scheme

n circles centering on each interest point with the radius r. Note that the minimum distance between any two points should be larger than r in order to avoid too much region is overlapped between two ROIs. Regarding to n, it should not be too large since more ROIs can degrade the robustness of our hash. A proper selection of the above parameters can be gained through experimental trials. The satisfactory results are shown in Section 3.

Global and Local Feature Extraction. The global and local features are extracted by performing FMT because of its robust principles, while the previous image hashing scheme [1] can neither capture the local image information nor provide the discrimination of small malicious content changes. For the sake of simplicity, we denote n ROI and LL subband as $R_i(x, y), i = 1, ..., n+1$, where R_{n+1} is LL subband containing global feature. In each $R_i(x, y)$, FMT is firstly performed to transform $R_i(x, y)$ into Fourier-polar domain $R_i(\rho, \theta)$. Along the θ-axis, we calculate

$$\overline{m}_{i\rho} = \frac{1}{360} \sum_{\theta=1}^{360} R_i(\rho, \theta) \tag{1}$$

where $\rho = 1, ..., \tilde{r}$. Therefore, for each $R_i(x, y)$ we get a feature vector $M_i = [\overline{m}_{i1}, \overline{m}_{i2}, ..., \overline{m}_{i\rho}, ..., \overline{m}_{i\tilde{r}}]$. Consequently, given n ROIs, the total features can be represented as $\mathbf{V} = [M_1, M_2, ..., M_i, ..., M_{n+1}]$. The first n vectors are local feature, while the last one is the global feature. It is worthy to clarify that $\tilde{r}$ may not be equal to r. The former $\tilde{r}$ is the radius in polar coordinate, while r denotes the radius of ROI in image coordinate. In the case of r is too large or too small to fit the size of hash, bilinear interpolation is used in polar coordinate to generate a satisfactory $\tilde{r}$.

Image Hash Generation. Given $n + 1$ feature vectors, the vectors are reformed as the matrix

$$\mathbf{V}_{n+1,\tilde{r}} = [M_1, M_2, ..., M_n, M_{n+1}]^\top = \begin{bmatrix} \overline{m}_{11} & \overline{m}_{12} & \cdots & \overline{m}_{1\tilde{r}} \\ \overline{m}_{21} & \overline{m}_{22} & \cdots & \overline{m}_{2\tilde{r}} \\ \vdots & \vdots & \ddots & \vdots \\ \overline{m}_{n,1} & \overline{m}_{n,2} & \cdots & \overline{m}_{n,\tilde{r}} \\ \overline{m}_{n+1,1} & \overline{m}_{n+1,2} & \cdots & \overline{m}_{n+1,\tilde{r}} \end{bmatrix} \tag{2}$$

where the first n rows are from n corresponding ROIs, and the last row is from the coarse represented whole image. In fact, the local modification in the image is indicated in horizontal direction of $\mathbf{V}$, whereas a corresponding effect can be reflected in vertical direction. Specifically, we propose a *column-wise combination method* to generate image hash. For each column in $\mathbf{V}$, we calculate the weighted sum

$$h_j = \sum_{i=1}^{n+1} w_{ij} \overline{m}_{ij} \tag{3}$$

where w_{ij} is pseudo-random weight number generated by a secret key k. The final hash vector is denoted as $\mathbf{H} = [h_1, h_2, ..., h_j, ..., h_{\tilde{r}}]$ and the $\tilde{r}$ is actually the size of final hash. Note that each element in $\mathbf{H}$ is affected by all ROIs and the coarse represented image. Even only one ROI feature vector or the coarse image feature is changed, the weighted sum of each column vector will be changed simultaneously. Therefore, all elements in $\mathbf{H}$ are changed as long as one ROI is changed. Finally, $\mathbf{H}$ can be quantized and represented as binary digit by using Gray code if necessary.

3 Evaluations

For the results presented in this section, the following settings are chosen: besides the standard test images, a natural image database [4] is used and all images are cropped to 512×512 and converted into 8-bit gray-scale, 3-level CDF wavelet transform is applied to downsample the image. To achieve a good robustness, $r = 64$ and $n = 5$ ROIs are selected after our simulations. The hash vector contains $\tilde{r} = 32$ elements. All vectors in decimal representation are quantized in the range $[0, 255]$ and are represented by using 8-bit Gray code if necessary. The Mean Square Error (MSE) and normalized Hamming distance are used as measure matrix of decimal hashes and binary hashes, respectively.

3.1 Evaluation of Robust and Discriminative Capability

The robustness of our scheme actually depends on the invariance of ROI construction. Given an original image, we generate 80 similar versions by manipulating the original image according to a set of CPOs listed in Table 1. Regarding to CCOs, 16×16 zero blocks are used to randomly replace the same size blocks in the original image. Figure 2 shows some examples of ROI construction under CPOs and CCOs. It can be observed that the ROIs almost cover the same regions in each similar image, which means *the interest point based ROIs can be robustly captured even if the image is strongly processed.* 80 hashes are calculated from those CCO images. Another 80 hashes are also generated by using 80 natural images which are randomly selected from the image database. Furthermore, the same processes are done by using other two novel existing schemes, FMT-based image hashing [1] and histogram-based image hashing [2].

Table 1. Types and Parameters of Content Preserving Operations

CPOs	Parameters	CPOs	Parameters
JPEG	Quality factor 10:10:100	Gaussian filter	Filter mask size 2:1:11
JPEG2000	Ratio 0.02:0.02:0.2	Average filter	Filter mask size 2:1:11
Gaussian noise	Variance 0.01:0.01:0.1	Median filter	Filter mask size 2:1:11
Speckle noise	Variance 0.01:0.01:0.1	Rescaling	Percentage 0.8:0.04:1.2

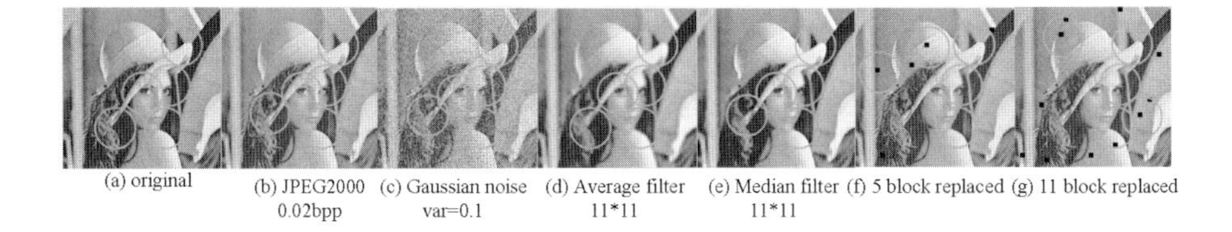

Fig. 2. Examples of ROI constructions under CPOs and CCOs

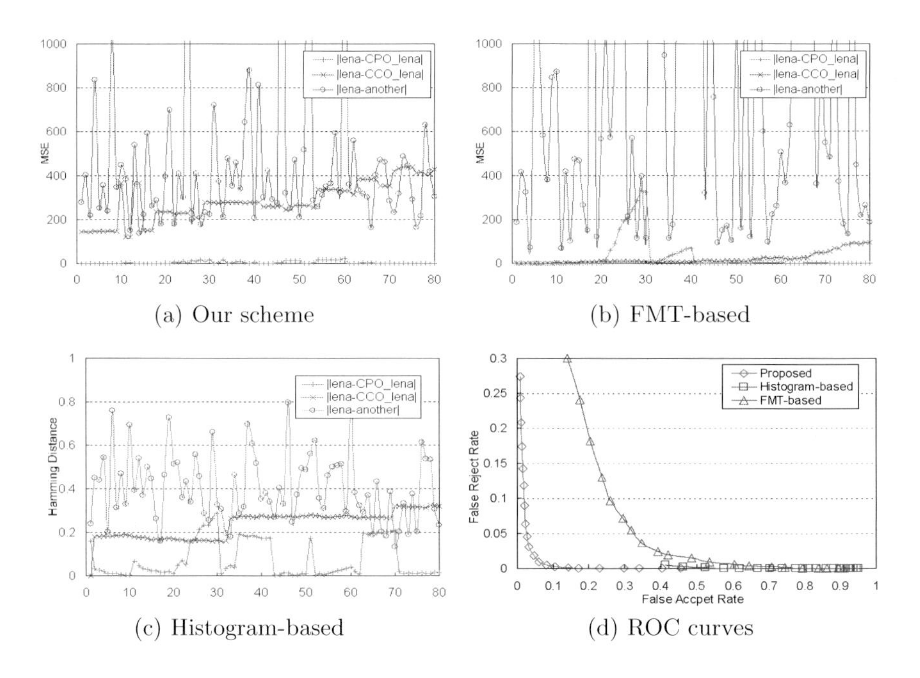

Fig. 3. Comparison of Robust and discriminative capability

The above results, as well as the hashes under CPOs, generated by performing the different image hashing schemes are depicted in Figure 3(a-c). It can be observed that our scheme obtains a very clear gap of hashes between CPOs and CCOs. That is, the proposed scheme performs a better discriminative capability under different kinds of image content operations. On the other hand, the results of other two schemes shown in Figure 3(b) and (c) are mixed and confused. The ambiguity between CPOs and CCOs will cause an inaccurate authentication result.

3.2 Statistical Analysis via ROC Curves

A detailed statistical comparison by using ROC curves is presented here. Firstly two image sets are constructed under CPOs and CCOs, respectively. 500 images are independently operated by the following CPOs: 1) JPEG with QF=20, 2)

JPEG2000 with compression ratio=0.05, 3) Gaussian noise with $\sigma = 0.03$, 4) Speckle noise with $\sigma = 0.03$, 5) Gaussian filter with $\sigma = 1$ and 5×5 mask, 6) Average filter and median filter with 5×5 masks, respectively, 7) Rescaling with 120%. Regarding to the CCO image set, $10, 20, ...80$ 16×16 blocks in original image are randomly replaced by zero blocks with the same size. Therefore, $500 \times 8 = 4000$ similar versions and also 4000 different versions are generated. The False Reject Rate (FRR) and False Accept Rate (FAR) are recorded as follows;

$$FRR(\varepsilon_1) = Probability(D(h(I,k), h(I_{simi}, k)) < \varepsilon_1) \tag{4}$$

$$FAR(\varepsilon_2) = Probability(D(h(I,k), h(I_{diff}, k)) > \varepsilon_2) \tag{5}$$

where $D(*)$ is the normalized hamming distance between two binary hash strings.

For fixed ε_1 and ε_2, the probabilities are computed as (4) and (5) by comparing the original and CPO/CCO images. We use $\varepsilon_2 = 2\varepsilon_1$, where ε_1 is varied in the range $[0.1, 0.3]$. The corresponding ROC curves of three different schemes are shown in Figure 3(d). Obviously, it is indicated that both the FRR and FAR of our scheme are much lower than other two schemes. Specifically, given a fixed ε_1 and ε_2, our scheme can distinguish CPO/CCO images with a much higher accuracy. Whereas the results from other two schemes are not satisfied especially for the histogram-based scheme which has a quite confused result.

4 Conclusion

In this paper, we have proposed a discriminative image hashing scheme based on ROI. The simulation results show that our scheme is not only resistent to various content preserving manipulations, but also discriminative to content changing operations. A statistical analysis via ROC curves has been also demonstrated to confirm the performance capability. The further work will focus on enhancing the robustness against geometric operations of our scheme. Moreover, the theoretical security analysis as in [5] will be further considered.

References

1. Swaminathan, A., Mao, Y., Wu, M.: Robust and secure image hashing. IEEE Transactions on Information Forensics and Security 1(2), 215–230 (2006)
2. Xiang, S., Kim, H., Huang, J.: Histogram-based image hashing scheme robust against geometric deformations. In: MM&Sec 2007: Proceedings of the 9th workshop on Multimedia & security, pp. 121–128. ACM, New York (2007)
3. Monga, V., Evans, B.: Perceptual image hashing via feature points: performance evaluation and tradeoffs. IEEE Transactions on Image Processing 15(11), 3453–3466 (2006)
4. Olmos, A., Kingdom, F.: Mcgill calibrated colour image database (2004), http://tabby.vision.mcgill.ca
5. Koval, O., Voloshynovskiy, S., Beekhof, F., Pun, T.: Security analysis of robust perceptual hashing. In: SPIE, vol. 6819 (2008)

Transformational Breathing between Present and Past: Virtual Exhibition System of the Mao-Kung Ting

Chun-Ko Hsieh[1,3], Xin Tong[2], Yi-Ping Hung[1], Chia-Ping Chen[1], Ju-Chun Ko[1], Meng-Chieh Yu[1], Han-Hung Lin[1], Szu-Wei Wu[1], Yi-Yu Chung[1], Liang-Chun Lin[1], Ming-Sui Lee[1], Chu-Song Chen[4], Jiaping Wang[2], Quo-Ping Lin[3], and I-Ling Liu[3]

[1] Graduate Institute of Networking and Multimedia, National Taiwan University, No. 1, Sec. 4, Roosevelt Road, Taipei, 10617 Taiwan
[2] Microsoft Research Asia, 5F, Beijing Sigma Center,49 Zhichun Road, Beijing, P.R. China
[3] National Palace Museum, 221 Chih-shan Rd.,Sec.2, Taipei 111, Taiwan
[4] Institute of Information Science, Academic Sinica, 128 Academia Road, Section 2, Nankang, Taipei 115, Taiwan
d94944001@ntu.edu.tw, xtong@microsoft.com, hung@csie.ntu.edu.tw, cpchen@iis.sinica.edu.tw, d94944002@ntu.edu.tw, d95944008@ntu.edu.tw, stanley538@gmail.com, 54ways@gmail.com, drizztbest@gmail.com, r96922003@ntu.edu.tw, Jiapw@microsoft.com, mslee@csie.ntu.edu.tw, song@iis.sinica.edu.tw, jameslin@npm.gov.tw, iliu@npm.gov.tw

Abstract. The Mao-Kung Ting is one of the most precious artifacts in the National Palace Museum. Having five-hundred-character inscription cast inside, the Mao-Kung Ting is regarded as a very important historical document, dating back to 800 B.C.. Motivated by revealing the great nature of the artifact and interpreting it into a meaningful narrative, we have proposed an innovative Virtual Exhibition System to facilitate communication between the Mao-Kung Ting and audiences. Consequently, we develop the Virtual Exhibition system into the following scenarios: "Breathing through the History" and "View-dependent display".

Keywords: Mao-Kung Ting, de-/weathering simulation technique, view dependent display.

1 Introduction

Museums have the generosity of spirit to share exquisite artifacts with the global audiences. With that spirit, the research teams have combined multiple technologies to interpret an invaluable Chinese artifact as an interactive artwork. How to reveal the great nature of the artifact? The Mao-Kung Ting has been selected, as it is a ritual

S. Boll et al. (Eds.): MMM 2010, LNCS 5916, pp. 707–712, 2010.

bronze vessel dating back to 800 B.C. Especially, the Mao-Kung Ting has been weathered for thousands of years [1]. For the reasons given above, the team sought to reconstruct a 3D model of Mao-Kung Ting, and develop interactive applications, which are “Breathing through the History” and ”View-dependent display.”

2 Related Wok

2.1 De-/Weathering

There are several natural influences that cause real-world surfaces to exhibit dramatic variation over the course of time. Some methods can simulate the de-/weathering effects[2],[3],[4],[5]. To model the de-/weathered appearance of the Mao-Kung Ting model easily and convincingly, we refer to a visual simulation technique called “appearance manifolds” [6].

2.2 Breath-Based Biofeedback

Many techniques, such as Optoelectronic Plethysmography, Ultra Wide Band, Respiratory Inductive Plethysmography, and Heart Rate Variability, are available for detecting respiration status [7], [8].Ultra Wide Band (UWB) radar is applied in variety of settings for remote measuring of heart activities and respiration of users [9]. UWB do not need any detectors or markers attached to bodies of users; therefore, we chose that technique in our system.

3 System Architecture

There are two architectures for the Virtual Exhibition System. One supports the scenario “Breathing though the History”; the other supports the scenario “View-dependent display”.

3.1 Breathing through the History

Breath detection module: The UWB is the major component of the breath detection module. (Fig. 1a).

De-/Weathering Display Module: The process begins with the detection of breath of users, followed by the de-/weathering algorithm activated in the 3D Mao-Kung Ting model. Then, the de-/weathering appearance of the Mao-Kung Ting will change according to the respiration status of the users, and be displayed on the Interactive Multi-resolution Tabletop (i-m-Top) [6].

3.2 View-Dependent Display

The system consists of two major components, described below. (Fig. 1b).

Fig. 1. (a) The architecture has the UWB Breath Detector set underneath the Interactive Multi-resolution Tabletop (i-m-Top). (b) The architecture is composed of a stereo camera and an Interactive Multi-resolution Tabletop system.

1) Interactive Multi-resolution Tabletop: the tabletop system is employed as a display monitor and multimedia server.

2) Stereo Camera System: the device is utilized to detect foreground objects and the positions of the user's head and the handheld device.

4 Implementation

We generate the 3D model firstly, and then the implementation can be divided into the following parts.

4.1 De-/Weathering Simulation

First, we had to prepare a weathered material sample and capture its BRDF at a single instant of time. (Fig. 2a) The sample must contain spatial variations, which depicts different degrees of weathering and can further be analyzed to acquire spatial and temporal appearance properties for synthesizing the weathering process. We tried to simulate the weathering process of bronze, and synthesized a piece of weathered sample according to the current study in Mao-Kung Ting.

Then we captured spatially-variant BRDF from each surface point on the flat sample using a linear light source device.(Fig. 2b) [10], and we fitted parameters of the isotropic Ward model [11] for each point to form a 7D appearance space defined by reflectance features.

It is typical for sample points to have a dense distribution in the appearance space. Hence we are able to construct an appearance manifold, which is a neighborhood graph among these sample points, by connecting each point to its k nearest neighbors and pruning the outliers in the graph.

After constructing the appearance manifold that approximates a subspace of weathered surface points for the material, the user identifies two sets of points to present the most weathered and least weathered appearances respectively. We then defined the degree of weathering for each point in the appearance manifold according to its relative distance between the two sets.

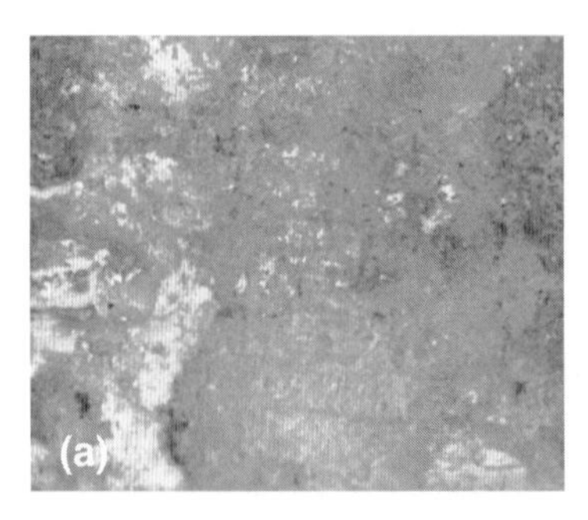

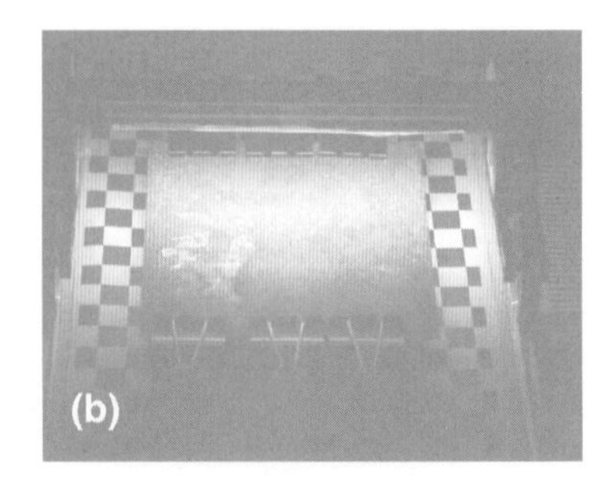

Fig. 2. (a) The weathered bronze sample. (b) The spatially-variant BRDF data capturing device.

Consequently, a weathering degree map has been obtained by replacing the appearance value of each sample point with its corresponding degree value in the appearance manifold.

Assuming we have distribution of large scale time-variant weathering degrees over the surface, which may be generated by manual specification or using existing visual simulation techniques, weathering and de-weathering appearances on a 3D model can be synthesized with the computed appearance manifold and degree map.

We used a multi-scale texture synthesis technique, like the one proposed in "appearance manifolds" [6], to synthesize the time-variant appearance sequence frame by frame. Specifically, the progress of synthesis includes three steps for each frame: 1) Initialize the degree values of each pixel by extrapolation from the appearance of the preceding frame. 2) Interpolate the initial degree values according to their geodesic distance along the shortest path to the set of most weathered points in the appearance manifold. 3) We needed to consider the neighborhood information on the material sample and incorporate changes in texture characteristic over time to avoid undesirable repetition of texture patterns. The initial appearance frame is finally refined by synthesis with multiple neighborhood scales.

4.2 Breath Detection

Whenever a user walks toward the UWB, the device begins to detect breath of the user. Then, the UWB data will be transmitted to system via Bluetooth; afterwards, the collected data will be analyzed by the system. These data also illustrate how human breathing impacts on the change of the Mao-Kung Ting during its aging process.

4.3 View-Dependent Display

The View-dependent display aims to provide users with the effect of visual fidelity, so that the displayed virtual artwork will be adjusted according to different view angles of the user. (Fig. 3)

Implementation processes of the view-dependent display are as following. First, the viewpoint of the user is estimated according to the position of the user's head. Since the interaction is designed for a single user, in a simple environment, a camera can capture a human object via foreground detection. Codebook Model is used for foreground detection. Then, according to the human features in the human object, such as face or hair, the position of the user's head will be identified in the image.

In this system, the stereo camera, with two optical lenses, contains distance information in pixels between camera and object in an image based on triangle theory. We can acquire a 3D coordinate with a camera from stereo image pair (I_{left} , I_{right}) by pixels:

$$p_{Camera(i,j)}(x,y,z) = \Gamma(I_{left}, p_{(i,j)}, I_{right}) \qquad \forall p_{(i,j)} \in I_{left} \tag{1}$$

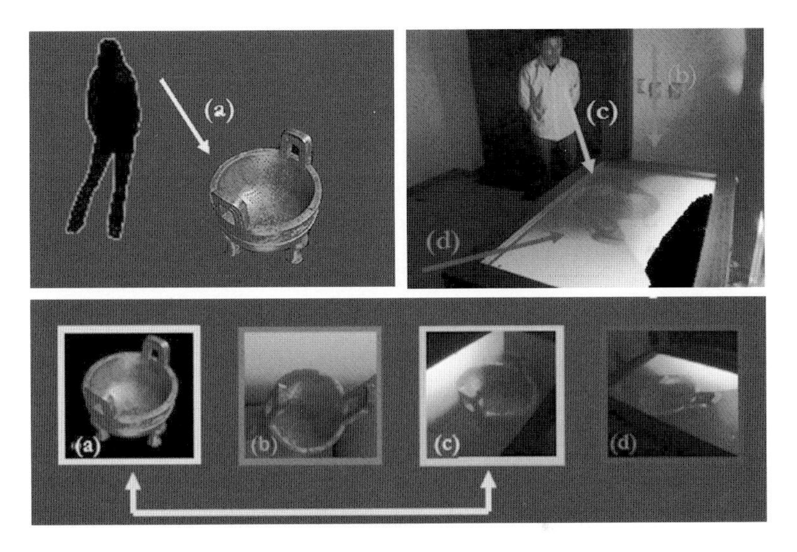

Fig. 3. The left image shows that the user stands in front the real MKT. The right image shows that different view points in front of the display table. The bottom image shows that the user's view (c) is almost as real as seen by a visitor in front of the MKT (a).

Nevertheless, a 3D coordinate acquired from the original position must be transferred to a display-centered coordinate. Therefore, we set the calibration board aligned with the surface of the tabletop. Through calibration, the extrinsic parameter [$\mathbf{R}_c$ | $\mathbf{t}_c$] is acquired from the camera. $\mathbf{T}_c$ is the vector for the distance from camera to the center of the calibration board. According to the Eq. 1, we can also transfer a displayed-centered coordinate to a camera-centered one.

$$P_{Camera}[X,Y,Z] = [R_c \mid \vec{t}_c] P_{imTop}[X,Y,Z] \tag{2}$$

$P_{camera}[X,Y,Z]$ is the set of $\mathbf{p}_{camera}(x,y,z)$. Actually, a partial 3D coordinate relative to the camera can be acquired from the stereo camera. Therefore, we can use the inverse function of [$\mathbf{R}_c$ | $\mathbf{t}_c$] to acquire a 3D coordinate relative to surface of the tabletop.

$$P_{imTop}[X,Y,Z] = [R_c^T \mid -R_c^T \vec{t}_c] P_{Camera}[X,Y,Z] \tag{3}$$

Since the camera detects the position of the user's head, and a 3D coordinate relative to the surface of the tabletop acquired, the effect of view-dependent display is accomplished.

5 Conclusions

Our artwork provides a virtual exhibition system to lead museum visitors to experience the unique features of an ancient artifact, Mao-Kung Ting. We have illustrated how technologies strongly support the design considerations of systems, such as de-/weathering technology, breathing based biofeedback technology, computer vision technology, and view-dependent display technology, which play a vital role in our systems. The demo video can be found at: http://ippr.csie.ntu.edu.tw/MKT/MKT.html

References

[1] Collection of National Palace Museum, http://www.npm.gov.tw/en/collection/selections_02.htm?docno=194&catno=19&pageno=2

[2] Dorsey, J., Hanrahan, P.: Modeling and rendering of metallic patinas. In: SIGGRAPH 1996, pp. 387–396 (1996)

[3] Dorsey, J., Edelman, A., Jensen, H.W., Legakis, J., Pedersen, H.K.: Modeling and rendering of weathered stone. In: SIGGRAPH 1999, pp. 225–234 (1999)

[4] Georghiades, A.S., Lu, J., Xu, C., Dorsey, J., Rushmeier, H.: Observing and transferring material histories. Tech. Rep. 1329, Yale University (2005)

[5] Gu, J., Tu, C.I., Ramamoorthi, R., Belhumeur, P., Matusik, W., Nayar, S.: Time-varying surface appearance: Acquisition, modeling and rendering. ACM Transaction on Graphics 25(3) (2006)

[6] Wang, J., Tong, X., Lin, S., Wang, C., Pan, M., Bao, H., Guo, B., Shun, H.-Y.: Appearance Manifolds for Modeling Time-Variant Appearance of Materials. ACM Transaction on Graphics 25(3) (2006)

[7] Yu, M.-C., Hung, Y.-P., Chang, K.-J., Chen, J.-S., Hsu, S.-C., Ko, J.-C., Ching-Yao: Multimedia Feedback for Improving Breathing Habit. In: 1st IEEE International Conference on Ubi-media Computing (U-Media 2008), Lanzhou University, China, July 15-16 (2008)

[8] Hoyer, D., Schmidt, K., Bauer, R., Zwiener, U., Kohler, M., Luthke, B., Eiselt, M.: Nonlinear analysis of heart rate and respiratory dynamics. IEEE Eng. Med. Biol. Mag. 16(1), 31–39 (1997)

[9] Immoreev, I.Y.: Practical Application of Ultra-Wideband Radars. Ultrawideband and Ultrashort Impulse Signals, Sevastopol, Ukraine, September 18-22 (2006)

[10] Gardner, A., Tchou, C., Hawkins, T., Debevec, P.: Linear light source reflectometry. ACM Trans. Graph. 22(3) (2003)

[11] Ward, G.J.: Measuring and modeling anisotropic reflection. In: SIGGRAPH 1992, pp. 265–272 (1992)

Learning Cooking Techniques from YouTube

Guangda Li, Richang Hong, Yan-Tao Zheng,
Shuicheng Yan, and Tat-Seng Chua

National University of Singapore, Singapore
{g0701808,eleyans}@nus.edu.sg, {hongrc,chuats}@comp.nus.edu.sg,
yantaozheng@gmail.com

Abstract. Cooking is a human activity with sophisticated process. Underlying the multitude of culinary recipes, there exist a set of fundamental and general cooking techniques, such as cutting, braising, slicing, and sauntering, etc. These skills are hard to learn through cooking recipes, which only provide textual instructions about certain dishs. Although visual instructions such as videos are more direct and intuitive for user to learn these skills, they mainly focus on certain dishes but not general cooking techniques. In this paper, we explore how to leverage YouTube video collections as a source to automatically mine videos of basic cooking techniques. The proposed approach first collects a group of videos by searching YouTube, and then leverages the trajectory bag of words model to represent human motion. Furthermore, the approach clusters the candidate shots into motion similar groups, and selects the most representative cluster and shots of the cooking technique to present to the user. The testing on 22 cooking techniques shows the feasibility of our proposed framework.

Keywords: Cooking techniques, video mining, YouTube.

1 Introduction

Recipes are the natural solution for people to learn how to cook, but cooking is not just about recipes. Underneath the recipe for various culinary dishes, cooking involves a set of fundamental and general techniques, including cutting, filleting, and roasting, etc. These basic cooking skills are hard to learn in cooking recipes, as they only provide textual instructions about certain dish. If user is presented by visual tutorial such as videos for these basic skills, it will be more direct and intuitive for them to understand, such as the examples given in Figure 1.

In recent years, millions of video on the web, make them a source for people to learn some basic cooking techniques. However, web video search engine cannot be automatically applied to visually demonstrate these basic cooking techniques, due to the following reasons. First, search results usually contain irrelevant videos, because textual metadata associated with the video in term of title, tags or surrounding text is not sufficiently accurate to locate videos about cooking techniques. Second, even if the best related video is identified, only part of the video presents this cooking technique.

S. Boll et al. (Eds.): MMM 2010, LNCS 5916, pp. 713–718, 2010.

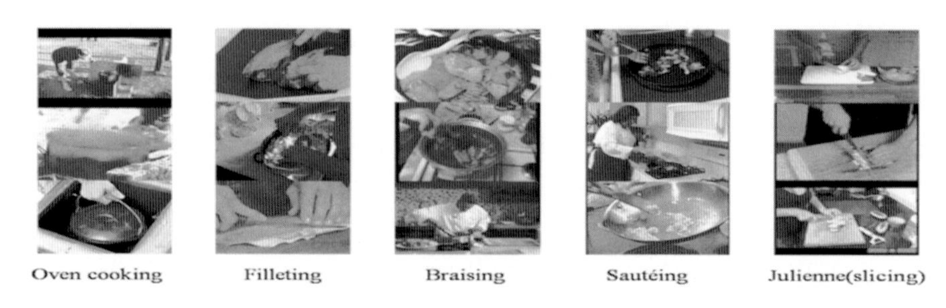

Fig. 1. Same type of cooking technique has similar motion pattern

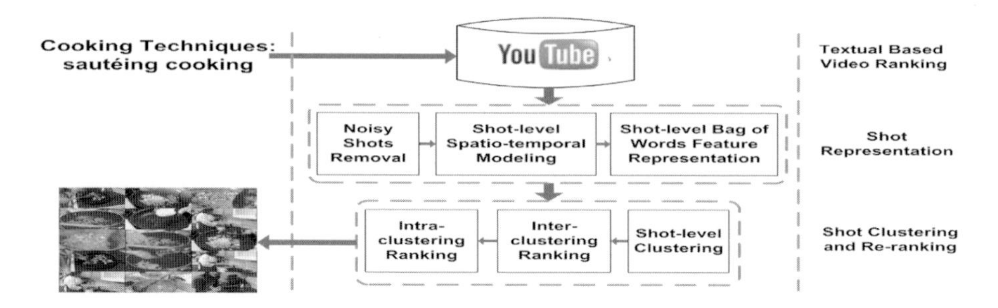

Fig. 2. Overall framework

In this paper, we explore how to mine video tutorials of basic cooking techniques from the noisy web video collection. Our target is to generate a group of representative video shots containing the desired cooking techniques,such as in figure 1. The premise is simple: video shots of the same cooking technique tend to share similar human motion patterns. There are three steps to learn cooking technique patterns. First, a cooking technique, such as "sauteing cooking", is taken as search keyword to submitted to YouTube[1] to find a group of relevant but noisy videos. Then we segment each video into shots. Second, video shots are represented by spatio-temporal bag of words feature. Noisy shots are removed based a set of heuristic rules. Then the cooking technique patterns are mined in an unsupervised fashion. The graph clustering method is utilized to learn the human motion patterns of cooking techniques. Then the most representative cluster is identified by ranking clusters based on their cluster density and scatter. Finally, the most representative shot within this cluster is identified based on its representativeness. The overall framework is given in Figure 2.

2 Related Work

To some extent, our work is to establish multimedia dictionary of cooking techniques. Li et al. [9] and Wang et al. [10] aimed to establish general multimedia

[1] YouTube Website: www.youtube.com

dictionary by leveraging community contributed generated images. However, there are very few works done in cooking domain. Shibata et al. [3] defined the internal structure of cooking videos as three steps: preparation, sauting and dishing up. Then they proposed a framework for acquiring object models of foods from predefined preparation step. Linguistic, visual and audio modalities are utilized to do further object recognition. Hamada et al. [4] considered cooking motions and appearances of foods to be visually important in a cooking video, which contribute to assemble abstract videos. The above two works mainly focus on applying multi-modality features on single video, but the redundant information from different videos is neglected. For example, different cooking videos may involve similar cooking techniques, as shown in Figure 1. With the advance of computer vision, especially in motion analysis, some works have been conducted on extracting spatio-temporal features to represent a certain kind of motion. Dollar et al. [5] used cuboids to extend the 2D local interest points, representing not only along the spatial dimensions but also in the temporal dimension. Ju et al. [2] proposed to model the spatio-temporal context information in a hierarchical way, where three levels of motion context information are exploited. This method outperforms most of others. For this reason, we model the shots by a number of motion features, specifically the sift-based trajectory.

3 Cooking Technique Learning

3.1 Shot Representation

Following most existing video processing systems [11], we take shot as the basic content unit. After segmenting video into shots, the next step is to model shots within a video by spatial-temporal features. We utilize the trajectory transition descriptor proposed in [8] to characterize the dynamic properties of trajectories within a video shot. Each shot is represented by N trajectory transition descriptors. However, it is infeasible to extract the trajectory transition descriptors for all shots of a video. For example, an 8 minutes MPEG format video which is converted from FLV format downloaded from YouTube will expand to 50M. And the overall number of shots for one video will be above one hundred. To minimize the computational workload, we have to identify some shot candidates from the whole video for the trajectory transition descriptor extraction. Due to our observation, some shots make no contribution for cooking motion techniques discovering. And Hamada et al. [4] pointed that in cooking video, shots with faces are usually less important than shots containing objects and motion. These shots are deemed to be noisy and will not contribute to the mining of cooking techniques. We, therefore, only take the rest of the shots for subsequent processing. Then we construct a trajectory vocabulary with 1000 words by hierarchical K-means algorithm over the sampled trajectory transition descriptors, and assign each trajectory transition descriptor to its closest (in the sense of Euclidean distance) trajectory word.

3.2 Constructing Match Shot Graph

After performing trajectory bag of words model on each shot, this set of shots are transformed to a graph representation by their shot similarity, in which the vertexes are shots. The edges connecting vertexes is quantified by its length. For shots i and j, its length is defined as follows:

$$d_{ij} = \frac{\|d_i - d_j\|}{|d_i| * |d_i|} \quad (1)$$

3.3 Ranking Clusters

As the distance between any two shots for a certain cooking technique has been established, we use clustering to expose different aspects of a cooking technique. Since we do not have a priori knowledge of the number of clusters, the k-means like clustering technique is unsuitable here. Therefore, we use mutual *knn* method to perform the clustering, which connect each shot with k nearest neighbors by their similarity. This is of great advantage over other clustering methods because we can simply control the number k of nearest neighbors to each shots but not the number of clusters. Then all these clusters compete for the chance of being selected to be shown to the user. We use the following criteria to compute a cluster's ranking score, similar to what was done in [6]:

$$RC_k = \frac{\sum_{j \in k, 0<m<K, m \neq k} |S_i - S_j|}{\sum_{j=1}^{n} |S_j - \overline{S_k}|} \quad (2)$$

numerator is inter-cluster distance (the average distance between shots within the cluster and shots outside of the cluster), and denominator is the intra-cluster distance (the average distance between shots within the cluster). A higher ratio indicates that the cluster is tightly formed and shows a motion coherent view, while a lower ratio indicates that the cluster is noisy and may not be motion coherent, or is similar to other clusters.

3.4 Learning Representative Shots

Given the result of clusters ranking, we rank the shots within a cluster to find the most representative shot. The representativeness score RS for shot j is calculated as follow:

$$RS_j = \frac{1}{|S_j - \overline{S_k}|} \quad (3)$$

A higher score indicates that the shot is tighter to the center of this cluster, while a lower score indicate that the shot is far from the center of this cluster.

4 Experiment

To validate the effectiveness of our proposed framework, we assembled a collection of videos posted on YouTube from May 2009 to June 2009, by YouTube API. Based on the cooking category in Wikipedia[2] and manually checking the availability of videos for each concept in YouTube, we got 22 cooking concepts. Then we select the top 20 videos for each cooking technique query.

[2] http://en.wikipedia.org/wiki/Category

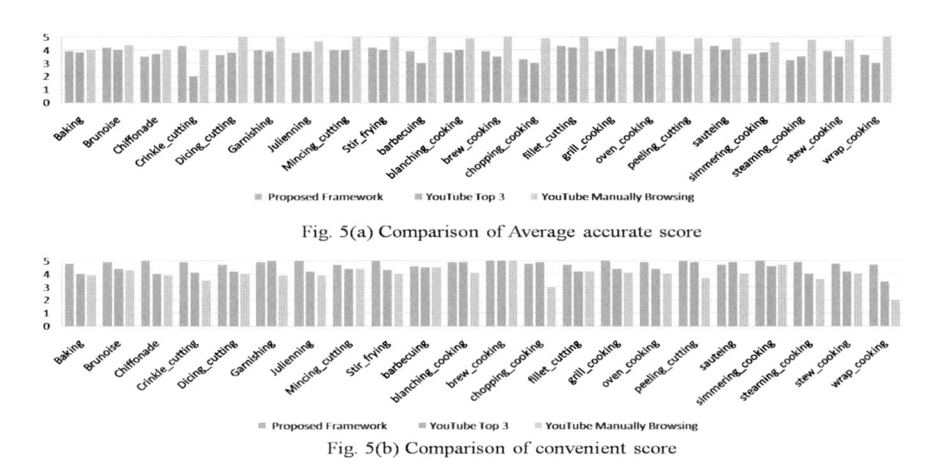

Fig. 5(a) Comparison of Average accurate score

Fig. 5(b) Comparison of convenient score

Fig. 3. Same type of cooking technique has similar motion pattern

4 evaluators are involved in our evaluations. Each evaluator was first presented the manually selected shot demonstration as in Figure 1 and the textual explanation for a certain cooking technique. The expectation is that evaluators can understand what a certain cooking technique goes on. After that, the evaluators were shown the top 3 ranked clusters and the top three ranked shots in the first cluster. The evaluators were also asked to search YouTube using cooking technique concepts respectively. They were asked to give satisfaction and convenience score range from 1 to 5 to indicate their preference.

We compare users' satisfaction score for the proposed system and two different strategies of using YouTube. We can see that in Figure 3(a), the overall satisfaction about accuracy of the proposed framework is better than only screening the top 3 videos from YouTube, but not as good as manually browsing and picking a retrieved videos from YouTube. This is because any automatic system cannot perform better than manual methods. In Figure 3(b), the comparison shows that user deemed our proposed our system was more convenient for them to learn the cooking techniques than the other two strategies. This is because evaluators

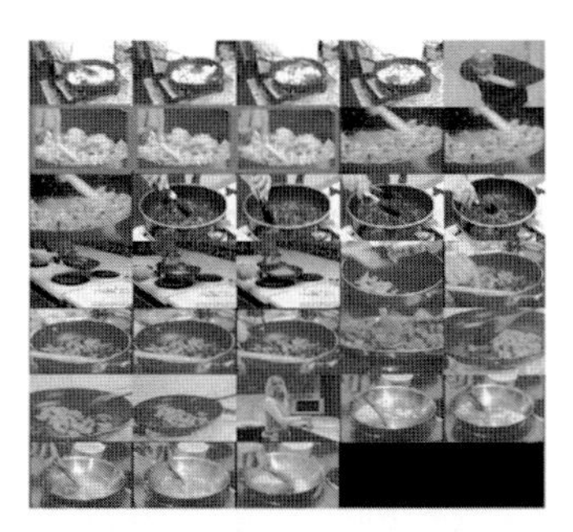

Fig. 4. Same type of cooking technique has similar motion pattern

need to click several times on retrieved videos to discover what they want using YouTube. However, our proposed framework eases the learning workload.

Figure 4 shows the generated group for two cooking techniques. The top three ranked shots are marked with red square. We can observe that some shots are from the same video, which could be caused by two reasons. First, the shot segmentation tool is sensitive to scene change, so continuous shots should be merged. Second, trajectory features for some shots are not discriminative enough for the graph clustering method.

5 Conclusion and Further Work

In this paper, we explore how to leverage YouTube video collections as a source to automatically mine videos of basic cooking techniques. The proposed approach take a simple premise: video shots of the same cooking technique tend to reveal similar human motion patterns. The future works are to mine cooking techniques on web videos utilizing the motion features and the embedded rich metadata, to facilitate the multimedia Question-Answering (QA) system as described in [1], [8] and [7].

References

1. Chua, T.-S., Hong, R., Li, G., Tang, J.: From Text Question-Answering to Multimedia QA. To appear in ACM Multimedia Workshop on Large-Scale Multimedia Retrieval and Mining, LS-MMRM (2009)
2. Sun, J., Wu, X., Yan, S., Cheong, L.F., Chua, T.-S., Li, J.: Hierarchical Spatio-Temporal Context Modeling for Action Recognition. In: Proceedings of IEEE International Conference on Computer Vision and Pattern Recognition (CVPR), Miami, Florida, US (2009)
3. Shibata, T., Kato, N., Kurohashi, S.: Automatic object model acquisition and object recognition by integrating linguistic and visual information. In: ACM'MM (2007)
4. Hamada, R., Satoh, S., Sakai, S.: Detection of Important Segments in Cooking Videos. In: Proceedings of the IEEE Workshop on Content-Based Access of Image and Video Libraries (Cbaivl 2001), p. 118 (2001)
5. Dollar, P., Rabaud, V., Cottrell, G., Belongie, S.: Behavior recognition via sparse spatio-temporal features. In: VS-PETS 2005, pp. 65–72 (2005)
6. Kennedy, L.S., Naaman, M.: Generating diverse and representative image search results for landmarks. In: Proceeding of the 17th international Conference on World Wide Web, Beijing, China, pp. 297–306 (2008)
7. Li, G., Ming, Z., Li, H., Chua, T.-S.: Video Reference: Question Answering on YouTube. In: ACM'MM (2009)
8. Hong, R., Tang, J., Tan, H.-K., Yan, S., Ngo, C.-W., Chua, T.-S.: Event Driven Summarization for Web Videos. In: ACM Multimedia Workshop on Social Media, WSM (2009)
9. Li, H., Tang, J., Li, G., Chua, T.-S.: Word2Image: Towards Visual Interpretation of Words. In: ACM'MM 2008 (2008)
10. Wang, M., Yang, K., Hua, X.-S., Zhang, H.-J.: Visual Tag Dictionary: Interpreting Tags with Visual Words. In: ACM Multimedia Workshop on Web-Scale Multimedia Corpus (2009)
11. Tang, S., Li, J.-T., et al.: TRECVID 2008 High-Level Feature Extraction By MCG-ICT-CAS. In: TRECVID Workshop, Gaithesburg, USA (2008)

Adaptive Server Bandwidth Allocation for Multi-channel P2P Live Streaming

Chen Tang, Lifeng Sun, and Shiqiang Yang

Key Laboratory of Media and Networking, MOE-Microsoft
Tsinghua National Laboratory for Information Science and Technology
Dept. of Computer Sci. & Tech., Tsinghua Univ., Beijing 100084, China

Abstract. With the growing popularity of P2P live streaming, more and more channels have been set up, however, this poses new challenges to P2P technology, such as stronger dynamic characteristic, high switching delay. In this paper, we focus on the server, which can allocate different bandwidth to meet the needs of each type of peers, especially the switching peers. On a close study of the basic tradeoff between the switching peers' and overall performance, we propose new peers first strategy and adaptive bandwidth allocation, and get a satisfactory result. Our results show that the full and rational utilization of server bandwidth is of great help for improving the performance of multi-channel P2P live streaming.

Keywords: P2P live streaming, multi-channel, new peers first, adaptive bandwidth allocation.

1 Introduction

During recent years, many researches on P2P and P2P live streaming have done, and won a tremendous success, such as data-driven/mesh-based P2P streaming protocol [1], GridMedia [2], iGridMedia [3], and Ration [4]. More and more practical P2P live streaming systems have put to use, and the largest supportable number of online people has also surpass the million mark, e.g., PPLive [5].

As P2P live streaming has got an extensive application and rapid development, it has entered the multi-channel era. However, this poses new challenges. As we know, the cooperation and coordination between peers is a huge advantage of P2P network, but this is inevitably impacted by the separation of different channels. High switching delay is another problem in multi-channel. On the other hand, the peer scheduling in P2P network, which is the focus of the past P2P research, doesn't go very far towards solving these problems.

Aim at the above-mentioned problems, we refocus on the server. We recognize that there are various peers in each channel. Then, we can provide QoS guarantee for different peers, through the server scheduling algorithm. In this paper, we mainly discuss the channel churn in multi-channel. A new-peers-first strategy, which prioritizes direct access of new peers to the server, has been implemented and evaluated. We reduce the switching delay by about 50%. But the basic tradeoff between the switching peers' and overall performance need be considered. Then we propose adaptive bandwidth allocation strategy.The whole study is based on pull-push protocol [2] and Ration algorithm [4].

S. Boll et al. (Eds.): MMM 2010, LNCS 5916, pp. 719–724, 2010.

The remainder of this paper is organized as follows. In Section 2, we implement and evaluate two server connecting strategies. In Section 3, we hash over new peers first strategy and adaptive bandwidth allocation strategy. The simulation results are given in Section 4. In section 5, we conclude this paper.

2 Performance Analysis of Server Connecting Strategy

2.1 The Role of the Server

As we know, there are dedicated streaming servers as the video source data providers in the P2P live streaming, which are different from the normal peers. Because the streaming media transmissions consume a lot of network bandwidth resource, the servers are indispensable. In the multi-channel system, the server is in response to the request peer, offers large amounts of streaming media data to the peer in each channel. We should change the role of the server, from passive response to active control. Through the analysis of known server bandwidth allocation, we reallocate server bandwidth to the peers with the different requirements, in order to utilize the limited bandwidth more rationally.

2.2 Server Connecting Strategy

The first factor to be considered is the server connecting strategy, which decides what kinds of peer can be connected to the server. Normally, the neighbor search of a node is random, so the connecting between two nodes doesn't need to be controlled. But the server is particular, the choice of its neighbors can affect the performance of the entire network. So the connecting between the server and the peers is not random, but complies with some rules.

iGridMedia [3] propose **rescue connection** to guarantee data transfer delay. In this strategy, once an absent packet is about to pass the deadline, the peer will directly request this packet from the server through the rescue connection established between the peer and the server. So video can play with the guarantee-delay. Another strategy is opposite, which limits the number of peers connected to the server. These peers are called first-level peers, because the server provides data to them directly. In this strategy, the video source data are sent to the first-level peers at first, and then spread to the entire network through the cooperation and coordination between peers.

Here, we define term *Peer Resource Index (PRI)*. It is defined as the ratio of the total peer upload capacity $\sum_{i=1}^{n} u_i$ to the minimum bandwidth resource demand (i.e, streaming rate r times the viewer number n), that is., PRI $=\sum u_i/nr$. Other performance indexes are as follows: **the peer quality**——the ratio of filling buffer, **the channel quality**——the quality sum of all peers in the channel, **the channel delay**——the average transfer delay of all packets in the channel.

The comparison of these two strategies is given below. Fig. 1 and Fig. 2 show that the rescue connection reduces the data transfer delay effectively, but the first-level peers structure is more helpful to improve quality. When PRI is 1.4, the difference becomes smaller. We can discuss the reasons in detail. The rescue connection is equivalent to allocating most of the bandwidth to send the rescue

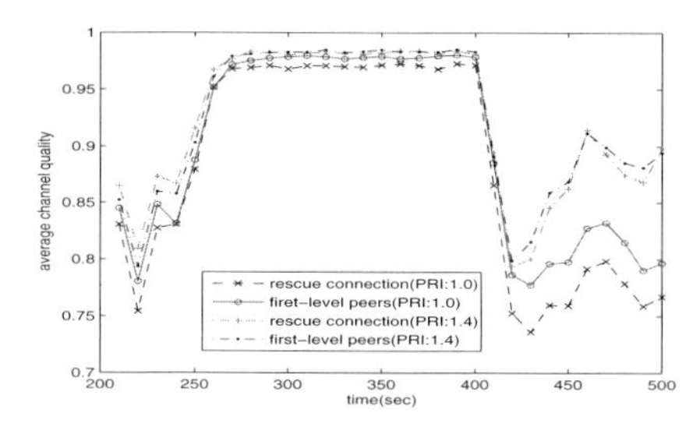

Fig. 1. Average channel quality with respect to PRI and server connecting strategy

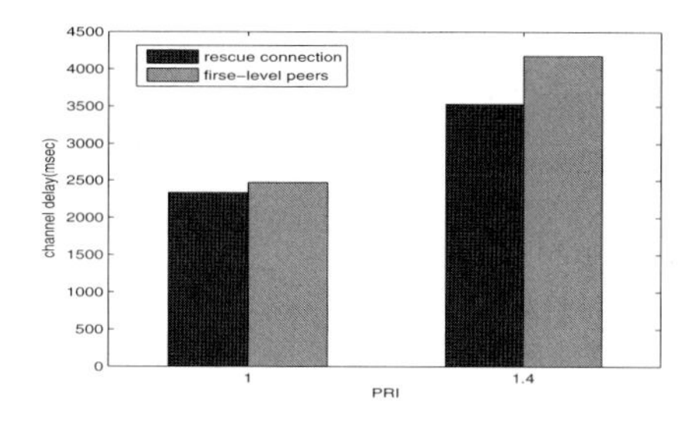

Fig. 2. Channel delay with respect to PRI and server connecting strategy

data, but not the newest data, especially when the bandwidth is tight. And the data from other peers is limited, so the peer quality sharply declines. On the other hand, the video data transmission needs to go through first-level peers, so the transfer delay is increased in the first-level peers' structure.

To sum up, we can know the effect of server bandwidth allocation. Especially when the bandwidth is tight, we can't ensure all performances to be good enough. Therefor, active control of server bandwidth allocation is important to adapt for the heterogeneous and dynamic characteristic of actual network.

3 Adaptive Server Bandwidth Allocation for Peer Diversity

Through the above analysis, we realize the importance of server bandwidth allocation. So we should continue to discuss how to allocate bandwidth to the peers connected to the server. In traditional P2P network, there are no differences between the peers for the server. But in fact, there are various peers in each channel, such as the switching peers. In this section, we will discuss how to control server bandwidth allocation to reduce the switching delay.

3.1 New Peers First Strategy

At first, let's analyze what should be done for the switching peer. Once a peer begins to switch channel, it should be disconnected from the original neighbor and take some time to find new neighbor in the new channel. Then, the switching peer receives data from its neighbor, but won't play video until the buffer count is more than a given ratio (such as 80%) of the total size of its playback buffer. So the search of the new neighbor and the rebuffering lead to the high switching delay. With respect to these two causes, we note the specificity of first-level peers. The first-level peers receive data from the server, and are easy to connect to the server. In other words, it solves these two problems. Based on this, we propose *new peers first strategy*——select the switching peers to be first-level peers.

The advantage of new peers first strategy is obvious. But it leads to new problem: the other peers have to reduce the data request to the server. Moreover, in P2P network, the impact of a peer is not isolated, but will be passed on to the neighbors and spread to the entire network finally. In other words, if the ratio of the switching peers in first-level peers is reduced, the overall performance will increase, however, the switching peers' performance will decline. The priority is so crucial that we should set it carefully. At section 4, we will show the evaluation results of new peers first strategy with different priorities.

3.2 Adaptive Bandwidth Allocation

As previously analyzed, new peers first strategy impacts the overall performance, and the adjustment of the priority should reach a compromise between new peers' and overall performance. Now, the question boils down to how calculate the ratio of the switching peers in first-level peers to meet the switching peers' needs as much as possible with ensuring the other peers' performance.

Ration [4] derives the relationship among streaming quality q, server bandwidth usage s, and the number of peers n in each channel c as follows:

$$q^c = \gamma^c (s^c)^{\alpha^c} (n^c)^{\beta^c}. \quad (1)$$

where γ,α,β can be estimated with least squares algorithms by using historical data, and the result indicates that $0<\alpha<1$ and $-1<\beta<0$.

We can also apply this formula to the switching peers. Let q_n be the switching peers' quality, n_n be the number of the switching peers' quality, s_n be the server bandwidth allocated to the switching peers, q_o,n_o,s_o be for the other peers. Through the training of historical data, we can get γ_n,α_n,β_n,γ_o,α_o,β_o. Our aim is to obtain the maximum of the switching peers' quality, with guaranteeing the other peers' quality. Such an objective can be formally represented as follows:

$$max\{q_n\} = max\{\gamma_n (s_n)^{\alpha_n} (n_n)^{\beta_n}\} = max\{\gamma_n (U * r_n)^{\alpha_n} (n_n)^{\beta_n}\}. \quad (2)$$

subject to

$$U = s_n + s_o = U * r_n + s_o, 0 \leq s_n \leq B_n, 0 \leq s_o \leq B_o, q_o \geq Q.$$

where the objective function $B = (\gamma n^{\beta})^{-\frac{1}{\alpha}}$ denoting the maximal server capacity requirement, that achieves q=1. As given value, U is the total server bandwidth, Q is the minimum of the other peers' quality what we should guarantee.

To solve this problem, we notice $\alpha>0$, so q and s are positively correlated. Besides, n can be got from the regular reporting. In other words, if we want to maximize q_n, we should increase r_n. We can calculate the extremes of r_n in three aspects, and take the minimum of them to guarantee the other peers' quality. Then ,our complete strategy is summarized in Table 1.

This adjustment isn't always carried out, but is triggered by some conditions. In practice, we receive the peers' information report regularly. Unless the quality of the switching peers or the other peers is reduced by more than 10%, we won't adjust the ratio. Because (1) is an empirical formula, the accuracy is limited. Therefore, the ratio do not need to change within some errors. This strategy can balance new peers' and overall performance well, through the server bandwidth rational allocation to the two types of peers. The simulation results will also be present at section 4.

Table 1. Adaptive Bandwidth Allocation Strategy

1 Collect the historical data with peer heartbeat messages, then we can predict n_n and n_o with ARIMA(0,2,1) model, and estimate γ_n,α_n,β_n, γ_o,α_o,β_o as mentioned above [4].
2 Calculate the minimum of s_o:$s_{omin} = (\frac{Q}{\gamma_o n_o^{\beta_o}})^{1/\alpha_o}$.Then $r_1 = 1 - s_{omin}/U$.
3 Calculate the maximum of s_n:$s_{nmax} = (\frac{1}{\gamma_n n_n^{\beta_n}})^{1/\alpha_n}$.Then $r_2 = s_{nmax}/U$.
4 Calculate the maximum of the ratio of the switching peer in first-level peers.Then $r_3 = n_n/n_f$,where n_f is the number of first-level peers.
5 Calculate r_n as $r_n = min\{r_1, r_2, r_3\}$.

4 Experimental Evaluations

Based on an implemented event-driven packet-level simulator coded in C++ and the multi-channel alteration, we implement the above strategies, conduct a series of simulations in this section.[1] The basic parameters of the simulation network are as follows: The default streaming rate is set to 500kbps, PRI is set to 1. The default neighbor count is 15, the default request window size is 20 seconds, the buffer size is 35 seconds, the simulation time is 500s, the channel number is 3, the peers' number of each channel is 300, so the server bandwidth is about 450Mbps. We set node-to-node latency matrix and the peers' bandwidth distribution according to the actual Internet. And for each point in the figures, we average the results by repeating 10 runs with different random seeds.

Fig. 3 shows that the rise speed of the switching peers' quality has a marked increase——about 50%, by new peers first strategy. But, as mentioned above, the other peers' quality is reduced. Because the switching peers account for only 10%, the decline is limited. We will do further testing in the following.

In Fig. 4 and Fig. 5, we raise the ratio of the switching peers in the network. The results show that the overall performances are reduced by new peers first

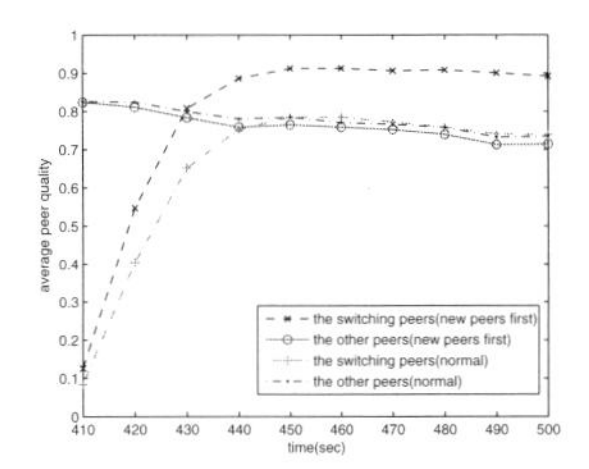

Fig. 3. Average peer quality with respect to the peers' type and whether to use new peers first strategy

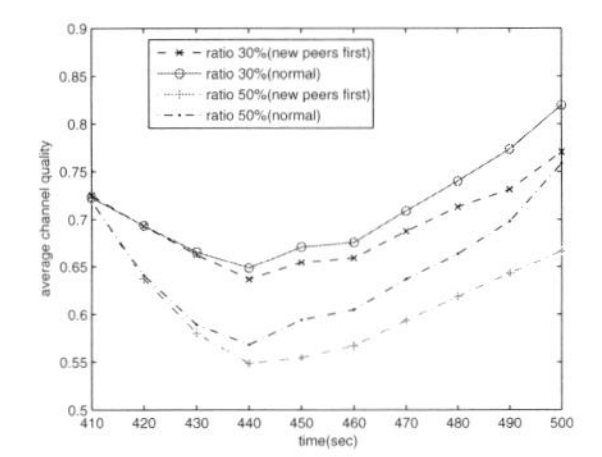

Fig. 4. Average channel quality with respect to the ratio of the switching peers in the network and whether to use new peers first strategy

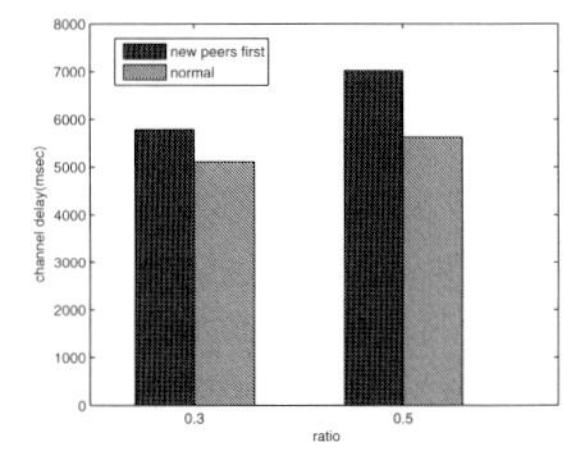

Fig. 5. Channel delay with respect to the ratio of the switching peers in the network and whether to use new peers first strategy

[1] The simulator is available online for free downloading at `http://media.cs.tsinghua.edu.cn/~zhangm`

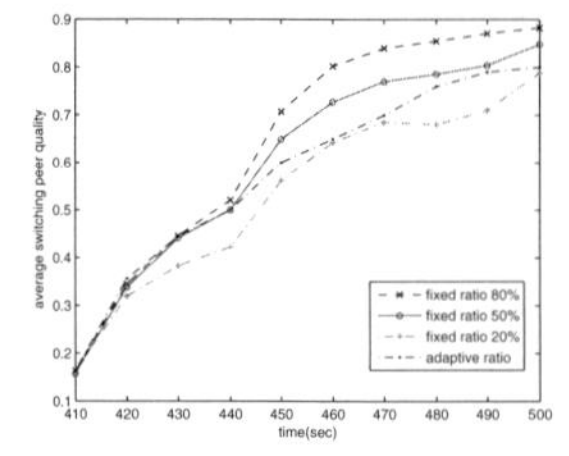

Fig. 6. Average switching peer quality with respect to the ratio of the switching peers in first-level peers

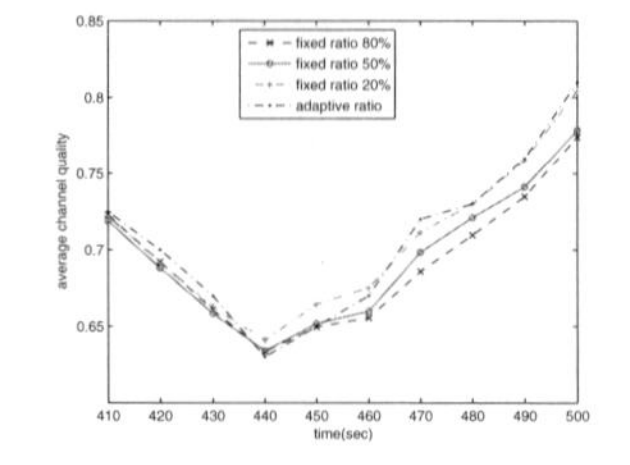

Fig. 7. Average channel quality with respect to the ratio of the switching peers in first-level peers

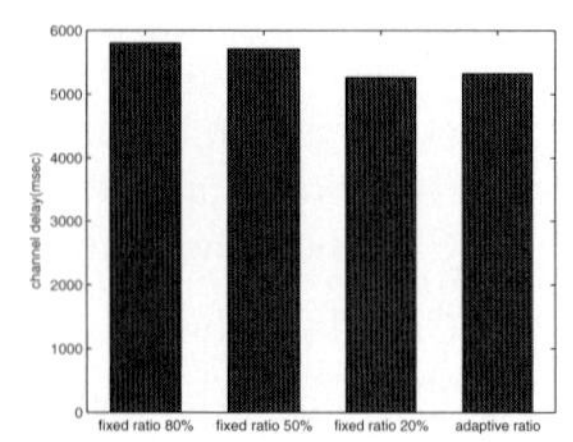

Fig. 8. Channel delay with respect to the ratio of the switching peers in first-level peers

strategy. Besides, the larger the ratio is, the more obvious the decline is. So we can't use new peers first strategy directly to guarantee the overall performances.

Fig. 6 , Fig. 7 and Fig. 8 indicate the ratio of the switching peers in first-level peers is a key factor.It has a positive correlation with the switching peers' performance, but negatively related to overall performance. So we can know the advantage of adaptive bandwidth allocation: without affecting overall performance, adaptive bandwidth allocation can maximize the switching peers' performance.

5 Conclusion and Future Work

In this paper, new peers first strategy and adaptive bandwidth allocation strategy are proposed to reduce the high switching delay. And the work of this paper can guide us to allocate the server bandwidth more fully and rationally. For future work, we can record the users' behavior to adjust adaptive bandwidth allocation strategy to adapt to the characteristics of the actual Internet.

Acknowledgments. This work was supported by NSFC under Grant No. 60503063 and No.60833009, and 863 program under Grant No.2009AA01Z328.

References

1. Zhang, X., Liu, J., Li, B., Yum, Y.S.: Coolstreaming/donet: A data-driven overlay network for live media streaming. In: Proc. IEEE INFOCOM'05. (2005) 2102–2111
2. Zhang, M., Luo, J., Zhao, L., Yang, S.: A peer-to-peer network for live media streaming - using a push-pull approach. In: Proceedings of the 13th annual ACM international conference on Multimedia, Singapore (2005) 287–290
3. Zhang, M., Sun, L., Xi, X., Yang, S.: igridmedia: Providing delay-guaranteed peer-to-peer live streaming service on internet. In: Proc. IEEE Globecom '08, New Orleans, USA (2008) 1–5
4. Wu, C., Li, B., Zhao, S.: Multi-channel live p2p streaming: Refocusing on servers. In: Proc. IEEE INFOCOM'08, Phoenix, USA (2008) 1355–1363
5. PPLive: Pplive (2009) `http://www.pplive.com/`.

Feature Subspace Selection for Efficient Video Retrieval

Anuj Goyal, Reede Ren, and Joemon M. Jose

Dept. of Computing Science, University of Glasgow, Glasgow, G12 8RZ, UK
{anuj,reede,jj}@dcs.gla.ac.uk

Abstract. The curse of dimensionality is a major issue in video indexing. Extremely high dimensional feature space seriously degrades the efficiency and the effectiveness of video retrieval. In this paper, we exploit the characteristics of document relevance and propose a statistical approach to learn an effective sub feature space from a multimedia document collection. This involves four steps: (1) density based feature term extraction, (2) factor analysis, (3) bi-clustering and (4) communality based component selection. Discrete feature terms are a set of feature clusters which smooth feature distribution in order to enhance the discrimination power; factor analysis tries to depict correlation between different feature dimensions in a loading matrix; bi-clustering groups both components and factors in the factor loading matrix and selects feature components from each bi-cluster according to the communality. We have conducted extensive comparative video retrieval experiments on the TRECVid 2006 collection. Significant performance improvements are shown over the baseline, PCA based K-mean clustering.

1 Introduction

Video retrieval attracts great interest from both industry and academic fields. However, existing retrieval systems have a high computational complexity. This is due to two reasons. First, a video document consists of many media modalities such as audio track, textual tags and visual frames. Second, video contents and associated semantics have no direct correlation with low-level features. Moreover, Wang [10] asserts that retrieval is not a simple discrimination on local features but a measurement of uncertainties among possible relevant documents. This indicates that the noise in the feature space will result in extra complexity in the measurement of document relevance and degrade retrieval performance. The optimization on video document representation is therefore essential to improve the effectiveness as well as the efficiency of a video retrieval system.

In this paper, we exploit techniques from statistical information retrieval to learn an efficient feature subspace from media collections. These techniques are: (1) dimension based density normalization; (2) factor analysis in the normalized feature space; and (3) bi-clustering for subspace allocation. This is because the theory of information retrieval has already developed many hypothesis on the feature distribution. These knowledge has not been used by traditional dimensionality reduction methods such as principle component analysis (PCA).

S. Boll et al. (Eds.): MMM 2010, LNCS 5916, pp. 725–730, 2010.

Moreover, retrieval is a statistical decision based on the difference of feature distributions in both document collection and queries [11]. Statistical information from document collections may facilitate the creation of a better feature space. We hence start our work from density normalization by projecting continuous distributed features to a set of discrete variables called feature terms. This projection will maximize the discrimination between documents in a collection. We then use factor analysis to compute the correlation between dimensions in the feature term space and get the loading matrix. To make groups in loading matrix we propose to apply bi-clustering on it. From each bi-cluster we select the component which has minimum communality as a feature subspace for document representation and for relevance computation.

The remainder of this paper is organized as follows. Section 2 brings a brief overview about the literature of textual term and feature subspace selection in content-based video retrieval. Our approach for feature subspace selection is presented in Section 3. Experiment configuration and evaluation results are stated in Section 4. Discussion and conclusion are found in Section 5.

2 Related Work

In this paper, we try to exploit techniques in statistical information retrieval for feature subspace selection. Many essential issues require explanation, such as document representation and relevance estimation.

2.1 Term Distribution

As an important part of term weighting, text term distribution has been well addressed to justify text retrieval models [3,1]. Many hypothesis have been proposed to simulate a general term distribution. Harter *et al.* [3] declare that a term should follow a 2-Poisson distribution, since term appearance is a Boolean random phenomena with a low average arrival rate. This model is extended by Margulis *et al.* [8] who test N-Poisson distributions. The authors hypothesize that N-Poisson might have provided a more precise estimation than a 2-Poisson hypothesis, if a term actually followed a Poisson-like distribution. Several class numbers from two to seven are evaluated on real document collections, although no optimized solution is reached. Amati *et al.* argue that the joint probability of multiple terms is so small that a simple uniform distribution is good enough for the term distribution modeling.

In multimedia retrieval, several approaches have been proposed to extract term-like media features, such as high-level features and SIFT-based local features. Although the distribution of these term-like media features has not been well studied, it is interesting to exploit effective textual models for multimedia documents.

2.2 Feature Subset Selection

Feature subset selection (FSS) is an important optimization approach for multimedia retrieval [4,2]. This technique aims to select the most effective feature components in the document representation without losing performance. The key operation of FSS is to estimate the discriminative power of a feature component. A multi-layer perceptron network is proposed in [6] to classify variables into two groups, effective and ineffective, where a stepwise discrimination is used as input. Principal Component Analysis (PCA) is also widely used to find an optimal solution in data representation. As will be shown later, these methods have many disadvantages.

In this paper, we use factor analysis (FA) for un-supervised selection of feature term components which overcomes the shortcomings of PCA. Furthermore, we apply bi-clustering on a loading matrix to group different feature components. Bi-clustering can separate factor subgroups more efficiently, because unlike K-means clustering, it is flexible to choose any dimensions of a feature as well as any combination of factors, from the loading matrix.

3 Methodology

In this section, we describe our method for feature subspace selection, including density normalization, FA, bi-clustering and feature components selection.

3.1 Density Normalization

Relevance is the core idea behind IR. This measurement is not a distance but a probability on content similarity. As Zhai *et al.* argued [11], relevance is closely associated with distribution density of documents in a collection. Normalizing feature distributions is therefore an effective method to enhance the discrimination between documents and a query. According to the hypothesis of uniform term distribution, we project a document collection to a new feature space, in which documents are sparsely distributed with equal distribution density. For the convenience, a discrete space (feature term space) is used. The extraction of a *feature term* is a projection from a multiple valued N-dimensional variable to an integer, *i.e.* clustering which assigns class labels to data samples. This projection can be symbolized as a function $\hat{f} : [0, K]^N \rightarrow \{0, 1, \ldots, M-1\} \sim \{0, 1\}^M$, where K denotes the range of a feature and M the number of classes. We regard these integers as *feature terms*. In one-dimensional case, $N = 1$.

For a collection D, the frequency of a feature term f_t is the times that a feature falls into a given value interval $t \in [0, M)$ (Equation 1).

$$f_t = |D_t|, D_t = \{d|\hat{f}(d) = t, d \in D\} \tag{1}$$

where d is a document in D. The probability of a *feature term* t is,

$$p(t) = \frac{f_t}{\sum_{i=0}^{M-1} f_i} \tag{2}$$

There are many approaches available to complete the projection from a feature to feature terms and is compared in [9]. We propose the usage of maximized information entropy because of the robustness.

$$Entropy_s(M) = -\frac{1}{\sqrt{M-1}} \sum_{i=0}^{M-1} p(t_i) \log(p(t_i)) \tag{3}$$

After the computation of feature terms, we employ factor analysis to select discriminant components.

3.2 Factor Analysis

We randomly sample video frames from the media collection. A matrix F is therefore generated, in which each row contains feature terms from a visual frame. Factor analysis is applied on the covariance matrix of F which generates the loading matrix Λ.

3.3 Bi-clustering

We try three methods to cluster the loading matrix. The component clustering only considers the overall distance between two components. This distance measurement makes the similarity in factor patterns questionable, as we think about factor combinations as well. In factor clusters, we have a group of different factors which behave almost the same for all components. However, it will miss some factor combinations that behave similarly only for some components, due to the constraint that all objects in a cluster should contain all components. To overcome these problems, we turn to bi-clustering over the loading matrix. Bi-clustering is a two-way data analysis and aims to find subgroups of rows and columns, which are as similar as possible to each other and as different as possible to the rest. The BiMax algorithm [7] is used for bi-clustering the loading matrix Λ.

3.4 Communality Based Feature Selection

We select the component with minimum communality from each bi-cluster for efficient representation. This is because components with minimum communality have minimum variance in common with other feature components and are therefore more discriminative than other components.

4 Experiment and Results

The TRECVid 2006 collection is used for evaluation, including 160 hours news videos and 24 content-based queries (Topic 173-196). For each query topic, from seven to eleven images are used as query examples and a ground truth is provided as a ranked list of 65 to 775 relevant shots. We use the state-of-art of PCA-based K-mean clustering [5] as the baseline. Results from the baseline are denoted by Run 1 in Table.1.

Table 1. rel_ret@1000(r_r@1000), MAP and P@20 for different experiments

		80% Feature Selected						60% Feature Selected					
		r_r@1000		MAP		P@20		r_r@1000		MAP		P@20	
RUN	Description	EH	HT	EH	HT	EH	HT	EH	HT	EH	HT	EH	HT
1	PCA + KMeans Centers(Baseline)	331	160	.0043	.0007	.0417	.0093	340	133	.0057	.0003	.0561	.0063
2	All feature components	384	199	.0066	.0009	**.0625**	**.0146**	same	bec.	no	feat.	sel.	
3	FA + Bi-Clust + Max. Comm.	335	162	.0059	.0004	.0458	.0125	331	123	.0067	.0005	.0583	.0063
4	FA + Bi-Clust Centers	360	170	.0074	.0005	.0500	.0083	347	135	.0057	.0003	.0396	.0063
5	FA + Bi-Clust + Min. Comm.	**389**	208	.0083	**.0012**	**.0625**	.0104	351	199	.0074	.0010	.0583	.0125
6	Density Norm. + FA + Bi-Clust + Min. Comm. (Proposed)	382	**256**	**.0109**	.0010	**.0625**	.0042	345	**256**	**.0100**	**.0013**	**.0646**	.0104

Density normalization is carried out on the entire TRECVid 2006 collection. Keyframes are sampled every ten visual frames and we compute feature distribution across the sample collection. Since, factor analysis is of high computational complexity, we randomly selected 100 frames from the collection for factor analysis and bi-clustering.

Two MPEG-7 visual features, edge histogram (EH) (80 components) and homogeneous texture (HT) (62 components) are extracted, as both of features are of high dimensionality. The number of components are decided by the number of bi-clusters, as we select one component from each bi-cluster. In addition, the number of bi-clusters can be changed by users. For the convenience, we fixed the number of selected components to a given ratio of the original size, *i.e.* 80% and 60% respectively. The Euclidean distance between feature terms is used to calculate the dissimilarity between query examples and keyframes. The top 1000 shots that are of minimum distance from any query example will be returned as query results.

In Table 1, six runs are stated: Run-1 represents the baseline, Run-2 uses all feature components without any component selection; Run-3, Run-4 Run-5 and Run-6 are experiments with factor analysis and bi-clustering, but with different configurations. Run-3 tests the maximum communality components in bi-clusters; Run-4 uses components nearest to bi-cluster center; Run-5 selects the minimum communality components from each bi-cluster. Run-3/4/5 show the effectiveness of factor analysis and bi-clustering in feature subspace learning but work on the original low-level features, that is without density normalization. Run-6 is our proposed approach which combines density normalization, factor analysis, bi-clustering and minimum communality based component selection. Run-6 proves the effectiveness of density normalization. Run-1 and 4 highlight the performance difference between PCA and FA based methods. In all configurations, both of low-level feature and both of the given ratio of components, factor analysis performs better than PCA. Run-3 and 5 verify the assumption that component with minimum communality is the most discriminating component in a bi-cluster. Experimental results strongly supports this assumption, as Run-5 significantly outperforms Run-3.

5 Conclusion

In this paper, we propose a statistical strategy to facilitate feature subset selection. The highlight of this work is the exploitation of the hypothesis from statistical information retrieval, which adapt a traditional feature selection scheme to the application of video retrieval. Experimental results show that our approach outperforms PCA-based K-mean clustering.

Acknowledgments

This research was supported by the EC under contract SALERO (FP6-027122) and SEMEDIA (FP6-045032).

References

1. Amati, G., Rijsbergen, C.J.V.: Probabilistic models of information retrieval based on measuring the divergence from randomness. ACM Transactions on Information Systems (TOIS) 20(4), 357–389 (2002)
2. Collins, R., Liu, Y., Leordeanu, M.: Online selection of discriminative tracking features. IEEE Trans. PAMI 27(10), 1631–1643 (2005)
3. Harter, S.: A probabilistic approach to automatic keyword indexing, part i on the distribution of speciality words in a technical literature. Journal of the ASIS 26, 197–216 (1975)
4. Jiang, W., Er, G., Dai, Q., Gu, J.: Similarity-based online feature selection in content-based image retrieval. IEEE Transactions on Image Processing 15(3), 702–712 (2006)
5. Jolliffe, I.: Principal Component Analysis. Springer, New York (1986)
6. Lin, T.-S., Meador, J.: Statistical feature extraction and selection for ic test pattern analysis. IEEE International Symposium on Circuits and Systems 1, 391–394 (1992)
7. Madeira, S., Oliveira, A.: Biclustering algorithms for biological data analysis: a survey. IEEE/ACM Transactions on Computational Biology and Bioinformatics 1(1), 24–45 (2004)
8. Margulis, E.: N-poisson document modelling. In: SIGIR 1992, pp. 177–189. ACM Press, New York (1992)
9. Ren, R., Jose, J.M.: Query generation from multiple media examples. In: 7th International Workshop on Content-Based Multimedia Indexing, pp. 138–143 (2009)
10. Wang, J.: Mean-variance analysis: A new document ranking theory in information retrieval. In: Boughanem, M., et al. (eds.) ECIR 2009. LNCS, pp. 4–16. Springer, Heidelberg (2009)
11. Zhai, C., Lafferty, J.: A risk minimization framework for information retrieval. Inf. Process. Manage. 42(1), 31–55 (2006)

A Novel Retrieval Framework Using Classification, Feature Selection and Indexing Structure

Yue Feng[1], Thierry Urruty[2], and Joemon M. Jose[1]

[1] University of Glasgow, Glasgow, UK
[2] University of Lille 1, Lille, France
yuefeng@dcs.gla.ac.uk, thierry.urruty@lifl.fr, jj@dcs.gla.ac.uk

Abstract. In this paper, we propose a framework to consider both the efficiency and effectiveness to achieve the trade-off in performance of Content Based Image Retrieval (CBIR). This framework includes: (i) concept based classification to classify images into different semantic concept groups and narrows down the search domain in retrieval; (ii) Feature selection model to analysis the relationship between queries and concept classes to reduce feature dimension; (iii) Multidimensional vector space indexing structure for real-time access to reduce the retrieval cost. In our experiments, we study the efficiency and the effectiveness of our method using one public collection and compared with one of state of the art methods.

1 Introduction

A good image retrieval model can not only help users achieve efficient and effective organisation of large image databases but also helps to bridge the gap between raw low level image features and high level semantic concepts. Bag of words [5] recently received a considerable amount of focus due to the benefit of extracting semantic information. Tirilly etal. [8] extract a lexicon of 50-100 visual concepts and employ these concepts in query-by-example video retrieval. They claimed that (1) semantic concepts were effective and efficient in content-based video retrieval and (2) a large concept lexicon was decisive to retrieval performance. Later, some large lexicons of visual concepts were developed and showed a good performance in broadcasting news video retrieval [2]. However, this new approach also introduces many challenges. The availability of visual concepts relies on the effectiveness of related detectors. The detection precision of some complex specified concepts is below 0.3 [7]. This indicates that a Boolean matching of these concepts may lead to a faulty decision at a high probability. In addition, too many concepts significantly increase the difficulty and the cost in the development of concept detectors. A small, generic, and reliable concept set might out-perform a large but incredible visual lexicon in content-based video retrieval.

The indexing based retrieval model is famous for its fast access into the data, but most of these methods are not efficient enough for multimedia data, which

S. Boll et al. (Eds.): MMM 2010, LNCS 5916, pp. 731–736, 2010.

contains high dimensionality and high level semantics. Existing indexing techniques perform well for some databases and poorly for others. The performance of the algorithms generally depends on the workload and sequential scan remains an efficient search strategy for similarity search.

Motivated by the above problems of (i) solving the semantic gap in image retrieval and (ii) improving the effectiveness without reducing too much efficiency, we propose a new retrieval model containing a classification, a feature selection and an indexing method. This process mainly aims at optimising the query processing time without reducing precision too much.

The rest of this paper is structured as follows. Section 2 presents the details of our proposed retrieval framework. The experimental results on TRECVid2008 corpus are shown in Section 3. Finally, Section 4 concludes the paper.

2 Methodology

The proposed methodology begins from image concept classification to group images into different concepts, then feature selection are followed to reduce the feature dimensions thereby reducing the processing cost. The indexing structure then constructs a geometrical representation of the data in the vector space for each category. Finally, a similarity measure is applied for retrieval.

2.1 Generic Image Concept Classification

Use of low level features can not give satisfactory retrieval results in many cases, especially when the high level concepts in the user's mind are not easily expressible in terms of the low level features. Given the conclusion that different images can be classified into different concept groups based on their coarse scene information [4], we used gist feature in classification.

The original SVM is designed for binary classification, which is hard to apply on the images with multiple concepts. To solve this problem, we downgrade the multiclass problem to a set of binary problems. First, a set of binary classifiers, each was trained to separate one class from the rest, is built. Each classifier represents one image concept. In other words, n hyperplanes are constructed, where n is the number of pre-defined class. Each hyperplane separates one class from the others using the following decision function. Thus, a new data point x is classified into the class with the largest decision function, $argmax_k f_k(x)$.

$$f(x) = sgn\left(\sum_{i=1}^{N} \alpha_i^0 y_i x_i \cdot x + b_0\right) \quad (1)$$

Different from the results of the traditional SVM, where each document is classified into one class and given a single label, we used a ranking label to represent the classification result. The multiple label is calculated using the distance of an image to each class.

2.2 Feature Selection

To represent the images after classification, we used four MPEG-7 standard visual features [3], which are colour structure (CS), colour layout (CL), homogeneous texture (HT) and edge histogram (EH), containing 410 dimension.

Considering (i) different queries contain different contents, the performance of each feature for different content is mostly not the same; (ii) the processing time will then be dominated by the dimension. Thus, using features irrelevant to the problem can seriously hamper the accuracy and increase the expensiveness.

In this paper, we applied a class specific feature selection method to find the best features for each different class. First, it uses SVM to evaluate each feature's potential with respect to each class. Second, we compute the F1-measure of each class given by the SVM classifier using different features, where the best feature is one with maximum F1-measure,

$$F1_c(f) \;=\; \frac{2 \times p_c(f) \times r_c(f)}{p_c(f) \;+\; r_c(f)} \qquad (2)$$

where, $p_c(f)$, $r_c(f)$ and $F1_c(f)$, the precision, recall and the F1-measure of a class $c \in \Omega$ for the low level visual feature $f \in \{CL,\; CS,\; EH,\; HT\}$ respectively, where Ω is the whole set of classes.

2.3 Indexing Structure

Given classification and feature extraction results, an indexing structure method is applied to (i) improve the retrieval efficiency and (ii) reduce the data storage volume.

To construct the indexing structure, we used a random projection clustering algorithm that is adapted to high dimensional data space. This clustering algorithm combines the ideas from random projection techniques and density-based clustering. More detailed can be found in [10]. All clusters are then indexed in a pyramid based indexing structure. The indexing structure construction and its performance are more detailled in [9].

2.4 Retrieval Methodology

Given a query topic with multiple query examples, the distance between each query and each concept category is first computed. Thus, a subset S^k_Ω of the whole collection with lowest distance can be chosen for each query example. We denote k, the number of selected classes, where $k \leq card(\Omega)$.

Given this selected subset S^k_Ω, we first retrieve the top K nearest neighbors of the query using the estimated "best" features of this subset. Thus, the retrieval algorithm only need access to the number of k selected classes and uses a number of α "best" features, instead of the whole collection with all features.

We denote $R(f, S^k_\Omega)$, the relevance of a low level feature f with respect to the subset S^k_Ω. This relevance of each visual feature is computed based on the

F1-measure $F1_c$ of each class c obtained as presented in Section 2.2 with the following formula:

$$R(f, S_{\Omega}^{k}) = (\sum_{c=0}^{k} F1_c(f)) \quad (3)$$

In order to measure the similarity amongst images inside the subset S_{Ω}^{k}, the "best" visual features are combined for ranking using distance measures recommended by Mpeg-7 [3].

This measure delivers a ranked list of images in the selected categories. It puts the relevant documents in the top of the ranked list, minimizing the time the user has to invest on acquiring the results.

3 Experiments

Our experiments are based TRECVid 2008 [6] collection, which contains 35,000 video shots corresponding to 730,000 none annotated key-frames.

In this experiment, TRECVid 2007 collection is used for training, while TRECVid 2008 collection are for testing and querying. According to the semantic of the 48 topics in TRECVid 2008 , 6 generic and 1 specific concept categories are pre-defined. Given the annotation ground truth of TRECVid 2007, 700 relevant key-frames are formed 7 categories for training purpose. For the testing set, 730,000 extracted key-frames are used. Finally, the performance is evaluated by TRECVid evaluation tool.

3.1 Experiments on TRECVid 2008 Collection

In this experiment, we evaluate our performance by comparing with (1) a baseline and (2) a state of the art method in TRECVid 2008 competition, where the baseline methods uses sequential scan with 7 classes (k) and 4 features (α).

Figures 1a and 1b show the *P@20* and *P@100* with respect to the number k of classes used. Figure 1a highlights the fact that selecting more than one class results in a huge increase in precision whatever α is. Except for $\alpha = 3$, selecting 2 classes is overall the best for the parameter k and using $\alpha = 4$ provides the best precision results. Meanwhile, similar observations and conclusions can be made in Figure 1b. In addition, the results from this figure are more stable compared to Figure 1a. It can be clearly observed that the more features selected, the better precision it achieved. In addition, the precision of using 2 classes is equivalent to those using more classes.

Figure 1c presents the number of relevant results retrieved with respect to k. The same observation from Figure 1a and Figure 1b can be made. However, this figure highlights that the baseline results are not necessarily the best. Using 2 to 5 classes and all features will improve the results slightly.

Figure 1d shows the average query processing time for one query image over the whole 48 topics. These results show that the time is almost linear. We also found that using an indexing structure helps save at least 20% of the time.

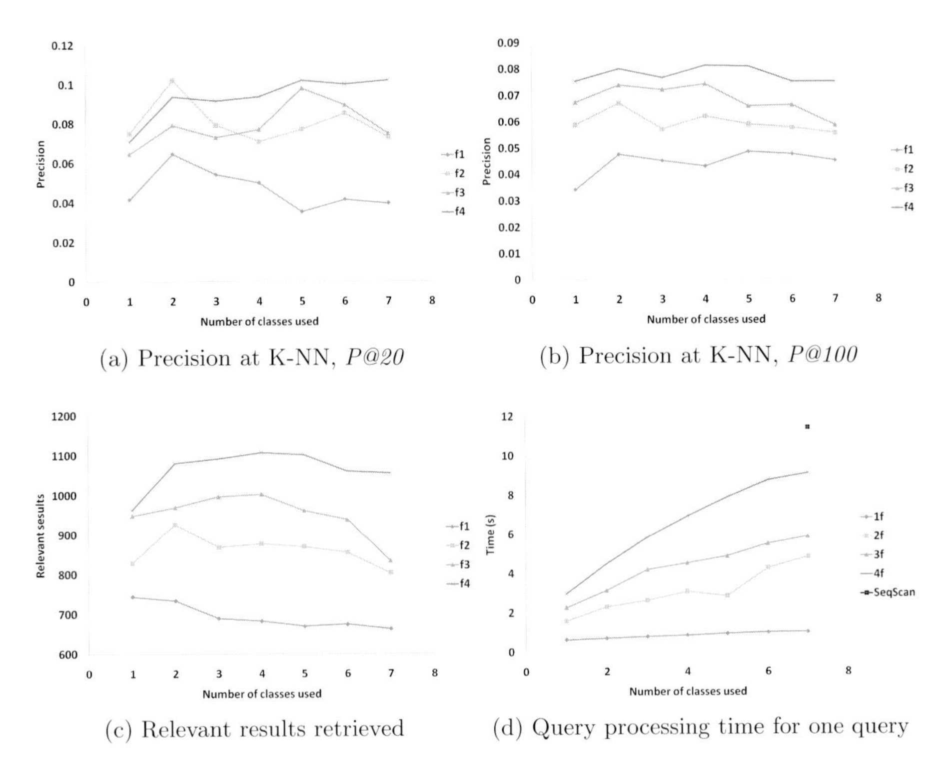

(a) Precision at K-NN, *P@20*

(b) Precision at K-NN, *P@100*

(c) Relevant results retrieved

(d) Query processing time for one query

Fig. 1.

Table 1 presents the results from the baseline and IBM's well performed method [1] in TRECVid 2008 competition and the proposed method. The IBM gives the best performance in terms of *P@100* and the baseline has the highest in relevant results retrieved. However, the time is 600+ and 9.1 seconds per query, respectively, which is not viable for online query processing. Comparing with the proposed results, using 2 classes with the 2 best features is a good choice.

Table 1. Selection of different values for α and k for TRECVid collection

(α, k)	P@100	Relevant	Time (s)
(1, 2)	0.09	744	0.7
(2, 2)	***0.13***	***926***	***2.3***
(3, 4)	0.15	1002	4.5
(4, 4)	0.16	1107	6.9
(4, 7)	0.15	1055	9.1
Baseline	0.15	1055	11.38
IBM	0.23	712	600+

Although it decreased *P@100* and the number of relevant results retrieved by 45% and 60%, it also reduced the processing time by 99.6% and 75% compared with IBM and the baseline, respectively. Thus, selecting 2 classes with 2 best features is the best trade off between precision and time.

4 Conclusion

In this paper, we discussed the possibility of combining classification, feature selection and an indexing structure in an application of CBIR. Our experiment on one real video collection showed the performance of our proposal, that is, to solve the semantic gap and to achieve the trade off between effectiveness and efficiency.

References

1. Tesic, J., Xie, L., Yan, R., Jiang, W., Natsev, A., Smith, J.R., Merler, M.: IBM research trecvid-2008 video retrieval system. In: TRECVid 2008, USA (2008)
2. Hauptmann, A.: Towards a large scale concept ontology for broadcast video. In: Enser, P.G.B., Kompatsiaris, Y., O'Connor, N.E., Smeaton, A., Smeulders, A.W.M. (eds.) CIVR 2004. LNCS, vol. 3115, pp. 674–675. Springer, Heidelberg (2004)
3. Martinez, J.M.: Mpeg-7 overview (2004), `http://www.chiariglione.org/mpeg/standards/mpeg-7/mpeg-7.htm`
4. Oliva, A., Torralba, A.: Modeling the shape of the scene: A holistic representation of the spatial envelope. Computer Vision 42(3), 145–175 (2001)
5. Sivic, J., Zisserman, A.: Video Google: A text retrieval approach to object matching in videos. In: ICCV 2003, October 2003, vol. 2, pp. 1470–1477 (2003)
6. Smeaton, A.F., Over, P., Kraaij, W.: Evaluation campaigns and trecvid. In: MIR 2006 workshop, NY, USA, pp. 321–330 (2006)
7. Snoek, C.G.M., Worring, M., van Gemert, J.C., Geusebroek, J.-M., Smeulders, A.W.M.: The challenge problem for automated detection of 101 semantic concepts in multimedia. In: ACM MULTIMEDIA 2006, pp. 421–430. ACM, New York (2006)
8. Tirilly, P., Claveau, V., Gros, P.: Language modeling for bag-of-visual words image categorization. In: CIVR 2008, pp. 249–258. ACM, New York (2008)
9. Urruty, T., Djeraba, C., Jose, J.M.: An efficient indexing structure for multimedia data. In: Proceedings of ACM MIR 2008, Vancouver, Canada. ACM, New York (2008)
10. Urruty, T., Djeraba, C., Simovici, D.A.: Clustering by random projections. In: Perner, P. (ed.) ICDM 2007. LNCS (LNAI), vol. 4597, pp. 107–119. Springer, Heidelberg (2007)

Fully Utilized and Low Design Effort Architecture for H.264/AVC Intra Predictor Generation*

Yiqing Huang, Qin Liu, and Takeshi Ikenaga

Graduate School of Information, Production and Systems, Waseda University, Japan
nestastam@ruri.waseda.jp

Abstract. Fully exploiting the spatial feature of image makes H.264/AVC standard superior in intra prediction part. However, when hardware is considered, full support of all intra modes will cause high design effort, especially for large image size. In this paper, we propose a low design effort solution for intra predictor generation, which is the most significant part in intra engine. Firstly, one parallel processing flow is given out, which achieves 37.5% reduction of processing time. Secondly, a fully utilized predictor generation architecture is given out, which saves 77.5% cycles of original one. With 30.11k gates at 200MHz, our design can support full-mode intra prediction for real-time processing of 4k×2k@60fps.

Keywords: Intra Prediction, H.264/AVC, Hardware Architecture.

1 Introduction

The latest video coding standard H.264/AVC provides superior coding performance to previous ones. The improvement of H.264/AVC is mainly due to the introduction of many new techniques. However, it also brings about complexity problem. In H.264/AVC, spatial and temporal information is fully utilized to achieve high compression ratio. There are totally 7 inter modes with different block size for motion estimation. In terms of spatial feature, nine 4×4, four 16×16 luminance intra modes are introduced. The final best modes is decided among all these inter and intra modes by analyzing rate distortion cost. So, the overall complexity is quite significant considering the real-time encoding process.

In hardware field, [1] firstly gives out one 4-stage real-time encoder and intra prediction (IP) engine is arranged in one single stage due to the huge complexity. For the whole IP engine, the most significant part is the intra predictor generation. As listed in [2], in one 4×4 sized sub-block, there are totally 30 cycles required for generating predictors of all intra 4×4 prediction modes (I4MB) and 10 cycles for intra 16×16 modes (I16MB). Since sixteen 4 × 4 sub-blocks exist in one MB, the total cycles will around 640. Although fast algorithms such as [3] [4] can achieve reduction of candidate intra mode to some extent, full support of all

* This work was support by CREST, JST and GCOE Program.

S. Boll et al. (Eds.): MMM 2010, LNCS 5916, pp. 737–742, 2010.

modes in hardware is a must to keep the video quality. In the worst case, all the prediction modes are required for the system. Moreover, the minimum required frequency (Req_Freq) for predictor generation will determine the design effort for the whole engine. Here, the Req_Freq is defined as Eq. 1. The cyc_per_MB is the required processing clock cycles for one MB and fps is the frames to be encoded every second. The frm_w and frm_h are the width and height of each frame. According to Eq. 1, When specification is extended to Full HD (1080p) or 4k×2k@60fps, the existing sequential way in [2] will cause extreme high design effort(1.24GHz), which is impossible to be accomplished.

$$Req\_Freq = cyc\_per\_MB \times fps \times \frac{frm\_w \times frm\_h}{256} \quad (1)$$

In our paper, we propose a low design effort intra predictor generation method. Firstly, by analyzing the data dependency, one 2-block parallel processing flow is proposed, which saves 37.5% processing time. Secondly, one dedicated fully utilized hardware architecture is given out, which simultaneously generates predictors of all modes. So, the number of processing cycle is further reduced.

The rest of paper is organized as follows. In section 2, the data dependency problem in existing work is analyzed and the proposed parallel processing flow is given out. In section 3, the feature and data flow of proposed hardware engine are analyzed in detail. Section 4 shows the experimental results and comparisons with other works. This paper concludes with section 5.

2 Parallel Processing Flow for Intra Predictor Generation

In literature [2], the whole intra predictor generation is based on the 4×4 sub-block scale. One 16×16 MB is separated into sixteen 4×4 sub-blocks. The processing flow is based on the raster scan order because of the data dependency problem. Each mode requires 4 clock cycles. There are some bubbles between two sub-blocks. In [2], the bubble period is fully utilized by inserting predictor generation of intra 16×16 modes. The I16MB modes are also organized in 4×4 sub-block scale. In fact, such kind of processing order is not a must and parallel scheme can be achieved lossless.

Fig. 1 is our proposed processing flow. We use circle marked with number to indicate each stage. Firstly, for current MB in process, the original '16-stage' based flow is optimized into '10-stage' way. So, about 37.5% processing time is reduced. From Fig. 1, it is also obvious that our proposal is a lossless optimization toward original raster scan order. In the first MB, 4×4_blk1 and 4×4_blk2 are individually processed in two stages. In the following part, predictor generation is in the form of 2-block scale. For example, when handle 4×4_blk3 in [2], the predictor generation of 4×4_blk5 can be executed together with 4×4_blk3 with no quality loss. Also, the last two stages of current MB are handled together with first two stages of the next MB, as shown in the top of Fig. 1. Full hardware utilization can be achieved during the whole intra predictor generation process.

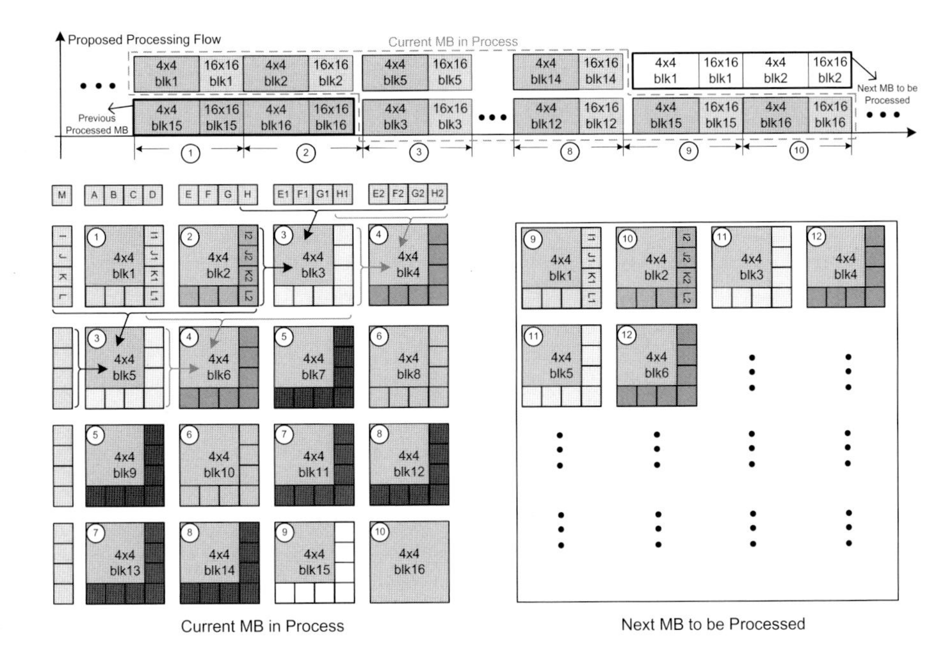

Fig. 1. Proposed Processing Flow

3 Fully Utilized Intra Predictor Generation Architecture

Previous section gives out one 2-block based parallel flow and 37.5% processing time is reduced. However, such adoption is also not enough for low effort design.

In [2], except horizontal and vertical modes (only use 1 cycle), the required cycles for rest I4MB or I16MB mode are in the period of 4, which means that the 16 predictors of one 4×4 sub-block can be obtained after 4 clock cycles. So, the total cycles for generating all luminance predictors of I4MB and I16MB modes are 640. When this structure is extended into Full HD or 4k×2k@60fps, the design effort will be increased to 157MHz and 1.24GHz, which is beyond maximum work frequency (55MHz). In fact, data reuse can be achieved among nine I4MB modes. Table. 1 demonstrates the calculation of first four predictors of I4MB mode. To simplify the description, the shift operations for generating final result are omitted. It is shown that value of some predictors within same I4MB mode or across different I4MB modes are the same. For example, we use bold fonts to mark the predictors with value (A+2B+C). It is obvious that three I4MB modes have this value. In fact, for one 4×4 sub-block, this value occurs 7 times among five I4MB modes. So, many operations are wasted. In our design, we fully enable the data reuse among all I4MB modes and propose one fully utilized predictor generation engine, as shown in Fig. 2. Two pipeline stages are inserted to output the results. Large multiplexors in original design are replaced with several small multiplexors. Predictors of all the I4MB and I16MB modes can be obtained after two cycles. The details are given in following parts.

Table 1. Predictors of I4MB Modes in 4×4 Sub-block

Pred(y,x)	V	H	DC	DDL	DDR	VR	HD	VL	HU
Pred(0,0)	A	I	Z	**A+2B+C**	I+2M+A	M+A	I+M	A+B	J+I
Pred(0,1)	B	I	Z	B+2C+D	M+2A+B	A+B	I+2M+A	B+C	K+2J+I
Pred(0,2)	C	I	Z	C+2D+E	**A+2B+C**	B+C	M+2A+B	C+D	K+J
Pred(0,3)	D	I	Z	D+2E+F	B+2C+D	C+D	**A+2B+C**	D+E	L+2K+J

Z=L+K+J+I+A+B+C+D, V: Vertical, H: Horizontal,
DDL: Diagonal Down Left, VR: Vertical Right, VL: Vertical Left,
DDR: Diagonal Down Right, HD: Horizontal Down, HU: Horizontal Up

For I4MB modes (except DC mode), the predictors within one 4×4 sub-block can be obtained by configuring Fig.2 into Fig.3. The bold blue arrow is the selected path. The input of Fig.3 is the left, up and up-right pixels of current sub-block (for instance, A to H, I to L, and M for 4×4_blk0 in Fig. 1). The output result after two clock cycles can be traced with selected path. From Fig.3, it is shown that predictors from O1 to O8 equal to the input values; and these values are output at the 1st pipeline stage together with O9 to O20 in Fig.3. For rest predictors (O21 to O33), they are output and stored at 2nd stage.

For I16MB modes, the horizontal and vertical modes can be easily implemented by our structure in Fig.2. As for I16MB plane mode, we can also generate all 16 predictors of one 4×4 sub-blocks by using 2 cycles. As defined in standard, Eq.2 is the calculation of plane predictor in each position (Pred(y,x)), where a, b, c are constant value for one MB and they can be calculated based on Eq.3 to Eq.4. Pel(-1,15) and Pel(15,-1) are pixels from previous MBs. The UR_w and LC_w are sum of weighted differences of upper row and left column, respectively. So, we change Eq.2 to Eq.5 to realize plane mode. The Sd are the seed value depending on the location of 4×4 sub-block. There are four seed value namely Sd1 to Sd4 listed in Eq.6 to Eq.7. Each Sd is for one column of 4×4 sub-blocks. For example, Sd1 is used for blk1, blk5, blk9, blk13; Sd3 is used for blk3, blk7, blk11, blk15. The difference of blk5 to blk1 is only 4c and this value can be

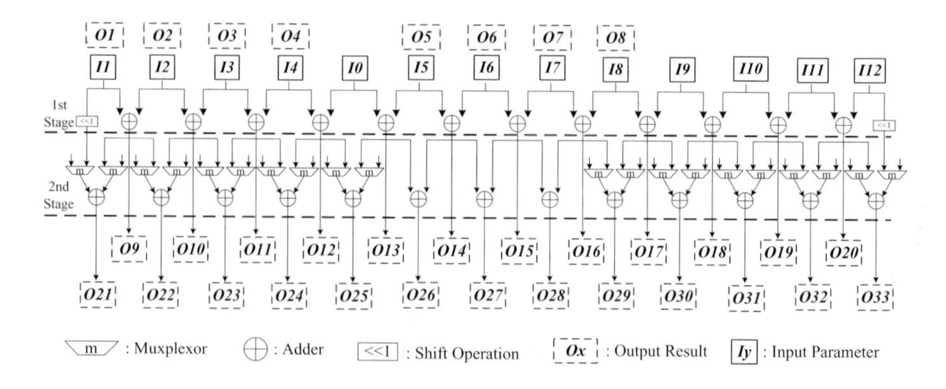

Fig. 2. Proposed Predictor Generation Engine

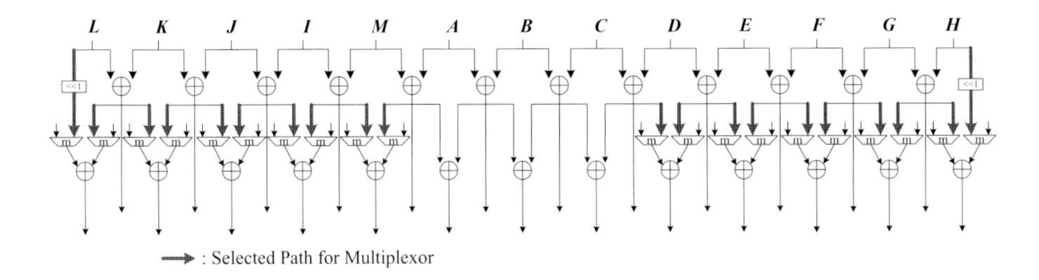

Fig. 3. Proposed Architecture for I4MB Modes

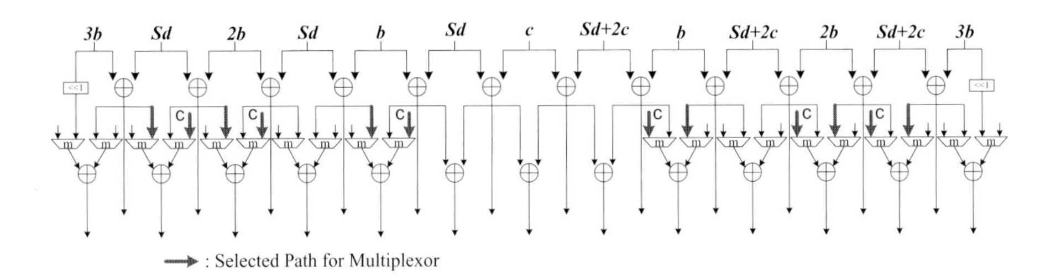

Fig. 4. Proposed Architecture for I16MB Plane Mode

supplemented during the shift operation. The configuration and input data of I16MB plane mode are shown in Fig.4. The output can be easily deduced.

For I16MB and I4MB DC modes, the results are average of upper and left pixels. The difference is the width and height of two DC modes, which decides the number of required pixels. By combining our structure with compressor tree structure, it is easy to realize DC modes. Also, With our structure, the four chroma 8×8 modes can be implemented with analogy.

$$Pred(y,x) = (a + b \times (x-7) + c \times (y-7) + 16) >> 5 \tag{2}$$

$$a = 16 \times Pel(-1,15) + 16 \times Pel(15,-1) \tag{3}$$

$$b = (5 \times UR_w + 32) >> 6, \qquad c = (5 \times LC_w + 32) >> 6 \tag{4}$$

$$Pred(y,x) = ((Sd + x \times b) + y \times c) >> 5 \tag{5}$$

$$Sd1 = a - 7b - 7c + 16, \qquad Sd2 = a - 3b - 7c + 16 \tag{6}$$

$$Sd3 = a + b - 7c + 16, \qquad Sd4 = a + 5b - 7c + 16 \tag{7}$$

4 Experimental Result

The proposed structure is synthesized with TSMC 0.18um technology. The result is shown in Table. 2. Since two parallel engines are used in our design, the hardware cost of our design is larger than previous one. However, considering the whole encoder design, 30k gates is not a significant value. The merit of our

Table 2. Experimental Result and Comparison

Design	Technology	Gate Count	Max Freq.	Max Spec.
[2]	0.18um	12945	55MHz	SDTV@31fps
ours	0.18um	30112	200MHz	4kx2k@60fps

Table 3. Comparison of Processing Cycles for One 4×4 Sub-block

Design	I4MB DC	I4MB rest	I16MB DC	I16MB rest	Total	Req_Freq for 4k×2k@60fps
[2]	4	26	4	6	40	1.24GHz
ours	1	2	3	3	9	175MHz

architecture is very obvious. The whole architecture is highly pipelined with simple and regular structure. The maximum work frequency is about 4 times than previous design. Also, as shown in Table.3, for one 4×4 sub-block, instead of using 40 clock cycles for all the modes in I4MB and I16MB, our architecture only needs 9 cycles, which saves 77.5% cycles. Moreover, the related design effort is greatly reduced when extending to higher specification. For example, by extending structure in [2] to 4k×2k@60fps, the minimum required frequency (Req_Freq) will become 1.24GHz. By using our structure with parallel scheme, only 175MHz is required to fulfill the throughput of 4k×2k@60fps real-time processing.

5 Conclusion Remarks

One fully utilized and low design effort engine for H.264/AVC intra predictor generation is proposed. Firstly, the data dependency problem in the conventional flow is analyzed and one parallel flow is given out, which achieve 37.5% reduction in processing time. Secondly, the proposed architecture can generate predictors of all I4MB and I16MB modes with only 22.5% cycles of previous design. Based on our parallel processing flow and fully utilized architecture, within 200MHz, our design can achieve real-time intra predictor generation for 4k×2k@60fps.

References

1. Huang, Y., et al.: A 1.3tops h.264/avc single-chip encoder for HDTV applications. In: ISSCC 2005, Feburary 2005, pp. 128–130.
2. Huang, Y., Hsieh, B., Chen, T., Chen, L.: Analysis, Fast Algorithm, and VLSI Architecture Design for H.264/AVC Intra Frame Coder. IEEE Trans. on CSVT 15(3), 378–401 (2005)
3. Pan, F., Lin, X., Rahardja, S., Lim, K.: Fast Mode Decision Algorithm for Intraprediction in H.264/AVC Video Coding. IEEE Trans. on CSVT 15(7), 813–822 (2005)
4. Tian, G., Zhang, T., Ikenaga, T., Goto, S.: A Fast Hybrid Decision Algorithm for H.264/AVC Intra Prediction Based on Entropy Theory. In: Huet, B., et al. (eds.) MMM 2009. LNCS, vol. 5371, pp. 85–95. Springer, Heidelberg (2009)

A Database Approach for Expressive Modeling and Efficient Querying of Visual Information

Ahmed Azough[1], Alexandre Delteil[1], Mohand-Said Hacid[2], and Fabien DeMarchi[2]

[1] Orange Labs, 38-40 rue du Général Leclerc, 92130 Issy-les-Moulineaux, France
[2] Université de Lyon, Batiment Nautibus, 8 boulevard Niels Bohr, 69622 Villeurbanne, France

Abstract. With the rapid development of multimedia technologies, the number of available multimedia resources is always increasing and the need for efficient multimedia modeling, indexing and retrieval techniques is growing. In this work, we propose a hierarchical, hybrid and semistructured data model for representing video data. Based on this model, we build a declarative, rule-based, constraint query language enabling to infer and to retrieve spatial, temporal or semantic relationships from information represented in the model and to intentionally specify relationships among objects and events. We introduce the concept of temporal and spatial frame of reference which allows to simultaneously locate video contents according to multiple spatiotemporal environments in real world. Our model and query language are extensible, application independent, expressive enough and quite suitable for multimedia information retrieval.

1 Introduction

Video contents need to be indexed so that answers to queries can be quickly computed. Studies have shown that most user queries are expressed using high level (i.e. semantic) concepts. However, structured and accurate annotation is still lacking for the vast majority of multimedia documents. The stratification approach [1] was one of the first video indexing schemes for video documents. This approach consists of associating each element of interest to an interval called stratum by specifying several levels of description. Hacid et al [2] have extended the stratification concept by defining temporal cohesion. Temporal cohesions allow a set of time segments to be associated with the same description. However, this work was limited to temporal modeling and did not take into account spatial modeling. In [3], an extension of AVIS system to spatial dimension is presented. However, little semantics can be inferred and stored in the data model. BilVideo system was presented in [4] as an original query system allowing to represent spatial, temporal and semantic information of objects in video documents. Nevertheless, the systems deals only with the spatial properties corresponding to the coordinates of objects on the screen. Objects and events might be spatially positioned following several frame references (e.g. soccer players should be located according to their position in the playfield in addition to their position on the screen).

In this work we propose a hierarchical, hybrid (i.e. low-level and high-level) and semistructured data model for representing video data. Based on this model, a declarative, rule-based, constraint query language that has a clear operational fixpoint and set-theoretical semantics is presented. The data model allows to attach, in multiple granularities, a segment of space or time to a set of objects, events and relations through

S. Boll et al. (Eds.): MMM 2010, LNCS 5916, pp. 743–748, 2010.

attribute value pairs describing their spatiotemporal positioning following multiple frames of reference.

2 Data Modeling

Our video data model and the underlying query language provide wider and more expressive spatiotemporal description. Some basic concepts relevant to our formal model are defined in this section.

Definition 1 (Spatio-temporal Frames of Reference). *A frame of reference (FoR) is a coordinate system used to measure the position of objects in it. It consists of an origin, a set of axis and a variable on each axis.*

Several frames of reference are used in this work. Among spatial *FoR*s, we use "Geographic" for placing objects in cities or countries..., "soccer playfield" for positioning players on play zones, "Residences" for locating people is different buildings... Among temporal *FoR*s, we use "Calendar" for using date, "soccer time" for locating the soccer actions following the soccer timing...

Definition 2 (Temporal Constraint). *An atomic temporal constraint is a formula of the form t Θ t' or t $\leq$ c where t and t' are variables, c is a constant and Θ is one of* $=, \leq, <, \neq, \geq, >$. *A complex temporal constraint is a boolean combination built from (atomic or complex) constraints by using logical connectives.*

Definition 3 (Duration). *A Duration is a pair $(c(t), tF)$ where $c(t)$ is a temporal constraint (t is a variable), and tF is a temporal frame of reference.*

For a given duration d, we denote $d.tmp\_const$ its temporal constraint and $p.for$ its temporal frame of reference.

Definition 4 (Spacial Constraint). *An atomic spatial constraint on a n-tuple of variables $(x_1, ..., x_n) \in \mathbb{R}^n$ is a formula of the form $f(x_1, \ldots, x_n)\ \Theta\ 0$, where n is called the dimension of the constraint and Θ is one of* $=, \leq, <, \neq, \geq, >$. *A complex spatial constraint is a boolean combination built from (atomic or complex) constraints by using logical connectives.*

Spatial constraints can be referred to by predefined nouns like names of cities, countries, etc. $city('London')$ is an example of spatial constraint.

Definition 5 (Position). *A position is a pair $(c(x_1, \ldots, x_n), sF)$ where $c(x_1, \ldots, x_n)$ is a spatial constraint and sF is a spatial frame of reference.*

For a given position p, we denote $p.spc\_const$ its spatial constraint and $p.for$ its spatial frame of reference.

Most of the time, people refer to video content using descriptors such as objects, events, attributes and relations. In order to build the data model of the video database, we assume the existence of the following countably infinite and disjoint sets:

- video identifiers : $\mathcal{ID}_{vid} = \{vid_1, vid_2, ...\}$
- object identifiers : $\mathcal{ID}_{obj} = \{oid_1, oid_2, ...\}$
- event identifiers : $\mathcal{ID}_{evt} = \{eid_1, eid_2, ...\}$
- object types : $\mathcal{C}_{obj} = \{c_1, c_2, ...\}$
- event types : $\mathcal{E}_{evt} = \{e_1, e_2, ...\}$
- spatial FoRs : $\mathcal{SFOR} = \{sF_1, sF_2, ...\}$,
- temporal FoRs : $\mathcal{TFOR} = \{tF_1, tF_2, ...\}$,
- relations : $\mathcal{R} = \{R_1, R_2, ...\}$,
- attributes : $\mathcal{A} = \{A_1, A_2, ...\}$,
- (atomic) constants : $\mathcal{D} = \{d_1, d_2, ...\}$.

Definition 6 (Video Location). *A video location l is defined as a triple $(vid, c_t(t), c_s(x, y))$ where vid is a video identifier, $c_t(t)$ is a temporal constraint (see definition 2) representing a temporal segment of video vid, and $c_s(x, y)$ is a spatial constraint (see definition 4) representing an area of the 2D screen.*

For a given video location l, we denote $l.vid$ the video identifier, $l.tmp\_const$ the temporal constraint and $l.spc\_const$ the spatial constraint.

Definition 7 (Attribute Value set). *Let A be an attribute in $\mathcal{A}$. We define an attribute value set associated with A as a set of tuples (val_i, l_i) where val_i is a value [2] and l_i is the video location where the attribute A_i gets as value val_i.*

Example :

- Given a person $Alex$ and his attribute $spatial\ location$ associated with a value set $\{[(city(London), Geographic), (vid_5, t \in [5min, 15min],$ "$x \in [10, 100], y \in [20, 300])$"], $[(city(Paris), Geographic), (vid_4, t \in [4min, 10min], ANY)]\}$. This means that from the 5th minute to the 15th of video vid_5, $Alex$ was in $London$ and his position on the screen was $\{x \in [10, 100], y \in [20, 300])\}$. Whereas from the 4th minute to the 10th of video vid_4, $Alex$ was in $Paris$. the keyword ANY means that his position on the screen was ignored.

Figure 1 depicts the schema of the adopted data model. Each content (object or event) is associated with an attribute $video\_locations$ that shows the positions of the content within the video stream, but also to attributes $space\_location$ and $time\_location$ that enables positioning it following other FoRs. Let X be a content (object or event) and $X.spc\_loc$ the set of space locations where X is located. We denote by $X.spc\_loc_{(sF)}$ the subset of $X.spc\_loc$ containing only space locations calculated according to the spatial FoR sF. Similarly, $X.tmp\_loc_{(tF)}$ refers to the subset of $X.tmp\_loc$ whose elements are time locations calculated according to the temporal FoR tF.

Example Let us consider the following event: *Monday, March 10th 2009, in Liverpool, the Liverpool Football player "Gerrard" scores a goal against the Real-Madrid goalkeeper "Casillas" in the 27th minute of the game.*

A simple database representing the previous event can be described based on our data model as follows:

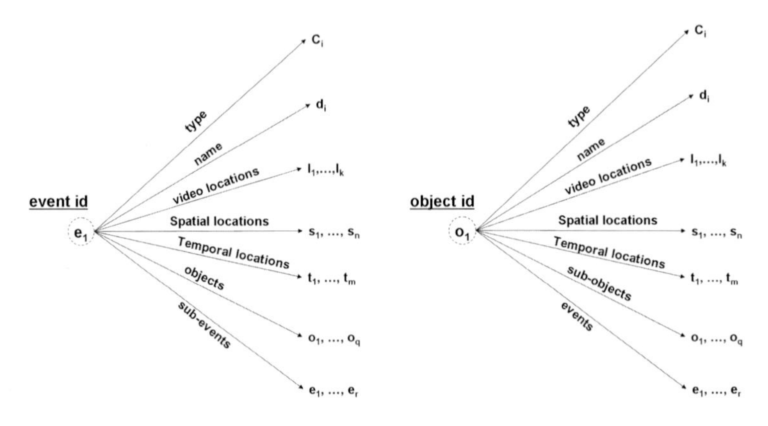

Fig. 1. The proposed data model for representing video content

$e =$ (*eid*, [type={"scoring goal"}, *vid_loc*={(v_5, t∈[02:31,03:00],ANY)}
spc_loc={[(area(penalty zone), 'soccer playfield'),(v_5,t∈[02:31,02:59],ANY)]},
tmp_loc={[(t>27:31,t<27:59, 'soccer time'),(v_5, t∈[02:31,02:59],ANY)]},
objects={o_1, o_2, o_3}]).
$o_1 =$ (oid_1, [*type*={"soccer ball"},
vid_loc={(v5, t∈[02:30,03:00],"x=450,y=250")},
spc_loc={[("goal box",soccer playfield),(v_5, t∈[02:31,03:00], ANY)]},
$o_2 =$ (oid_2, [*type*="soccer player",*name*="Gerrard",
vid_loc={(v_5,t∈[02:26,03:00],"x∈[475,525], y∈[350,450]")},
spc_loc={[("x=45,y=13",'soccer playfield'),(v5, t∈[02:31,03:40],ANY)]},
$o_3 =$ (oid_3, [*type*="soccer Goalkeeper",*name*="Casillas",
vid_loc={(v_5,t∈[02:26,03:00],"x∈[325,375],y∈[300,400"])},
spc_loc={ [(x=40,y=1,soccer playfield),(v_5, t∈[02:31,02:40],ANY)]},

3 Rule-Based Query Language

In this work, a declarative rule based language is used. It allows to reason with objects, events and spatiotemporal constraints specified using the video data model. Queries can refer to both semantic and low level visual layers. They can be specified in fine granularity with the possibility to retrieve only the part of the video where the conditions given in the query are satisfied. Spatial, temporal and semantic conditions are specified as predicates which make it easier and more intuitive to formulate complex query conditions.

Definition 8 (Predicate Symbol). *In order to define the set of predicate symbols, we assume that each relation P in* **R** *of arity n is associated with a predicate symbol* P *of arity n, that two unary predicate symbols* Event *and* Object *represent respectively the events and objects classes, and that unary predicate symbols* Spacial_location *and* Temporal_location *refer respectively to Spacial_location and Temporal_location attributes.*

The query "*List all the segments of films where Tom Cruise appears and the roles he was playing*" can be expressed by the following rule:

$$q(V,T_1,T_2,R) \leftarrow Object(O), O.name = \text{"Tom Cruise"}, (R,L) \in O.role \\ L \in O.vid\_loc, L.vid = V, L.tmp\_const \Rightarrow t \in [T_1,T_2],$$

The query "*List all the athletes that has participated in the 100m sprint running during the olympic games of Beijing*" can be written as follows:

$$q(N) \leftarrow Event(E_1), Event(E_2), E_1.type = \{\text{"}100m\ spring\ running\text{"}\}, \\ E_2.name = \{\text{"Beijing Olympic Games"}\}, E_1 \in E_2.sub\_events, \\ Object(O), O \in E_1.objects, O.type = \{\text{"athlete"}\}, N = O.name,.$$

The query "*List the events that has occurred at January* 1st*, 1945 and the videos where they are filmed*" is expressed as:

$$q(E,V) \leftarrow Event(E), Duration(D), Vid\_Loc(L), (D,L) \in E.tmp\_loc_{(Calendar)}, \\ D.tmp\_const \Rightarrow \{t = \text{"1945-01-01"}\}, V = L.vid$$

The query "*List the video sequences shot at April the* 1st *2009 in Buckingham Palace where the president Barack Obama appears on the left of the Queen Elizabeth*" is expressed by:

$$q(I) \leftarrow Object(O_1), Object(O_2), I = L_1.tmp\_const, V = L_1.vid, \\ O_1.name = \{Barak\ Obama\}, O_2.name = \{Elizabeth\ II\}, \\ (P_1,L_1) \in O_1.spc\_loc_{(\text{"}Residences\text{"})}, (P_2,L_2) \in O_2.spc\_loc_{(\text{"}Residences\text{"})} \\ P_1.spc\_const \Rightarrow palace('Buckingham'), P_2.spc\_const \Rightarrow palace('Buckingham'), \\ L_1.spc\_const.sup_{(x)} < L_2.spc\_const.inf_{(x)}, \\ (D_1,L_1) \in O_1.tmp\_loc_{(\text{"}Calendar\text{"})}, (D_2,L_2) \in O_2.tmp\_loc_{(\text{"}Calendar\text{"})} \\ D_1.tmp\_const \Rightarrow \{t = \text{"2009-04-01"}\}, D_2.spc\_const \Rightarrow \{t = \text{"2009-04-01"}\}$$

Let's consider a multi-camera surveillance system installed in a metro station, with a camera fixed on the check point and sending a video stream V_c, and a second camera fixed on the hall of the station and sending a video stream V_h. We can express the query "*list the people entering the station hall without crossing the check point*" by:

$$q(O) \leftarrow Object(O), (V_h, I_h, ANY) \in O.vid\_loc, \\ not((V_c, I_c, ANY) \in O.vid\_loc, sup(I_c) < Inf(I_h)),$$

4 Implementation

The designed framework, given in figure 2, is composed of three major packages that are the "Automatic Content Extractor", the "Manual Annotator", and the "Research Engine". The "Automatic Content Extractor" allows for classifying video contents into three levels; features level, mid level corresponding to elementary objects, and high level corresponding to events and abstract concepts. The "Manual Annotator" allows users to annotate video objects and instantiate databases at the three levels. Finally, the "Research Engine" enables for querying the multimedia database at the three levels based on the language presented in this paper. The "Research Engine" is composed of a query interpreter written in Java on the top of the deductive object oriented semantic engine based on FLORID [5] and an image based retrieval system based on QBIC[6].

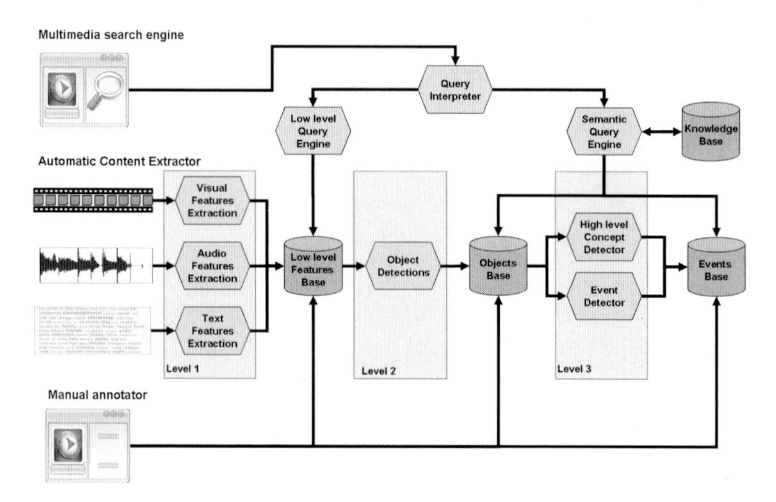

Fig. 2. The proposed framework for annotating, extracting and retrieving video content

5 Conclusion

In this paper, a novel framework, for modeling, indexing and querying semantic objects and events from video documents has been presented. The framework is based on a novel data model enabling spatial, temporal, and semantic modeling of events and objects occurring in video documents.

We believe that our model is one of the first proposals combining objects, events and relations for specifying semantics of video data specified by multiple spatial and temporal frames of reference. We adopt an object oriented approach to associate each object (or event) in the database to its corresponding positions and durations where it occurs. This allows for considerably reducing the time of answering queries and thus augmenting the efficiently of the research engine.

References

1. Chua, T.-S., Ruan, L.-Q.: A video retrieval and sequencing system. ACM Trans. (1995)
2. Hacid, M.-S., Decleir, C., Kouloumdjian, J.: A database approach for modeling and querying video data. IEEE Trans. on Knowl. and Data Eng. 12(5), 729–750 (2000)
3. Koprulu, M., Cicekli, N.K., Yazici, A.: Spatio-temporal querying in video databases. Inf. Sci. Inf. Comput. Sci. 160(1-4), 131–152 (2004)
4. Dönderler, M.E., Şaykol, E., Arslan, U., Ulusoy, O., Güdükbay, U.: Bilvideo: Design and implementation of a video database management system. Multimedia Tools Appl. (2005)
5. Kifer, M., Lausen, G., Wu, J.: Logical foundations of object-oriented and frame-based languages. J. ACM 42(4), 741–843 (1995)
6. Flickner, M., Sawhney, H., Niblack, W., Ashley, J., Huang, Q., Dom, B., Gorkani, M., Hafner, J., Lee, D., Petkovic, D., Steele, D., Yanker, P.: Query by image and video content: The qbic system. Computer (1995)

A Multiple Instance Approach for Keyword-Based Retrieval in Un-annotated Image Database

Jun Jiao[1], Chen Shen[2], Bo Dai[3], and Xuan Mo[4]

[1] Computer Science & Technology Department, Nanjing University, Nanjing, China
failedjj@gmail.com
[2] School of Software and Electronics, Peking University, Beijing, China
scv119@gmail.com
[3] Institute of Automation, Chinese Academy of Sciences, Beijing, China
daobo08@gucas.ac.cn
[4] Department of Automation, Tsinghua University, Beijing, China
blademamama@gmail.com

Abstract. In image retrieval, if user can describe their query concepts by keywords, search results can be returned efficiently and precisely by matching query keywords with text annotation in image databases. However, even if the query keyword is given, keyword-based retrieval can not be applied directly in an image database without any text annotation. The development of Web mining and searching techniques has enabled us to search images in Web by keywords. Thus, we can search the query keywords given by user through Web to obtain example images, and then find those images relevant to user's query in image database with the help of these example images. In order to improve the image retrieval performance, we adopt multiple instance learning when calculating the similarity between example images and images in database. Experiments validate that our method can effectively improve the retrieval performance in un-annotated image database.

Keywords: Image Retrieval, Keyword-Based Retrieval, Multiple Instance Learning.

1 Introduction

With the rapid development in multimedia devices and Internet, the past decade has seen an enormous increase in digital images that people need to manage during work and everyday life.

To make describing image search queries much easier for the user, keyword-based image retrieval techniques emerged and attracted the attention of researchers. In previous research, keyword-based image retrieval techniques usually require image databases to be annotated with text information. However, due to the gap between semantic words and visual features, query-based methods usually can not be applied to an image database without any text annotation. Hoi

S. Boll et al. (Eds.): MMM 2010, LNCS 5916, pp. 749–754, 2010.

and Lyu [4] proposed a method to implement keyword-based image retrieval in un-annotated image database through web mining and search techniques. Nevertheless, the method that they used to collect related images often tends to pick out many images irrelevant to user's query, thus heavily affecting the image retrieval performance. In addition, after obtaining the example images, the CBIR method that they adopted is also so simple that more sophisticated method can be used to enhance the precision of the search result.

In this paper, based on [4], we proposed a new keyword-based retrieval approach to un-annotated image database. Compared with [4], our method has two distinct advantages: (1) Our method to collect relevant images through Web can lead to example images with fewer irrelevant images. (2) In the CBIR phase, we treat each image as a bag of instances and adopted one-class SVM [1][6][3] to improve the image retrieval performance.

This paper is organized as follows: Section 2 presents our method to collect relevant images and image retrieval in detail. Section 3 reports the experiment results. Section 4 concludes.

2 The Proposed Method

The method is composed of two steps: (1) Collecting relevant images. (2) Content based image retrieval. Both steps will be explained in detail in the following subsections.

2.1 Collecting Relevant Images

In this step, we first obtain user's query keywords, then use these keywords to collect relevant images in the Web. We adopted the image collecting method used in WEBSEIC to collect images from Web. [7]. This method consists of two steps: (1) Filter out irrelevant images by text information (2) Pick out relevant images by visual information.

We first use a keyword-based webpage search technique to obtain a search result of user's query. The set of top ranked webpages in the search result is denoted as U. All the images in U are then extracted, denoted as I, and I_i represents the ith image in I. For each image I_i in I, we extract text evidence from its description tags including filenames, ALT attribute of IMG tag and anchors, connect and turn them into feature vectors represented by $d_i = (w_{d_i 1}, w_{d_i 2}, ..., w_{d_i t})$; text surrounding each image is also extracted and turned into feature vectors $s_i = (w_{s_i 1}, w_{s_i 2}, ..., w_{s_i t})$. User's query keywords are represented as $q = (w_{q1}, w_{q2}, ..., w_{qt})$. Binary weight is used for all vectors. Given the query q, the conditional probability of observing d_i or s_i can be calculated by equation(1), in which e_i represents d_i or s_i.

$$P(e_j|q) = \frac{\sum_{k=1}^{t} w_{jk} w_{qk}}{\sqrt{\sum_{k=1}^{t} w_{jk}} \sqrt{\sum_{k=1}^{t} w_{qk}}} \tag{1}$$

We use equation (2) to combine probability of both d_i and s_i calculated by equation(1) to obtain the conditional probability of observing image I_i when given the user's query q, in which Pd_{jq} and Ps_{jq} represent the probability of d_i and s_i calculated by equation(1) respectively.

$$P(I_j|q) = \eta \times [1 - (1 - Pd_{jq})(1 - Ps_{jq})] \quad (2)$$

We calculate the conditional probability of observing each image in I, then rank all the images in I according to its probability value and pick out the images with high probability of being observed.

Let $C = \{x_1, x_2, ..., x_{|c|}\}$ represent the possibly relevant images obtained previously. For each image C_i in C, we extract its visual feature and represent it as m-dimensional feature vector $D_i = (d_{i1}, d_{i2}, ..., d_{im})$. Let z denote a point in visual feature space, we use equation(3) to calculate its density, in which $|C|$ represents the number of images in C.

$$Density(z) = \sum_{i=1}^{|C|} e^{-\sum_{j=1}^{m} |z_j - x_{ij}|^2} \quad (3)$$

For all images in C, we use algorithm described in algorithm(1) to filter out those irrelevant images.

Algorithm 1. Filter out irrelevant images using visual information

---

Parameter:

N: When number of images is less than N, the algorithm will terminates.

Input:

C: a set of possibly relevant images.

Output:

P: a set of relevant images

Process:

1. Step 1: For each image C_i in C, extract its visual feature and represent it as feature vector D_i.
2. Step 2: Calculate each image's density using equation(3).
3. Step 3: Rank all the images according to their density
4. Step 4: Remove the half of images with lower density out of C
5. Step 5: If the number of remaining images is less than N, algorithm terminates, let P represent the set of remaining images. Otherwise goto step 2.

2.2 Content-Based Image Retrieval

For each image in the relevant images set P and image database, we use Blobworld [2] to turn it into several image regions. Let l denote the total number of regions extracted from all images in P, we regard each image region as an instance and represent the set of all the regions as $X = \{X_1, X_2, ..., X_l\}$, in which X_i is the feature vector of a region. For each image D_i in image database D, we represent the set of instances extracted from it as $Y_{D_i} = \{Y_{D_i}^1, ..., Y_{D_i}^{l_i}\}$, in which l_i denotes the number of regions extracted from image D_i.

One-class SVM[1] is then adopted to explore the regions related to user's query. Since the relevant instances share great similarity in visual feature, they lie close to each other and form a relatively dense region in the feature space. This can be formulated into the following optimization problem:

$$\min_{R\in\mathbb{R},\zeta\in\mathbb{R}^l,c\in F} R^2 + \frac{1}{vl}\sum_i \zeta_i \tag{4}$$

subject to

$$\|\phi(X_i) - c\|^2 \leq R^2 + \zeta_i,\ \zeta_i \geq 0\ and\ i = 1, ..., l \tag{5}$$

By solving the above optimization problem, we can calculate the similarity score between a region $Y_{D_i}^j$ of the image D_i in database and the regions $\{X_1, X_2, ..., X_l\}$ of example images using the following equation[6]:

$$f(Y_{D_i}^j) = R^2 - \sum_{i,j} \alpha_i\alpha_j k(X_i, X_j) + 2\sum_i \alpha_i k(X_i, Y_{D_i}^j) - k(Y_{D_i}^j, Y_{D_i}^j) \tag{6}$$

For each D_i in image database D, we set the score of D_i to the maximum similarity score of its regions:

$$score(D_i) = \max\{f(Y_{D_i}^j)|1 \leq j \leq l_k\} \tag{7}$$

We rank all the images in D according to their score calculated by equation (9), and return the top images with highest score to user as image retrieval results.

3 Experiment

3.1 Experiment Settings

In order to test the performance of proposed method, we randomly gathered 2000 images from COREL image data set to form an image database, which are composed of 20 different classes, with 100 images per class. We extracted 64 dimensional Color Texture Moment(CTM)[5] and 64 dimensional color histogram, and combined these two features together to represent each image.

Due to its fame and excellent performance, Google is chosen as the search engine to implement key-word based webpage search. For each query given by user, we pick out the top 50 pages in the webpage search result returned by Google and extracted all the images contained in these webpages. The number of images remained after filtering irrelevant images using textual information is 100, and the finally remained example images are 20.

For the kernel of one-class SVM, we adopted Gaussian kernel due to its good ability to capture non-linearity:

$$k(X, Y) = e^{-\|X-Y\|^2/2\sigma^2} \tag{8}$$

10 classes out of the total 20 classes are chosen to be used as query to test the performance. The 10 classes chosen are: *Dessert, Dolphin, Firework, Green Leaf, Lion, Night Owl, Puma, Rare Car, Rhinoceros, Wave.* We use their class name as the query keyword directly.

3.2 Experiment Result

We compared our proposed method with two other methods: (1)Method1: the method proposed in [4], which uses another kind of image collecting method and single-instance technique in CBIR. (2)Method2: the method that adopted our image collecting algorithm, but, in CBIR, uses the same single-instance technique as Method1. The precision in top 20, 50 and 100 images of the image retrieval results are used to evaluate these three methods. The experiment results are presented in Table(1).

From Tabel 1, we can see that the image retrieval precision on *Dessert*, *Night Owl* and *Rare Car* are very low. Through analysis, we attribute this phenomenon to the following two reasons: (1) The limited ability of the feature extraction method that we adopted. On *Dessert* and *Rare Car*, the related images are composed of various color, and it is really difficult to distinguish these images from others using color histogram. Therefore, although our method can improve the quality of example images and performance of CBIR, it cannot improve the image retrieval precision on these two classes. (2) On all these three classes, the images that we collected through Web are very different from the images in the database, thus leading to the poor performance of CBIR.

Since Method 1 and Method 2 share the same algorithm during CBIR phase, their performance are determined by the image collecting method that they used. On all other 7 classes, the precision of Method 2 is higher than Method 1. This validates that our method of collecting example images from Web can lead to example images with better quality. The performance of our proposed method

Table 1. Retrieval precision of three methods in top 20, 50 and 100 images

	Top 20			Top 50			Top 100		
Data	Method1	Method2	Our	Method1	Method2	Our	Method1	Method2	Our
Dessert	0.00	0.00	0.00	0.04	0.05	0.05	0.04	0.06	0.04
Dolphin	0.10	0.13	**0.43**	0.08	0.10	**0.38**	0.10	0.12	**0.25**
Firework	0.20	0.24	**0.31**	0.16	0.20	**0.27**	0.14	0.18	**0.21**
Green Leaf	0.11	0.13	**0.25**	0.07	0.07	**0.16**	0.06	0.05	**0.13**
Lion	0.35	0.44	**0.46**	0.29	0.38	**0.40**	0.24	0.30	**0.34**
Night Owl	0.02	0.00	0.00	0.02	0.02	0.03	0.02	0.03	0.03
Puma	0.29	0.45	**0.54**	0.25	0.38	**0.44**	0.19	0.25	**0.31**
Rare Car	0.05	0.05	0.03	0.02	0.02	0.03	0.03	0.03	0.03
Rhinoceros	0.33	0.41	**0.44**	0.26	0.33	**0.36**	0.17	0.23	**0.28**
Wave	0.40	0.52	0.50	0.30	0.34	**0.40**	0.22	0.25	**0.29**
Average	0.185	0.227	**0.296**	0.149	0.189	**0.252**	0.121	0.150	**0.191**

is also better than Method 2, which validates that our use of multiple-instance method in CBIR can effectively improve the image retrieval precision.

4 Conclusion and Future Work

The core idea of this paper is to improve the previous keyword-based image retrieval method for un-annotated image databases by using new image collecting method and multiple-instance image retrieval. Experiments results show that, on most classes, our method can improve the image retrieval performance effectively. However, due to the simple feature extracting method and query keyword that we used, our method cannot improve the retrieval precision on some classes, and would not deteriorate the performance either. By using more sophisticated feature extracting method and query keyword, our method will achieve even better performance.

References

1. Scholkopf, B., Platt, J.C., Shawe, J.T., Smola, A.J., Williamson, R.C.: Estimating the support of a high-dimensional Distribution. Neural Computation 13, 1443–1471 (2001)
2. Carson, C., Belongie, S., Greenspan, H., Malik, J.: Blobworld: image segmentation using expectation-maximization and its application to image querying. IEEE Transactions on Pattern Analysis and Machine Intelligence 24, 1026–1038 (2002)
3. Zhang, C.C., Chen, X., Chen, M., Chen, S.C., Shyu, M.L.: A Multiple Instance Learning Approach for Content Based Image Retrieval Using One-Class Support Vector Machine. In: Proc. IEEE International Conference on Multimedia and Expo., pp. 1142–1145 (2005)
4. Hoi, C.H., Lyu, M.R.: Web image learning for searching semantic concepts in image databases. In: Proc. the 13th International World Wide Web Conference, pp. 406–407 (2004)
5. Yu, H., Li, M.J., Zhang, H.J., Feng, J.F.: Color texture moments for content-based image retrieval. In: Proc. International Conference on Image Processing, vol. 3, pp. 929–932 (2002)
6. Chen, Y.Q., Zhou, X.S., Huang, T.S.: One-class SVM for learning in image retrieval. In: Proc. International Conference on Image Processing, vol. 1, pp. 34–37 (2001)
7. Zhou, Z.H., Dai, H.B.: Exploiting image contents in Web search. In: Proc. the 20th International Joint Conferences on Artificial Intelligence, pp. 2928–2933 (2007)

On the Advantages of the Use of Bitstream Extraction for Video Summary Generation

Luis Herranz and José M. Martínez

Video Processing and Understanding Lab, Escuela Politécnica Superior
Universidad Autónoma de Madrid, 28049 Madrid, Spain
{luis.herranz,josem.martinez}@uam.es

Abstract. Video summaries are compact representations used in search and video retrieval. As content itself, summaries are stored and distributed using a suitable video coding format, such as H.264/AVC. The generation of the bitstream of the summary usually consist of the decoding of the input bitstream and the encoding of the selected frames (i.e. transcoding). This approach can be computationally very demanding and unsuitable if the summary must be generated on demand with low delay. This paper analyzes the advantages of an alternative approach using bitstream extraction instead of transcoding. Experimental results show that the bitstream can be generated much faster than with transcoding and without any loss of quality.

1 Introduction

Browsing and retrieval of video content is a time consuming task. Video summarization techniques[5] try to provide the user with a compact representation containing enough information to get a quick idea of what happens in the video. Most modalities of video summaries, such as storyboards, video skims and fast playbacks, are built by selecting some frames from the input sequence.

Some applications, such as personalized and interactive summarization, may require the summary to be generated on demand, as the content of the summary is customized after the interaction with the user or with his/her personal profile. One drawback of conventional summarization approaches is the generation of the bitstream itself, which is based on the transcoding of the input sequence after selecting the frames of the summary. Thus, the inherent complexity of transcoding may lead to a high delay. This aspect of the whole summarization process is barely considered, but it may become critical in this applications. For long summaries, such as video skims, low delay generation is a very challenging problem, and very efficient approaches are required.

Transcoding is frequently used in (non content-based) bitstream adaptation to constrained bitrate conditions. Many transcoding architectures have been proposed in this context[1,4,6], trading off rate-distortion performance and efficiency. An alternative approach for bitstream adaptation is scalable video coding. In this case the bitstream is arranged into different layers with increasing bitrate and quality. An important advantage is that the adaptation to a given bitrate

S. Boll et al. (Eds.): MMM 2010, LNCS 5916, pp. 755–760, 2010.

is very simple and consists of the extraction of the appropriate layers for the required bitrate.

Similarly, bitstream extraction can be used for summary generation together with an alternative model for representing video summaries[2,3]. However, this approach is studied only from the efficiency point of view, with experiments comparing extraction and transcoding processing times. In this paper we remark other inherent advantages of this approach, such as better efficiency and rate-distortion performance and controlled drift, in contrast to the conventional approach based on transcoding.

The rest of the paper is organized as follows. Section 2 briefly describes the stages of a summarization system. Sections 3 and 4 describe architectures for bitstream generation based on transcoding and extraction. Finally, Section 5 presents some experimental simulations and Section 6 draws the conclusions.

2 Bitstream Generation of Video Summaries

Every summarization system has two different stages: analysis and generation. The analysis stage, using the term *analysis* in a wide sense, includes all the processes addressed to characterize and represent the content in order to remove semantic redundancies, and the selection of the frames to be included in the summary. Feature extraction, shot boundary detection, high level structuring, keyframe selection, clustering or optimization algorithms are examples of operations that can be included in the analysis stage. Most works in video summarization[5] deal with this stage, but the generation stage is barely studied.

However, in many scenarios, the generation of the bitstream is critical for efficient summarization. The generation stage obtains the coded bitstream of the summary from the input bitstream, once the sequence has been analyzed and the frames to be included in the summary have been determined.

The analysis and generation stages are connected by some kind of summary description with the frames to be included in the summary. Both analysis and generation are not required to be done at the same time, as the summary description can be stored as metadata.

3 Architecture Based on Transcoding

Fig. 1a shows a conventional summarization architecture using a transcoder for the generation stage. It can be obtained cascading a decoder and an encoder, which is the approach used in most of video summarization systems. Besides, it is simple to separate the analysis from the coding, as the analysis stage usually works with uncompressed frames as basic units for summarization.

The transcoder shown in Fig. 1a has all the stages of a conventional decoder (entropy decoding, dequantization, inverse transform and motion compensation) and a conventional closed-loop encoder (motion estimation and compensation, transform, quantization and entropy coding). The link between them is the frame

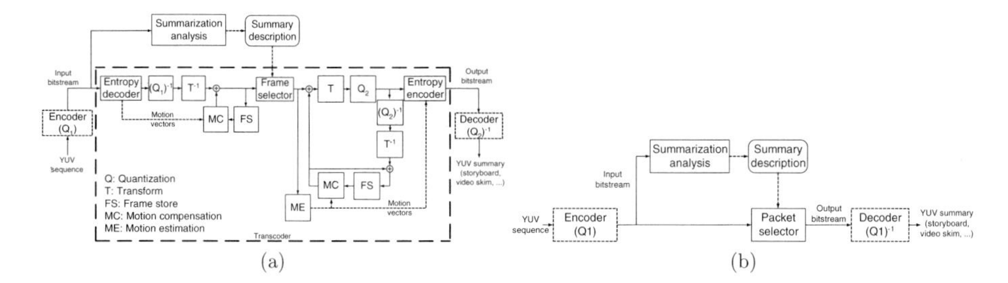

(a) (b)

Fig. 1. Summarization architectures: of a) transcoding based, b) extraction based

selector, which is also the entry point for summarization. Only frames belonging to the summary are encoded into the summary bitstream.

Summary generation using transcoding is very inefficient, specially for long summaries, such as video skims. The generation delay depends on these factors and the number of frames to be included in the summary. The most time consuming part of the whole process is motion estimation. However, limited search ranges or simplified search algorithms lead to a degradation of the rate-distortion performance. In addition, transcoding suffers from an inherent drawback related to the additional quantization (Q_2) introduced by the transcoder. A first loss of quality occurred before transcoding, when the input sequence was lossy encoded with a first quantization (Q_1). When comparing transcoding architectures, the decoder-encoder cascade with full range search is the optimal architecture which gives the best end to end rate-distortion performance, and it is used as reference in most transcoding algorithms[1,4,6].

4 Architecture Based on Extraction

If the coding format is the same for both input and output bitstreams, an alternative approach for the generation of the bitstream of the summary is bitstream extraction (see Fig. 1b). Similar to the extraction approach used in adaptation, extraction for summarization is guided by the summary description[2]. The whole transcoder is replaced by an extractor which consist of a packet selector. The packet selector selects only the packets containing the compressed frames and discards those not required.

This approach has two inherent advantages. Extraction is a very simple operation which requires few resources and that can be done very efficiently. Besides, the frames themselves are not modified, so the quality of each frame is the same as in the input bitstream. Compared to transcoding, there is no quality degradation due to an additional quantization stage. However, the selection of frames is not so arbitrary as in the transcoding based architecture, as frames are not independent (i.e. they are related by prediction), and coding structure must be taken into account. For this reason, we use the model described in [3] to represent the summaries for extraction.

5 Experimental Results

We compared experimentally the efficiency and rate-distortion performance of transcoding and extraction approaches, in the context of H.264/AVC with hierarchical B-frames. We implemented an extractor for H.264/AVC, using a simple bitstream description generated by the encoder. The transcoder used is a cascade of the JM 12.4 decoder and encoder. The optimal transcoding architecture in terms of rate-distortion performance is the cascade of decoder and decoder with full range search. However, it is computationally very intensive. To trade off quality and efficiency five variations were tested combining full and EPZS search and size of the search window (0, 8 or 64 pixels): FULL64, FULL8, EPZS64, EPZS8 and ZERO (only zero vectors are evaluated). Due to the large number of possible summaries, we consider only the original sequence, which is not exactly a summary, but in practice it gives a good measure of the efficiency and rate-distortion performance of both approaches. Due to the unavailability of long test sequences coded in an uncompressed format (e.g. YUV), required to measure PSNR, we encoded the sequence *stefan* (300 frames of 352x288 pixels) with the JM 12.4 encoder, with dyadic structures, full search (64 pixel window size) and different values of GOP size and quantization parameter. For each test, the transcoder uses the same GOP size and the same quantization parameter as the encoder.

5.1 Rate-Distortion Performance

Fig. 2a and Fig. 2b show the rate-distortion curves for all the approaches and the impact of quantization parameter and GOP size. As expected, extraction outperforms transcoding, as the quality is not degraded by an additional quantization stage. Besides, the quality of transcoding degrades more with longer GOPs, suggesting that requantization affects more to motion predicted frames, and that the quantization error is propagated and accumulated in other intercoded frames. In the transcoding experiments, FULL64 has the best quality, with EPZS64 close to it. Even using only intracoded frames (i.e. GOP=1 in Fig. 2b) the quality is still degraded by transcoding.

The effect of motion estimation is better shown in Fig. 3a. Although the transcoder uses a closed-loop drift-free architecture, a progressive loss of quality within the GOPs is evident in the plots. The degradation propagates backwards (the intracoded frame is the last frame in the GOP) until a new intracoded frame is found. This degradation is higher for longer GOPs, and the mean PSNR decreases, as shown in Fig. 2b. For extraction, the absence of requantization avoids this problem, and the only quality loss is due to the source encoder.

5.2 Efficiency

Fig. 3b shows the mean frames per second obtained in the case of the processing of the whole sequence. The experiments were performed on an Intel Core 2 CPU at 1.8 GHz. With this implementation, extraction is clearly faster than transcoding, with a large gap between them. Although simplified architectures and optimized implementations can greatly improve the performance of transcoding,

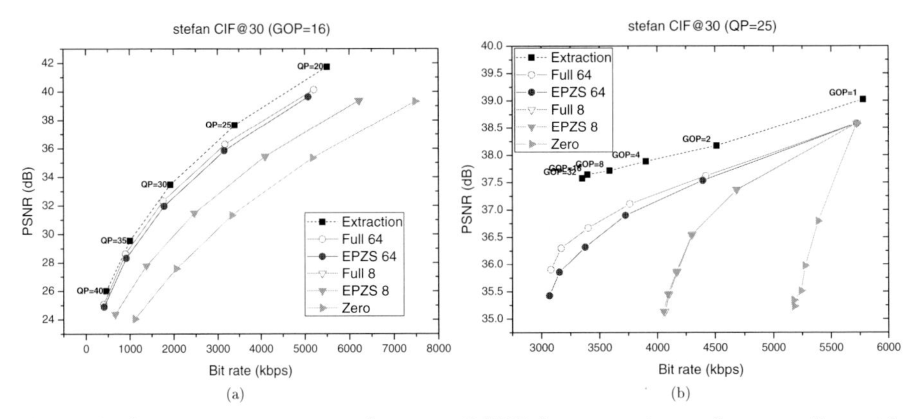

Fig. 2. Performance comparison of average PSNR for extraction and transcoding with a) different values of, b) different GOP sizes

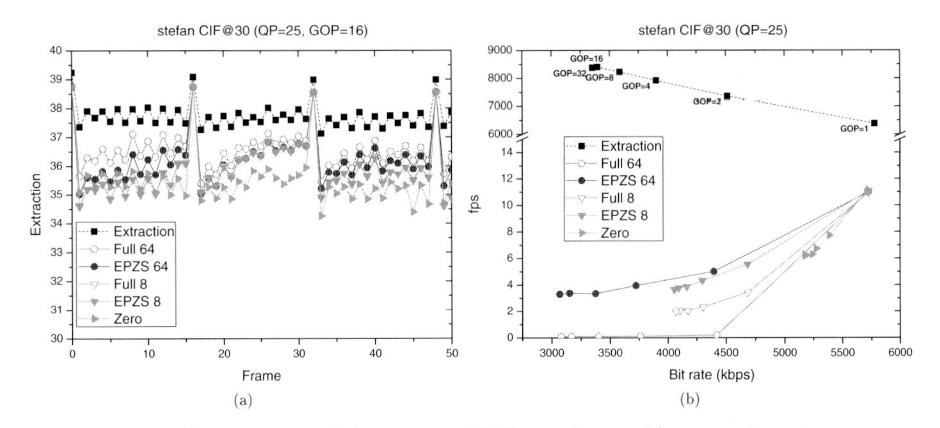

Fig. 3. Comparison of a) average PSNR per frame, b) processing time

extraction seems to remain as the best option when high efficiency in the generation is required. For extraction, the only factor having some noticeable influence in the performance is the bitrate. An inverse linear trend is observed, with slower processing as packets become larger.

The efficiency of transcoding is highly related to the motion estimation strategy, ranging from 15% of the total transcoding time for ZERO to 98% with FULL64, in the worst cases. EPZS and small search areas speed up the transcoding, although still below real time processing. In contrast to extraction, transcoding is faster for shorter GOPs, as motion estimation is used in fewer frames, and particularly fast when only intracoding is used.

However, there are simplified architectures for transcoding, more efficient, although having worse rate-distortion performance than the decoder-encoder cascade with full search. Open-loop architectures such as requantization[6,1] are computationally efficient, but they suffer from drift. A cascaded pixel-domain transcoder (CPDT) consists of a concatenation of a decoder and a simplified encoder, which reuses motion information from the decoder, with an efficiency comparable to ZERO but with a quality significantly lower than the decoder-encoder cascade for H.264/AVC bitstreams[4].

6 Conclusions

Extraction is an interesting alternative to transcoding for the generation of the bitstream in video summarization. This approach has two inherent advantages: efficiency, derived from the simplicity of the adaptation method, and quality preservation, in contrast to transcoding which introduces an additional loss of quality. In this paper we have studied the architectures of both approaches along with an experimental comparison for H.264/AVC. The size of the GOP is an important parameter that has effect not only on the precision of the summarization analysis, but also on the rate-distortion performance and efficiency of the generation process.

Acknowledgment

This work is partially supported by the Spanish Government under project TEC2007-65400 (SemanticVideo) and by the Comunidad de Madrid under project P-TIC-0223-0505 (PROMULTIDIS).

References

1. De Cock, J., Notebaert, S., Van de Walle, R.: A novel hybrid requantization transcoding scheme for H.264/AVC. In: Proc. 9th International Symposium on Signal Processing and Its Applications, ISSPA 2007, pp. 1–4 (2007)
2. Herranz, L.: Integrating semantic analysis and scalable video coding for efficient content-based adaptation. Multimedia Systems 13(2), 103–118 (2007)
3. Herranz, L., Martínez, J.M.: An integrated approach to summarization and adaptation using H.264/MPEG-4 SVC. Signal Processing: Image Communication 24(6), 499–509 (2009)
4. Lefol, D., Bull, D., Canagarajah, N., Redmill, D.: An efficient complexity-scalable video transcoder with mode refinement. Signal Processing: Image Communication 22(4), 421–433 (2007)
5. Truong, B.T., Venkatesh, S.: Video abstraction: A systematic review and classification. ACM Trans. Multimedia Comput. Commun. Appl. 3(1), 3 (2007)
6. Xin, J., Lin, C.-W., Sun, M.-T.: Digital video transcoding. Proceedings of the IEEE 93(1), 84–97 (2005)

Image Clustering via Sparse Representation

Jun Jiao[1], Xuan Mo[2], and Chen Shen[3]

[1] Computer Science & Technology Department, Nanjing University, Nanjing, China
failedjj@gmail.com
[2] Department of Automation, Tsinghua University, Beijing, China
blademamama@gmail.com
[3] School of Software and Electronics, Peking University, Beijing, China
scv119@gmail.com

Abstract. In recent years, clustering techniques have become a useful tool in exploring data structures and have been employed in a broad range of applications. In this paper we derive a novel image clustering approach based on a sparse representation model, which assumes that each instance can be reconstructed by the sparse linear combination of other instances. Our method characterizes the graph adjacency structure and graph weights by sparse linear coefficients computed by solving ℓ^1-minimization. Spectral clustering algorithm using these coefficients as graph weight matrix is then used to discover the cluster structure. Experiments confirmed the effectiveness of our approach.

Keywords: Image Clustering, Spectral Clustering, Sparse Representation.

1 Introduction

Clustering algorithms are widely used in data mining and pattern recognition problems. The goal of clustering is to determine the intrinsic grouping in a set of data. Spectral clustering[1][7][13] algorithms have been successfully used in computer vision. Compared to the traditional algorithms , spectral clustering has many fundamental advantages. It is very simple to implement and can be solved efficiently by standard linear algebra methods. In addition, results obtained by spectral clustering often outperform the traditional approaches. However, the success of spectral clustering depends heavily on the choice of the similarity measure, but this choice is generally not treated as part of the learning problem. Thus, time-consuming manual feature selection and weighting is often a necessary precursor to the use of spectral methods[5].

Recently, several works have considered methods to relieve this burden by incorporating prior knowledge into the metric, either in the setting of K-means clustering[4][9] or spectral clustering[10][12]. In this paper, we consider a complementary approach to use ℓ^1 graph to construct the similarity matrix needed by spectral clustering. This method is based on the assumption that each instance can be reconstructed from the sparse linear combination of other instances. We calculate a ℓ^1 graph by solving ℓ^1 minimization problems. Weights of this graph

S. Boll et al. (Eds.): MMM 2010, LNCS 5916, pp. 761–766, 2010.

are then used as the weight matrix for spectral clustering. Experiments result shows that our spectral clustering method achieves excellent performance in image clustering.

The rest of this paper is organized as follows. Section 2 briefly introduces some related works. Section 3 presents our method to calculate the ℓ^1 graph. Section 4 reports the experiment results on synthetic data set. Section 5 concludes.

2 Related Works

2.1 Graph Construction

Given n instances $\mathbf{x}_1, ..., \mathbf{x}_N \in \mathbb{R}^d$, previous graph based algorithms usually construct a weighted graph with n nodes in the following way:(1)Constructing graph adjacency: Two nodes $\mathbf{x}_i$ and $\mathbf{x}_j$ are connected in the graph if they are considered to be close. (2)Calculating graph weight. Weights in the graph are used to reflect how strong two nodes are related.

An obvious drawback of this method is that the calculation of graph structure and weights is divided into two different steps. The structure of the graph has already been fixed after the first step. Therefore, the calculation of graph weights in the second step will be heavily constrained by this fixed graph structure.

2.2 Spectral Clustering

Spectral clustering[1][7][13] methods arise from concepts in spectral graph theory. The basic idea is to construct a weighted graph from the initial data set where each node represents a pattern and each weighted edge simply takes into account the similarity between two patterns. In this framework the clustering problem can be seen as a graph cut problem, which can be tackled by means of the spectral graph theory. The core of this theory is the eigenvalue decomposition of the Laplacian matrix of the weighted graph obtained from data. In fact, there is a close relationship between the second smallest eigenvalue of the Laplacian and the graph cut.

3 The Proposed Algorithm

Our proposed clustering algorithm is composed of three steps: (1) Constructing graph W for spectral clustering. (2) Solving the k smallest eigenvectors from W. (3) Using K-means to cluster the instances with eigenvectors as features. These three steps will be introduced in detail in the following subsections.

3.1 Constructing Graph

The method for constructing sparse graph that we adopted in this paper was firstly proposed in [11] for semi-supervised learning. We first introduce some notations: Given n instances $\mathbf{x}_1, ..., \mathbf{x}_N \in \mathbb{R}^d$ and $X = [\mathbf{x}_1, ..., \mathbf{x}_N]$ is the column matrix of all instances, our aim is to construct a sparse weighted graph G to characterize the relationship between instances. In order to obtain a sparse representation of a new instance $\mathbf{y}$, one obvious way is to solve the following ℓ^0-norm optimization problem:

$$\hat{\mathbf{a}} = \arg\min \|\mathbf{a}\|_0, s.t. X\mathbf{a} = \mathbf{y}, \tag{1}$$

where $\|\cdot\|_0$ counts the number of nonzero entries in a vector. However, the problem of optimizing equation(2) is NP-hard: there is no algorithm for solving it more efficiently than enumerating all subsets of $\mathbf{a}$, and it is difficult even to approximate as well[3][8]. To overcome this issue, Wright et al.[8] proposed to solve the following ℓ^1-minimization problem instead:

$$\hat{\mathbf{a}} = \arg\min \|\mathbf{a}\|_1, s.t. X\mathbf{a} = \mathbf{y}, \tag{2}$$

It has been proved that if the $\mathbf{a}_0$ sought is sparse enough, the solution of the ℓ^0- minimization problem is equal to the solution of equation(3), and equation(3) can be solved in polynomial time by standard linear programming methods[8].

Yan and Wang [11] used equation(3) to represent each instance in the form of the linear combination of other instances. For each instance $\mathbf{x}_i$, set $X_i = X \setminus \mathbf{x}_i = [\mathbf{x}_1, ..., \mathbf{x}_{i-1}, \mathbf{x}_{i+1}, ..., \mathbf{x}_N]$, then the reconstruction weight $\hat{\mathbf{a}}_i$ for $\mathbf{x}_i$ can be calculated by solving the following ℓ^1-minimization problem:

$$\hat{\mathbf{a}}_i = \arg\min \|\mathbf{a}_i\|_1, s.t. X_i\mathbf{a}_i = \mathbf{x}_i, \tag{3}$$

Let $\mathbf{a}_i^j$ denote the jth entry of $\mathbf{a}_i$, the ℓ^1 graph is then determined in the following way: a direct edge is placed from node $\mathbf{x}_i$ to $\mathbf{x}_j$, iff $\mathbf{a}_i^j \neq 0$, and the weight W_{ij} is set to $|\mathbf{a}_i^j|$. Due to the fact that ℓ^1- minimization automatically leads to a sparse representation, the graph obtained here is a sparse graph. It is also important to note that the graph obtained is a directed graph. In order to make it symmetric for spectral clustering, we transform the weight matrix W into $(W^T + W)/2$ and use it as the weight matrix for clustering.

3.2 Spectral Clustering

After calculating the sparse representation for each instance, we use the classical spectral clustering algorithm[1] with weight matrix W to discover the cluster structure. We first calculate the normalized matrix $L = D - W$, where D is a diagonal matrix with $D_{ii} = \sum_{j=1}^{n} W_{ij}$. Then the eigen-vectors of L is solved to obtain the first k eigenvectors $\mathbf{v}_1, ..., \mathbf{v}_k$. Each row of eigenvectors is then normalized and regarded as a representation of the corresponding instance. K-means clustering algorithm is finally employed to cluster the rows of eigenvectors.

The main advantages of our algorithm are as follow: (1) Compared with many ℓ^2 based clustering algorithm, the ℓ^1 minimization employed in our algorithm can lead to a sparse representation. (2) Our clustering method can determine both the graph adjacency and weight in ℓ^1 optimization, while most of the

Algorithm 1. Image Clustering via Sparse Representation

Input:

X: A set of real valued instances $X = [\mathbf{x}_1, ..., \mathbf{x}_N], \mathbf{x}_i \in \mathbf{R}^d$.

Parameter:

k: The number of clusters

Process:

1. Normalize the instances to have unit ℓ^2 norm.
2. For each instance $\mathbf{x}_i$, solve equation(4) to obtain $\mathbf{a}_i$
3. If $\mathbf{a}_i^j \neq 0$, set the weight $W_{ij} = |\mathbf{a}_i^j|$, $1 \leq i, j \leq N$
4. $W = (W^T + W)/2$
5. $D_i \leftarrow \sum_{j=1}^{N} W_{ij}, \quad D \leftarrow diag\{D_i\}_i$
6. $L = D - W$
7. Compute the first k eigenvectors $\mathbf{v}_1, ..., \mathbf{v}_k$ of L, then form these eigenvectors into $V = [\mathbf{v}_1, ..., \mathbf{v}_k] \in \mathbb{R}^{n \times k}$
8. Normalize each row of V: $V_{ij} = V_{ij}/(\sum_k v_{ik}^2)^{\frac{1}{2}}$
9. Each row of V is considered as an instance and cluster these instances into k clusters with K-means clustering algorithm.

previous clustering algorithms separate them into two steps. (3) While many other clustering algorithms are very sensitive to the parameters, our algorithm is parameter free. The performance of traditional spectral clustering is heavily related to the choice of σ in Heat Kernel.

4 Experiment

In this section, we present the results of applying our algorithm on different kinds of image data set. To validate that our algorithm can achieve better clustering performance and is very stable, we compare our algorithm with the classical clustering algorithm proposed in [1], which adopted Heat Kernel as the similarity measure.

We tested our algorithm on the famous COREL image data set. 6 classes of images are picked out from Corel image data set, where each image class is composed of 100 images. Among these 6 classes, we chose 2, 3, 4, 5 and 6 classes seperately to form six data sets with different number of clusters. Each image is represented in the combination of 64 dimensional color histogram and 64 dimensional Color Texture Moment [6]. Clustering Accuracy is used to measure the the clustering performance[2]. Various σ for Hear Kernel are used to carry out the experiment, and the accuracy is calculated by repeating the experiment 100 times, then averaging the accuracy. The results are presented in Figure 1.

We can see from Figure 1 that our method, which is parameter free , performs better than the spectral clustering method using Gaussian function, which is sensitive to the parameter σ.

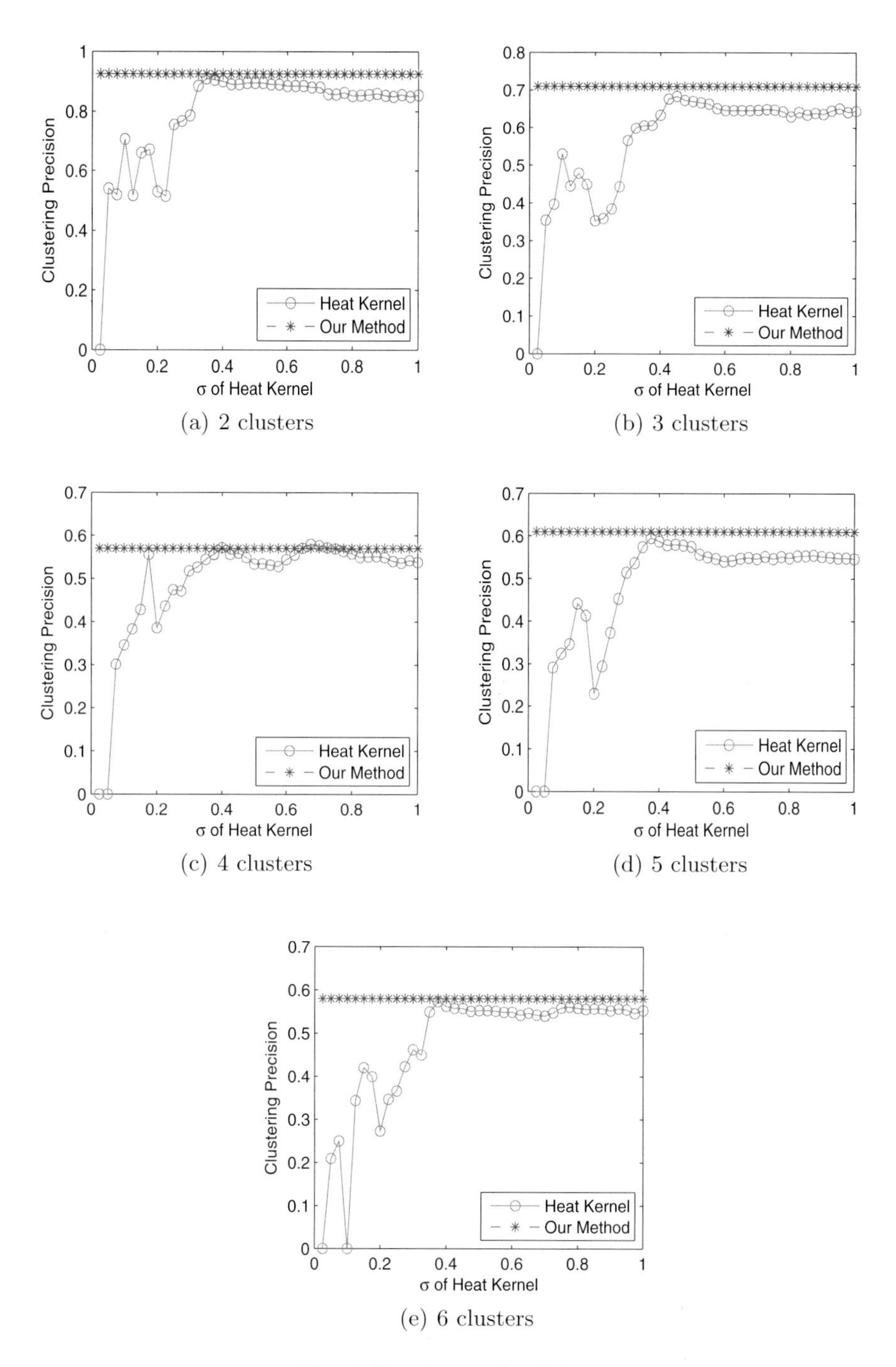

(a) 2 clusters

(b) 3 clusters

(c) 4 clusters

(d) 5 clusters

(e) 6 clusters

Fig. 1. Results on Corel data set

5 Conclusion

In this paper, we proposed a novel image clustering approach based on a sparse representation model. Similarity matrix are obtained from neighborhoods information by solving ℓ^1-minimization programming problems, then spectral clustering algorithm is applied to the similarity matrix obtained. Our method is parameter free and can lead to a sparse representation of similarity matrix. In the experiments, we show that our algorithm can be used to cluster images effectively.

References

1. Ng, A., Jordan, M., Weiss, Y.: On spectral clustering: analysis and an algorithm. In: Advances in Neural Information Processing Systems, vol. 14, pp. 849–856 (2002)
2. Cai, D., He, X., Han, J.: Document clustering using locality preserving indexing. IEEE Transactions on Knowledge and Data Engineering 17, 1624–1637 (2005)
3. Amaldi, E., Kann, V.: On the approximability of minimizing nonzero variables or unsatisfied relations in linear systems. Theoretical Computer Science 209, 237–260 (1998)
4. Xing, E.P., Ng, A.Y., Jordan, M.I., Russell, S.: Distance metric learning, with application to clustering with side-information. In: Proceedings of Neural Information Processing Systems, vol. 15, pp. 505–512 (2002)
5. Bach, F.R., Jordan, M.I.: Learning Spectral Clustering, With Application To Speech Separation. Journal of Machine Learning Research 7, 1963–2001 (2006)
6. Yu, H., Li, M.J., Zhang, H.J., Feng, J.F.: Color texture moments for content-based image retrieval. In: Proc. International Conference on Image Processing, vol. 3, pp. 929–932 (2002)
7. Shi, J., Malik, J.: Normalized cuts and image segmentation. IEEE Transactions on Pattern Analysis and Machine Intelligence 22, 888–905 (2000)
8. Malik, J., Ganesh, A., Yang, A., Ma, Y.: Robust face recognition via sparse representation. IEEE Transactions on Pattern Analysis and Machine Intelligence 31, 210–227 (2008)
9. Wagstaff, K., Cardie, C., Rogers, S., Schroedl, S.: Constrained K-means clustering with background knowledge. In: Proc. the Eighteenth International Conference on Machine Learning, pp. 577–584 (2001)
10. Kamvar, S.D., Klein, D., Manning, C.D.: Spectral learning. In: Proc. the Eighteenth International Joint Conference on Artificial Intelligence, pp. 561–566 (2003)
11. Yan, S.C., Wang, H.: Semi-supervised Learning by Sparse Representation. In: Proc. SIAM Data Mining Conference, pp. 792–801 (2009)
12. Yu, S.X., Shi, J.: Grouping with bias. In: Advances in Neural Information Processing Systems (2002)
13. Weiss, Y.: Segmentation using eigenvectors: a unifying view. In: Proc. of the Seventh International Conference on Computer Vision, vol. 2, p. 975 (1999)

A Parameterized Representation for the Cartoon Sample Space

Yuehu Liu, Yuanqi Su, Yu Shao, and Daitao Jia

Institute of Artificial Intelligence and Robotics,
Xi'an Jiaotong University, Xi'an, P.R. China, 710049
liuyh@mail.xjtu.edu.cn

Abstract. In this paper, the cartoon sample space and its organization are presented. The samples consisting the faces and the corresponding facial cartoons drawn by artists, are parameterized according to the proposed cartoon face model, forming the painting parameter definition (PPD) and the reference shape definition (RSD). And tow conversions constitute the maps between RSD and PPD. Using the cartoon sample space, the parametric representation and organization of the samples across different styles can be resolved in a unified framework, which forms the basis for automatic cartoon face modeling and facial cartoon rendering based on the sample learning.

1 Introduction

To capture and render the characteristics of human face is a fundamental and yet challenging work in computer vision and computer graphics. Various models have been proposed for rendering the human face with different artistic styles such as the line drawing, portraits, cartoons, and so on. Their effectiveness has been proved, however, the automatic capturing and setting of model parameters are still problems difficult to be solved. As shown in Figure 1, different artistic styles show their difference in the uses of color schema, shapes, combination rules and so on. Previous works have not taken the various diversities led by different artistic style into consideration[1][2][3][4][5].

In this paper, we try to supply an unified framework to represent and generate facial cartoons with different artistic styles. To comprise the diversities led by artistic styles, a layered cartoon face model is proposed. A group of facial cartoons drawn by different artist are collected, and subsequently parameterized with the layered cartoon face model, resulting in the cartoon sample space. Given photos of face, the corresponding cartoon can automatically be generated in assigned style by learning from the cartoon sample space.

2 The Layered Cartoon Face Model

The layered cartoon face model is organized in a shape-driven way. It contains the reference shape layer, representing shape of the original face, and the painting

S. Boll et al. (Eds.): MMM 2010, LNCS 5916, pp. 767–772, 2010.

Fig. 1. Some portraits and facial cartoons drawn by existing systems

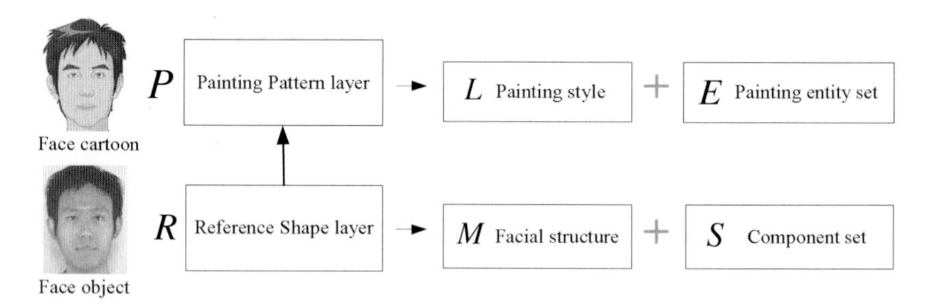

Fig. 2. The layered cartoon face model

pattern layer, representing renderings information of a specific style for facial cartoon as shown in Figure 2. Besides these, some extra information is added, including the gender, age and artistic style.

The reference shape layer is denoted as $R = (M, S)$, representing the shape information for facial components in the original face. The facial structure records the location of components on face, while the set of components includes $l - brow$ for left eyebrow, $r - brow$ right eyebrow, $l - eye$ left eye, $r - eye$ right eye, $nose$ for nose, $mouth$ for mouth, and $f - form$ for contour of face, which are used as super index as in formula (1).

$$S = \{S^{f-form}, S^{l-brow}, S^{r-brow}, S^{l-eye}, S^{r-eye}, S^{nose}, S^{mouth}\} \tag{1}$$

The painting pattern $P = (L, E)$ consists of the layout L and the renderings E of all components as in formula (2). The painting style L records the relative size, direction and location of different E^{cp}. Rendering E^{cp} for component cp is a set of sequentially implemented painting entities. Each painting entity is defined by the rendering shape, the rendering element and rendering rule.

$$E = \{E^{f-form}, E^{l-brow}, E^{r-brow}, E^{l-eye}, E^{r-eye}, E^{nose}, E^{mouth}\} \tag{2}$$

3 Parameterized Cartoon Sample Space

A group of typical faces and the corresponding facial cartoons are collected, and subsequently parameterized by the layered model, resulting in the parameterized set $\mathcal{D} = \{(R^k, P^k), k = 1 \cdot \cdot N\}$, named the cartoon sample space. The sample space can be decomposed into subspace $\mathcal{D}^{cp} = \{(R^{k,cp}, P^{k,cp}), k = 1 \cdot \cdot N\}$ according to the facial component cp, including left eyebrow, right eyebrow, left eye, right eye, nose, mouth and facial form. For example, all nose samples form the subspace $\mathcal{D}^{nose}$.

The parameters of the sample are recorded with the painting parameter definition noted as PPD, and the reference shape definition (RSD), they represent the painting pattern and the reference shape of the sample respectively. Research on photometric facial synthesis tells that new facial image can be synthesized by replacing facial component such as eyes, eyebrows, nose or facial form. In the same way, splitting face cartoon sample into components makes it possible to generate desired personalized facial cartoon from limited samples. Unified parameterizations and its organization are suitable for multi-style cartoon sample space, which is extensible and independent with the artistic style.

3.1 The Painting Parameter Definition – PPD

The PPD is comprised of the parameterized painting pattern of the sample, including the painting style, the rendering shape of the painting entity, the rendering elements and their rendering rules, which are defined in formula (3).

$$P = (L, E^{cp}|_{cp=f-form,l-brow,r-brow,l-eye,r-eye,nose,mouth}) \tag{3}$$

with $E^{cp} = \left(E^{cp}_{shape}, E^{cp}_{element}, E^{cp}_{rule}\right)$

E^{cp}_{shape} is the rendering shape, which describes the geometrical shape of the painting entity. For rendering a painting entity with specific artistic style, the painting entity is decomposed into some rendering elements such as drawing lines, filling region with specific styles. These rendering elements are abstracted,

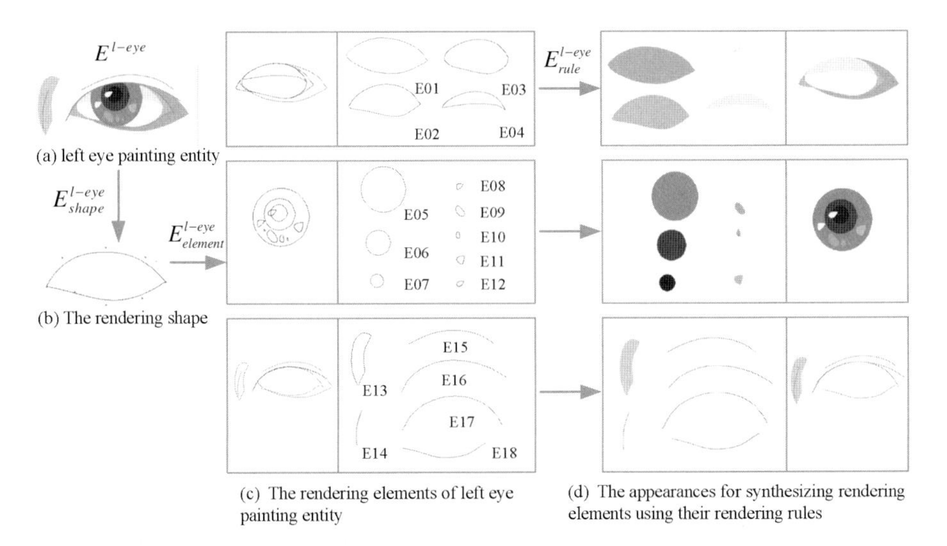

Fig. 3. An example for the painting parameter definition

and classifies into two categories: the curve element for defining an open outline and the region element for closed region. $E^{cp}_{element}$ records a group of the rendering elements composed the painting entity E^{cp}.

E^{cp}_{rule} denotes the rendering rules of rendering element, which decides the appearance and rendering method of the element. For the curve element, The private properties include color, degree, weight, line style, brush type, and brush hight, while the private properties of the region element contain color, boundary, and filling style. Their public properties have layer, visibility, rendering order which set the relation between different elements. These properties make up the rendering rules, determine the rendering method of the painting entity.

These rendering rules also represent the drawing algorithms and the artistic style of the facial cartoon. The cartoon rendering engine can learn them to synthesis multi-style facial cartoon without human interaction.

An instance of the left-eye painting entity E^{l-eye} is showed in Figure 3. Its rendering shape E^{l-eye}_{shape} is a curve connected by 8 feature points, there are 18 rendering elements: $E01$ to $E13$ are the region elements, and $E14$ to $E18$ the curve elements. Each element corresponds with a rendering rule, for example, the rendering rule of the region element $E02$ is (Element ID=E02, Layer No=60, Visible=true, Rendering order=2, Fill color=(220,122,75), Boundary={(84,556),(104,529), ...,}, Fill style=monochrome), and the curve element $E18$ is (Element ID=E18, Layer No=60, Visible=true, Rendering order=5, Line color=(137,69,34), Line degree=3, Line wight=5, Brush tyle=solid, Control point weight=0, Line style=line, Brush hight=32).

3.2 The Reference Shape Definition – RSD

The RSD records the reference shape parameters of the sample, which includes the facial structure and the facial components denoted as formula (4).

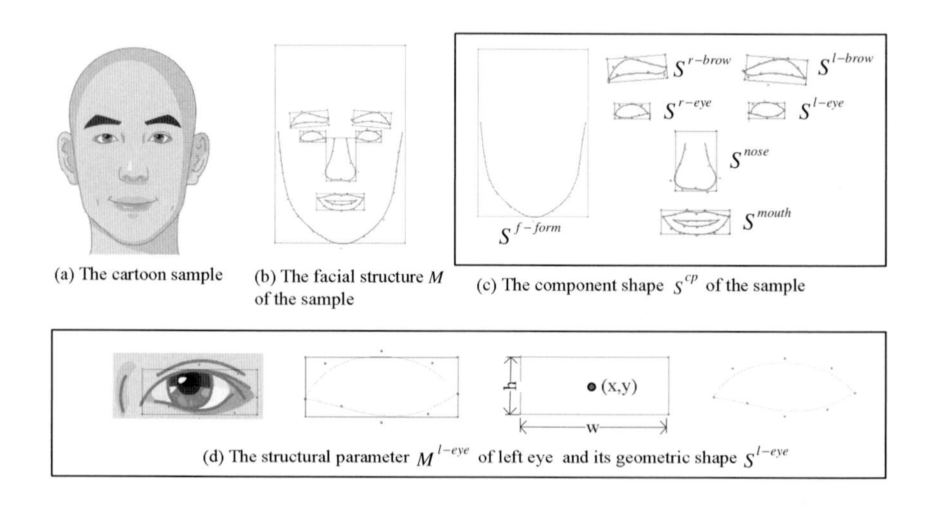

(a) The cartoon sample (b) The facial structure M of the sample (c) The component shape S^{cp} of the sample

(d) The structural parameter M^{l-eye} of left eye and its geometric shape S^{l-eye}

Fig. 4. The RSD of the cartoon sample

$$R = (M^{cp}, S^{cp}|_{cp=f-form,l-brow,r-brow,l-eye,r-eye,nose,mouth}) \quad (4)$$

A new notation, M^{cp} defines structural parameters of the facial component, which is composed by a 5-tuple $M^{cp} = (x, y, w, h, \theta)$, (x, y) is center coordinate of component, w component width, h component height and θ direction with respect to horizon. S^{cp} is the geometric shape of the facial component. The facial structure $M = \{M^{cp}\}$ gives the arrangement of facial components on a face, and can be calculated by the contour features of the face.

Figure 4 illustrates the facial structure of a sample and its component shapes. The RSD of left-eye component sample S^{l-eye} and its structural parameter M^{l-eye} are shown in Figure 4(d).

3.3 Two Conversions from *R* to *P*

There are two kinds of mapping between the reference shape R and the painting pattern P: one is the linear mapping Φ from the facial structure M to the layout L, called the shape conversion; Another is the non-linear mapping Ψ between the facial component S to the painting entity E, named the rendering conversion.

Given an input face, its reference shape parameters (M, S) can be automatically extracted and calculated[6]. The painting pattern parameters of the corresponding cartoon can be obtained by using the conversions and learning samples, denoted as formula (5).

$$\begin{aligned} L &= \Phi\,(M; \mathcal{D}) \\ E &= \Psi\,(S; \mathcal{D}) \end{aligned} \quad (5)$$

For the shape conversion Φ, suppose the layout L is the convex combination of those in sample space as formula (6), the coefficient α_k represents the weight of each sample, which are constraint to $\alpha_k \geq 0$, $\sum_{k=1}^{N} \alpha_k = 1$. These coefficients can be calculated by using the corresponding facial structure and those in the sample space, the layout can be determined for a specific facial cartoon[7].

$$L = \sum_{k=1}^{N} \alpha_k L^k \quad (6)$$

The rendering conversion Ψ is used to calculate the painting entity E. Since the facial components are relatively independent, Ψ can be simplified to each component denoted as Ψ^{cp}. Ψ^{cp} is determined as follows: Firstly, select the most appropriate component sample from the sample subspace to meet the given facial component shape. Secondly, calculate the difference between component shape and the selected sample. Final, a compensation process is used by warping the entity according to the shape difference[7].

4 Conclusion

A cartoon face model and the sample space are proposed which help the representation and generation of multi-style cartoons in a unified framework. For an input face object, with the help of sample space and two corresponding conversions, the cartoon rendering engine can automatically generate facial cartoons

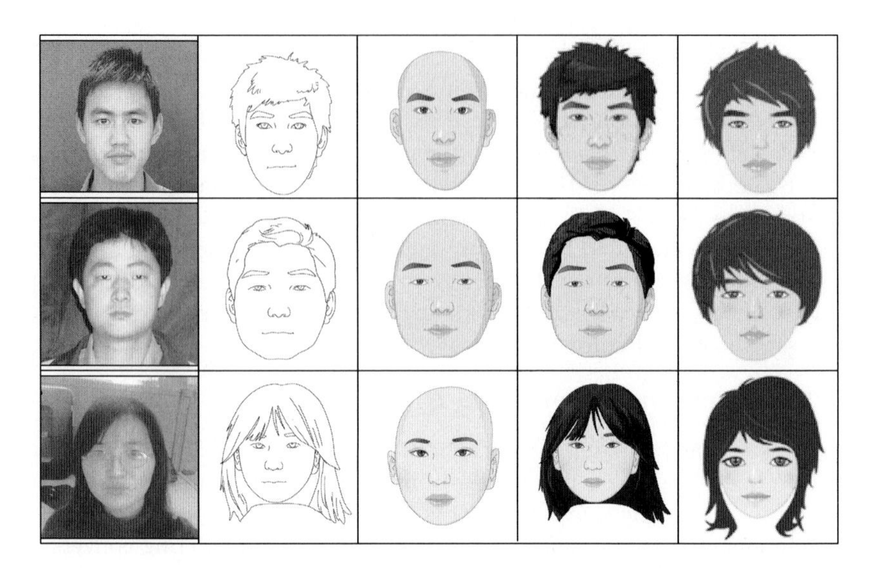

Fig. 5. Some generated facial cartoons with different artistic styles

with different artistic style, some are shown in Figure 5. Experimental results verify the effectiveness of the proposed parameterized representation of the cartoon sample space.

Acknowledgment. The project was supported by the Research Fund for the Doctoral Program of Higher Education of China (No.20060698025), and the NSF of China(No.60775017).

References

1. Thorisson, K.R.: ToonFace: A System for Creating and Animating Interactive Cartoon Faces, Learning and Common Sense Section Technical Report (1996)
2. Chen, H., Liu, Z., Rose, C., et al.: Example-Based Composite Sketching of Human Portraits. In: The 3rd Int'l symposium on Non-photorealistic aniamtion and rendering, pp. 95–101 (2004)
3. MSN cartoon, `http://cartoon.msn.com.cn`
4. Chen, H., Zheng, N.N., Liang, L., et al.: PicToon: A personalized Image-based Cartoon System. In: ACM international conference on Multimedia, pp. 171–178 (2002)
5. Tominaga, H., Fujiwara, M., Murakami, T., et al.: On KANSEI facial image processing for computerized facial caricaturing system PICASSO. In: IEEE International Conference on SMC (1999)
6. Colmenarez, A.J., Xiong, Z., Huang, T.S.: Facial Analysis from Continuous Video with Applications to Human-computer Interface. Kluwer Academic Publishers, Boston (2004)
7. Liu, Y., Su, Y., Wu, Z., Yang, Y.: NatureFace: A Cartoon Face Producer for Mobile Content Service. In: IAPR Conference on Pattern Recognition: WMMP 2008, USA (2008)

Enhancing Seeker-Bars of Video Players with Dominant Color Rivers

Klaus Schoeffmann and Laszlo Boeszoermenyi

Institute of Information Technology (ITEC), University of Klagenfurt
Universitätsstr. 65-67, 9020 Klagenfurt, Austria
{ks,laszlo}@itec.uni-klu.ac.at

Abstract. Interactive navigation through a video is a simple way for a user to get a quick overview of its content and to find interesting scenes. Although common video players provide only poor navigation facilities – in comparison to real video search applications – they are often employed by users due to their simplicity. We present a tool using a similarly simple interaction method but enabling much more efficient navigation.

1 Introduction

Video navigation is known as the interactive process of navigating through a single video, usually to get an overview of its content and to locate potentially interesting scenes. An example application enabling video navigation is a common video player that provides a seeker-bar for random access. Many users employ such video players for video search as these tools provide an intuitive user interface which is easy to use for non-experts as well. For instance, the popular YouTube player still uses a simple seeker-bar for navigation. We present a video navigation tool that provides an extended seeker-bar which can visualize the progression of dominant colors throughout the whole video. Our extension allows navigation like a usual seeker-bar but shows a colorized content abstraction in the background, which can be helpful for many search tasks. While there is already existing work on that general idea, we further improved this concept in several ways:

1. We use not only one dominant color (DC) per frame but rather several ones (e.g. 1st, 2nd, 3rd DC, a.s.o.), which allows a user to derive more useful information from the visualization. For example, in a news video we can easily locate segments showing a black-dressed anchorman in front of an orange background and distinguish them from orange commercials (Fig. 2).
2. Our seeker-bar provides an *overview visualization* for the whole video as well as a *detailed visualization* for a user-defined segment. The overview visualization contains a *zoom window* specifying the clipping for the detailed visualization (Fig. 1). A user can interactively change the size and location of the zoom window and easily navigate through videos of any duration.
3. The user can restrict the visualization to a spatial region. For example, in a quiz show this allows to find all the different questions or answers, which are always presented at the same spatial position in same or similar colors.

S. Boll et al. (Eds.): MMM 2010, LNCS 5916, pp. 773–775, 2010.

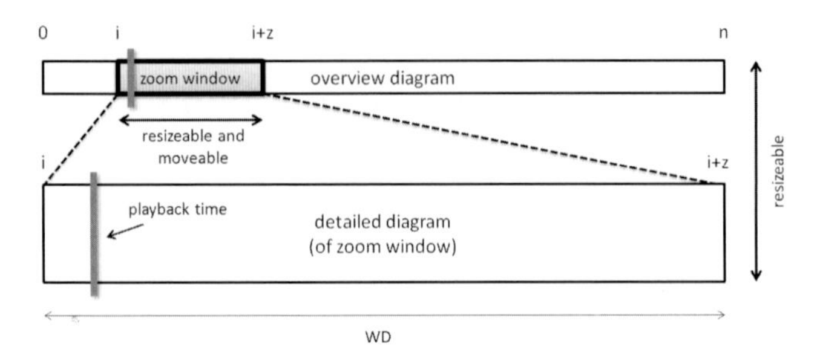

Fig. 1. Interaction model of the enhanced seeker-bar for a video with n frames [1]

2 Dominant Color Rivers

For every frame of the video we quantize the color values to a 64-bin histogram, extract the five-most dominant colors, and visualize these colors in a vertical line. This vertical line consists of sub-lines, one for each dominant color. The height of each sub-line is determined by the proportional number of pixels related to that dominant color. The adjacent visualization of such composed vertical lines results in a diagram showing the progression of dominant colors throughout the video. We use that visualization as an interactive visual time-line for navigation. The vertical position of a dominant color is set according to the bin number in the color histogram, where the smallest bin is drawn at the bottom and the highest bin is drawn at the top. This ensures a smooth "river-like" visualization of one dominant bin in temporal sequence, since the relative position of one particular bin stays constant over time.

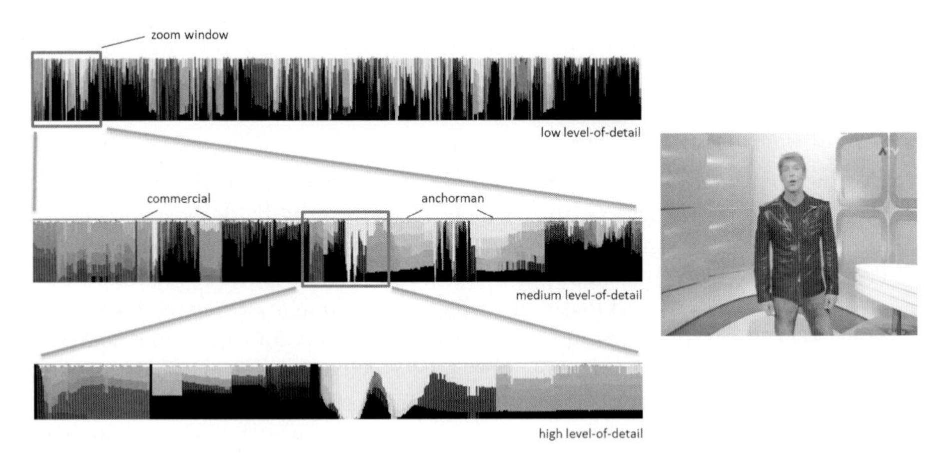

Fig. 2. Dominant color rivers (detailed diagram of Fig 1 only) visualized for a news video at several levels of detail. While the first visualization shows the entire video (full zoom window), the second one shows a small range from the beginning, and the third one shows an even smaller segment, as defined by the zoom windows through the user.

Fig. 2 shows an example of the enhanced seeker-bar for a news video at several levels of detail. The visual seeker-bar helps users at navigation by giving a rough content abstraction instead of a (relative) time information only. For example, if the user knows that a news anchorman appears several times in a video, wearing black clothes and usually in front of an orange background, the user can easily detect those events by typical color patterns in the visualization (compare Fig 2). On playback the visualization smoothly scrolls along with the playback time, showing the current time position at the center of the diagram.

The user can also restrict the visualization to a particular spatial area by drawing a rectangle directly on the display area. This allows a user to concentrate on colors of small areas which would not be shown in the visualization due to their minor proportional extend in comparison to the entire frame. For example, in a video of a quiz show with dark colors in general this helps to concentrate on colors of areas showing questions/answers only (see Fig. 3).

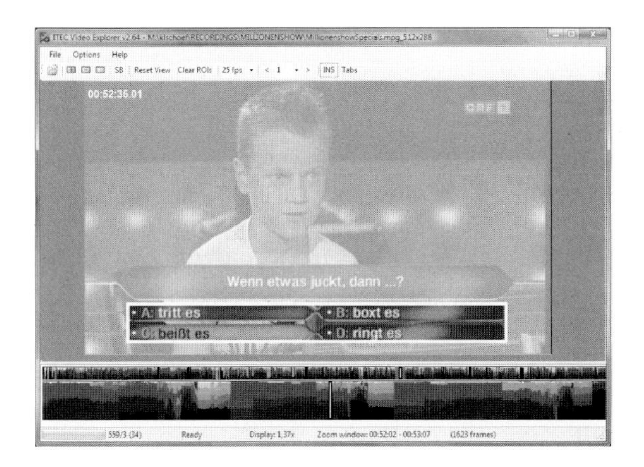

Fig. 3. Restricting dominant color visualization to a spatial area (white rectangle)

3 Conclusions

Our enhancement to seeker-bars allows simple and intuitive navigation in a video and enables more efficient search than common seeker-bars do. The visualization requires a short preceding content analysis step. The method could be used as an alternative to common seeker-bars, especially in systems which process videos before users consume them. Such an example system is YouTube, for instance.

Reference

1. Schoeffmann, K., Boeszoermenyi, L.: Video browsing using interactive navigation summaries. In: Proceedings of the 7th International Workshop on Content-Based Multimedia Indexing, Chania,Crete. IEEE, Los Alamitos (2009)

AdVR: Linking Ad Video with Products or Service

Shi Chen[1], Jinqiao Wang[1], Bo Wang[1], Ling-yu Duan[2], Qi Tian[3], and Hanqin Lu[1]

[1] National Laboratory of Pattern Recognition, Institute of Automation, Beijing, China
{schen,jqwang,bwang,luhq}@nlpr.ia.ac.cn
[2] Institute of Digital Media, School of EE & CS, Peking University, Beijing, China
lingyu@pku.edu.cn
[3] University of Texas at San Antonio, San Antonio, TX, USA
qitian@cs.utsa.edu

Abstract. This demonstration (AdVR: *Ad Video to service Recommendation*), attempts to build a semantic linking between ad video and relevant products or service by progressive search on E-commerce websites. We firstly extract representative images to summarize the video. Then we search visually similar product images by Spectral Hashing, rank the contextually textual information by tag aggregation, and refine the results by textual re-search. Finally, several products relevant to the ad will be recommended to users. Experiments on popular E-commerce websites demonstrate the attractiveness and effectiveness of our approach which semantically links video content with web services.

Keywords: Ads Analysis, Ads Recommendation, Video Search.

1 Introduction

Ad video is a very popular advertising form to promote goods, services, and ideas. The proliferation of user generated content (UGC) websites brings high potential to develop advertising industry on web and especially to find a viewer-friendly and advertiser beneficial solution to launch ads. There has been some work about contextual advertising [4], such as less intrusive insertion of relevant ad video in streams by content matching, but the inherent semantics of video ads is much less exploited. It will be very useful and potential to do ads recommendation by linking ad videos with related products promoted on websites.

In this demonstration, we present an effective and progressive way to semantically represent ad video and link it with online products or service towards ads recommendation in a style of cross-media. An example of AdVR is shown in Figure1. At first, we analyze video ads to capture the subset of representative images about advertised products. Then we try to collect relevant images and textual information from the Internet through visual matching. After that, we rank the contextually text information of visual search results by tag aggregation and refine the results with textual re-search. Finally, we parse the textual tags of top matched products and make more meaningful online recommendation based on the re-search results from the Web.

S. Boll et al. (Eds.): MMM 2010, LNCS 5916, pp. 776–778, 2010.

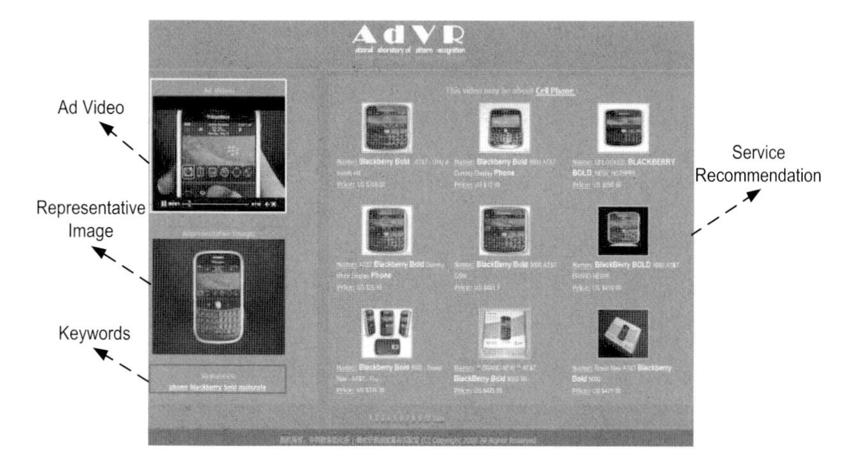

Fig. 1. An example of AdVR

2 System Overview

Figure 2 shows the overall framework of AdVR, which contains offline and online parts. In the offline part, we collect products' information from shopping websites including eBay and Amazon, build feature and tag indexes to make preparations for the online part. The online part is composed of five modules: video analysis, visual search, tag aggregation, textual re-search and service recommendation.

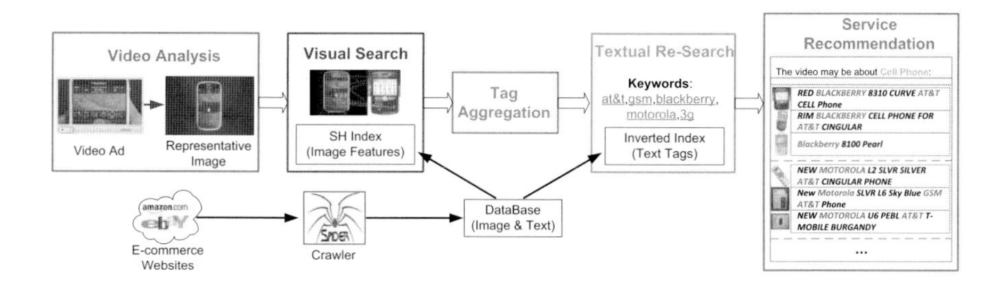

Fig. 2. Overall framework of AdVR

3 System Implementation

We detect the presence of FMPI (*Frames Marked with Product Images*) to identify a commercial video [2]. An FMPI image can be regarded as a kind of document image involving graphics, images, and texts and is usually used to highlight the advertised products. We cluster the gist descriptor of all FMPI images and simply select one as the representative image to represent the video.

We extract several local (SURF, Shape-Context, Geometric-Blur) and global features (Colour histogram, Grid Gabor) for all images in database and the representative FMPI images. We use a Naïve Bayes Nearest Neighbor classifier to search visually similar images. With the help of Spectral Hashing, we can search the similar images quickly. In addition, we dynamically select and adjust features' weights by entropy theory to combine these features to get better results.

Then we use the products' textual information (tags) to refine the search results. With the help of the K-lines-based tag aggregation algorithm [3], the visual search results can be classified by tags. We remain the top classes of them to select keywords for textual re-search. We execute text-based search to find more precise and semantically relevant products. Finally, with useful product tags (names, prices and other descriptions), we give users rich and detailed recommendation based on the textual research results. Users will see images and tags of relevant products and can also find detailed information of each product by its link on E-commercial websites.

4 Experiments and Performance

We built up a large dataset of product images and tags from popular shopping websites. We selected 13 popular categories of ad videos for experiments. Our product information database contains about 40,000 items. Mean Average Precision (MAP) is employed to evaluate our results. The MAPs of all videos in the three steps (visual search, tag aggregation and textual re-search) of AdVR are: 0.26, 0.28, and 0.4. In all, our progressive linking scheme can improve the semantic understanding of ads step by step and give quite precise recommendation of products at last.

5 Conclusion

This demonstration presents an online product/service recommendation system by linking ad video with relevant product information across the Web. Various techniques are combined, such as video analysis, image retrieval and multimodal fusion. It can semantically analyze ad videos and provide users product information relevant to the advertisements. We will improve the performance of recommendation by improving the matching precision and carefully sorting the tags of products.

References

1. Weiss, Y., Torralba, A., Fergus, R.: Spectral Hashing. In: Proc. Advances in Neural Information Processing Systems (2008)
2. Duan, L.Y., Wang, J., Zheng, Y., Jin, J.S., Lu, H., Xu, C.: Segmentation, categorization, and identification of commercials from tv streams using multimodal analysis. In: Proc. ACM MM 2006, pp. 202–210 (2006)
3. Fischer, I.: New Method for Spectral Clustering. Technical Report No.IDSIA-12-04, Hebrew University, Israel
4. Mei, T., Hua, X.S., Li, S.: Contextual In-Image Advertising. In: Proc. ACM Multimedia, pp. 439–448 (2008)

Searching and Recommending Sports Content on Mobile Devices

David Scott[1], Cathal Gurrin[1,2], Dag Johansen[2], and Håvard Johansen[2]

[1] Dublin City University, Dublin, Ireland
[2] Untiversity of Tromsø, Norway
{david.scott,cgurrin}@computing.dcu.ie, {haavardj,dag}@cs.uit.no

Abstract. As users rely more on mobile devices to access to video and web information we must adapt current technologies and develop new ones to support multimodal device access to digital archives. In this work we report on a prototype video retrieval system for TV sports content that utilizes sports summarization and personalization to deliver a multi-modal user experience.

1 Introduction

In recent years we have witnessed a revolution in how people locate and access video content. Viewers are no longer restricted to accessing videos through a broadcaster's TV schedule. Now, they can access video content on-demand using digital video recorders, online video archives and increasingly from their mobile devices. A survey conducted by the Nielson company shows that the employment rate of mobile video access has increased with 52.2% from 2008 to 2009 [1]. This massive surge is set to continue as more and more content is migrated to mobile devices.

At present, most video content available to mobile devices is simply delivered via modified front-ends to web archives such as YouTube. There has been little effort to adapt these services specifically for access by mobile devices. Although mobile devices have improved much recently in the way they handle user input (e.g. presence of qwerty keyboard, touch screens, etc.) it is still more challenging for a user to access content using a mobile device compared to stationary devices. As such, it remains necessary to minimize user input.

To support multimodal device access to a digital video archive, we will describe DAVVI, a video search and retrieval system for sports video content that utilizes automatic sports summarization and recommendation technologies to support both mobile and desktop device access.

2 Prototype System Description

Our current sports video retrieval prototype has an archive of several months of Norwegian and UK football content. It supports both web and mobile access. There are a number of innovative aspects of the system, including:

S. Boll et al. (Eds.): MMM 2010, LNCS 5916, pp. 779–781, 2010.

- Automatic event segmentation and tagging of events in sports matches.
- Real-time, automatic generation of result video documents (not lists of clips) as the unit of retrieval.
- Personalization and recommendation of sports content based on a stored user profile, primarily to support mobile access.
- Identification of a number of key event types, such as: goal, corner, free, yellow card, and penalty.
- Multiple bitrates streaming to support multimodal device access.

The most innovative aspects of this demo include sports segmentation with keyframe extraction, the recommendation of content, and the video documents result generation, each of which we will now discuss.

The event segmentation technique for sports content has received a lot of attention recently using purely visual processing techniques, for example [2] uses an SVM to combine evidences from a number of key attributes that suggest an important event taking place, however the accuracy is still only in the region of 70%. By utilizing external sources of evidence we have been able to surmount this problem and achieve near 100% accuracy. However, this poses another research challenge, that of accurately identifying the optimal start and end of an event, which ideally should be defined by the specific user query. For example, one needs to define where a goal event begins and ends and not simply to return a specific length result. The external source of evidence we utilize was football match statistics trawled from the web, and this gives us an accurate timestamp of the actual event taking place (to the second). Once this is known, the event segmentation technique, based on [2], determined where the event started (build up to the event) and where the event ended (moved on to another separate event). Two segmentation approaches were employed, manual and automatic, and we are currently evaluating the effectiveness of both.

In the *manual segmentation* approach, each event type is segmented using a predefined length (non-optimized to the event) of a number of seconds. Yellow cards for example were identified as short event and produced ten second segments, but goal events defined as more important events and are allocated thirty second segments.

The *automatic segmentation* technique optimizes the length of the event segment, which can for example be shorter for mobile devices and longer for stationary devices. The starting point is defined by an increase in activity of crowd noise, fast visual motion and camera panning, which was also further reinforced by the parsed online match statistics. End points are determined as crowd excitement reduction, less camera movement and less player movement.

The unit of retrieval for this prototype is not individual video clips, but a generated video document that is comprised of a number of high ranked clips that best match a user's information need and profile. On the desktop device the result is a custom generated result video document, however the component clips are also retuned and can be manipulated by the user (e.g., reordered, deleted, etc.) to improve the resulting video document, which itself may be shared with buddies. However on the mobile device, taking into account the screen size limitation, the result of a query is only a custom generated result video document. For example, a result may contain all the goals scored by 'Ronaldo' after a corner kick in last season's football matches.

For recommendation purposes, content can be recommended on both mobile and desktop devices. The desktop prototype allows multi-user login, complex query formation, saving of playlists, recommendation of content to buddies and the creation of user profiles based on content access. To support mobile device access, the mobile version is bound to a single user, supports simple user query formation (e.g. free-text queries) and is mostly focused on utilizing the user profile to proactively seek out and recommend interesting content to the user (e.g. based on recent activity of a favorite player, or viewing habits of buddies, which is a form of collaborative filtering). The user profile is maintained by employing both explicit and implicit feedback, with explicit feedback only employed on the desktop interface. Finally, hybrid filtering is employed to recommend content.

Figure 1 shows screenshots of both the desktop and the mobile versions of the DAVVI prototype sports search engine. In the mobile interface you can see the text search box and a listing of the most recent video documents generated.

Fig. 1. Interface to the Desktop and Mobile DAVVI

3 Future Work/Conclusions

In DAVVI we have developed a prototype sports event centered search engine which supports multimodal device access. For future work, an extensive user study is planned along with continuing to improve underlying algorithms and recommendation engines. Process migration to allow seamless video playback when swopping between different interaction devices is under development (e.g. changing devices from desktop to mobile would not affect.

References

1. The Nielsen Company. Television, Internet and Mobile Usage in the U.S. A2/M2 Three Screen Report, 1st Quarter 2009. Web publication (May 2009), http://blog.nielsen.com/nielsenwire/online_mobile/americans-watching-more-tv-than-ever/
2. Sadlier, D., O'Connor, N.: Event Detection in Field Sports Video using Audio-Visual Features and a Support Vector Machine. IEEE Transactions on Circuits and Systems for Video Technology 15(10), 1225–1233 (2005)

Extended CBIR via Learning Semantics of Query Image

Chuanghua Gui, Jing Liu, Changsheng Xu, and Hanqing Lu

Institute of Automation, Chinese Academy of Sciences, Beijing 100190, China
{chgui,jliu,csxu,luhq}@nlpr.ia.ac.cn

Abstract. This demo presents a web image search engine via learning semantics of query image. Unlike traditional CBIR systems which search images according to visual similarities, our system implements an extended CBIR (ExCBIR) which returns both visually and semantically relevant images. Given a query image, we first automatically learn its semantic representation from those visual similar images, and then combine the semantic representation and their visual properties to output the searching result. Considering that different visual features have variously discriminative power under a certain semantic context, we give more confidence to the feature whose result images are more consistent on semantics. Experiments on a large-scale web images demonstrate the effectiveness of our system.

Keywords: Web image search, semantic learning, feature selection.

1 Introduction

With the prevalence of the Internet and digital cameras, more and more digital images access to our life. Accordingly, the capabilities to efficient and effective image retrieval have become increasingly important and necessary.

Generally, there are two types of image retrieval approaches. One is the content-based image retrieval (CBIR), as a way to search visually similar images given a query image. As we know, the semantic gap between the low-level visual description and the high-level semantics has become a major obstacle to CBIR. Some work introduces relevance feedback from the users to get better results [1]. However, as images are described only with visual features, it is hard to ensure the relevance from semantic perspective. That is, it cannot fundamentally bridge the semantic gap.

Text-based image retrieval (TBIR) is the other approach. TBIR requires annotating each web image in advance and searches semantically relevant images given a text query. Due to the query polysemy, the results always contain multiple topics and they are mixed together. To attack it, Jing et al. [2] proposed to cluster the resulted images according to their semantics. They first identified several key phrases related to a given query, and assigned all the result images to the corresponding phrases. Ding et al. [3] further improved IGroup by clustering the key phrases into semantic clusters. However, both methods tend to make users puzzled because too many phrases are given, which make the results really diverse.

In this demo, we extend the traditional CBIR by learning the semantic representation of the query image and return the results similar both on visual appearance and semantic meanings. Given a query image, we first perform CBIR to obtain some

S. Boll et al. (Eds.): MMM 2010, LNCS 5916, pp. 782–785, 2009.

visually similar image sets corresponding to different visual features separately. Then we learn a semantic vector for each set and weightedly combine them into a semantic representation of the query image according to different confidence of each feature. Specially, the semantic consistence of each set is used to measure the confidence. More semantic consistence indicates that the feature is more appropriate to describe the query image. Finally, we present a co-search process by designing a vision-and-semantics combined formulation to search relevant images.

2 System Implementation

The system is composed of three main steps: basic CBIR, semantic learning and co-search, which are illustrated in Fig. 1.

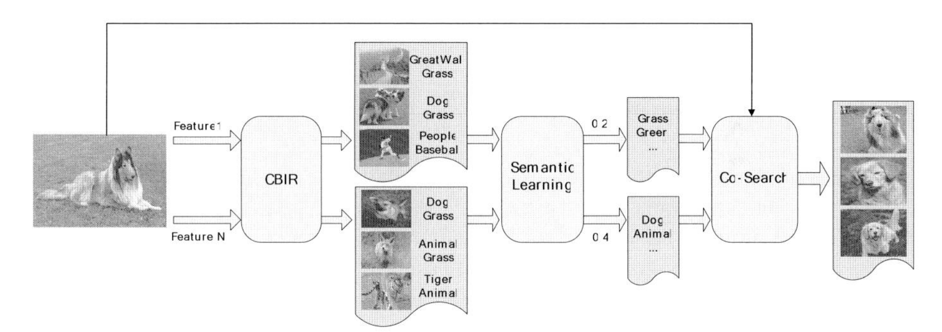

Fig. 1. Illustration of the system framework

2.1 CBIR

We perform CBIR to obtain some visually similar image sets corresponding to different visual features. Local Sensitive Hashing (LSH) [4] is used to speed up the matching process. LSH uses a hash function h to divide the database images into many bins. All images with the same output value of h are placed in a single bin. We execute the hash function to a query image and map it into a bin by its output. Only the images in the same bin as the query image are regarded as the results.

2.2 Semantic Learning

For each obtained image set by CBIR, we first mine representative keywords from the title, URL, anchor text and surrounding text of each image as candidates. Then, the *tf-idf* strategy is applied to rank these candidates and the top ones are selected as one preliminary semantic representation. To all kinds of visual feature, the same learning processes are performed and we combine the learned representations into the final semantic representation of the query image. During the combination, we will consider different confidences of the visual features. The semantic consistence among the textual information of each image set is used to evaluate the confidence of the visual feature. Specially, the more consistent the result set is, the more appropriate the corresponding feature is to describe the query image. We express the textual information of images with word vectors and map each vector as a point into the semantic space, and then compute the intensity of all the points as the measure of the semantic consistence.

2.3 Co-search

After obtaining the semantic vector and the semantic consistence of each feature, we apply a linear fusion to combine the ranked lists of TBIR and CBIR, which is formulated as:

$$S_{final} = S_{TBIR} + \alpha * S_{CBIR} \tag{1}$$

where α is a parameter to leverage the rate of the semantic similarity and the visual similarity. That is, the bigger α gives more confidence on the visual similarity while less on the semantic relevance, and vice versa. S_{TBIR} is the semantic relevance between the learned semantic representation of the query image and the parsed textual information of images. S_{CBIR} is the visual similarity, which is defined as:

$$S_{CBIR} = \sum c_i S_{f_i} \tag{2}$$

where c_i is the semantic consistence of the *i-th* feature.

3 Performance and Evaluation

Our system indexed 8 million web images in total. These images were crawled from Google and Flickr. Three types of visual features are extracted, including 144-dimensional Color Correlogram, 24-dimensional Polynomial Wavelet Tree and 36-dimensional Color Histogram. The parsed textual information, i.e., title, URL, ALT tag, anchor text and surrounding text is also indexed. To evaluate our system, ten participants are asked to search in a traditional CBIR system and our system with arbitrary queries they liked. The mean average precision (MAP) is employed, which is widely used by the image retrieval community. Experimental results of 50 query images show that our system (MAP=0.27) performs much better than traditional CBIR system (MAP=0.16). Fig.2 presents some search examples by CBIR and our system respectively.

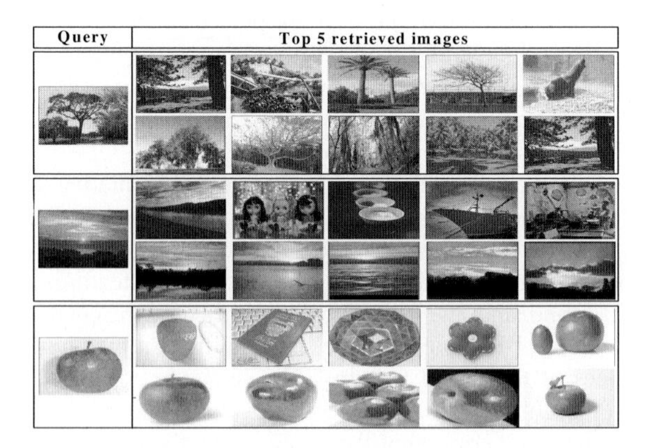

Fig. 2. The search result list by traditional CBIR (given in the 1-st, 3-rd and 5-th rows) and our method (given in the 2-nd, 4-th and 6-th rows) respectively

References

1. Rui, Y., Huang, T.S., Mehrotra: Relevance feedback: A power tool in interactive content-based image retrieval. IEEE Trans. Circuits and Systems for Video Technology (1998)
2. Jing, F., Wang, C., Yao, Y., Zhang, L., Ma, W.Y.: IGroup: Web Image Search Results Clustering. ACM Multimedia (2006)
3. Ding, H., Liu, J., Lu, H.: Hierarchical Clustering-Based Navigation of Image Search Results. ACM Multimedia (2008)
4. Datar, M., Immorlica, N.: Locality-sensitive hashing scheme based on p-stable distributions. In: The 20th annual symposium on Computational geometry (2004)

A Gesture-Based Personal Archive Browser Prototype

ZhenXing Zhang[1,2], Cathal Gurrin[1], Hyowon Lee[1], and Denise Carthy[2]

[1] Centre for Digital Video Processing, Dublin City University, Ireland
[2] Biomedical Diagnostics Inistiture, Dublin City University, Ireland
{zzhang,cgurrin,hlee}@computing.dcu.ie, denise.carthy@dcu.ie

Abstract. As personal digital archives of multimedia data become more ubiquitous, the challenge of supporting multimodal access to such archives becomes an important research topic. In this paper we present and positively evaluate a prototype gesture-based interface to a large personal media archive which operates on a living room TV using a Nintendo Wiimote for interaction.

1 Personal Media Archives in the Livingroom Environment

The increasing trend recently of people becoming content creators and not just consumers poses a challenge for organising and accessing the resulting personal archives of multimedia data. We have also noted a recent trend towards the integration of personal content management technologies into the enjoyment-oriented (lean-back) environment of the living-room, for example DVR functionality or WWW access on the TV. In this paper, we are concerned with the integration of personal content organisation facilities into the livingroom TV environment, which poses a number of challenges because it needs to be performed by non-expert users, with a remote control in a distractive (lean-back) environment, and not at a desktop computer with use of a keyboard and mouse (the typical lean-forward environment).

Previous research into managing personal media archives on a desktop device does not directly transfer to the living room environment, for reasons such as user interaction support, difficulty of querying and even device processor speed. It is our conjecture, however, that a living-room TV acts as a natural focal point for accessing personal media archives, and that taking into account the significant limitations and challenges of developing for such an environment is essential to successfully deploy multimedia content organisation technologies. Indeed initial work in the area by Lee et al. [1] suggests that simplicity of interaction is crucial for the livingroom environment, more crucial than in any other digital media domain and that ultimately, this simplicity of interaction determines the success or otherwise of any new applications. The challenge therefore is to marry the competing requirements of supporting complex digital multimedia archive organization technologies with the simplicity of interaction required when developing for the livingroom environment.

In this work we describe and demonstrate an interactive TV application, for the livingroom, which employs a gesture-based interface (using a Nintendo Wiimote) to manage a large archive of personal media gathered using a Microsoft Sensecam[2].

Studies on interactive TV interaction highlight the special characteristics of the livingroom environment and they show design implications and guidelines for a

S. Boll et al. (Eds.): MMM 2010, LNCS 5916, pp. 786–788, 2010.

technology operating in such a context. Characteristics such as use of a remote control as input device, increased viewing distance and enjoyment oriented (not the complete focus of task-orientation) usage scenarios prevail. Based on these studies, previous research and our own experiences of developing information retrieval systems for lean-back devices [1], we have compiled a set of guidelines for developing interactive multimedia applications for the livingroom, or any lean-back, environment:

- *Represent complex digital multimedia objects visually*. Complex multimedia objects, such as photo collections, video archives or HDM archives need to be visually represented and easily manipulatable on screen.
- *Minimise user input where possible* and proactively recommend content or support information seeking via a small number of frequently used features.
- *Engage the user* with simple, low-overhead interaction methodologies, that are enjoyable to use, easy to learn and engaging in a distracting environment.

2 A Prototype Gesture Based Diary Interface

Following the three guidelines, we developed a gesture-based browsing interface to an archive of Microsoft Sensecam [2] images operating on a living room (40 inch) TV, using a Nintendo Wiimote. Two weeks (about 50,000 photos or 3,500 per day) of Sensecam data, gathered by one wearer, was employed for this experiment, which we feel to be a good example of a challenging personal archive. The Sensecam images were tagged with date/time and location, as is standard for digital photos. How the three guidelines (above) impacted on the prototype is now illustrated:

- *Represent complex digital multimedia objects visually*. A HDM archive is an enormous repository of data and as such it needs to be summarised and visually easy to browse and interpret on any device. We used the event segmentation technique of Doherty et al. [3] to organize each day's images into a set of about thirty discrete events, which utilized visual processing of temporal image dissimilarity coupled with and analysis of Sensecam sensor data. A keyframe was automatically selected to represent each event based on an automatic analysis of its visual significance within that event.
- The prototype *minimized user input* by providing a diary-style calendar interface as the key access mechanism. A user could select next/previous days (a simple Wiimote gesture) and also select next/previous event (another simple gesture). Upon selecting an event, the event playback began which cycled through the images comprising that event at a fixed speed. The speed of this playback (from pause to fast-forward/fast-rewind) was user controlled by twisting the Wiimote as if one is twisting a dial or a knob.
- The prototype engages the user with *low overhead and low learning time* interaction methodologies that users found both easy and enjoyable to use.

A user evaluation was carried out of this prototype gesture based interface (Fig 1) with six novice users and the sensecam owner. All users received five minutes training on both interfaces. The prototype was compared to a similar interface on a desktop device with mouse and keyboard (lean-forward) interaction. Six tasks (four known-item and two ad-hoc search) were allocated to each participant (on alternate interfaces) and ordered to as to avoid bias. The average time taken for known item search was 77 seconds for the gesture interface and 73 seconds for the desktop interface, which was similar. For the ad-hoc task however, users of the gesture interface found 50% more relevant images than the desktop interface under the same time constraints. For the Sensecam owner (with a good knowledge of the data), the gesture interface was significantly better at finding known items with little difference on the ad-hoc search. In a qualitative examination, users found the gesture interface to be more satisfactory, efficient, productive and easier to recover from error when compared to the desktop interface which was more comfortable and easier to learn.

In conclusion, we have presented a set of guidelines for developing interactive search/browsing systems in a livingroom environment and evaluated a prototype system adhering to these guidelines. We found that tailoring the underlying algorithms to suit the limitations of the target environment can result in an equivalent (or better) performing system than an equivalent desktop implementation. Since the integration of more personal content organisation technologies into the livingroom environment is likely, these guidelines and results are a valuable initial contribution.

Fig. 1. The prototype gesture-based interface showing playback from the seventh event of the 12th April 2009, which took place in early afternoon in Dublin, Ireland

References

1. Lee, H., Ferguson, P., Gurrin, C., Smeaton, A.F., O'Connor, N., Park, H.S.: Balancing the Power of Multimedia Information Retrieval and Usability in Designing Interactive TV. In: Proceedings of uxTV 2008, Mountain View, CA, October 22-24 (2008)
2. Hodges, S., Williams, L., Berry, E., Izadi, S., Srinivasan, J., Butler, A., Smyth, G., Kapur, N., Wood, K.: SenseCam: A retrospective memory aid. In: Dourish, P., Friday, A. (eds.) UbiComp 2006. LNCS, vol. 4206, pp. 177–193. Springer, Heidelberg (2006)
3. Doherty, A.R., Smeaton, A.F.: Automatically Segmenting Lifelog Data into Events. In: WIAMIS 2008 - 9th International Workshop on Image Analysis for Multimedia Interactive Services, Klagenfurt, Austria (May 2008)

E-learning Web, Printing and Multimedia Format Generation Using Independent XML Technology

Alberto González Téllez

Universidad Politécnica de Valencia,
Dept. De Informática de Ssitemas y Computadores,
Camino de Vera, s/n,
46022 Valencia, Spain
agt@disca.upv.es

Abstract. E-learning content delivery through the Internet is nowadays a very active topic in academy, industry and research. Common tools used in presential learning are not adequate in the new context then new tools are required in order to produce, manage and deliver learning content in a way as much independent and flexible as possible. We present a set of tools that allows independent e-learning production and one-source-several-format delivery using vendor independent XML languages, particularly Docbook and SMIL.

Keywords: E-learning, Delivering Formats, XML, Docbook, SMIL.

1 Introduction

E-learning content delivery through the Internet by means of Learning Management Systems (LMS) is made using several presentation formats, particularly printing (PDF), web (HTML) and multimedia. It is then very convenient to separate content from presentation as much as possible. Furthermore, liking content to a particular application or software vendor is also not desirable in order to be able to easily move content among different LMS. Organizations can in that way interchange content and choose the more satisfactory platform at any time without any lease.

We propose a set of tools build around the open and vendor independent standards Docbook [1] and SMIL [2]. Docbook is a veteran XML compliant language oriented to electronic book production with a complete set of format generation XSLT style sheets. In order to be productive an author oriented editor is required. Unfortunately we have not found yet an open solution and we rely on the commercial editor XXE (XMLmind XML Editor). It is an affordable general author oriented XML editor, fully customizable in its Professional Edition. SMIL (Synchronized Multimedia Integration Language) is also an XML compliant language oriented to multimedia composition. It is just and specification from the W3C and then a particular platform has to be chosen in order to make it operative. At present we are using RealPlayer from Realnetworks as the target player. This requires that media elements included in an SMIL composition have to be transcoded to Realnetworks formats, particular Realaudio and Realvideo. This can be performed by means of the free tool Real Producer

S. Boll et al. (Eds.): MMM 2010, LNCS 5916, pp. 789–792, 2010.

Basic. In spite of being a proprietary platform RealPlayer is multiplatform (Windows, Linux and MasOS) and free. It is also open through the Helix project.

The actual need to use proprietary authoring tools do not affect content because it can be easily move to other environments as far as they comply with Docbook and SMIL specs.

2 E-learning, Web and Printing Format Generation

Docbook is a very powerful open and independent product that allows unlimited customization. In our academic environment we are interested in producing several delivery formats and in content reuse both applied to our centralized pool of learning content. In order to achieve this we have developed a Docbook markup and style sheets customization [3], combined with an XXE customization in order to get a user friendly integration. In this way we are able to generate the following documents without any content replication:

- Theory modules in IMS Content Package format and exams in IMS QTI format able to be imported in our Sakai based e-learning platform named PoliformaT.
- Class presentation in chunked HTML format that include a summary of the class notes with text font size adjusted.
- Class notes and exercise collections (with and without solutions) in high quality PDF format.
- Printed test and open question exams. Test exams can be automatically reordered at the question and response options levels.

We make extensive use of XInclude, very conveniently implemented in XXE, particularly to produce printed exams and to reuse content in related subjects.

The Docbook based publishing process is depicted in figure 1.

3 Production and Integration of Multimedia Compositions

The possibility of learning content delivery from an LMS, that in our University started in 2006 with PoliformaT, combined with the startup in 2008 of the streaming service PoliTube, have pushed us to develop a technique to enrich our static text+image documents with multimedia presentations.

In [4][5] we proposed an authoring environment to produce SMIL presentations based on free and open software. We also proposed an integration mechanism of these presentations in HTML modules published in PoliformaT. We do not have proper streaming service for SMIL delivery and then the integration in PoliformaT performs the download of the whole presentation before playing. That requires an undesirable wait time but the advantage is the interactive capabilities that SMIL offers (i.e. asynchronous time navigation with time links index). An alternative that avoids waiting before playing is to convert SMIL presentations to pure video and to publish it in PoliTube. The disadvantage with this option is that interactive capabilities are lost.

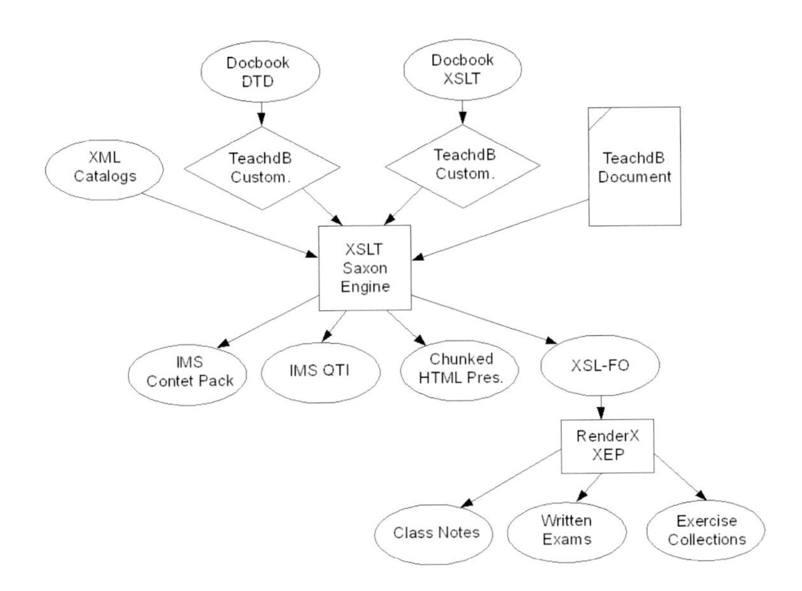

Fig. 1. E-learning publishing process

Following the previous considerations we are extending our previous work with the following tasks:

- Design of an automatic generation tool to produce SMIL compositions from a set of clips and presentation layout design.
- Production of MP4 video version of SMIL compositions and integration of both MP4 and SMIL into our modules published in PoliformaT.

The SMIL producer tool requires the Java Runtime Environment to run the tool that is implemented in Java and rely on JAXP to generate SMIL. It also needs RealProducer (Basic or Plus) to transcode presentation clips and RealPlayer to play the resulting SMIL composition.

Authors have to specify the following item sequences to the tool:

- Clips to be included in the presentation in a format supported by RealProducer.
- Video or audio narration clips.
- Time navigation index entries and how they map with the clip sequence.

Based on the author inputs and the time information available in real media formats, the SMIL producer generates a SMIL file and a folder that includes all the media required transcoded in RealPlayer supported formats, particularly realvideo, realaudio and realtext.

We have developed several presentation templates (figure 2), depending on the type of video camera used to capture presenter video, the version of RealProducer (Basic of Plus) available, the type of narration (voice or video) and the number of time link index entries.

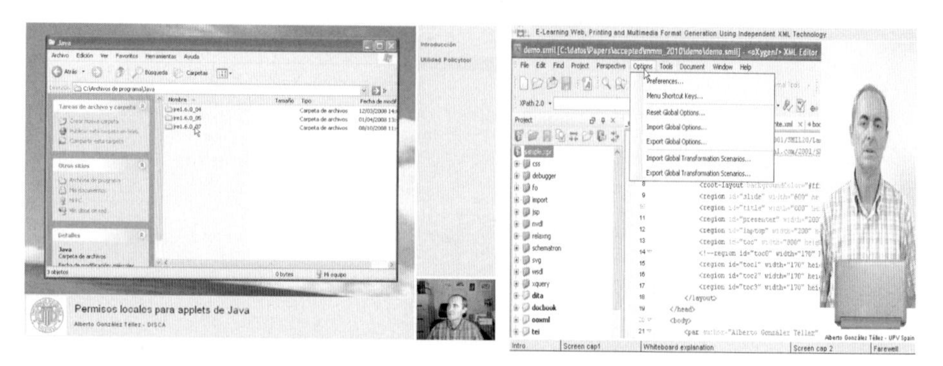

Fig. 2. SMIL composition templates

To produce an MP4 version able to be delivered by PoliTube, we capture the presentation played by RealPlayer with a screen video capture utility (i.e. CamStudio) and encode it into an MP4 container using a video encoder utility (i.e. MediaCoder).

The integration in PoliformaT HTML modules of both versions of multimedia presentations is made by including every presentation as an image. The image title is a link that opens an HTML page that embeds the corresponding MP4 video published in PoliTube adjusting video size for the best visual quality. The image is in fact a Java applet that performs the download of the SMIL version allowing local playing with good interactive response. This second alternative requires RealPlayer, Java and Java security permissions properly configured. We have developed a Java application to transparently configure Java security.

A virtual demonstration of the techniques proposed can be found in [6] in both MP4 video and SMIL formats.

References

1. Docbook main sites, `http://docbook.sourceforge.net`
2. SMIL site, `http://www.w3.org/AudioVideo/`
3. González, A.: Teaching Document Production and Management with Docbook. In: Proceeding of the II International Conference on Web Information Systems and Technologies, Setubal, pp. 188–195 (2006)
4. Gonzalez, A.: Authoring Multimedia Learning Material using Open Standards. Interactive Technology and Smart Education 4(4), 192–199 (2007)
5. SMIL presentation of [4],
`http://www.disca.upv.es/agonzale/mtel2007/mtel2007.html`
6. Virtual demonstration,
`http://www.disca.upv.es/agonzale/mmm2010/demo.html`

Dynamic Video Collage*

Yan Wang[1,2], Tao Mei[1], Jingdong Wang[1], and Xian-Sheng Hua[1]

[1] Microsoft Research Asia, Beijing 100190, P.R. China
[2] University of Science and Technology of China, Hefei 230027, P.R. China
grapeot@mail.ustc.edu.cn, {tmei,jingdw,xshua}@microsoft.com

Abstract. This demo presents a video visualization technique named Dynamic Video Collage (DVC). By selecting representative frames, extracting their regions of interest, constructing collages with the gradually coming frames, and displaying the collages via animations, we aim to mimic the storytelling of the video content by dynamically presenting a series of collages. In contrast to our previous work on Video Collage [1], DVC is able to show video dynamics in a more visually appealing way.

1 Introduction

With the rapid development of digital recording devices and internet video services, video data are explosively increasing. One way to get a quick glance of the main content of such amount of video data is video presentation, which aims at providing a compact summary with important information. There are two categories of video presentation methods: *static summary* and *video skimming*. Static summary approaches usually use a single image to present a video sequence, e.g., Video Collage [1] and Video Synopsis [2]. Although compact and easy for transmission and storage, static summary is unable to show video dynamics. On the other hand, video skimming uses a short video sequence to summarize the highlight segments. Although the dynamic summary contains enough time-involving information, it remains a challenge how to make the summary visually appealing as well as show the overall story.

We present in this paper a new video presentation approach, called Dynamic Video Collage (DVC), which represents a tradeoff between the compactness and visual pleasure of static summarization and the dynamics of video skimming. On one hand, it is compact and visually pleasing, like static summary, by generating a collage with overlaid keyframes. On the other, it can also show dynamics by gradually displaying a series of collages, like video skimming. To generate a DVC, when processing a video, we first select the representative frames (also called keyframes) from a part of the video which has been processed, and create a photo collage based on these frames [3]. Once a new keyframe is detected, the collage is automatically and gradually updated with the new keyframe, in an overlay manner with the most important part of the keyframe shown. Such

* This work was performed when Yan Wang was a visiting student at Microsoft Research Asia.

S. Boll et al. (Eds.): MMM 2010, LNCS 5916, pp. 793–795, 2010.

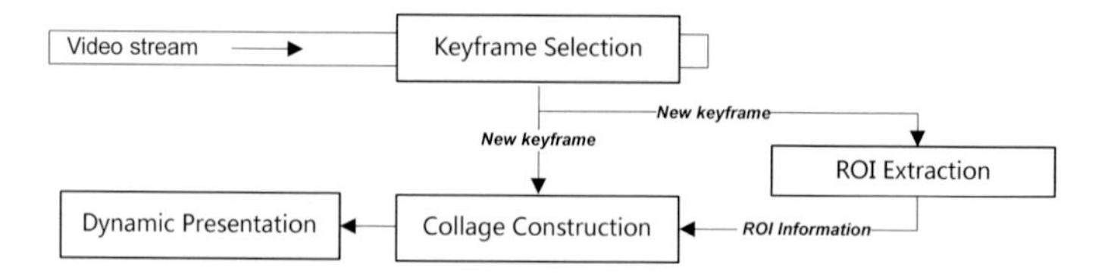

Fig. 1. Approach overview for generating a Dynamic Video Collage (DVC)

dynamic procedure (i.e., the process of keyframe detection and the changes of collage) is displayed in an animation. By DVC, users can get the main content as well as the video dynamics in a visually appealing and attractive way.

Our work is highly motivated from Video Collage [1] which generates a collage image to present a video sequence, and Picture Collage [3] which presents an image collection using a picture collage in an overlay manner. However, DVC is different from Video Collage in that it can show video dynamics and it constructs the collage in an overlay manner rather than in seamless fusion way; DVC is also different from Picture Collage in that it can update the collage with the increasing processed keyframes and thus can preserve the story of video sequence.

2 Approach

DVC has the following basic properties. First, the displayed frames should be representative with temporal structure maintained so as to preserve the informative video content. Second, the collage should be compact enough for efficient browsing. Third, it is expected to have the ability to present video dynamics so as to make the presentation more impressive. Figure 1 shows the approach overview for generating a DVC. Given a video sequence, we first perform shot detection to select a set of representative frames (i.e., keyframes). Once a keyframe is detected, the region-of-interest (ROI) is extracted and fed into the collage construction module to update the collage. When the collage construction module receives a new keyframe, it will update the collage with this keyframe. The changing procedure of the collage is gradually displayed in an animation manner. The process of DVC generation is like storytelling of the whole video sequence. The video dynamics are represented by the gradually enlarged collages which are generated by the incrementally detected keyframes.

ROI extraction. To make the collage compact, only the ROI of each keyframe is made visible in DVC. We adopt a fuzzy growing process to extract a saliency map for each keyframe [4]. By using two different fuzzy growing thresholds, we can get a bi-level ROI map.

Collage construction. To make the collage compact and visually appealing, we adopt an overlay collage in [3], which constructs a collage with saliency maximization, blank space minimization, saliency ratio balance, and orientation diversity from a collection of images. Collage construction is formulated as a Maximum a Posteriori (MAP) problem in which the corresponding solution infers a set of

Fig. 2. An example of "dynamic video collage" for a home video. The upper parts indicate the collages, while the lower parts indicate the detected keyframes. The highlighted keyframes are the new ones integrated to the collages. Please note that (A) is with 14 keyframes, while (B) is with 15.

state variables indicating the positions, layers, and orientations. We employed the Morkov chain Monte Carlo (MCMC) algorithm [5] for problem optimization.

3 Demonstrations

Figure 2 shows an example of DVC generated from a home video. The screen of DVC is divided into two parts: collage and keyframe list. In (A), DVC generates a collage with 14 keyframes, while in (B), with a new keyframe, DVC is able to update the collage smoothly and in an animation way. Although the collages in the DVC are compact, the ROI of each keyframe is visible. In this way, DVC presents a new visualization technique for video with impressive results.

References

1. Wang, T., Mei, T., Hua, X.-S., Liu, X., Zhou, H.-Q.: Video collage: A novel presentation of video sequence. In: IEEE International Conference on Multimedia & Expo. Beijing, China, pp. 1479–1482 (2007)
2. Rav-Acha, A., Pritch, Y., Peleg, S.: Making a long video short: Dynamic video synopsis, pp. 435–441 (2006)
3. Wang, J., Sun, J., Quan, L., Tang, X., Shum, H.Y.: Picture collage. In: Proceedings of IEEE Conference on Computer Vision and Pattern Recognition, pp. 347–354 (2006)
4. Ma, Y.-F., Zhang, H.-J.: Contrast-based image attention analysis by using fuzzy growing. In: Proceedings of ACM Multimedia, pp. 374–381 (2003)
5. Andrieu, C., Doucet, A., Jordan, M.I.: An introduction to mcmc for machine learning. Machine Learning 50, 5–43 (2003)

VDictionary: Automatically Generate Visual Dictionary via Wikimedias

Yanling Wu[1,2], Mei Wang[2,3], Guangda Li[2], Zhiping Luo[2], Tat-Seng Chua[2], and Xumin Liu[1]

[1] College of Information Engineering, Capital Normal University, Beijing, China
liuxumin@126.com
[2] School of Computing, National University of Singapore, Singapore
{wuyanlin,wangmei,luozhipi,chuats}@comp.nus.edu.sg, g0701808@nus.edu.sg
[3] School of Computer Science and Technique, Donghua University, China

Abstract. This paper presents a novel system to automatically generate visual explanation by exploiting the visual information in Wikimedia Commons and the automatic image labeling techniques. Sample images and the sub object based training data are obtained from Wikimedia Commons. Then propose an image labeling algorithm to extract salient semantic sub object. Each sub object is assigned to a semantic label. In this way, different semantic-level visual references are provided in our system.

1 Introduction

Compared to text, visual information, such as image, contains richer information. Recently, many online visual dictionaries leverage the visual world [1] for their application purpose. Visual dictionary is designed to quickly answer questions when you know what something looks like but not what it is called, or when you know the word but cannot picture the object. Fig. 1 illustrates an example about object "car" from the famous website *Visual Dictionary Online* [1]. However, the textual definitions are always developed by professional experts in such websites, so that they need much time and labors, which greatly impair the scalability and completeness.

Region-based image annotation algorithms have been used to support automatic visual dictionary construction. However, their performances may largely depend on the image segmentation techniques, which are very fragile and erroneous process [2]. In addition, how to select salient sub objects of a given image for annotating is a complicated problem. Most image annotation algorithms heavily rely on the training data [6]. It is difficult to obtain a region-based annotation training set.

Nowadays, some photo websites, such as Wikimedia [3], are popular in daily life. Fig. 2 illustrates an example of "transport" hierarchy obtained from Wikimedia Commons. As can be seen, information from Wikimedia Commons has the following characteristics: 1) Textual concepts and images are well organized by professional users; 2) Images in a given object category is relatively "pure";

S. Boll et al. (Eds.): MMM 2010, LNCS 5916, pp. 796–798, 2010.

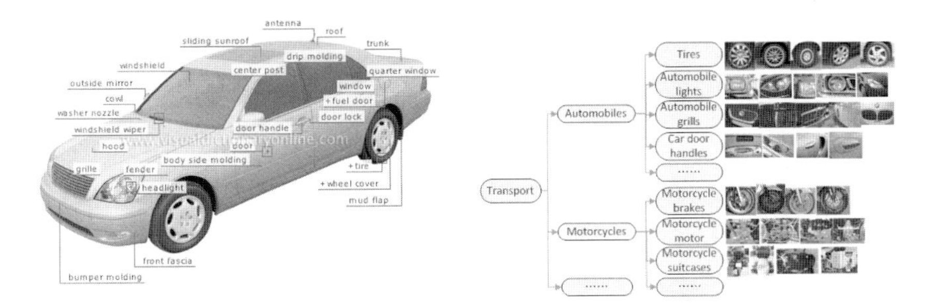

Fig. 1. The dictionary online

Fig. 2. The transport hierarchy from Wikimedia

and 3) Visual information of the sub objects, which can be used as the training data for automatic sub object annotation is also provided by Wikipedia.

In this paper we present a novel system to automatically generate visual explanation. This system utilizes web source Wikimedia Commons and statistical image labeling techniques.

2 The Highlight of the System

Fig. 3 shows the framework of the proposed system. First, we feed each given object in a dictionary into Wikimedia Commons, and then obtain sample images S in this category and the images in its subcategory as the sub object annotation training data T. We employ SIFT detector and descriptor [4] to determine salient local point and compute the description respectively. It can be observed that dense salient feature points often appear at the location where interest object parts exist. We then exploit mean shift clustering (MSC) [5] to automatically generate sub objects for the image in S, and then exploit k-NN classifier to assign label for each sub object based on T.

Automatic Sub Object Generation: Since the location of the meaningful sub object is usually at the place where the feature points clusters into a dense area. MSC is always used to find dense regions in the data space. Therefore, to obtain meaningful image representation, we employ MSC to find the semantic patches where local salient regions are densely distributed based on their geometric positions. After clustering, we can obtain a set of cluster centers. By Defining two thresholds α_1 and α_2. α_1 is used to refine clusters, α_2 is used to filter the isolate local feature points. Taking the cluster center as the center, the scale of the farthest points as the radius, we can obtain the image patches, where interest sub objects always exist.

Automatic Semantic Region Generation: We assign the sub object label for each patch as the sub objects have been obtained. For a given training set T, we use k-NN classifier to assign the sub object label for each sub object patch.

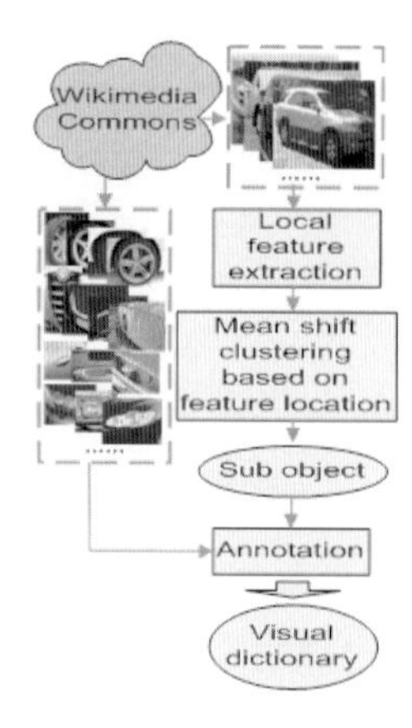

Fig. 3. System Framework

Fig. 4. System interface

Demonstration Interface: Fig. 4 shows the interface of the system. It consists of two panels: control panel and image panel. In the control panel, the image selected button "Browse" allows the user to select an example image or a given object to be explained. To click the "Show Image" button, the example image or the selected representative image for the object will be presented in the image panel. By clicking "Sub Object Generation" the visual panel will present the image on which the semantic parts are marked. While clicking "Show Annotation" the textual annotation of each sub object will be labeled. In the image panel, interactive zooming is supported to show the detailed information of the image.

3 Conclusion

This paper presented an automatic visual dictionary system. It can help users find visual answers automatically.

References

1. Visual Dictionary Online, http://visual.merriam-webster.com/
2. Gao, Y., Fan, J., Luo, H., Xue, X., Jain, R.: Automatic Image Annotation by Incorporating Feature Hierarchy and Boosting to Scale up SVM Classifiers. ACM Multimedia (2006)
3. Wikimedia, http://commons.wikimedia.org/wiki/Main_Page
4. Lowe, D.G.: Distinctive image features fromscale-invariant keypoints. International Journal of Computer Vision 60, 91–110 (2004)
5. Wang, P., Lee, D., Gray, A., Rehg, J.: Fast mean shift with accurate and stable convergence. In: International Conference on Artificial Intelligence and Statistics, pp. 604–611 (2007)
6. Wang, M., Hua, X.S., Tang, J., Hong, R.: Beyond Distance Measurement: Constructing Neighborhood Similarity for Video Annotation. IEEE Transactions on Multimedia 11(3), 465–476 (2009)

Video Reference: A Video Question Answering Engine

Lei Gao[1], Guangda Li[2], Yan-Tao Zheng[2], Richang Hong[2], and Tat-Seng Chua[2]

[1] Peking University
[2] National University of Singapore
{gaoleiss,yantaozheng}@gmail.com, {hongrc,chuats}@comp.nus.edu.sg,
g0701808@nus.edu.sg

Abstract. Community-based question answering systems have become very popular for providing answers to a wide variety of "how-to" questions. However, most such systems present only textual answers. In many cases, users would prefer visual answers such as videos which are more intuitive and informative. The Video Reference system is proposed as a solution to the above problem. It automatically extracts videos from YouTube[1] as a video reference responding to a textual question. The demo shows results on real questions sampled from Yahoo! Answering.

Keywords: Video question answering, YouTube.

1 Introduction

Community-based question answering, such as Yahoo! Answers[2] , have become more and more popular on the web. People seek answers of a variety of "how-to" questions or "what about" questions by either searching for similar questions on their own or waiting for other users to answer. However, even when the best answer is presented, user may still have difficulty in grasping the knowledge, since textual answers are often too complicated to follow, especially for "how-to" questions and event-based question [1] [4]. Here, we present a demo named Video Reference to present visual answers such as videos, which will be more direct and intuitive for user to understand. This demo is based on the techniques proposed in [3]. It shows how textual analysis, visual analysis, opinion voting and content redundancy can be fused together to return the best video answer to a given natural language style question.

In recent years, millions of video on the web, make them a source for people to learn some basic cooking techniques. However, web video search engine cannot be automatically applied to visually demonstrate these basic cooking techniques, due to the following reasons. First, search results usually contain irrelevant videos, because textual metadata associated with the video in term of title, tags or surrounding text is not sufficiently accurate to locate videos about cooking techniques. Second, even if the best related video is identified, only part of the video presents this cooking technique.

[1] YouTube: www.youtube.com
[2] Yahoo! Answers: http://answers.yahoo.com/

S. Boll et al. (Eds.): MMM 2010, LNCS 5916, pp. 799–801, 2010.

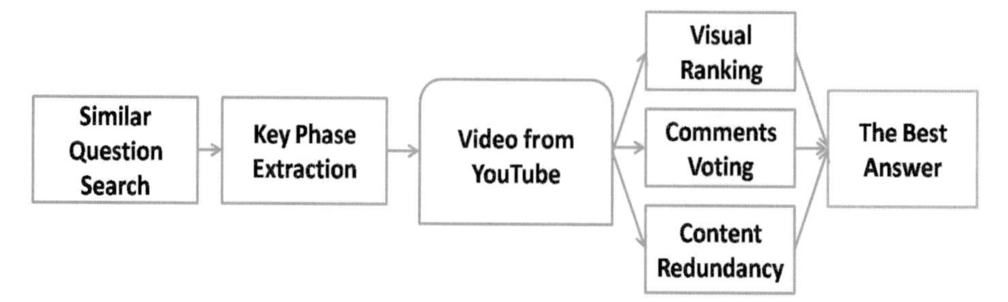

Fig. 1. The overall system architecture of Video Reference

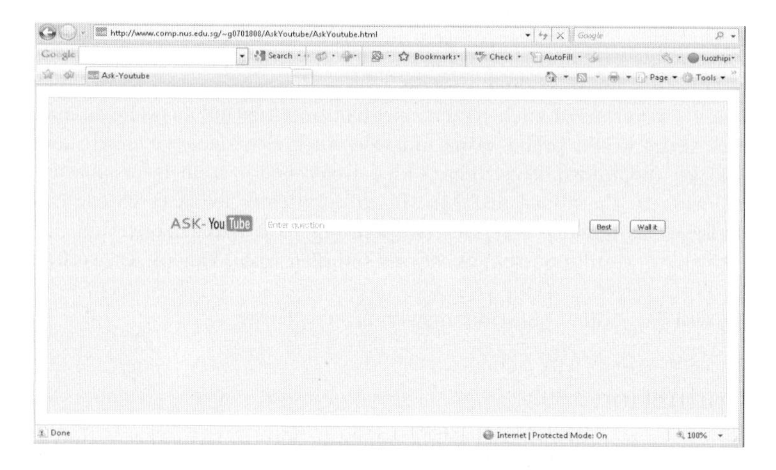

Fig. 2. Video Reference System interface: The input is a question

2 System Overview

The system automatically extracts video answers from online videos in respond to a textual question. The overall system architecture is presented in Figure 1. It is composed of three modules for the identification of video answers from YouTube.

Question Analysis: Given the original question, text-based similar question search finds questions that express similar information needs from a large archive of user generated questions in Yahoo! Answers. Since web video search engines do not perform well with verbose query, we then parse the question into phrases and identify the most informative phrases as query.

Content Analysis: Category filtering is first conducted on training images based on the key concepts identified from the question. We then adopt an extended version of k nearest neighbor classifier to classify the presence of question-related visual concepts in videos based on an adaptive vocabulary tree method, as proposed in [2].

Comments Voting: While visual information presents the relevance, positive comments of videos can reveal the video's popularity. Thus we use opinion analysis as a tool to indicate video's popularity by analyzing the past viewer's

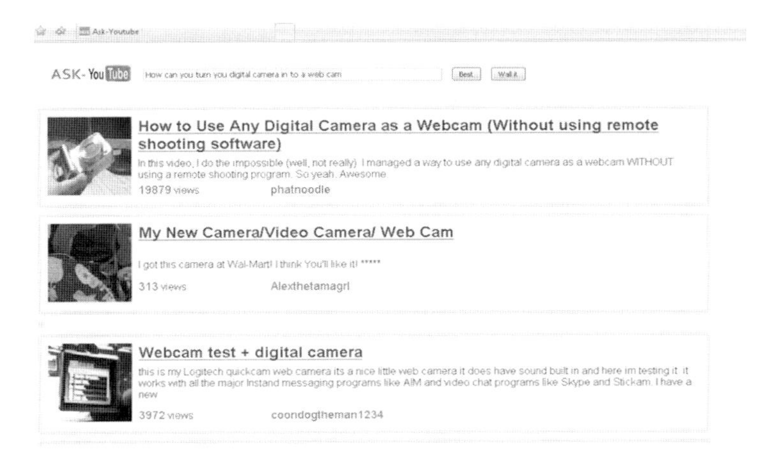

Fig. 3. System response to a question: "How can you turn you digital camera into a web cam"

comments. The result of visual re-ranking, opinion voting, and redundancy analysis are then fused to identify the precise video answer using the adapted Bayes rule. The re-ranked video is shown to be more accurate and relevant than those retrieved solely from YouTube keywords search [2]. A snapshot of the system interface is showed in figure 2.

The demo presents results on real questions sampled from Yahoo! Answering. At the current stage, we only tested the system on questions from Electronic device domain. One example result is given in figure reffig:result. In future, we will demonstrate the system with more questions from amending cars, home DIY, cooking, fitness domains, and so on.And we will incorporate video tag analysis techniques into the system, which are similar to work done in [5] and [6].

References

1. Chua, T.-S., Hong, R., Li, G., Tang, J.: From Text Question-Answering to Multimedia QA. To appear in ACM Multimedia Workshop on Large-Scale Multimedia Retrieval and Mining (2009)
2. Song, J.: Scalable image retrieval based on feature forest. In: ACCV (2009)
3. Li, G., Ming, Z., Li, H., Chua, T.-S.: Video Reference: Question Answering on YouTube. In: ACM'MM (2009)
4. Hong, R., Tang, J., Tan, H.-K., Yan, S., Ngo, C.-W., Chua, T.-S.: Event Driven Summarization for Web Videos. In: ACM Multimedia Workshop on Social Media (2009)
5. Li, H., Tang, J., Li, G., Chua, T.-S.: Word2Image: Towards Visual Interpretation of Words. In: ACM'MM 2008 (2008)
6. Wang, M., Yang, K., Hua, X.-S., Zhang, H.-J.: Visual Tag Dictionary: Interpreting Tags with Visual Words. In: ACM Multimedia Workshop on Web-Scale Multimedia Corpus (2009)
7. Tang, S., Li, J.-T., et al.: TRECVID 2008 High-Level Feature Extraction By MCG-ICT-CAS. In: TRECVID Workshop, Gaithesburg, USA (2008)

Author Index

Printing: Mercedes-Druck, Berlin
Binding: Stein+Lehmann, Berlin